Physiology and Pathology of Channels in the Nervous System

Physiology and Pathology of Channels in the Nervous System

Editor: Kelli Packer

AMERICAN
MEDICAL PUBLISHERS
www.americanmedicalpublishers.com

AMERICAN
MEDICAL PUBLISHERS
www.americanmedicalpublishers.com

Cataloging-in-Publication Data

Physiology and pathology of channels in the nervous system / edited by Kelli Packer.
p. cm.
Includes bibliographical references and index.
ISBN 979-8-88740-378-6
1. Nervous system--Pathophysiology. 2. Nervous system--Diseases--Pathophysiology.
3. Ion channels. 4. Neurotransmitters. 5. Neurology. I. Packer, Kelli.
RC347 .P49 2023
616.8--dc23

American Medical Publishers,
41 Flatbush Avenue,
1st Floor, New York,
NY 11217, USA

ISBN 979-8-88740-378-6 (Hardback)

Contents

Preface

This book has been a concerted effort by a group of academicians, researchers and scientists, who have contributed their research works for the realization of the book. This book has materialized in the wake of emerging advancements and innovations in this field. Therefore, the need of the hour was to compile all the required researches and disseminate the knowledge to a broad spectrum of people comprising of students, researchers and specialists of the field.

Single membrane channels are created through combination of two distinct membrane protein families, pannexins and connexins. These channels function as aqueous pores absorbent to small molecules and ions, to permit the diffusional interchange between the intracellular and extracellular environments. Ion fluxes across the plasma membrane in all mammalian cells are controlled by these specific membrane proteins. The channels create functional oligomeric complexes by interacting with ancillary proteins that contribute to health maintenance and cellular homeostasis. Membrane channels having varied ion selectivity modulate membrane excitability and potential. The formation of action potential triggers muscle contraction, among other cellular functions. These proteins are targets of many drugs ranging from antiepileptics to analgesics because of their role in the pathophysiology of numerous human diseases. Thus, ion channel dysfunction causes the onset of a variety of human diseases known as channelopathies, such as metabolic, cardiovascular, neural and autoimmune diseases. This book aims to shed light on the physiology and pathology of channels in the nervous system. It presents researches and studies performed by experts across the globe. This book will serve as a reference to a broad spectrum of readers.

At the end of the preface, I would like to thank the authors for their brilliant chapters and the publisher for guiding us all-through the making of the book till its final stage. Also, I would like to thank my family for providing the support and encouragement throughout my academic career and research projects.

Editor

Pannexin2 oligomers localize in the membranes of endosomal vesicles in mammalian cells while Pannexin1 channels traffic to the plasma membrane

Daniela Boassa[1]*, Phuong Nguyen[1], Junru Hu[1], Mark H. Ellisman[1,2] and Gina E. Sosinsky[1,2]*

[1] National Center for Microscopy and Imaging Research, Center for Research in Biological Systems, University of California, San Diego, La Jolla, CA, USA
[2] Department of Neurosciences, University of California, San Diego, La Jolla, CA, USA

***Correspondence:**
Daniela Boassa and Gina E. Sosinsky, National Center for Microscopy and Imaging Research, School of Medicine, University of California, San Diego, BSB 1000, 9500 Gilman Drive, MC 0608, La Jolla, CA 92093-0608, USA
e-mail: dboassa@ucsd.edu; gsosinsky@ucsd.edu

Pannexin2 (Panx2) is the largest of three members of the pannexin proteins. Pannexins are topologically related to connexins and innexins, but serve different functional roles than forming gap junctions. We previously showed that pannexins form oligomeric channels but unlike connexins and innexins, they form only single membrane channels. High levels of Panx2 mRNA and protein in the Central Nervous System (CNS) have been documented. Whereas Pannexin1 (Panx1) is fairly ubiquitous and Pannexin3 (Panx3) is found in skin and connective tissue, both are fully glycosylated, traffic to the plasma membrane and have functions correlated with extracellular ATP release. Here, we describe trafficking and subcellular localizations of exogenous Panx2 and Panx1 protein expression in MDCK, HeLa, and HEK 293T cells as well as endogenous Panx1 and Panx2 patterns in the CNS. Panx2 was found in intracellular localizations, was partially N-glycosylated, and localizations were non-overlapping with Panx1. Confocal images of hippocampal sections immunolabeled for the astrocytic protein GFAP, Panx1 and Panx2 demonstrated that the two isoforms, Panx1 and Panx2, localized at different subcellular compartments in both astrocytes and neurons. Using recombinant fusions of Panx2 with appended genetic tags developed for correlated light and electron microscopy and then expressed in different cell lines, we determined that Panx2 is localized in the membrane of intracellular vesicles and not in the endoplasmic reticulum as initially indicated by calnexin colocalization experiments. Dual immunofluorescence imaging with protein markers for specific vesicle compartments showed that Panx2 vesicles are early endosomal in origin. In electron tomographic volumes, cross-sections of these vesicles displayed fine structural details and close proximity to actin filaments. Thus, pannexins expressed at different subcellular compartments likely exert distinct functional roles, particularly in the nervous system.

Keywords: pannexin channels, ATP signaling, tetracysteine tag, miniSOG, correlated light and electron microscopy, electron tomography, intercellular communication, connexin

INTRODUCTION

The pannexins (Panx1, Panx2, and Panx3) are unique entities in the "connexin-like" family of proteins. These three membrane proteins are topologically similar to innexins (invertebrate gap junction proteins) (Panchin et al., 2000; Locovei et al., 2006a) with some sequence similarities, but function as single membrane channels (pannexons) in mammals (Locovei et al., 2006a). As shown in **Figure 1A**, the hallmarks of this folding topology are the four transmembrane segments, conserved cysteine residues found in the two extracellular loops, and the cytoplasmic domains comprised of the amino and carboxy termini and the connecting loop between transmembrane segments two and three. Based on these similarities to connexins, the transmembrane segments most likely are α-helical. However, unlike connexins, pannexins contain four strictly conserved cysteines instead of six (Ambrosi et al., 2010) and an N-linked glycosylation site in one of the two extracellular loops (N254 for Panx1, putative site N86 for Panx2, and N71 for Panx3) (Boassa et al., 2007; Penuela et al., 2013). Pannexins oligomerize to form channel structures also called pannexons.

Functionally, Panx1 and Panx3 have both been shown to perform a role in paracrine intercellular signaling. The ubiquitous Panx1 forms an ATP release channel that can interact with specific P2 purinergic receptors as part of an ATP signaling pathway (Locovei et al., 2006b; Pelegrin and Surprenant, 2006), but also can be differentially modulated by voltage and K^+ (Wang et al., 2014). Panx1 is mechanosensitive, a factor important in calcium wave propagation (Locovei et al., 2006b). Several studies have established that Panx1 channels are part of adaptive or inflammasome responses in the immune system (Kanneganti et al., 2007; Silverman et al., 2008; Schroder et al., 2010; Maslieieva and Thompson, 2014). Panx1 channels play an important role in the nervous system since they are

Abbreviations: Panx1, pannexin1; Panx2, pannexin2; CNS, central nervous system; EM, electron microscopy; CLEM, correlated lig and electron microscopy.

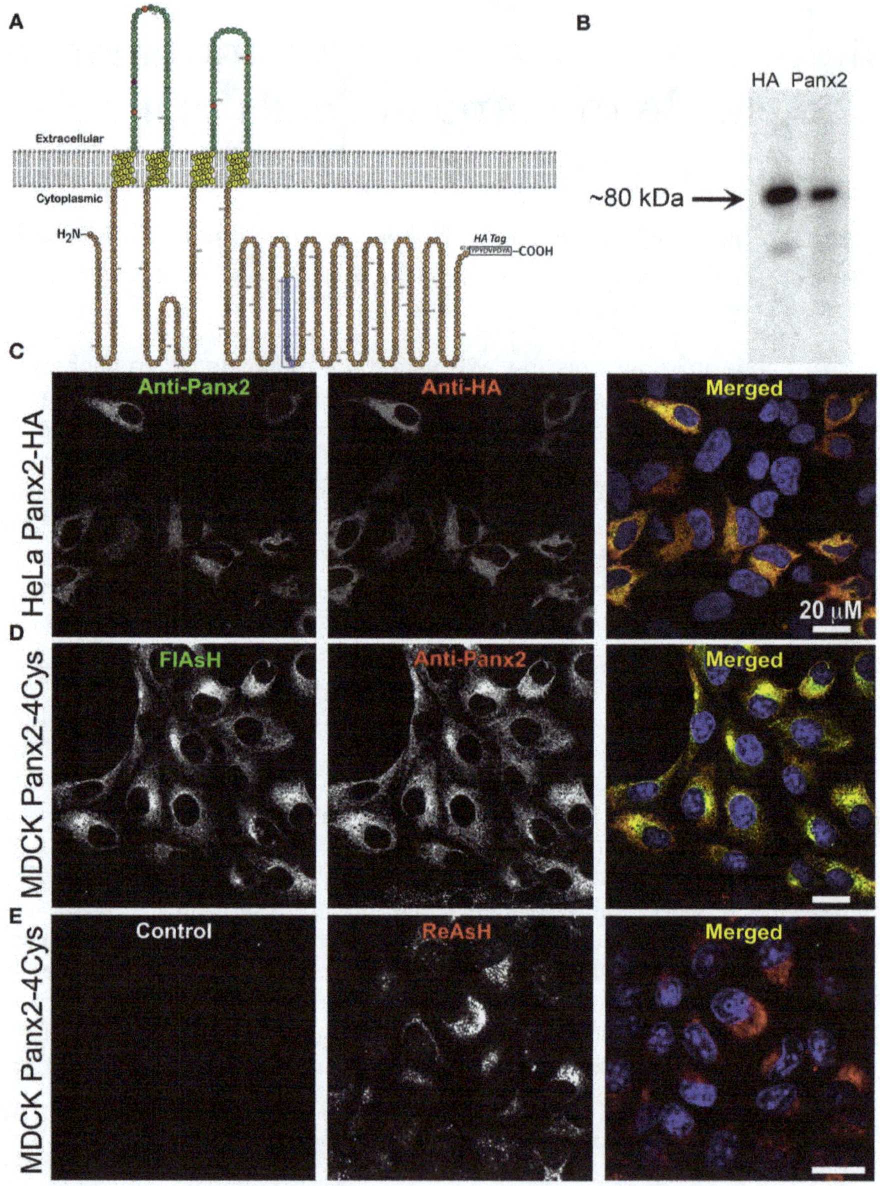

FIGURE 1 | Exogenous Panx2 is localized intracellularly in stable expressing cell lines. (A) Schematic of the membrane topology of a Panx2 monomer. The Panx2 peptide sequence used to generate the carboxy terminal antibody used in this study is indicated in blue as well as an appended HA tag. **(B)** Western blot of a cell lysate from HeLa cells stably expressing Panx2-HA. Both the anti-Panx2 (polyclonal) and anti-HA (monoclonal) antibodies detect a ~80-kDa band corresponding to the molecular mass of Panx2-HA. **(C–E)** Images are single plane confocal micrographs. **(C)** In this HeLa cell line, immunofluorescence demonstrates almost complete overlap of anti-Panx2 (left image) and anti-HA antibodies (middle image). Anti-Panx2 is displayed in green, anti-HA in red and DAPI is blue in the merged images for **(C)** (right column). **(D)** MDCK cells stably expressing Panx2-4Cys stained with FlAsH (green) and the anti-Panx2 antibody (red) also show almost complete overlap of the two fluorescent signals. **(E)** In this panel, no primary antibody was used in the initial incubations (control, left image), while ReAsH staining (red) confirmed the presence of Panx2-4Cys (middle image).

expressed in significant levels in neurons and astrocytes (Vogt et al., 2005; Weickert et al., 2005; Ray et al., 2006; Huang et al., 2007; Bargiotas et al., 2011; Cone et al., 2013). During ischemia and subsequent anoxic depolarization, Panx1 channels in neurons are activated by NMDA receptors and open releasing ATP and glutamate (Thompson et al., 2008). This activation of neuronal Panx1 channels can occur in response to nitric oxide or Src activation of NMDA receptors, as recently reviewed by Thompson (2014).

Panx3, which is more similar to Panx1 than to Panx2 in terms of molecular mass and its channel characteristics, localizes to the plasma membrane as the fully glycosylated membrane channel (Penuela et al., 2009); Panx3 is thought to be expressed mostly embryonically or only in skin or cartilage (Panchin et al., 2000;

Bruzzone et al., 2003; Baranova et al., 2004), although published immunoblot analyses showed weak bands with a second higher molecular weight species present in a broad range of tissues (Penuela et al., 2007; Langlois et al., 2014). It has been proposed that Panx3 channels in cartilage switch the chondrocyte cell fate from proliferation to differentiation by regulating the intracellular ATP/cAMP levels (Iwamoto et al., 2010).

Because combinations of connexin isoforms can form heteromeric hemichannels or heterotypic intercellular channels, early research investigated whether the same was true for pannexins after they were first identified. Bruzzone et al. (2003) originally reported that co-injection of Panx1 and Panx2 mRNA in unpaired and paired Xenopus oocytes resulted in functional hemichannels and intercellular channels, respectively with properties distinct from Panx1 homomeric hemichannels and intercellular channels. However, a subsequent papers by Naus' group (Lai et al., 2009; Le Vasseur et al., 2014) reported *only* intracellular localizations of Panx2 in mammalian cells. A different study concluded that Panx2 is found at the plasma membrane when co-expressed with Panx1 in NRK or HEK 293T cells (Penuela et al., 2009), however, within light microscopic resolution, it is unclear if in these overlapping areas of staining, Panx1 and Panx2 make heteromeric channels or form mixed populations of homomeric channels. We previously showed that preparations of Panx1 and Panx2 purified from baculovirus infected Sf9 cells made stable homomeric functional channels but unstable heteromeric channels (Ambrosi et al., 2010). Presumably an oligomerization mis-match occurs because Panx1 formed a homomeric hexamer while homomeric Panx2 pannexons were octameric (Ambrosi et al., 2010). We also found that the individual characteristics of Panx1 and Panx2 homomeric channel openings when expressed in Xenopus oocytes were different from each other, again suggesting that these two isoforms create two different kinds of pannexons (Ambrosi et al., 2010). A recent study demonstrated that when Panx1 and Panx2 channels expressed in Xenopus oocytes were stimulated to a putative open state, there was significantly less membrane currents or Yo-Pro dye uptake of Panx2 channels as compared to Panx1 channels (Hansen et al., 2014). The authors reasoned that either Panx2 channels required different physiological conditions from Panx1 to open or Panx2 is expressed at low levels at the plasma membrane.

As described in this study, using differential labeling and imaging approaches in immortalized tissue culture cells we observed that Panx1 and Panx2 channels had different sub-cellular localizations. Subsequently, we addressed this question using light microscopic imaging of endogenous pannexins in native brain tissue complemented by correlated light and electron microscopic studies (CLEM) using EM compatible genetically encoded probes that allow investigation of the distribution of Panx2 at significantly higher resolution than conventional fluorescence microscopy. We report here that Panx1 and Panx2 were differentially localized both in neurons and astrocytes in the adult mouse brain. Recombinant protein expression in different cell lines confirmed these observations of segregated Panx1 and Panx2 sub-cellular localizations. Previously, our group and others showed that Panx1 is fully N-glycosylated and transported to the cell membrane (Boassa et al., 2007, 2008; Penuela et al., 2007). In contrast, we present data here that Panx2 has an intracellular localization in the membrane of cytoplasmic endosomal vesicles and exists as a partially-glycosylated species. The resolution provided by electron microscopy suggests that Panx2 pannexons could operate as vesicular channels that are in transport to the cell membrane.

MATERIALS AND METHODS

ANTIBODIES AND REAGENTS

Below we provide antibody identification numbers in The Antibody Registry, http://antibodyregistry.org/ for the antibodies used in this study.

Pannexin antibodies

Panx1 and Panx2 antibodies were generated against peptides using sequences in the N-terminus (Panx1 mouse monoclonal; N-terminus (LKEPTEPKFKGLRLE characterization fully described in Cone et al. (2013) and the C-terminus (Panx2 rabbit polyclonal; (EPPVVKRPRKKMKWI, amino acids 420–434 **Figure 1A**). These peptides and anti-peptide antibodies were custom produced and purified by Abgent, Inc. (San Diego, CA). These antibodies recognize invariant sequences in rodent and human pannexins. The chicken anti-Panx1 antibody with a distal C-terminal epitope was provided by Dr. Gerhard Dahl (Locovei et al., 2006a) and our characterizations of it are documented in Cone et al. (2013). Panx2 antibodies were used at dilutions of 1:15,000 for western blots and 1:250–1:500 for immunofluorescence experiments.

Anti-HA antibodies

In this study, we used a monoclonal HA tag antibody (Sigma Aldrich, St. Louis, MO Catalog Number H9658, Antibody Registry ID AB_260092) for both immunofluorescence and western blots. Dilutions used were 1:10,000 for western blots and 1:250–1:500 for immunofluorescence experiments.

Cellular/Organellar markers

The following antibodies were used as markers of subcellular compartments: anti-Rab4 (BD Biosciences Catalog Number 610888, Antibody registry ID AB_398205), anti-Clathrin (BD Biosciences Catalog Number 610499, Antibody registry ID AB_397865), anti-p47a/AP3M1 (BD Biosciences Catalog #610900, AB_10015260) and anti-adaptin β (BD Biosciences Catalog #610381, Antibody Registry ID AB_397765). We used anti-GFAP antibodies (Advanced Immunochemical Incorporated Catalog #031223—Lot 1gf, Antibody registry ID AB_2314538). Antibodies were used at 1:250–500 dilutions.

DNA CONSTRUCTS

All constructs in this paper use amino acid sequences for rat pannexins (Uniprot accession codes P60571 and E0X643 for Panx2 and Panx1, respectively). The schematic of the Panx2 sequence shown in **Figure 1A** was made using the web-based program Protter (Omasits et al., 2014) http://wlab.ethz.ch/protter/start/. The rat Panx2 construct used in this study is described more in detail in Ambrosi et al. (2010). Methods for development and characterization of tagged Panx1 and Panx2 stably expressing mammalian cell lines are described in Boassa et al. (2007) and Boassa et al. (2010).

CELL LINES, TRANSFECTION, AND TRANSDUCTION METHODS

MDCK, HeLa, and HEK 293T cells were cultured in Dulbecco's modified Eagle's medium (Mediatech, Inc., Manassas, VA) supplemented with 10% FBS in a 37°C incubator with 10% CO_2. Transient transfections were carried out using Lipofectamine 2000 reagent (Life Technologies, Carlsbad, CA) following the manufacturer's protocol and 0.5 micrograms of tagged/untagged Panx2 DNA subcloned into a pcDNA3.1 vector. The transfection mixture was added to the cells plated onto 6-well dishes containing poly-d-lysine-coated coverslips or onto poly-d-lysine-coated glass bottom culture dishes (MatTek, Ashland, MA) for CLEM. Cells were incubated for 4–6 h after which the transfection medium was removed and replaced with regular growth medium. Transfected cells grew for 24–48 h post-transfection, then were fixed for either light microscopy or CLEM (both fixation conditions described below) and prepared for fluorescent imaging.

Transductions were carried out using a retroviral system according to the protocols from the Nolan laboratory (www.stanford.edu/group/nolan). Untagged or tagged-Panx2 was subcloned into a modified version of the mammalian retroviral vector pCLNCX-Hygro (originally obtained from Imgenex, La Jolla, CA). Experiments were conducted on stably untagged and tagged Panx2-expressing cell lines generated by transduction followed by selection with the antibiotic hygromycin (Gibco-BRL, Life Technologies).

WESTERN BLOTS FOR TISSUE CULTURE CELLS

MDCK, HeLa, and HEK 293T cells stably expressing untagged or tagged Panx2 proteins grown on Petri dishes were washed three times with Hanks' balanced salt solution pre-warmed at 37°C. Proteins were extracted from cells in SDS buffer containing 4% beta-mercaptoethanol, 1 mM phenylmethylsulfonyl fluoride, and a protease inhibitor mixture (Sigma Aldrich). Whole cell lysates were separated by SDS-PAGE on a 4–20% Tris-Glycine gel (Life Technologies) and then electrophoretically transferred to a PVDF membrane (Millipore, Bedford, MA). Membranes were blocked overnight in 5% non-fat dry milk made in 0.5% Tween-20 in PBS (PBS-T) and incubated with primary antibodies for 1 h at room temperature. Membranes were then washed in PBS-T to remove excess primary antibody, incubated in horseradish peroxidase-linked secondary antibodies in 5% non-fat dry milk in PBS-T for 1 h at room temperature and finally rinsed 3 times for 10 min each in PBS-T. To visualize proteins, the membrane was processed for enzyme-linked chemiluminescence using an ECL Kit (Amersham Biosciences).

ANALYSIS OF POST-TRANSLATIONAL MODIFICATIONS OF PANX1 AND PANX2 EXPRESSING TISSUE CULTURE CELLS

Whole cell lysates from HEK 293T, HeLa and MDCK cells stably expressing Panx2-HA, Panx2-WT, or Panx1-WT proteins were incubated with 10 units of calf intestinal alkaline phosphatase (CIAP) for 5 h at 37°C; or with 1500 units of *N*-glycosidase F (PNGase F) (New England Biolabs, Beverly, MA) for 5 h at 37°C at pH 7.5 following procedures we used in Boassa et al. (2007) or 50 μM pan-caspase inhibitor Z-VAD-OMe-FMK for 5 h at 37°C (Millipore, Billerica, MA) following a procedures described in Penuela et al. (2014). Samples were boiled for 5 min, and western blotted as described above.

IMMUNOFLUORESCENCE AND CONFOCAL IMAGING

Tissue culture cells plated on poly-d-lysine coated coverslips were fixed in 4% paraformaldehyde in phosphate-buffered saline (PBS) for 15 min, washed in PBS, permeabilized in 0.1% Triton X-100, and blocked in 1% BSA, 50 mM glycine and 2% normal serum. The primary antibodies were mixed in blocking buffer diluted five-fold in PBS, and applied for 1 h at room temperature. The secondary antibodies were diluted in the same buffer, and applied for 1 h at room temperature. Confocal immunofluorescence images (1024 × 1024 pixels) were acquired on the Olympus Fluoview 1000 laser scanning confocal microscope using a 60X oil immersion objective with numerical aperture 1.42.

METHOD FOR COLOCALIZATION ANALYSIS

For each set of images where we were interested in quantifying overlaps of fluorescence signals, a Manders' coefficient for colocalization was calculated using the JACoP plugin (Bolte and Cordelieres, 2006), a colocalization analysis tool for ImageJ (rsb.info.nih.gov/ij). Here, we used the Manders' coefficient as a measure of how much the pixel intensities in one channel match with pixel intensities in the same location in the other channel. The Manders' coefficient quantifies the degree of overlapping of two fluorophores and varies for 0–1, where 0 corresponds to non-overlapping and 1 reflects 100% colocalization. The average Manders' coefficient and its standard deviation were analyzed for multiple images of each sample, taking the number of cells in each image into consideration (i.e., Manders' coefficient for each image is weighted by the number of cells).

STATISTICAL ANALYSIS

We used the Prism software package (GraphPad, La Jolla CA) to test for statistical significance levels between samples. In particular, an unpaired t-test for pair-wise comparisons with Welch's correction for unequal standard deviations and with a significance cutoff set to $\alpha = 0.05$ was applied to the Manders' coefficients of our co-localization studies.

PREPARATION OF RODENT BRAIN TISSUE FOR IMMUNOHISTOCHEMISTRY AND WESTERN BLOTS

All experiments involving vertebrate animals conform to the National Institutes of Health *Guide for the Care and Use of Laboratory Animals* and were approved by the Institutional Animal Care and Use Committee of the University of California San Diego. The animal welfare assurance number is A3033-01.

Immunohistochemistry

Animals were fully anesthetized with an intraperitoneal injection of pentobarbital, and oxygenated Ringer's solution at 37°C containing xylocaine and heparin was perfused transcardially for 3 min followed by 4% paraformaldehyde for 10 min. The brain was removed and post-fixed in 4% paraformaldehyde overnight at 4°C. Sagittal sections were cut on a Leica vibratome at a thickness of 50 microns and stored at −20°C in cryoprotectant solution (30% glycerol, 30% ethylene glycol in PBS) until processed for immunohistochemistry. Free-floating tissue sections were blocked with 4% normal donkey serum, 1% bovine serum albumin, 1% cold water fish gelatin, 0.5% Triton X-100, and 50 mM glycine in PBS for 1 h at room temperature. Primary antibodies were mixed in blocking buffer diluted five-fold in PBS, and then applied overnight at 4°C. FITC, Rhodamine RedX, and Cy5 conjugated donkey secondary antibodies (Jackson ImmunoResearch Laboratories, Inc., West Grove, PA) were applied for 2.5 h at room temperature. The immunolabeled tissue samples were carefully mounted as flat as possible using gelvatol as mounting medium. Fluorescence imaging was performed on an Olympus Fluoview 1000 laser scanning confocal microscope using a 40X oil immersion objective with numerical aperture 1.3, or a 60X oil immersion objective with numerical aperture 1.42.

Western blots

Animals were fully anesthetized with an intraperitoneal injection of pentobarbital, and oxygenated Ringer's solution at 37°C containing xylocaine and heparin was perfused transcardially. The brain was dissected and brain lysates were then prepared by tissue homogenization in RIPA buffer containing 1X complete protease inhibitor cocktail (Roche). The volume of buffer was adjusted with the weight of tissue (2–3 mg per 10 μl). Tissue and cell debris was removed by centrifugation. The lysate was boiled for 5 min in 1X SDS sample buffer containing 4% beta-mercaptoethanol and 1 mM phenylmethylsulfonyl fluoride, and then loaded on a 4–20% Tris-Glycine gel (Life Technologies), and transferred to a PVDF membrane (Millipore, Bedford, MA). Membranes were blocked overnight in 5% non-fat dry milk and 0.5% Tween-20 in PBS, and incubated for 1 h with primary antibody at room temperature. After extensive washes in 0.5% Tween-20 in PBS, membranes were incubated for 1 h with secondary antibody at room temperature. Western blots were developed with an ECL kit (Amersham Biosciences).

TETRACYSTEINE LABELING, PHOTOOXIDATION OF DIAMINOBENZIDINE (DAB) AND ELECTRON MICROSCOPY (EM)

We used protein fusions with either the tetracysteine tag (FLNCCPGCCMEP) liganded with FlAsH or ReAsH for fluorescence imaging (Martin et al., 2005) or the weakly fluorescent MiniSOG (Shu et al., 2011), both developed by Roger Y. Tsien's laboratory for *in situ* protein localization. The ReAsH-labeled Panx2-4Cys proteins, or the Panx2-MiniSOG fusion proteins were photooxidized in the presence of DAB and oxygen for protein staining for EM as described in several of our publications (Boassa et al., 2007, 2008, 2013).

EM TOMOGRAPHY

MDCK cells stably expressing Panx2-4Cys were stained with ReAsH-EDT_2, fixed, photooxidized, and processed for electron microscopy (EM) as described in Boassa et al. (2013). Half micron sections were cut from Durcupan embedded specimens using a diamond knife and then coated with carbon on both sides. Colloidal gold particles (10 nm diameter) were deposited on each side to serve as fiducial markers. For reconstruction, three tilt series of images were recorded at regular tilt (angular increments of 2° from −60° to +60° increments) with a JEOL 4000EX intermediate high-voltage electron microscope operated at 400 kV. The specimens were irradiated before initiating a tilt series to limit anisotropic specimen thinning during image collection. Tilt series were recorded using a 4 × 4 k custom high resolution slow-scan CCD camera system delivering 25% contrast at Nyquist. A rough alignment for each tilt series was calculated using the IMOD package (Kremer et al., 1996). Fine alignment of projections and combination of the three tilt series data for the final 3D reconstruction were performed using the TxBR reconstruction package (Lawrence et al., 2006). Reconstructed volumes were viewed and vesicles segmented using the automated thresholding algorithm in Amira (FEI Visualization Sciences Group, Burlington, MA). Tomogram animations were generated using Amira. Single particle 2D class averages of 2D cross-sections through vesicle tomographic slices were obtained using the EMAN2 software package (Tang et al., 2007), which was also used to measure vesicle diameters. All image data, tomograms, segmentations and full-resolution movies for the tomogram presented in **Figure 7** can be accessed for downloading in the Cell Centered Database (http://www.ccdb.org) (CCDB Microscopy Product ID:48720; Project ID:2010) and Cell Image Library (www.cellimagelibrary.org).

RESULTS

In this study, we examined the sub-cellular distributions of Panx2 oligomers and how these localizations differ from those of Panx1 channels. In previous publications, we had established that Panx1 channels are found at the plasma membrane and in intracellular stores characteristic of anterograde protein trafficking (Boassa et al., 2007, 2008; Dolmatova et al., 2010). Published studies of sub-cellular localizations of Panx2 using fluorescence microscopy are contradictory. Some studies indicated Panx2 distribution both at the cell membrane and intracellular locations (Penuela et al., 2009; Swayne et al., 2010) and others reported that Panx2 was only in the cytoplasm (Lai et al., 2009; Wicki-Stordeur et al., 2013; Le Vasseur et al., 2014). Thus, as part of this study we extended experiments to electron microscopic imaging in order to gain resolution and cellular context.

ANTI-PANX2 ANTIBODY LOCALIZATIONS OF EXOGENOUSLY EXPRESSED TAGGED PANX2 AND ANTIBODY VALIDATION

As a first step, we developed a polyclonal antibody against a carboxy terminus peptide composed of amino acids 420–434 (see blue amino acids and box in the Panx2 topology model in **Figure 1A**) and cell lines stably expressing Panx2 with either an HA tag or a tetracysteine domain (4Cys) appended to the C-terminus. These genetic epitope tags served both as aids in localization and as validation tests for our anti-Panx2 antibody. As demonstrated in **Figure 1B**, staining of western blots of the same cell lysates from a HeLa cell line stably expressing Panx2-HA predominantly contained a band at about 80 kDa identified by both a monoclonal anti-HA antibody and our polyclonal anti-Panx2 antibody. It is worth noting that Penuela et al. (2014) have reported that Panx2 exhibited caspase cleavage with caspase 3 or caspase 7 treatment resulting in a band of ~36 kDa. For our constructs, the Panx2 caspase cleavage site is located N-terminally to our antibody epitope and the HA tag, so it is possible that bands of a size ~36-38 kDa in molecular mass could represent a caspase cleaved Panx2. Bands greater than 80 kDa should indicate Panx2 post-translational modifications.

Using this Panx2 antibody and epitope tagged Panx2, subcellular localizations were investigated by confocal microscopy with Panx2-expressing HeLa or MDCK cells. Confocal immunofluorescence images of Panx2-HA -expressing HeLa cells (**Figure 1C**) showed almost complete overlap of FITC (Anti-Panx2, **Figure 1C** left image) and Rhodamine RedX (Anti-HA, **Figure 1C**, middle image) fluorescence when merged (Panx2 = green; HA = red, right image), as indicated by the yellow signal. Nuclei were stained with DAPI (blue) for better identification of cells. As an additional validation, MDCK cells stably expressing Panx2-4Cys were stained with anti-Panx2 antibodies and the FlAsH reagent that specifically binds to the 4Cys tag (Gaietta et al., 2002). In the merged image in **Figure 1D** (right image), the labeling from the green FlAsH (left image) and red anti-Panx2 (middle image) were coincident as displayed by the intense yellow color. Following the same antibody labeling procedure, but using a solution lacking the Panx2 primary antibody resulted in images devoid of fluorescent signal (**Figure 1E**, left image, control). Images of ReAsH labeling showed intracellular fluorescence indicating the presence of Panx2-4Cys (**Figure 1E**, middle, right images) and the specificity of our anti-Panx2 antibody.

PANX2 PROTEIN SIZES IN BRAIN TISSUE DIFFERS FROM MODEL CELL LINES

We confirmed that our polyclonal Panx2 antibody was specific for Panx2 by incubating this antibody with the immunizing peptide, which effectively abolished western blot signal detection (**Figure 2A**). Comparison of western blots of rat and mouse brain lysates using our anti-Panx2 antibody revealed a weak band similar in size to the expected ~74 kDa we observed in western blots of HeLa cells stably expressing exogenous untagged Panx2 wild type proteins (**Figure 2B**). A faster migrating strong band at about 45 kDa was also detected as well as 2–4 higher molecular mass bands. These lower bands were not due to cross-reactivity of the Panx2 antibodies with Panx1 as immunoblotting of lysates from endogenously expressing Panx1 MDCK cells did not show any bands (data not shown). Lower bands could result from caspase cleavage occurring during tissue collection. However, to prevent caspase-dependent cleavage we treated our lysates with the pan-caspase inhibitor ZVAD (**Figure 2C**). No differences were detected in the banding pattern when compared to untreated samples suggesting that the presence of lower bands detectable either by Anti-HA or Anti-Panx2 antibody in some preparations could be the result of non-specific protein degradation.

POST-TRANSLATIONAL MODIFICATIONS OF PANX2 IN TISSUE CULTURE CELLS

While Panx1 and Panx3 have N-linked glycosylation sites that allow for trafficking to the plasma membrane (Boassa et al., 2007; Penuela et al., 2007), Panx2 has a putative glycosylation site at N86 that has been reported to exhibit partial glycosylation (Penuela et al., 2009) where the band shift is far less than a fully glycosylated form (as compared with Panx1). In **Figure 2C** we show immunoblot analysis of lysates from stably expressing Panx2-HA MDCK, HEK 293T and HeLa cells treated either with

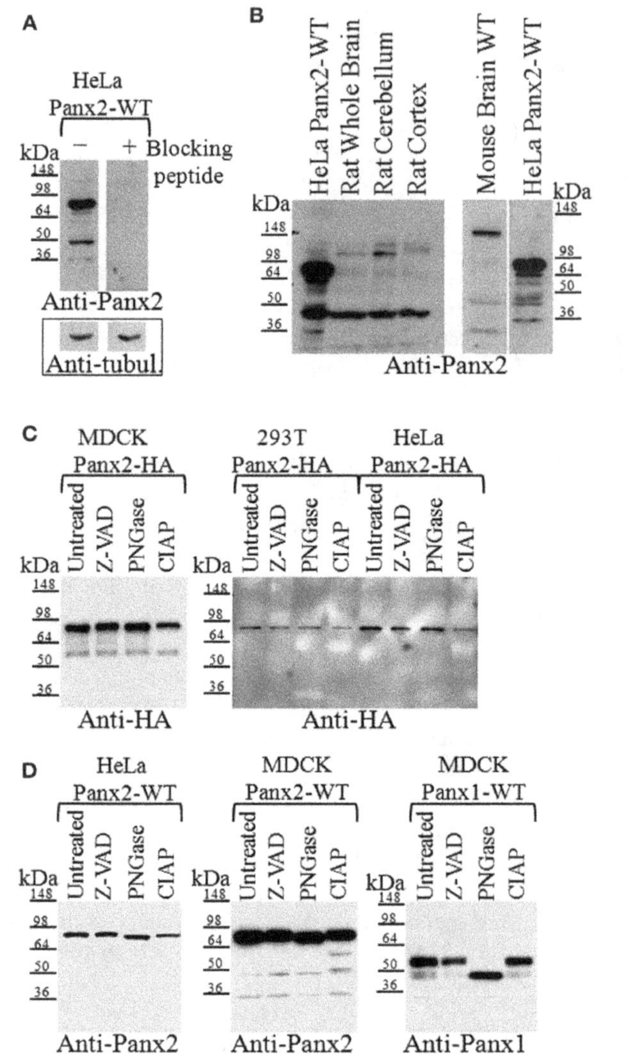

FIGURE 2 | Western blot analysis of Panx2 in tissue and post-translational modifications in tissue culture cells. (A) Incubation of the Panx2 antibody with the immunizing peptide prior to western blotting eliminated the labeling of untagged Panx2-WT proteins stably expressed in HeLa cells. Western blots for alpha-tubulin levels shown below the Panx2 immunoblots served as loading controls for these lysates. **(B)** The expected mass of Panx2 monomers based on their amino acid sequence is ~74 kDa and western blot analysis of HeLa cell lysates stably expressing untagged rat Panx2 showed a major band at about this molecular mass (left and right hand lanes). Some lower bands were also observed at about 45 kDa. In tissue lysates from rat and mouse brains, a weaker ~74 kDa band was observed as well as few faster migrating bands that may represent caspase cleaved Panx2 species. **(C)** Cell lysates from three cell lines, MDCK, HEK 293T, and HeLa, each stably expressing HA-tagged Panx2, were treated with either Z-VAD-OMe-FMK (Z-VAD) to inhibit caspase-dependent cleavage, PNGase to reduce glycosylation, or CIAP for protein dephosphorylation. We did not see any significant shift in the bands indicating that at least in these cell types, HA-tagged Panx2 is not highly phosphorylated or glycosylated. **(D)** However, cell lysates from HeLa, and MDCK, each stably expressing untagged Panx2-WT revealed a slight shift in banding pattern following PNGase treatment. As positive control, lysates from MDCK cells stably expressing untagged Panx1-WT showed significant band shifts after PNGase treatment. Western blots shown here are representative of at least three independent experiments per group.

PNGase to determine its N-linked glycosylation status or with CIAP for dephosphorylation. We found that both treatments did not significantly shift the banding pattern, indicating that in our Panx2-HA samples glycosylation and phosphorylation were not detectable. However, lysates from Panx2-WT (untagged) proteins stably expressed in HeLa or MDCK cells revealed a small shift in the bands following PNGase treatment, similarly to what has been previously shown (Penuela et al., 2009) suggesting that the untagged Panx2 proteins exists both as a non-glycosylated core protein (GLY0) and a high mannose-type glycoprotein (GLY1) while the Panx2-HA exists only as a non-glycosylated core protein (**Figure 2D**). Interestingly, following CIAP treatment two new bands appeared at about 60 and 50 kDa in the lysates from Panx2-WT (untagged) proteins, although this seemed to vary among cell lines. As a positive control, lysates from MDCK cells stably expressing Panx1-WT (untagged) proteins showed a change in banding pattern following PNGase treatment (**Figure 2D**).

CELLULAR DISTRIBUTIONS OF PANX1 AND PANX2 ARE NON-OVERLAPPING IN CELL CULTURES AND BRAIN TISSUE

While Panx2 oligomers have been reported to traffic to the plasma membrane when co-expressed with Panx1 (Penuela et al., 2009), we found that co-expressing untagged Panx1 and Panx2 in various cell lines resulted in non-overlapping populations of Panx1 and Panx2. As shown in **Figure 3A**, when co-expressed in three different cell lines (MDCK, HEK 293T, and HeLa cells), Panx1 had strong plasma membrane fluorescence while Panx2 labeling was localized into intracellular compartments. For these experiments we used untagged, wild type Panx1 and Panx2, and immunolabeling was performed with a monoclonal Panx1 antibody and the polyclonal Panx2 antibody described in the previous section. We also obtained the same results in experiments with myc and HA tagged versions of Panx1 and Panx2, respectively, expressed in these cell types (data not shown) and also in our initial experiments using a chicken antibody provided by Gerhard Dahl (Locovei et al., 2006a; Cone et al., 2013).

In order to quantify the overlap between Panx1 and Panx2 fluorescence, we calculated the Manders' coefficient for these single plane confocal images (**Figure 3B**). Manders' coefficients for colocalization for HEK 293T and MDCK cell images (number of analyzed images were 8 and 11, respectively) were below 0.05 (5% colocalization), while a similar analysis of 7 HeLa cell confocal images had an average Manders' coefficient of about 0.2 (20% colocalization).

Because previous studies based on *in situ* hybridization of Panx2 mRNA have indicated that the CNS is enriched in Panx2 (Bruzzone et al., 2003; Vogt et al., 2005; Zappala et al., 2007; Bargiotas et al., 2011), we looked for the segregation and overlap of Panx1 and Panx2 within this tissue. Triple labeling of 50 μm thick mouse brain sections with Panx1, Panx2, and glial fibrillary acidic protein (GFAP) antibody, a specific marker for astrocytes, demonstrated that the two Panx isoforms localize at different subcellular compartments in both astrocytes and neurons. In CA1 hippocampus (**Figure 4**), we observed that Panx2 was highly localized to pyramidal cell somata while Panx1 was distributed in cell bodies as well as in axons (*see enlargement of yellow and blue boxed area in Panel A*). In neuronal cell somas, Panx2 labeling did not overlap with Panx1 and was mainly perinuclear

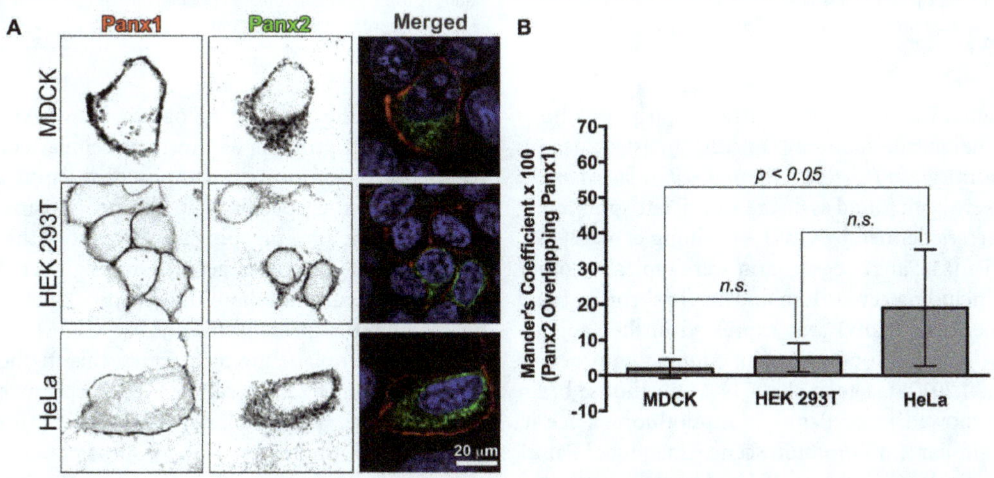

FIGURE 3 | Recombinant Panx1 and Panx2 have non-overlapping cellular distributions in various cell lines. (A) MDCK, HEK 293T, and HeLa cells were co-transfected with wild type untagged Panx1 and Panx2. Images are single confocal slices. In the single channel confocal images (left and middle columns), the color table has been inverted for better visualization of the individual fluorescence channels and in particular, vesicle populations. Panx1 was labeled with a mouse monoclonal antibody (Cy5 secondary antibody detection, shown in red in the merged image) while our polyclonal C-terminal Panx2 antibody was used to stain Panx2 (FITC secondary detection, shown in green in the merged image). In all three cell types, Panx1 was found mostly in the plasma membrane (left column, red color in the "Merged" images in the right hand column) while Panx2 remained in intracellular locations (middle column, green color in the Merged images in the right hand column). In the right hand column images ("Merged"), the nuclei are stained with DAPI (blue). We calculated a Manders' colocalization coefficient in order to quantify the degree of overlap of Panx1 and Panx2 in these cell lines. The graph of average percent overlap (as indicated by the Manders' coefficient × 100) shown in **(B)** demonstrates that there is very little overlap (20% and below) between Panx1 and Panx2 in these three cell types with significantly higher overlap in HeLa cells. The numbers of images analyzed were 8, 7, and 11 for HEK 293T cells, HeLa cells and MDCK cells, respectively. Error bars indicate standard deviations. (n.s., not significant).

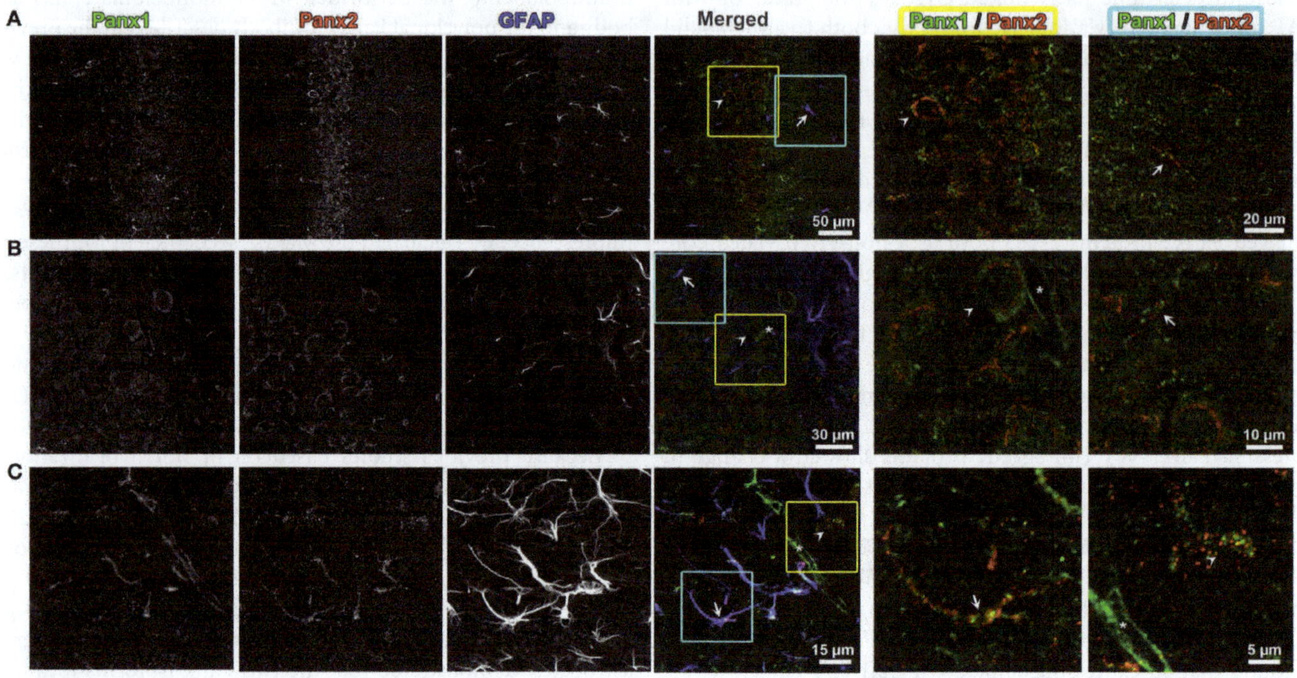

FIGURE 4 | Endogenous Panx1 and Panx2 show non-overlapping signals in hippocampal neurons and astrocytes. Confocal images of the CA1 field of the hippocampus in the adult mouse brain are shown in **(A–C)**. Images in **(A,B)** are single confocal planes while **(C)** represents maximum intensity projections of confocal image stacks to better show the astrocyte morphology. The tissue has been immunolabeled for Panx1 (left black and white image column), Panx2 (middle black and white image column) and the astrocytic marker, GFAP (right black and white image column). In the color image column labeled "Merged" (middle column of figure), the three black and white images were superimposed with the Panx1, Panx2, and GFAP labelings displayed in green, red and blue, respectively. Three-fold enlargements of two areas highlighted by yellow and cyan boxes in each of these merged images are displayed in the two far right columns (*cyan = far right column, yellow = column left of right hand column*). These enlargements only display the Panx1 (green) and Panx2 (red) signal to better show the cellular segregation of Panx1 and Panx2 in CA1 pyramidal cells and astrocytes. Example astrocytes are indicated by arrows, neurons by arrowheads and capillaries by an asterisk. Note that in astrocytes and neurons there is little overlap in populations of Panx1 and Panx2. Also, the capillaries contain Panx1, but not Panx2.

while the Panx1 signal was localized to intracellular populations with little plasma membrane staining (*Panel B*). In astrocytes (*as exemplified by the astrocyte in Panel 4C blue and yellow boxed area*), Panx1 and Panx2 were both found as discrete punctate spots along the astrocytic processes defined by GFAP (Bushong et al., 2002). These patterns of Panx1/Panx2 segregation were typical of other areas of the brain including cerebellum and cerebral cortex (data not shown). Furthermore, Panx1 was expressed in the capillaries and blood vessels throughout the brain, confirming previous reports (Burns et al., 2012; Gaete et al., 2014). Our mouse brain stainings typically showed more Panx1 immunofluorescence in blood vessels than similar immunofluorescence imaging of Panx1 in rat brain blood vessels (Cone et al., 2013). However, we noted in Cone et al. (2013) that the level of vessel staining was dependent on the primary antibody used for labeling. More importantly and relevant to this study was that we never found Panx2 immunofluorescence in cerebral blood vessels.

CORRELATED LIGHT AND EM (CLEM) IMAGING OF PHOTOOXIDIZED PANX2-miniSOG REVEALS PANX2 IN INTRACELLULAR COMPARTMENTS

In order to investigate the localization of Panx2 at high resolution, we took advantage of EM imaging in combination with protein specific staining to obtain increased spatial resolution of the labeled proteins within their cellular context. Specifically, we used a fusion protein composed of Panx2 and miniSOG for highlighting the specific distribution of Panx2. MiniSOG is a portion of the phototropin-2 protein from the plant *Arabidopsis thaliana* that has been genetically engineered for the specific purpose of using its diaminobenzidine (DAB) reactive properties to identify protein fusions by CLEM (Shu et al., 2011). Unlike immunolabeling techniques, this method has the advantage that strong fixatives can be used to optimally preserve cellular ultrastructure, a particularly important feature when analyzing membrane bound intracellular compartments. Panx2-miniSOG was transiently expressed in HeLa cells. The miniSOG tag is weakly fluorescent as shown in **Figure 5A** left image (highlighted by the white arrowheads), and following strong illumination with blue light it generates enough singlet oxygen to catalyze the deposition of osmophilic DAB polymers onto the tagged Panx2. **Figure 5A** middle and right images shows before and after photooxidation transmitted light micrographs. The black arrowheads identify the same cells as in **Figure 5A** left image (white arrowheads) that expressed the fluorescent Panx2-miniSOG. These transmitted light micrographs are useful for monitoring the photooxidation process, correlating the appearance of the

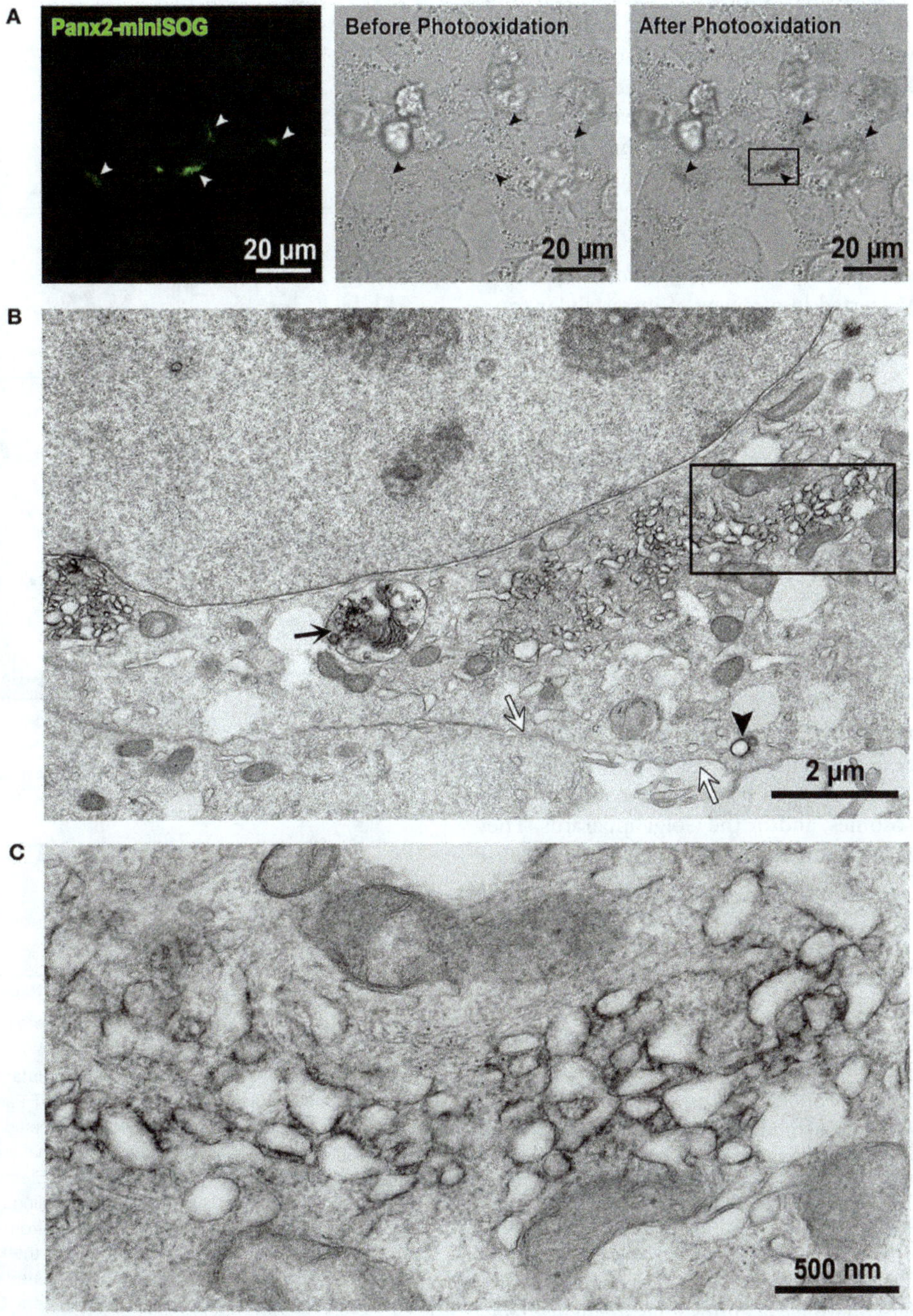

FIGURE 5 | EM imaging reveals intracellular localizations of Panx2-miniSOG in membrane bound structures. (A) Confocal image from Panx2-miniSOG transiently expressed in HeLa cells (*left image*, white arrowheads point to miniSOG fluorescence). DIC images of the same area before (*middle image*) and after photooxidation (*right image*) with black arrowheads highlighting the same areas as in the fluorescence image. **(B)** Displayed is a low magnification EM thin section (corresponding to the boxed area in the right hand post-photooxidation DIC image in **A**). The black arrow points to a Panx2-miniSOG containing lysosome while the black arrowhead indicates a stained Panx2-miniSOG vesicle. Note that the plasma membrane is unstained (white arrows) and thus, contains no Panx2-miniSOG channels. **(C)** A higher four-fold magnification view of the boxed area in **(B)** revealed Panx2-miniSOG in intracellular tubulo-vesicular membranous compartments and minimal ER localizations.

optically visible DAB reaction product with the detected miniSOG intrinsic fluorescence, and for tracking the same areas in EM images. The osmium/DAB complex then serves as an electron dense label for EM. In the electron micrograph presented in **Figure 5B**, Panx2-miniSOG appears dark within intracellular membrane bound structures. This area is a higher magnification view of the boxed portion of cytoplasm between the nucleus and plasma membrane in **Figure 5A** right image. Within this region, labeled structures include elongated membranous portions, individual vesicles (black arrowhead) and a degradation compartment (black arrow) is readily apparent. In contrast, staining is not observed at the level of the plasma membrane (white arrows). A four-fold enlargement of the boxed area in **Figure 5B**, displayed in **Figure 5C**, highlights the cytoplasmic protrusions from the Panx2 oligomer seen in these intracellular membrane cross-sections.

Panx2 channels colocalize in early or recycling endosomes

D'Hondt et al. (2011) asserted in a review of intracellular functions of pannexin channels that pannexons may serve an endoplasmic reticulum (ER) Ca^{2+} release function. Their hypothesis was based mainly on a previous report that Panx1-EGFP stably overexpressed in human prostate cancer epithelial LNCaP cells had a fluorescent ER localization pattern overlapping with BODIPY-Brefeldin A fluorescence (Vanden Abeele et al., 2006) that correlated with Ca^{2+} leakage from the ER with tagged and untagged Panx1 ER expression. While we originally found that double immunolabeling of Panx2 and the ER marker protein calnexin in Panx2 transfected tissue culture cells revealed diffuse intracellular staining and significant overlap with calnexin (data not shown), suggesting that Panx2 localized in the ER, the CLEM data shown in **Figures 5, 7** revealed instead that Panx2 is localized in the membrane of intracellular vesicles suggestive of endosomal systems and not in the ER. Endosomes represent sorting compartments within cells that serve to move proteins from/to the plasma membrane, lysosomes, and/or the Golgi apparatus. They are characterized as early, late and recycling endosomes. Because we saw no evidence for anterograde trafficking of Panx2 to the plasma membrane and a previous publication reported Panx2 being localized to the endolysomal system (Wicki-Stordeur et al., 2013), we performed immuno-colocalizations between Panx2 and endosomal constituent proteins used as markers for various endosomal compartments (**Figure 6A**). Four endosomal proteins were chosen for our colocalization experiments: Clathrin (endocytic vesicles), Adaptin β (protein sorting from the TGN and endosomes as well as clathrin-mediated endocytosis) (Boehm and Bonifacino, 2002), Rab4 (early and recycling endosomes) and p47A (vesicles derived from the Golgi). p47A is also known as AP3M1, a rat homolog of clathrin-associated adaptor proteins that interacts with the tyrosine-based sorting signal in the trans-Golgi network (Dell'Angelica et al., 1997). The monomeric GTPase Rab4 is associated with early endosomes, regulates recycling vesicle formation and provides for vesicle sorting, before it reaches lysosomes for degradation (Mellman, 1996; Mohrmann et al., 2002).

We expressed Panx2-WT (untagged) in HEK 293T cells and as expected, all four vesicular markers showed diffuse cytoplasmic fluorescence distributions. Overlap of Panx2-WT and vesicular markers is indicated by the yellow color in the merged **Figure 6A** single plane confocal images. For Clathrin and Adaptin β, this percent overlap (Manders' Coefficient × 100) was low at ~10–19% ($n = 107$ and $n = 131$ cells, respectively), whereas the percent overlap between Panx2 and Rab4 signals was statistically significantly higher at ~60% ($n = 119$) and ~48% for p47A and Panx2 ($n = 144$ cells) (**Figure 6B**). Thus, the high percentage overlap indicated that Panx2 is most likely localized to early or recycling endosomes.

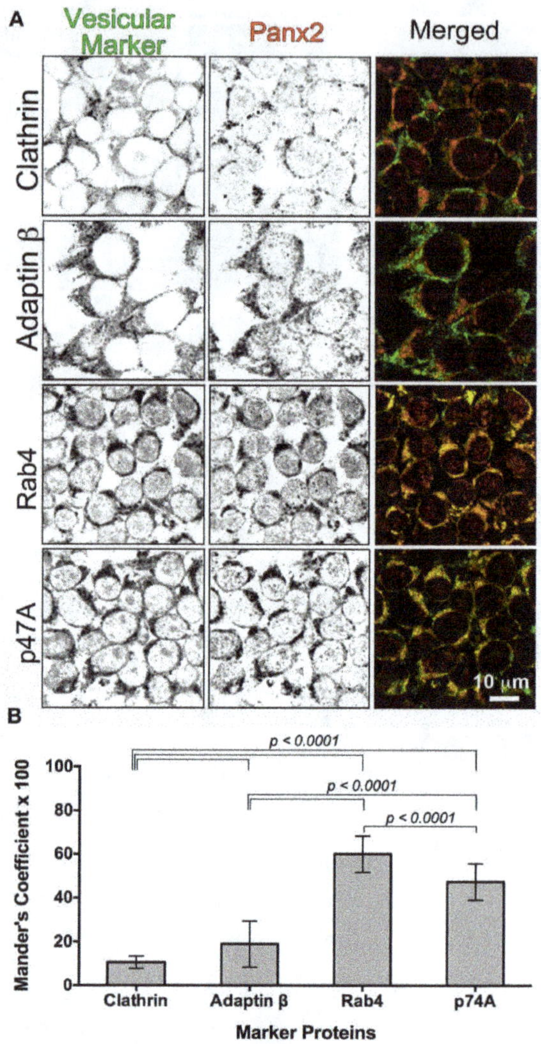

FIGURE 6 | Colocalization of Panx2 with cellular markers indicates early endosomal sorting. In **(A)**, untagged WT Panx2 (red) and four organellar markers (green) were co-immunolabeled with the polyclonal anti-C terminal Panx2 antibody and a monoclonal antibody specific to each organellar proteins. The four protein markers, Clathrin, Adaptin β, Rab4 and p47A label clathrin-coated vesicles, early/recycling endosomes, early endosomes, and degradation vesicles derived from the Golgi, respectively. Both the single plane confocal images **(A)** and the Percent Colocalization (Manders' Coefficient × 100) graph in **(B)** indicated that Panx2 overlaps with early endosomes. Sample sizes for Manders' Coefficient calculation: 107 (clathrin), 119 (rab4), 131 (adaptin β) and 144 (p47A) cells. The errors bars indicate standard deviations.

3D visualization by electron tomography demonstrated that intracellular vesicles contain Panx2 channels

In order to obtain a higher resolution view of Panx2 channels within intracellular vesicles, we used a fusion protein of Panx2 with the current generation of tetracysteine tag (Martin et al., 2005). Tetracysteine tags (4Cys) are small (~2 kDa), but require an exogenous biarsenical ligand for visualization. In tissue culture cells, diffusion of this ligand and binding to 4Cys is efficient and the red fluorescent biarsenical ReAsH ligand gives good photooxidation-driven DAB reaction product. Since DAB deposition is a localized reaction, we have found that DAB/osmium labeling better defines macromolecular complexes the closer the reaction occurs to the protein. We stably expressed Panx2-4Cys in MDCK cells, a cell line we have previously found to be particularly amenable to CLEM imaging and suffer less from over-expression of connexin superfamily proteins than HEK 293T or HeLa cells (Boassa et al., 2007, 2010).

As displayed in the fluorescence image in **Figure 7A**, the ReAsH labeling was localized intracellularly. This area was photooxidized and the sample dishes processed for EM. Using triple tilt electron tomography of 0.5 μm thick sections from these photooxidized areas, we obtained a tomographic volume of a portion of the cell containing vesicles (**Supplemental Movie**). The volumetric slab is shown in **Figure 7B.** Three vesicles are shown in cross section in a single slice from the tomogram (**Figure 7C**). Two of these are highlighted by cyan arrows while many actin filaments are found in close proximity (yellow arrow) or apposition in these vesicles (yellow arrowhead in **Figure 7D**). This is not surprising as Ohashi et al., showed that actin filaments regulated by cortactin was required for segregation of early endosomes in the perinuclear area, inducing movement of each endosome toward the cell center and preventing fusion events (Ohashi et al., 2011). Automated segmentation based on intensity thresholding using Amira was used to outline the sphere-like vesicles (colored in gold) and the proteinaceous coat (colored in blue) that surrounded the membrane boundary of these vesicles (**Figure 7D**).

Significant structural details were seen in the vesicle cross-section in this and other similar tomograms. A higher magnification view of one of the vesicles is displayed in **Figure 7E** and striations perpendicular to the membrane bilayer were apparent. We measured the diameters of these vesicles from the cross-sections in the tomogram. The vesicle population was variable with average ~88 ± 12.9 nm standard deviation (range 54.8–109.2 nm, $n = 46$ vesicles). The vesicles were not always spherical as we found the average ellipticity was 1.12 ± 0.15 in these cross-sectional views. Thus, sub-tomogram averaging would not have yielded a 3D consistent vesicle structure. Instead, we selected 128 × 128 pixel areas of vesicle membrane cross-sections from 2D slices in the full-resolution tomogram and used class averaging (Tang et al., 2007) to enhance any repeating substructure. In **Figure 7F**, we present three class averages out of 16. The left hand average has less defined repeating membranous substructure, but shows the thicker layer of staining on the cytosolic side of the vesicle, most likely due to the extensive Panx2 cytoplasmic domains and tag position. As defined by the staining, the membrane thickness was ~3.5–4 nm. The middle and right hand averages showed more variations in the membrane bound region and at the cytosolic surface. The dark striations we interpret as stain within the channel pore. The blue box represents an area of about 8.0 × 10.0 nm, dimensions that are reasonable for a Panx2 channel based on our own publication (Ambrosi et al., 2010) and the measured dimensions of connexin hemichannels from crystallographic analyses (Maeda et al., 2009; Oshima et al., 2011).

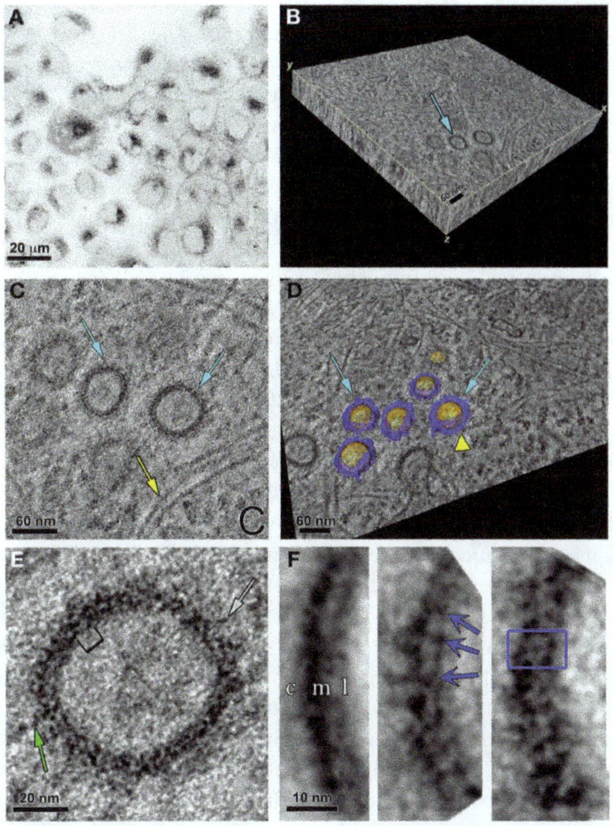

FIGURE 7 | Specific labeling of Panx2-4Cys and electron tomography highlight Panx2 distribution in the membrane of intracellular vesicles. (A) MDCK cells stably expressing Panx2-4Cys were stained with ReAsH-EDT$_2$, fixed and imaged with a confocal microscope before photooxidation. The image is displayed with an inverted color table (black is the highest fluorescence signal, white is no signal). **(B)** 3D representation of the EM tomographic volume as a 3D slab. Cyan arrow points to vesicles containing Panx2-4Cys oligomers. **(C)** Single X-Y slice from the tomogram showing Panx2-4Cys labeling in vesicle cross-sections (cyan arrows). Note the actin filaments in close proximity to the Panx2-4Cys-labeled vesicles (yellow arrow). **(D)** Automated segmentation of Panx2 vesicles (cyan arrows): yellow represents the lipid bilayer of the vesicle, while blue highlights the large domains of the Panx2 oligomers protruding from the vesicle membrane into the cytoplasm. The yellow colored arrowhead indicates an actin filament in close apposition to a vesicle. Based on comparisons with connexin channels, these protrusions most likely are from the large cytoplasmic domains of Panx2-4Cys. **(E)** Single slice of the tomogram showing a subarea of a Panx2-4Cys containing vesicle. Here stain-excluding areas (protein and lipid) are white and stain is black. The green arrows point to stained fine protrusions from the cytoplasmic surface of the vesicles. The bracket indicates an area bounded by stain that is ~8 nm. By analogy with connexin hemichannels, this distance should correspond to approximately the diameter through the membrane bound portion of Panx2 channels. **(F)** Three representative class averages obtained by single particle analysis of boxed cross-sectional areas. In the left hand average, c, cytosolic side; m, membrane; l, lumen of vesicle. Note the thicker cytosolic staining due to the deposition of DAB/osmium to the side containing the 4Cys/ReAsH tag. The blue arrows in the middle average and box in right hand average highlight substructure defined by stain from the Panx2 pore within the membrane bound portion (right hand average: inside box dimensions = ~ 8 × 10 nm).

DISCUSSION

In contrast to the wealth of literature on the functionality of Panx1 and Panx3 as putative ATP channels, the functional role of Panx2 channels remains unresolved. Penuela et al. (2014) postulated that Panx2 may have ATP release functions similar to Panx1 because of a putative C-terminal caspase3 and caspase7 cleavage site analogous to a Panx1 caspase site that causes ATP release upon cleavage (Chekeni et al., 2010; Dourado et al., 2014). Nonetheless, it is clear from the data presented in this study and previously published studies by other laboratories (Lai et al., 2009; D'Hondt et al., 2011; Wicki-Stordeur et al., 2013; Le Vasseur et al., 2014) that Panx2 trafficking behavior and endolysosomal localizations are markedly different from Panx1 and Panx3 (Boassa et al., 2007; Penuela et al., 2007).

SUBCELLULAR ANALYSIS OF PANX2 AND PANX1 IN MODEL CELL SYSTEMS AND BRAIN TISSUE SHOWED SIMILAR SEGREGATION PATTERNS

We used immunofluorescence imaging to confirm that Panx2 and Panx1 trafficked to intracellular compartments and the plasma membrane, respectively. The highly segregated populations of Panx1 and Panx2 in our light micrographs indicated that heteromeric combinations usually do not form. Our results are contrary to the co-immunoprecipitation and light microscopy imaging experiments of Penuela et al. (2009) and are confirmatory of our earlier hypothesis that Panx1/Panx2 channels are unstable *in vitro* (Ambrosi et al., 2010). Thus, it is unlikely that heteromeric channels would be found in native, unperturbed organ systems.

Because conventional fluorescence confocal microscopy does not provide sufficient resolution to examine details of subcellular compartments below the diffraction limit and provides information just based on signal from the probe, we performed EM on tissue culture cells expressing Panx2 C-terminally tagged with either miniSOG or 4Cys and reacted such that Panx2 could be easily identified within the cytoplasm at nanometer resolution (**Figures 5**, **7**). We found strong staining in vesicle associated compartments, consistent with endolysosome distributions previously identified by Wicki-Stordeur et al. (2013). Sites for both endocytotic recognition and endolysosomal targeting sequences have also been identified in the Panx2 sequence using consensus sequence analysis programs (Boyce et al., 2014). Two endolysosomal targeting sequences were identified in the intracellular loop (residue 153–158) and a distal part of the C-terminus (amino acids 647–652) while an area of the C-terminus sequence contained an endocytotic recognition sequence at amino acids 376–379. Rarely do we see Panx2 EM localizations in the ER as suggested by co-localization studies that solely used light microscopy for imaging (D'Hondt et al., 2011).

The endosomal colocalization experiments we presented indicated that expression patterns were consistent with early endosomal vesicle populations. The fact that we observed vesicle distributions in different cell types as well as in transient transfections and stable Panx2 expressing cell lines confirmed that this mode of trafficking is not an expression artifact. While early endosomes have typically been associated with endocytotic events (Jovic et al., 2010), early and recycling endosomes are now thought to act as entry portals, sorting stations, and signaling platforms. This mechanism provides an alternative route in biosynthetic pathways from traditional anterograde vesicular trafficking (De Matteis and Luini, 2008; Gould and Lippincott-Schwartz, 2009). The vesicular membrane cross-sections in our tomographic analysis contained Panx2 channel structures within the vesicle membrane and with distinct extensions into the cytosol (**Figure 7**) that are consistent with what is known about Panx2 protein topology. These images are very different from CLEM images of Panx1 channels we previously published in Boassa et al. (2007). Interestingly, endosomal cargo recycling typically involves the actin cytoskeleton (Gautreau et al., 2014), consistent with our observations of close associations of actin filaments in the tomograms.

In the present study, we observed strong non-overlapping Panx1 and Panx2 expression in both astrocytes and neurons in the adult mouse brain (**Figure 4**). While the Panx2 monomer has a similar molecular folding, it is much larger (~74 kDa; 674 amino acids) than Panx1 (~48 kDa; 426 amino acids) or Panx3 (~45 kDa; 392 amino acids). Its sequence is more distantly related to Panx1 and Panx3 than those proteins are to each other (Baranova et al., 2004). The mRNA for Panx2 is found in significant quantities in the CNS, both in astrocytes and neurons (Baranova et al., 2004; Bargiotas et al., 2011). Panx1 and Panx2 have a widespread and similar mRNA distribution, but Panx1 and Panx2 mRNA levels are inversely regulated during the development of the rat brain with significantly increase in Panx2 expression during post-natal development (Vogt et al., 2005). Small amounts of mRNA expression have been demonstrated in other tissues such as thyroid, prostate, liver and heart (Bruzzone et al., 2003). At the protein level, in cochlea immunolocalizations, Panx2 labeling was restricted to the basal cells in the stria vascularis and was also detected in the spiral ganglion neurons, but no overlapping labeling for Panx1 and Panx2 was observed (Wang et al., 2009).

Several studies have attempted to define a relationship between Panx2 within brain tissue and conditions that would up-regulate it. A gene array analysis showed an overall reduction of Panx2 in gliomas and a direct correlation was observed between Panx2 expression and post-diagnosis survival in patients (Litvin et al., 2006). Human Panx2 DNA is located in the same chromosomal regions that are implicated in human gliomas (Ino et al., 1999; Oskam et al., 2000; Hu et al., 2004) and exogenous Panx2 transfected into rat C6 glioma cells displayed a flattened morphology and increased cell–cell contacts and significantly reduced *in vitro* oncogenicity parameters (Lai et al., 2009). Zappala et al. (2007) showed that Panx2 was exclusively expressed by neurons of normoxic astrocytes–neurons primary co-cultures. Conversely, after 20 min of hypoxia, Panx2 was strongly expressed also in astrocytes. The *de novo* expression of Panx2 in astroglial cells after ischemia was proposed to play a role in the neuroprotection after ischemia (Zappala et al., 2007). A recent study showed that hippocampal seizure activity induced by a seizuregenic protocol using Co^{2+} up-regulated Panx2 mRNA and protein expression (Mylvaganam et al., 2010).

POST-TRANSLATIONAL MODIFICATIONS OF PANX2

Several post-translational modifications are possible based solely on analysis of the Panx2 sequence (Penuela et al., 2014). During brain development Panx2 has been described to be dynamically expressed over the course of post-natal hippocampal neurogenesis *in vivo* and *in vitro* with changes observed between intracellular and plasma membrane localizations associated with S-palmitoylation. Specifically, Panx2 is palmitoylated and localized intracellularly in neural progenitor cells and depalmitoylated and at the plasma membrane in mature neurons (Swayne et al., 2010). These authors hypothesized that S-palmitoylation may regulate Panx2 intracellular sorting and protein-membrane interactions during neural development. Contrary to Panx1 and Panx3 where a complex glycosylated product (whose band on a western blot we originally named GLY2) (Boassa et al., 2007) was expressed at the plasma membrane, Panx2 was reported to exist in a high mannose form (GLY1) that exhibits only slight band shifts from the non-glycosylated form (GLY0) (Penuela et al., 2009). In this study, we have confirmed the existence of two forms for untagged Panx2 proteins. However, the tagged Panx2-HA exists only as a non-glycosylated core protein showing no change in the banding pattern after PNGase treatment; similarly dephosphorylation by alkaline phosphatase was ineffective indicating minimal or no post-translational modifications.

CONCLUSIONS

Taken together our studies demonstrate that Panx2 is differently trafficked from Panx1. This endolysosomal trafficking mechanism implies that Panx2 has a different regulation mechanism than the other pannexins. While future studies may or may not establish ATP release as its functional role, significant Panx2 expression levels in the CNS and its concentration in the membrane of intracellular vesicles might suggest that latent Panx2 channels may require an external stimulus or insult to the cell that causes transport to the cell surface in fulfillment of its functional role.

ACKNOWLEDGMENTS

We thank Gerhard Dahl for the gift of his chicken polyclonal 4515 anti-Panx1 antibody and Masako Terada for her expert assistance in electron tomography data acquisition, processing and visualization. This work was started under NSF grant MCB-0543934 and completed with support from NIH grants GM072881 and GM065937 (Gina E. Sosinsky), GM086197 (Roger Y. Tsien) and AHA Grant 10SDG2610281 (Daniela Boassa). The work presented here was conducted at the National Center for Microscopy and Imaging Research at San Diego, which is supported by NIH Grant GM103412 awarded to Dr. Mark Ellisman.

SUPPLEMENTARY MATERIAL

Supplemental Movie | Animation showing sequences of (1) one tilt series used for the triple tilt series tomography, (2) X-Y slices through the triple tilt tomogram and (3) automated segmentation of six vesicles (*gold = membrane; blue is proteinaceous material extending from the membrane into the cytoplasm*). The small black dots at the top and bottom of the tomogram and in the tilt series are gold beads deposited on the top and bottom of the section that are used for fiducial alignment, beam and section distortion corrections in the 3D reconstruction process (see Materials and Methods). Data, metadata, segmentation and the full resolution movie for this tomogram can be downloaded from the Cell Centered Database (CCDB MPID:48720 PID:2010) as noted in the Materials and Methods section.

REFERENCES

Ambrosi, C., Gassmann, O., Pranskevich, J. N., Boassa, D., Smock, A., Wang, J., et al. (2010). Pannexin1 and Pannexin2 channels show quaternary similarities to connexons and different oligomerization numbers from each other. *J. Biol. Chem.* 285, 24420–24431. doi: 10.1074/jbc.M110.115444

Baranova, A., Ivanov, D., Petrash, N., Pestova, A., Skoblov, M., Kelmanson, I., et al. (2004). The mammalian pannexin family is homologous to the invertebrate innexin gap junction proteins. *Genomics* 83, 706–716. doi: 10.1016/j.ygeno.2003.09.025

Bargiotas, P., Krenz, A., Hormuzdi, S. G., Ridder, D. A., Herb, A., Barakat, W., et al. (2011). Pannexins in ischemia-induced neurodegeneration. *Proc. Natl. Acad. Sci. U.S.A.* 108, 20772–20777. doi: 10.1073/pnas.1018262108

Boassa, D., Ambrosi, C., Qiu, F., Dahl, G., Gaietta, G., and Sosinsky, G. (2007). Pannexin1 channels contain a glycosylation site that targets the hexamer to the plasma membrane. *J. Biol. Chem.* 282, 31733–31743. doi: 10.1074/jbc.M702422200

Boassa, D., Berlanga, M. L., Yang, M. A., Terada, M., Hu, J., Bushong, E. A., et al. (2013). Mapping the subcellular distribution of alpha-synuclein in neurons using genetically encoded probes for correlated light and electron microscopy: implications for Parkinson's disease pathogenesis. *J. Neurosci.* 33, 2605–2615. doi: 10.1523/JNEUROSCI.2898-12.2013

Boassa, D., Qiu, F., Dahl, G., and Sosinsky, G. (2008). Trafficking dynamics of glycosylated pannexin 1 proteins. *Cell Commun. Adhes.* 15, 119–132. doi: 10.1080/15419060802013885

Boassa, D., Solan, J. L., Papas, A., Thornton, P., Lampe, P. D., and Sosinsky, G. E. (2010). Trafficking and recycling of the connexin43 gap junction protein during mitosis. *Traffic* 11, 1471–1486. doi: 10.1111/j.1600-0854.2010.01109.x

Boehm, M., and Bonifacino, J. S. (2002). Genetic analyses of adaptin function from yeast to mammals. *Gene* 286, 175–186. doi: 10.1016/S0378-1119(02)00422-5

Bolte, S., and Cordelieres, F. P. (2006). A guided tour into subcellular colocalization analysis in light microscopy. *J. Microsc.* 224, 213–232. doi: 10.1111/j.1365-2818.2006.01706.x

Boyce, A. K., Prager, R. T., Wicki-Stordeur, L. E., and Swayne, L. A. (2014). Pore positioning: current concepts in Pannexin channel trafficking. *Channels (Austin).* 8, 110–117. doi: 10.4161/chan.27287

Bruzzone, R., Hormuzdi, S. G., Barbe, M. T., Herb, A., and Monyer, H. (2003). Pannexins, a family of gap junction proteins expressed in brain. *Proc. Natl. Acad. Sci. U.S.A.* 100, 13644–13649. doi: 10.1073/pnas.2233464100

Burns, A. R., Phillips, S. C., and Sokoya, E. M. (2012). Pannexin protein expression in the rat middle cerebral artery. *J. Vasc. Res.* 49, 101–110. doi: 10.1159/000332329

Bushong, E. A., Martone, M. E., Jones, Y. Z., and Ellisman, M. H. (2002). Protoplasmic astrocytes in CA1 stratum radiatum occupy separate anatomical domains. *J. Neurosci.* 22, 183–192.

Chekeni, F. B., Elliott, M. R., Sandilos, J. K., Walk, S. F., Kinchen, J. M., Lazarowski, E. R., et al. (2010). Pannexin 1 channels mediate 'find-me' signal release and membrane permeability during apoptosis. *Nature* 467, 863–867. doi: 10.1038/nature09413

Cone, A. C., Ambrosi, C., Scemes, E., Martone, M. E., and Sosinsky, G. E. (2013). A comparative antibody analysis of pannexin1 expression in four rat brain regions reveals varying subcellular localizations. *Front. Pharmacol.* 4:6. doi: 10.3389/fphar.2013.00006

Dell'Angelica, E. C., Ohno, H., Ooi, C. E., Rabinovich, E., Roche, K. W., and Bonifacino, J. S. (1997). AP-3: an adaptor-like protein complex with ubiquitous expression. *EMBO J.* 16, 917–928. doi: 10.1093/emboj/16.5.917

De Matteis, M. A., and Luini, A. (2008). Exiting the Golgi complex. *Nat. Rev. Mol. Cell Biol.* 9, 273–284. doi: 10.1038/nrm2378

D'Hondt, C., Ponsaerts, R., De Smedt, H., Vinken, M., De Vuyst, E., De Bock, M., et al. (2011). Pannexin channels in ATP release and beyond: an unexpected rendezvous at the endoplasmic reticulum. *Cell. Signal.* 23, 305–316. doi: 10.1016/j.cellsig.2010.07.018

Dolmatova, E., Baum, J., Boassa, D., Sosinsky, G., and Duffy, H. (2010). Discovery of a novel cardiac myocyte complex involved In cell-cell signaling following myocardial infarction: the role of pannexin1 channels. *Am. J. Physiol. Heart Circ. Physiol.* 303, H1208–H1218. doi: 10.1016/j.hrthm.2010.09.043

Dourado, M., Wong, E., and Hackos, D. H. (2014). Pannexin-1 is blocked by its C-terminus through a delocalized non-specific interaction surface. *PLoS ONE* 9:e99596. doi: 10.1371/journal.pone.0099596

Gaete, P. S., Lillo, M. A., and Figueroa, X. F. (2014). Functional role of connexins and pannexins in the interaction between vascular and nervous system. *J. Cell. Physiol.* 229, 1336–1345. doi: 10.1002/jcp.24563

Gaietta, G., Deerinck, T. J., Adams, S. R., Bouwer, J., Tour, O., Laird, D. W., et al. (2002). Multicolor and electron microscopic imaging of connexin trafficking. *Science* 296, 503–507. doi: 10.1126/science.1068793

Gautreau, A., Oguievetskaia, K., and Ungermann, C. (2014). Function and regulation of the endosomal fusion and fission machineries. *Cold Spring Harb. Perspect. Biol.* 6:a016832. doi: 10.1101/cshperspect.a016832

Gould, G. W., and Lippincott-Schwartz, J. (2009). New roles for endosomes: from vesicular carriers to multi-purpose platforms. *Nat. Rev. Mol. Cell Biol.* 10, 287–292. doi: 10.1038/nrm2652

Hansen, D. B., Ye, Z. C., Calloe, K., Braunstein, T. H., Hofgaard, J. P., Ransom, B. R., et al. (2014). Activation, permeability, and inhibition of astrocytic and neuronal large pore (hemi)channels. *J. Biol. Chem.* 289, 26058–26073. doi: 10.1074/jbc.M114.582155

Hu, J., Jiang, C. C., Ng, H. K., Pang, J. C., Tong, C. Y., and Chen, S. Q. (2004). A genome-wide allelotype study of primary and corresponding recurrent glioblastoma multiforme in one patient. *Chin. Med. J.* 117, 456–458.

Huang, Y., Grinspan, J. B., Abrams, C. K., and Scherer, S. S. (2007). Pannexin1 is expressed by neurons and glia but does not form functional gap junctions. *Glia* 55, 46–56. doi: 10.1002/glia.20435

Ino, Y., Silver, J. S., Blazejewski, L., Nishikawa, R., Matsutani, M., Von Deimling, A., et al. (1999). Common regions of deletion on chromosome 22q12.3-q13.1 and 22q13.2 in human astrocytomas appear related to malignancy grade. *J. Neuropathol. Exp. Neurol.* 58, 881–885. doi: 10.1097/00005072-199908000-00010

Iwamoto, T., Nakamura, T., Doyle, A., Ishikawa, M., De Vega, S., Fukumoto, S., et al. (2010). Pannexin 3 regulates intracellular ATP/cAMP levels and promotes chondrocyte differentiation. *J. Biol. Chem.* 285, 18948–18958. doi: 10.1074/jbc.M110.127027

Jovic, M., Sharma, M., Rahajeng, J., and Caplan, S. (2010). The early endosome: a busy sorting station for proteins at the crossroads. *Histol. Histopathol.* 25, 99–112.

Kanneganti, T. D., Lamkanfi, M., Kim, Y. G., Chen, G., Park, J. H., Franchi, L., et al. (2007). Pannexin-1-mediated recognition of bacterial molecules activates the cryopyrin inflammasome independent of Toll-like receptor signaling. *Immunity* 26, 433–443. doi: 10.1016/j.immuni.2007.03.008

Kremer, J. R., Mastronarde, D. N., and McIntosh, J. R. (1996). Computer visualization of three-dimensional image data using IMOD. *J. Struct. Biol.* 116, 71–76. doi: 10.1006/jsbi.1996.0013

Lai, C. P., Bechberger, J. F., and Naus, C. C. (2009). Pannexin2 as a novel growth regulator in C6 glioma cells. *Oncogene* 28, 4402–4408. doi: 10.1038/onc.2009.283

Langlois, S., Xiang, X., Young, K., Cowan, B. J., Penuela, S., and Cowan, K. N. (2014). Pannexin 1 and pannexin 3 channels regulate skeletal muscle myoblast proliferation and differentiation. *J. Biol. Chem.* 289, 30717–30731. doi: 10.1074/jbc.M114.572131

Lawrence, A., Bouwer, J. C., Perkins, G., and Ellisman, M. H. (2006). Transform-based backprojection for volume reconstruction of large format electron microscope tilt series. *J. Struct. Biol.* 154, 144–167. doi: 10.1016/j.jsb.2005.12.012

Le Vasseur, M., Lelowski, J., Bechberger, J. F., Sin, W. C., and Naus, C. C. (2014). Pannexin 2 protein expression is not restricted to the CNS. *Front. Cell. Neurosci.* 8:392. doi: 10.3389/fncel.2014.00392

Litvin, O., Tiunova, A., Connell-Alberts, Y., Panchin, Y., and Baranova, A. (2006). What is hidden in the pannexin treasure trove: the sneak peek and the guesswork. *J. Cell. Mol. Med.* 10, 613–634. doi: 10.1111/j.1582-4934.2006.tb00424.x

Locovei, S., Bao, L., and Dahl, G. (2006a). Pannexin 1 in erythrocytes: function without a gap. *Proc. Natl. Acad. Sci. U.S.A.* 103, 7655–7659. doi: 10.1073/pnas.0601037103

Locovei, S., Wang, J., and Dahl, G. (2006b). Activation of pannexin 1 channels by ATP through P2Y receptors and by cytoplasmic calcium. *FEBS Lett.* 580, 239–244. doi: 10.1016/j.febslet.2005.12.004

Maeda, S., Nakagawa, S., Suga, M., Yamashita, E., Oshima, A., Fujiyoshi, Y., et al. (2009). Structure of the connexin 26 gap junction channel at 3.5 A resolution. *Nature* 458, 597–602. doi: 10.1038/nature07869

Martin, B. R., Giepmans, B. N., Adams, S. R., and Tsien, R. Y. (2005). Mammalian cell-based optimization of the biarsenical-binding tetracysteine motif for improved fluorescence and affinity. *Nat. Biotechnol.* 23, 1308–1314. doi: 10.1038/nbt1136

Maslieieva, V., and Thompson, R. J. (2014). A critical role for pannexin-1 in activation of innate immune cells of the choroid plexus. *Channels (Austin).* 8, 131–141. doi: 10.4161/chan.27653

Mellman, I. (1996). Endocytosis and molecular sorting. *Annu. Rev. Cell Dev. Biol.* 12, 575–625. doi: 10.1146/annurev.cellbio.12.1.575

Mohrmann, K., Gerez, L., Oorschot, V., Klumperman, J., and Van Der Sluijs, P. (2002). Rab4 function in membrane recycling from early endosomes depends on a membrane to cytoplasm cycle. *J. Biol. Chem.* 277, 32029–32035. doi: 10.1074/jbc.M203064200

Mylvaganam, S., Zhang, L., Wu, C., Zhang, Z. J., Samoilova, M., Eubanks, J., et al. (2010). Hippocampal seizures alter the expression of the pannexin and connexin transcriptome. *J. Neurochem.* 112, 92–102. doi: 10.1111/j.1471-4159.2009.06431.x

Ohashi, E., Tanabe, K., Henmi, Y., Mesaki, K., Kobayashi, Y., and Takei, K. (2011). Receptor sorting within endosomal trafficking pathway is facilitated by dynamic actin filaments. *PLoS ONE* 6:e19942. doi: 10.1371/journal.pone.0019942

Omasits, U., Ahrens, C. H., Muller, S., and Wollscheid, B. (2014). Protter: interactive protein feature visualization and integration with experimental proteomic data. *Bioinformatics* 30, 884–886. doi: 10.1093/bioinformatics/btt607

Oshima, A., Tani, K., Toloue, M. M., Hiroaki, Y., Smock, A., Inukai, S., et al. (2011). Asymmetric configurations and N-terminal rearrangements in connexin26 gap junction channels. *J. Mol. Biol.* 405, 724–735. doi: 10.1016/j.jmb.2010.10.032

Oskam, N. T., Bijleveld, E. H., and Hulsebos, T. J. (2000). A region of common deletion in 22q13.3 in human glioma associated with astrocytoma progression. *Int. J. Cancer* 85, 336–339. doi: 10.1002/(SICI)1097-0215(20000201)85:3<336::AID-IJC7>3.0.CO;2-9

Panchin, Y., Kelmanson, I., Matz, M., Lukyanov, K., Usman, N., and Lukyanov, S. (2000). A ubiquitous family of putative gap junction molecules. *Curr. Biol.* 10, R473–R474. doi: 10.1016/S0960-9822(00)00576-5

Pelegrin, P., and Surprenant, A. (2006). Pannexin-1 mediates large pore formation and interleukin-1beta release by the ATP-gated P2X7 receptor. *EMBO J.* 25, 5071–5082. doi: 10.1038/sj.emboj.7601378

Penuela, S., Bhalla, R., Gong, X. Q., Cowan, K. N., Celetti, S. J., Cowan, B. J., et al. (2007). Pannexin 1 and pannexin 3 are glycoproteins that exhibit many distinct characteristics from the connexin family of gap junction proteins. *J. Cell Sci.* 120, 3772–3783. doi: 10.1242/jcs.009514

Penuela, S., Bhalla, R., Nag, K., and Laird, D. W. (2009). Glycosylation regulates pannexin intermixing and cellular localization. *Mol. Biol. Cell* 20, 4313–4323. doi: 10.1091/mbc.E09-01-0067

Penuela, S., Gehi, R., and Laird, D. W. (2013). The biochemistry and function of pannexin channels. *Biochim. Biophys. Acta* 1828, 15–22. doi: 10.1016/j.bbamem.2012.01.017

Penuela, S., Lohman, A. W., Lai, W., Gyenis, L., Litchfield, D. W., Isakson, B. E., et al. (2014). Diverse post-translational modifications of the pannexin family of channel-forming proteins. *Channels (Austin).* 8, 124–130. doi: 10.4161/chan.27422

Ray, A., Zoidl, G., Wahle, P., and Dermietzel, R. (2006). Pannexin expression in the cerebellum. *Cerebellum* 5, 189–192. doi: 10.1080/14734220500530082

Schroder, K., Zhou, R., and Tschopp, J. (2010). The NLRP3 inflammasome: a sensor for metabolic danger? *Science* 327, 296–300. doi: 10.1126/science.1184003

Shu, X., Lev-Ram, V., Deerinck, T. J., Qi, Y., Ramko, E. B., Davidson, M. W., et al. (2011). A genetically encoded tag for correlated light and electron microscopy of intact cells, tissues, and organisms. *PLoS Biol.* 9:e1001041. doi: 10.1371/journal.pbio.1001041

Silverman, W., Locovei, S., and Dahl, G. (2008). Probenecid, a gout remedy, inhibits pannexin 1 channels. *Am. J. Physiol. Cell Physiol.* 295, C761–C767. doi: 10.1152/ajpcell.00227.2008

Swayne, L. A., Sorbara, C. D., and Bennett, S. A. (2010). Pannexin 2 is expressed by postnatal hippocampal neural progenitors and modulates neuronal commitment. *J. Biol. Chem.* 285, 24977–24986. doi: 10.1074/jbc.M110.130054

Tang, G., Peng, L., Baldwin, P. R., Mann, D. S., Jiang, W., Rees, I., et al. (2007). EMAN2: an extensible image processing suite for electron microscopy. *J. Struct. Biol.* 157, 38–46. doi: 10.1016/j.jsb.2006.05.009

Thompson, R. J. (2014). Pannexin channels and ischaemia. *J. Physiol.* doi: 10.1113/jphysiol.2014.282426. [Epub ahead of print].

Thompson, R. J., Jackson, M. F., Olah, M. E., Rungta, R. L., Hines, D. J., Beazely, M. A., et al. (2008). Activation of pannexin-1 hemichannels augments aberrant bursting in the hippocampus. *Science* 322, 1555–1559. doi: 10.1126/science.1165209

Vanden Abeele, F., Bidaux, G., Gordienko, D., Beck, B., Panchin, Y. V., Baranova, A. V., et al. (2006). Functional implications of calcium permeability of the channel formed by pannexin 1. *J. Cell Biol.* 174, 535–546. doi: 10.1083/jcb.200601115

Vogt, A., Hormuzdi, S. G., and Monyer, H. (2005). Pannexin1 and Pannexin2 expression in the developing and mature rat brain. *Brain Res. Mol. Brain Res.* 141, 113–120. doi: 10.1016/j.molbrainres.2005.08.002

Wang, J., Ambrosi, C., Qiu, F., Jackson, D. G., Sosinsky, G., and Dahl, G. (2014). The membrane protein Pannexin1 forms two open-channel conformations depending on the mode of activation. *Sci. Signal.* 7, ra69. doi: 10.1126/scisignal.2005431

Wang, X. H., Streeter, M., Liu, Y. P., and Zhao, H. B. (2009). Identification and characterization of pannexin expression in the mammalian cochlea. *J. Comp. Neurol.* 512, 336–346. doi: 10.1002/cne.21898

Weickert, S., Ray, A., Zoidl, G., and Dermietzel, R. (2005). Expression of neural connexins and pannexin1 in the hippocampus and inferior olive: a quantitative approach. *Brain Res. Mol. Brain Res.* 133, 102–109. doi: 10.1016/j.molbrainres.2004.09.026

Wicki-Stordeur, L. E., Boyce, A. K., and Swayne, L. A. (2013). Analysis of a pannexin 2-pannexin 1 chimeric protein supports divergent roles for pannexin C-termini in cellular localization. *Cell Commun. Adhes.* 20, 73–79. doi: 10.3109/15419061.2013.791681

Zappala, A., Li Volti, G., Serapide, M. F., Pellitteri, R., Falchi, M., La Delia, F., et al. (2007). Expression of pannexin2 protein in healthy and ischemized brain of adult rats. *Neuroscience* 148, 653–667. doi: 10.1016/j.neuroscience.2007.06.028

2

Pannexin 1 regulates bidirectional hippocampal synaptic plasticity in adult mice

Alvaro O. Ardiles[1]*, Carolina Flores-Muñoz[1], Gabriela Toro-Ayala[1], Ana M. Cárdenas[1], Adrian G. Palacios[1], Pablo Muñoz[1,2], Marco Fuenzalida[3], Juan C. Sáez[1,4] and Agustín D. Martínez[1]*

[1] Centro Interdisciplinario de Neurociencia de Valparaíso, Facultad de Ciencias, Instituto de Neurociencia, Universidad de Valparaíso, Valparaíso, Chile
[2] Escuela de Medicina, Facultad de Medicina, Universidad de Valparaíso, Valparaíso, Chile
[3] Centro de Neurobiología y Plasticidad Cerebral, Instituto de Fisiología, Universidad de Valparaíso, Valparaíso, Chile
[4] Departamento de Fisiología, Pontificia Universidad Católica de Chile, Santiago, Chile

***Correspondence:**
Alvaro O. Ardiles and Agustín D. Martínez, Centro Interdisciplinario de Neurociencia de Valparaíso, Facultad de Ciencias, Instituto de Neurociencia, Universidad de Valparaíso, Avenida Gran Bretaña 1111, Playa Ancha, Valparaíso, 2360102, Chile
e-mail: alvaro.ardiles@cinv.cl; agustin.martinez@uv.cl

The threshold for bidirectional modification of synaptic plasticity is known to be controlled by several factors, including the balance between protein phosphorylation and dephosphorylation, postsynaptic free Ca^{2+} concentration and NMDA receptor (NMDAR) composition of GluN2 subunits. Pannexin 1 (Panx1), a member of the integral membrane protein family, has been shown to form non-selective channels and to regulate the induction of synaptic plasticity as well as hippocampal-dependent learning. Although Panx1 channels have been suggested to play a role in excitatory long-term potentiation (LTP), it remains unknown whether these channels also modulate long-term depression (LTD) or the balance between both types of synaptic plasticity. To study how Panx1 contributes to excitatory synaptic efficacy, we examined the age-dependent effects of eliminating or blocking Panx1 channels on excitatory synaptic plasticity within the CA1 region of the mouse hippocampus. By using different protocols to induce bidirectional synaptic plasticity, Panx1 channel blockade or lack of Panx1 were found to enhance LTP, whereas both conditions precluded the induction of LTD in adults, but not in young animals. These findings suggest that Panx1 channels restrain the sliding threshold for the induction of synaptic plasticity and underlying brain mechanisms of learning and memory.

Keywords: pannexin 1, hippocampus, LTD, LTP, synaptic plasticity, NMDA receptors, mice

INTRODUCTION

Pannexins constitute a family of integral membrane proteins with moderately homologous sequences, but with significant similarity in transmembrane topology when compared to connexins and innexins, which are the classic gap junction channel proteins in vertebrates and invertebrates, respectively (Panchin et al., 2000). The following three members compose the Pannexin family: pannexin 3, which is ubiquitously expressed in many vertebrate tissues, pannexin 1 (Panx1) and pannexin 2, which are abundantly expressed in the brain (Bruzzone et al., 2003). Panx1 has been the most widely studied Pannexin family member thus far and, although originally identified as a gap junction-forming protein, it has been consistently shown to form functional non-junction channels that mediate the exchange of molecules between the cytoplasm and the extracellular space (MacVicar and Thompson, 2010). In fact, Panx1 forms a non-selective channel with high conductance (Bao et al., 2004), through which charged molecules such as ATP have been observed to pass (Bao et al., 2004; Locovei et al., 2006). Nevertheless, permeability to ATP remains controversial (Romanov et al., 2012). It is interesting to note, however, that Panx1 interacts with purinergic receptors mediating the release of ATP towards the extracellular space (Pelegrin and Surprenant, 2006; Locovei et al., 2007). In this regard, mounting evidence has indicated that ATP and purinergic signaling are involved in diverse functions in the central nervous system (CNS), such as neurotransmission, neuromodulation and inflammation (Fields and Stevens, 2000). In this sense, Panx1 channels could also be important contributors to such processes (Silverman et al., 2009; Prochnow et al., 2012).

In the CNS, Panx1 mRNA is expressed in a development-dependent manner in the cerebellum, cerebral cortex, olfactory bulb and hippocampus, exhibiting high levels in embryonic and young systems, and declining in adult tissue (Vogt et al., 2005). At the cellular level, Panx1 channels have been identified in astrocytes, pyramidal cells and interneurons (Vogt et al., 2005; Huang et al., 2007; Zoidl et al., 2007). In neurons, Panx1 channels have been detected in postsynaptic densities, where they co-localize with glutamate receptors and scaffold proteins such as postsynaptic density protein 95 (PSD-95; Zoidl et al., 2007). However, the functional significance of the postsynaptic localization of Panx1 channels is not completely understood. The absence of Panx1 was recently shown to increase long-term potentiation (LTP) along with behavioral alterations, including increases in anxiety as well as spatial and recognition impairments, suggesting that Panx1 modulates synaptic plasticity and is needed for proper learning (Prochnow et al., 2012). In the present report, we extended these studies to other forms of synaptic plasticity. We found that blockade of Panx1 channels or lack of Panx1 causes a shift from long-term depression (LTD) to LTP. These results suggest that Panx1 channels could regulate the sliding threshold for excitatory synaptic plasticity, hence increasing the likelihood of synaptic strengthening elicited by conditioning stimulation.

MATERIALS AND METHODS

ANIMALS

The experiments were carried out in 1- to 12 month-old male C57BL/6 or Panx1$^{-/-}$ mice. The generation of Panx1$^{-/-}$ (KO) mice has been described previously (Anselmi et al., 2008). Mice were housed at 21 ± 1°C at constant humidity (55%) and in a 12/12 h dark–light cycle, with a light phase from 08:00 AM to 08:00 PM. Food and water were provided *ad libitum*. The use and care of the animals were approved by the Ethics and Animal Care Committee of Universidad de Valparaíso.

ELECTROPHYSIOLOGY

Hippocampal slices were prepared as previously reported (Ardiles et al., 2012). Animals were sacrificed under deep halothane anesthesia, and their brains were quickly removed. The hippocampus was removed and sectioned into 300–400 μm thick slices by using a vibratome (Vibratome 1000 plus, Ted Pella Inc., CA, USA). The slices were transferred and maintained for 1 h at room temperature in normal artificial cerebrospinal fluid (ACSF). Normal ACSF was similar to the dissection buffer, except that sucrose was replaced by 124 mM NaCl, $MgCl_2$ was lowered to 1 mM, and $CaCl_2$ was raised to 2 mM. All recordings were collected in a submersion-recording chamber perfused with ACSF (30 ± 0.5°C; 2 ml/min). Field excitatory postsynaptic potentials (FPs) were evoked by stimulating Schaffer collaterals with 0.2 ms pulses delivered through concentric bipolar stimulating electrodes (FHC), and were recorded extracellularly in CA1 stratum radiatum. Baseline responses were recorded by using half-maximum stimulation intensity at 0.033 Hz. Basal synaptic transmission was assayed by determining input–output relationships from FPs generated by gradually increasing the stimulus intensity; the input was the peak amplitude of the fiber volley (FV), and the output was the initial slope of FP. Paired-pulse facilitation was elicited by using an interstimulus interval range between 50–500 ms. Long-term potentiation was induced by high frequency stimulation (HFS; Tang et al., 1999) consisting of a single tetanus of 100 Hz or by theta burst stimulation (TBS; Ardiles et al., 2012), consisting of four theta epochs composed by 10 trains of four pulses (at 100 Hz) delivered at 5 Hz. Long-term depression was induced by low frequency stimulation (LFS; Ardiles et al., 2012) or by paired-pulse low frequency stimulation (PP-LFS; Ardiles et al., 2012), both consisting of 900 pulses at 1 Hz. Long-term depression also was induced by 900 pulses at 5 and 10 Hz (Cui et al., 2013). These protocols were delivered following 20 min of baseline transmission. For chemical-LTD, slices were superfused for 5 min with 30 μM NMDA. Panx1 channels were blocked with 200 μM probenecid. Glial glutamate transporter was blocked with 10 μM DL-TBOA. D-AP5, NMDA and DL-TBOA were obtained from Tocris (Bristol, UK). Probenecid was obtained from Molecular Probes (Eugene, OR, USA).

IMMUNOBLOTTING

After electrophysiological recordings, hippocampal slices were quickly frozen and processed for Western blotting as previously described (Ardiles et al., 2012). Samples were homogenized in ice-cold lysis buffer (150 mM NaCl, 10 mM Tris-Cl, pH 7.4, EDTA 2 mM, 1% Triton X-100 and 0.1% SDS), supplemented with a protease and phosphatase inhibitor cocktail (Thermo Scientific, Rockford, IL, USA) by using a homogenizator. Protein samples were centrifuged twice for 5 min at 14,000 revolutions per minute (rpm) (4°C). Protein concentration was determined with the BCA Protein Assay Kit (Thermo Scientific, Rockford, IL, USA). For synaptic proteins, 40 μg of protein per lane were resolved by 10% SDS-PAGE, followed by immunoblotting on PVDF membranes (Merck Millipore, Merck KGaA, Darmstadt, Germany) with mouse anti-β-actin (Santa Cruz Biotechnology, Dallas, TX, USA) and rabbit anti-Panx1 (CT-395; Penuela et al., 2007). Band intensities were visualized by enhanced chemiluminiscence kit (ECL, Pierce, Thermo Scientific, Rockford, IL, USA) and the intensity of each band was scanned and densitometrically quantified through the use of Image J software (NIH, Bethesda, MD, USA).

BIOTINYLATION

Surface biotinylation was performed in acute hippocampal slices as previously reported with minor modifications (Lee et al., 2003; Thomas-Crusells et al., 2003). Slices were briefly preincubated in ACSF at 30°C for 1 h, washed twice with ice-cold ACSF and then incubated with sulfo-NHS-SS-Biotin (Thermo Scientific; 1 mg/ml in ACSF) for 45 min on ice with gentle rotation. Excess biotin was removed by means of two brief washes with 10 mM lysine (in ACSF) and two ACSF washes. Slices were then lysed in 500 μl of lysis buffer, centrifuged at 14,000 rpm for 5 min at 4°C and supernatants were discarded. Pellets were resuspended in lysis buffer and biotinylated cell-surface proteins were precipitated with high capacity neutravidin agarose resin (Thermo Scientific, Rockford, IL, USA) and the mixture was rotated overnight at 4°C. After several washes with lysis buffer, precipitates were collected by centrifugation (14,000 rpm for 1 min) and detected by immunoblot.

QUANTITATIVE REVERSE TRANSCRIPTION POLYMERASE CHAIN REACTION

Total RNA was extracted from brain tissue by using the SV Total RNA Isolation System (Promega Corp., Madison, WI, USA) according to the manufacturers' instructions. RNA concentration was determined spectrophotometrically at 260 nm. One microgram of RNA was incubated with 0.5 μg of oligo-dT primer and reverse transcriptase (ImProm-II Reverse Transcription; Promega), according to the manufacturer's specifications.

Levels of Panx1 RNA were analyzed by qRT-PCR using the following primers: Forward 5′-GCCAGAGAGTGGAGTTCAAAGA-3′; Reverse 5′-CATTAGCAGGACGGATTCAGAA-3′. qRT-PCRs were performed in a total volume of 20 μl consisting of 100 ng of cDNA, 10 μl of Kapa Sybr®Fast Master mix (Kapa Biosystems) and 0.4 μl of each primer under the following cycling conditions: uracil DNA glycosylase (UDG) activation at 50°C for 2 min, polymerase activation at 95°C for 5 min, followed by 40 cycles of denaturation at 95°C for 10 s, annealing at 55°C for 30 s, extension at 72°C for 30 s, followed by a melt curve of 95°C for 15 s, 55°C for 15 s and 95°C for 15 s. Equal loading of cDNA was confirmed by amplification of the cyclophilin with the following primers: Forward 5′- AGGTCCTGGCATCTTGTCCAT-3′ and Reverse 5′-GAACCGTTTGTGTTTGGTCCA-3′. The purity of the amplified products was checked with a dissociation curve.

STATISTICS

All data are presented as mean ± standard error of the mean (SEM). Data analysis was carried out using the Prism software (GraphPad Software Inc.). Statistical comparisons were performed by means of a two-tailed t-test and one-way ANOVA, followed by a Tukey's or repeated measures/two-way ANOVA and subsequently by Bonferroni's *post hoc* tests.

RESULTS

PANX1 RNA AND PROTEIN LEVELS DECREASE IN THE BRAIN WITH AGE

Early expression analyses have revealed a widespread distribution of Panx1 mRNA in many areas of the CNS, including the olfactory bulb, cerebral cortex, hippocampus and cerebellum (Bruzzone et al., 2003), exhibiting changes that occur in a time-dependent manner that is, Panx1 mRNA is expressed at high levels in embryonic and young postnatal whole brain, but declines during adulthood (Vogt et al., 2005). Accordingly, quantitative-RT-PCR analyses of samples from different brain areas revealed a decrease in Panx1 mRNA levels for adult (9–12 months) wild type (WT) mice compared to young (1 month) WT mice (**Figure 1A**). Consistent with this, we detected almost similar levels of reduction in total Panx1 protein levels in the adult cerebral cortex, hippocampus and cerebellum compared to Panx1 levels found in the same brain areas of young WT animals (**Figure 1C**). Because Panx1 channels are localized mainly in the plasma membrane where they exert their main functions, we decided to investigate whether the reduction in total Panx1 protein levels in WT adult mice would also result in similar levels of reduction in the membrane compartment. We observed even larger levels of reduction in protein Panx1 levels in the plasma membrane of adult hippocampus tissue compared to young (about 60% of reduction; **Figures 1D,E**). As a control for the process of biotinylation of plasma membrane proteins we analyzed the levels of biotinylation of Connexin43 (Cx43), which is another membrane protein that form hemichannels mainly in the astroglial cells (Dermietzel et al., 1989). We observed that Cx43 levels were slightly increased during adulthood (**Figures 1D,E**), consistent with the importance of this proteins in astroglial function. These results suggest that reduction in the plasma membrane levels of

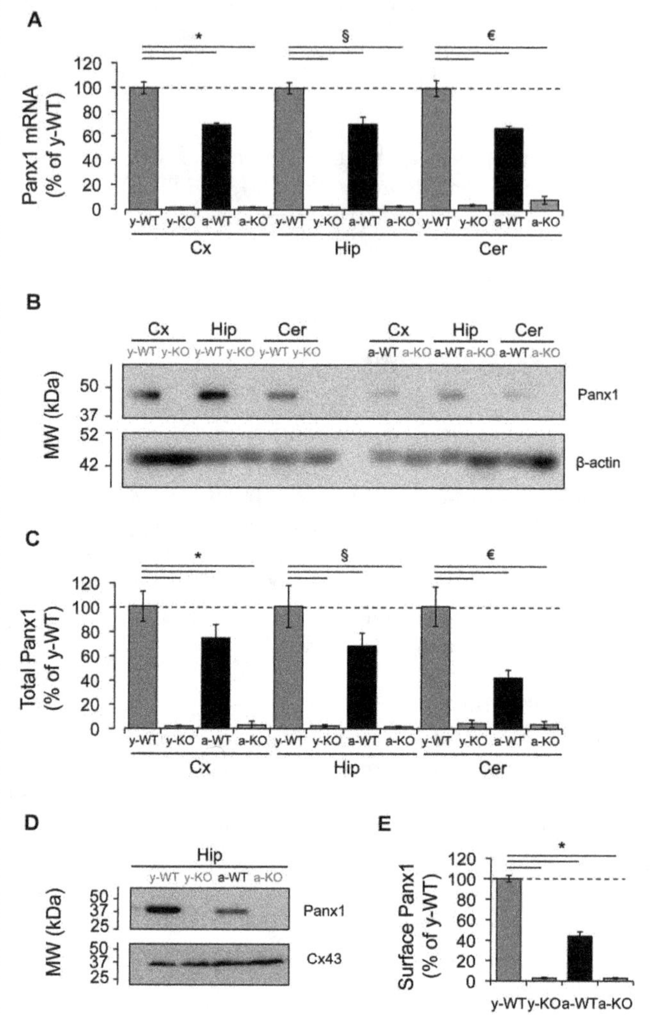

FIGURE 1 | Decreased expression of Panx1 transcript and protein in the adult mice brain. (A) Quantitative RT-PCR analysis of relative abundance of Panx1 mRNA in the cerebral cortex (Cx), hippocampus (Hip), and cerebellum (Cer) from young wild type (y-WT, gray), young Panx1 knockout (y-KO, blue), adult wild type (a-WT, black) and adult Panx1 knockout mice (a-KO, green). The transcript values were normalized to the levels of cyclophilin (Cyp1) ($N = 4$). **(B)** Representative Western blot showing the expression levels of Panx1 protein in homogenates of cerebral cortex (Cx), hippocampus (Hip) and cerebellum (Cer) of young or adult animals. β-actin expression in samples was used for loading control. **(C)** Quantification of Panx1 protein expression by densitometry analysis of bands from four independent Western blots ($N = 4$), including the one shown in **(B)**. Values were normalized to β-actin loading control. **(D)** Western blots of plasma membrane biotinylated proteins of hippocampal slices probed with anti-Panx1 and anti-Cx43 antibodies. **(E)** Quantification of protein expression by densitometry analysis of Panx1 bands from three independent Western blots ($N = 3$) like the one shown in **(D)**. All data are plotted as mean ± SEM related to results of y-WT animals. Statistical differences were calculated using 2 way-ANOVA, followed by *post hoc* Bonferroni's test. * $p < 0.05$ vs. y-WT Cx; § $p < 0.05$ vs. y-WT Hip; € $p < 0.05$ vs. y-WT Cer.

Panx1 was a specific process not affected by the biotinylation procedure. Finally, both Panx1 mRNA and protein were absent in Panx1$^{-/-}$ (KO) mice (**Figure 1**). These data confirm previously reported age-dependent expressions of Panx1 in the CNS and encourages us to evaluate whether Panx1 channels play a role in synaptic functions and how they may be affected by age.

PANX1 CHANNELS REGULATE EXCITATORY SYNAPTIC TRANSMISSION IN ADULT BUT NOT IN YOUNG BRAIN

There is some evidence for the participation of Panx1 channels in synaptic activity. For example, prolonged activation of Panx1 channels triggered by increased NMDAR activity leads to aberrant ionic currents and abnormal neuronal bursting, which contribute to epileptiform activity (Thompson et al., 2008). In line with the latter, lack of Panx1 improves the outcome of kainic acid-induced status epilepticus in juvenile mice (Santiago et al., 2011), whereas both Panx1 blockade and the $P2X_7$ receptor silencing increase susceptibility to pilocarpine-induced seizures in adult mice (Kim and Kang, 2011). Moreover, the absence of Panx1 leads to increased excitability in the adult hippocampus (Prochnow et al., 2012). In order to evaluate the role of Panx1 channels in synaptic transmission at different ages, we recorded synaptic activity in Schaffer collateral-CA1 synapses (Sc-CA1) evoking FP (**Figure 2A**). We noted in young animals (1 month) that FP slopes and FV amplitudes evoked by different stimulation intensities were indistinguishable between slices from young WT (y-WT) and KO (y-KO) mice, suggesting that lack of Panx1 does not affect excitability (**Figures 2A3,A4**). Consequently, input-output curves showed no difference in basal transmission between young transgenic mice and WT littermates (**Figure 2A5**). However, we found a significant increase in FP slopes in slices obtained from adult KO mice (a-KO) without significant changes in FV amplitude for high stimulation intensities compared to slices of adult WT (a-WT, **Figures 2A3**, **4**). Similar results were obtained in slices from a-WT incubated with the Panx1 channel blocker probenecid (a-WT+Pbncd) as compared to untreated slices (**Figures 2A3,A4**). Input-output curves revealed small but statistically significant differences between a-WT+Pbncd and a-KO compared to a-WT mice, showing that when the same number of axons was recruited after stimulation (same FV value) there was a slightly larger FP in slices from a-WT+Pbncd and a-KO mice (**Figure 2A5**). These results suggest that the elimination of Panx1 signaling increases the release of transmitters, or increases postsynaptic responsiveness. To explore the first possibility, we measured paired pulse facilitation (PPF) as a function of presynaptic activity and found no differences between the groups (**Figure 2B**). Although these results indicate that the absence of Panx1 does not affect the release probability, we cannot rule out that Panx1 channels contribute to presynaptic functions. In fact, it has been suggested that the absence of Panx1 could reduce ATP catabolism and increase extracellular adenosine levels, therefore increasing glutamate release (Prochnow et al., 2012). Consequently, in slices incubated with a glial glutamate transporter blocker (TBOA, 10 μM) to increase glutamate availability at synapse, we observed larger FP slopes in slices from a-KO and a-WT+Pbncd compared to slices from a-WT (**Figures 2C1,C2**). This indirectly suggests that Panx1 channels could contribute to controlling the probability of glutamate release. However, whether the blockade or absence of Panx1 affects postsynaptic mechanisms remains to be determined.

BLOCKADE OR LACK OF PANX1 CHANNELS FACILITATES LTP AND OBLITERATES LTD ONLY IN ADULT ANIMALS

It is believed that memories are stored by modifications in synaptic connections between neurons. Two types of activity-dependent synaptic modifications are LTP and LTD, which have received considerable support to be considered as cellular memory mechanisms (Bliss and Collingridge, 1993). In the hippocampus, both LTP and LTD can be elicited at the same synapses by different frequencies of stimulation, where high frequencies induce LTP and low frequencies induce LTD (Dudek and Bear, 1992, 1993). The most accepted model states that high and low frequencies of presynaptic stimulation evoke different levels of postsynaptic calcium influx through an NMDAR-dependent pathway leading to the activation of phosphatases or kinases, respectively (Lisman, 1989; Lüscher and Malenka, 2012). As suggested by Prochnow et al. (2012), Panx1 stabilizes synaptic plasticity and is needed for learning. However, this study was restricted to the role of Panx1 in LTP in adult mice, without considering the analysis of other types of synaptic plasticity. Hence, we decided to evaluate the role of Panx1 channels in bidirectional synaptic plasticity with age by using different protocols and frequency stimulations to induce LTP and LTD in the Sc-CA1 pathway.

First, we used two protocols that we have used before to reveal age-dependent changes in LTP and LTD (Ardiles et al., 2012). We observed that strong conditioning stimulation (4TBS; 4 × 100 Hz) elicited potentiation in both young and adult animals. However, significantly enhanced LTP was exhibited by a-KO compared with the other groups (**Figure 3A**). On the other hand, reliable LTD was elicited with a PP-LFS protocol (1 Hz, 15 min) in brain slices of both y- and a-WT and in brain slices from y-KO mice (**Figure 3B**). However, this stimulation resulted in transient LTD in a-WT+Pbncd that decayed back to baseline after an hour. Long-term depression however, was absolutely absent in a-KO mice, where this protocol elicited an enduring potentiation (**Figure 3B**). These data suggest that lack of Panx1 impairs NMDAR-dependent plasticity in adults but not in young animals. To further confirm the possible effects of Panx1 blockade on LTP, we used a stimulation protocol that normally induces transient and non-sustained LTP (**Figure 4**). We observed that one TBS (1 × 100 Hz; TBS1) protocol induced transient LTP in WT slices, which returned to baseline levels after 20–30 min of applying the protocol (**Figure 4A**). Interestingly, the same conditioning stimulus (TBS1) produced significantly greater potentiation levels in slices from KO mice, which remained stable after 20–30 min compared to WT slices (**Figures 4A,B**). When a second TBS (TBS2) was delivered 35 min after the application of the first TBS (TBS1), we observed an increase in the synaptic response in both WT and KO slices. However, this potentiation was maintained for a longer time only in KO slices, while it fell close to baseline levels in WT slices (**Figure 4A**). In the presence of probenecid, which was added 20 min before the application of TBS2, we observed a robust potentiation in WT slices, reaching similar levels than those observed in KO slices, which remained stable for an hour after TBS2 application (**Figure 4B**). To rule out non-specific effects of probenecid and/or the participation of other Pannexins expressed in neurons, such as Panx2, slices from KO animals were incubated with probenecid

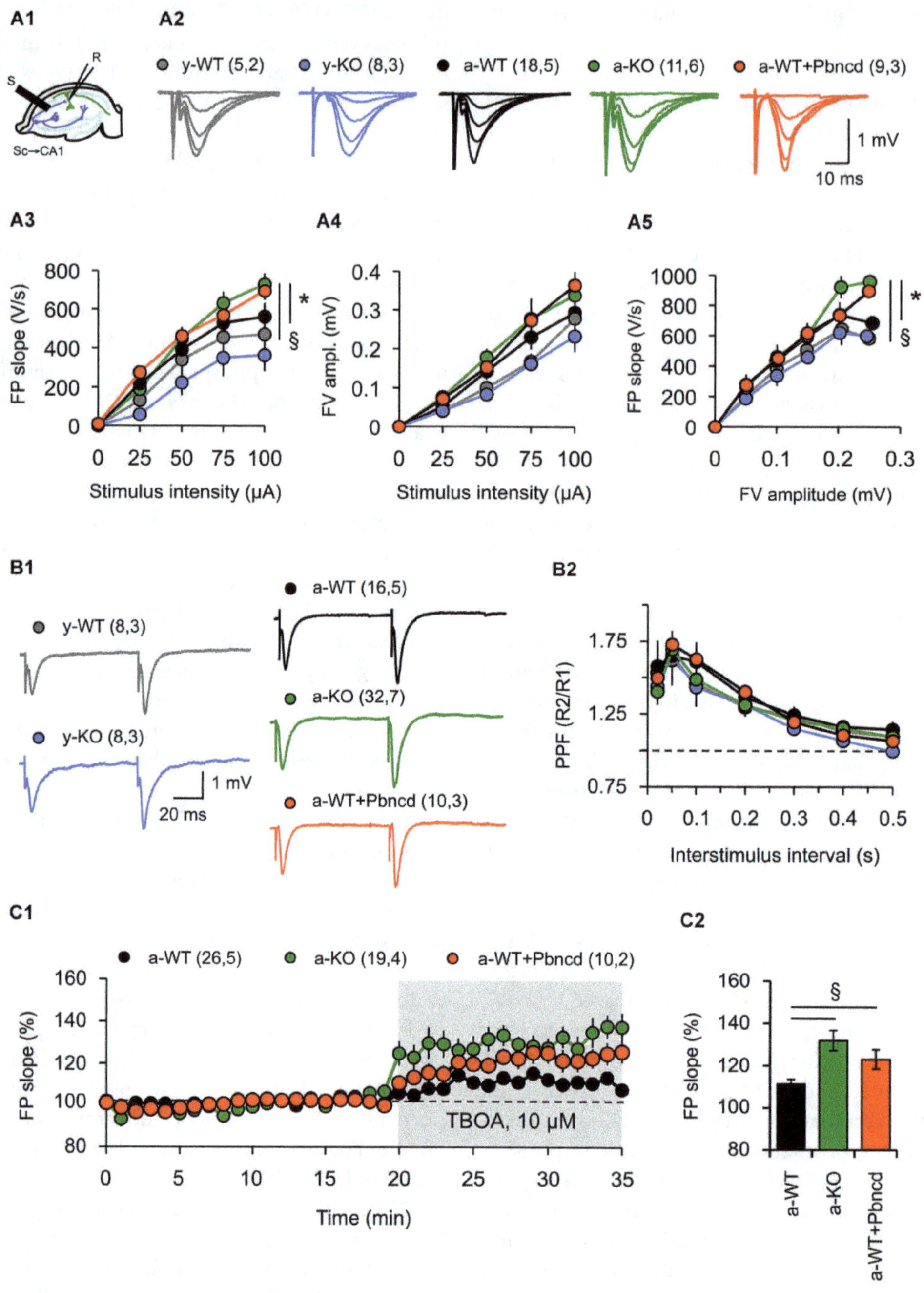

FIGURE 2 | Absence and blockade of Panx1 channels alter hippocampal synaptic transmission in adult, but not in young mice. **(A1)** Cartoon depicting stimulus-record electrode configuration to record synaptic activity in Sc-CA1 synapse in hippocampal slices. **(A2)** representative FP at different stimulus intensities for young wild type (y-WT, gray line), young Panx1 knockout (y-KO, blue line), adult wild type (a-WT, black line), adult wild type plus probenecid 100 μM (a-WT+Pbncd, dotted line) and adult Panx1 knockout mice (a-KO, green line). **(A3–A5)** Input-output curves showing the relationship between FP slope **(A3)**, fiber volley amplitude **(A4)** and stimulus intensity; and fiber volley amplitude and FP slope **(A5)**. An increased FP slope was observed in a-WT+Pbncd and a-KO mice compared to either y-WT or a-WT mice. **(B1)** Representative FP traces at interstimulus intervals of 100 ms. **(B2)** Paired-pulse facilitation (PPF) of the FP at various interstimulus intervals. No significant differences were observed between WT and KO mice. **(C1)** Absence and blockade of Panx1 channels significantly increased glutamate release and spillover in adult animals. DL-TBOA (TBOA 10, μM) was perfused after 20 min of basal transmission. **(C2)** Averaged increments in basal synaptic transmission induced by TBOA. The values in parentheses indicate the number of hippocampal slices (left) and the number of animals (right) used. All data are plotted as mean ± SEM. Statistical differences were calculated using ANOVA, followed by *post hoc* Bonferroni's test. Asterisks indicate statistical significance of the observed differences. *$p < 0.05$ vs. y-WT; § $p < 0.05$ vs. a-WT.

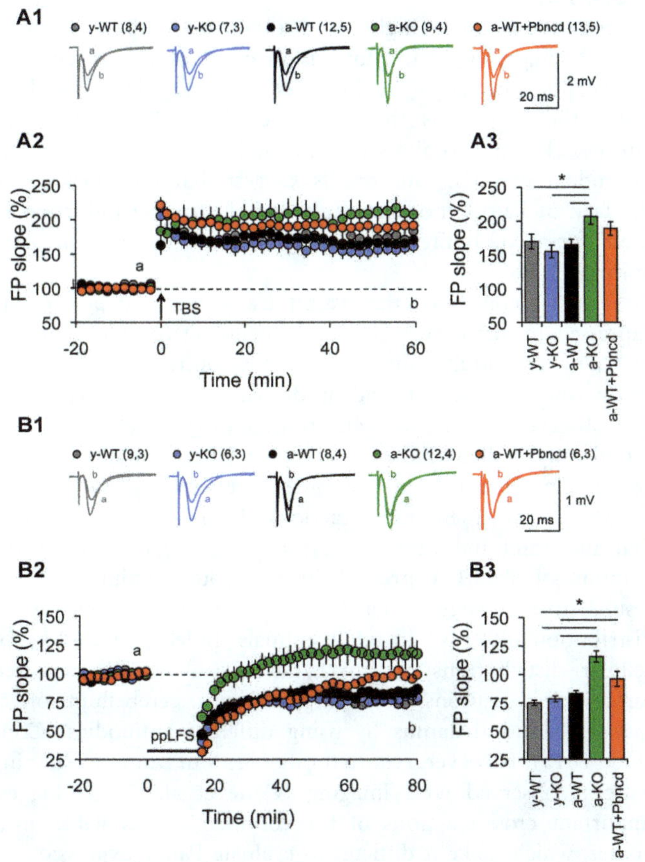

FIGURE 3 | Increased long term potentiation (LTP) and absent long term depression (LTD) in the Schaffer collateral–CA1 pathway from adult Panx1 knock-out and WT mice treated with Pbncd. (A1) Representative traces of FPs recorded 1 min before (a) and 60 min after (b) TBS. **(A2)** LTP obtained in slices from young wild type (y-WT, gray line), young Panx1 knockout (y-KO, blue line), adult wild type (a-WT, black line), adult wild type plus probenecid 100 μM (a-WT+Pbncd, red line) and adult Panx1 knockout mice (a-KO, green). Long-term potentiation was induced by the delivery of TBS at the time indicated by the arrow. **(A3)** Magnitude average of LTP determined as responses between 50 and 60 min after TBS. Long-term potentiation was significantly different for a-KO mice compared to y-WT, y-KO and a-WT mice. **(B1)** Representative traces of FPs recorded 1 min before (a) and 60 min after (b) PP-LFS. **(B2)** Long-term depression obtained in slices from y-WT (gray line), y-KO (blue line), a-WT (black line), a-WT plus probenecid 100 μM (a-WT+Pbncd, dotted line) and a-KO hippocampal slices (green line). Long-term depression was induced by the delivery of PP-LFS at the time indicated by the line. **(B3)** Magnitude average of LTD determined as responses between 50 and 60 min after PP-LFS. Long-term depression was absent in a-WT+Pbncd and in a-KO mice. The values in parentheses indicate the number of hippocampal slices (left) and the number of animals (right) used. All data are plotted as mean ± SEM. Statistical differences were calculated using ANOVA, followed by *post hoc* Tukey test. Asterisks indicate statistical significance of the observed differences (*$p < 0.05$).

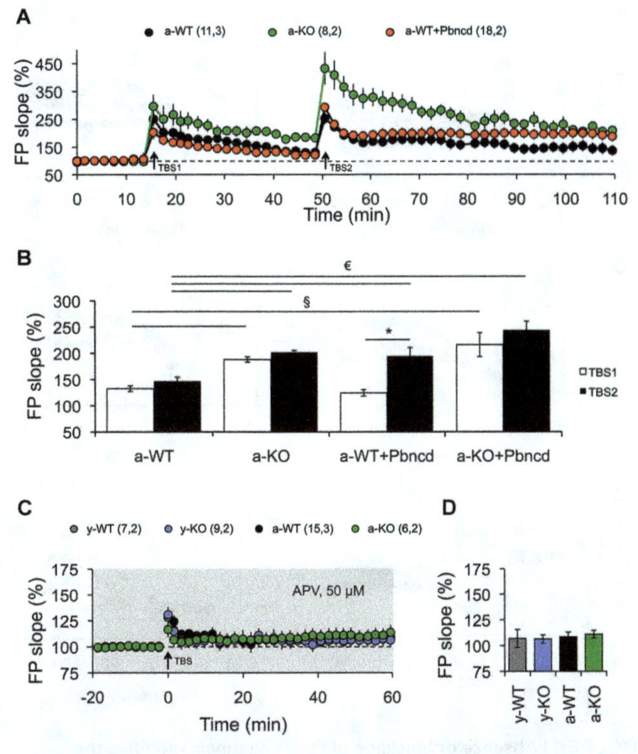

FIGURE 4 | Blockade or absence of Panx1 channels facilitates the induction of LTP. (A) Long-term potentiation induced by one TBS (TBS_1). TBS1 induced a stable LTP in adult Panx1 knockout (a-KO, green square), but transient LTP in adult wild type (a-WT, black circle), which returned to baseline after 30 min. A second TBS (TBS2) applied to the same synapses elicited significantly more potentiation in a-KO but not in a-WT mice. The blockade of Panx1 with Pbncd added 20 min before the application of the TBS2, induced a significant enhancement of potentiation in a-WT (a-WT+Pbncd, red circle). **(B)** Magnitude averages of LTP were determined as responses between 30 and 35 min after the first TBS (open bar) and 50 and 60 min after the second TBS (filled bar). **(C)** Long-term potentiation induced by TBS was completely blocked by the incubation of APV 50 μM. **(D)** Magnitude average of LTP determined as responses between 50 and 60 min after TBS. The values in parentheses indicate the number of hippocampal slices (left) and the number of animals (right) used. *$p < 0.05$ for TBS1 vs. TBS2; §$p < 0.05$ between TBS1; €$p < 0.05$ between TBS2. All data are plotted as mean ± SEM.

to perform similar experiments as those described above. We found that probenecid did not modify LTP magnitudes compared to untreated KO slices (**Figure 4B**). These results suggest that probenecid affects LTP only in WT animals, implying that its effects on LTP were specific over Panx1 channels. Consistent with the results described thus far using preconditioning stimuli, we observed that a high frequency stimulation protocol (100 Hz, 1 s; **Figure 5A**), which is a standard protocol to induce LTP, overall reiterates the same results and clearly shows that the elimination of Panx1 or its blockage potentiates LTP independently of the protocol used to induce it. These protocols induced NMDAR-dependent LTP, since the increase of LTP was completely blocked by APV (**Figures 4C,D**). These findings suggest that Panx1 could constrain the synaptic strengthening induced by conditioning stimulation, and when Panx1 is blocked or absent, a change in the threshold of synaptic plasticity occurs in adult animals.

To further test this prediction, we then investigated how synaptic plasticity induced by different frequencies of stimulation is affected by blockage or deletion of Panx1. Reliable LTD was

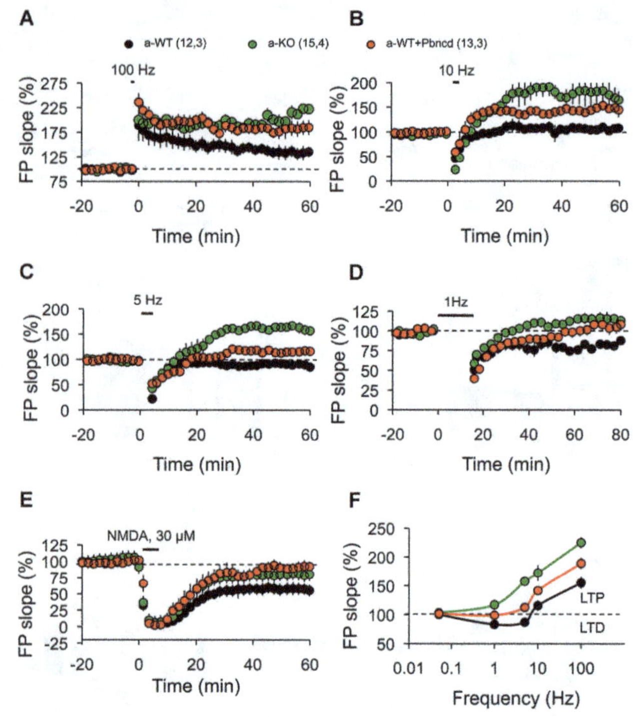

FIGURE 5 | Absence or blockade of Panx1 channels modifies the threshold for inducing excitatory hippocampal synaptic plasticity. (A) One tetanus (100 Hz, 1 s) induced significantly more potentiation in adult wild type plus Pbncd (a-WT+Pbncd, white circle) and adult Panx1 knockout (a-KO, green circle) compared to adult wild type (a-WT, black) mice. **(B)** 10 Hz stimulation for 1.5 min evoked significant potentiation in a-WT+Pbncd and a-KO mice compared to a-WT mice. **(C)** 5 Hz stimulation for 3 min produced potentiation in a-WT+Pbncd and a-KO mice, whereas it produced a slight depression in a-WT. **(D)** 1 Hz stimulation for 15 min induced a reliable LTD in WT mice, whereas it elicited a mild potentiation in a-WT+Pbncd and a-KO mice. **(E)** 30 μM of NMDA for 5 min induced LTD in all groups. **(F)** Summary of data for synaptic plasticity at different frequencies of stimulation. The values in parentheses indicate the number of hippocampal slices (left) and the number of animals (right) used. All data are plotted as mean ± SEM.

elicited with a standard LFS protocol (900 pulses, 1 Hz) in slices from a-WT animals, but it was absent in a-KO animals (**Figure 5D**). Long-term depression was also significantly reduced in a-WT+Pbncd, in which LTD decayed quickly back to baseline after an hour (**Figure 5D**). This absence of LTD observed in a-KO and in a-WT+Pbncd was also evident when using other frequencies of stimulation (5–10 Hz) to induce LTD (**Figure 5C**). On the contrary, the protocols of 900 pulses at 10 Hz (**Figure 5B**) and 5 Hz (**Figure 5C**) to induce LTD in a-KO slices evoked a prolonged form of potentiation instead of depression (**Figures 5B,C**). Despite the absence of LTD elicited with conditioning stimulation, the application of NMDA (30 μM during 5 min) induced chemical-LTD in slices from all animals groups, although a-WT showed a slightly, but not significantly greater depression of the synaptic responses compared to a-WT+Pbncd and a-KO (**Figure 5E**), thus indicating that the lack or blockade of Panx1 also affects the induction of this form of LTD.

DISCUSSION

Recent studies have highlighted the participation of Panx1 in LTP, suggesting that Panx1 is involved in this mechanism of synaptic plasticity (Prochnow et al., 2012). Here, we have tested the hypothesis that the blockade of Panx1 not only affects LTP, but also alters LTD in the hippocampus in an age-dependent manner. Our results suggest that the blockade or the lack of Panx1 modifies the threshold for the induction of these forms of bidirectional synaptic plasticity only in adult animals.

First, we confirmed that Panx1 transcripts and protein are expressed in different brain areas, including the hippocampus of young and adult animals. However, contrary to a previous study where an age-dependent decrease in Panx1 expression was observed by *in situ* hybridization (Vogt et al., 2005), we found only a slight reduction in total Panx1 protein levels in adults compared to young animals. Interestingly, by comparing the plasma membrane fractions (**Figure 1E**) we observed that the adult hippocampus expresses less than 40% of the amount of Panx1 expressed in the young animals. These results strongly suggest that there is an important cellular redistribution of Panx1 in adult animals. Indeed, multiple subcellular distributions of Panx1 have been recently reported for principal neurons in the hippocampus, cerebellum, olfactory bulb and thalamus by using different antibodies (Cone et al., 2013). However, reported differences in immuno-staining patterns observed with imaging (Cone et al., 2013) suggest important cross-reactions of the commercially available antibodies, which make it difficult to evaluate Panx1 expression by immunolocalization.

Second, we found that upon high intensity stimulation, synaptic transmission in the adult hippocampus increases in conditions where Panx1 is blocked or absent (**Figure 2**), indicating that Panx1 regulates excitatory synaptic transmission. Moreover, our findings that the blockage of the glial glutamate transporter significantly increases excitatory transmission in both a-KO mice and a-WT+Pbncd support a previously suggested mechanism, in which Panx1 modulates a presynaptic component of the synaptic transmission by increasing glutamate release and spillover (Prochnow et al., 2012). However, since we observed a greater FP slope in a-KO and a-WT+Pbncd animals compared to a-WT mice (**Figure 2A**), we cannot rule out a postsynaptic contribution, for example, in the mobilization of glutamate receptors.

Third, in agreement with a previous report (Prochnow et al., 2012) we observed increased LTP in the Schaffer Collateral–CA1 pathway of a-KO and a-WT+Pbncd animals (**Figure 3A**), whereas prolonged LTP was observed in a-KO slices induced with protocols that in WT mice elicited LTD (**Figure 3B**). This seems to indicate that a metaplastic shift and therefore a modification in the threshold for synaptic plasticity would occur. In fact, a shift in the LTP/LTD induction threshold produced by the blockage or lack of Panx1 was evident after plotting the responses to the different frequency stimulations. In this regard, a moderate and strong shift to the right in the frequency-response curve was observed for a-KO and a-WT+Pbncd slices, respectively (**Figure 5F**). These effects were restricted to adult animals in agreement with the

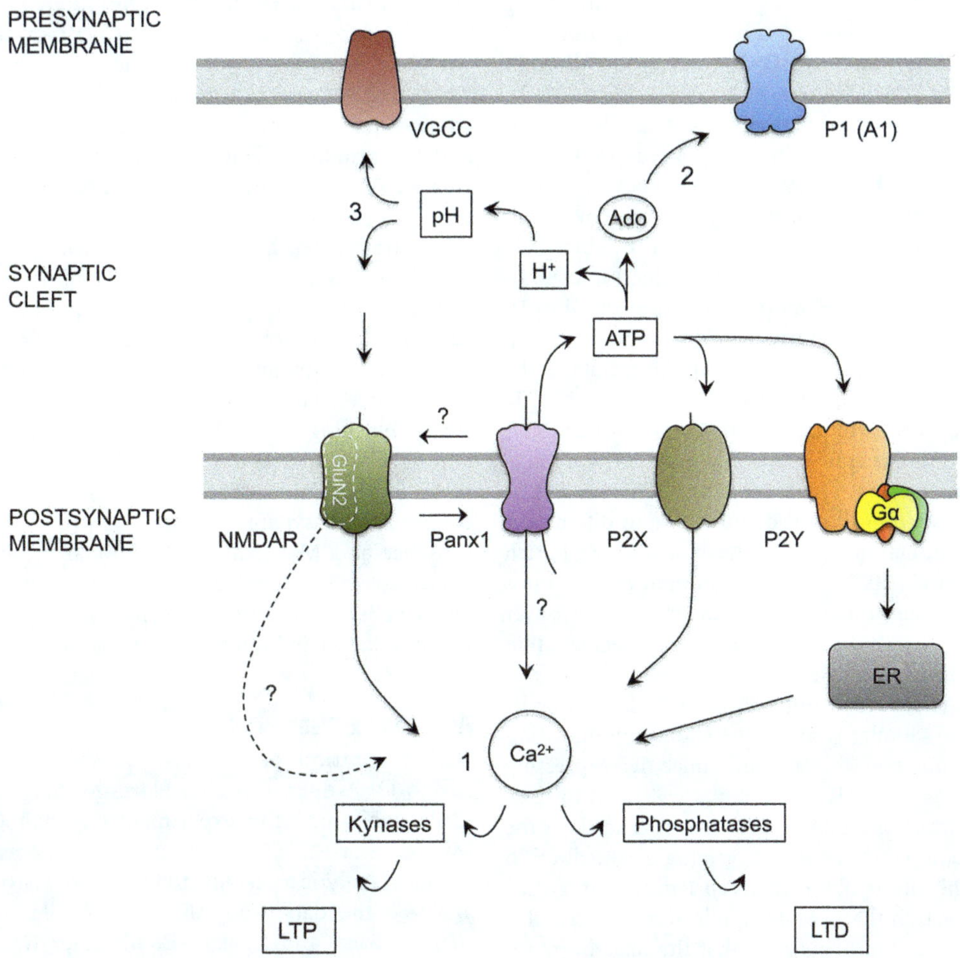

FIGURE 6 | Hypothesis for the role of Panx1 channels in the regulation of excitatory synaptic plasticity. (1) Calcium Hypothesis. Upon neuronal activity, NMDAR activation trigger calcium influx into dendritic spine. Depending on the kinetics and magnitude of the calcium concentration increments, kinases or phosphatases are activated promoting the insertion or remotion of AMPARs mediating LTP and LTD respectively (Lüscher and Malenka, 2012). NMDAR activation, probably through depolarization of post-synaptic membrane, activates Panx1 channels (Thompson et al., 2008) producing more calcium influx and ATP release by Panx1 channels. ATP released by Panx1 channels may activate ionotropic (P2X) and metabotropic (P2Y) purinergic receptors. P2X depolarizes the membrane and allows calcium entry, whereas P2Y controls calcium release from intracellular stores such as endoplasmic reticulum (ER) through the activation of G protein (Gα), therefore both receptors may contribute to the increase in the cytosolic calcium concentration (Collo et al., 1996; Yamazaki et al., 2003; Wang et al., 2004; Abbracchio et al., 2009). In addition, Panx1 channels could interact with specific NMDAR subunits exerting a modulatory effect over these receptors. We also speculate that Panx1 channels could facilitate the localization or function of specific NMDAR subunit in the post-synaptic membrane, therefore in the absence of Panx1, a greater activation of kynases instead phosphatases could be due to a change in the composition of GluN2 subunits, leading a change in the kinetics and calcium influx through NMDARs. **(2)** Activation of pre-synaptic Adenosine receptors Hypothesis. In the synapses, ATP release through Panx1 channels could be converted into adenosine (Ado), which in turn can activate P1 purinergic receptors (adenosine receptors). The activation of A1 adenosine receptor (A1), located in the presynaptic terminals reduces the release of glutamate. Therefore, in the absence of Panx1 the depletion of extracellular ATP and adenosine promotes an increase in the neurotransmitter release (Prochnow et al., 2012). **(3)** Regulation of synaptic cleft pH Hypothesis. Extracellular ATP hydrolysis also generates protons and phosphate that produce a decrease in the extracellular pH (Vroman et al., 2014). This acidification could inhibit both, voltage gated calcium channels (VGCC) present in the presynaptic terminal, and NMDAR on postsynaptic membrane, reducing glutamate release and NMDA receptor activation. The absence of Panx1 channels or their inhibition prevents the release of ATP, stops the production of phosphate buffer and produce the alkalization of the synaptic cleft, which increase the release probability and the activation of NMDARs.

idea that different synaptic plasticity mechanisms could operate in young and adult animals (Foster, 1999).

It has been well established that aging is associated to a shift in synaptic plasticity, favoring a decrease in synaptic transmission and in the ability to induce LTP (reviewed in Kumar, 2011). These changes are already observed in middle-aged rodents (Rex et al., 2005; Fouquet et al., 2011). Nevertheless, we did not observe differences in a-WT animals, but found that the opposite situation occurs in a-KO mice, where transmission and LTP (but not LTD) seem to be favored in adulthood. These data suggest that Panx1 plays a pivotal role in the balance of plasticity mechanisms during aging. In this regard, it has been postulated that a shift in synaptic plasticity induction that promotes LTD over LTP during aging results from changes in expression, subunit composition and posttranslational modification of NMDAR, causing alterations in calcium influx (Foster, 1999; Kumar, 2011; Magnusson, 2012). Mounting evidence has shown differential roles of GluN2 subunits in LTP and LTD induction. For instance, early reports found that GluN2A/B antagonists block LTP, whereas GluN2C/D antagonists block LTD (Hrabetova and Sacktor, 1997; Hrabetova et al., 2000). More recent studies have shown that the selective blockade of GluN2B-containing NMDAR abolishes the induction of LTD, but not of LTP, whereas the preferential inhibition of GluN2A subunits prevents LTP without affecting LTD (Liu et al., 2004). On the other hand, the overexpression of GluN2B facilitates the synaptic potentiation induced by stimulation within the 10–100 Hz range (Tang et al., 1999), whilst the overexpression of GluN2A abolishes the induction of synaptic depression in the 3–5 Hz range (Cui et al., 2013). In addition, age-dependent differences in the expression of GluN2 subunits can also contribute to differences in the LTP and LTD induction observed during aging (Berberich et al., 2005; Bartlett et al., 2007), as GluN2 subunit expression is critically modified during development (Hestrin, 1992; Monyer et al., 1994; Sans et al., 2000). In this sense, we can speculate that the effects of the blockade or absence of Panx1 observed in our data could change the subunit composition of NMDARs or its active state (**Figure 6**). Similar shifts in the sliding modification threshold have been observed in transgenic mice overexpressing the GluN2B subunit of NMDAR (Tang et al., 1999) and in mice lacking calcineurin, which is a phosphatase that is crucial for the LTD mechanism (Zeng et al., 2001), suggesting a contribution of these factors to the effects observed when Panx1 is absent or inhibited. The observation that NMDA application induces LTD in both adult WT and KO mice suggests that the machinery for the expression of LTD is intact, but the signal to trigger this form of synaptic plasticity could be altered. These observations raised the question of how the blockade or absence of Panx1 modifies the synaptic activation of NMDAR and how this activation leads to an increase in LTP in detriment of LTD. Prochnow et al. (2012) suggest that the effects that they observed in excitability and LTP could be explained by a reduction in the catabolism of ATP released by Panx1, leading to a depletion of adenosine in the synaptic cleft and therefore facilitating the activation of NMDARs (**Figure 6**). Moreover, a recent report by Vroman et al. (2014), shows that extracellular ATP hydrolysis inhibits retinal synaptic transmission between horizontal cells and photoreceptors, by changing pH in synaptic cleft. ATP released by Panx1 channels located on horizontal cell dendrites, is hydrolyzed by ecto-ATPase into inosine, phosphate and protons ions acidifying the synaptic cleft. This change in the pH inhibits voltage dependent calcium channels present in the cone synaptic terminal, causing reduction in glutamate release by the cones. On the contrary, the closing of Panx1 channels prevents the release of ATP, and therefore stops the production of phosphate buffer and leads to alkalization of the synaptic cleft. A similar mechanism could occur in hippocampal Sc-CA1 synapse, since the opening of voltage gated calcium channels (VGCC) depends on both intracellular and extracellular pH (Tombaugh and Somjen, 1997) and hippocampus synaptic membranes present strong ecto-ATPase and ecto-ATPDase activity (Kukulski et al., 2004). Additionally, NMDARs also are modulated by extracellular pH (Gottfried and Chesler, 1994), where an increase in extracellular pH facilitates the activation, whereas a decrease in extracellular pH inhibits NMDAR. Thus, when Panx1 channels are blocked, a decrease in ATP release and the resulting alkalization could enhance synaptic strength (**Figure 6**). We believe that additional mechanisms could take place (**Figure 6**). One possible explanation is that the calcium concentration reached during LFS protocols through NMDARs in KO mice is greater than that in WT mice, enhancing kynase activation over phosphatase activation; nonetheless additional experiments are necessary to test this hypothesis. Regardless of the precise mechanism, the present findings emphasize a new role of Panx1 in excitatory hippocampal synaptic plasticity, where simply blocking Panx1 shifts the threshold balance between LTP and LTD. We have studied the effects of Panx1 on bidirectional synaptic plasticity in the Sc-CA1 pathway, because it is the most comprehensive synaptic model for NMDAR-dependent plasticity (Lüscher and Malenka, 2012). Similar synaptic plasticity processes have been described in other brain areas, as in neocortical and cerebellar synapses (Malenka and Bear, 2004). However, whether Panx1 is important to these brain areas remains to be elucidated.

ACKNOWLEDGMENTS

Author contributions: Alvaro O. Ardiles designed the study; Alvaro O. Ardiles, Carolina Flores-Muñoz, and Gabriela Toro-Ayala performed the experiments; Adrian G. Palacios, Pablo Muñoz, Ana M. Cárdenas, and Juan C. Sáez contributed new reagents/analytic tools; Alvaro O. Ardiles, Carolina Flores-Muñoz, analyzed the data; and Alvaro O. Ardiles, Marco Fuenzalida, Juan C. Sáez, and Agustín D. Martínez wrote the paper. This work was funded by FONDECYT 3130759 to Alvaro O. Ardiles, CORFO 13IDL2-18271, CONICYT ICM- P09-022-F, ANILLO-ACT1104 and FONDECYT 1130855 to Agustín D. Martínez. The authors are especially thankful to Dr. Hannah Monyer for kindly providing the Panx1 KO mice, and Drs. Dale Laird and Silvia Penuela for generously gift the CT-395 antibody.

REFERENCES

Abbracchio, M. P., Burnstock, G., Verkhratsky, A., and Zimmermann, H. (2009). Purinergic signalling in the nervous system: an overview. *Trends Neurosci.* 32, 19–29. doi: 10.1016/j.tins.2008.10.001

Anselmi, F., Hernandez, V. H., Crispino, G., Seydel, A., Ortolano, S., Roper, S. D., et al. (2008). ATP release through connexin hemichannels and gap junction transfer of second messengers propagate Ca2+ signals across the inner ear. *Proc. Natl. Acad. Sci. U S A* 105, 18770–18775. doi: 10.1073/pnas.0800793105

Ardiles, A. O., Tapia-Rojas, C. C., Mandal, M., Alexandre, F., Kirkwood, A., Inestrosa, N. C., et al. (2012). Post-synaptic dysfunction is associated with spatial and object recognition memory loss in a natural model of Alzheimer's disease. *Proc. Natl. Acad. Sci. U S A* 109, 13835–13840. doi: 10.1073/pnas.1201209109

Bao, L., Locovei, S., and Dahl, G. (2004). Pannexin membrane channels are mechanosensitive conduits for ATP. *FEBS Lett.* 572, 65–68. doi: 10.1016/j.febslet.2004.07.009

Bartlett, T. E., Bannister, N. J., Collett, V. J., Dargan, S. L., Massey, P. V., Bortolotto, Z. A., et al. (2007). Differential roles of NR2A and NR2B-containing NMDA receptors in LTP and LTD in the CA1 region of two-week old rat hippocampus. *Neuropharmacology* 52, 60–70. doi: 10.1016/j.neuropharm.2006.07.013

Berberich, S., Punnakkal, P., Jensen, V., Pawlak, V., Seeburg, P. H., Hvalby, Ø., et al. (2005). Lack of NMDA receptor subtype selectivity for hippocampal long-term potentiation. *J. Neurosci.* 25, 6907–6910. doi: 10.1523/jneurosci.1905-05.2005

Bliss, T. V., and Collingridge, G. L. (1993). A synaptic model of memory: long-term potentiation in the hippocampus. *Nature* 361, 31–39. doi: 10.1038/361031a0

Bruzzone, R., Hormuzdi, S. G., Barbe, M. T., Herb, A., and Monyer, H. (2003). Pannexins, a family of gap junction proteins expressed in brain. *Proc. Natl. Acad. Sci. U S A* 100, 13644–13649. doi: 10.1073/pnas.2233464100

Collo, G., North, R. A., Kawashima, E., Merlo-Pich, E., Neidhart, S., Surprenant, A., et al. (1996). Cloning OF P2X5 and P2X6 receptors and the distribution and properties of an extended family of ATP-gated ion channels. *J. Neurosci.* 16, 2495–2507.

Cone, A. C., Ambrosi, C., Scemes, E., Martone, M. E., and Sosonsky, G. E. (2013). A comparative antibody analysis of Pannexin1 expression in four rat brain regions reveals varying subcellular localizations. *Front. Pharmacol.* 4:6. doi: 10.3389/fphar.2013.00006

Cui, Z., Feng, R., Jacobs, S., Duan, Y., Wang, H., Cao, X., et al. (2013). Increased NR2A:NR2B ratio compresses long-term depression range and constrains long-term memory. *Sci. Rep.* 3:1036. doi: 10.1038/srep01036

Dermietzel, R., Traub, O., Hwang, T. K., Beyer, E., Bennett, M. V., Spray, D. C., et al. (1989). Differential expression of three gap junction proteins in developing and mature brain tissues. *Proc. Natl. Acad. Sci. U S A* 86, 10148–10152. doi: 10.1073/pnas.86.24.10148

Dudek, S. M., and Bear, M. F. (1992). Homosynaptic long-term depression in area CA1 of hippocampus and effects of N-methyl-D-aspartate receptor blockade. *Proc. Natl. Acad. Sci. U S A* 89, 4363–4367. doi: 10.1073/pnas.89.10.4363

Dudek, S. M., and Bear, M. F. (1993). Bidirectional long-term modification of synaptic effectiveness in the adult and immature hippocampus. *J. Neurosci.* 13, 2910–2918.

Fields, R. D., and Stevens, B. (2000). ATP: an extracellular signaling molecule between neurons and glia. *Trends Neurosci.* 23, 625–633. doi: 10.1016/s0166-2236(00)01674-x

Foster, T. C. (1999). Involvement of hippocampal synaptic plasticity in age-related memory decline. *Brain Res. Rev.* 30, 236–249. doi: 10.1016/s0165-0173(99)00017-x

Fouquet, C., Petit, G. H., Auffret, A., Gaillard, E., Rovira, C., Mariani, J., et al. (2011). Early detection of age-related memory deficits in individual mice. *Neurobiol. Aging* 32, 1881–1895. doi: 10.1016/j.neurobiolaging.2009.11.001

Gottfried, J. A., and Chesler, M. (1994). Endogenous H+ modulation of NMDA receptor-mediated EPSCs revealed by carbonic anhydrase inhibition in rat hippocampus. *J. Physiol.* 478, 373–378.

Hestrin, S. (1992). Developmental regulation of NMDA receptor-mediated synaptic currents at a central synapse. *Nature* 357, 686–689. doi: 10.1038/357686a0

Hrabetova, S., and Sacktor, T. C. (1997). Long-term potentiation and long-term depression are induced through pharmacologically distinct NMDA receptors. *Neurosci. Lett.* 226, 107–110. doi: 10.1016/s0304-3940(97)00252-8

Hrabetova, S., Serrano, P., Blace, N., Tse, H. W., Skifter, D. A., Jane, D. E., et al. (2000). Distinct NMDA receptor subpopulations contribute to long-term potentiation and long-term depression induction. *J. Neurosci.* 20:RC81.

Huang, Y., Grinspan, J. B., Abrams, C. K., and Scherer, S. S. (2007). Pannexin1 is expressed by neurons and glia but does not form functional gap junctions. *Glia* 55, 46–56. doi: 10.1002/glia.20435

Kim, J.-E., and Kang, T.-C. (2011). The P2X7 receptor-pannexin-1 complex decreases muscarinic acetylcholine receptor-mediated seizure susceptibility in mice. *J. Clin. Invest.* 121, 2037–2047. doi: 10.1172/JCI44818

Kukulski, F., Sévigny, J., and Komoszyński, M. (2004). Comparative hydrolysis of extracellular adenine nucleotides and adenosine in synaptic membranes from porcine brain cortex, hippocampus, cerebellum and medulla oblongata. *Brain Res.* 1030, 49–56. doi: 10.1016/j.brainres.2004.09.044

Kumar, A. (2011). Long-term potentiation at CA3-CA1 hippocampal synapses with special emphasis on aging, disease and stress. *Front. Aging Neurosci.* 3:7. doi: 10.3389/fnagi.2011.00007

Lee, H. K., Takamiya, K., Han, J.-S., Man, H., Kim, C.-H., Rumbaugh, G., et al. (2003). Phosphorylation of the AMPA receptor GluR1 subunit is required for synaptic plasticity and retention of spatial memory. *Cell* 112, 631–643. doi: 10.1016/s0092-8674(03)00122-3

Lisman, J. (1989). A mechanism for the Hebb and the anti-Hebb processes underlying learning and memory. *Proc. Natl. Acad. Sci. U S A* 86, 9574–9578. doi: 10.1073/pnas.86.23.9574

Liu, L., Wong, T. P., Pozza, M. F., Lingenhoehl, K., Wang, Y., Sheng, M., et al. (2004). Role of NMDA receptor subtypes in governing the direction of hippocampal synaptic plasticity. *Science* 304, 1021–1024. doi: 10.1126/science.1096615

Locovei, S., Scemes, E., Qiu, F., Spray, D. C., and Dahl, G. (2007). Pannexin1 is part of the pore forming unit of the P2X(7) receptor death complex. *FEBS Lett.* 581, 483–488. doi: 10.1016/j.febslet.2006.12.056

Locovei, S., Wang, J., and Dahl, G. (2006). Activation of pannexin 1 channels by ATP through P2Y receptors and by cytoplasmic calcium. *FEBS Lett.* 580, 239–244. doi: 10.1016/j.febslet.2005.12.004

Lüscher, C., and Malenka, R. C. (2012). NMDA receptor-dependent long-term potentiation and long-term depression (LTP/LTD). *Cold Spring Harb. Perspect. Biol.* 4:a005710. doi: 10.1101/cshperspect.a005710

MacVicar, B. A., and Thompson, R. J. (2010). Non-junction functions of pannexin-1 channels. *Trends Neurosci.* 33, 93–102. doi: 10.1016/j.tins.2009.11.007

Magnusson, K. R. (2012). Aging of the NMDA receptor: from a mouse's point of view. *Future Neurol.* 7, 627–637. doi: 10.2217/fnl.12.54

Malenka, R. C., and Bear, M. F. (2004). LTP and LTD: an embarrassment of riches. *Neuron* 44, 5–21. doi: 10.1016/j.neuron.2004.09.012

Monyer, H., Burnashev, N., Laurie, D. J., Sakmann, B., and Seeburg, P. H. (1994). Developmental and regional expression in the rat brain and functional properties of four NMDA receptors. *Neuron* 12, 529–540. doi: 10.1016/0896-6273(94)90210-0

Panchin, Y., Kelmanson, I., Matz, M., Lukyanov, K., Usman, N., and Lukyanov, S. (2000). A ubiquitous family of putative gap junction molecules. *Curr. Biol.* 10, R473–R474. doi: 10.1016/s0960-9822(00)00576-5

Pelegrin, P., and Surprenant, A. (2006). Pannexin-1 mediates large pore formation and interleukin-1 beta release by the ATP-gated P2X7 receptor. *EMBO J.* 25, 5071–5082. doi: 10.1038/sj.emboj.7601378

Penuela, S., Bhalla, R., Gong, X.-Q., Cowan, K. N., Celetti, S. J., Cowan, B. J., et al. (2007). Pannexin 1 and pannexin 3 are glycoproteins that exhibit many distinct characteristics from the connexin family of gap junction proteins. *J. Cell Sci.* 120, 3772–3783. doi: 10.1242/jcs.009514

Prochnow, N., Abdulazim, A., Kurtenbach, S., Wildförster, V., Dvoriantchikova, G., Hanske, J., et al. (2012). Pannexin1 stabilizes synaptic plasticity and is needed for learning. *PLoS One* 7:e51767. doi: 10.1371/journal.pone.0051767

Rex, C. S., Kramár, E. A., Colgin, L. L., Lin, B., Gall, C. M., and Lynch, G. (2005). Long-term potentiation is impaired in middle-aged rats: regional specificity and reversal by adenosine receptor antagonists. *J. Neurosci.* 25, 5956–5966. doi: 10.1523/jneurosci.0880-05.2005

Romanov, R. A., Bystrova, M. F., Rogachevskaya, O. A., Sadovnikov, V. B., Shestopalov, V. I., and Kolesnikov, S. S. (2012). The ATP permeability of pannexin 1 channels in a heterologous system and in mammalian taste cells is dispensable. *J. Cell Sci.* 15, 5514–5523. doi: 10.1242/jcs.111062

Sans, N., Petralia, R. S., Wang, Y. X., Blahos, J. 2nd, Hell, J. W., and Wenthold, R. J. (2000). A developmental change in NMDA receptor-associated proteins at hippocampal synapses. *J. Neurosci.* 20, 1260–1271.

Santiago, M. F., Veliskova, J., Patel, N. K., Lutz, S. E., Caille, D., Charollais, A., et al. (2011). Targeting pannexin1 improves seizure outcome. *PLoS One* 6:e25178. doi: 10.1371/journal.pone.0025178

Silverman, W. R., de Rivero Vaccari, J. P., Locovei, S., Qiu, F., Carlsson, S. K., Scemes, E., et al. (2009). The pannexin 1 channel activates the inflammasome in neurons and astrocytes. *J. Biol. Chem.* 284, 18143–18151. doi: 10.1074/jbc.M109.004804

Tang, Y. P., Shimizu, E., Dube, G. R., Rampon, C., Kerchner, G. A., Zhuo, M., et al. (1999). Genetic enhancement of learning and memory in mice. *Nature* 401, 63–69. doi: 10.1038/43432

Thomas-Crusells, J., Vieira, A., Saarma, M., and Rivera, C. (2003). A novel method for monitoring surface membrane trafficking on hippocampal acute slice preparation. *J. Neurosci. Methods* 23, 159–166. doi: 10.1016/s0165-0270(03)00050-5

Thompson, R. J., Jackson, M. F., Olah, M. E., Rungta, R. L., Hines, D. J., Beazely, M. A., et al. (2008). Activation of pannexin-1 hemichannels augments aberrant bursting in the hippocampus. *Science* 322, 1555–1559. doi: 10.1126/science.1165209

Tombaugh, G. C., and Somjen, G. C. (1997). Differential sensitivity to intracellular pH among high- and low- threshold Ca2+ currents in isolated rat CA1 neurons. *J. Neurophysiol.* 77, 639–653.

Vogt, A., Hormuzdi, S. G., and Monyer, H. (2005). Pannexin1 and Pannexin2 expression in the developing and mature rat brain. *Brain Res. Mol. Brain Res.* 141, 113–120. doi: 10.1016/j.molbrainres.2005.08.002

Vroman, R., Klaassen, L. J., Howlett, M. H., Cenedese, V., Klooster, J., Sjoerdsma, T., et al. (2014). Extracellular ATP hydrolysis inhibits synaptic transmission by increasing ph buffering in the synaptic cleft. *PLoS Biol.* 12:e1001864. doi: 10.1371/journal.pbio.1001864

Wang, Y., Haughey, N. J., Mattson, M. P., and Furukawa, K. (2004). Dual effects of ATP on rat hippocampal synaptic plasticity. *Neuroreport* 15, 633–636. doi: 10.1097/00001756-200403220-00012

Yamazaki, Y., Kaneko, K., Fujii, S., Kato, H., and Ito, K. (2003). Long-term potentiation and long-term depression induced by local application of ATP to hippocampal CA1 neurons of the guinea pig. *Hippocampus* 13, 81–92. doi: 10.1002/hipo.7999

Zeng, H., Chattarji, S., Barbarosie, M., Rondi-Reig, L., Philpot, B. D., Miyakawa, T., et al. (2001). Forebrain-specific calcineurin knockout selectively impairs bidirectional synaptic plasticity and working/episodic-like memory. *Cell* 107, 617–629. doi: 10.1016/s0092-8674(01)00585-2

Zoidl, G., Petrasch-Parwez, E., Ray, A., Meier, C., Bunse, S., Habbes, H. W., et al. (2007). Localization of the pannexin1 protein at postsynaptic sites in the cerebral cortex and hippocampus. *Neuroscience* 146, 9–16. doi: 10.1016/j.neuroscience.2007.01.061

3

Regulation of gap junction channels by infectious agents and inflammation in the CNS

Paul Castellano[1,2] and Eliseo A. Eugenin[1,2]*

[1] Public Health Research Institute (PHRI), New Jersey Medical School, Rutgers The State University of New Jersey, Newark, NJ, USA
[2] Department of Microbiology and Molecular Genetics, New Jersey Medical School, Rutgers The State University of New Jersey, Newark, NJ, USA

Correspondence:
Eliseo Eugenin, Department of Microbiology and Molecular Genetics, New Jersey Medical School, Rutgers The State University of New Jersey, 225 Warren Street, Newark, NJ 07103, USA
e-mail: eliseo.eugenin@rutgers.edu

Gap junctions (GJs) are conglomerates of intercellular channels that connect the cytoplasm of two or more cells, and facilitate the transfer of ions and small molecules, including second messengers, resulting in metabolic and electrical coordination. In general, loss of gap junctional communication (GJC) has been associated with cellular damage and inflammation resulting in compromise of physiological functions. Recently, it has become evident that GJ channels also play a critical role in the pathogenesis of infectious diseases and associated inflammation. Several pathogens use the transfer of intracellular signals through GJ channels to spread infection and toxic signals that amplify inflammation to neighboring cells. Thus, identification of the mechanisms by which several infectious agents alter GJC could result in new potential therapeutic approaches to reduce inflammation and their pathogenesis.

Keywords: hemichannel, astrocytes, HIV, microglia, oligodendrocytes

INTRODUCTION

Gap junction (GJ) channels are formed by connexins (Cxs), a family of proteins with more than 21 members in humans (for review comparing mouse and human Cxs, see Willecke et al., 2002). Each channel is formed by two hemichannels which are hexamers of homologous subunit proteins, termed Cxs (For reviews on structure and function, see Bennett et al., 1991; Saez et al., 2003). Unopposed hemichannels (uHCs) can be formed by one (homomeric connexons) or several (heteromeric) types of Cxs. GJ channels can be formed by two identical, homotypic, or different, heterotypic, subunits of hemichannels. These multiple combinations enable channels formed by different Cxs to vary in their biophysical properties and permeability (reviewed Harris and Bevans, 2001). The large internal diameter of the pore is around 12 A°, and allows ions and intracellular messengers to diffuse between connected cells, including inositol trisphosphate (IP_3), calcium, cyclic nucleotides, metabolites, toxic molecules, neurotransmitters, viral peptides, and electrical signals (reviewed Saez et al., 2003; Bennett and Zukin, 2004). Through the diffusion of these second messengers, GJs coordinate several physiological functions including electrotonic properties, secretion of glucose, uptake and diffusion of glutamate and other neurotransmitters (reviewed Bennett et al., 1991; Saez et al., 2003). uHCs can also be opened on the cell surface, providing autocrine and paracrine communication systems. Some of the molecules released from the cytoplasm into the extracellular space through the opening of uHC are ATP, prostaglandin E_2 (PGE_2), glutamate, aspartate and ions (reviewed Wang et al., 2013; Lohman and Isakson, 2014).

Within the CNS, Cxs are highly expressed in all cells including brain microvascular endothelial cells, astrocytes, oligodendrocyters, microglia, and neurons (**Figure 1**). Cx43 and Cx30 are the main Cxs found in astrocytes, while neurons mostly express Cx36, Cx30.2, and Cx45. Oligodendrocytes express Cx29, Cx32, Cx31.3, Cx45, and Cx47, and upon activation microglia express Cx32, Cx36, and Cx43 (For a detailed review of Cx expression and function in CNS parenchyma under normal physiologic conditions and diseased conditions, see Eugenin et al., 2012). Thus, alterations in Cx expression and gap junctional communication (GJC) have a large impact upon CNS function, including response to injury, behavior, synaptic and blood brain barrier (BBB) stability (Eugenin et al., 2012). In this review we will describe how several pathogens dysregulate Cx expression in the CNS.

PARENCHYMAL GJC IS SHUTDOWN IN RESPONSE TO MOST INFECTIOUS AGENTS AND ASSOCIATED INFLAMMATION

It is accepted that astrocytic GJC contribute to the stability of neuronal networks by favoring metabolic and electrical cooperation between connected cells. However, upon release of classical inflammatory mediators, including IL-1β (John et al., 1999; Duffy et al., 2000), NO (Bolaños and Medina, 1996), ATP (Meme et al., 2004), TGF-β (Reuss et al., 1998, 2000), endothelins (Giaume et al., 1992), and acidification (H^+ and lactic acid) (Abudara et al., 2001), expression of Cxs and GJC is reduced or totally shutdown. With respect to infectious agents, it is known that *S. aureus* infection or lipopolysaccharide (LPS) treatment of astrocytes reduces Cx43 and Cx36 expression and GJC. Herpes simplex

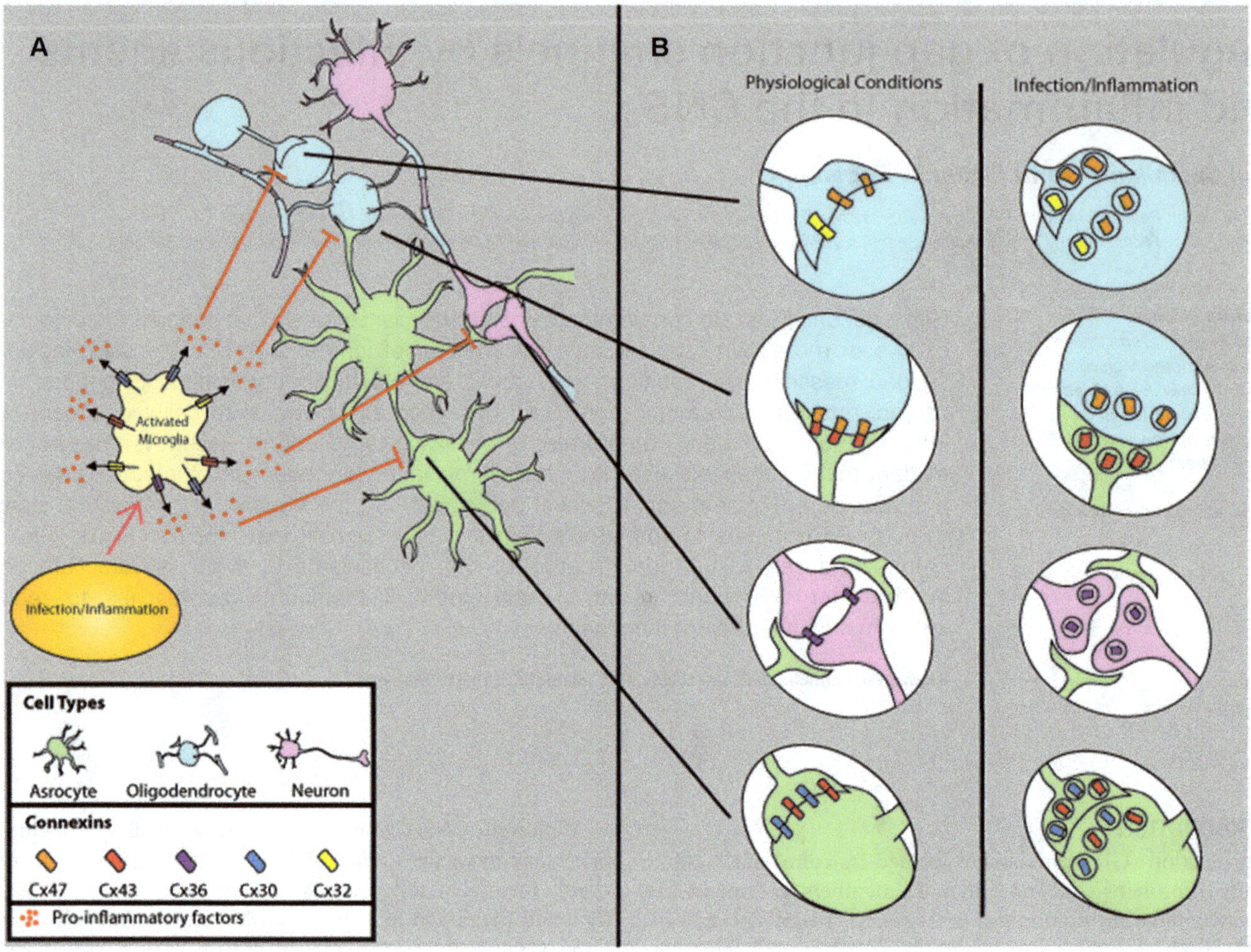

FIGURE 1 | Comparison of gap junction connectivity under physiological and infectious/inflammatory conditions. (A) Normally, all cells of the CNS are coupled by GJ to coordinate several physiological functions, including metabolic and electrical coordination. However, upon infection/inflammation GJC is reduced by a mechanism that involves microglia activation and release of pro-inflammatory factors such as PGEs, ATP, neurotransmitters, cytokines, and chemokines. **(B)** Corresponds to the changes in GJ synaptic connections in response to inflammation and infectious agents. Most of these factors induce internalization, degradation, and reduction of GJC.

virus-2 (HSV-2) reduces expression of connexins and GJC by direct tyrosine phosphorylation of Cx43 (Crow et al., 1990; Filson et al., 1990; Fischer et al., 2001; Musée et al., 2002; Koster-Patzlaff et al., 2007, 2009; Karpuk et al., 2011). Bovine papillomavirus type 4 E8, when bound to ductin, causes loss of GJC in primary fibroblast (Faccini et al., 1996; Ashrafi et al., 2000). Swine Flu virus down regulates endothelial Cx43 expression by a extracellular-signal-regulated kinase and c-Jun N-terminal kinase dependent mechanism (Hsiao et al., 2010). Borna virus infection of the CNS reduces Cx43 and Cx36 expression (Koster-Patzlaff et al., 2007, 2009). Reduction of Cx36 expression in Huntington's disease (Petrasch-Parwez et al., 2004) and global reduction of Cx43, Cx32, and Cx47 in Balo's disease (Masaki et al., 2012) are prominent features in these disorders.

As described above, several disorders and pathogens reduce Cx expression and GJC. In the next sections we will describe how dysregulation of connexin containing channels participate in the pathogenesis of several CNS diseases and the potential role of infectious diseases in the physiological functions mediated by GJC. Our hypothesis is that infectious agents and associated inflammation through GJ impairment can impede upon neuronal myelination, plasticity, migration, and cellular differentiation.

Cx EXPRESSION AND GJC IN PARENCHYMAL CELLS ARE DOWN REGULATED BY INFECTIOUS AGENTS AND ASSOCIATED INFLAMMATION

Cx EXPRESSION AND GJC ARE ESSENTIAL FOR NEURONAL PLASTICITY

GJ channels contribute to coordinated electrical signaling by serving as a low resistance bridge to transfer electrical signals that influence the receptor components of receptor neurons. The receptor composition of neuronal post-synaptic membranes is highly plastic and influenced by the frequency and strength of signaling from effector neurons. GJs containing Cx36 in hippocampal neurons contribute to neuronal plasticity by participating in synchronizing sub-threshold oscillatory activity essential for long term potentiation (Placantonakis et al., 2006) (for review of electrical activity in Cx KO models see Söhl et al., 2004). Furthermore, close proximity of GJs and NMDA receptors allows for the reciprocal regulation of both synapses (Pereda et al., 2003; Rash et al., 2004). Loss of Cx36 GJC in knockout mice did not involve changes in the slope of excitatory post synaptic potentials (EPSPs) in the hippocampus, but was likely due to increased NR2A/NR2B ratios of NMDA receptor subunits (Wang and Belousov, 2011) indicating that electrical synapses have a major role in regulation of glutamatergic synaptic plasticity in the hippocampus. Borna disease virus infection induces long term reduction of Cx36 mRNA and protein (for up to 8 weeks) in neurons of infected and even in uninfected regions of the hippocampus, which is accompanied by release of pro-inflammatory cytokines that further induce neuronal death (Koster-Patzlaff et al., 2009). LPS also reduces Cx36 expression and GJC in the hippocampus and neocortex (Dobrenis et al., 2005), and Huntington's disease mouse models had reduced Cx36 expression in the retina (Petrasch-Parwez et al., 2004). Therefore, pathological conditions that reduce Cx36 GJC and expression, such as Borna disease virus (Koster-Patzlaff et al., 2009), Huntington's disease (Petrasch-Parwez et al., 2004), and inflammation (Dobrenis et al., 2005) have the potential to hinder synaptic plasticity.

As indicated previously, Cx43 and Cx30 are the predominant Cxs expressed in astrocytes, and form GJs that serve critical roles in plasticity, extracellular synaptic metabolite recycling, neuronal excitability, immune activation, inflammation, and BBB integrity (reviewed Eugenin et al., 2012). Electrophysiological recordings of the somatosensory barrel cortex in astrocyte-specific conditional Cx43 KO mice do not show increased amplitude of low frequency potentials (LFP) observed in control mice indicating astrocyte Cx43 is essential for LTP (Han et al., 2014). LPS treatment or *S. aureus* infection reduces GJC in astrocytes for up to 16 h post-infection (Karpuk et al., 2011), followed by reduction in Cx43 expression after 24 h (Esen et al., 2007) indicating that these inflammatory factors have profound effects upon synaptic plasticity and subsequent congnitive functioning.

Abnormal GJC and Cx expression may also influence cognitive function beyond plasticity. Conditional Cx43 KO mice displayed a reduction in environment exploration and increased anxiety based on performance in slit and open-field observations (Han et al., 2014), suggesting that Cx43 dysfunction due to bacterial infection or inflammation may contribute to abnormal cognitive functioning. Moreover, juvenile mice bearing a missense mutation of oligodendrocyte Cx47 also displayed increased anxiety in open field observations, and Cx30/Cx47 double knockouts had severe motor impairments (Han et al., 2014). Therefore, cognitive and motor functions can become impaired during inflammation due to reduced Cx expression and GJC. Our proposal is that infectious agents and subsequent inflammation in the CNS compromises Cx expression and GJC, and repeated bouts of inflammation can have profound effects upon anxiety, learning, and memory.

Cx EXPRESSION AND GJC ARE ESSENTIAL FOR STEM CELL MIGRATION AND DIFFERENTIATION

Damage to the CNS from traumatic injury, inflammation, or resection requires the cellular repopulation of neurons to infected areas, but neuronal communication is never fully repaired. Multipotential stem cells have the ability to repopulate damaged parenchyma, but inflammation reduces their ability to migrate and differentiate where needed. The adult CNS has been identified to contain stem cells in the subventricular zone (SVZ), and attempts have been made to utilize the multipotential nature of these cells to repair and repopulate areas of CNS damage (for a review of neurogenesis and migration involving the SVZ, see Brazel et al., 2003). Stem cells of the CNS require Cx expression for differentiation into mature cells (Yang et al., 2005; Hartfield et al., 2011; Lemcke et al., 2013) by allowing glial Cx43 expressing cells to provide a scaffold for directing migration (Miragall et al., 1997; Elias et al., 2007; Cina et al., 2009; Kunze et al., 2009; Marins et al., 2009). In adult mice, neuronal progenitors have been identified as originating from the anterior portion of the SVZ (Doetsch et al., 1997). These progenitors travel through the rostral migratory stream (RMS) and give rise to neurons in the olfactory bulb (Lois and Alvarez-Buylla, 1994). A study attempting to identify proliferating cells of the RMS and olfactory bulb identified that BrdU positive cells correlated with lower levels of Cx43 expression in adult mice (Miragall et al., 1997). However, it cannot be assumed that the BrdU positive cells are actively replicating, and may have possibly already received the signaling needed for differentiation into their final cell fate. This correlated with a study in which knockdown of Cx43 in a neural differentiation culture model had significantly reduced proliferation and differentiation (Lemcke et al., 2013), and neurospheres obtained from Cx36 knockdown had reduced numbers of differentiated neurons and coupling (Hartfield et al., 2011).

Interestingly, Cx43 appeared to form a scaffold in the RMS/Cx43/Brdu$^+$ study (Miragall et al., 1997), which may correlate with the dependence of Cx43 for migration. Repopulation of hippocampal neurons and glia occurs from the migration and differentiation of radial glia (RG)-like precursors that are dependent upon Cx43 and Cx30, as shown through conditional KO and viral ablation models in which there were reductions in Prox1$^+$ cells, NeuN and Ki67 staining in the dentate gyrus (Kunze et al., 2009). Therefore, repopulation of hippocampal parenchyma depends upon Cx43 and Cx30 expression, suggesting that chronic inflammation or pathogenesis in this region could impede migration and repopulation after injury or inflammation by a Cx dependent mechanism.

Cx43 is essential for neuronal migration in the developing brain, and fetal infection/inflammation that reduces Cx43 expression can have devastating consequences. Using a conditional KO model in which the c-terminal tail of Cx43 was deleted in nestin$^+$ cells demonstrated that neurons did not migrate to the cortical plate, and halted at the intermediate zone (Cina et al., 2009). Proper migration depends on surface expression of Cx, but not GJC, as seen using a mutant model in which the conserved tyrosine in the third transmembrane domain of Cx43 and Cx26 was deleted. This deletion prevents the opening of GJ channels, but Cx participation in surface adhesion and migration is conserved (Elias et al., 2007). Therefore neuronal migration depends on both surface expression and an intact c-terminal tail of Cx43, but not GJC. Reduced expression of Cx43 during inflammation of the neonatal brain could have permanent effects from impaired neuronal migration, such as ganglion cell layer deterioration observed in congenital ocular toxoplasmosis from *T. gondii* infection (Safar et al., 1995), and reduced white matter area and atrophy seen in adult brains after fetal influenza infection (Fatemi et al., 2008). We propose that reducing inflammation from infectious agents or maintaining Cx expression within the CNS will increase progenitor cell migration and neuronal differentiation by maintaining the Cx scaffold needed for migration and also by allowing differentiation signals to correctly influence progenitors.

GJC IS ESSENTIAL FOR PROPER MYELINATION

Myelin is essential for rapid neuronal signaling by contributing to saltatory conduction. Without myelination of axons, depolarization dissipates in a distance dependent manner and signals cannot be conveyed to post-synaptic membranes. Loss of myelination is a hallmark of various diseases that result in motor and cognitive deficit, such as multiple sclerosis (MS), amyotropic lateral sclerosis, and Alzheimer's disease, but the contribution of infectious agents upon demyelination has not been examined.

Oligodendrocytes are responsible for myelinating neurons in the CNS, and astrocytes provide metabolic support of this process. Homotypic Cx47 and Cx32 GJs couple oligodendrocytes to themselves, and heteromeric Cx47/Cx43 GJs couple oligodendrocytes to astrocytes (**Figure 1**; Rash et al., 2001). Double KO models of Cx47/Cx32 or Cx47/Cx43 have severe myelin abnormalities and die within 90 days (Tress et al., 2012; May et al., 2013), but single knockouts of Cx32 or Cx47 do not result in premature death or alterations in GJC (Menichella et al., 2003). This indicates there may be compensatory function when one Cx is disrupted, and both forms of Cxs need to be down regulated in order for a pathogenic phenotype to arise. In line with the KO model, Experimental Autoimmune Encephalitis (EAE) mouse models induced by intraperitoneal injection with recombinant myelin oligodendrocyte glycoprotein (MOG) had reduced levels of Cx32 and Cx47 during relapsing stages of inflammation (Eugenin et al., 2012; Markoullis et al., 2012). Further evidence using an inducible double KO model indicates lack of astrocytic Cx43 and Cx30 produces widespread myelinic edema and vacuolization, accompanied by hippocampal CA1 region-specific pathology (Lutz et al., 2009). Therefore, localized acute neuroinflammation may have long term effects upon myelination by downregulating Cxs essential for maintaining oligodendrocyte integrity.

Acute inflammation can also induce demyelination through the activation of microglia that secrete toxic mediators. Microglia are resident phagocytic cells of the CNS that migrate to points of infection and injury to engulf debris, clear infection, and amplify inflammation, but they are not always beneficial in eliminating neuropathogenesis without inducing severe damage to myelin. LPS activated microglia reduce the production of myelin basic protein (MBP), and induce oligodendrocyte progenitor cell (OPC) death by releasing TNFα (Pang et al., 2010) through a mechanism mediated by glutamate that opens uHCs (Takeuchi et al., 2006). *In vivo* activation of oligodendrocyte TNFα receptor induces severe myelin vacuolization and death (Akassoglou et al., 1998). Interestingly, in the above study, OPC death did not occur until after 16 h of LPS treatment, *in vitro*. However, after 48 h nearly all OPCs died (Pang et al., 2010), indicating acute inflammation has the capability of reducing remyelination with devastating consequences long after initial infection/inflammation if microglia are not deactivated. Therefore, TNFα induces the release of glutamate through uHCs in microglia (Takeuchi et al., 2006) during chronic or relapsing/remitting inflammation (**Figure 1**) in the CNS will result in demyelination.

Aside from inducing OPC death, microglial release of TNFα (Pang et al., 2010) may indirectly induce demyelination by dysregulation of GJC (Karpuk et al., 2011) and water content in astrocytes (Sharma et al., 2010). Direct LPS injection into mice spinal cords induces continued demyelination over 30 days with significant down regulation of astrocytic Cx30, Cx43, and Aquaporin-4 (AQP4) after 3 days (Sharma et al., 2010). In post-mortem brain slices of patients with Balo's disease, a demyelinating condition similar to MS, immunostaining revealed reductions in Cx43, Cx32, Cx37, and AQP4 in concentric lesions, but no reactivity of anti-Cx or anti-AQP4 antibodies were detected (Masaki et al., 2012), suggesting that down regulation of these membrane proteins was likely to contribute to the neuro-pathogenesis of disease but may not be the cause. AQP4 is closely associated with astrocyte GJs (Sharma et al., 2010), and its expression is necessary for proper water balance in the CNS (Alexander et al., 2008; Rama Rao et al., 2014; Wu et al., 2014). Therefore, demyelinating diseases result in down regulation of Cxs and AQP4 that are essential for maintaining myelin integrity.

Reduced expression of AQP4 induces astrocyte swelling (Alexander et al., 2008) and retraction of endfeet (Alvestad et al., 2013) associated with increased brain edema in rat models of hepatic encephalopathy (Jayakumar et al., 2014), which correlated with depolymerization of actin cytoskeleton in cultured human astrocytes (Nicchia et al., 2005). In agreement, whole cell patch clamp of *S. aureus* infected astrocytes have increased membrane capacitance (Karpuk et al., 2012). Therefore, localized acute neuroinflammation may have long term effects upon myelination by downregulating Cxs essential for maintaining oligodendrocyte integrity. In contrast to Balo's disease, brain slices from human MS patients had increased expression of Cx43 likely due to astrogliosis, but similar reduced expression of Cx32 in chronic active lesions (Eugenin et al., 2012). Therefore, decreased Cx32 expression due to inflammation may be sufficient to impair myelination in humans, and the pathology of MS differs from Balo's disease.

We propose Cx expression and GJC between oligodendrocytes are essential for myelination, and reducing inflammation rapidly may decrease damage to myelin. As stated above, reduced inflammation will also help reduce the pathogenic effects upon plasticity, progenitor cell migration, and cellular differentiation.

Cx EXPRESSION AND GJC ARE ENHANCED IN IMMUNE CELLS TO RESOLVE PATHOGENESIS IN THE CNS

ACTIVATED MICROGLIA AMPLIFY INFLAMMATION AND ADVERSELY AFFECT NEURONS BY RELEASING TOXIC MEDIATORS THROUGH uHCS

In all CNS diseases, immune cells play a critical role in controlling inflammation. However, unresolved inflammation in the CNS can have devastating consequences, including impaired myelination (Akassoglou et al., 1998; Markoullis et al., 2012) sensorimotor deficits (Han et al., 2014), and loss of signaling that contributes to learning and memory (Koster-Patzlaff et al., 2009; Wang and Belousov, 2011). Activated microglia release pro-inflammatory factors, such as ATP (Orellana et al., 2013) and glutamate (Takeuchi et al., 2006), through uHCs that serve to amplify inflammation with devastating effects if left unchecked. Neuronal beading, a sign of neurotoxicity, was induced by activated microglial release of glutamate from Cx32 uHCs (Takeuchi et al., 2005, 2006). Treatment of cultured neurons with either IL-1β or TNFα reduced the number of neuronal processes and expression of microtubule-associated protein 2. There was also an increase of intracellular and extracellular glutamate production and mitochondrial release of glutaminase, which collectively induced neuronal death after 3 days (Ye et al., 2013). TNFα mediated reduction of excitatory amino acid transporter (EAAT) and glutamate toxicity is also attributed to neuronal death in Japanese encephalitis virus (JEV; Chen et al., 2012), West Nile virus (WNV; Blakely et al., 2009), and Sindbis virus (Carmen et al., 2009). Therefore, release of pro-inflammatory factors through uHCs from activated microglia can have devastating affects upon CNS parenchyma.

PERIPHERAL IMMUNE INVASION IS NECESSARY FOR CLEARANCE OF INFLAMMATION IN THE CNS, AND IS HIGHLY DEPENDENT ON Cx EXPRESSION AND GJC

Immune responses in the CNS can be broken down into three general phases: first, the secretion of chemotactic factors from damaged areas; second, the migration of immune cells to the point of injury, inflammation, or pathogen invasion; and third, the clearance of debris, infectious agents and recovery. Cx expression and GJC are critical for each of these phases and for generating an adaptive immune response. In monocytes, inflammatory signals induce Cx43 expression allowing migration across the BBB (Eugenin et al., 2003). This is a key step that allows population of the CNS with phagocytic antigen presenting cells that are essential for reducing inflammation (Akassoglou et al., 1998).

Macrophages and dendritic cells (DCs) play an essential role in generating and adaptive immune response for clearing bacterial and fungal infections through phagocytosis, antigen presentation, and release of factors that reduce inflammation. Cx43 is essential for phagocytic activity in peritoneal macrophages as seen in a mouse model implementing immunostaining and confocal microscopy of Cx43, LAMP-2, and dyna beads (Anand et al., 2008), and paralleled phagocytic mechanisms involving Cx43 in monocyte-derived macrophages in the CNS are likely. Phagocytosis serves to clear debris from damaged parenchyma, followed by antigen processing and presentation. A study using CX3CR1^{GFP+} mice (to endogenously label antigen presenting cells) immunized with MOG (to induce neuroinflammation), found CD11c^{+} DCs abundantly infiltrated CNS parenchyma, were positive for myelin antigen after 21 days, and participated in cross-presentation with T-cells (Sosa et al., 2013). Antigen cross-presentation at immunological synapses is critical for adaptive immune response, and membrane bound Cx43 is a necessary component of supramolecular activation clusters (Mendoza-Naranjo et al., 2011). Coupled cells are able to share antigens and trigger response in cytotoxic T lymphocytes even when some cells were never exposed directly to a pathogen (Neijssen et al., 2005). Using a mimetic peptide targeting the extracellular portion of Cx43, DC to T-cell (Elgueta et al., 2009) and DC to DC (Sosa et al., 2013) cross presentation and activation is inhibited *in vitro*, indicating Cx43 is necessary for efficient adaptive immune response in the CNS.

Collectively, these data suggest that immune cells utilize GJs to enhance adaptive immune responses and clearance of inflammation from infectious agents in the CNS.

SEVERAL PATHOGENS USE GJC TO SPREAD INFECTION AND INFLAMMATION

In general, parenchymal GJC is shut down in the CNS upon initial inflammation (Koster-Patzlaff et al., 2007, 2009; Karpuk et al., 2011) and followed by reduction in Cx expression (Esen et al., 2007). In a mouse model of *S. aureus* infection, astrocyte GJC was dramatically reduced near the margins of abcesses after 3 days of infection, but there were no significant changes in protein expression of Cx43 or Cx30. At this same timepoint opening of uHC was evident as compared with uninfected contralateral brain slices (Karpuk et al., 2012), indicating that early infection reduces GJC and opens uHCs in astrocytes. Interestingly, the timecourse for uHC channel opening in astrocytes was similar to the timecourse of uHC channel opening induced by amyloid-β treatment. Neurotoxicity induced by amyloid-β was associated with uHC opening (Panx1 and Cx43) and subsequent release of glutamate and ATP (Orellana et al., 2011).

Release of toxic factors through uHCs during infection promotes inflammation and induces neurotoxicity. *In vivo* ATP imaging of rat spinal cord after laminectomy and weight drop injury found a dramatic reduction in the amount of ATP released around the site of injury in Cx43 KO or knockdown models (Huang et al., 2012; O'Carroll et al., 2013), with correlative improvement in Bousso mouse scale (BMS) score for locomotion. However, complete recovery in action potential propagation was not achieved in Cx43 KO mice after 7 days (Huang et al., 2012), which indicates astrocytic Cx43 uHC release of ATP during inflammation is not the only contributor to neuronal damage. Moreover, the release of ATP acts as a chemoattractant for activated, pro-inflammatory microglia that further amplify injury through increased release of toxic factors (for a review of microglia chemoattraction see Chekeni and Ravichandran, 2011). Therefore, infectious agents amplify inflammation from the release of toxic mediators through uHC that attract microglia and amplify inflammation. Thus, we propose that the use of uHC blockers during CNS inflammation will reduce neuronal damage and improve prognosis.

PARTICIPATION OF GJ AND uHC IN BYSTANDER KILLING BY SEVERAL PATHOGENS

Bystander killing is the process by which an infected cell induces the death of a neighboring uninfected cell. Bystander killing occurs between microglia-microglia (Ribot et al., 2007), astrocytes-astrocytes (Eugenin and Berman, 2007), astrocytes-neurons (Loov et al., 2012), and astrocytes-endothelial cells (Eugenin et al., 2011), and evidence suggests this process is GJ mediated (Eugenin and Berman, 2007; Eugenin et al., 2011). However, the exact mechanisms involving bystander cell death observed in pathogenic conditions of the CNS are not completely understood.

GJs amplify injury and induce bystander cell death during metabolic stress. Bcl2 is an anti-apoptotic mitochondrial protein, and can provide resistance to cell death when transfected into a glioma cell line. Using a Cx43 double-transfectant (Bcl-2^{+}Cx43^{+}) and selecting for varying degrees of expression of each protein revealed that bystander cell death due to metabolic stress was in direct proportion to the number and density of GJs with less resistant neighbors (Lin et al., 1998). Therefore, the ability to resist apoptosis from high expression of Bcl-2 did not protect cells from death if a weaker neighbor was compromised, as long as there was active GJ coupling. This was repeated with variations of stress in mono-cultures and mixed cultures, and the same conclusion was drawn (Lin et al., 1998).

Interestingly, HIV may have the ability to control GJC for its own benefit by developing resistance to cell death in cultured human astrocytes, while inducing the death of uninfected bystanders in a GJ dependent manner (Eugenin and Berman, 2013). Bystander cell death was reduced using GJ blockers, as did blocking IP_3 or cytochrome c (CytC) mediated signaling. CytC is a pro-apoptotic signaling molecule that mediates caspase activation by participating in apoptosome formation, and microinjection of CytC in cultured astrocytes induces 100% apoptosis. However, microinjection of CytC into HIV infected astrocytes did not induce cell death, suggesting that HIV infection protects astrocytes from apotopsis (Eugenin and Berman, 2013). Therefore, HIV infected astrocytes are protected from cell death while killing uninfected neighbors by a GJ dependent mechanism. This is consistent with the low level of chronic inflammation seen in patients with HIV associated neurocognitive disorder (HAND), and indicates HIV may have the ability to control GJs for its own benefit in cell death resistance and dissemination.

The HIV protective effect on astrocytes may parallel cell death resistance in gliomas by a common mechanism involving CytC and Bcl-2. CytC is a key component of the mitochondrial electron transport chain (ETC), and donates an electron to complex IV for the final step in H^{+} transport into the intermembrane space. An early step of apoptosis induced by metabolic stress is the release of CytC from the inner mitochondrial membrane into the cytoplasm, where it interacts with Apaf-1 and activates apoptosome formation. The exact components that selectively induce release of mitochondrial components are currently under investigation, but a series of outer mitochondrial membrane (OMM) proteins have been identified as capable of allowing the release of CytC. Bcl-2 associated X (Bax) protein is a pro-apoptotic cytosolic member of the Bcl-2 family that translocates to the OMM and is associated with CytC release and apoptosis (for reviews of CytC induced apoptosis and the relation with OMM proteins, see Kilbride and Prehn, 2013; Renault et al., 2013). However, in drug resistant gliomas Bax does not translocate, and overexpression of Bcl-2 conveys resistance to staurosporine induced apoptosis (Murphy et al., 2000). Interestingly, resistance to Temozolomide (TMZ), an anti-tumor pharmaceutical studied for treating gliomas, is influenced by Cx43 expression and correlates with mitochondrial alterations including release of CytC and increased Bcl-2/Bax ratios. Knockdown of Cx43 using shRNA in a LN229 glioma cell line increases Bax expression, reduces co-localization of CytC with mitochondria, and conveys reduced TMZ resistance which increases tumor susceptibility to pharmaceutical treatment *in vitro* (Murphy et al., 2000, 2012). Therefore, there is a mechanism by which Cx43 influences mitochondrial signaling pathways involving apoptosis, and HIV may support Cx43 expression and gap junctional intercellular communication (GJIC) to convey a protective effect.

Use of the retroviral beta-galactosidase at gag (BAG) vector and Herpes Simplex Virus Thymidine Kinase (HSV-tk) has expanded our knowledge of the relationship between GJIC and bystander cell death. Several studies validate the use of this retroviral vector in drug delivery intending to preserve brain tissue while treating gliomas, transfected a retroviral BAG vector into rat brain inoculated with a C6 glioma cell line (Short et al., 1990). Vector insertion into glioma cells was much greater than normal brain tissue because only a low level of retroviral incorporation into host DNA is possible in non-replicating cells, while glioma cells are extensively labeled since they rapidly divide (Miller et al., 1990). HSV-tk transfection combined with glanciclovir (GCV) treatment, an antiretroviral pharmaceutical, reduced medullablastoma xenograft tumor volume (Rosolen et al., 1998), but complete eradication was not shown. However, we propose this technique can be used to hunt tumor cells in the CNS for drug delivery after resection.

Combining HSV-tk therapy with induced expression of Cx43 could have promise as a treatment for malignant gliomas after resection. Resection is not always successful in eradicating gliomas because glial tumor cells are able to disseminate and intersperse themselves in brain parenchyma far from the original tumor mass, making resection ineffective in reducing reemergence. Neural stem cells (NSCs) engineered to secrete IL-12, and bone marrow derived stem cells (BMSCs), have shown the ability to track migrating tumor cells (Ehtesham et al., 2002; Nakamizo et al., 2005), thereby allowing the targeting of interspersed tumor cells. Glioma stem cell tumorspheres have low GJC, and inducing the expression of Cx43 inhibited self-renewal and invasiveness (Yu et al., 2011). Moreover, bystander cell apoptosis is increased in proportion to GJ connectivity (Lin et al., 1998). Therefore, combining HSV-tk therapy with induced expression of Cx43 should be a more effective eliminator of undetectable malignant gliomas after resection. Using a rat model in which C6 glioma cells transfected to express Cx43 were injected into the caudate nucleus, followed by injection of BMSC-tk cells into the tumor site and GCV intraperitoneal injection, animals were able to live six times longer and reduced tumor volume by a quarter of the original size was observed (Huang et al., 2010). Therefore, bystander killing of tumor cells induced by the expression of Cx43 shows promise in combination with current therapies.

CONCLUSIONS

We have outlined the involvement of Cx expression and GJC in neuronal plasticity, myelination, migration, stem cell differentiation, and discussed the devastating effects inflammation and infectious diseases can incur on these processes. Inflammation induced by disease is amplified by the opening of uHCs in microglia, and is followed by peripheral immune invasion into the CNS to resolve the damage. Adaptive immune responses following inflammation in the CNS are also dependent on Cx expression and GJC, and the mechanisms of adaptive immune response beyond antigen cross-presentation are an active field of study. In contrast, GJs are also used by some pathogens to induce bystander killing, and this phenomena has been explored to treat malignant gliomas in a novel HSV-tk/Cx43 "hunter-killer" tactic after resection. Thus, further studies to control GJC and uHCs can improve the detrimental effects of infectious diseases and associated inflammation in the CNS.

ACKNOWLEDGMENTS

We would like to thank the Alfred P. Sloan Foundation Minority fellowship (to Paul Castellano). This work was supported by the National Institutes of Mental Health grant, MH096625, and PHRI funding (to Eliseo A. Eugenin). We thank Stephani Velasquez, Nancy Ruel, Kelsey A. Miller for advice and corrections of the manuscript. We also thank Stephanie Castellano for assisting in the design and assembly of **Figure 1**. The authors had no financial interest.

REFERENCES

Abudara, V., Jiang, R. G., and Eyzaguirre, C. (2001). Acidic regulation of junction channels between glomus cells in the rat carotid body. Possible role of [Ca (2+).](i). *Brain Res.* 916, 50–60. doi: 10.1016/s0006-8993(01)02862-1

Akassoglou, K., Bauer, J., Kassiotis, G., Pasparakis, M., Lassmann, H., Kollias, G., et al. (1998). Oligodendrocyte apoptosis and primary demyelination induced by local TNF/p55TNF receptor signaling in the central nervous system of transgenic mice: models for multiple sclerosis with primary oligodendrogliopathy. *Am. J. Pathol.* 153, 801–813. doi: 10.1016/s0002-9440(10)65622-2

Alexander, J. J., Jacob, A., Cunningham, P., Hensley, L., and Quigg, R. J. (2008). TNF is a key mediator of septic encephalopathy acting through its receptor, TNF receptor-1. *Neurochem. Int.* 52, 447–456. doi: 10.1016/j.neuint.2007.08.006

Alvestad, S., Hammer, J., Hoddevik, E. H., Skare, O., Sonnewald, U., Amiry-Moghaddam, M., et al. (2013). Mislocalization of AQP4 precedes chronic seizures in the kainate model of temporal lobe epilepsy. *Epilepsy Res.* 105, 30–41. doi: 10.1016/j.eplepsyres.2013.01.006

Anand, R. J., Dai, S., Gribar, S. C., Richardson, W., Kohler, J. W., Hoffman, R. A., et al. (2008). A role for connexin43 in macrophage phagocytosis and host survival after bacterial peritoneal infection. *J. Immunol.* 181, 8534–8543. doi: 10.4049/jimmunol.181.12.8534

Ashrafi, G. H., Pitts, J. D., Faccini, A., McLean, P., O'Brien, V., Finbow, M. E., et al. (2000). Binding of bovine papillomavirus type 4 E8 to ductin (16K proteolipid), down-regulation of gap junction intercellular communication and full cell transformation are independent events. *J. Gen. Virol.* 81(Pt. 3), 689–694. doi: 10.1099/vir.0.014746-0

Bennett, M. V., and Zukin, R. S. (2004). Electrical coupling and neuronal synchronization in the Mammalian brain. *Neuron* 41, 495–511. doi: 10.1016/s0896-6273(04)00043-1

Bennett, M. V., Barrio, L. C., Bargiello, T. A., Spray, D. C., Hertzberg, E., and Saez, J. C. (1991). Gap junctions: new tools, new answers, new questions. *Neuron* 6, 305–320. doi: 10.1016/0896-6273(91)90241-q

Blakely, P. K., Kleinschmidt-DeMasters, B. K., Tyler, K. L., and Irani, D. N. (2009). Disrupted glutamate transporter expression in the spinal cord with acute flaccid paralysis caused by West Nile virus infection. *J. Neuropathol. Exp. Neurol.* 68, 1061–1072. doi: 10.1097/nen.0b013e3181b8ba14

Bolaños, J. P., and Medina, J. M. (1996). Induction of nitric oxide synthase inhibits gap junction permeability in cultured rat astrocytes. *J. Neurochem.* 66, 2091–2099. doi: 10.1046/j.1471-4159.1996.66052091.x

Brazel, C. Y., Romanko, M. J., Rothstein, R. P., and Levison, S. W. (2003). Roles of the mammalian subventricular zone in brain development. *Prog. Neurobiol.* 69, 49–69. doi: 10.1016/s0301-0082(03)00002-9

Carmen, J., Rothstein, J. D., and Kerr, D. A. (2009). Tumor necrosis factor-alpha modulates glutamate transport in the CNS and is a critical determinant of outcome from viral encephalomyelitis. *Brain Res.* 1263, 143–154. doi: 10.1016/j.brainres.2009.01.040

Chekeni, F. B., and Ravichandran, K. S. (2011). The role of nucleotides in apoptotic cell clearance: implications for disease pathogenesis. *J. Mol. Med. (Berl)* 89, 13–22. doi: 10.1007/s00109-010-0673-7

Chen, C. J., Ou, Y. C., Chang, C. Y., Pan, H. C., Liao, S. L., Chen, S. Y., et al. (2012). Glutamate released by Japanese encephalitis virus-infected microglia involves TNF-alpha signaling and contributes to neuronal death. *Glia* 60, 487–501. doi: 10.1002/glia.22282

Cina, C., Maass, K., Theis, M., Willecke, K., Bechberger, J. F., and Naus, C. C. (2009). Involvement of the cytoplasmic C-terminal domain of connexin43 in neuronal migration. *J. Neurosci.* 29, 2009–2021. doi: 10.1523/jneurosci.5025-08.2009

Crow, D. S., Beyer, E. C., Paul, D. L., Kobe, S. S., and Lau, A. F. (1990). Phosphorylation of connexin43 gap junction protein in uninfected and Rous sarcoma virus-transformed mammalian fibroblasts. *Mol. Cell. Biol.* 10, 1754–1763.

Dobrenis, K., Chang, H. Y., Pina-Benabou, M. H., Woodroffe, A., Lee, S. C., Rozental, R., et al. (2005). Human and mouse microglia express connexin36 and functional gap junctions are formed between rodent microglia and neurons. *J. Neurosci. Res.* 82, 306–315. doi: 10.1002/jnr.20650

Doetsch, F., Garcia-Verdugo, J. M., and Alvarez-Buylla, A. (1997). Cellular composition and three-dimensional organization of the subventricular germinal zone in the adult mammalian brain. *J. Neurosci.* 17, 5046–5061.

Duffy, H. S., John, G. R., Lee, S. C., Brosnan, C. F., and Spray, D. C. (2000). Reciprocal regulation of the junctional proteins claudin-1 and connexin43 by interleukin-1beta in primary human fetal astrocytes. *J. Neurosci.* 20, 597–603. doi: 10.1167/iovs.08-2606

Ehtesham, M., Kabos, P., Gutierrez, M. A., Chung, N. H., Griffith, T. S., Black, K. L., et al. (2002). Induction of glioblastoma apoptosis using neural stem cell-mediated delivery of tumor necrosis factor-related apoptosis-inducing ligand. *Cancer Res.* 62, 7170–7174. doi: 10.1038/nn.3019

Elgueta, R., Tobar, J. A., Shoji, K. F., De Calisto, J., Kalergis, A. M., Bono, M. R., et al. (2009). Gap junctions at the dendritic cell-T cell interface are key elements for antigen-dependent T cell activation. *J. Immunol.* 183, 277–284. doi: 10.4049/jimmunol.0801854

Elias, L. A., Wang, D. D., and Kriegstein, A. R. (2007). Gap junction adhesion is necessary for radial migration in the neocortex. *Nature* 448, 901–907. doi: 10.1038/nature06063

Esen, N., Shuffield, D., Syed, M. M., and Kielian, T. (2007). Modulation of connexin expression and gap junction communication in astrocytes by the gram-positive bacterium S. aureus. *Glia* 55, 104–117. doi: 10.1002/glia.20438

Eugenin, E. A., and Berman, J. W. (2007). Gap junctions mediate human immunodeficiency virus-bystander killing in astrocytes. *J. Neurosci.* 27, 12844–12850. doi: 10.1523/jneurosci.4154-07.2007

Eugenin, E. A., and Berman, J. W. (2013). Cytochrome c dysregulation induced by HIV infection of astrocytes results in bystander apoptosis of uninfected astrocytes by an IP and calcium-dependent mechanism. *J. Neurochem.* 127, 644–651. doi: 10.1111/jnc.12443

Eugenin, E. A., Basilio, D., Saez, J. C., Orellana, J. A., Raine, C. S., Bukauskas, F., et al. (2012). The role of gap junction channels during physiologic and pathologic conditions of the human central nervous system. *J. Neuroimmune Pharmacol.* 7, 499–518. doi: 10.1007/s11481-012-9352-5

Eugenin, E. A., Clements, J. E., Zink, M. C., and Berman, J. W. (2011). Human immunodeficiency virus infection of human astrocytes disrupts blood-brain barrier integrity by a gap junction-dependent mechanism. *J. Neurosci.* 31, 9456–9465. doi: 10.1523/JNEUROSCI.1460-11.2011

Eugenin, E. A., Branes, M. C., Berman, J. W., and Saez, J. C. (2003). TNF-alpha plus IFN-gamma induce connexin43 expression and formation of gap junctions between human monocytes/macrophages that enhance physiological responses. *J. Immunol.* 170, 1320–1328. doi: 10.4049/jimmunol.170.3.1320

Faccini, A. M., Cairney, M., Ashrafi, G. H., Finbow, M. E., Campo, M. S., and Pitts, J. D. (1996). The bovine papillomavirus type 4 E8 protein binds to ductin and causes loss of gap junctional intercellular communication in primary fibroblasts. *J. Virol.* 70, 9041–9045.

Fatemi, S. H., Reutiman, T. J., Folsom, T. D., Huang, H., Oishi, K., Mori, S., et al. (2008). Maternal infection leads to abnormal gene regulation and brain atrophy in mouse offspring: implications for genesis of neurodevelopmental disorders. *Schizophr. Res.* 99, 56–70. doi: 10.1016/j.schres.2007.11.018

Filson, A. J., Azarnia, R., Beyer, E. C., Loewenstein, W. R., and Brugge, J. S. (1990). Tyrosine phosphorylation of a gap junction protein correlates with inhibition of cell-to-cell communication. *Cell Growth Differ.* 1, 661–668.

Fischer, N. O., Mbuy, G. N., and Woodruff, R. I. (2001). HSV-2 disrupts gap junctional intercellular communication between mammalian cells in vitro. *J. Virol. Methods* 91, 157–166. doi: 10.1016/s0166-0934(00)00260-3

Giaume, C., Cordier, J., and Glowinski, J. (1992). Endothelins inhibit junctional permeability in cultured mouse astrocytes. *Eur. J. Neurosci.* 4, 877–881. doi: 10.1111/j.1460-9568.1992.tb00198.x

Han, Y., Yu, H. X., Sun, M. L., Wang, Y., Xi, W., and Yu, Y. Q. (2014). Astrocyte-restricted disruption of connexin-43 impairs neuronal plasticity in mouse barrel cortex. *Eur. J. Neurosci.* 39, 35–45. doi: 10.1111/ejn.12394

Harris, A. L., and Bevans, C. G. (2001). Exploring hemichannel permeability in vitro. *Methods Mol. Biol.* 154, 357–377. doi: 10.1385/1-59259-043-8:357

Hartfield, E. M., Rinaldi, F., Glover, C. P., Wong, L. F., Caldwell, M. A., and Uney, J. B. (2011). Connexin 36 expression regulates neuronal differentiation from neural progenitor cells. *PLoS One* 6:e14746. doi: 10.1371/journal.pone.0014746

Hsiao, H. J., Liu, P. A., Yeh, H. I., and Wang, C. Y. (2010). Classical swine fever virus down-regulates endothelial connexin 43 gap junctions. *Arch. Virol.* 155, 1107–1116. doi: 10.1007/s00705-010-0693-1

Huang, C., Han, X., Li, X., Lam, E., Peng, W., Lou, N., et al. (2012). Critical role of connexin 43 in secondary expansion of traumatic spinal cord injury. *J. Neurosci.* 32, 3333–3338. doi: 10.1523/JNEUROSCI.1216-11.2012

Huang, Q., Liu, X. Z., Kang, C. S., Wang, G. X., Zhong, Y., and Pu, P. Y. (2010). The anti-glioma effect of suicide gene therapy using BMSC expressing HSV/TK combined with overexpression of Cx43 in glioma cells. *Cancer Gene Ther.* 17, 192–202. doi: 10.1038/cgt.2009.64

Jayakumar, A. R., Tong, X. Y., Curtis, K. M., Ruiz-Cordero, R., Abreu, M. T., and Norenberg, M. D. (2014). Increased toll-like receptor 4 in cerebral endothelial cells contributes to the astrocyte swelling and brain edema in acute hepatic encephalopathy. *J. Neurochem.* 128, 890–903. doi: 10.1111/jnc.12516

John, G. R., Scemes, E., Suadicani, S. O., Liu, J. S., Charles, P. C., Lee, S. C., et al. (1999). IL-1beta differentially regulates calcium wave propagation between primary human fetal astrocytes via pathways involving P2 receptors and gap junction channels. *Proc. Natl. Acad. Sci. U S A* 96, 11613–11618. doi: 10.1073/pnas.96.20.11613

Karpuk, N., Burkovetskaya, M., and Kielian, T. (2012). Neuroinflammation alters voltage-dependent conductance in striatal astrocytes. *J. Neurophysiol.* 108, 112–123. doi: 10.1152/jn.01182.2011

Karpuk, N., Burkovetskaya, M., Fritz, T., Angle, A., and Kielian, T. (2011). Neuroinflammation leads to region-dependent alterations in astrocyte gap junction communication and hemichannel activity. *J. Neurosci.* 31, 414–425. doi: 10.1523/JNEUROSCI.5247-10.2011

Kilbride, S. M., and Prehn, J. H. (2013). Central roles of apoptotic proteins in mitochondrial function. *Oncogene* 32, 2703–2711. doi: 10.1038/onc.2012.348

Koster-Patzlaff, C., Hosseini, S. M., and Reuss, B. (2007). Persistent Borna Disease Virus infection changes expression and function of astroglial gap junctions in vivo and in vitro. *Brain Res.* 1184, 316–332. doi: 10.1016/j.brainres.2007.09.062

Koster-Patzlaff, C., Hosseini, S. M., and Reuss, B. (2009). Loss of connexin36 in rat hippocampus and cerebellar cortex in persistent Borna disease virus infection. *J. Chem. Neuroanat.* 37, 118–127. doi: 10.1016/j.jchemneu.2008.10.004

Kunze, A., Congreso, M. R., Hartmann, C., Wallraff-Beck, A., Huttmann, K., Bedner, P., et al. (2009). Connexin expression by radial glia-like cells is required for neurogenesis in the adult dentate gyrus. *Proc. Natl. Acad. Sci. U S A* 106, 11336–11341. doi: 10.1073/pnas.0813160106

Lemcke, H., Nittel, M. L., Weiss, D. G., and Kuznetsov, S. A. (2013). Neuronal differentiation requires a biphasic modulation of gap junctional intercellular communication caused by dynamic changes of connexin43 expression. *Eur. J. Neurosci.* 38, 2218–2228. doi: 10.1111/ejn.12219

Lin, J. H., Weigel, H., Cotrina, M. L., Liu, S., Bueno, E., Hansen, A. J., et al. (1998). Gap-junction-mediated propagation and amplification of cell injury. *Nat. Neurosci.* 1, 494–500. doi: 10.1038/2210

Lohman, A. W., and Isakson, B. E. (2014). Differentiating connexin hemichannels and pannexin channels in cellular ATP release. *FEBS Lett.* 588, 1379–1388. doi: 10.1016/j.febslet.2014.02.004

Lois, C., and Alvarez-Buylla, A. (1994). Long-distance neuronal migration in the adult mammalian brain. *Science* 264, 1145–1148. doi: 10.1126/science.8178174

Loov, C., Hillered, L., Ebendal, T., and Erlandsson, A. (2012). Engulfing astrocytes protect neurons from contact-induced apoptosis following injury. *PLoS One* 7:e33090. doi: 10.1371/journal.pone.0033090

Lutz, S. E., Zhao, Y., Gulinello, M., Lee, S. C., Raine, C. S., and Brosnan, C. F. (2009). Deletion of astrocyte connexins 43 and 30 leads to a dysmyelinating phenotype and hippocampal CA1 vacuolation. *J. Neurosci.* 29, 7743–7752. doi: 10.1523/JNEUROSCI.0341-09.2009

Marins, M., Xavier, A. L., Viana, N. B., Fortes, F. S., Froes, M. M., and Menezes, J. R. (2009). Gap junctions are involved in cell migration in the early postnatal subventricular zone. *Dev. Neurobiol.* 69, 715–730. doi: 10.1002/dneu.20737

Markoullis, K., Sargiannidou, I., Schiza, N., Hadjisavvas, A., Roncaroli, F., Reynolds, R., et al. (2012). Gap junction pathology in multiple sclerosis lesions and normal-appearing white matter. *Acta Neuropathol.* 123, 873–886. doi: 10.1007/s00401-012-0978-4

Masaki, K., Suzuki, S. O., Matsushita, T., Yonekawa, T., Matsuoka, T., Isobe, N., et al. (2012). Extensive loss of connexins in Balo's disease: evidence for an auto-antibody-independent astrocytopathy via impaired astrocyte-oligodendrocyte/myelin interaction. *Acta Neuropathol.* 123, 887–900. doi: 10.1007/s00401-012-0972-x

May, D., Tress, O., Seifert, G., and Willecke, K. (2013). Connexin47 protein phosphorylation and stability in oligodendrocytes depend on expression of Connexin43 protein in astrocytes. *J. Neurosci.* 33, 7985–7996. doi: 10.1523/JNEUROSCI.5874-12.2013

Meme, W., Ezan, P., Venance, L., Glowinski, J., and Giaume, C. (2004). ATP-induced inhibition of gap junctional communication is enhanced by interleukin-1 beta treatment in cultured astrocytes. *Neuroscience* 126, 95–104. doi: 10.1016/s0306-4522(04)00232-5

Mendoza-Naranjo, A., Bouma, G., Pereda, C., Ramirez, M., Webb, K. F., Tittarelli, A., et al. (2011). Functional gap junctions accumulate at the immunological synapse and contribute to T cell activation. *J. Immunol.* 187, 3121–3132. doi: 10.4049/jimmunol.1100378

Menichella, D. M., Goodenough, D. A., Sirkowski, E., Scherer, S. S., and Paul, D. L. (2003). Connexins are critical for normal myelination in the CNS. *J. Neurosci.* 23, 5963–5973. doi: 10.1523/JNEUROSCI.0325-09.2009

Miller, D. G., Adam, M. A., and Miller, A. D. (1990). Gene transfer by retrovirus vectors occurs only in cells that are actively replicating at the time of infection. *Mol. Cell. Biol.* 10, 4239–4242.

Miragall, F., Albiez, P., Bartels, H., de Vries, U., and Dermietzel, R. (1997). Expression of the gap junction protein connexin43 in the subependymal layer and the rostral migratory stream of the mouse: evidence for an inverse correlation between intensity of connexin43 expression and cell proliferation activity. *Cell Tissue Res.* 287, 243–253. doi: 10.1007/s004410050749

Murphy, K. M., Ranganathan, V., Farnsworth, M. L., Kavallaris, M., and Lock, R. B. (2000). Bcl-2 inhibits Bax translocation from cytosol to mitochondria during drug-induced apoptosis of human tumor cells. *Cell Death Differ.* 7, 102–111. doi: 10.1038/sj.cdd.4400597

Murphy, N., Cowley, T. R., Richardson, J. C., Virley, D., Upton, N., Walter, D., et al. (2012). The neuroprotective effect of a specific P2X(7) receptor antagonist derives from its ability to inhibit assembly of the NLRP3 inflammasome in glial cells. *Brain Pathol.* 22, 295–306. doi: 10.1111/j.1750-3639.2011.00531.x

Musée, J., Mbuy, G. N., and Woodruff, R. I. (2002). Antiviral agents alter ability of HSV-2 to disrupt gap junctional intercellular communication between mammalian cells in vitro. *Antiviral Res.* 56, 143–151. doi: 10.1016/s0166-3542(02)00106-7

Nakamizo, A., Marini, F., Amano, T., Khan, A., Studeny, M., Gumin, J., et al. (2005). Human bone marrow-derived mesenchymal stem cells in the treatment of gliomas. *Cancer Res.* 65, 3307–3318. doi: 10.2106/JBJS.K.01291

Neijssen, J., Herberts, C., Drijfhout, J. W., Reits, E., Janssen, L., and Neefjes, J. (2005). Cross-presentation by intercellular peptide transfer through gap junctions. *Nature* 434, 83–88. doi: 10.1038/nature03290

Nicchia, G. P., Srinivas, M., Li, W., Brosnan, C. F., Frigeri, A., and Spray, D. C. (2005). New possible roles for aquaporin-4 in astrocytes: cell cytoskeleton and functional relationship with connexin43. *FASEB J.* 19, 1674–1676. doi: 10.1096/fj.04-3281fje

O'Carroll, S. J., Gorrie, C. A., Velamoor, S., Green, C. R., and Nicholson, L. F. (2013). Connexin43 mimetic peptide is neuroprotective and improves function following spinal cord injury. *Neurosci. Res.* 75, 256–267. doi: 10.1016/j.neures.2013.01.004

Orellana, J. A., Shoji, K. F., Abudara, V., Ezan, P., Amigou, E., Saez, P. J., et al. (2011). Amyloid beta-induced death in neurons involves glial and neuronal hemichannels. *J. Neurosci.* 31, 4962–4977. doi: 10.1523/JNEUROSCI.6417-10.2011

Orellana, J. A., Montero, T. D., and von Bernhardi, R. (2013). Astrocytes inhibit nitric oxide-dependent Ca(2+) dynamics in activated microglia: involvement of ATP released via pannexin 1 channels. *Glia* 61, 2023–2037. doi: 10.1002/glia.22573

Pang, Y., Campbell, L., Zheng, B., Fan, L., Cai, Z., and Rhodes, P. (2010). Lipopolysaccharide-activated microglia induce death of oligodendrocyte progenitor cells and impede their development. *Neuroscience* 166, 464–475. doi: 10.1016/j.neuroscience.2009.12.040

Pereda, A., O'Brien, J., Nagy, J. I., Bukauskas, F., Davidson, K. G., Kamasawa, N., et al. (2003). Connexin35 mediates electrical transmission at mixed synapses on Mauthner cells. *J. Neurosci.* 23, 7489–7503. doi: 10.1073/pnas.0804793105

Petrasch-Parwez, E., Habbes, H. W., Weickert, S., Lobbecke-Schumacher, M., Striedinger, K., Wieczorek, S., et al. (2004). Fine-structural analysis and connexin expression in the retina of a transgenic model of Huntington's disease. *J. Comp. Neurol.* 479, 181–197. doi: 10.1002/cne.20327

Placantonakis, D. G., Bukovsky, A. A., Aicher, S. A., Kiem, H. P., and Welsh, J. P. (2006). Continuous electrical oscillations emerge from a coupled network: a study of the inferior olive using lentiviral knockdown of connexin36. *J. Neurosci.* 26, 5008–5016. doi: 10.1523/jneurosci.0146-06.2006

Rama Rao, K. V., Verkman, A. S., Curtis, K. M., and Norenberg, M. D. (2014). Aquaporin-4 deletion in mice reduces encephalopathy and brain edema in experimental acute liver failure. *Neurobiol. Dis.* 63, 222–228. doi: 10.1016/j.nbd.2013.11.018

Rash, J. E., Pereda, A., Kamasawa, N., Furman, C. S., Yasumura, T., Davidson, K. G., et al. (2004). High-resolution proteomic mapping in the vertebrate central nervous system: close proximity of connexin35 to NMDA glutamate receptor clusters and co-localization of connexin36 with immunoreactivity for zonula occludens protein-1 (ZO-1). *J. Neurocytol.* 33, 131–151. doi: 10.1023/b:neur.0000029653.34094.0b

Rash, J. E., Yasumura, T., Dudek, F. E., and Nagy, J. I. (2001). Cell-specific expression of connexins and evidence of restricted gap junctional coupling between glial cells and between neurons. *J. Neurosci.* 21, 1983–2000. doi: 10.1111/j.1460-9568.2008.06431.x

Renault, T. T., Teijido, O., Antonsson, B., Dejean, L. M., and Manon, S. (2013). Regulation of Bax mitochondrial localization by Bcl-2 and Bcl-x(L): keep your friends close but your enemies closer. *Int. J. Biochem. Cell Biol.* 45, 64–67. doi: 10.1016/j.biocel.2012.09.022

Reuss, B., Hertel, M., Werner, S., and Unsicker, K. (2000). Fibroblast growth factors-5 and -9 distinctly regulate expression and function of the gap junction protein connexin43 in cultured astroglial cells from different brain regions. *Glia* 30, 231–241. doi: 10.1002/(sici)1098-1136(200005)30:3<231::aid-glia3>3.0.co;2-1

Reuss, B., Dermietzel, R., and Unsicker, K. (1998). Fibroblast growth factor 2 (FGF-2) differentially regulates connexin (cx) 43 expression and function in astroglial cells from distinct brain regions. *Glia* 22, 19–30. doi: 10.1002/(sici)1098-1136(199801)22:1<19::aid-glia3>3.0.co;2-3

Ribot, E., Bouzier-Sore, A. K., Bouchaud, V., Miraux, S., Delville, M. H., Franconi, J. M., et al. (2007). Microglia used as vehicles for both inducible thymidine kinase gene therapy and MRI contrast agents for glioma therapy. *Cancer Gene Ther.* 14, 724–737. doi: 10.1038/sj.cgt.7701060

Rosolen, A., Frascella, E., di Francesco, C., Todesco, A., Petrone, M., Mehtali, M., et al. (1998). In vitro and in vivo antitumor effects of retrovirus-mediated herpes simplex thymidine kinase gene-transfer in human medulloblastoma. *Gene Ther.* 5, 113–120. doi: 10.1038/sj.gt.3300559

Saez, J. C., Berthoud, V. M., Branes, M. C., Martinez, A. D., and Beyer, E. C. (2003). Plasma membrane channels formed by connexins: their regulation and functions. *Physiol. Rev.* 83, 1359–1400. doi: 10.1016/j.cellsig.2010.03.005

Safar, E. H., Abd-el Ghaffar, F. M., Saffar, S. A., Makled, K. M., Habib, K. S., el Abiad, R., et al. (1995). Incidence of Toxoplasma and Toxocara antibodies among out-patients in the Ophthalmic Research Institute, Egypt. *J. Egypt. Soc. Parasitol.* 25, 839–852.

Sharma, R., Fischer, M. T., Bauer, J., Felts, P. A., Smith, K. J., Misu, T., et al. (2010). Inflammation induced by innate immunity in the central nervous system leads to primary astrocyte dysfunction followed by demyelination. *Acta Neuropathol.* 120, 223–236. doi: 10.1007/s00401-010-0704-z

Short, M. P., Choi, B. C., Lee, J. K., Malick, A., Breakefield, X. O., and Martuza, R. L. (1990). Gene delivery to glioma cells in rat brain by grafting of a retrovirus packaging cell line. *J. Neurosci. Res.* 27, 427–439. doi: 10.1002/jnr.490270322

Söhl, G., Odermatt, B., Maxeiner, S., Degen, J., and Willecke, K. (2004). New insights into the expression and function of neural connexins with transgenic mouse mutants. *Brain Res. Brain Res. Rev.* 47, 245–259. doi: 10.1016/j.brainresrev.2004.05.006

Sosa, R. A., Murphey, C., Ji, N., Cardona, A. E., and Forsthuber, T. G. (2013). The kinetics of myelin antigen uptake by myeloid cells in the central nervous system during experimental autoimmune encephalomyelitis. *J. Immunol.* 191, 5848–5857. doi: 10.4049/jimmunol.1300771

Takeuchi, H., Jin, S., Wang, J., Zhang, G., Kawanokuchi, J., Kuno, R., et al. (2006). Tumor necrosis factor-alpha induces neurotoxicity via glutamate release from hemichannels of activated microglia in an autocrine manner. *J. Biol. Chem.* 281, 21362–21368. doi: 10.1074/jbc.m600504200

Takeuchi, H., Mizuno, T., Zhang, G., Wang, J., Kawanokuchi, J., Kuno, R., et al. (2005). Neuritic beading induced by activated microglia is an early feature of neuronal dysfunction toward neuronal death by inhibition of mitochondrial respiration and axonal transport. *J. Biol. Chem.* 280, 10444–10454. doi: 10.1074/jbc.m413863200

Tress, O., Maglione, M., May, D., Pivneva, T., Richter, N., Seyfarth, J., et al. (2012). Panglial gap junctional communication is essential for maintenance of myelin in the CNS. *J. Neurosci.* 32, 7499–7518. doi: 10.1523/JNEUROSCI.0392-12.2012

Wang, N., De Bock, M., Decrock, E., Bol, M., Gadicherla, A., Vinken, M., et al. (2013). Paracrine signaling through plasma membrane hemichannels. *Biochim. Biophys. Acta* 1828, 35–50. doi: 10.1016/j.bbamem.2012.07.002

Wang, Y., and Belousov, A. B. (2011). Deletion of neuronal gap junction protein connexin 36 impairs hippocampal LTP. *Neurosci. Lett.* 502, 30–32. doi: 10.1016/j.neulet.2011.07.018

Willecke, K., Eiberger, J., Degen, J., Eckardt, D., Romualdi, A., Guldenagel, M., et al. (2002). Structural and functional diversity of connexin genes in the mouse and human genome. *Biol. Chem.* 383, 725–737. doi: 10.1515/bc.2002.076

Wu, Q., Zhang, Y. J., Gao, J. Y., Li, X. M., Kong, H., Zhang, Y. P., et al. (2014). Aquaporin-4 mitigates retrograde degeneration of rubrospinal neurons by facilitating edema clearance and glial scar formation after spinal cord injury in mice. *Mol. Neurobiol.* doi: 10.1007/s12035-013-8607-3. [Epub ahead of print].

Yang, S. R., Cho, S. D., Ahn, N. S., Jung, J. W., Park, J. S., Jo, E. H., et al. (2005). Role of gap junctional intercellular communication (GJIC) through p38 and ERK1/2 pathway in the differentiation of rat neuronal stem cells. *J. Vet. Med. Sci.* 67, 291–294. doi: 10.1292/jvms.67.291

Ye, L., Huang, Y., Zhao, L., Li, Y., Sun, L., Zhou, Y., et al. (2013). IL-1beta and TNF-alpha induce neurotoxicity through glutamate production: a potential role for neuronal glutaminase. *J. Neurochem.* 125, 897–908. doi: 10.1111/jnc.12263

Yu, J., Wu, J., Bagchi, I. C., Bagchi, M. K., Sidell, N., and Taylor, R. N. (2011). Disruption of gap junctions reduces biomarkers of decidualization and angiogenesis and increases inflammatory mediators in human endometrial stromal cell cultures. *Mol. Cell. Endocrinol.* 344, 25–34. doi: 10.1016/j.mce.2011.04.011

4

Role of connexin channels in the retinal light response of a diurnal rodent

Angelina Palacios-Muñoz[1], Maria J. Escobar[2], Alex Vielma[1], Joaquín Araya[1], Aland Astudillo[2], Gonzalo Valdivia[1], Isaac E. García[1], José Hurtado[1,3], Oliver Schmachtenberg[1], Agustín D. Martínez[1] and Adrian G. Palacios[1,3]*

[1] *Facultad de Ciencias, Centro Interdisciplinario de Neurociencia de Valparaíso, Universidad de Valparaíso, Valparaíso, Chile*
[2] *Departamento de Electrónica, Universidad Técnico Federico Santa María, Valparaíso, Chile*
[3] *Instituto de Sistemas Complejos de Valparaíso, Valparaíso, Chile*

***Correspondence:**
Adrian G. Palacios, Facultad de Ciencias, Centro Interdisciplinario de Neurociencia de Valparaíso, Universidad de Valparaíso, Pasaje Harrington 287, Valparaíso, Chile
e-mail: adrian.palacios@uv.cl

Several studies have shown that connexin channels play an important role in retinal neural coding in nocturnal rodents. However, the contribution of these channels to signal processing in the retina of diurnal rodents remains unclear. To gain insight into this problem, we studied connexin expression and the contribution of connexin channels to the retinal light response in the diurnal rodent *Octodon degus* (degu) compared to rat, using *in vivo* ERG recording under scotopic and photopic light adaptation. Analysis of the degu genome showed that the common retinal connexins present a high degree of homology to orthologs expressed in other mammals, and expression of Cx36 and Cx43 was confirmed in degu retina. Cx36 localized mainly to the outer and inner plexiform layers (IPLs), while Cx43 was expressed mostly in cells of the retinal pigment epithelium. Under scotopic conditions, the b-wave response amplitude was strongly reduced by 18-β-glycyrrhetinic acid (β-GA) (−45.1% in degu, compared to −52.2% in rat), suggesting that connexins are modulating this response. Remarkably, under photopic adaptation, β-GA increased the ERG b-wave amplitude in degu (+107.2%) while reducing it in rat (−62.3%). Moreover, β-GA diminished the spontaneous action potential firing rate in ganglion cells (GCs) and increased the response latency of ON and OFF GCs. Our results support the notion that connexins exert a fine-tuning control of the retinal light response and have an important role in retinal neural coding.

Keywords: retina, physiology, neural coding, multi-electrode array (MEA), connexins

INTRODUCTION

It is well-accepted that gap junction channels are major components of the nervous system that mediate both electrical and metabolic coupling between neurons and glial cells (Demb and Pugh, 2002; Connors and Long, 2004; Sohl et al., 2005; Bloomfield and Volgyi, 2009). Electrical synapses are formed by gap junctions, which are composed of connexins subunits (Guldenagel et al., 2000; Sohl et al., 2000; Willecke et al., 2002). Many studies of the last decade have revealed the widespread expression of connexins in the nervous system, including the retina (Kar et al., 2012; Saez and Leybaert, 2014) of mammals with different life styles, e.g., primates, guinea pig, ground squirrel, rabbit, rat and mouse. Cx36, which forms channels with small unitary conductance and low voltage sensitivity, is expressed in rodent retina (Al-Ubaidi et al., 2000) in cone and rod photoreceptors, bipolar cells (BCs), AII amacrine cells and ganglion cells (GCs) (Feigenspan et al., 2001, 2004; Guldenagel et al., 2001; Mills et al., 2001; Deans et al., 2002; DeVries et al., 2002; Hidaka et al., 2002, 2004; Lee et al., 2003; Degen et al., 2004; Han and Massey, 2005; Schubert et al., 2005; Dedek et al., 2006; O'Brien et al., 2012). Other connexins are also expressed in retinal neurons. Cx45 has been detected in subpopulations of BCs, amacrine cells and GCs (Lin et al., 2005; Maxeiner et al., 2005; Dedek et al., 2006), and Cx57 in horizontal cells (HCs) (Massey et al., 2003; Hombach et al., 2004; Ciolofan et al., 2007). Moreover, Cx43 is expressed in Müller cells (Janssen-Bienhold et al., 1998; Johansson et al., 1999; Kihara et al., 2006) and cells of the retinal pigment epithelium (Janssen-Bienhold et al., 1998).

The functional contribution of each retinal connexin type has only partly been unveiled. For example, deletion of the Cx36 gene in mouse reduces the b-wave of the electroretinogram (ERG) under scotopic conditions (Guldenagel et al., 2001; Maxeiner et al., 2005), and eliminates ON-center GC responses (Deans et al., 2002). Previous studies have shown that retinal gap junctions are regulated by several factors, including light, circadian rhythm as well as neuromodulators such as nitric oxide and dopamine (Bloomfield and Volgyi, 2009). These factors may induce post-translational modifications of Cxs. For example, under photopic conditions, phosphorylation of mouse Cx36 modulates ON BCs and AII amacrine cell synapses (Kothmann et al., 2009). Limited evidence is available regarding the importance of retinal connexins in nocturnal vs. diurnal species. Lee et al. (2003) have shown that Cx36 is expressed in the retina of guinea pig, a crepuscular rodent. More information has been obtained from studies in zebrafish, a diurnal teleost fish, in which coupling between photoreceptors is controlled by phosphorylation of Cx35, an ortholog of mammalian Cx36, which is part of the mechanisms that control light and dark adaptation cycles (Li et al., 2009). To further understand the role of connexins in the retina of diurnal species, we studied the retinal expression of connexins Cx36 and Cx43 and the general contribution of connexins to retinal light responses in *Octodon degus* (degu), a crepuscular diurnal rodent (Ardiles et al., 2013) that presents a high percentage of cone photoreceptors (30%) (Jacobs et al., 2003) with different spectral sensitivities (500 nm M-cones and 360 nm UV S-cones) (Chavez et al., 2003). The results were compared to rat, a standard nocturnal model with a low percentage of cones (1–3%). We found that general blockage of connexin channels with 18-β-glycyrrhetinic acid (β-GA) (Xia and Nawy, 2003; Pan et al., 2007) *in vivo* and *in vitro* had similar effects on the scotopic light response, but opposing results under photopic adaptation in both rodent species, supporting a differential role of connexins in the retinal cone pathways of diurnal vs. nocturnal species.

MATERIALS AND METHODS

ANIMALS

Adult male and female *Octodon degus* and *Sprague dawley* rats were maintained in the animal facility of the Universidad de Valparaiso, at 20–25°C on a 12-h light-dark cycle, with access to food and water *ad-libitum.* All experiments were approved by the bioethics committee of the Universidad de Valparaiso, in accordance with the bioethics regulation of the Chilean Research Council (CONICYT), and had an approved animal welfare assurance (NIH A5823-01).

WESTERN BLOTS

After ERG experiments, the animals were deeply anesthetized with halothane and decapitated. The right eye (control) was removed immediately and opened along the *ora serrata.* The lens and vitreous humor were removed and the isolated retinas were homogenized in ice-cold lysis buffer containing 50 mM Tris-HCl pH 7.4, 150 mM NaCl, 0.1% Triton X-100, 0.6 mM PMSF and a cocktail of proteases (Sigma P8340) and phosphatase inhibitors (Sigma P5726), at 4°C. The homogenates were centrifuged at 13,000 g for 20 min at 4°C. The supernatant was collected and the total protein contents were determined according to the method of Bradford. Aliquots of tissue samples corresponding to 40 μg of total protein were heated to 100°C for 5 min with Laemmli sample buffer (300 mM Tris-HCl pH 6.8, 50% glycerol, 10% SDS, 500 mM DTT, 0.05% β-mercaptoethanol, 0.01% bromophenol blue) and loaded onto 10% polyacrylamide gels. The proteins were blotted to polyvinyl difluoride membrane (Amersham Biosciences, Buckinghamshire, UK) and blocked for 1 h at room temperature in PBS with 0.2% Tween-20 and 8% non-fat dry milk. The membrane was incubated overnight at 4°C with a mouse monoclonal anti-Cx35/36 antibody (MAB 3045, Chemicon; dilution 1:1000); or a rabbit polyclonal anti-Cx43 antibody (C-6219, Sigma; dilution: 1:300) in blocking solution. The membrane was rinsed in PBS with 0.2% Triton X-100 (four times, 15 min each time) and incubated for 2 h at room temperature with a 1:1000 dilution of horseradish peroxidase-conjugated goat anti-mouse or anti-rabbit IgG antibody (Jackson Immuno Research Laboratories). Thereafter, the blot was washed in PBS with 0.2% Triton X-100 and bound immunoglobulins were visualized with SuperSignal West Pico chemiluminescent substrate (Pierce Biotechnology, Rockford, IL, USA) on Kodak BioMax Light film.

IMMUNOHISTOCHEMISTRY

For immunohistochemistry, eyes were fixed in 4% paraformaldehyde (PFA) in PBS (0.1 M, pH 7.4) for 10 min. A small hole was cut out with a fine needle through the *ora serrata,* and the eye was immersed again in 4% PFA (2 h at 4°C) followed by three washes in PBS. For cryoprotection, the fixed eye was immersed in sequentially increasing concentrations of sucrose in PBS (10%, 20%, 1 h in each concentration and finally in 30% at 4°C overnight). Thereafter, the eye was frozen in tissue freezing medium (Tissue-Tek OCT Compound, Sakura Finetek Europe). Frozen sections of 15 μm thickness were obtained with a cryostat (Leica CM 1900, Germany) at −20°C, and mounted on poly-L-lysine coated microscope slides. To wash out the cryomatrix, the sections were rinsed four times with PBS plus 0.3% Triton X-100 (TBS). Nonspecific binding was blocked by incubating the sections for 1 h at room temperature in TBS containing 10% normal goat serum (NGS). Retinal sections were incubated overnight at 4°C with the primary antibodies diluted in the blocking solution: Mouse monoclonal anti-Cx35/36 (MAB 3045, Chemicon; dilution 1:200), or rabbit polyclonal anti-Cx43 (C-6212, Sigma; dilution 1:300). The sections were washed four times with TBS for 15 min and incubated for 2 h at room temperature with the secondary antibodies diluted in blocking solution; Cy3-conjugated affinity pure Goat anti-mouse IgG antibody (113-166-072, Jackson Immuno Research Lab; dilution: 1.1000) or Cy2-conjugated affinity pure Goat anti-rabbit IgG antibody (111-226-047, Jackson Immuno Research Lab; dilution 1:1000). Some slices were incubated for 30 min with propidium iodide (dilution 1:8000 in TBS) to stain nuclei. The retinal sections were mounted in Fluomount (Dako Industries, Carpenteria, CA, USA). In control experiments, non-specific binding was tested by omitting the primary antibody. The slices were imaged using a spinning disk confocal microscope Olympus (BX-DSU, Olympus, Japan) and captured using an ORCA-2 camera (Hamamatsu Photonics, Japan). Images were acquired and processed using the Cell-R program (Olympus Soft Imaging Solutions, Germany) and processed using AutoQuantX 2.2.2 (Media Cybernetics, USA) deconvolution software.

BIOINFORMATICS

Multiple alignments followed by neighbor joining and bootstrap analyses were performed to align protein sequences of different connexins genes from human, guinea pig, degu and rat. Connexins of the α, β, γ, and δ groups were included in the dendrogram. Gap junction orthologs tend to group together whereas paralog sequences are further apart (Volgyi et al., 2013a). All sequences for degu shown in **Figure 1** were extracted from PUBMED (Gene). The species and protein sequence IDs used for this analysis are: *Homo Sapiens* Cx43 NP_000156.1,

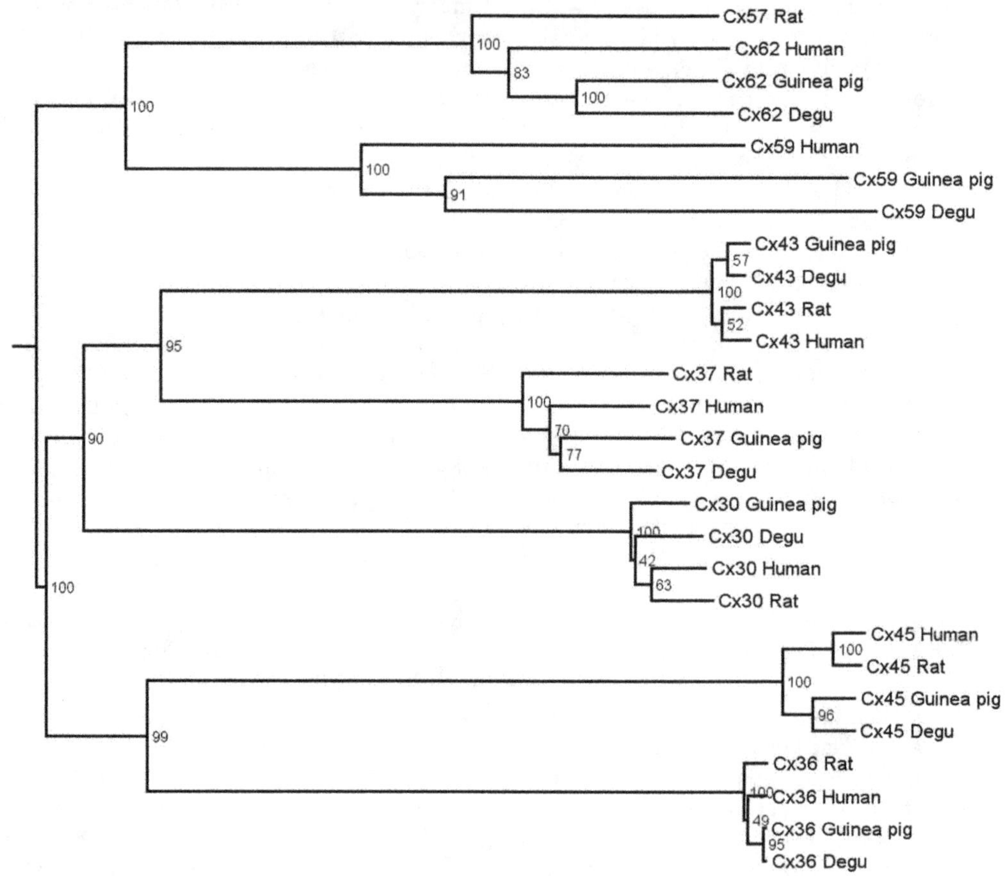

FIGURE 1 | Phylogenetic tree of gap junctions present in the retina of different mammals. Multiple alignments with subsequent neighbor-joining and bootstrap analysis were performed on protein sequences of human, guinea pig, degu, and rat. Connexin orthologs tend to group together whereas the paralog sequences are further apart.

Cx37 NP_002051.2, Cx59 NP_110399.2, Cx62 NP_115991.1, Cx30 NP_001103689.1, Cx45 NP_005488.2, Cx36 NP_065711.1; *Cavia porcellus* Cx43 NP_001166219.1, Cx37 XP_003471533.1, Cx59 XP_005008590.1, Cx62 XP_005008615.1, Cx30 XP_00500 7068.1, Cx45 XP_003465784.1, Cx36 XP_003475711.1; *Octodon degus* Cx37 XP_004641059.1, Cx59 XP_004648949.1, Cx62 XP_ 004649057.1, Cx30 XP_004639430.1, Cx45 XP_004633888.1, Cx36 XP_004643177.1, CX43 XP_004630289.1; *Rattus norvegicus* Cx43 NP_036699.1, Cx37 NP_067686.1, Cx57 NP_00116 6979.1, Cx30 NP_445840.1, Cx45 NP_001078850.1, Cx36 NP_062154.1.

IN VIVO ERG

All procedures and the optical stimulation apparatus of the ERG system have been described previously (Chavez et al., 2003; Peichl et al., 2005). In brief, the a-wave ERG corresponds to the photoreceptor response and the b-wave to the depolarization of ON BCs (Brown, 1968). For scotopic conditions, animals were pre-adapted for 2 h to total darkness, and for photopic conditions, they were adapted to constant background light for 20 min with a quartz tungsten lamp (150 W) producing an illumination of 240 $\mu W/cm^2 sr$ at the cornea. After halothane induction, animals were anesthetized with ketamine (Troy laboratories, Smithfield, Australia) and xylazine (Bayer SA, Brazil) and maintained during an experiment at 32°C with a thermoregulated bed pad. After local anesthesia with lidocaine (1%) and atropine sulfate (1%), a silver/silver chloride ring electrode was placed at the cornea and a subcutaneous platinum electrode on the skin was used as reference. A xenon lamp (LBLS-509 Sutter Instruments, Novato, CA, USA), with a monochromator (1200 lines mm^{-1} grating, ORIEL, Stratford, CT, USA) was used to produce a 500 nm narrowband (20 nm half-bandwidth) light stimulus. Optical isolation from secondary monochromator wavelength emissions was obtained by using a RG500 long-pass filter. Flash duration (10 ms under scotopic and 30 ms under photopic conditions) and flash intervals (15 or 1 s for scotopic and photopic recordings, respectively) were controlled by an electronic shutter (Uniblitz, Vincent Associates, Rochester, NY, USA) operated by custom software. The intensity range spanned between 0.065 and 65.65 photons/μm^2 for scotopic conditions, and between 4.3 and 215 photons/μm^2 under photopic conditions. ERG light-evoked responses were amplified with an AC/DC amplifier (A-M Systems, Model 3000, Carlsbourg, WA, USA), band-pass filtered between 1 and 100 Hz and digitalized with an A/D interface (CB-68LP, National Instruments, Austin, TX, USA). Each ERG corresponds to the average of 20–30 flashes.

MULTI-ELECTRODE RECORDING, ANIMAL PREPARATION, AND VISUAL STIMULATION

A multi-electrode array (MEA USB-256, Multichannel Systems GmbH, Germany) for *in vitro* isolated retina experiments was used to record action potential firing from a population of GCs. For MEA experiments, the animals were dark-adapted and deeply anesthetized with halothane before decapitation. Under dim red light, both eyes were enucleated and one of the retinas was diced into quarters while the other was stored in AMES medium in the dark for further experiments. For recordings, one piece of retina was mounted (GCs down) onto a dialysis membrane placed into a ring device mounted in a traveling (up/down) cylinder, which was moved to contact the electrode surface of the MEA recording array. Visual stimuli were generated by a custom software created with PsychoToolbox (Matlab) on a MiniMac Apple computer and projected onto the retina with a LED projector (PLED-W500, Viewsonic, USA) equipped with an electronic shutter (Vincent Associates, Rochester, USA) and connected to an inverted microscope (Lens 4×, Eclipse TE2000, NIKON, Japan). The image was conformed by 380 × 380 pixels, each covering 5 μm^2. Since rodents are dichromatic (green and blue/UV cones), in our experiments with checkerboard stimuli only the B (blue) and G (green) beams of the projector were used, while the R (red) channel was used for signal synchronization. For the measurement of GC receptive fields (RF), a checkerboard stimulus with a bin size of 100 µm was used at rate of 60 fps. Optical density filters in the optical path were used to control final light intensity. A CCD camera (Pixelfly, PCO, USA) attached to the microscope was used for online visualization and calibration of the light stimuli projected onto the recording array. With the use of checkerboard as stimulus only the photopic condition was tested.

DRUG APPLICATIONS

The general connexin channel blocker, 18-β-glycyrrhetinic acid (β-GA, G10105, Sigma) was dissolved in ethanol (EtOH) and aliquots were diluted in PBS (0.1 M, pH 7.4). Final concentrations of EtOH were less than 0.2% in the β-GA solution. For *in vivo* ERG experiments, intravitreal injections of β-GA (10 µl of final solutions), were performed after local anesthesia (see below and Delgado et al., 2009), using a fine needle through the *ora serrata*. The volume of the rat and degu vitreous was estimated as 0.15 ml (Naarendorp and Williams, 1999) and 0.22 ± 0.02 ml, respectively, and used to calculate the final concentration of β-GA in the eye (Moller and Eysteinsson, 2003). The ERG a- and b-wave were assessed 30 min after β-GA injection from the same eye recorded previously as control. The effects of the gap junction blocker remained unaltered for several hours (data not shown). Moreover, the injection of 0.2% EtOH in PBS (10 µl) did not produce changes in the ERG amplitude (data not shown). For *in vitro* multi-electrode recording of isolated retina, β-GA (50 µM) was added to the perfusion solution 30 min before GC recording.

DATA ANALYSIS AND CELL CLASSIFICATION

In ERG recordings, peak amplitudes of a- and b-wave were calculated by fitting the peak of the response to a polynomial function of fifth order using IGOR Pro Software, subtracting the averaged baseline. Results are shown as mean ± s.e.m, and statistical significance was evaluated by a paired Student's *t*-test, using Graph Pad InStat software (La Jolla, CA, USA). For MEA experiments, MC_Rack software (MultichannelSystems) served for signal visualization. After each experiment, the software Offline Sorter (Plexon, TX, USA) was used for spike sorting and Neuroexplorer (NexTechnologies, Madison, WI, USA) for further statistical analysis. Moreover, a custom software was used to calculate the spike-triggered average (STA) associated with the corresponding linear RF of each GC. As a result of the STA processing, a spatiotemporal volume characterizing the dynamics of the cellular response is obtained. In the time domain, the spatiotemporal RF is formed by 18 images corresponding to 300 ms before the spike event. The pixel with the maximal deviation from the mean was detected and a 2D Gaussian fit was adjusted in order to estimate the center of the RF. The temporal profile of the GC was obtained extracting for each frame the value of the pixel located at the position of the RF center. A STA was computed for each GC under control and β-GA conditions. Either ON or OFF cell characterization was obtained from the temporal profiles of the estimated RF. Following a procedure similar to (Chichilnisky and Kalmar, 2002; Field et al., 2007), temporal cell profiles were used for cell classification in terms of ON/OFF type. Furthermore, using principal component analysis (PCA), the temporal profiles were projected on a space with a lower dimensionality, and the classification was finally performed using k-means. Additionally, the time-to-peak parameter was measured for each temporal profile, which refers to the time of the temporal profile, previous to the occurrence of a spike, with the highest deviation being positive (GC ON) or negative (GC OFF).

RESULTS

HIGH DEGREE OF HOMOLOGY BETWEEN RETINAL CONNEXINS OF DEGU AND OTHER MAMMALS AND EXPRESSION OF Cxs36 AND 43 IN DEGU RETINA

Figure 1 shows the estimated phylogenetic tree for the main connexins present in vertebrate retinas. Compared to other mammals, degu shows similarities in terms of protein sequences of the main retinal connexins. This result extends the observations of Völgyi for a large number of vertebrates (Volgyi et al., 2013a). We then studied the expression of Cx36 and Cx43, the most common neuronal and glial connexins, respectively. Cx36 immunoblotting of degu and rat retina homogenates revealed a band with an apparent molecular mass of about 36 kDa (**Figure 2A**). The specificity of the antibody was confirmed with rat brain cortex homogenate as positive control and heart extract from rat as negative control, due to the absence of Cx36 from this tissue. Retinal cryosections from degu and rat (**Figures 2B,C**) presented an intense Cx36 immunolabeling as bright dots along the OPL, where the terminals of the photoreceptors are located. Consistent with previous reports (Feigenspan et al., 2001; Frank et al., 2010) we detected weak Cx36 immunoreactivity in the OPL of rat retina, also occasionally covered by blood vessels with strong non-specific labeling (**Figure 2C**, white arrowhead). On the other hand, immunolocalization of Cx36 was both strong and dense within the inner plexiform layer (IPL) in both species (**Figures 2B,C**).

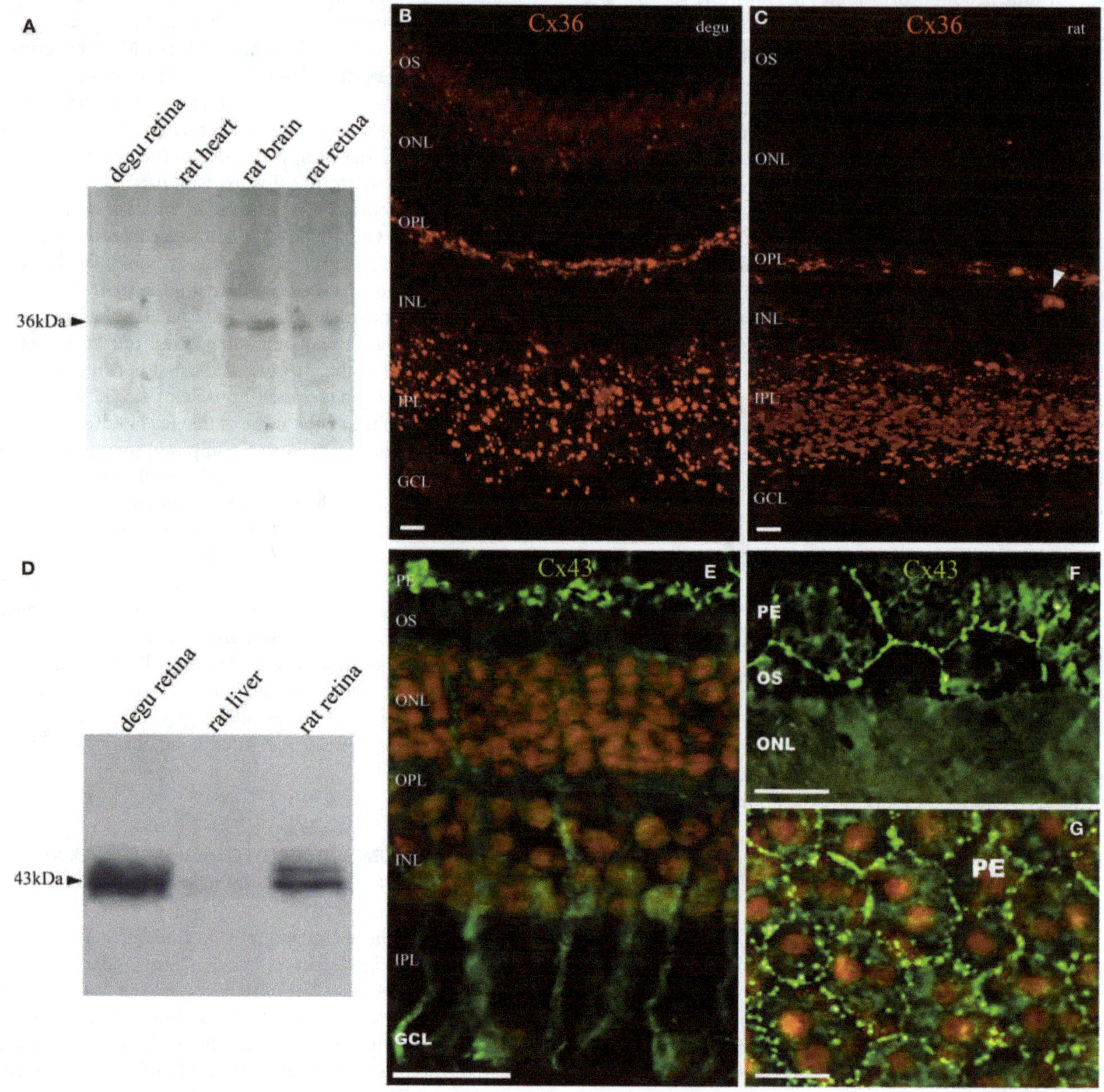

FIGURE 2 | Expression and immunolocalization of Cx36 and Cx43 in the retina of degu. (A) Immunodetection of Cx36 (arrowhead) in retinal extracts of degu and rat, and brain cortex as positive control. No immunostaining was visible in heart extracts of rat. **(B,C)** Fluorescent micrograph showing Cx35/36 immunolabeling in degu and rat retina. Immunolocalization of Cx36 along the OPL and in the IPL was observed as intensely fluorescent puncta. Less frequently, Cx36 labeling was observed in the GCL, where puncta were localized to the margins of cell bodies. The arrowhead points to a blood vessel with strong non-specific labeling **(D)** Immunodetection of Cx43 in retinal extracts of degu and rat. Cx43 shows the expected electrophoretic mobility around 43 kDa. **(E)** Fluorescent micrograph of retinal Cx43 immunoreactivity in degu. The immunoreactive puncta of Cx43 were clearly visible between pigment epithelial cells (top) and possibly in glial cells present in the inner retinal layers. **(F,G)** Fluorescence micrographs of pigment epithelium layer cells showing abundant Cx43 gap junction plaques in cellular appositions. The cell nuclei were counterstained with propidium iodide staining. PE, pigment epithelial cells; OS, photoreceptor outer segment; ONL, outer nuclear layer; OPL, outer plexiform layer; INL, inner nuclear layer; IPL, inner plexiform layer; GCL, ganglion cell layer. Scale bars: 20 μm **(B,C,F,G)** and 50 μm **(E)**.

The Cx43 Western blots of degu retina homogenates show the characteristic band profile around 43 kDa (**Figure 2D**), which is similar to that observed in rat heart extracts, a tissue known to express high levels of this protein. No immunostaining was visible in rat liver extracts, consistent with the lack of expression of Cx43 in liver. Immunolocalization of Cx43 in degu retina showed staining in cells reminiscent of retinal Müller cells (**Figure 2E**) and at the borders of pigment epithelium cells (PE), revealing the prominent hexagonal array of these cells (**Figures 2F,G**). No immunolabeling of Cx43 was observed in other layers of degu retina. These observations are consistent with previous studies that showed a lack of Cx43 expression in retinal neurons, but a high expression in pigment epithelial cells and glial elements of several vertebrate retinas, including human (Coca-Prados et al., 1992; Janssen-Bienhold et al., 1998).

EFFECT OF β-GA ON THE B-WAVE OF THE ERG

β-GA was used to investigate the contribution of gap junction channels to the generation of visual responses in degu and rat. Representative ERG response patterns are shown in **Figure 3** for both species before and after injection of β-GA. Under dark-adapted conditions, β-GA produced a reduction of the ERG b-wave amplitude by 45.1 % ($n = 6, p = 0.036$) in degu and 52.2 % in rat ($n = 4, p = 0.037$) for high intensity stimuli (**Figures 3A,B; Table 1**). This effect was only appreciable in degu at high stimulation intensities. However, under photopic light conditions, the amplitude of the ERG b-wave increased by 107.2 % ($n = 4, p = 0.006$) in degu, while it was reduced by 62.3% ($n = 4, p = 0.032$) in rat (**Figures 3C,D; Table 1**). These results indicate that β-GA produced a similar global effect on both species under scotopic conditions, but opposite effects under photopic adaptation.

GANGLION CELL RECEPTIVE FIELDS UNDER CONTROL AND β-GA CONDITIONS

Although we found that inhibition of connexins channels modulates the ERG b-wave, reflecting altered responses of ON BCs, the output of visual signals from the retina may not be affected in the same way by connexin inhibition. Therefore, we studied the function of connexins in degu retina at the level of GC action potentials (spikes), using a multi-electrode array for *in vitro* experiments under photopic conditions. **Figure 4** shows an example of a RF (left) and its temporal profile (right) for an ON (top) and OFF (bottom) GC recorded in degu retina. **Table 2** shows the distribution of ON/OFF GC types for control and β-GA conditions. Under control conditions, around 80% of GCs match type OFF, a result consistent with the literature for guinea pig and rat (Peichl, 1989; Zaghloul et al., 2003). The estimation of the linear RF allowed us to detect two main changes concerning the number of GCs and their response time with β-GA (**Table 2**). We observed the number of characterized GCs with a valid RF decreased under β-GA treatment for both ON (12.7%) and OFF (11.1%) type GCs. **Figure 5A** shows the time profile of all the GCs with a valid RF according to the criteria stated in the Materials and Methods Section for control and β-GA conditions. Fitting the time profiles with splines and using PCA and k-means, GCs clustered into five functional groups, differing in the polarity, shape and time-to-peak of their action potentials. In the presence of β-GA, the control clusters remained in place, but showed overall larger response latencies, resulting in temporal profiles with slower dynamics compared to control. The difference between the response times was parameterized using time-to-peak values (**Figure 4**), revealing that under both control and β-GA conditions, OFF cells are significantly faster than ON cells. Globally, the increment observed in the time-to-peak value for the two ON and OFF classes of GCs under β-GA treatment was 30.1 ms (46.42%, Welch's t-test p-value < 0.005) for ON and 13.2 ms (27.2%, Welch's t-test p-value < 0.005) for OFF type GCs (**Table 2**). Specifically, the effect of connexin channel blockage on the time course of the RF of different cell types is shown in **Figure 5**. **Figure 5B** shows how β-GA treatment affected the different functional GCs types in different ways. Some of the ON GCs increased their time-to-peak interval, reaching up to 117 ms compared to a mean of 64.61 ms (control). **Figure 5C** shows the comparison of time-to-peak latencies for the different clusters shown in **Figure 5A**. Evidently, a certain type of ON GC (blue cluster) is more affected by β-GA than others.

Another important outcome to analyze is the number of spikes evoked by checkerboard stimulus experiments. In response to β-GA treatment, we observed a decrease in the number of evoked spikes of ON GCs from 8.2 (±7.3 SD, $n = 115$) to 4.4 (±4.2 SD, $n = 48$) spikes per second or OFF GCs from 6.2 (±5.6 SD, $n = 265$) to 3.3 (±3.1 SD, $n = 152$) spikes per second. This result suggests that in degu the increase of the photopic b-wave under β-GA treatment is not reflected by an increase in the mean GC spike rate.

DISCUSSION

The analysis of protein sequence relationships for the principal classes of retinal connexins of degu, guinea pig, rat and human based on the classification proposed by Cruciani and Mikalsen (2006) and discussed in Volgyi et al. (2013a) shows that most common retinal connexins are present in the degu genome and that they are closely related to other mammalian species. The phylogenetic tree in **Figure 1** suggests a high degree of homology of the degu Cx36 ortholog with guinea pig, rat and human Cx36 (Volgyi et al., 2013b). Therefore, it can be expected that the anti-Cx36 rat antibody used here cross-reacts with degu Cx36. Indeed, the Cx36 antibody labeled a band with an electrophoretic mobility similar to rat Cx36 in degu retinal extracts (**Figure 2A**). In addition, localization of Cx36 in degu retina is mostly similar to previous findings in others rodents (**Figures 2B,C**). One exception is that Cx36 shows higher expression in the OPL compared to rat retina where less Cx36-positive puncta are observed, consistent with previous findings in nocturnal rodents like the rat (**Figure 2C**) (Frank et al., 2010). In addition, rat and degu retinas show a distributed Cx36 staining in the IPL throughout the OFF and ON sublayers, although labeling was a little more concentrated in the latter (**Figure 2B**), similar to previous findings in other nocturnal rodents. The punctate Cx36 staining pattern found in degu was very similar to the closely related guinea pig. In the latter, an intense staining is observed throughout the OPL, where Cx36 forms homologous gap junctions between neighboring cone-cone and rod-rod photoreceptors and forms heterologous gap junctions between cone and rod photoreceptors (Lee et al., 2003; Feigenspan et al., 2004). The higher density of Cx36 in the OPL of degu retina might be due to the presence of a significantly higher number of cones reaching a peak density of 50.000 mm^{-2} (Jacobs et al., 2003) compared to 7.000 mm^{-2} in rats (Jacobs et al., 2001). The IPL of degu retina also shows punctuate and intense staining of Cx36, particularly in the ON sublayer (**Figure 2B**). The localization of Cx36 immunoreactivity in the IPL is in good agreement with other mammalian retinas, where Cx36 was described in somata and dendrites of retinal amacrine AII and GCs (Feigenspan et al., 2001; Mills et al., 2001; Hidaka et al., 2002). Interestingly, in *Gallus gallus*, a diurnal bird, Cx36 is

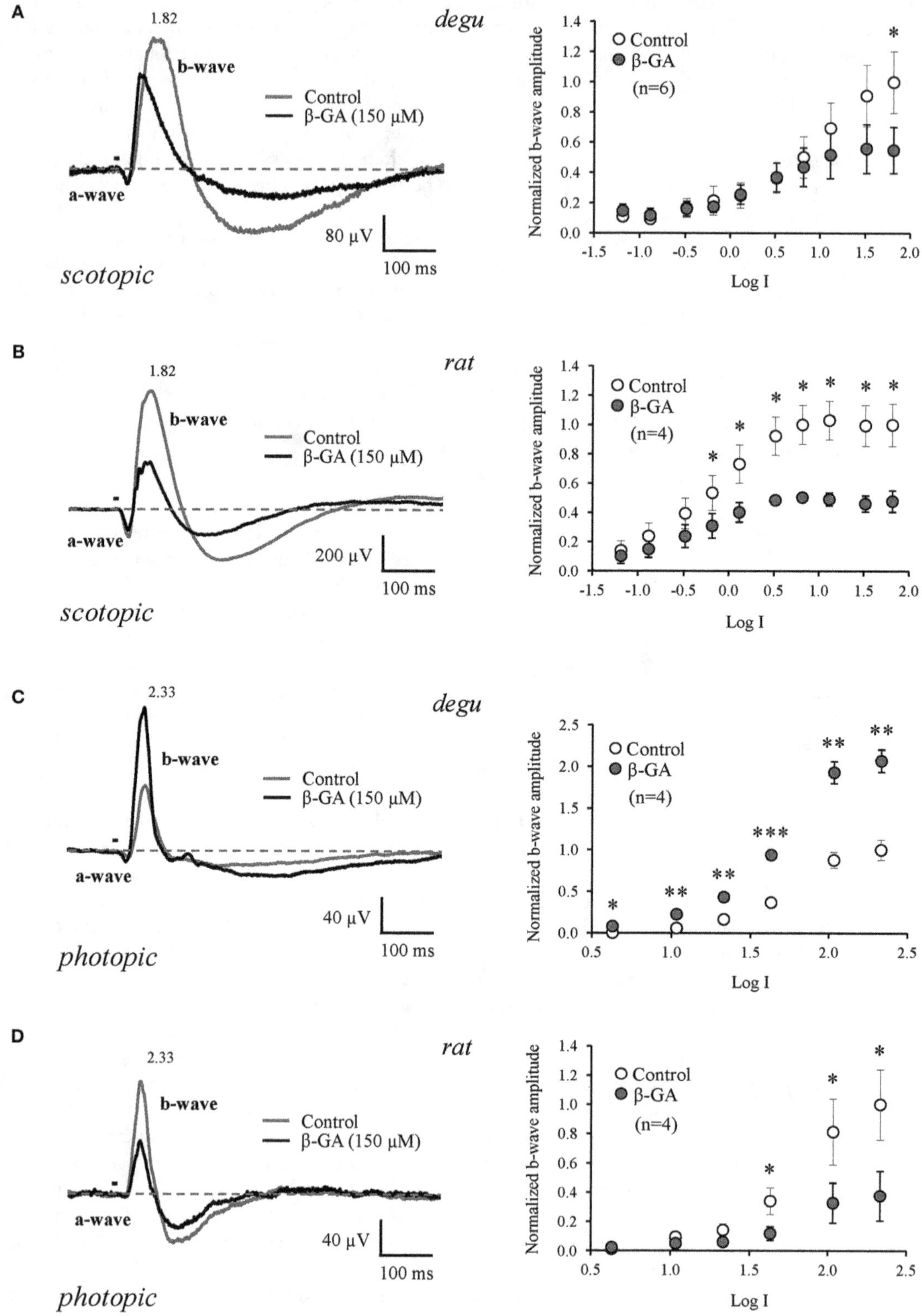

FIGURE 3 | Differential effect of connexin channel inhibition on the amplitude of the ERG b-wave in the retina of degu and rat during dark and light adapted conditions. Left: representative ERG tracers obtained from degu **(A–C)** and rat **(B–D)** eyes under control conditions or after treatment with β-GA (150 μM), at the maximum intensity used. Bars indicate the stimulus duration ($\lambda = 500$ nm). Dotted lines indicate baseline level. Recordings were realized under dark **(A,B)** and light **(C,D)** adapted conditions. Right: intensity-response functions under dark and light conditions before and after treatments with β-GA. Asterisks represent the statistical significance with respect to control (Paired *t*-test, $^{*}p < 0.05$; $^{**}p < 0.01$; $^{***}p < 0.001$).

Table 1 | Summary of the effects of β-GA on the ERG parameters (mean ± s.e.m.) in degu and rat.

ERG-SCOTOPIC ADAPTED						
Log intensity	**b-wave amplitude (μV) degu (*n* = 6)**			**b-wave amplitude (μV) rat (*n* = 4)**		
	Control	**β-GA**		**Control**	**β-GA**	
−0.88	9.8 ± 2.0	12.4 ± 4.7	–	136.3 ± 45.9	86.0 ± 39.1	–
0.52	38.4 ± 11.4	38.9 ± 10.0	–	526.7 ± 72.9	276.2 ± 79.9*	↓
1.82	105.1 ± 23.4	57.8 ± 16.0*	↓	569.7 ± 78.5	272.4 ± 88.85*	↓
ERG-PHOTOPIC ADAPTED						
	b-wave amplitude (μV) degu (*n* = 4)			**b-wave amplitude (μV) rat (*n* = 4)**		
	Control	**β-GA**		**Control**	**β-GA**	
1.03	2.8 ± 1.2	10.7 ± 1.5**	↑	6.5 ± 2.1	3.5 ± 3.2	–
1.63	17.4 ± 1.9	44.2 ± 2.2***	↑	23.2 ± 6.3	8.1 ± 3.9*	↓
2.33	47.2 ± 7.1	97.7 ± 7.8**	↑	68.0 ± 16.4	25.7 ± 14.2*	↓

*Asterisks show statistically significant differences (Paired t-test; *$p < 0.05$; **$p < 0.01$; ***$p < 0.001$). Arrows indicate the effect tendency.*

localized mainly in the OFF sublamina of the IPL (Kihara et al., 2008), suggesting that Cx36 is present in distinct retinal circuits in the mature *Gallus gallus* retina. The different distribution of Cx36 may be due to the fact that *Gallus gallus* has phylogenetically evolved an essentially rodless retina, whereas degu, although diurnal, preserves an important rod pathway consistent with the results of our study discussed below.

It is well-known that in the IPL of mammalian retina, Cx36 and Cx45 are the principal neuronal gap junction proteins, forming: "bi-homotypic" gap junction channels, with Cx45 coupling to Cx45 and Cx36 coupling to Cx36 (Li et al., 2008); or "heterotypic" gap junctions channels (Maxeiner et al., 2005; Dedek et al., 2006) between mouse AII amacrine and ON cone BC. Both Cx36 and Cx45 have an important role in visual signal transmission in the primary rod-to-cone circuit in the IPL of mammalian retina. It can be assumed that they have similar functions because it is possible to functionally replace Cx45 by Cx36 in the retina of mice (Frank et al., 2010). Furthermore, the expression and distribution of Cx43 in degu retina (**Figures 2D–G**) were largely similar to findings in other mammals, with staining in pigment epithelial and glial cells, Müller cells and astrocytes (Janssen-Bienhold et al., 1998; Johansson et al., 1999; Kihara et al., 2006).

We confirm previous findings showing that connexins contribute to the b-wave generated by ON BCs at the level of the OPL, where synapses between photoreceptors, bipolar and HCs are found. Under scotopic adaptation, the rat presented higher sensitivity to light compared to degu, as expected for a nocturnal animal (**Table 1**). However, in our full field stimulus at the highest stimulus intensity (1.82 log), the b-wave decreased in both species after β-GA application. The effect of β-GA on the ERG b-wave of rat started at lower stimulus intensities compared to degu, consistent with a robust scotopic rod system. The main scotopic rod pathway involves rods, rod ON BCs, amacrine cells and ON cone BCs; the last connection consisting of electrical synapses (Deans et al., 2002). Therefore, the blockage of Cx36 gap junctions is expected to decrease the response of the ON pathway, consistent with our findings. Along these lines of evidence, in mice, targeted disruption of Cx36 or Cx45 affects primarily the transmission of the rod pathway (b-wave) in dark-adapted retinas (Maxeiner et al., 2005). However, we cannot discard the possible contribution of Cx36 gap junctions between rods and cones in this mechanism (Deans et al., 2002), since β-GA is blocking connexin channels indiscriminately in the whole retina.

A somehow different picture emerges under photopic conditions, where the b-wave is driven essentially by the ON cone BC pathway (Deans et al., 2002), and does not involve Cx36 gap junction channels with AII cells. Under control conditions and full field stimulation, both rodent types display similar b-wave responses. However, β-GA increased the maximum amplitude of the b-wave in degu by 107%, whereas in rat it decreased the b-wave by 62% when measured at the highest light intensity (2.33 log, **Table 1**). In degu, β-GA increased the b-wave across nearly all stimulus intensities, whereas in rat β-GA reduced the b-wave only at the highest stimulus intensity. For rat, our results are similar to the ones observed in the Cx36$^{-/-}$ mice (Guldenagel et al., 2001), suggesting that in both nocturnal species a similar pathways is involved, probably because their retinas are dominated by the rod pathway. Similar effects were observed in the retina of Cx45$^{-/-}$ mice under photopic stimulation (Maxeiner et al., 2005). On the other hand, our results in degu are consistent with findings in the diurnal goldfish, where meclofenamic acid (MFA) a general gap junction blocker, also increased the photopic b-wave (Kim and Jung, 2012), suggesting that this mechanism could be present across diurnal vertebrate species.

We can only speculate about a possible explanation for the opposite effects on the photopic b-wave observed here after connexin channel blockage. Under photopic conditions, the neural circuit of the OPL involves inhibitory HCs that might play a critical role. It has been shown that HCs of type A and type B are electrically coupled, supporting an inhibitory feedback mechanism that reduces the response of photoreceptors. Interestingly, cats, rabbits and guinea pigs, the latter closely related to degu, present A and B type HCs in their retinas, while rat, mouse and gerbil have only type B HCs (Peichl and Gonzalez-Soriano, 1994; Sohl et al., 2005). Type A HCs in mammals receive mainly cone input and have a large integration area, up to 2 mm wide. In

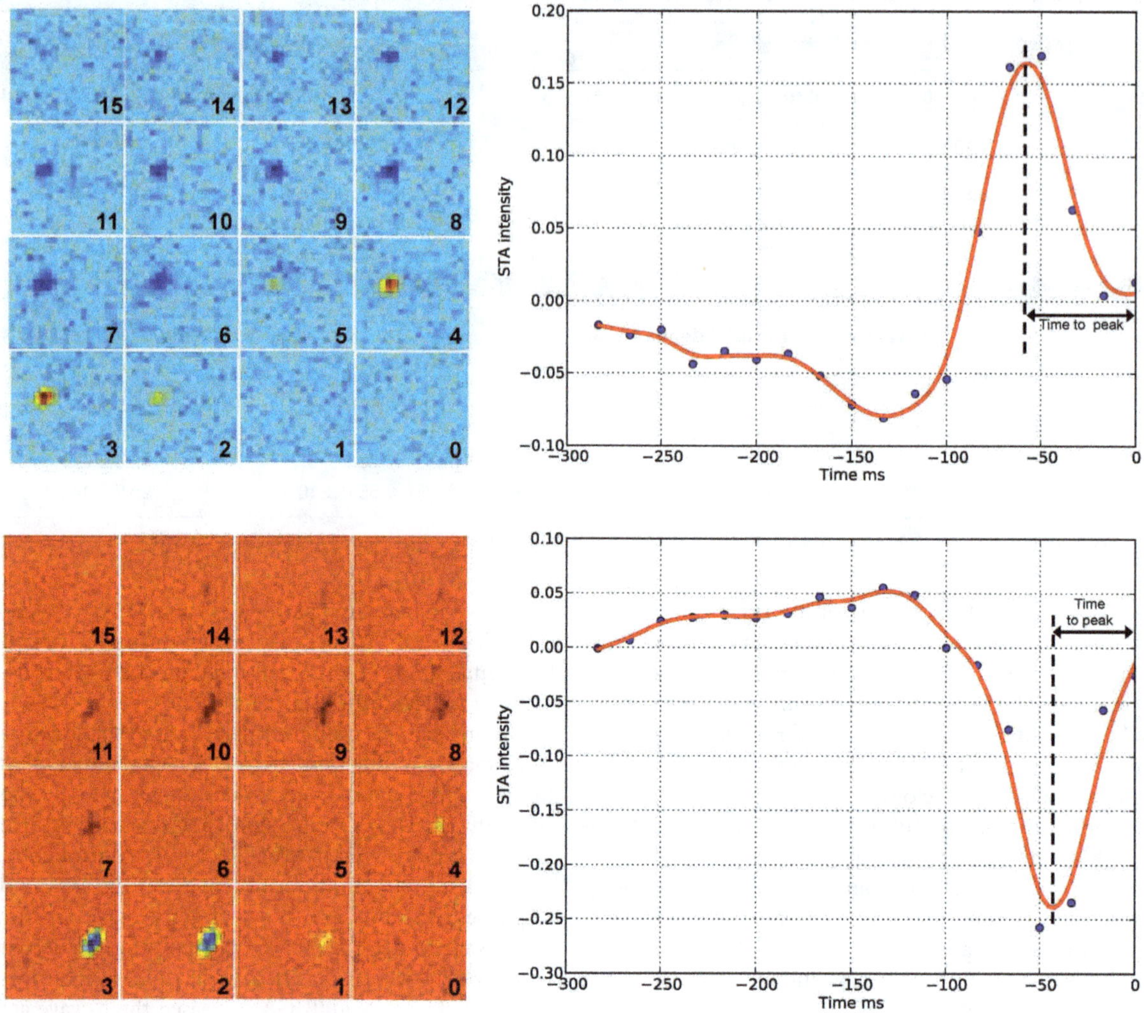

FIGURE 4 | Spike-triggered average (left) and temporal profile (right) under control conditions. The left panel shows the result of the STA algorithm on the response of GCs to checkerboard stimuli. Each frame has a number indicating the order before the spike time (zero represents the spike occurrence time point). The complete sequence covers −250 ms before the spike time with a time resolution of 16.67 ms. Blue/red represents low/high intensities, respectively. Top, example of an ON GC (STA estimated from $n = 2882$ spikes); bottom, example of an OFF GC (STA estimated from $n = 9263$ spikes). In both cases, the light intensity before the spike occurrence is either a decrease (top) or an increase of the mean intensity of the stimuli. The right panel shows the temporal profile of the zone of the receptive field with the highest response. Time zero represents the spike occurrence. Time-to-peak represents the time between the spike occurrence and the maximum peak of the stimulus. Each blue point corresponds to a time panel figure on the left. The red line is a spline interpolation of the curve defined by the blue dots.

Table 2 | Distribution of ON and OFF retinal GCs in degu retina estimated according to their RF, using STA and checkerboard stimulation.

Cell type	Control A ($n = 3$) (%)	Control B ($n = 2$) (%)	βGA (50 μM) ($n = 2$) (%)	Control A time-to-peak	Control B time-to-peak	βGA (50 μM) time-to-peak
ON	240 (20.1)	180 (20.6)	23 (23.0)	67.36 ± 12.06	64.81 ± 8.41	94.9 ± 19.3
OFF	955 (79.9)	693 (79.4)	77 (77.0)	48.60 ± 8.13	48.34 ± 7.44	61.5 ± 12.1
Total	1195 (100)	873 (100)	100 (100)			

*Control A corresponds to three different experiments (including the two in Control B) under normal conditions and represents a total of 1195 GCs. Control B and β-GA corresponds to two experiments under control ($n = 873$ GCs) and β-GA conditions (50 μM; $n = 100$ GCs). Time-to-peak corresponds to the interval between the stimulus presentation and the spike occurrence. The last two columns show the statistics of time-to-peak values (in ms, see **Figure 5**) for each GC type.*

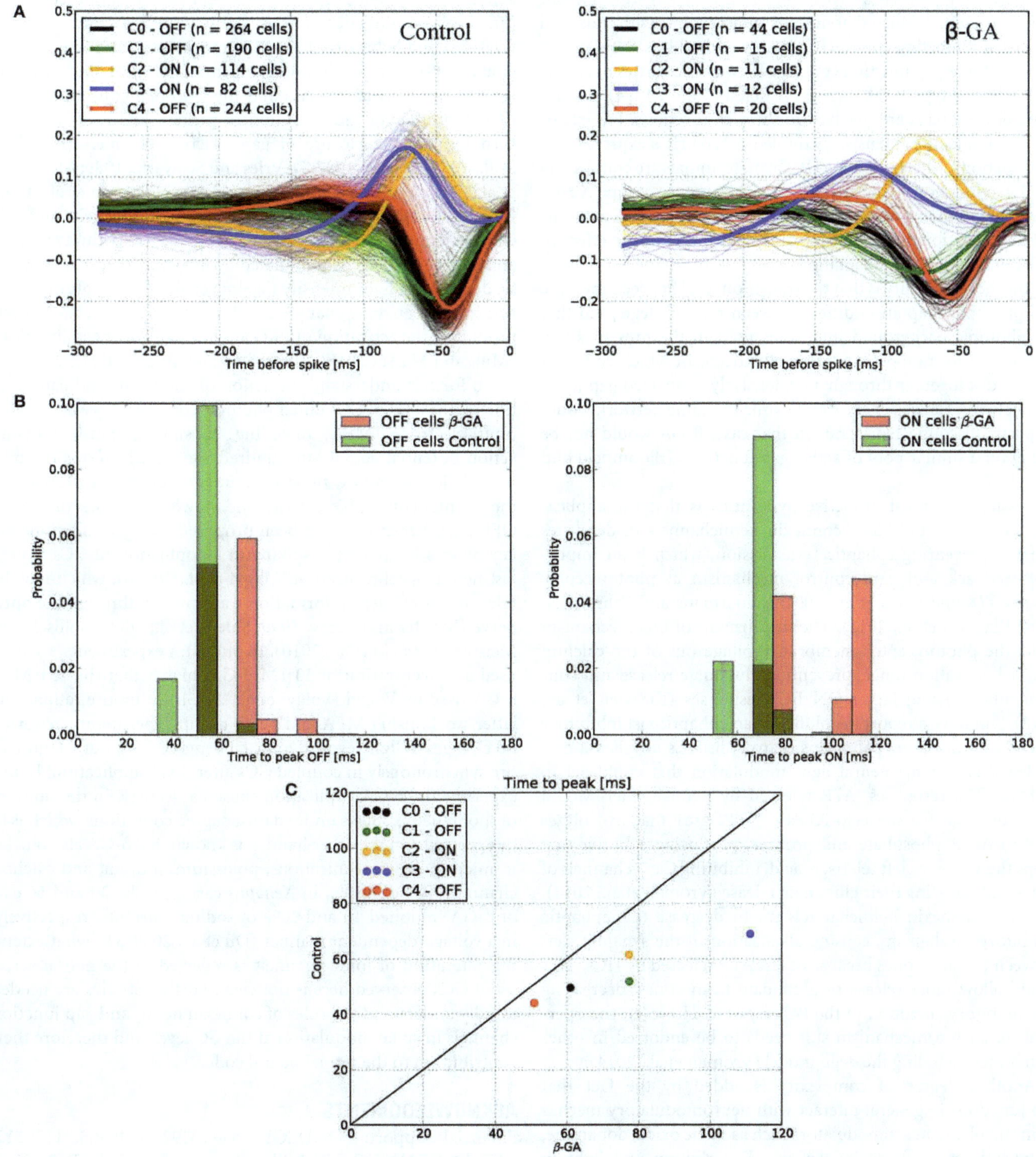

FIGURE 5 | Functional ganglion cell characterization under control and β-GA conditions. Using STA analysis, ganglion cells were classified and clustered into five different groups according to their ON or OFF preference and response timing. **(A)** The dynamics of the response timing is represented for both control (left) and β-GA conditions (right). β-GA decreased the number of ganglion cells with a valid RF. **(B)** Distribution of time-to-peak responses under β-GA and control conditions for ON and OFF ganglion cells. Under β-GA treatment, the response latency in the temporal response profile increased compared to control. For both ON and OFF cells, the latencies observed in the β-GA condition presented a larger standard deviation compared to controls. ON cells were more affected, allowing to fit a bimodal distribution (right). **(C)** Comparison of time-to-peak vs. time-to-zero-cross for the same conditions described in **(B)**. Comparison of time-to-peak for control and β-GA treatment for the same cluster allows concluding that every cell type was affected by β-GA, in particular the ON type labeled as C3. The strait line indicates zero difference between β-GA and control.

comparison, type B HCs have a smaller integration area of about 1 mm, an axon terminal that makes contact mainly with rods, and dendrites contacting cones. HCs in cats and rabbits show strong low-resistance gap junction coupling between dendrites of type A cells, whereas type B cells are less coupled (Vaney, 1994). Several types of connexins are involved in these connections. For example, in rabbits, the extensively coupled A-type HCs express Cx50 gap junctions, whereas B-type HCs axon terminals are coupled via Cx57 gap junctions (Cha et al., 2012). The absence of type A HCs from rat and their presence in degu supports a plausible anatomical and functional mechanism to explain the opposite effect of β-GA under photopic adaptation.

Another possibility is that the transition level between mesopic and photopic adaptation differs between rat and degu, and that the photopic background intensities used in the present study correspond to mesopic conditions for degu, in which rods and cones work together through an extensively connected gap junction network (Jacobs et al., 2003), while the same network would be partly disconnected in rat. In that case, β-GA would not be acting on a similar pool of active gap junctions (Bloomfield and Volgyi, 2009).

Possibly the most attractive hypothesis is that the application of β-GA could block connexin hemichannels in dendrites of HCs, preventing ephaptic transmission, which is an important feedback and gain control mechanism at photoreceptor synapses (Kamermans et al., 2001; Kamermans and Fahrenfort, 2004; Klaassen et al., 2011). The impairment of this mechanism shifts the photoreceptor membrane voltage out of the calcium channel activation range, preventing glutamate release and consequently allowing larger ON BCs responses (Klaassen et al., 2012). Similarly, a report in goldfish of an ephaptic fast inhibitory mechanism involving pannexins (Panx1) in HCs unveiled a new feedback circuit for neural light modulation that could act to increase BCs responses. ATP released by hemichannels would be a substrate for the ecto-ATPase NTPDase1 that hydrolyzes ATP to AMP, phosphate and protons, producing a buffer that keeps the synaptic cleft relatively acid, inhibiting Ca^{2+} channels of cones and reducing their glutamate release (Vroman et al., 2014). Blocking pannexin hemichannels might decrease this ephaptic inhibitory mechanism, causing alkalization of the synaptic cleft between HC and cones because of less ATP released by HCs. This would allow more release of glutamate from cones, decreasing the inhibitory surround of the BC response. However, the operation of such a mechanism still needs to be endorsed in other vertebrates including the degu model (Vroman et al., 2014).

Another degree of complexity is added by the fact that gap junctions in general interact with neuromodulatory mechanisms involving neuromodulators such as nitric oxide, dopamine, acetylcholine, regulated by the circadian rhythm and light to change retinal light sensitivity (Deans et al., 2002; Bloomfield and Volgyi, 2009). In this context, in rabbits electrical coupling of HCs decreases under scotopic and photopic conditions, but increases under mesopic adaptation (Xin and Bloomfield, 1999). In mouse, scotopic conditions were shown to increase coupling between HCs (Pandarinath et al., 2010). Furthermore, in rabbit and mouse, gap junction coupling of alpha-GCs is modulated by light. This increase in coupling is not reflected by an increase of spontaneous activity, but an increase in the correlated spike activity (Hu et al., 2010). In HCs and AII amacrine cells, gap junction coupling depends on light as well as on dopamine (DeVries and Schwartz, 1989; Bloomfield et al., 1997; Baldridge et al., 1998). The activation of dopamine receptors increases the activity of adenylate cyclase and the cytosolic concentration of cAMP, producing the activation of protein kinase A, which in turn induces the closure of gap junction channels (Piccolino et al., 1984; Lasater, 1987; DeVries and Schwartz, 1989; McMahon et al., 1989; Witkovsky and Schutte, 1991; Hampson et al., 1992; Bloomfield et al., 1997). Moreover, AII amacrine cells and ON cone BC coupling, a fundamental part of the rod pathway (Veruki and Hartveit, 2002), is modulated by nitric oxide, whose synthesis depends on light intensity (Vielma et al., 2012). This molecule activates the enzyme guanylate cyclase and increases cGMP levels producing the activation of PKG and the closure of gap junctions (Mills and Massey, 1995; Bloomfield and Volgyi, 2009).

To further understand the action of connexins and gap junctions at the level of GCs under photopic conditions, we performed multi-electrode (MEA) recordings to simultaneously measure action potential activity in hundreds of GCs. In terms of spike modulation by β-GA, these experiments revealed a decrease in the spontaneous spike activity and slower responses of ON or OFF GCs. Recently, it has been proposed that gap junctions are key players in concerted spiking in a population of GCs trough fast neural synchronization (Volgyi et al., 2013b), which may be relevant for efficient information transmission through the optic nerve (Meister and Berry, 1999; Shlens et al., 2008; Gollisch and Meister, 2010; Hu et al., 2010). In our MEA experiments, we have used a concentration of 50 μM β-GA, higher that the 25 μM of β-GA used by Völgyi (Volgyi et al., 2013b) in mouse retina. The latter work, using MEA and patch clamp experiments, reported no changes in light-evoked ionic currents of GCs, but a failure to fire synchronously in coupled GCs after β-GA application. In our experiments, β-GA application caused a decrease in the number of spontaneous spikes under dark-adapted conditions, which is to be expected. On the other hand, it is known that β-GA also blocks, in micromolar concentrations, potassium, sodium, and calcium channels. For example, in *Xenopus* cardiac cells, 30 and 50 μM of β-GA inhibited 10 and 20% of sodium currents, respectively, in a voltage-dependent manner (Du et al., 2009). To what extend this alteration of ionic channels is reflected in the neural activity of GCs observed here is not clear. Further studies are needed to fully understand the roles of connexin hemi- and gap junction channels in spike modulation at the GC level, and therefore their contribution to the retinal neural code.

ACKNOWLEDGMENTS

Financial support: FONDECYT #1110292, 1140403, 1120513, ANR-47 CONICYT, and Millennium Institute ICM-P09-022-F. GV is a thesis student from the Biochemistry program, Instituto de Química, Pontificia Universidad Católica de Valparaíso. We thank Ximena Baez for technical help in some of the experiments and the reviewers for valuable suggestions.

REFERENCES

Al-Ubaidi, M. R., White, T. W., Ripps, H., Poras, I., Avner, P., Gomes, D., et al. (2000). Functional properties, developmental regulation, and chromosomal localization of murine connexin36, a gap-junctional protein expressed preferentially in retina and brain. *J. Neurosci. Res.* 59, 813–826. doi: 10.1002/(SICI)1097-4547(20000315)59:6<3C813::AID-JNR14>3E3.0.CO;2-#

Ardiles, A. O., Ewer, J., Acosta, M. L., Kirkwood, A., Martinez, A. D., Ebensperger, L. A., et al. (2013). Octodon degus (Molina 1782): a model in comparative biology and biomedicine. *Cold Spring Harb. Protoc.* 2013, 312–318. doi: 10.1101/pdb.emo071357

Baldridge, W. H., Vaney, D. I., Weiler, R. (1998). The modulation of intercellular coupling in the retina. *Semin. Cell Dev. Biol.* 9, 311–318. doi: 10.1006/scdb.1998.0235

Bloomfield, S. A., and Volgyi, B. (2009). The diverse functional roles and regulation of neuronal gap junctions in the retina. *Nat. Rev. Neurosci* 0.10, 495–506. doi: 10.1038/nrn2636

Bloomfield, S. A., Xin, D., and Osborne, T. (1997). Light-induced modulation of coupling between AII amacrine cells in the rabbit retina. *Vis. Neurosci.* 14, 565–576. doi: 10.1017/S0952523800012220

Brown, K. T. (1968). The electroretinogram: its components and their origins. *Vision Res.* 8, 633–677. doi: 10.1016/0042-6989(68)90041-2

Cha, J., Kim, H. L., Pan, F., Chun, M. H., Massey, S. C., and Kim, I. B. (2012). Variety of horizontal cell gap junctions in the rabbit retina. *Neurosci. Lett.* 510, 99–103. doi: 10.1016/j.neulet.2012.01.010

Chavez, A. E., Bozinovic, F., Peichl, L., and Palacios, A. G. (2003). Retinal spectral sensitivity, fur coloration, and urine reflectance in the genus octodon (rodentia): implications for visual ecology. *Invest. Ophthalmol. Vis. Sci.* 44, 2290–2296. doi: 10.1167/iovs.02-0670

Chichilnisky, E. J., and Kalmar, R. S. (2002). Functional asymmetries in ON and OFF ganglion cells of primate retina. *J. Neurosci.* 22, 2737–2747.

Ciolofan, C., Lynn, B. D., Wellershaus, K., Willecke, K., and Nagy, J. I. (2007). Spatial relationships of connexin36, connexin57 and zonula occludens-1 in the outer plexiform layer of mouse retina. *Neuroscience* 148, 473–488. doi: 10.1016/j.neuroscience.2007.06.003

Coca-Prados, M., Ghosh, S., Gilula, N. B., and Kumar, N. M. (1992). Expression and cellular distribution of the alpha 1 gap junction gene product in the ocular pigmented ciliary epithelium. *Curr. Eye Res.* 11, 113–122. doi: 10.3109/02713689209000061

Connors, B. W., and Long, M. A. (2004). Electrical synapses in the mammalian brain. *Annu. Rev. Neurosci.* 27, 393–418. doi: 10.1146/annurev.neuro.26.041002.131128

Cruciani, V., and Mikalsen, S. O. (2006). The vertebrate connexin family. *Cell. Mol. Life Sci.* 63, 1125–1140. doi: 10.1007/s00018-005-5571-8

Deans, M. R., Volgyi, B., Goodenough, D. A., Bloomfield, S. A., and Paul, D. L. (2002). Connexin36 is essential for transmission of rod-mediated visual signals in the mammalian retina. *Neuron* 36, 703–712. doi: 10.1016/S0896-6273(02)01046-2

Dedek, K., Schultz, K., Pieper, M., Dirks, P., Maxeiner, S., Willecke, K., et al. (2006). Localization of heterotypic gap junctions composed of connexin45 and connexin36 in the rod pathway of the mouse retina. *Eur. J. Neurosci.* 24, 1675–1686. doi: 10.1111/j.1460-9568.2006.05052.x

Degen, J., Meier, C., Van Der Giessen, R. S., Sohl, G., Petrasch-Parwez, E., Urschel, S., et al. (2004). Expression pattern of lacZ reporter gene representing connexin36 in transgenic mice. *J. Comp. Neurol.* 473, 511–525. doi: 10.1002/cne.20085

Delgado, L. M., Vielma, A. H., Kahne, T., Palacios, A. G., and Schmachtenberg, O. (2009). The GABAergic system in the retina of neonate and adult Octodon degus, studied by immunohistochemistry and electroretinography. *J. Comp. Neurol.* 514, 459–472. doi: 10.1002/cne.22023

Demb, J. B., and Pugh, E. N. (2002). Connexin36 forms synapses essential for night vision. *Neuron* 36, 551–553. doi: 10.1016/S0896-6273(02)01062-0

DeVries, S. H., Qi, X., Smith, R., Makous, W., and Sterling, P. (2002). Electrical coupling between mammalian cones. *Curr. Biol.* 12, 1900–1907. doi: 10.1016/S0960-9822(02)01261-7

DeVries, S. H., and Schwartz, E. A. (1989). Modulation of an electrical synapse between solitary pairs of catfish horizontal cells by dopamine and second messengers. *J. Physiol.* 414, 351–375.

Du, Y., Zhang, S., Wu, H., Zou, A., Lei, M., Cheng, L., et al. (2009). Glycyrretinic acid blocks cardiac sodium channels expressed in Xenopus oocytes. *J. Ethnopharmacol.* 125, 318–323. doi: 10.1016/j.jep.2009.06.016

Feigenspan, A., Janssen-Bienhold, U., Hormuzdi, S., Monyer, H., Degen, J., Sohl, G., et al. (2004). Expression of connexin36 in cone pedicles and OFF-cone bipolar cells of the mouse retina. *J. Neurosci.* 24, 3325–3334. doi: 10.1523/JNEUROSCI.5598-03.2004

Feigenspan, A., Teubner, B., Willecke, K., and Weiler, R. (2001). Expression of neuronal connexin36 in AII amacrine cells of the mammalian retina. *J. Neurosci.* 21, 230–239. doi: 10.1002/cne.20759

Field, G. D., Sher, A., Gauthier, J. L., Greschner, M., Shlens, J., Litke, A. M., et al. (2007). Spatial properties and functional organization of small bistratified ganglion cells in primate retina. *J. Neurosci.* 27, 13261–13272. doi: 10.1523/JNEUROSCI.3437-07.2007

Frank, M., Eiberger, B., Janssen-Bienhold, U., de Sevilla Muller, L. P., Tjarks, A., Kim, J. S., et al. (2010). Neuronal connexin-36 can functionally replace connexin-45 in mouse retina but not in the developing heart. *J. Cell Sci.* 123, 3605–3615. doi: 10.1242/jcs.068668

Gollisch, T., and Meister, M. (2010). Eye smarter than scientists believed: neural computations in circuits of the retina. *Neuron* 65, 150–164. doi: 10.1016/j.neuron.2009.12.009

Guldenagel, M., Ammermuller, J., Feigenspan, A., Teubner, B., Degen, J., Sohl, G., et al. (2001). Visual transmission deficits in mice with targeted disruption of the gap junction gene connexin36. *J. Neurosci.* 21, 6036–6044.

Guldenagel, M., Sohl, G., Plum, A., Traub, O., Teubner, B., Weiler, R., et al. (2000). Expression patterns of connexin genes in mouse retina. *J. Comp. Neurol.* 425, 193–201. doi: 10.1002/1096-9861(20000918)425:2<193::AID-CNE3>3E3.0.CO;2-N

Hampson, E. C., Vaney, D. I., and Weiler, R. (1992). Dopaminergic modulation of gap junction permeability between amacrine cells in mammalian retina. *J. Neurosci* 0.12, 4911–4922.

Han, Y., and Massey, S. C. (2005). Electrical synapses in retinal ON cone bipolar cells: subtype-specific expression of connexins. *Proc. Natl. Acad. Sci. U.S.A.* 102, 13313–13318. doi: 10.1073/pnas.0505067102

Hidaka, S., Akahori, Y., and Kurosawa, Y. (2004). Dendrodendritic electrical synapses between mammalian retinal ganglion cells. *J. Neurosci.* 24, 10553–10567. doi: 10.1523/JNEUROSCI.3319-04.2004

Hidaka, S., Kato, T., and Miyachi, E. (2002). Expression of gap junction connexin36 in adult rat retinal ganglion cells. *J. Integr. Neurosci.* 1, 3–22. doi: 10.1142/S0219635202000025

Hombach, S., Janssen-Bienhold, U., Sohl, G., Schubert, T., Bussow, H., Ott, T., et al. (2004). Functional expression of connexin57 in horizontal cells of the mouse retina. *Eur. J. Neurosci.* 19, 2633–2640. doi: 10.1111/j.0953-816X.2004.03360.x

Hu, E. H., Pan, F., Volgyi, B., and Bloomfield, S. A. (2010). Light increases the gap junctional coupling of retinal ganglion cells. *J. Physiol.* 588, 4145–4163. doi: 10.1113/jphysiol.2010.193268

Jacobs, G. H., Calderone, J. B., Fenwick, J. A., Krogh, K., and Williams, G. A. (2003). Visual adaptations in a diurnal rodent, Octodon degus. *J. Comp. Physiol. A Neuroethol. Sens. Neural Behav. Physiol.* 189, 347–361. doi: 10.1007/s00359-003-0408-0

Jacobs, G. H., Fenwick, J. A., and Williams, G. A. (2001). Cone-based vision of rats for ultraviolet and visible lights. *J. Exp. Biol.* 204, 2439–2446.

Janssen-Bienhold, U., Dermietzel, R., and Weiler, R. (1998). Distribution of connexin43 immunoreactivity in the retinas of different vertebrates. *J. Comp. Neurol.* 396, 310–321.

Johansson, K., Bruun, A., and Ehinger, B. (1999). Gap junction protein connexin43 is heterogeneously expressed among glial cells in the adult rabbit retina. *J. Comp. Neurol.* 407, 395–403.

Kamermans, M., and Fahrenfort, I. (2004). Ephaptic interactions within a chemical synapse: hemichannel-mediated ephaptic inhibition in the retina. *Curr. Opin. Neurobiol.* 14, 531–541. doi: 10.1016/j.conb.2004.08.016

Kamermans, M., Fahrenfort, I., Schultz, K., Janssen-Bienhold, U., Sjoerdsma, T., and Weiler, R. (2001). Hemichannel-mediated inhibition in the outer retina. *Science* 292, 1178–1180. doi: 10.1126/science.1060101

Kar, R., Batra, N., Riquelme, M. A., and Jiang, J. X. (2012). Biological role of connexin intercellular channels and hemichannels. *Arch. Biochem. Biophys.* 524, 2–15. doi: 10.1016/j.abb.2012.03.008

Kihara, A. H., Mantovani de Castro, L., Belmonte, M. A., Yan, C. Y., Moriscot, A. S., and Hamassaki, D. E. (2006). Expression of connexins 36, 43, and 45 during postnatal development of the mouse retina. *J. Neurobiol.* 66, 1397–1410. doi: 10.1002/neu.20299

Kihara, A. H., Paschon, V., Cardoso, C. M., Higa, G. S., Castro, L. M., Hamassaki, D. E., et al. (2008). Connexin36, an essential element in the rod pathway, is highly expressed in the essentially rodless retina of Gallus gallus. *J. Comp. Neurol.* 512, 651–663. doi: 10.1002/cne.21920

Kim, D. Y., and Jung, C. S. (2012). Gap junction contributions to the goldfish electroretinogram at the photopic illumination level.

Korean J. Physiol. Pharmacol. 16, 219–224. doi: 10.4196/kjpp.2012.16.3.219

Klaassen, L. J., Fahrenfort, I., and Kamermans, M. (2012). Connexin hemichannel mediated ephaptic inhibition in the retina. *Brain Res.* 1487, 25–38. doi: 10.1016/j.brainres.2012.04.059

Klaassen, L. J., Sun, Z., Steijaert, M. N., Bolte, P., Fahrenfort, I., Sjoerdsma, T., et al. (2011). Synaptic transmission from horizontal cells to cones is impaired by loss of connexin hemichannels. *PLoS Biol* 9:e1001107. doi: 10.1371/journal.pbio.1001107

Kothmann, W. W., Massey, S. C., and O'Brien, J. (2009). Dopamine-stimulated dephosphorylation of connexin 36 mediates AII amacrine cell uncoupling. *J. Neurosci.* 29, 14903–14911. doi: 10.1523/JNEUROSCI.3436-09.2009

Lasater, E. M. (1987). Retinal horizontal cell gap junctional conductance is modulated by dopamine through a cyclic AMP-dependent protein kinase. *Proc. Natl. Acad. Sci. U.S.A.* 84, 7319–7323. doi: 10.1073/pnas.84.20.7319

Lee, E. J., Han, J. W., Kim, H. J., Kim, I. B., Lee, M. Y., Oh, S. J., et al. (2003). The immunocytochemical localization of connexin 36 at rod and cone gap junctions in the guinea pig retina. *Eur. J. Neurosci.* 18, 2925–2934. doi: 10.1046/j.1460-9568.2003.03049.x

Li, H., Chuang, A. Z., and O'Brien, J. (2009). Photoreceptor coupling is controlled by connexin 35 phosphorylation in zebrafish retina. *J. Neurosci.* 29, 15178–15186. doi: 10.1523/JNEUROSCI.3517-09.2009

Li, X., Kamasawa, N., Ciolofan, C., Olson, C. O., Lu, S., Davidson, K. G., et al. (2008). Connexin45-containing neuronal gap junctions in rodent retina also contain connexin36 in both apposing hemiplaques, forming bihomotypic gap junctions, with scaffolding contributed by zonula occludens-1. *J. Neurosci.* 28, 9769–9789. doi: 10.1523/JNEUROSCI.2137-08.2008

Lin, B., Jakobs, T. C., and Masland, R. H. (2005). Different functional types of bipolar cells use different gap-junctional proteins. *J. Neurosci.* 25, 6696–6701. doi: 10.1523/JNEUROSCI.1894-05.2005

Massey, S. C., O'Brien, J. J., Trexler, E. B., Li, W., Keung, J. W., Mills, S. L., et al. (2003). Multiple neuronal connexins in the mammalian retina. *Cell Commun. Adhes.* 10, 425–430. doi: 10.1080/cac.10.4-6.425.430

Maxeiner, S., Dedek, K., Janssen-Bienhold, U., Ammermuller, J., Brune, H., Kirsch, T., et al. (2005). Deletion of connexin45 in mouse retinal neurons disrupts the rod/cone signaling pathway between AII amacrine and ON cone bipolar cells and leads to impaired visual transmission. *J. Neurosci.* 25, 566–576. doi: 10.1523/JNEUROSCI.3232-04.2005

McMahon, D. G., Knapp, A. G., and Dowling, J. E. (1989). Horizontal cell gap junctions: single-channel conductance and modulation by dopamine. *Proc. Natl. Acad. Sci. U.S.A.* 86, 7639–7643. doi: 10.1073/pnas.86.19.7639

Meister, M., and Berry, M. J. 2nd. (1999). The neural code of the retina. *Neuron* 22, 435–450. doi: 10.1016/S0896-6273(00)80700-X

Mills, S. L., and Massey, S. C. (1995). Differential properties of two gap junctional pathways made by AII amacrine cells. *Nature* 377, 734–737. doi: 10.1038/377734a0

Mills, S. L., O'Brien, J. J., Li, W., O'Brien, J., and Massey, S. C. (2001). Rod pathways in the mammalian retina use connexin 36. *J. Comp. Neurol.* 436, 336–350. doi: 10.1002/cne.1071

Moller, A., and Eysteinsson, T. (2003). Modulation of the components of the rat dark-adapted electroretinogram by the three subtypes of GABA receptors. *Vis. Neurosci.* 20, 535–542. doi: 10.1017/S0952523803205071

Naarendorp, F., and Williams, G. E. (1999). The d-wave of the rod electroretinogram of rat originates in the cone pathway. *Vis. Neurosci.* 16, 91–105. doi: 10.1017/S0952523899161054

O'Brien, J. J., Chen, X., Macleish, P. R., O'Brien, J., and Massey, S. C. (2012). Photoreceptor coupling mediated by connexin36 in the primate retina. *J. Neurosci.* 32, 4675–4687. doi: 10.1523/JNEUROSCI.4749-11.2012

Pan, F., Mills, S. L., and Massey, S. C. (2007). Screening of gap junction antagonists on dye coupling in the rabbit retina. *Vis. Neurosci.* 24, 609–618. doi: 10.1017/S0952523807070472

Pandarinath, C., Bomash, I., Victor, J. D., Prusky, G. T., Tschetter, W. W., and Nirenberg, S. (2010). A novel mechanism for switching a neural system from one state to another. *Front. Comput. Neurosci.* 4:2. doi: 10.3389/fncom.2010.00002

Peichl, L. (1989). Alpha and delta ganglion cells in the rat retina. *J. Comp. Neurol.* 286, 120–139. doi: 10.1002/cne.902860108

Peichl, L., Chavez, A. E., Ocampo, A., Mena, W., Bozinovic, F., and Palacios, A. G. (2005). Eye and vision in the subterranean rodent cururo (*Spalacopus cyanus, Octodontidae*). *J. Comp. Neurol.* 486, 197–208. doi: 10.1002/cne.20491

Peichl, L., and Gonzalez-Soriano, J. (1994). Morphological types of horizontal cell in rodent retinae: a comparison of rat, mouse, gerbil, and guinea pig. *Vis. Neurosci.* 11, 501–517. doi: 10.1017/S095252380000242X

Piccolino, M., Neyton, J., and Gerschenfeld, H. M. (1984). Decrease of gap junction permeability induced by dopamine and cyclic adenosine 3':5'-monophosphate in horizontal cells of turtle retina. *J. Neurosci.* 4, 2477–2488.

Saez, J. C., and Leybaert, L. (2014). Hunting for connexin hemichannels. *FEBS Lett.* 588, 1205–1211. doi: 10.1016/j.febslet.2014.03.004

Schubert, T., Maxeiner, S., Kruger, O., Willecke, K., and Weiler, R. (2005). Connexin45 mediates gap junctional coupling of bistratified ganglion cells in the mouse retina. *J. Comp. Neurol.* 490, 29–39. doi: 10.1002/cne.20621

Shlens, J., Rieke, F., and Chichilnisky, E. (2008). Synchronized firing in the retina. *Curr. Opin. Neurobiol.* 18, 396–402. doi: 10.1016/j.conb.2008.09.010

Sohl, G., Guldenagel, M., Traub, O., and Willecke, K. (2000). Connexin expression in the retina. *Brain Res. Brain Res. Rev.* 32, 138–145. doi: 10.1016/S0165-0173(99)00074-0

Sohl, G., Maxeiner, S., and Willecke, K. (2005). Expression and functions of neuronal gap junctions. *Nat. Rev. Neurosci.* 6, 191–200. doi: 10.1038/nrn1627

Vaney, D. I. (1994). Patterns of Neuronal Coupling in the Retina. *Prog. Retin. Eye Res.* 13, 301–355. doi: 10.1016/1350-9462(94)90014-0

Veruki, M. L., and Hartveit, E. (2002). Electrical synapses mediate signal transmission in the rod pathway of the mammalian retina. *J. Neurosci.* 22, 10558–10566.

Vielma, A. H., Retamal, M. A., and Schmachtenberg, O. (2012). Nitric oxide signaling in the retina: what have we learned in two decades? *Brain Res.* 1430, 112–125. doi: 10.1016/j.brainres.2011.10.045

Volgyi, B., Kovacs-Oller, T., Atlasz, T., Wilhelm, M., and Gabriel, R. (2013a). Gap junctional coupling in the vertebrate retina: variations on one theme? *Prog. Retin. Eye Res.* 34, 1–18. doi: 10.1016/j.preteyeres.2012.12.002

Volgyi, B., Pan, F., Paul, D. L., Wang, J. T., Huberman, A. D., and Bloomfield, S. A. (2013b). Gap junctions are essential for generating the correlated spike activity of neighboring retinal ganglion cells. *PLoS ONE* 8:e69426. doi: 10.1371/journal.pone.0069426

Vroman, R., Klaassen, L. J., Howlett, M. H., Cenedese, V., Klooster, J., Sjoerdsma, T., et al. (2014). Extracellular ATP hydrolysis inhibits synaptic transmission by increasing ph buffering in the synaptic cleft. *PLoS Biol.* 12:e1001864. doi: 10.1371/journal.pbio.1001864

Willecke, K., Eiberger, J., Degen, J., Eckardt, D., Romualdi, A., Guldenagel, M., et al. (2002). Structural and functional diversity of connexin genes in the mouse and human genome. *Biol. Chem.* 383, 725–737. doi: 10.1515/BC.2002.076

Witkovsky, P., and Schutte, M. (1991). The organization of dopaminergic neurons in vertebrate retinas. *Vis. Neurosci.* 7, 113–124. doi: 10.1017/S0952523800010981

Xia, Y., and Nawy, S. (2003). The gap junction blockers carbenoxolone and 18beta-glycyrrhetinic acid antagonize cone-driven light responses in the mouse retina. *Vis. Neurosci.* 20, 429–435. doi: 10.1017/S0952523803204089

Xin, D., and Bloomfield, S. A. (1999). Dark- and light-induced changes in coupling between horizontal cells in mammalian retina. *J. Comp. Neurol.* 405, 75–87. doi: 10.1002/(SICI)1096-9861(19990301)405:1%3C75::AID-CNE6%3E3.0.CO;2-D

Zaghloul, K. A., Boahen, K., and Demb, J. B. (2003). Different circuits for ON and OFF retinal ganglion cells cause different contrast sensitivities. *J. Neurosci.* 23, 2645–2654.

5

Hemichannels in neurodegenerative diseases: Is there a link to pathology?

Megan Bosch[1] and Tammy Kielian[2]*

[1] *Department of Pharmacology and Experimental Neuroscience, University of Nebraska Medical Center, Omaha, NE, USA*
[2] *Department of Pathology and Microbiology, University of Nebraska Medical Center, Omaha, NE, USA*

Correspondence:
Tammy Kielian, Department of Pathology and Microbiology, University of Nebraska Medical Center, 985900 Nebraska Medical Center, Omaha, NE 68198-5900, USA
e-mail: tkielian@unmc.edu

Although originally considered a structural component of gap junctions, connexin hemichannels (HCs) are now recognized as functional entities capable of influencing metabolic gradients within the CNS, allowing direct communication between the intra- and extracellular milieus. Besides connexins, HCs can also be formed by pannexins, which are not capable of gap junction assembly. Both positive and negative effects have been attributed to HC activity in the context of neurodegenerative diseases. For example, HCs can exert neuroprotective effects by promoting the uptake of neurotoxic molecules, whereas chronic HC opening can disrupt molecular gradients leading to cellular dysfunction and death. The latter scenario has been suggested for multiple neurodegenerative disorders, including Alzheimer's disease (AD) and more recently, lysosomal storage disorders, which are the focus of this perspective. Currently available evidence suggests a complex role for HCs in neurodegenerative disorders, which sets the stage for future studies to determine whether targeting HC action may improve disease outcomes.

Keywords: connexin, hemichannels, neurodegeneration, lysosomal storage disease, Alzheimer's disease

INTRODUCTION

Hemichannels (HCs) are composed of six connexin (Cx) subunits that assemble into hexameric pores that traffic to the plasma membrane where they can remain uncoupled or pair with adjacent HCs on neighboring cells to form gap junction channels (Contreras et al., 2003). Besides Cxs, HCs can also be formed by pannexins (Panx), which are not capable of gap junction assembly. This Perspective focuses on HC involvement in neurodegenerative diseases; therefore, gap junction activity will not be discussed here but has been the subject of several excellent reviews related to CNS disorders (Orellana et al., 2009; Eugenin et al., 2012). HCs are permeable to small hydrophilic molecules, such as ATP, Ca^{2+}, glutamate, glucose, and glutathione, which are critical for CNS homeostasis by maintaining ionic and metabolic gradients and can also control autocrine/paracrine signaling (Retamal et al., 2007; Kielian, 2008; Rouach et al., 2008; Sánchez et al., 2009; Schalper et al., 2010; Orellana et al., 2011a; Bennett et al., 2012; Fiori et al., 2014).

Currently, a total of 10 Cx and 2 Panx isoforms have been reported to be expressed in the brain (Giaume et al., 2013). The repertoire of Cx protein expression is distinct among various CNS cell types; however, the functional impact of these differences remains to be fully defined. For example, astrocytes primarily express Cx43 and Cx30, as well as, Cx26, Cx40, Cx45 and Cx46; microglia utilize Cx43, Cx36, and Cx32; and neurons express Cx36, Cx26, Cx45, and Cx57 (Rouach et al., 2002; Mika and Prochnow, 2012). Studies have shown that a selective pattern of Cx HC expression may orchestrate extracellular signaling networks between various glial cell types and/or neurons in a homotypic or heterotypic fashion (Giaume et al., 2013; May et al., 2013). Panx expression has been reported in neurons, astrocytes, and more recently microglia (Orellana et al., 2009, 2011b; Sáez et al., 2013b); however, an outstanding question is whether Cx/Panx HC function differs within various CNS cell subpopulations. For example, it is becoming clear that astrocytes exhibit regional heterogeneity and it will be interesting to determine whether this is associated with differences in the molecular composition of Cx/Panx HCs and/or sensitivity to HC opening.

Originally considered a structural component of gap junctions, strong evidence has emerged to support a role for HCs in maintaining cellular and tissue homeostasis, by allowing cells to directly communicate with their surrounding microenvironment and relay signals via the release of molecules that activate extracellular receptors in an autocrine/paracrine manner (Orellana et al., 2009). In the last decade, several reports have linked inflammation and dysregulated HC activity in the CNS. The summation of this work suggests that HCs have a dual role in regulating molecular gradient homeostasis in the context of neurodegenerative diseases. On the one hand, transient HC activity has been suggested to be protective during normal physiologic states as well as acute insults or inflammation. For example, astrocyte HCs have been shown to promote glucose uptake, which is considered a second glucose-lactate pathway for neurons. In addition, HCs could release lactate, which could also be beneficial for neurons (Rouach et al., 2008; Giaume et al., 2013). Proinflammatory cytokines (i.e., IL-1β and TNF-α) are known to activate glia and promote

HC opening, which may serve as a means to propagate glial activation or neuron activity by the release of bioactive molecules acting in an autocrine/paracrine manner. Conversely, sustained HC opening during chronic neurodegenerative diseases may promote disease progression by perturbing metabolic gradients and the exaggerated release of toxic molecules to induce cell death (**Figure 1**).

HC INVOLVEMENT IN NEURODEGENERATIVE DISEASES

A common denominator linking neurodegenerative diseases and HC opening is neuroinflammation, characterized by microglial and astrocyte activation and secretion of inflammatory mediators. Microglia, the resident macrophages of the CNS parenchyma, phagocytose cellular debris and produce a wide array of proinflammatory molecules (Kettenmann et al., 2011; Lyman et al., 2014). These mediators, which include cytokines (i.e., IL-β, IL-6, and TNF-α), reactive oxygen and nitrogen species, glutamate, and neurotrophic factors enhance cell mobility, phagocytosis, and can be neuroprotective when carefully regulated (Kielian, 2008; Mika and Prochnow, 2012). Under physiological conditions, transient astrocyte activation plays a pivotal role in maintaining neurotransmitter levels at the tripartite synapse and metabolite trafficking. However, during neuroinflammation, chronic proinflammatory mediator release may cause dramatic and potentially detrimental alterations in the way microglia and astrocytes communicate and regulate homeostasis via HC opening (**Figure 1**). The end result is a milieu that can lead to cellular toxicity or dysfunction, which over time, can manifest as cognitive and/or motor decline depending on the CNS site affected (Finn et al., 2011; Orellana et al., 2011a; Xiong and Kielian, 2013).

HC ACTIVITY IN LYSOSOMAL STORAGE DISEASES

Lysosomal storage diseases (LSDs) encompass a large group of inherited metabolic disorders characterized by the accumulation of storage material within lysosomes. Collectively, LSDs afflict 1 out of every 6700 live births and approximately 75% of the LSDs currently identified impact CNS function (Meikle et al., 1999; Sands and Haskins, 2008). Although a relatively new area of investigation, recent studies have reported perturbed HC activity in two distinct LSDs, namely Juvenile Neuronal Ceroid Lipofuscinosis (JNCL) and Niemann-Pick type C (NPC).

JNCL is caused by a mutation in the *CLN3* gene that most commonly spans exons 7–8 (Janes et al., 1996; Cotman et al., 2002; Drack et al., 2013). Brains of JNCL patients at autopsy as well as JNCL mouse models have shown that areas of activated microglia and astrocytes correlate with regions of neuron loss, along with elevated levels of IL-1β and ceramide, the latter representing a key lipid mediator involved in inflammation and apoptosis (Pontikis et al., 2004; Mencarelli and Martinez-Martinez, 2013; Xiong and Kielian, 2013). A recent study from our laboratory using primary microglia from the $CLN3^{\Delta ex\,7/8}$ mouse model of JNCL demonstrated that when challenged with "danger signals" elevated in the brains of JNCL patients (i.e., ceramide and neuron lysate), $CLN3^{\Delta ex\,7/8}$ microglia released significantly more proinflammatory mediators compared to wild type cells, which remained largely non-responsive (Xiong and Kielian,

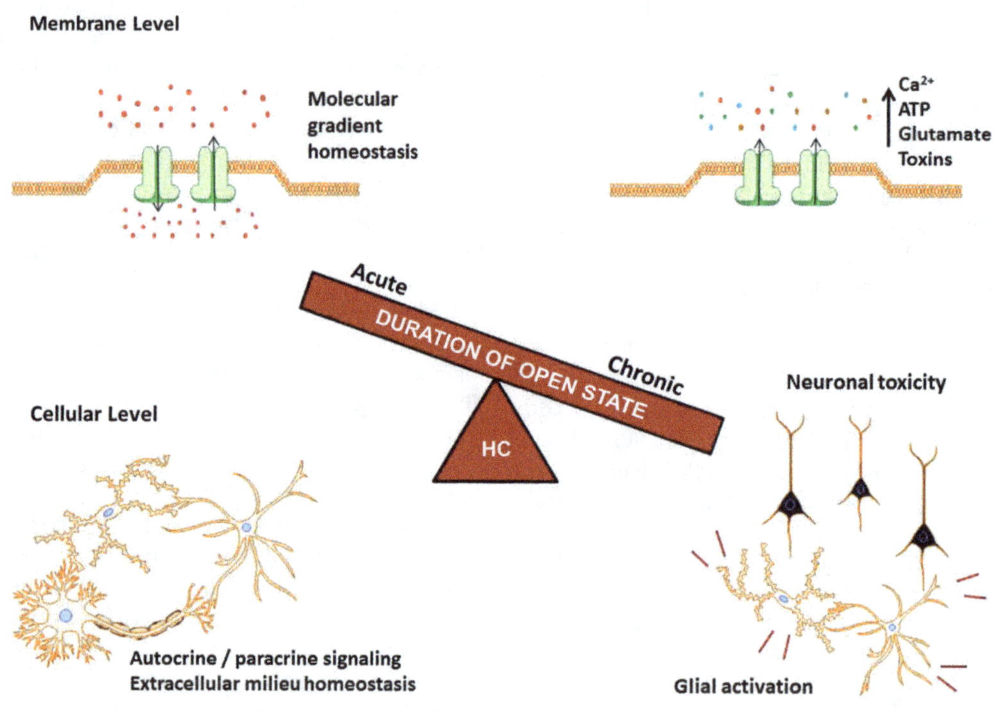

FIGURE 1 | Balance of hemichannel (HC) regulation in the CNS. HC opening is under tight control, where acute activity helps to maintain molecular gradients, cell homeostasis, and autocrine/paracine signaling. A slight disruption of this balance, as can occur during neurodegenerative diseases and inflammatory signals, causes increased HC opening. Chronic HC activity can result in the ablation of vital molecular gradients, toxin release into the extracellular milieu, and potentiation of glial activation and neuron death.

2013). Furthermore, CLN3$^{\Delta ex7/8}$ microglia displayed increased HC opening, which was associated with elevated glutamate and ATP release. Glutamate accumulation can cause neuronal excitotoxicity and a role for glutamate excitotoxicity in JNCL progression has been previously reported (Finn et al., 2011).

Another recent study from our laboratory revealed a transient increase in astrocyte HC activity in disease-affected regions of the CLN3$^{\Delta ex7/8}$ mouse brain as early as postnatal day 30, which significantly preceded neuron loss that is not evident until 6–8 months of age (Burkovetskaya et al., 2014). However, this increase was transient, since CLN3$^{\Delta ex7/8}$ astrocyte HC function began to decline at postnatal day 60, eventually falling below levels observed in wild type mice by postnatal day 90, suggesting a progressive decline in astrocyte function at later stages of disease. Treatment of CLN3$^{\Delta ex7/8}$ mice with the HC inhibitor INI-0602, a blood-brain barrier permeable derivative of carbenoxolone (Takeuchi et al., 2011), significantly reduced lysosomal storage material accumulation in specific brain regions. In addition, astrocyte gap junction communication was significantly elevated in CLN3$^{\Delta ex7/8}$ mice, which was predicted to occur via HC closure, although this was not apparent in acute brain slices *ex vivo* (Burkovetskaya et al., 2014). Nonetheless, aberrant HC activity in astrocytes and microglia may contribute to neuron loss in JNCL, particularly when considering that glial activation predates neuron death by several months in this mouse model (Pontikis et al., 2004). Unresolved questions are whether changes in HC function are responsible for the brain metabolic disturbances reported in JNCL and whether HC involvement extends to other CNS cell types (i.e., neurons and microglia).

NPC is caused by a mutation in the *NPC1* or *NPC2* genes, with the former being most common. NPC1 and NPC2 are required for cholesterol clearance and their absence causes the accumulation of cholesterol and other lipids in lysosomes (Rosenbaum and Maxfield, 2011). Similar to JNCL, neuroinflammation has been implicated in NPC pathology (Baudry et al., 2003). However, a recent study by Sáez et al. (2013a) suggested that the increase in HC activity observed in NPC might not be related to neuroinflammation *per se* but rather, the mutation itself (Sáez et al., 2013a). Specifically, primary astrocyte cultures from NPC$^{-/-}$ mice displayed increased HC activity under baseline conditions compared with cells from wild type and NPC$^{+/-}$ animals. In addition, acute hippocampal slices from NPC$^{-/-}$ mice at postnatal day 2 revealed increased astrocyte HC activity that could be blocked using the general HC blocker La^{3+} and Cx43 antibody. These data imply the involvement of Cx43 HCs and suggest that dysfunctional HCs occur at the earliest phase of NPC disease. It remains to be determined whether this HC activity represents an attempt by astrocytes to regain homeostasis in the context of NPC mutation or whether HC opening sets the stage for downstream neuropathology. Nevertheless, the available evidence supports a role for HCs in two distinct LSDs that have devastating consequences on the CNS (Finn et al., 2011; Sáez et al., 2013a; Burkovetskaya et al., 2014).

ALZHEIMER'S DISEASE AND HC FUNCTION

Alzheimer's disease (AD) is currently the leading cause of dementia in adults over the age of 65 years (Tiiman et al., 2013). Hallmark symptoms of AD include memory impairment, loss of abstract thought and language skills, and alterations in personality (Welander et al., 2009). Pathologically, AD is characterized by the extracellular accumulation of amyloid-beta (Aβ) peptide into senile plaques and the intracellular hyperphosphorylation of tau protein into neurofibrillary tangles. These aggregates cause neuronal damage, generation of reactive oxygen and nitrogen species, neuroinflammation, and defects in cell-cell communication (Selkoe, 2001; Small and Duff, 2008; Quintanilla et al., 2012; von Bernhardi and Eugenin, 2012). The underlying mechanisms that elicit plaque and tangle accumulation in AD remain elusive.

Using an APP/PS1 mouse model of AD, Mei et al. (2010) observed increased Cx43 and Cx30 expression in astrocyte processes invading the plaque core and Cx43 immunoreactivity has also been associated with plaques in human AD tissues (Nagy et al., 1996; Mei et al., 2010). Along with increased Cx expression, Aβ has been reported to increase HC activity in neurons, astrocytes, and microglia (Orellana et al., 2011b). Similar to the series of events reported in JNCL above, Aβ elicits a proinflammatory response in resident glial cells typified by inflammatory cytokine, glutamate, and ATP release, which subsequently triggers HC opening in neighboring neurons. It is postulated that neuron HC opening is one factor responsible for neuron death that can further enhance neuroinflammation and propagate the neurodegenerative process.

Indeed, HC involvement in AD was also demonstrated by Takeuchi et al. (2011), where treatment of APP/PS1 mice with the HC inhibitor INI-0602 improved cognitive function (Takeuchi et al., 2011). In addition, INI-0602 blocked neurotoxic glutamate release from activated microglia both *in vitro* and *in vivo*. Although still in the relatively early stage of exploration, these findings suggest that targeting HCs could prove to be beneficial in combating AD symptoms and progression.

CURRENT CHALLENGES IN THE FIELD OF HC BIOLOGY

Since the discovery that HCs can exert functional activity, significant efforts have been made to characterize their roles in neurodegenerative diseases; however, many limitations still exist. One primary issue is the availability of reagents that can specifically block HC action. Many of the pharmacological inhibitors currently used to study HCs also affect gap junctions, which must be considered when these compounds are used experimentally. In addition, many widely used inhibitors are not selective for a particular HC type, which leaves the identification of HC protein composition in question. A potential solution is the use of Cx-specific antibodies or Cx/Panx mimetic peptides that can selectively block HC permeability (Sáez et al., 2013a). A second difficulty when studying Cx HCs is the methods used to evaluate their activity, since it is challenging to distinguish between Cx gap junction and HC action. For example, HCs are often reported by increases in either Cx immunoreactivity in brain tissues or Western blots. However, both gap junctions and HCs are comprised of Cxs making it problematic to discriminate between the two. Functionally, the primary method used to measure HC activity is ethidium uptake. However, based on its low molecular weight, numerous types of open channels are permeable to ethidium, which calls to caution how HC ethidium uptake data is interpreted. Here the combination of ethidium uptake coupled with a Cx/Panx selective inhibitor can begin to narrow the action to Cx/Panx channels; however, for the reasons mentioned above, identification of HC vs. gap junction channels remains an area of debate. Perhaps the best means to quantifying HC activity is single channel electrophysiology. This approach is feasible in cultured cells; however, it is significantly more challenging in brain slices, as reflected in a study by Kang et al. (2008) where electrophysiological evidence of HC activity was obtained in only 18 out of 700 recordings (Kang et al., 2008). Regardless of these technical limitations, careful consideration should be given to select the best model system to examine HC function. Many studies characterizing the functional roles of HCs have been performed in reconstituted liposomes or cell cultures (Fiori et al., 2014). While these methods provide useful insights, they are not able to recapitulate *in vivo* environments. Live tissue slices are better representative of *in vivo* events and have provided more functional data; however, this approach also has its limitations. Specifically, upon excision, the brain slice begins to slowly deteriorate due to cell damage inflicted during the cutting process concomitant with microglial activation, which may impact the results obtained. To ensure tissue viability for an extended period of time, slices are continuously bathed with artificial cerebrospinal fluid which is equilibrated with CO_2 and care should be taken to evaluate cells that are positioned well below the cut surface to avoid signals from damaged regions. Future research avenues could be directed towards discovering new minimally invasive methods for studying HCs *in vivo* such as two-photon microscopy or channel tracers that can be imaged using functional magnetic resonance imaging (MRI) or single photon emission computed topography (SPECT; Kielian, 2008).

OUTSTANDING QUESTIONS AND CONCLUSIONS

Significant progress has been made towards elucidating the functional role of HCs in neurodegenerative diseases; however, many unanswered questions remain. The first is whether HC dysfunction is an active contributor to disease pathology or merely a consequence. In the case of AD, neuroinflammation indirectly increases HC channel activity in multiple cell types (Orellana et al., 2011a; Quintanilla et al., 2012). Numerous reports by others also support a similar relationship where HC activation and neuroinflammation act as co-factors, where one event influences the progression of the other. If correct, then therapeutically targeting HCs would only potentially slow disease progression but not be curative. In contrast, a direct connection between HC activity and inflammation does not appear to exist in NPC. Specifically, increased HC activity in neurons, astrocytes, and microglia was not induced by proinflammatory stimulators, suggesting that HC action precedes neuroinflammation in NPC. A similar case for changes in HC activity predating neuroinflammation is seen in JNCL, since HC dysfunction was observed early, whereas overt inflammation is not evident until later stages of disease (Xiong and Kielian, 2013; Burkovetskaya et al., 2014). The similarities between these two LSDs suggest that a core response may be operative that is not tied to inflammation *per se*. Nonetheless, as the intensity of the inflammatory response increases with advancing disease, it is likely that inflammatory mediators will impact HC activity, perhaps in a fashion that has already been described in AD and glial cell culture models (**Figure 1**). The complexity of HC composition combined with distinct expression patterns on various CNS cell types suggest that new insights are bound to emerge to account for this diversity in HCs in terms of CNS homeostasis and pathology.

ACKNOWLEDGMENTS

This work was supported by the National Institutes of Health National Institute of Neurological Disorders and Stroke (NINDS) R21NS084392-01A1, Bee For Battens, The Saoirse Foundation, and the UNMC Dean's Pediatric Research Fund (to Tammy Kielian).

REFERENCES

Baudry, M., Yao, Y., Simmons, D., Liu, J., and Bi, X. (2003). Postnatal development of inflammation in a murine model of Niemann-Pick type C disease: immunohistochemical observations of microglia and astroglia. *Exp. Neurol.* 184, 887–903. doi: 10.1016/s0014-4886(03)00345-5

Bennett, M. V., Garre, J. M., Orellana, J. A., Bukauskas, F. F., Nedergaard, M., and Saez, J. C. (2012). Connexin and pannexin hemichannels in inflammatory responses of glia and neurons. *Brain Res.* 1487, 3–15. doi: 10.1016/j.brainres.2012.08.042

Burkovetskaya, M., Karpuk, N., Xiong, J., Bosch, M., Boska, M. D., Takeuchi, H., et al. (2014). Evidence for Aberrant astrocyte hemichannel activity in Juvenile Neuronal Ceroid Lipofuscinosis (JNCL). *PLoS One* 9:e95023. doi: 10.1371/journal.pone.0095023

Contreras, J. E., Saez, J. C., Bukauskas, F. F., and Bennett, M. V. (2003). Gating and regulation of connexin 43 (Cx43) hemichannels. *Proc. Natl. Acad. Sci. U S A* 100, 11388–11393. doi: 10.1073/pnas.1434298100

Cotman, S. L., Vrbanac, V., Lebel, L. A., Lee, R. L., Johnson, K. A., Donahue, L. R., et al. (2002). Cln3(Deltaex7/8) knock-in mice with the common JNCL mutation exhibit progressive neurologic disease that begins before birth. *Hum. Mol. Genet.* 11, 2709–2721. doi: 10.1093/hmg/11.22.2709

Drack, A. V., Miller, J. N., and Pearce, D. A. (2013). A novel c.1135_1138delCTGT mutation in CLN3 leads to juvenile neuronal ceroid lipofuscinosis. *J. Child Neurol.* 28, 1112–1116. doi: 10.1177/0883073813494812

Eugenin, E. A., Basilio, D., Saez, J. C., Orellana, J. A., Raine, C. S., Bukauskas, F., et al. (2012). The role of gap junction channels during physiologic and pathologic conditions of the human central nervous system. *J. Neuroimmune Pharmacol.* 7, 499–518. doi: 10.1007/s11481-012-9352-5

Finn, R., Kovacs, A. D., and Pearce, D. A. (2011). Altered sensitivity of cerebellar granule cells to glutamate receptor overactivation in the Cln3(Deltaex7/8)-knock-in mouse model of juvenile neuronal ceroid lipofuscinosis. *Neurochem. Int.* 58, 648–655. doi: 10.1016/j.neuint.2011.02.003

Fiori, M. C., Reuss, L., Cuello, L. G., and Altenberg, G. A. (2014). Functional analysis and regulation of purified connexin hemichannels. *Front. Physiol.* 5:71. doi: 10.3389/fphys.2014.00071

Giaume, C., Leybaert, L., Naus, C. C., and Saez, J. C. (2013). Connexin and pannexin hemichannels in brain glial cells: properties, pharmacology and roles. *Front. Pharmacol.* 4:88. doi: 10.3389/fphar.2013.00088

Janes, R. W., Munroe, P. B., Mitchison, H. M., Gardiner, R. M., Mole, S. E., and Wallace, B. A. (1996). A model for Batten disease protein CLN3: functional implications from homology and mutations. *FEBS Lett.* 399, 75–77. doi: 10.1016/s0014-5793(96)01290-2

Kang, J., Kang, N., Lovatt, D., Torres, A., Zhao, Z., Lin, J., et al. (2008). Connexin 43 hemichannels are permeable to ATP. *J. Neurosci.* 28, 4702–4711. doi: 10.1523/JNEUROSCI.5048-07.2008

Kettenmann, H., Hanisch, U. K., Noda, M., and Verkhratsky, A. (2011). Physiology of microglia. *Physiol. Rev.* 91, 461–553. doi: 10.1152/physrev.00011.2010

Kielian, T. (2008). Glial connexins and gap junctions in CNS inflammation and disease. *J. Neurochem.* 106, 1000–1016. doi: 10.1111/j.1471-4159.2008.05405.x

Lyman, M., Lloyd, D. G., Ji, X., Vizcaychipi, M. P., and Ma, D. (2014). Neuroinflammation: the role and consequences. *Neurosci. Res.* 79, 1–12. doi: 10.1016/j.neures.2013.10.004

May, D., Tress, O., Seifert, G., and Willecke, K. (2013). Connexin47 protein phosphorylation and stability in oligodendrocytes depend on expression of Connexin43 protein in astrocytes. *J. Neurosci.* 33, 7985–7996. doi: 10.1523/JNEUROSCI.5874-12.2013

Mei, X., Ezan, P., Giaume, C., and Koulakoff, A. (2010). Astroglial connexin immunoreactivity is specifically altered at beta-amyloid plaques in β-amyloid precursor protein/presenilin1 mice. *Neuroscience* 171, 92–105. doi: 10.1016/j.neuroscience.2010.08.001

Meikle, P. J., Hopwood, J. J., Clague, A. E., and Carey, W. F. (1999). Prevalence of lysosomal storage disorders. *JAMA* 281, 249–254. doi: 10.1001/jama.281.3.249

Mencarelli, C., and Martinez-Martinez, P. (2013). Ceramide function in the brain: when a slight tilt is enough. *Cell. Mol. Life Sci.* 70, 181–203. doi: 10.1007/s00018-012-1038-x

Mika, T., and Prochnow, N. (2012). Functions of connexins and large pore channels on microglial cells: the gates to environment. *Brain Res.* 1487, 16–24. doi: 10.1016/j.brainres.2012.07.020

Nagy, J. I., Li, W., Hertzberg, E. L., and Marotta, C. A. (1996). Elevated connexin43 immunoreactivity at sites of amyloid plaques in Alzheimer's disease. *Brain Res.* 717, 173–178. doi: 10.1016/0006-8993(95)01526-4

Orellana, J. A., Froger, N., Ezan, P., Jiang, J. X., Bennett, M. V., Naus, C. C., et al. (2011a). ATP and glutamate released via astroglial connexin 43 hemichannels mediate neuronal death through activation of pannexin 1 hemichannels. *J. Neurochem.* 118, 826–840. doi: 10.1111/j.1471-4159.2011.07210.x

Orellana, J. A., Saez, P. J., Shoji, K. F., Schalper, K. A., Palacios-Prado, N., Velarde, V., et al. (2009). Modulation of brain hemichannels and gap junction channels by pro-inflammatory agents and their possible role in neurodegeneration. *Antioxid. Redox Signal.* 11, 369–399. doi: 10.1089/ars.2008.2130

Orellana, J. A., Shoji, K. F., Abudara, V., Ezan, P., Amigou, E., Saez, P. J., et al. (2011b). Amyloid β-induced death in neurons involves glial and neuronal hemichannels. *J. Neurosci.* 31, 4962–4977. doi: 10.1523/JNEUROSCI.6417-10.2011

Pontikis, C. C., Cella, C. V., Parihar, N., Lim, M. J., Chakrabarti, S., Mitchison, H. M., et al. (2004). Late onset neurodegeneration in the Cln3-/- mouse model of juvenile neuronal ceroid lipofuscinosis is preceded by low level glial activation. *Brain Res.* 1023, 231–242. doi: 10.1016/j.brainres.2004.07.030

Quintanilla, R. A., Orellana, J. A., and von Bernhardi, R. (2012). Understanding risk factors for Alzheimer's disease: interplay of neuroinflammation, connexin-based communication and oxidative stress. *Arch. Med. Res.* 43, 632–644. doi: 10.1016/j.arcmed.2012.10.016

Retamal, M. A., Froger, N., Palacios-Prado, N., Ezan, P., Saez, P. J., Saez, J. C., et al. (2007). Cx43 hemichannels and gap junction channels in astrocytes are regulated oppositely by proinflammatory cytokines released from activated microglia. *J. Neurosci.* 27, 13781–13792. doi: 10.1523/jneurosci.2042-07.2007

Rosenbaum, A. I., and Maxfield, F. R. (2011). Niemann-Pick type C disease: molecular mechanisms and potential therapeutic approaches. *J. Neurochem.* 116, 789–795. doi: 10.1111/j.1471-4159.2010.06976.x

Rouach, N., Avignone, E., Meme, W., Koulakoff, A., Venance, L., Blomstrand, F., et al. (2002). Gap junctions and connexin expression in the normal and pathological central nervous system. *Biol. Cell.* 94, 457–475. doi: 10.1016/s0248-4900(02)00016-3

Rouach, N., Koulakoff, A., Abudara, V., Willecke, K., and Giaume, C. (2008). Astroglial metabolic networks sustain hippocampal synaptic transmission. *Science* 322, 1551–1555. doi: 10.1126/science.1164022

Sáez, P. J., Orellana, J. A., Vega-Riveros, N., Figueroa, V. A., Hernandez, D. E., Castro, J. F., et al. (2013a). Disruption in connexin-based communication is associated with intracellular Ca(2)(+) signal alterations in astrocytes from Niemann-Pick type C mice. *PLoS One* 8:e71361. doi: 10.1371/journal.pone.0071361

Sáez, P. J., Shoji, K. F., Retamal, M. A., Harcha, P. A., Ramirez, G., Jiang, J. X., et al. (2013b). ATP is required and advances cytokine-induced gap junction formation in microglia in vitro. *Mediators Inflamm.* 2013:216402. doi: 10.1155/2013/216402

Sánchez, H. A., Orellana, J. A., Verselis, V. K., and Saez, J. C. (2009). Metabolic inhibition increases activity of connexin-32 hemichannels permeable to Ca2+ in transfected HeLa cells. *Am. J. Physiol. Cell Physiol.* 297, C665–C678. doi: 10.1152/ajpcell.00200.2009

Sands, M. S., and Haskins, M. E. (2008). CNS-directed gene therapy for lysosomal storage diseases. *Acta Paediatr. Suppl.* 97, 22–27. doi: 10.1111/j.1651-2227.2008.00660.x

Schalper, K. A., Sanchez, H. A., Lee, S. C., Altenberg, G. A., Nathanson, M. H., and Saez, J. C. (2010). Connexin 43 hemichannels mediate the Ca2+ influx induced by extracellular alkalinization. *Am. J. Physiol. Cell Physiol.* 299, C1504–C1515. doi: 10.1152/ajpcell.00015.2010

Selkoe, D. J. (2001). Alzheimer's disease results from the cerebral accumulation and cytotoxicity of amyloid beta-protein. *J. Alzheimers Dis.* 3, 75–80.

Small, S. A., and Duff, K. (2008). Linking Abeta and tau in late-onset Alzheimer's disease: a dual pathway hypothesis. *Neuron* 60, 534–542. doi: 10.1016/j.neuron.2008.11.007

Takeuchi, H., Mizoguchi, H., Doi, Y., Jin, S., Noda, M., Liang, J., et al. (2011). Blockade of gap junction hemichannel suppresses disease progression in mouse models of amyotrophic lateral sclerosis and Alzheimer's disease. *PLoS One* 6:e21108. doi: 10.1371/journal.pone.0021108

Tiiman, A., Palumaa, P., and Tougu, V. (2013). The missing link in the amyloid cascade of Alzheimer's disease—metal ions. *Neurochem. Int.* 62, 367–378. doi: 10.1016/j.neuint.2013.01.023

von Bernhardi, R., and Eugenin, J. (2012). Alzheimer's disease: redox dysregulation as a common denominator for diverse pathogenic mechanisms. *Antioxid. Redox Signal.* 16, 974–1031. doi: 10.1089/ars.2011.4082

Welander, H., Franberg, J., Graff, C., Sundstrom, E., Winblad, B., and Tjernberg, L. O. (2009). Abeta43 is more frequent than Abeta40 in amyloid plaque cores from Alzheimer disease brains. *J. Neurochem.* 110, 697–706. doi: 10.1111/j.1471-4159.2009.06170.x

Xiong, J., and Kielian, T. (2013). Microglia in juvenile neuronal ceroid lipofuscinosis are primed toward a pro-inflammatory phenotype. *J. Neurochem.* 127, 245–258. doi: 10.1111/jnc.12385

6

Extracellular gentamicin reduces the activity of connexin hemichannels and interferes with purinergic Ca^{2+} signaling in HeLa cells

Vania A. Figueroa[1,2] *, Mauricio A. Retamal[2], Luis A. Cea[1], José D. Salas[2], Aníbal A. Vargas[1], Christian A. Verdugo[2], Oscar Jara[3], Agustín D. Martínez[3] and Juan C. Sáez[1,3] *

[1] Departamento de Fisiología, Facultad de Ciencias Biológicas, Pontificia Universidad Católica de Chile, Santiago, Chile
[2] Centro de Fisiología Celular e Integrativa, Facultad de Medicina, Clínica Alemana Universidad del Desarrollo, Santiago, Chile
[3] Instituto Milenio, Centro Interdisciplinario de Neurociencia de Valparaíso, Universidad de Valparaíso, Valparaíso, Chile

***Correspondence:**
Vania A. Figueroa, Centro de Fisiología Celular e Integrativa, Facultad de Medicina, Clínica Alemana Universidad del Desarrollo, Avenida Las Condes 12438, Lo Barnechea, 7710162 Santiago, Chile
e-mail: vaniafigueroa@udd.cl or;
Juan C. Sáez, Departamento de Fisiología, Facultad de Ciencias Biológicas, Pontificia Universidad Católica de Chile, Alameda 340, 8331150 Santiago, Chile
e-mail: jsaez@bio.puc.cl

Gap junction channels (GJCs) and hemichannels (HCs) are composed of protein subunits termed connexins (Cxs) and are permeable to ions and small molecules. In most organs, GJCs communicate the cytoplasm of adjacent cells, while HCs communicate the intra and extracellular compartments. In this way, both channel types coordinate physiological responses of cell communities. Cx mutations explain several genetic diseases, including about 50% of autosomal recessive non-syndromic hearing loss. However, the possible involvement of Cxs in the etiology of acquired hearing loss remains virtually unknown. Factors that induce post-lingual hearing loss are diverse, exposure to gentamicin an aminoglycoside antibiotic, being the most common. Gentamicin has been proposed to block GJCs, but its effect on HCs remains unknown. In this work, the effect of gentamicin on the functional state of HCs was studied and its effect on GJCs was reevaluated in HeLa cells stably transfected with Cxs. We focused on Cx26 because it is the main Cx expressed in the cochlea of mammals where it participates in purinergic signaling pathways. We found that gentamicin applied extracellularly reduces the activity of HCs, while dye transfer across GJCs was not affected. HCs were also blocked by streptomycin, another aminoglycoside antibiotic. Gentamicin also reduced the adenosine triphosphate release and the HC-dependent oscillations of cytosolic free-Ca^{2+} signal. Moreover, gentamicin drastically reduced the Cx26 HC-mediated membrane currents in *Xenopus laevis* oocytes. Therefore, the extracellular gentamicin-induced inhibition of Cx HCs may adversely affect autocrine and paracrine signaling, including the purinergic one, which might partially explain its ototoxic effects.

Keywords: aminoglycosides, connexins, Cx26, dye uptake, membrane current, intracellular calcium

INTRODUCTION

Two cells in close contact can exchange metabolites, second messengers and ions through gap junction channels (GJCs; Goldberg et al., 2004). Consequently, GJCs are key elements for diverse coordinated physiological responses of cell communities in most organs. Each GJC is made by the serial docking of two hemichannels (HCs), each one contributed by one of two adjacent cells. In turn, each HC is composed of six protein subunits called connexins (Cxs; Sáez et al., 2003). Undocked HCs are cell membrane channels permeable to ions and small molecules, constituting a communication pathway between the cytoplasm and the extracellular environment (Sáez et al., 2005).

Several mutations in genes coding for Cxs cause congenital prelingual syndromic and non-syndromic hearing loss. About 50% of the hearing loss cases are congenital, being most frequently caused by mutations in the GJB2 gene that encodes Cx26 (Martínez et al., 2009). The remaining 50% of these cases is due to environmental causes, including bacterial and viral infections, acoustic trauma and ototoxic drugs (Resendes et al., 2001). However, it remains unknown whether Cx26 HCs participate in the etiology of acquired hearing loss.

According to the World Health Organization (WHO, 1994), ototoxic drugs are substances of various structures and classes that cause harmful effects in hearing and/or balance organs. This group of molecules includes the aminoglycosides, which are the most common and dangerous ototoxic drugs. Isolated from *Streptomyces* or *Micromonospora*, they are highly hydrophilic antibiotics and have two or more amino groups that confer them a basic nature (Begg and Barclay, 1995). Due to high efficacy and low cost, aminoglycosides are widely used in the treatment of infections caused by aerobic Gram-negative bacteria as well as some mycobacteria (González and Spencer, 1998). However, their use is limited due to negative side effects on kidney and cochlea (for recent reviews, see López-Novoa et al., 2011; Xie et al., 2011). In the inner ear, the ototoxicity induced by aminoglycosides culminates in

Abbreviations: CCCP, carbonyl cyanide 3-chlorophenylhydrazone; CPA, cyclopiazonic acid; Cxs, connexins; Etd, ethidium; GJCs, gap junction channels; HCs, hemichannels.

the destruction of the cochlear sensory hair cells (Huth et al., 2011).

The most extensively studied aminoglycoside is gentamicin. Among the proposed mechanisms for its ototoxicity and nephrotoxicity, it has been suggested the generation of reactive oxygen species and nitric oxide (Sha and Schacht, 1998; Basnakian et al., 2002; Hong et al., 2006; Choung et al., 2009). The routes of gentamicin uptake in the inner ear structures are not fully understood, but several mechanisms have been proposed, including mechanotransducer channels (METs) located on stereocilia of hair cells, endocytosis in the apical or basolateral membranes of the organ of Corti, TRP channels or adenosine triphosphate (ATP) receptors (Huth et al., 2011). Moreover, it has been reported that MET channel opening is required to induce gentamicin toxicity in hair cells, suggesting an intracellular toxicity mechanism (Alharazneh et al., 2011). In addition, early studies suggested that cochleotoxicity of gentamicin is an excitotoxic process involving the activation of NMDA receptors (Basile et al., 1996), which might also be explained by the generation of free radicals (Sha and Schacht, 1998). In contrast, aminoglycosides have been shown to block a variety of ionic channels, such as hair cell METs (Jaramillo and Hudspeth, 1993), acetylcholine receptors (Blanchet et al., 2000) and purinergic ionotropic channels (Bongartz et al., 2010), all of which might partially explain their toxic effects. Purinergic signaling is one of the main mechanisms of paracrine signaling in the cochlea and has been associated to activation of K^+ recycling in cochlear supporting cells, being fundamental for the normal functioning of this sensory organ (Zhu and Zhao, 2010).

More than a decade ago, Todt et al. (1999) demonstrated that extracellular gentamicin inhibits gap junctional electrical coupling through free radical production in isolated cochlear supporting cells. However, the possible effects of gentamicin on HCs remain unknown, and the mechanisms responsible for its effects on GJCs are still not fully understood. In this work, we show that gentamicin applied to the extracellular media reduces the functional activity of HCs in a Cx composition-independent way. However, the same gentamicin concentration has no effect on intercellular dye coupling through GJCs. Gentamicin also reduces the oscillations of Cx HC-dependent cytosolic Ca^{2+} signals elicited by extracellular ATP in HeLa-Cx26 cells. Moreover, we found that gentamicin, like other GJC/HC blockers, reduces ATP release in HeLa-Cx26 cells, triggered by divalent cation-free solution (DCFS) or by UTP, a purinergic receptor agonist. Since Cx HCs play a relevant role as cellular membrane pathways for autocrine/paracrine signaling including the purinergic signaling of the inner ear (Zhao et al., 2005), gentamicin-induced Cx HC dysfunction may adversely affect these pathways, partially explaining the ototoxicity induced by this antibiotic through an extracellular mechanism.

MATERIALS AND METHODS

REAGENTS

Ethidium (Etd) bromide, $LaCl_3$, adenosine triphosphate disodium (Na_2ATP), cyclopiazonic acid (CPA, sarcoplasmic reticulum Ca^{2+}-pump inhibitor), carbonyl cyanide 3-chlorophenylhydrazone (CCCP, H^+ ionophore and uncoupler of oxidative phosphorylation in mitochondria), U73122 [Phospholipase C (PLC) inhibitor], carbenoxolone (CBX) and streptomycin were obtained from Sigma-Aldrich (St. Louis, MO, USA). Fura-2-AM was obtained from Molecular Probes (Eugene, OR, USA) and gentamicin sulfate from Invitrogen Life Technologies (Carlsbad, CA, USA). Polyclonal anti-$P2Y_2$, -$P2Y_4$, and -$P2Y_6$ receptor antibodies, as well as goat anti-rabbit and anti-mouse secondary antibodies conjugated to horseradish peroxidase, were obtained from Santa Cruz Biotechnology Inc. (Santa Cruz, CA, USA). The mimetic peptide Gap26 (sequence: N-VCYDKSFPISHVR-C) was synthesized by Beijing SBS Genetech Co. Ltd. (Beijing, China).

GFP-TAGGED Cx26 CONSTRUCT IN pcDNA3.1

The rCx26-GFP fusion protein (rCx26-GFP) was generated as described previously by Jara et al. (2012). The coding region of rat Cx26 (rCx26, NM_001004099.1) was subcloned into pcDNA3.1/CT-GFP-TOPO (Invitrogen Life Technologies, Carlsbad, CA, USA), according to the manufacturer's instructions. The coding region of the construct was fully sequenced.

CELL CULTURE

HeLa-Parental cells were obtained from ATCC (CCL-2; ATCC, Rockville, MD, USA) and were stably transfected with the rCx26-GFP construct using Lipofectamine 2000 (Invitrogen Life Technologies, Carlsbad, CA, USA), according to the manufacturer's instructions. HeLa cells stably expressing rCx26 (NM_001004099.1) were kindly provided by Dr. Bruce Nicholson (Department of Biochemistry at the University of San Antonio, San Antonio, TX, USA), while HeLa cells stably expressing mouse Cx26 (mCx26, NM_008125.3), Cx43 (mCx43, NM_010288.3), or Cx45 (mCx45, NM_001159383.1) were kindly provided by Dr. Klaus Willecke from the LIMES Institute (Bonn Universität, Germany). All cell lines were grown at 37°C and 5% CO_2 in DMEM supplemented with 10% fetal bovine serum (GIBCO, Invitrogen), 100 U/ml penicillin, 100 μg/ml streptomycin sulfate, and 0.5 μg/ml puromycin to select transfected cells. HeLa rCx26-GFP cells were selected with 500 μg/ml G418. All cell lines were used 48 h after seeding, and all antibiotics used for selection and maintenance were not included during this period of time. Untransfected HeLa-Parental cells were used as control.

HC ACTIVITY

The HC activity was evaluated using the Etd uptake method as described (Figueroa et al., 2013). In brief, sub-confluent HeLa cells grown on glass coverslips were washed twice with recording solution [in mM: NaCl (148); KCl (5); $CaCl_2$ (1.8); $MgCl_2$ (1); glucose (5); HEPES (5), pH = 7.4] containing 5 μM Etd. Basal fluorescence intensity from the nucleus of each cell was recorded for 5 min, using an Olympus BX 51W1I upright microscope (Olympus America Inc., Center Valley, PA, USA). Next, cells were washed three times with (Ca^{2+}/Mg^{2+}-free) DCFS and the fluorescence intensities of the nuclei were recorded. At the end of each experiment, the Cx HC blocker La^{3+} was added to confirm HC mediated Etd uptake (Contreras et al., 2002; Retamal et al., 2007). Dye uptake of each cell was digitally photographed using a CCD monochrome camera (CFW-1310M; Scion; Frederick, MD, USA). Images were captured every 30 s (exposure time = 30 ms, gain = 0.5). Metafluor software (version 6.2R5, Universal Imaging Co., Downingtown, PA, USA) was used for data acquisition and off-line image analysis. The fluorescence intensity of at least 30 cells per experiment was averaged and plotted against time (expressed in minutes). Lastly, the slope, here called Etd uptake rate, was calculated using Microsoft Excel software and expressed as arbitrary units per minutes (AU/min).

hCx26 cRNA PREPARATION AND INJECTION INTO *Xenopus laevis* OOCYTES

The plasmid pOocyte-Cx26, containing human Cx26 cDNA (hCx26), was kindly provided by Dr. Guillermo Altenberg (Texas Tech University Health Sciences Center, Lubbock, TX, USA). cRNA coding for hCx26 was prepared as previously described (Figueroa et al., 2013). To reduce expression of endogenous Cx38, an antisense oligonucleotide directed against Cx38 was used. After the injection of the cRNA, Oocytes were maintained in Barth's solution [in mM: NaCl (88); KCl (1); $CaCl_2$ (5); $MgCl_2$ (0.8); HEPES (10), pH = 7.4] supplemented with 0.1 mg/ml gentamicin and 20 units/ml of penicillin-streptomycin.

MEMBRANE CURRENT VIA HCs

Dual whole cell voltage clamp recordings of *X. laevis* oocytes injected with hCx26 cRNA were carried out as described (Retamal et al., 2011) using a two electrode voltage clamp amplifier for oocytes (Warner Instruments, model OC-725C) connected to a digital-to analog converter (Molecular Devices, model DigiData 1440A). ND96 medium [in mM: NaCl (96); KCl (2); $CaCl_2$ (1.8); $MgCl_2$ (1); HEPES (10), pH = 7.4] was used as bath solution in all experiments. Recording pipettes were filled with 3 M KCl. For data acquisition and analysis, the pClamp 10 software was used. Currents were measured after 15 s rectangular voltage pulses, ranging from −60 to +40 mV, in 10 mV steps with a holding potential of −60 mV and 10 s intervals between pulses. Female *X. laevis* were obtained from the animal facility of Universidad de Chile, and the Commission of Bioethics and Biosafety of the Universidad del Desarrollo approved the experimental protocols.

INTRACELLULAR Ca^{2+} SIGNAL

The intracellular Ca^{2+} signal was evaluated as described (Figueroa et al., 2013). The intracellular Ca^{2+} signal was monitored in Fura-2-AM (5 μM) loaded HeLa cells grown on glass coverslips, and recording solution described above for dye uptake experiments was used. Fluorescence from regions of interest (ROI's) covering single Fura-2 loaded cells was determined at excitation wavelengths of 340 and 380 nm, while fluorescence emission was collected at 510 nm every 3 s using an Olympus BX 51W1I upright microscope. The intracellular Ca^{2+} signal was calculated as $R = F_{340\ nm}/F_{308\ nm}$, and the background was subtracted. Ca^{2+} transients were evoked by extracellular application of ATP and signals obtained were averaged, including at least 30 cells per experiment. Subsequently, the area under curve (AUC) and duration (measured as the time between the first increase in Ca^{2+} signal until the return to baseline) were calculated and represented graphically.

EXTRACELLULAR ATP MEASUREMENT

Adenosine triphosphate release from HeLa-rCx26 cells was determined using the ATP bioluminescence assay kit (Sigma) in combination with a spectrofluorometer (Jasco Corp., FP-63000, Tokyo, Japan). HeLa-rCx26 cells were seeded into 60 mm culture plates 24 h before each experiment or until they reached 70% confluence. For extracellular ATP measurements, the culture medium was removed and cells were washed twice with DCFS or Ca^{2+}/Mg^{2+}-containing solution. Then, cells were incubated for 5 min in 500 μl of DCFS or treated with 100 μM UTP in Ca^{2+}/Mg^{2+}-containing solution to induce ATP release. Subsequently, the 500 μl of extracellular solution were carefully collected to avoid damaging the cells, and ATP content was determined immediately using the luciferin/luciferase bioluminescence assay. To this end, 50 μl of the extracellular solution were mixed in the cuvette with 50 μl of luciferin/luciferase reagent, and the average light signal was measured for 10 s with a spectrofluorometer (Jasco Corp., FP-63000, Tokyo, Japan). The different reagents used in these experiments were diluted in 500 μl of DCFS or Ca^{2+}/Mg^{2+} solution resulting in the following concentrations: 200 μM gentamicin and 100 μM CBX. The ATP concentration was determined by using a luminescence standard curve, and data were normalized by the total cellular protein present in the plate. When CBX and gentamicin were used, they were included from the first wash at the indicated concentration.

WESTERN BLOT ANALYSES

Relative levels of proteins were assayed by Western blot analyses as described (Figueroa et al., 2013). Blots were incubated with primary polyclonal anti-$P2Y_2$, -$P2Y_4$, or -$P2Y_6$ antibodies overnight at 4°C, followed by five washes with TBS, 1% Tween-20 buffer. Then, they were incubated with goat anti-rabbit secondary antibodies and conjugated with horseradish peroxidase. The immunoreactivity was detected by electrogenerated chemiluminescence (ECL) using the SuperSignal kit (Pierce, Rockford, IL, USA) according to the manufacturer's instructions. Blots were also developed with α-tubulin, used as loading control.

DYE COUPLING

The functional state of GJCs was evaluated as described (Figueroa et al., 2013). Briefly, confluent HeLa cell cultures grown on coverslips were used in each experiment. Single cells were iontophoretically microinjected with a glass micropipette filled with 75 mM Etd in water or Lucifer yellow (LY, 5% w/v in 150 mM LiCl). Dye coupling index was calculated as the mean number of cells to which the dye spread occurred. All microinjections were performed in HCO_3^--free F-12 medium buffered with 10 mM HEPES (pH 7.4) containing 200 μM La^{3+} to avoid cell leakage of the microinjected dye via HCs. Fluorescent cells were observed using a Nikon inverted microscope equipped with epifluorescence illumination (Xenon arc lamp) and Nikon B filter to LY (excitation wavelength 450–490 nm; emission wavelength above 520 nm) and XF34 filter to Etd fluorescence (Omega Optical, Inc., Brattleboro, VT, USA). Photomicrographs were obtained using a CCD monochrome camera (CFW-1310M; Scion; Frederick, MD, USA). In all experiments, dye coupling was tested by injecting a minimum of 14 cells.

STATISTICS

Statistical analysis was performed using GraphPad Prism 5 software for Windows (GraphPad Software, San Diego, CA, USA). Data sets (means ± SEM) were compared using one-way analysis of variance (ANOVA) followed by a Tukey's post-test or Student's *t*-test for pair-wise comparisons.

RESULTS

GENTAMICIN INHIBITS HC ACTIVITY ELICITED BY EXTRACELLULAR DIVALENT CATION-FREE SOLUTION

The functional state of HCs was determined by measuring the cellular uptake of Etd. At a concentration of 5 μM, Etd diffuses across the cell membrane preferentially via HCs (Schalper et al., 2008). Once in the intracellular space, it binds to nucleic acids and emits fluorescence that is proportional to the HCs activity (Contreras et al., 2002; Schalper et al., 2008; Orellana et al., 2010). Etd uptake was measured during 5 min in HeLa-Parental and HeLa-rCx26 cells exposed to a saline solution with physiologic extracellular divalent cation concentrations (**Figure 1A**, Ca^{2+}/Mg^{2+}). In HeLa-rCx26 cells, the fluorescence intensity was slightly higher than in HeLa-Parental cells (**Figure 1B**, Ca^{2+}/Mg^{2+}), which do not express Cx26 HCs. In order to increase the open probability of rCx26 HCs, cells were bathed in DCFS (Jara et al., 2012). A rapid increase in fluorescence intensity was measured in HeLa-rCx26 but not in HeLa-Parental cells (**Figures 1A,B**, DCFS). Lastly, the application of 200 μM gentamicin to the bath solution reduced the DCFS-induced Etd uptake in HeLa-rCx26 cells to values similar to those recorded in the presence of physiologic extracellular Ca^{2+}/Mg^{2+} concentrations either in HeLa-rCx26 or -Parental cells (**Figure 1B**, gentamicin).

The basal Etd uptake rate of HeLa-rCx26 cells bathed with saline solution containing divalent cations was slightly higher than that of HeLa-Parental cells ($m_1 = 0.06$ AU/min for HeLa-rCx26 v/s $m_1 = 0.04$ AU/min for HeLa-Parental; **Figure 1B** Ca^{2+}/Mg^{2+} first 5 min and C), which might result from basal activity of rCx26 HCs. However, the Etd uptake and Etd uptake rate of HeLa-rCx26 cells was ~two fold higher when bathed with DCFS than in the presence of extracellular divalent cations (**Figures 1B,C**; $m_1 = 0.06$ AU/min v/s $m_2 = 0.13$ AU/min), whereas the Etd uptake rate of HeLa-Parental cells in DCFS remained indistinguishable from the rate measured in the presence of physiological

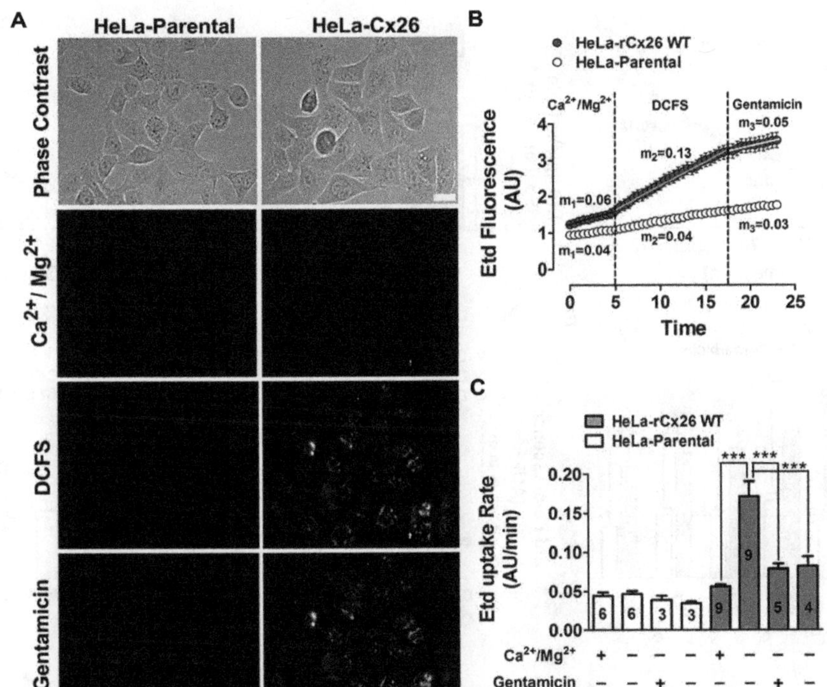

FIGURE 1 | Gentamicin blocks the ethidium uptake induced by a divalent cation-free solution (DCFS) in HeLa-Cx26 cells. The ethidium (Etd) uptake was evaluated in HeLa-Parental or -Cx26 cells bathed with divalent cation (Ca^{2+}/Mg^{2+}) or DCFS solution in time-lapse measurement experiments. **(A)** Representative fluorescent fields showing Etd florescence of HeLa-Parental or -rCx26 cells incubated in saline solution containing 5 μM Etd, under control conditions (Ca^{2+}/Mg^{2+}, 5 min), after exposure to DCFS (10 min) or after 5 min exposure to DCFS containing 200 μM gentamicin. Scale bar, 40 μm. **(B)** Representative time-lapse experiments showing Etd uptake in HeLa-Cx26 and -Parental cells under control conditions (Ca^{2+}/Mg^{2+}, first 5 min), after exposure to DCFS solution followed by the application of 200 μM gentamicin (last 5 min). m_1, m_2, m_3 = average slope. Measurements were taken every 30 s as fluorescence emission intensity of Etd bound to DNA and referred as fluorescence intensity expressed in arbitrary units (AUs). Each value corresponds to the mean ± SEM of at least 30 cells. **(C)** Etd uptake rate of Hela-Parental or -rCx26 cells measured under control conditions and after exposure to DCFS with or without 200 μM La^{3+} or 200 μM gentamicin. Data are presented as means ± SEM, the digit within each bar corresponds to the number of independent experiments under that condition. $***P < 0.001$.

concentrations of divalent cations (**Figures 1B,C**, Ca^{2+}/Mg^{2+} v/s DCFS). Then, the Etd uptake was drastically inhibited by 200 μM gentamicin (**Figure 1B**) and the Etd uptake rate values changed from m_2 = 0.13 AU/min to m_3 = 0.05 AU/min in Hela-rCx26 cells, but was not affected in HeLa-Parental cells (**Figure 1C**, DCFS + gentamicin). Similar results were obtained in HeLa-rCx26 cells bathed with DCFS and treated with 200 μM La^{3+} (**Figure 1C**), a widely used Cx HC blocker (Sáez et al., 2003). The gentamicin-induced Etd uptake rate inhibition in Hela-rCx26 cells was found to be reversible because it was restored by replacing the bath solution with gentamicin-free DCFS (**Figures 2A,B**). The gentamicin-induced inhibition of rCx26 HCs occurred in a concentration-dependent manner, being 133.4 ± 1.1 μM the concentration that induced 50% Etd uptake rate inhibition (IC50; **Figure 2C**). Gentamicin also inhibited Etd uptake in HeLa-rCx26GFP cells as well as in mCx26, mCx43, or mCx45 bathed in DCFS (**Figure 2D**). In addition, streptomycin (Strep 200 μM), another aminoglycoside antibiotic, also inhibited the Etd uptake of HeLa-rCx26 (**Figures 2E,F**), -mCx43 and -mCx45 cells in DCFS (**Figure 2F**). In HeLa-mCx45 cells, Etd uptake was reduced to values below the ones recorded under control condition (**Figure 2F**), suggesting that these HCs present a higher open probability under basal condition.

GENTAMICIN INHIBITS THE CELL MEMBRANE CURRENT MEDIATED BY Cx26 HCs

To test whether gentamicin inhibits Cx26 HCs, we recorded membrane currents generated by the application of rectangular command voltages under whole cell dual voltage clamp in *X. laevis* oocytes expressing hCx26. Endogenous oocyte Cx38 expression was inhibited by using specific antisense. Forty eight hours after the cRNA injection, oocytes were depolarized from −60 to +40 mV (10 mV steps) for 15 s (**Figure 3A**). Under control conditions (ND96 solution containing 1.8 mM Ca^{2+}), the activation of an outward current was evident (at voltages above 0 mV) followed by a tail current upon repolarization to −60 mV (**Figure 3A**). These currents were virtually absent in oocytes injected with Cx38 antisense oligonucleotide (**Figure 3B**), indicating this was mediated by activated hCx26 HCs. Then, cells were

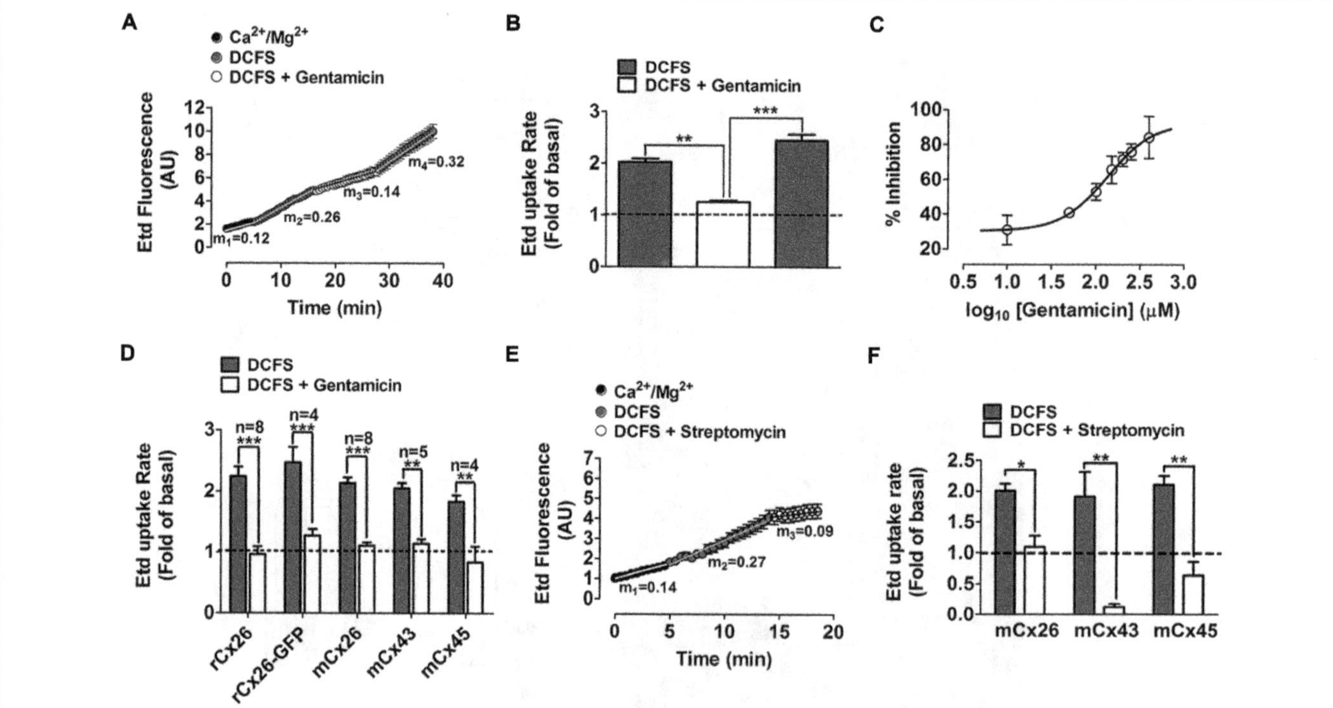

FIGURE 2 | Gentamicin blocks connexin26 hemichannels in a reversible- and concentration-dependent manner and also blocks connexin43 and connexin45 hemichanels. (A) Representative time-lapse experiment showing ethidium (Etd) uptake in cells transfected with rat connexin26 (HeLa-rCx26 cells) under control conditions (Ca^{2+}/Mg^{2+}), after exposure to DCFS, followed by 5 min exposure to DCFS containing 200 μM gentamicin and washout of gentamicin with DCFS (last 10 min), m_1, m_2, m_3, m_4 = average slope. **(B)** Average Etd uptake rate (experiments as shown in **A**); data were normalized to the control Etd uptake level (dotted line). **(C)** Concentration-response curve showing the inhibition of DCFS-induced Etd uptake prompted by different gentamicin concentrations in HeLa-rCx26 cells. Each point represents the mean ± SEM (n = 3). Values were normalized to the maximal response in the absence of gentamicin and were included in the Hill equation, IC_{50} = 133.4 ± 1.1 μM, R^2 = 0.95. **(D)** Etd uptake rates of HeLa cells expressing rat Cx26 (rCx26), mouse Cx26 (mCx26), rat Cx26-GFP (rCx26-GFP), mouse Cx43 (mCx43) or mouse Cx45 (mCx45), after exposure to DCFS followed by 5 min exposure to DCFS containing 200 μM gentamicin. Data were normalized to the Etd uptake value measured under control conditions (Ca^{2+}/Mg^{2+} solution, dotted line) and are presented as mean ± SEM (n = 3). **(E)** Representative time-lapse curve showing Etd uptake in HeLa-rCx26 cells in control conditions and after exposure to DCFS followed by 200 μM streptomycin, m_1, m_2, m_3 = average slope. **(F)** DCFS-induced Etd uptake normalized to control uptake in absence (gray bar) or presence (white bar) of 200 μM streptomycin of HeLa cells expressing rCx26, mCx43, or mCx45. Values recorded in at least 30 cells per experiment were included (n = 3). $^*P < 0.05$, $^{**}P < 0.01$, $^{***}P < 0.001$.

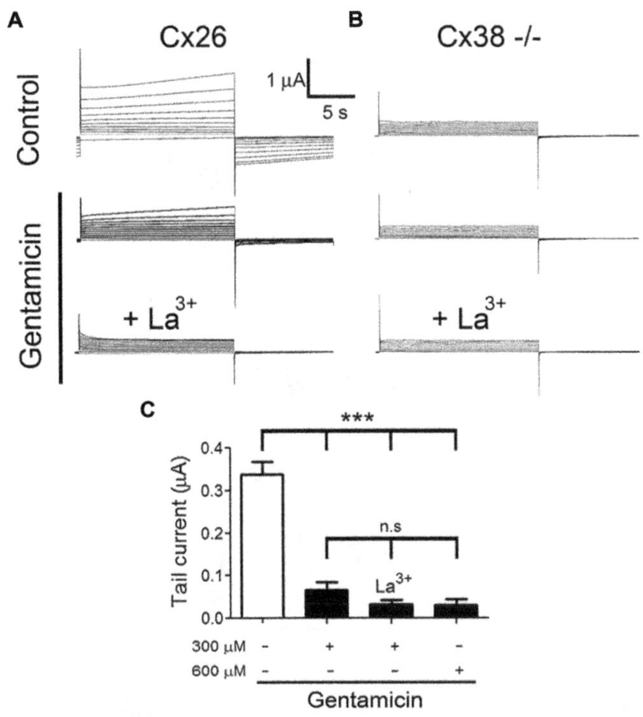

FIGURE 3 | Gentamicin inhibits the membrane current mediated by human connexin hemichannels. The membrane current of *Xenopus laevis* oocytes injected with connexin38 antisense oligonucleotide and human connexin (Cx26) cRNA was evaluated in whole cell modality using two electrodes. **(A)** Currents induced by depolarization from −60 to +40 mV (10 mV steps, for 15 s) under control condition (upper panel) or in the presence of 300 or 600 μM gentamicin (middle panel) and 300 μM gentamicin plus 200 μM La^{3+} (lower panel). **(B)** Currents induced by depolarization from −60 to +40 mV in oocytes non-injected with Cx26 cRNA under control conditions (upper panel) or in the presence of 300 or 600 μM gentamicin (middle panel) and 300 μM gentamicin plus 200 μM La^{3+} (lower panel). **(C)** Average maximal tail currents from 5 oocytes in the presence or absence of 300 or 600 μM gentamicin and 300 μM gentamicin plus 200 μM La^{3+}. $^{***}P < 0.005$ and n.s: not significant.

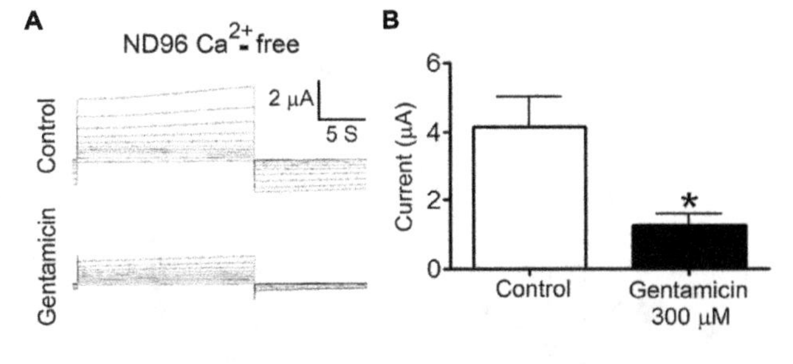

FIGURE 4 | Gentamicin inhibits the membrane current mediated by human connexin26 hemichannels in nominal Ca^{2+}-free solution. The membrane current of *Xenopus laevis* oocytes injected with connexin38 antisense oligonucleotide and human connexin26 cRNA was evaluated in whole cell modality using two electrodes in ND96 without Ca^{2+} and Mg^{2+}. **(A)** Currents induced by depolarization from −60 to +40 mV (10 mV steps, for 15 s) under control conditions (upper panel) or in the presence of 300 μM gentamicin (lower panel). **(B)** Average maximal currents at +40 mV from 5 oocytes in the presence or absence of 300 μM gentamicin. $^{*}P < 0.01$.

treated for 3–5 min with 300 μM gentamicin applied in the bath, and both the maximal and tail currents were drastically reduced (**Figure 3A**, middle) and completely abolished by the subsequent addition of La^{3+} (**Figure 3A**, bottom). To minimize contamination with endogenous currents, we measured the maximal current at −60 mV before the depolarization at +40 mV. It was observed that gentamicin-induced a concentration-dependent decrease of maximal tail current (**Figure 3C**). Under control conditions, the maximal tail current recorded was 0.34 ± 0.03 μA, and after the addition of 300 or 600 μM gentamicin it was 0.07 ± 0.02 and 0.03 ± 0.02 μA, respectively (**Figure 3C**). Similar results were obtained in presence of 200 μM La^{3+} in the extracellular solution (**Figure 3C**). In the absence of extracellular Ca^{2+}, outward currents generated with the different command voltages were more prominent than those recorded in bath solution containing 1.8 mM Ca^{2+} (**Figure 4A**). In the absence of extracellular Ca^{2+}, 300 μM gentamicin inhibited the membrane currents (**Figure 4A**) by ~70% (from 4.15 ± 0.87 to 1.28 ± 0.32 μA; **Figure 4B**).

GENTAMICIN REDUCES THE ATP-INDUCED Ca^{2+} SIGNAL IN HeLa-rCx26 CELLS

Since HCs have been suggested to play a major role in inner ear purinergic signaling (Zhao et al., 2005) and because inhibition of Cx43 HCs reduces the intracellular Ca^{2+} signals elicited by bradykinin (De Bock et al., 2012), we tested whether gentamicin affects the ATP-induced Ca^{2+} signals in HeLa-rCx26 cells.

Gentamicin was applied to the bath solution 5 min prior to stimulation with extracellular ATP in presence of 1.8 mM $[Ca^{2+}]_0$. Fura-2 loaded HeLa-rCx26-GFP cells were used to facilitate identification of cells expressing Cx26 and 10 μM ATP was applied in all experiments. **Figure 5** shows a representative experiment including five representative cells in each record. In all cells, the application of 10 μM ATP induced a fast initial increase in Ca^{2+} signal. In HeLa-Parental cells, small oscillations with ~5 s intervals were superimposed with a sustained Ca^{2+} signal increase that decayed progressively over time (**Figure 5A**). In contrast, in Hela-rCx26-GFP cells the oscillations were more evident than in parental cells (**Figure 5C**). The ATP induced-Ca^{2+} signal increases were slightly reduce by 200 μM gentamicin in HeLa-Parental cells (**Figure 5B**). However, gentamicin significantly reduced the ATP-induced Ca^{2+} signals in HeLa-rCx26-GFP cells (**Figure 5D**). Therefore, the most frequent effect of gentamicin was a reduction in oscillation frequency (e.g., in HeLa-rCx26-GFP), thus, from the first Ca^{2+} rise to the return to baseline, cells oscillated an average of 8.8 ± 0.4 times in control conditions and were reduced to 5.1 ± 0.4 by 200 μM gentamicin (100 cells randomly analyzed in seven independent experiments). The AUC and duration of the ATP-elicited Ca^{2+} signals were significantly reduced by gentamicin in HeLa rCx26-GFP, but the reductions observed in HeLa-Parental cells were not statistically significant (**Figures 5E,F**). We also measured the Ca^{2+} signal after 20 min incubation with gentamicin, and no significant differences with respect to the values recorded after 5 min treatment were found (data not shown). Similar results were obtained in HeLa rCx26-GFP after 10 min preincubation with 100 μM CBX, a HC/GJC blocker, or 20 min preincubation with 200 μM GAP-26 (**Figures 5E,F**), a mimetic

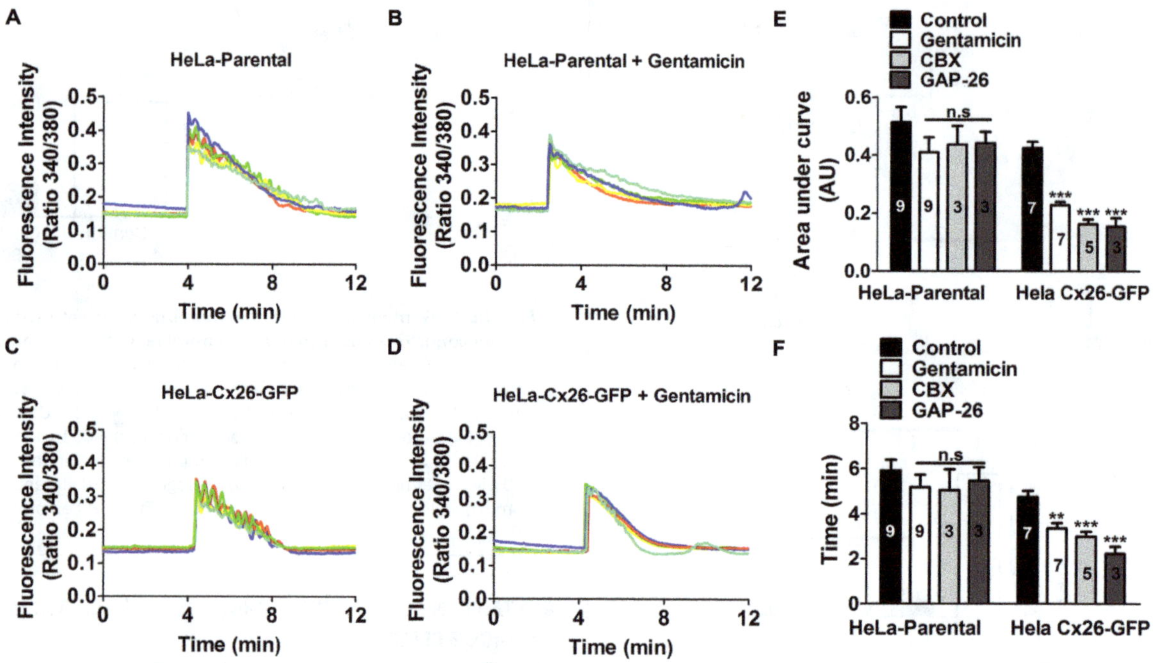

FIGURE 5 | Gentamicin and hemichannel blockers reduce the ATP-evoked Ca^{2+} signal in HeLa-rCx26GFP cells. HeLa-Parental cells or HeLa cells stably transfected with rCx26-GFP were loaded with Fura-2 and then stimulated with bath application of 10 μM ATP. **(A)** Time course of ATP-evoked intracellular Ca^{2+} signal under control conditions in HeLa-Parental cells. **(B)** Time course of ATP-evoked intracellular Ca^{2+} signal in HeLa-Parental cells pre-incubated for 5 min with 200 μM gentamicin. Five representative cells are shown in each case. **(C)** Time course of ATP-evoked intracellular Ca^{2+} signal under control conditions in HeLa-rC26-GFP cells. **(D)** Time course of ATP-evoked intracellular Ca^{2+} signal in HeLa-rC26-GFP cells pre-incubated for 5 min with 200 μM gentamicin. **(E)** Compiled data showing the amplitude of Ca^{2+} signal changes evoked by extracellular ATP (mean ± SEM) in HeLa parental and HeLa-rCx26GFP, measured as area under curve expressed as AU in control conditions or after 5 min preincubation with 200 μM gentamicin, 10 min with 100 μM carbenoxolone (CBX) or 20 min with 200 μM GAP-26. **(F)** Duration (time), from the first Ca^{2+} signal increase until the return to baseline (mean ± SEM) in HeLa-Parental and HeLa-rCx26GFP in control conditions or after 5 min preincubation with 200 μM gentamicin, 10 min with 100 μM CBX or 20 min with 200 μM GAP-26. For calculations of the average traces (as shown in A) 40 cells per experiment were included. The number of independent experiments is indicated in each bar. $^{**}P < 0.01$, $^{***}P < 0.001$.

peptide that blocks Cx26 HCs (Evans and Leybaert, 2007). However, these blockers did not have significant effect on parental cells (**Figures 5E,F**).

It is known that extracellular ATP activates P2Y receptors coupled to G proteins, which in turn can activate PLC-dependent pathways, generating IP_3 and thus promoting Ca^{2+} mobilization from the sarcoplasmic reticulum (Abbracchio et al., 2006). To determine whether this intracellular pathway is involved in the ATP-induced Ca^{2+} signal oscillations in HeLa-rCx26, we used a pharmacological approach. The possible involvement of PLC was tested using U73122 as inhibitor since, as mentioned above; PLC is a central component of the signal transduction mediated by activation of metabotropic purinergic receptors. To evaluate the participation of the endoplasmic reticulum in the ATP-induced Ca^{2+} signal oscillations, we used CPA, a sarcoplasmic-endoplasmic reticulum Ca^{2+}-ATPase pump inhibitor (Suzuki et al., 1992). We determined the possible involvement of mitochondria in the ATP-induced Ca^{2+} signal oscillations (Ishii et al., 2006) in HeLa-rCx26 using the protonophore CCCP, which inhibits the mitochondrial Ca^{2+} transport by collapsing the proton electrochemical gradient (Bygrave, 1978).

To identify an optimal concentration at which the inhibitory effect was reproducible, the effect of different concentrations of each blocker was tested in Fura-2 loaded HeLa rCx26-GFP cells. **Figure 6** shows concentration-response experiments in which each blocker exhibited a concentration-dependent inhibition of ATP-induced Ca^{2+} signals. Neither U73122 nor CCCP eliminated completely the Ca^{2+} signals, but they strongly reduced them (**Figures 6A,C,D**). However, 10 μM CPA inhibited completely the ATP-induced Ca^{2+} signals (**Figures 6B,D**). These data indicate that extracellular ATP is acting through metabotropic receptors, activating PLC and releasing Ca^{2+} from the reticulum. To confirm these results, we tested whether HeLa cells express metabotropic purinergic receptors. Western blot analyses for P2Y receptors contained in whole HeLa-Parental and HeLa-Cx26 cell lysates were performed. We used polyclonal antibodies against different types of P2Y receptors ($P2Y_2R$, $P2Y_4R$, and $P2Y_6R$) previously reported in HeLa cells (Okuda et al., 2003). Immunoreactive bands of ~34 and 95 kDa were detected for the three receptors in Hela-Parental cells and HeLa cells expressing rCx26 and rCx26-GFP, respectively (**Figure 7**). These bands were also detected in mouse brain (MB) lysate used as positive control. P2YRs receptors have been described to be 308–379 amino acid proteins with mass ranging from 41 to 53 kDa after glycosylation (D'Ambrosi et al., 2006). Bands with fast electrophoretic mobility were near the predicted molecular mass of

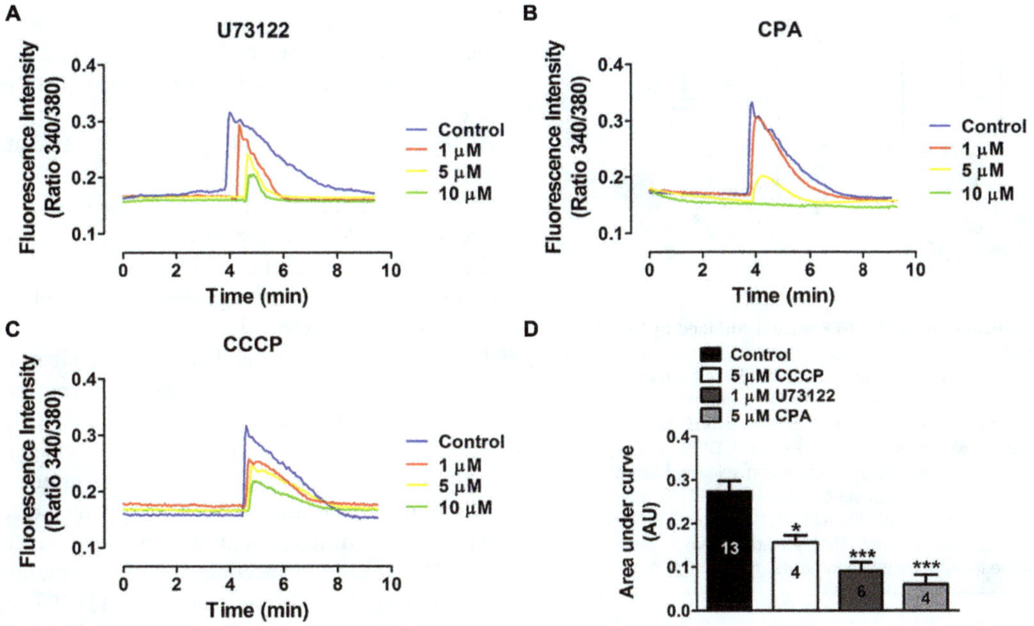

FIGURE 6 | Concentration-dependent effects of phospholipase C and intracellular Ca^{2+} stores inhibitors on the ATP-induced Ca^{2+} signal. The effect of different concentrations of **(A)** U73122, a phospholipase C (PLC) inhibitor, **(B)** CPA, an endoplasmic reticulum Ca^{2+}-ATPase pump inhibitor and **(C)** CCCP, a protonophore, was tested on the Ca^{2+} signal induced by 10 μM ATP. Representative average traces (including 40 cells per trace) measured in Fura-2 loaded HeLa-rCx26-GFP cells are shown. Cells were preincubated with each blocker during 10 min prior to stimulation with ATP. **(D)** Compiled data showing the amplitude of the ATP-evoked Ca^{2+} signals (mean ± SEM) measured as area under the curve and expressed as AU (mean ± SEM). Each trace represents the average from 40 cells per experiment. The number of independent experiments is indicated within each bar. $^{*}P < 0.05$, $^{***}P < 0.001$.

$P2Y_2$, $P2Y_4$, and $P2Y_6$ receptor protein subunits deduced from their cDNA sequence, being 42, 41 and 36 kDa, respectively (P41231, P51582, Q15077; SwissProtKB). These bands correspond to the monomeric forms while higher-order bands between 72 and 95 kDa may correspond to post-translational modified forms due to glycosylation (Sage and Marcus, 2002; Delbro et al., 2005), oligomeric forms of each receptor (D'Ambrosi et al., 2006, 2007) or heterodimerization between purinergic receptor subtypes (Yoshioka et al., 2001; Nakata et al., 2005; Suzuki et al., 2006). Notably, HeLa cells transfected with rCx26 or rCx26-GFP contained higher levels of the monomeric bands of each receptor protein than HeLa-Parental cells did (**Figures 7A,B**, 43 kDa). The functional expression of each receptor identified was also confirmed using a pharmacological approach with different purinergic agonists such as ATP, UTP, and UDP (data not shown).

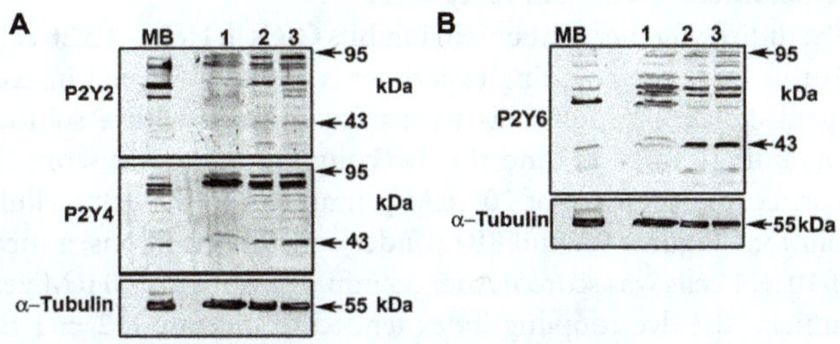

FIGURE 7 | Detection of $P2Y_2$, $P2Y_4$, and $P2Y_6$ receptors in HeLa-Parental and HeLa-Cx26 cells. Total homogenates of HeLa-Parental and HeLa cells expressing rCx26 or rCx26-GFP were resolved in sodium dodecyl sulfate polyacrylamide gel electrophoresis (12% SDS-PAGE). Membranes were probed with **(A)** anti-$P2Y_2$R and anti-$P2Y_4$R polyclonal or **(B)** anti-$P2Y_6$R antibodies. Protein loading was controlled with anti-α-tubulin monoclonal antibody. Bands were detected with secondary anti-rabbit or anti-mouse antiserum coupled to horseradish peroxidase by electrogenerated chemiluminescence (ECL). MB, mouse brain (tissue lysate); Lane 1, HeLa-Parental; Lane 2, HeLa-rCx26; Lane 3, HeLa-rCx26-GFP.

GENTAMICIN REDUCES ATP RELEASE TRIGGERED BY ACTIVATION OF P2Y RECEPTORS IN HeLa-rCx26 CELLS

Previous reports have shown that the expression of Cxs allows HeLa cells to release ATP in response to DCFS or purinergic receptor activation by UTP (Cotrina et al., 1998; De Vuyst et al., 2007). We studied the changes in extracelullar ATP using the luciferin-luciferase assay in response to DCFS to increase the open probability of rCx26 HCs in HeLa cells. In DCFS the amount of extracellular ATP was about 10-fold greater with respect to control conditions (Ca^{2+}/Mg^{2+}, **Figure 8A**). Additionally, the application of 200 μM gentamicin or 100 μM CBX reduced the DCFS-induced ATP release to values similar to those recorded in the presence of physiologic extracellular Ca^{2+}/Mg^{2+} concentrations (**Figure 8A**). We also measured the release of ATP mediated by activation of P2Y purinergic receptors by extracellular 100 μM UTP in HeLa rCx26 in the presence of extracelullar 1.8 mM Ca^{2+} and 1 mM Mg^{2+}. The ATP concentration increased about seven fold compared to basal

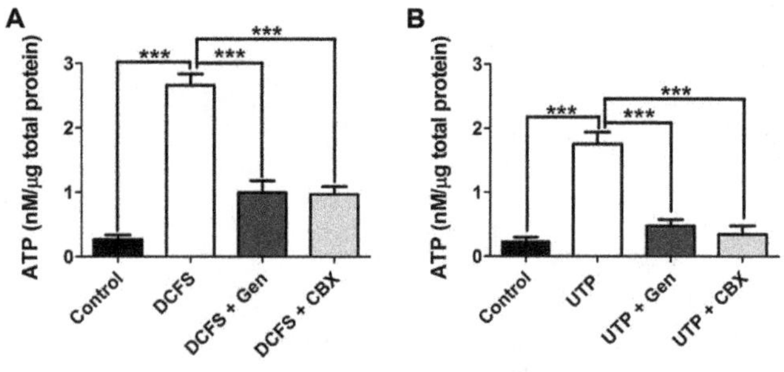

FIGURE 8 | Gentamicin reduces the ATP release induced by DCFS or UTP in HeLa-rCx26 cells. ATP release from HeLa-rCx26 cells in response DCFS and purinergic receptor stimulation by UTP. ATP released to the extracellular solution was determined with an ATP bioluminescence assay kit 5 min after stimulation. **(A)** ATP released by HeLa-rCx26 cells exposed to DCFS, DCFS + 200 μM gentamicin (DCFS + Gen), DCFS + 200 μM CBX (DCFS + CBX) or with Ca^{2+}/Mg^{2+} containing solution (control). **(B)** ATP released by HeLa-rCx26 stimulated with 100 μM UTP, 100 μM UTP + 200 μM gentamicin (UTP + Gen), 100 μM UTP +100 μM CBX (UTP + CBX) in Ca^{2+}/Mg^{2+} solution (control). The data are shown as means [ATP] ± SEM of three independent experiments. $***P < 0.001$.

release (without UTP, **Figure 8B**). Additionally the UTP-induced ATP release was drastically reduced by 200 μM gentamicin or 100 μM CBX (**Figure 8B**). These results suggest that gentamicin influences the amplitude, duration and shape of the ATP-induced Ca^{2+} signal, probably by decreasing the release of ATP through Cx HCs.

GENTAMICIN DOES NOT INHIBIT INTERCELLULAR GAP JUNCTIONAL COMMUNICATION IN HeLa-rCx26 CELLS

We studied whether gentamicin inhibits GJCs in HeLa-rCx26 cells. To this end, dye coupling experiments were performed in confluent cultures. Single cells were microinjected with a solution containing Etd or LY, and the dye coupling index was scored in absence and presence of 200 μM gentamicin in the extracellular solution (**Figures 9A** and **8B**). Under control conditions, a mean of 10 ± 1 cells was scored. After 35 min exposure to 200 μM gentamicin, the dye coupling index tended to increase (12 ± 1 Etd coupled cells), although this response was not statistically significant compared to that of the control value (14 injected cells per experiment, in three independent cultures). Similar results were obtained using LY (from 8 ± 1 cells increased to 9 ± 1 cells after treatment with gentamicin; **Figures 9A,B**). In HeLa cells transfected with mCx43 the Etd coupling was not significantly affected by gentamicin applied in the extracellular solution (6 ± 1 cells under control conditions and 7 ± 1 cells after treatment with gentamicin; **Figure 9B**). When gentamicin was included in the pipette to a final concentration of 200 uM with the Etd solution, no significant differences were observed in the number of coupled cells in cultures of rCx26 or mCx43 (**Figure 9B**), indicating that gentamicin not affect the dye coupling when applied intracellularly.

DISCUSSION

In this work, we found that the aminoglycoside antibiotic gentamicin applied extracellularly is a reversible Cx26 HC blocker that does not affect the functional state of gap junctions. Moreover, gentamicin drastically reduced both, the intracellular Ca^{2+} increase and the release of ATP induced by P2Y receptor activation by DCFS or UTP. Thus, the cellular toxicity of gentamicin might be explained partially by a perturbation of the autocrine/paracrine HC-dependent cell signaling, as in that mediated by extracellular ATP.

The addition of GFP to the C-terminus of rCx26 did not interfere with the HC blocking effect of gentamicin, suggesting that the C-terminal domain is not involved in gentamicin-induced inhibition of HCs. In addition, gentamicin also inhibited HCs formed by mCx43 or mCx45, which have longer C-terminus, indicating that its action is independent of the HC composition and the length of the C-terminus, a characteristic reported previously for other HC/GJC blockers (D'hondt et al., 2009; Wright et al., 2009). Our results differ from a previous study where rCx26 was shown not to form voltage-gated HCs in *Xenopus* oocytes and Neuro2A cells (González et al., 2006). The apparent discrepancy may be explained by differences in experimental procedures. González et al. (2006) used extracellular solution containing 1.8 mM Ca^{2+} and 0.8 mM Mg^{2+}. In contrast, we used DCFS, which increases the open probability of all Cx HCs so far studied (Sáez et al., 2003). In agreement with our findings, activation of rCx26 HCs expressed in HeLa cells has been previously induced by DCFS (Jara et al., 2012).

Aminoglycosides block several ionic channels including hair cell METs (Jaramillo and Hudspeth, 1993), ACh receptors (Blanchet et al., 2000; Amici et al., 2005) and purinergic ionotropic channels (Bongartz et al., 2010). In most of these channels, gentamicin acts as open channel blocker (Kroese et al., 1989; Bongartz et al., 2010). Here, gentamicin was shown to inhibit Cx26 HC open either by DCFS or depolarization of the cellular membrane. Both conditions are known to increase the open probability of Cx26 HCs (Sánchez et al., 2010; Jara et al., 2012), and under these conditions we observed the most potent inhibitory effect of gentamicin. Thus, our data suggest that gentamicin could also act as open HC blocker.

In organotypic mouse cochlea, long distance propagation of intercellular calcium waves occurred through ATP released via Cx30 and Cx26 HCs, whereas GJCs composed by the same Cxs allow the simultaneous diffusion of IP_3 across coupled cells (Anselmi et al., 2008). Moreover, ATP released through HCs located in cochlear supporting cells modulates the electromotility of the outer hair cells by activation of its ionotropic purinergic receptors that determine the cochlear sensitivity to sound stimulation in mammals (Zhao et al., 2005). Thus, altered purinergic signaling due to the blockade of HCs could have important consequences in auditory processing since purinergic transduction controls important aspects of the inner ear physiology, including the cochlear fluids homeostasis, essential for hearing (Housley and Gale, 2010). Relevant to this paracrine signaling, we found that gentamicin reduces the ATP-induced Ca^{2+} signal in HeLa-rCx26, which is in agreement with a previous report in which the ATP-induced Ca^{2+} transients of both HeLa-Cx26 and -Cx43 cells were inhibited by GAP26 and 18-βGA, two GJC/HC blockers. However, these compounds did not significantly affect the ATP-induced Ca^{2+} signal in HeLa-Parental cells (Verma et al., 2009).

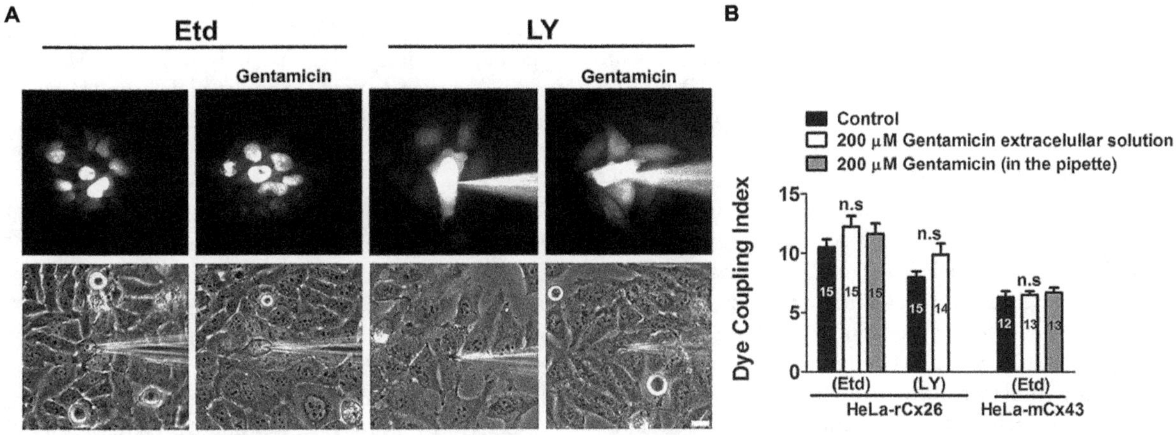

FIGURE 9 | Gentamicin does not inhibit the intercellular transfer of ethidium or Lucifer yellow between HeLa-rCx26 –mCx43 cells. Gap junctional communication, expressed as dye coupling index, was evaluated in confluent cultures of HeLa-rCx26 or –mCx43 cells. A single cell was microinjected with ethidium (75 mM Etd in water) or Lucifer yellow (5% LY w/v in 150 mM LiCl) and the number of cells to which Etd or LY was transferred in 2 min was scored. Gentamicin was applied extracellularly or was included in the pipette at a final concentration of 200 μM. **(A)** Photomicrographs showing a fluorescent field of HeLa-xCx26 cells microinjected with Etd or LY and transfer to adjacent cells under control conditions and in the presence of 200 μM gentamicin applied extracellularly 5 min before the assay. Scale bar, 20 μm. **(B)** Quantitative analysis of dye coupling between HeLa-rCx26 and HeLa-mCx43 cells under control conditions (control) or treated with 200 μM gentamicin in the extracellular solution or in the pipette, as indicated. Bar graphs summarize the results of three independent experiments for each condition (the number of microinjections is indicated within each bar). Values correspond to the mean ± SEM of fluorescent cells.

Our results demonstrate that activation of purinergic receptors by UTP increases the release of ATP in HeLa-rCx26 cells, which was inhibited by gentamicin and CBX. This is consistent with previous reports showing that the expression of Cxs allows HeLa cells to release ATP through HCs, as well as to mediate intercellular transfer of second messengers, such as IP_3, and thereby generate larger Ca^{2+} waves (Cotrina et al., 1998; Paemeleire et al., 2000; Beltramello et al., 2005). Therefore, gentamicin could alter the purinergic signaling by blocking Cx26 HCs and consequently would reduce the ATP release. However, the Cx26 HC permeability to Ca^{2+} (Sánchez et al., 2010; Fiori et al., 2012) might also be involved in the ATP-induced Ca^{2+} signal.

HeLa cells have been shown to express low levels of functional $P2X_7Rs$, which can be upregulated by pro-inflammatory cytokines, such as IFNγ (Welter-Stahl et al., 2009). However, and in agreement with Verma et al. (2009), our results in control conditions indicate that metabotropic purinergic receptors are the main mediators of the ATP-evoked Ca^{2+} signal in HeLa rCx26-GFP because they are strongly reduced by PLC inhibition and completely abrogated by sarcoplasmic-endoplasmic reticulum Ca^{2+}-ATPase inhibition. Accordingly, it has been shown that ATP-evoked Ca^{2+} signals in HeLa cells are mediated by metabotropic receptors and are not significantly different from those exhibited by HeLa cells pre-treated with oATP, a $P2X_7R$ antagonist (Okuda et al., 2003; Welter-Stahl et al., 2009). In spite of the fact that gentamicin could block $P2X_2Rs$ (Bongartz et al., 2010), the previously mentioned results rule out the possibility that gentamicin affects the ATP-evoked signals by blocking ionotropic purinergic receptors. In support of this interpretation, a previous study has shown that $P2X_2R$ is not expressed in HeLa cells (Welter-Stahl et al., 2009). On the other hand, intercellular Ca^{2+} signals elicited through photo stimulation of caged IP_3 propagate normally in cochlear organotypic cultures that lack $P2X_7Rs$ but fail to propagate in cultures with defective expression of Cx26 or Cx30 (Anselmi et al., 2008). Recently, Hu et al. (2012) reported that the expression of Cx26 increases in the cochlear lateral wall of rats 3 h after gentamicin administration, which could compensate for the loss of HC activity. It remains unknown how the overexpression of this protein is related to acquired hearing loss. In this line, we found that expression of Cx26 or Cx26-GFP enhanced the levels of the monomeric receptor. Changes in $P2Y_1R$ and $P2Y_4R$ levels in response to deletion of Cx43 in astrocytes have been described, and these changes in P2YRs also change the mode of propagation of intercellular Ca^{2+} signals (Suadicani et al., 2003). Moreover, other authors have suggested that Cx43 could interact directly with some purinergic receptors, including the $P2Y_1R$ (Iacobas et al., 2007). In light of the findings described above, we hypothesize that changes in the distribution of monomeric forms of purinergic receptors in response to Cx26 expression observed in the experiments described herein are due to direct interaction between Cx26 or Cx26-GFP with endogenously expressed purinergic receptors. However, future experiments are needed to address this issue. Our findings are relevant in the purinergic signaling context, because there is a growing amount of evidence showing that ATP released through HCs can trigger intercellular signaling, directly activating metabotropic and ionotropic purinergic receptors of cells in close contact, regulating both physiological and pathological processes in most tissues (Baroja-Mazo et al., 2013).

Finally, we found that gentamicin does not inhibit dye coupling via rCx26 or mCx43 GJCs when applied extracellularly. These results differ from those published by Todt et al. (1999) that showed inhibition of electrical gap junctional coupling induced by gentamicin in Hensen cells. Additionally, the inhibitory effect is time-dependent and the major effect was observed after 20 min treatment with gentamicin (Todt et al., 1999). In our experiments, cells were preincubated with gentamicin for only 5 min, and dye coupling was assessed during the following 30 min but no significant changes were observed. Relevant to this finding, it has been reported that some subtypes of HeLa cells are resistant to oxidative stress and grow at high oxygen concentrations due to potent antioxidant mechanisms (Campian et al., 2004). This feature might explain why we did not observe gentamicin-induced reduction in dye coupling as that observed in Hensen cells through an indirect mechanism involving production of free radicals and suppression by catalase (Todt et al., 1999). Another possibility could be that gentamicin requires access to the intracellular medium to block GJCs, however, we also performed experiments using gentamicin intracellularly through the microinjection pipette and we do not see significant changes in the dye coupling. Generally, an increase in gap junctional communication is associated with coordination of physiological responses (Ren et al., 1994; Mears et al., 1995; Curti and Pereda, 2004; Orellana et al., 2010; Johansen et al., 2011; Mhaske et al., 2013). In the present study, we did not find a statistical difference in the dye coupling index in presence of gentamicin in HeLa cells expressing rCx26 or Cx43. Therefore, we conclude that gentamicin applied extra or intracellularly does not change the dye coupling in HeLa transfected cells.

Recent studies have demonstrated that 24 h exposure to gentamicin 418, an antibiotic structurally and functionally similar to gentamicin, increases Cx43 phosphorylation and gap junction coupling in tubular proximal epithelial cell lines but decreases cell viability (Yao et al., 2010). Moreover, gain in HC activity has been associated with an increased incidence in cell death (Contreras et al., 2002; Orellana et al., 2010). Therefore, the gentamicin-induced closure might be linked to a deficient autocrine/paracrine signaling as discussed above, rather than the direct consequence of HC closure. Although gentamicin inhibits HCs and not GJCs, its use as HC blocker should be taken cautiously because of its pleiotropic effect on other membrane channels and its effect on free radical generation as discussed above.

Altogether, the data presented in this work suggest a new pharmacological target that could explain the deleterious side effects of gentamicin in the inner ear.

CONCLUSION

In the present article, we report that extracellular gentamicin inhibits the activity of Cx26 HCs expressed in HeLa cells and *Xenopus* oocytes in a concentration-dependent manner without affecting the intercellular coupling through Cx26 GJCs. This effect does not exclusively require the *C*-terminus of Cx26 because it is also observed in HCs with other Cxs composition presenting a much longer C-terminal than Cx26. Finally and as described for other HC blockers, gentamicin reduced the ATP-induced Ca^{2+} signals.

AUTHOR CONTRIBUTIONS

Vania A. Figueroa, Mauricio A. Retamal, Agustín D. Martínez, and Juan C. Sáez designed research; Vania A. Figueroa, Mauricio A. Retamal, Luis A. Cea, José D. Salas, Aníbal A. Vargas, Christian A. Verdugo, and Oscar Jara performed research; Vania A. Figueroa, Mauricio A. Retamal, Luis A. Cea, José D. Salas, Aníbal A. Vargas, Christian A. Verdugo analyzed data; and Vania A. Figueroa, Mauricio A. Retamal, Agustín D. Martínez, and Juan C. Sáez wrote the paper.

ACKNOWLEDGMENTS

We would like to thank Ms. Paola Fernández and Ms. Teresa Vergara for their technical support. This work was partially funded by FONDECYT projects 1120214 and Anillo ACT-1104 (to Mauricio A. Retamal and Agustín D. Martínez), 1111033 (to Juan C. Sáez), 3130577 (to Vania A. Figueroa) Chilean Science Millennium Institute (P09-022-F; to Juan C. Sáez, and Agustín D. Martínez) and CONICYT 24100132 (to Vania A. Figueroa), and FONDEF DO7I1086 (to Juan C. Sáez) grant.

REFERENCES

Abbracchio, M. P., Burnstock, G., Boeynaems, J. M., Barnard, E. A., Boyer, J. L., Kennedy, C., et al. (2006). International Union of Pharmacology LVIII: update on the P2Y G protein coupled nucleotide receptors: from molecular mechanisms and pathophysiology to therapy. *Pharmacol. Rev.* 58, 281–341. doi: 10.1124/pr.58.3.3

Alharazneh, A., Luk, L., Huth, M., Monfared, A., Steyger, P. S., Cheng, A. G., et al. (2011). Functional hair cell mechanotransducer channels are required for aminoglycoside ototoxicity. *PLoS ONE* 6:e22347. doi: 10.1371/journal.pone.0022347

Amici, M., Eusebi, F., and Miledi, R. (2005). Effects of the antibiotic gentamicin on nicotini- acetylcholine receptors. *Neuropharmacology* 49, 627–637. doi: 10.1016/j.neuropharm.2005.04.015

Anselmi, F., Hernández, V. H., Crispino, G., Seydel, A., Ortolano, S., Roper, S. D., et al. (2008). ATP release through connexin hemichannels and gap junction transfer of second messengers propagate Ca^{2+} signals across the inner ear. *Proc. Natl. Acad. Sci. U.S.A.* 105, 18770–18775. doi: 10.1073/pnas.0800793105

Baroja-Mazo, A., Barberà-Cremades, M., and Pelegrín, P. (2013). The participation of plasma membrane hemichannels to purinergic signaling. *Biochim. Biophys. Acta* 1828, 79–93. doi: 10.1016/j.bbamem.2012.01.002

Basile, A. S., Huang, J. M., Xie, C., Webster, D., Berlin, C., and Skolnick, P. (1996). *N*-methyl-D-aspartate antagonists limit aminoglycoside antibiotic-induced hearing loss. *Nat. Med.* 12, 1338–1343. doi: 10.1038/nm1296-1338

Basnakian, A. G., Kaushal, G. P., and Shah, S. V. (2002). Apoptotic pathways of oxidative damage to renal tubular epithelial cells. *Antioxid. Redox Signal.* 4, 915–924. doi: 10.1089/152308602762197452

Begg, E. J., and Barclay, M. L. (1995). Aminoglycosides-50 years on. *Br. J. Clin. Pharmacol.* 39, 597–603.

Beltramello, M., Piazza, V., Bukauskas, F. F., Pozzan, T., and Mammano, F. (2005). Impaired permeability to Ins(1,4,5)P3 in a mutant connexin underlies recessive hereditary deafness. *Nat. Cell Biol.* 7, 63–69. doi: 10.1038/ncb1205

Blanchet, C., Erostegui, C., Sugasawa, M., and Dulon, D. (2000). Gentamicin blocks ACh-evoked K^+ current in guinea-pig outer hair cells by impairing Ca^{2+} entry at the cholinergic receptor. *J. Physiol.* 525, 641–654. doi: 10.1111/j.1469-7793.2000.t01-1-00641.x

Bongartz, E. V., Rettinger, J., and Hausmann, R. (2010). Aminoglycoside block of P2X2 receptors heterologously expressed in *Xenopus laevis* oocytes. *Purinergic Signal.* 6, 393–403. doi: 10.1007/s11302-010-9204-9

Bygrave, F. L. (1978). Properties of energy-dependent calcium transport by rat liver microsomal fraction as revealed by initial-rate measurements. *Biochem. J.* 170, 87–91.

Campian, J. L., Qian, M., Gao, X., and Eaton, J. W. (2004). Oxygen tolerance and coupling of mitochondrial electron transport. *J. Biol. Chem.* 279, 46580–46587. doi: 10.1074/jbc.M406685200

Choung, Y. H., Taura, A., Pak, K., Choi, S. J., Masuda, M., and Ryan, A. F. (2009). Generation of highly-reactive oxygen species is closely related to hair cell damage in rat organ of corti treated with gentamicin. *Neuroscience* 161, 214–226. doi: 10.1016/j.neuroscience.2009.02.085

Contreras, J. E., Sánchez, H. A., Eugenín, E. A., Speidel, D., Theis, M., Willecke, K., et al. (2002). Metabolic inhibition induces opening of unapposed connexin43 gap junction hemichannels and reduces gap junctional communication in cortical astrocytes in culture. *Proc. Natl. Acad. Sci. U.S.A.* 99, 495–500. doi: 10.1073/pnas.012589799

Cotrina, M. L., Lin, J. H., Alves-Rodrigues, A., Liu, S., Li, J., Azmi-Ghadimi, H., et al. (1998). Connexins regulate calcium signaling by controlling ATP release. *Proc. Natl. Acad. Sci. U.S.A.* 95, 15735–15740. doi: 10.1073/pnas.95.26.15735

Curti, S., and Pereda, A. E. (2004). Voltage-dependent enhancement of electrical coupling by a subthreshold sodium current. *J. Neurosci.* 24, 3999–4010. doi: 10.1523/JNEUROSCI.0077-04.2004

D'Ambrosi, N., Iafrate, M., Saba, E., Rosa, P., and Volonté, C. (2007). Comparative analysis of P2Y4 and P2Y6 receptor architecture in native and transfected neuronal systems. *Biochim. Biophys. Acta* 1768, 1592–1599. doi: 10.1016/j.bbamem.2007.03.020

D'Ambrosi, N., Iafrate, M., Vacca, F., Amadio, S., Tozzi, A., Mercuri, N. B., et al. (2006). The P2Y4 receptor forms homo-oligomeric complexes in several CNS and PNS neuronal cells. *Purinergic Signal.* 4, 575–582. doi: 10.1007/s11302-006-9014-2

De Bock, M., Wang, N., Bol, M., Decrock, E., Ponsaerts, R., Bultynck, G., et al. (2012). Connexin43 hemichannels contribute to cytoplasmic Ca^{2+} oscillations by providing a bimodal Ca^{2+}-dependent Ca^{2+} entry pathway. *J. Biol. Chem.* 287, 12250–12266. doi: 10.1074/jbc.M111.299610

Delbro, D. S., Nylund, G., and Nordgren, S. (2005). Demonstration of P2Y purinergic receptors in the HT human colon cancer cell line. *Auton. Autacoid Pharmacol.* 25, 163–166. doi: 10.1111/j.1474-8673.2005.00345.x

De Vuyst, E., Decrock, E., De Bock, M., Yamasaki, H., Naus, C. C., Evans, W. H., et al. (2007). Connexin hemichannels and gap junction channels are differentially influenced by lipopolysaccharide and basic fibroblast growth factor. *Mol. Biol. Cell* 18, 34–46. doi: 10.1091/mbc.E06-03-0182

D'hondt, C., Ponsaerts, R., De Smedt, H., Bultynck, G., and Himpens, B. (2009). Pannexins, distant relatives of the connexin family with specific cellular functions? *Bioessays* 9, 953–974. doi: 10.1002/bies.200800236

Evans, W. H., and Leybaert, L. (2007). Mimetic peptides as blockers of connexin channel-facilitated intercellular communication. *Cell Commun. Adhes.* 6, 265–273. doi: 10.1080/15419060801891034

Figueroa, V., Sáez, P. J., Salas, J. D., Salas, D., Jara, O., Martínez, A. D., et al. (2013). Linoleic acid induces opening of connexin26 hemichannels through a PI3K/Akt/Ca^{2+}-dependent pathway. *Biochim. Biophys. Acta* 3, 1169–1179. doi: 10.1016/j.bbamem.2012.12.006

Fiori, M. C., Figueroa, V., Zoghbi, M. E., Saéz, J. C., Reuss, L., and Altenberg, G. A. (2012). Permeation of calcium through purified connexin26 hemichannels. *J. Biol. Chem.* 48, 40826–40834. doi: 10.1074/jbc.M112.383281

Goldberg, G. S., Valiunas, V., and Brink, P. R. (2004). Selective permeability of gap junction channels. *Biochim. Biophys. Acta* 1662, 96–101. doi: 10.1016/j.bbamem.2003.11.022

González, D., Gómez-Hernández, J. M., and Barrio, L. C. (2006). Species specificity of mammalian connexin-26 to form open voltage-gated hemichannels. *FASEB J.* 20, 2329–2338. doi: 10.1096/fj.06-5828com

González, L. S. III, and Spencer, J. P. (1998). Aminoglycosides: a practical review. *Am. Fam. Physician* 8,1811–1820.

Hong, S. H., Park, S. K., Cho, Y. S., Lee, H. S., Kim, K. R., Kim, M. G., et al. (2006). Gentamicin induced nitric oxide-related oxidative damages on vestibular afferents in the guinea pig. *Hear. Res.* 211, 46–53. doi: 10.1016/j.heares.2005.08.009

Housley, G., and Gale, J. (2010). Purinergic signalling in the inner ear-perspectives and progress. *Purinergic Signal.* 6, 151–153. doi: 10.1007/s11302-010-9196-5

Hu, P., Lai, R., and Xie, D. (2012). Gentamicin affects connexin26 expression in the cochlear lateral wall. *B-ENT* 8, 77–84.

Huth, M. E., Ricci, A. J., and Cheng, A. G. (2011). Mechanisms of aminoglycoside ototoxicity and targets of hair cell protection. *Int. J. Otolaryngol.* 2011:937861. doi: 10.1155/2011/937861

Iacobas, D. A., Suadicani, S. O., Iacobas, S., Chrisman, C., Cohen, M. A., Spray, D. C., et al. (2007). Gap junction and purinergic P2 receptor proteins as a functional unit: insights from transcriptomics. *J. Membr. Biol.* 217, 83–91. doi: 10.1007/s00232-007-9039-7

Ishii, K., Hirose, K., and Iino, M. (2006). Ca^{2+} shuttling between endoplasmic reticulum and mitochondria underlying Ca^{2+} oscillations. *EMBO Rep.* 4, 390–396. doi: 10.1038/sj.embor.7400620

Jara, O., Acuña, R., García, I. E., Maripillán, J., Figueroa, V., Sáez, J. C., et al. (2012). Critical role of the first transmembrane domain of Cx26 in regulating oligomerization and function. *Mol. Biol. Cell* 17, 3299–3311. doi: 10.1091/mbc.E11-12-1058

Jaramillo, F., and Hudspeth, A. J. (1993). Displacement-clamp measurement of the forces exerted by gating springs in the hair bundle. *Proc. Natl. Acad. Sci. U.S.A.* 90,1330–1334. doi: 10.1073/pnas.90.4.1330

Johansen, D., Cruciani, V., Sundset, R., Ytrehus, K., and Mikalsen, S. O. (2011). Ischemia induces closure of gap junctional channels and opening of hemichannels in heart-derived cells and tissue. *Cell. Physiol. Biochem.* 28, 103–114. doi: 10.1159/000331719

Kroese, A. B. A., Das, A., and Hudspeth, A. J. (1989). Blockage of the transduction channels of hair cells in the bullfrog's sacculus by aminoglycoside antibiotics. *Hear. Res.* 37, 203–218. doi: 10.1016/0378-5955(89)90023-3

López-Novoa, J. M., Quiros, Y., Vicente, L., Morales, A. I., and López-Hernández, F. J. (2011). New insights into the mechanism of aminoglycoside nephrotoxicity: an integrative point of view. *Kidney Int.* 1, 33–45. doi: 10.1038/ki.2010.337

Martínez, A. D., Acuña, R., Figueroa, V., Maripillán, J., and Nicholson, B. (2009). Gap-junction channels dysfunction in deafness and hearing loss. *Antioxid. Redox Signal.* 11, 309–322. doi: 10.1089/ars.2008.2138

Mears, D., Sheppard, N. F. Jr., Atwater, I., and Rojas, E. (1995). Magnitude and modulation of pancreatic beta-cell gap junction electrical conductance in situ. *J. Membr. Biol.* 146, 163–176. doi: 10.1007/BF00238006

Mhaske, P. V., Levit, N. A., Li, L., Wang, H. Z., Lee, J. R., Shuja, Z., et al. (2013). The human Cx26-D50A and Cx26-A88V mutations causing keratitis-ichthyosis-deafness syndrome display increased hemichannel activity. *Am. J. Physiol. Cell Physiol.* 304, C1150–C1158. doi: 10.1152/ajpcell.00374.2012

Nakata, H., Yoshioka, K., Kamiya, T., Tsuga, H., and Oyanagi, K. (2005). Functions of heteromeric association between adenosine and P2Y receptors. *J. Mol. Neurosci.* 26, 233–238. doi: 10.1385/JMN:26:2-3:233

Okuda, A., Furuya, K., and Kiyohara, T. (2003). ATP-induced calcium oscillations and change of P2Y subtypes with culture conditions in HeLa cells. *Cell Biochem. Funct.* 21, 61–68. doi: 10.1002/cbf.992

Orellana, J. A., Hernández, D. E., Ezan, P., Velarde, V., Bennett, M. V., Giaume, C., et al. (2010). Hypoxia in high glucose followed by reoxygenation in normal glucose reduces the viability of cortical astrocytes through increased permeability of connexin43 hemichannels. *Glia* 58, 329–343. doi: 10.1002/glia.20926

Paemeleire, K., Martin, P. E., Coleman, S. L., Fogarty, K. E., Carrington, W. A., Leybaert, L., et al. (2000). Intercellular calcium waves in HeLa cells expressing GFP-labeled connexin 43, 32, or 26. *Mol. Biol. Cell* 11, 1815–1827. doi: 10.1091/mbc.11.5.1815

Ren, P., de Feijter, A. W., Paul, D. L., and Ruch, R. J. (1994). Enhancement of liver cell gap junction protein expression by glucocorticoids. *Carcinogenesis* 15, 1807–1813. doi: 10.1093/carcin/15.9.1807

Resendes, B. L., Williamson, R. E., and Morton, C. C. (2001). At the speed of sound: gene discovery in the auditory system. *Am. J. Hum. Genet.* 69, 923–935. doi: 10.1086/324122

Retamal, M. A., Evangelista-Martínez, F., Leon-Paravic, C. G., Altenberg, G. A., and Reuss, L. (2011). Biphasic effect of linoleic acid on connexin46 hemichannels. *Pflügers Arch.* 461, 635–643. doi: 10.1007/s00424-011-0936-3

Retamal, M. A., Froger, N., Palacios-Prado, N., Ezan, P., Sáez, P. J., Sáez, J. C., et al. (2007). Cx43 hemichannels and gap junction channels in astrocytes are regulated oppositely by proinflammatory cytokines released from activated microglia. *J. Neurosci.* 27, 13781–13792. doi: 10.1523/JNEUROSCI.2042-07.2007

Sáez, J. C., Berthoud, V. M., Brañes, M. C., Martínez, A. D., and Beyer, E. C. (2003). Plasma membrane channels formed by connexins:their regulation and functions. *Physiol. Rev.* 83, 1359–1400. doi: 10.1152/physrev.00007.2003

Sáez, J. C., Retamal, M. A., Basilio, D., Bukauskas, F. F., and Bennett, M. V. L. (2005). Connexin-based gap junction hemichannels: gating mechanisms. *Biochim. Biophys. Acta* 1711, 215–224. doi: 10.1016/j.bbamem.2005.01.014

Sage, C. L., and Marcus, D. C. (2002). Immunolocalization of P2Y4 and P2Y2 purinergic receptors in strial marginal cells and vestibular dark cells. *J. Membr. Biol.* 185, 103–115. doi: 10.1007/s00232-001-0116-z

Sánchez, H. A., Mese, G., Srinivas, M., White, T. W., and Verselis, V. K. (2010). Differentially altered Ca^{2+} regulation and Ca^{2+} permeability in Cx26 hemichannels formed by the A40V and G45E mutations that cause keratitis ichthyosis deafness syndrome. *J. Gen. Physiol.* 136, 47–62. doi: 10.1085/jgp.201010433

Schalper, K. A., Palacios-Prado, N., Retamal, M. A., Shoji, K. F., Martínez, A. D., and Sáez, J. C. (2008). Connexin hemichannel composition determines the FGF-1-induced membrane permeability and free [Ca^{2+}]i responses. *Mol. Biol. Cell* 19, 3501–3513. doi: 10.1091/mbc.E07-12-1240

Sha, S. H., and Schacht, J. (1998). Are aminoglycoside antibiotics excitotoxic? *Neuroreport* 17, 3893–3895. doi: 10.1097/00001756-199812010-00023

Suadicani, S. O., De Pina-Benabou, M. H., Urban-Maldonado, M., Spray, D. C., and Scemes, E. (2003). Acute downregulation of Cx43 alters P2Y receptor expression levels in mouse spinal cord astrocytes. *Glia* 42, 160–171. doi: 10.1002/glia.10197

Suzuki, M., Muraki, K., Imaizumi, Y., and Watanabe, M. (1992). Cyclopiazonic acid, an inhibitor of the sarcoplasmic reticulum Ca^{2+}-pump, reduces Ca^{2+}dependent K^+ currents in guinea-pig smooth muscle cells. *Br. J. Pharmacol.* 107, 134–140. doi: 10.1111/j.1476-5381.1992.tb14475.x

Suzuki, T., Namba, K., Tsuga, H., and Nakata, H. (2006). Regulation of pharmacology by hetero-oligomerization between A1 adenosine receptor and P2Y2 receptor. *Biochem. Biophys. Res. Commun.* 351, 559–565. doi: 10.1016/j.bbrc.2006.10.075

Todt, I., Bgezahayo, A., Ernst, A., and Kolb, H.-A. (1999). Inhibition of gap junctional coupling in cochlear supporting cells by gentamicin. *Pflügers Arch.* 438, 865–867. doi: 10.1007/s004240051117

Verma, V., Hallett, M. B., Leybaert, L., Martin, P. E., and Evans, W. H. (2009). Perturbing plasma membrane hemichannels attenuates calcium signalling in cardiac cells and HeLa cells expressing connexins. *Eur. J. Cell Biol.* 88, 79–90. doi: 10.1016/j.ejcb.2008.08.005

Welter-Stahl, L., da Silva, C. M., Schachter, J., Persechini, P. M., Souza, H. S., Ojcius, D. M., et al. (2009). Expression of purinergic receptors and modulation of P2X7 function by the inflammatory cytokine IFNgamma in human epithelial cells. *Biochim. Biophys. Acta* 1788, 1176–1187. doi: 10.1016/j.bbamem.2009.03.006

World Health Organization (WHO). (1994). *Strategies for Prevention of Hearing Impairment from Ototoxic Drugs.* Report of a WHO informal consultation. WHO/PDH/95. 2. Geneva: World Health Organization.

Wright, C. S., van Steensel, M. A., Hodgins, M. B., and Martin, P. E. (2009). Connexin mimetic peptides improve cell migration rates of human epidermal keratinocytes and dermal fibroblasts in vitro. *Wound Repair Regen.* 2, 240–249. doi: 10.1111/j.1524-475X.2009.00471.x

Xie, J., Talaska, A. E., and Schacht, J. (2011). New developments in aminoglycoside therapy and ototoxicity. *Hear. Res.* 281, 28–37. doi: 10.1016/j.heares.2011.05.008

Yao, J., Huang, T., Fang, X., Chi, Y., Zhu, Y., Wan, Y., et al. (2010). Disruption of gap junctions attenuates aminoglycoside-elicited renal tubular cell injury. *Br. J. Pharmacol.* 160, 2055–2068. doi: 10.1111/j.1476-5381.2010.00860.x

Yoshioka, K., Saitoh, O., and Nakata, H. (2001). Heteromeric association creates a P2Y-like adenosine receptor. *Proc. Natl. Acad. Sci. U.S.A.* 98, 7617–7622. doi: 10.1073/pnas.121587098

Zhao, H. B., Yu, B., and Fleming, C. R. (2005). Gap junctional hemichannel-mediated ATP release and hearing controls in the inner ear. *Proc. Natl. Acad. Sci. U.S.A.* 102, 18724–18729. doi: 10.1073/pnas.0506481102

Zhu, Y., and Zhao, H. B. (2010). ATP-mediated potassium recycling in the cochlear supporting cells. *Purinergic Signal.* 2, 221–229. doi: 10.1007/s11302-010-9184-9

7

Neuronal involvement in muscular atrophy

Bruno A. Cisterna[1]*, Christopher Cardozo[2,3] and Juan C. Sáez[1,4]*

[1] Departamento de Fisiología, Pontificia Universidad Católica de Chile, Santiago, Chile
[2] Center of Excellence for the Medical Consequences of Spinal Cord Injury, James J. Peters Veterans Affairs Medical Center, Bronx, NY, USA
[3] Departments of Medicine and Rehabilitation Medicine, Icahn School of Medicine at Mount Sinai, New York, NY, USA
[4] Instituto Milenio, Centro Interdisciplinario de Neurociencias de Valparaíso, Universidad de Valparaíso, Valparaíso, Chile

***Correspondence:**
Bruno A. Cisterna and Juan C. Sáez, Departamento de Fisiología, Pontificia Universidad Católica de Chile, Alameda 340, 8331150 Santiago, Chile
e-mail: bcisterna@uc.cl; jsaez@bio.puc.cl

The innervation of skeletal myofibers exerts a crucial influence on the maintenance of muscle tone and normal operation. Consequently, denervated myofibers manifest atrophy, which is preceded by an increase in sarcolemma permeability. Recently, *de novo* expression of hemichannels (HCs) formed by connexins (Cxs) and other none selective channels, including $P2X_7$ receptors ($P2X_7Rs$), and transient receptor potential, sub-family V, member 2 (TRPV2) channels was demonstrated in denervated fast skeletal muscles. The denervation-induced atrophy was drastically reduced in denervated muscles deficient in Cxs 43 and 45. Nonetheless, the transduction mechanism by which the nerve represses the expression of the above mentioned non-selective channels remains unknown. The paracrine action of extracellular signaling molecules including ATP, neurotrophic factors (i.e., brain-derived neurotrophic factor (BDNF)), agrin/LDL receptor-related protein 4 (Lrp4)/muscle-specific receptor kinase (MuSK) and acetylcholine (Ach) are among the possible signals for repression for connexin expression. This review discusses the possible role of relevant factors in maintaining the normal functioning of fast skeletal muscles and suppression of connexin hemichannel expression.

Keywords: acetylcholine, hemichannels, connexins, trophic factors, electrical activity

INTRODUCTION

The control of skeletal muscle function by the nervous system has been of interest to researchers for more than 100 years in studies examining diverse aspects from the effects of mechanical loading to functions of specific molecular signals (Baldwin et al., 2013). The nervous system exerts control over skeletal muscles by two mechanisms: (1) neuromotor control, by which muscle contraction is initiated by nerve impulses generated in the brain cortex or the brainstem, depolarization of the sarcolemma and electromechanical coupling; and (2) neurotrophic control, which is independent of the electrical activity of motoneurons, and depends on the release of soluble factors from the nerve terminals of motor neurons at the neuromuscular junction (NMJ).

The importance of neural influences on skeletal muscle is evident from the rapid and severe muscular atrophy that occurs whenever there is loss of neural continuity (e.g., due to CNS injury, or the transection or compression of a nerve) (Tomanek and Lund, 1973; Zeman et al., 2009); the ensuing atrophy is considerably more rapid than that from other etiologies such as immobilization (Tomanek and Lund, 1974), cachexia (Dell, 2002; Tisdale, 2002), malnutrition (Morley, 2012), severe burns (Wu et al., 2010), aging (Demontis et al., 2013), dystrophies (Rahimov and Kunkel, 2013), and myasthenia gravis (Keesey, 2004; Ishii et al., 2005). Muscle atrophy results in great extent from accelerated turnover of proteins by the ubiquitin-proteasome pathway, often coupled with diminished rates of protein synthesis (Glass, 2003). Critical roles for signaling by myostatin, NF-kB, and FoxO1 and FoxO3A have been defined and have been reviewed in detail (Rüegg and Glass, 2011; Jackman et al., 2013).

Application of electrical stimulation to nerves to elicit muscle contractions can prevent or largely reverse muscle wasting due to paralysis indicating the critical role of muscle contraction in suppressing the signaling responsible for muscle atrophy (Dudley-Javoroski and Shields, 2008; Kim et al., 2008). There is also a vital role for presence of an intact lower motor neuron and NMJ, as demonstrated by findings of slowed muscle atrophy after spinal isolation (a variant of spinal cord injury, SCI) as compared to nerve transection (Hyatt et al., 2003). On the other hand, denervated muscle after temporary sensory nerve innervation, which provides support to the denervated muscle, improves functional recovery (Bain et al., 2001; Zhao et al., 2004).

One consequence of nerve transection is increased membrane permeability, reduced membrane potential, and increased membrane excitability. Most of these changes have recently been proposed to result from the *de novo* synthesis and insertion of connexin 39, 43 and 45 channels into the sarcolemma, which in turn have been found to mediate atrophy of fast skeletal muscle (Cea et al., 2013). This review compiles and discusses the information on the influence of the nervous system on skeletal muscles and their atrophy, and introduces the current state of knowledge regarding mechanisms by which the nervous system regulates skeletal muscle and its function.

MUSCULAR ATROPHY INDUCED BY DENERVATION

When muscle is denervated due to injury of lower motor neurons there ensues a flaccid paralysis and rapid atrophy with reduction in muscle mass, strength and myofiber diameter; apoptosis of myofiber occurs (Siu and Alway, 2005) together with loss of muscle fibers (Tews, 2002). Most reports indicate that as early as 7 days post-denervation there is a significantly decreased diameter of myofibers in mice (Bruusgaard and Gundersen, 2008; Cea et al., 2013), rats (Pellegrino and Franzini, 1963) and guinea-pigs (Tomanek and Lund, 1973).

It is well documented that the axonal stump that remains after nerve injury undergoes a degenerative process known as Wallerian degeneration (Salzer and Bunge, 1980). However, the axonal stump maintains some physiological activity on skeletal muscles for up to 1 day. Pioneering investigations in denervation showed that there is a direct relationship between the length of the axonal stump and time course of failure of the stump to transmit impulses to the muscle (Eyzaguirre et al., 1952; Gutmann et al., 1955; Birks et al., 1960). The axonal stump was demonstrated to retain the ability to generate spontaneous miniature end-plate potentials (MEPPs) and end-plate potentials (EPPs) that evoke muscle contraction for 8–10 h. Failure of the stump to generate MEPPs is preceded by a gradual decrease in their frequency, while EPPs fail abruptly (Miledi and Slater, 1970). In addition, it was established that the ability of the stump to transmit impulses is prolonged by about 45 min for each additional centimeter in the axonal stump, suggesting that there is a direct relationship between the length of the axonal stump and the transmission of impulses to the muscle (Miledi and Slater, 1970). Similarly, the axonal stump length also influences the onset of muscular disorders such as fibrillation and hypersensitivity to acetylcholine (Ach; Luco and Eyzaguirre, 1955). This finding suggests that there is transport and release of factor(s) from the axonal stump, which ultimately are depleted as axonal reserves are consumed by the myofibers. This idea was strengthened with a clinical observation of weakness and muscle atrophy after accidental overdose of vincristine, which blocks axonal transport (Maeda et al., 1987). Together, these observations indicate need for renewal of the synaptic machinery of the nerve endings responsible for the MEPPs and EPPs.

The onset of fibrillation potentials in denervated muscles (Luco and Eyzaguirre, 1955; Hník and Skorpil, 1962), usually coincides with the reduction of resting membrane potential (Ware et al., 1954; Thesleff, 1963; Albuquerque and Thesleff, 1968), and was one of the first changes described; it has been suggested that fibrillation is the result of membrane depolarization (Ware et al., 1954; Li, 1960; Gage et al., 1989). However, there is no conclusive evidence about the origin of these alterations or their interrelationship.

The membrane depolarization after denervation is associated with several changes in ion current, permeability and concentrations. During the first post-denervation week, there is an increase in intracellular Na^+ concentration, a decrease in intracellular K^+ concentration, and an increase in total calcium content, as well as increased Na^+ permeability (P_{Na+}) and Na^+ conductance, and decreased P_{K+} (Purves and Sakmann, 1974b; Picken and Kirby, 1976; Smith and Thesleff, 1976; Kotsias and Venosa, 2001). This can be explained in part by the massive expression of ion channels such as cardiac-type voltage-gated Na^+ channels (Sekiguchi et al., 2012), fetal acetyl choline receptor (Emmanouilidou et al., 2010) isoform and associated hypersensitivity to ACh (Rosenblueth and Luco, 1937), the tetrodotoxin (TTX) Na^+-resistant channels (Harris and Thesleff, 1971), hemichannels (HCs) formed by connexins (Cxs 39, 43, and 45; Cx HCs), pannexin1 (Panx1) channels, purinergic ionotropic $P2X_7$ receptors ($P2X_7$Rs), transient receptor potential, sub-family V, member 2 (TRPV2), all of which are channels that are found at high levels in the sarcolemma within the first week after denervation (Cea et al., 2013). There is also increased expression of the cardiac Ca^{2+} permeable dihydropyridine receptor isoform, but this occurs at 25 days post-denervation (Péréon et al., 1997), and thus is not directly related to the onset of resting membrane potential decay. Recently, it was demonstrated that absence of two Cxs (43 and 45) significantly decreased the loss of myofiber size in muscles studied at day 7 post-denervation and prevented activation of the p65 subunit of NF-kB and up regulation of pro-inflammatory cytokines (TNF-α and IL-1β) (Cea et al., 2013). This finding raises the question of whether *de novo* expression of Cxs is an upstream response to many of the changes observed in myofibers after denervation. If so, their expression and activation might be somehow regulated by the innervation state or activity of the myofibers.

MUSCULAR ATROPHY INDUCED BY SPINAL CORD INJURY

The disruption of continuity of the nervous system at the level of upper motor neurons may occur as a consequence of neurological conditions such as stroke, multiple sclerosis, or injury to the spinal cord (SCI) and acutely results in paralysis and atrophy of muscles; in SCI, affected muscles are those innervated by motor neurons arising from spinal cord segments below the level of the injury (Shields and Dudley-Javoroski, 2007; Qin et al., 2010). These neurological conditions result in diverse abnormalities, including spastic paralysis (Maynard et al., 1990; Sköld et al., 1999), weakness (Thomas et al., 1997), and extensor plantar responses. In SCI the lower motor neurons remain intact (Kaelan et al., 1988; Bjugn et al., 1997) but deterioration of motor neuron arborization and motor endplates occurs. Very heterogeneous NMJ subgroups (pre and post synaptic) are observed, with some present massive sprouting of nerve terminals, some loss of concentrated clusters of ACh receptors (AChRs) and others remaining intact (Burns et al., 2007); neuromuscular transmission is impaired (Ollivier-Lanvin et al., 2009). While denervation rapidly results in flaccid paralysis and late fibrillation, SCI initially presents as spinal shock and flaccid paralysis and is followed by the development over a period of several weeks or more of hyperreflexia and spasticity (Ditunno et al., 2004; Harris et al., 2006). Following SCI there ensues a brisk and extensive atrophy of skeletal muscle in mice, rats, and humans (Qin et al., 2010). In rodents, transection of the spinal cord, which results in complete loss of volitional activation of motor neurons arising below the anatomical level of the SCI, causes hindlimb muscle atrophy by as much as 40–60% (Ung et al., 2010; Wu et al., 2012). Similarly, muscle biopsy studies in humans suggest that following SCI, muscle fibers atrophy by 27 to 56% within the first 6–18 months after injury (Castro et al., 1999). These changes are associated with marked reductions in contractile force and fatigue resistance, loss of slow- and fast-twitch oxidative fibers, and diminished levels of enzymes for oxidative phosphorylation (Qin et al., 2010). It has been shown that in muscle from SCI rats studied at 56 days after the onset of paralysis, there are elevated sarcolemmal levels of Cxs 39, 43 and 45, and Panx 1 (Cea et al., 2013), which as noted above stimulates activation of the p65 subunit of NF-kB and drives muscle atrophy in muscle paralyzed by nerve transection. These findings suggest that elevation of membrane hemichannel expression may be involved in initiating muscle atrophy after SCI, and other neurological disorders that spare lower motor neurons such as stroke or multiple sclerosis.

ELECTRICAL STIMULATION REVERSES MUSCLE ATROPHY

Evidence supporting the view that the lack of neuromotor activity is responsible for the characteristics of muscle atrophy comes from the experiments with electrical stimulation (Hník et al., 1962; Salmons et al., 2005; Adami et al., 2007). After denervation, the early initiation functional electrical stimulation (FES) reverses the fibrillation potentials (Jones and Vrbová, 1970; Purves and Sakmann, 1974a), ACh hypersensitivity (Lomo and Rosenthal, 1972; Lomo and Westgaard, 1975), and TTX-resistant Na^+ channel expression (Award et al., 1965). However, it does not prevent loss of membrane potential (Squecco et al., 2009). In SCI, loss of muscle mass is also reversed by FES (Scremin et al., 1999). When used for extended periods, FES greatly increases muscle volume, strength and endurance, and the expression of slow myosin heavy chain isoforms (Dudley-Javoroski and Shields, 2008; Qin et al., 2010). It was recently demonstrated that muscle responds rapidly to FES with the greatest number of gene expression changes occurring within the first 1–3 days after initiating electrical stimulation (Ma et al., 2007; Wu et al., 2013). Genes regulated by FES included those for nicotinic AChRs (Adams and Goldman, 1998; Wu et al., 2013), suggesting remodeling of NMJs, possibly to support more efficient neuromuscular transmission. A striking difference between the effects of FES on chronically paralyzed muscle of rats or humans, and physical activity in the absence of SCI is the delayed and impaired upregulation of genes, supporting oxidative phosphorylation in response to FES (Rochester et al., 1995; Wu et al., 2013). Whether this reflects an effect of a non-physiologic pattern of neural and neuromuscular activation, improperly organized signaling molecules at the NMJ, or impaired mechanisms downstream of AChR activation, such as absence of key transcription factors or slowly reversible epigenetic modifications, is unclear. Eventually, more normal gene expression responses are induced in humans by FES, suggesting that the deficit, regardless of its cause, is reversible. Intriguingly, even brief periods of FES are sufficient to prevent muscle atrophy after SCI (Baldi et al., 1998; Kim et al., 2008), suggesting that ACh actions on skeletal muscle persist long after the membrane depolarization and contraction have ceased. The nature of these persistent effects remains unclear.

CONNEXINS AND SKELETAL MUSCLE

Connexins are membrane proteins that form poorly selective channels in the cellular membrane that are also called HCs or connexons. Classically, an HC forms an axially aligned complex with another HC present in the membrane of an adjacent cell to form an intercellular pore (gap junction channel), which directly connects the cytoplasm of adjoining cells (Sáez et al., 2005). More recently, HCs have been found to connect the intra and extra-cellular compartments, allowing transfer of ions such as Na^+ (Li et al., 2001), K^+ (Wallraff et al., 2006), and Ca^{2+} (Li et al., 2001; Sánchez et al., 2009; Schalper et al., 2010; Fiori et al., 2012), entry of nutrients such as glucose (Retamal et al., 2007), release of metabolic products such as glutathione (Rana and Dringen, 2007), as well as of autocrine and paracrine signals such as ATP (Stout et al., 2002), NAD^+ (Bruzzone et al., 2001), cADPR (Song et al., 2011), IP_3 (Gossman and Zhao, 2008), glutamate (Ye et al., 2003), and prostaglandin E_2 (Cherian et al., 2005).

Myoblasts express Cxs and form gap junctions that are essential for development during early and late stages of myogenesis (Constantin and Cronier, 2000). These gap junctions are likely to coordinate gene expression and metabolic responses among differentiating myoblasts (Kalderon et al., 1977; Dennis et al., 1981; Constantin and Cronier, 2000; Araya et al., 2003, 2004; von Maltzahn et al., 2004; Belluardo et al., 2005). In the terminal stage of myogenesis there is down regulation of Cx expression (Armstrong et al., 1983; Proulx et al., 1997; Constantin and Cronier, 2000), and the progressive decline of electrical coupling between myofibers (Dennis et al., 1981; Ling et al., 1992). Connexins are absent in normal skeletal muscle fibers but they have been detected in myofibers of adult muscles undergoing regeneration after injury (Araya et al., 2004; von Maltzahn et al., 2004; Belluardo et al., 2005) and in the sarcolemma of muscle fibers at 7 days post-denervation or 56 days after SCI (Cea et al., 2013). Studies of denervation atrophy in mice deficient for skeletal muscle Cx43 and Cx45 have also demonstrated important roles for these HCs in the signaling through which atrophy occurs. As noted above, the double knockout reduced denervation atrophy by ~70% for fast muscles at 7 days associated with complete inhibition of the activation of the p65 subunit of NF-kB, which as mentioned above, has been shown to be a critical regulator of denervation atrophy. To summarize, Cxs are expressed during myogenesis, when muscle cells are not innervated, disappear within few days after birth, when muscle cells are innervated, rapidly emerge after denervation or paralysis due to upper motor neuron injury, and mediate key signals responsible for denervation atrophy.

These findings strongly suggest that innervation and/or neuromuscular activity suppress Cx expression in the sarcolemma of adult. The mechanism(s) that controls the expression of Cxs in skeletal muscle is, however, unknown, and the only existing evidence relates to the process of myogenesis and points to microRNAs (miRNAs). Anderson et al. showed that miRNA-206 down-regulates Cx43 after birth (Anderson et al., 2006), and that this miRNA is up-regulated in turn by the myogenic transcription the factors myogenin and MyoD which promote myogenic commitment (Rao et al., 2006). In adulthood, miRNA-206 is

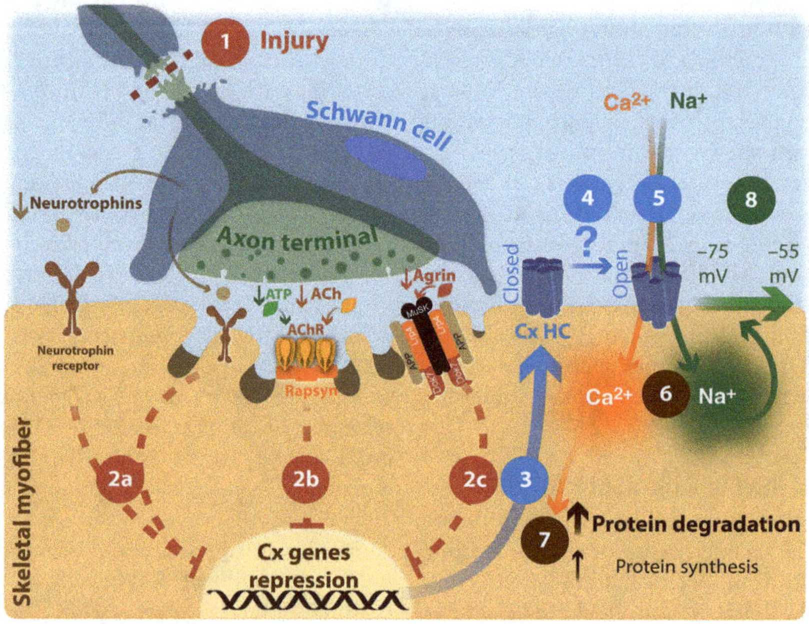

FIGURE 1 | Neuronal involvement in muscular atrophy: proposed model. **(1)** Injury of upper- or lower motor neuron. **(2)** Repression of connexin expression by motor neuron through an unknown mechanism is interrupted. **(2a) neurotrophin signaling**; **(2b) AChR signaling**; ACh: acetylcholine; AChR: acetylcholine receptor. ATP co-released with ACh increases AChR activity. **(2c) agrin/Lrp4/APP/MuSK signaling**; MuSK: Muscle-Specific Kinase; Dok-7: docking protein-7; Lpr4: LDL receptor-related protein 4; APP: amyloid precursor protein. **(3)** Hemichannels formed by connexins are expressed; Cx HC: connexin hemichannel. **(4)** Increases probability of opening of Cx HC by an unknown mechanism. **(5)** Entering Ca^{2+} and Na^{+} ions through non-selective ion channels such as Cx HC **(6)**, increased cytoplasmic resting concentrations of these ions. **(7)** Protein unbalance. **(8)** Reduction in the membrane potential is generated.

dramatically induced in a mouse model of ALS, and delays the disease progression and promotes regeneration of neuromuscular synapses (Williams et al., 2009). However, it is known that transcription factors responsible for up-regulation of miRNA-206, MyoD and myogenin, show an increase in expression within the first week after denervation (Nikolic et al., 2010) and SCI (Dupont-Versteegden et al., 1998), which raises questions about the importance of miRNAs as regulators of the expression of Cxs during adulthood, since as mentioned above, Cxs are up-regulated under conditions of loss of nerve continuity. This situation suggests the existence of other mechanisms that regulate expression of Cxs in the adult skeletal muscle (Oyamada et al., 2005). The above-mentioned findings that the expression of Cxs in skeletal muscles is inhibited after birth, long after innervation occurs, but at a time when a marked increase in neuromotor activity is required, indicates that Cx expression levels are most likely influenced by a neuromotor activity-related mechanism.

MECHANISMS THAT COULD MAINTAIN THE LOW EXPRESSION OF Cxs IN ADULT SKELETAL MYOFIBERS

The importance of Cxs throughout the life of the myofibers is rather well established, and there is considerable evidence that the neuromotor activity is related to their down-regulation in adulthood. However, the mechanism responsible for this regulation is not known. In this section we will discuss the possible mechanisms related to this issue (**Figure 1**).

ACETYLCHOLINE (ACh)

The most studied function of ACh is its role in the conversion of a neuron electrical signal into a chemical signal in the NMJs to produce a mechanical response in the muscle. However, there is also another function that is less well known, but equally important regarding the development and maintenance of the NMJ.

During postsynaptic differentiation, AChR clustering is initiated by a nerve-independent mechanism (Lin et al., 2001; Yang et al., 2001). The muscle-specific receptor tyrosine kinase (MuSK) together with Wnt ligands are involved in the prepatterning of adult AChRs, organizing them into concentrated clusters (Jing et al., 2009). From the time that a motor neuron innervates a myofiber, ACh through the cyclin-dependent kinase 5 (Cdk5) pathway disperses the AChR clusters that failed to be positioned with the nerve terminal (Fu et al., 2005; Lin et al., 2005), and remains as negative signal in the formation of AChR clusters in adulthood (Misgeld et al., 2005). The Cdk5-mediated regulation of AChR localization is poorly understood; we know however that the intermediate filament protein nestin interacts with Cdk5 and is required for ACh-induced association of p35, the co-activator of Cdk5, with the muscle membrane (Yang et al., 2011).

At level of the synaptic junction, blocking the release of the synaptic vesicle with botulinum neurotoxin (Kinder et al., 2001; Jirmanova et al., 1964), or blocking of AChR with alpha bungarotoxin (Ringel et al., 1975; Shen et al., 2006), also produces a rapid onset of atrophy. Moreover, in myasthenia gravis, a pathological condition characterized by the generation of anti-AChR antibodies, also show changes in skeletal muscles similar to those induced by denervation (Berrih-Aknin and Le Panse, 2014).

On the other hand, immobilization and denervation induce *de novo* expression of neuronal nicotinic α7 AChRs (α7AChRs) in myofibers, which is Ca^{2+} permeable (Dickinson et al., 2007), and contributes to neurotransmission (Tsuneki et al., 2003; Lee et al., 2014). Thus, α7AChRs together with aforementioned non-selective channels could contribute to the increase in intracellular Ca^{2+} that occurs after denervation.

AGRIN

Agrin is a proteoglycan released by the motor nerve terminal. This protein binds MuSK and its crucial co-receptors LDL receptor-related protein 4 (Lrp4) and amyloid precursor protein (APP; Choi et al., 2013). Agrin plays a positive role in post-synaptic differentiation by inducing and maintaining AChR clustering *in vitro* and *in vivo* (Nitkin et al., 1987; McMahan, 1990; Ferns et al., 1992; Ruegg et al., 1992; DeChiara et al., 1996; Gautam et al., 1996; Glass et al., 1996; Lin et al., 2001; Yang et al., 2001). Different studies showed that the prepatterning of AChR begins before innervation (Yang et al., 2000, 2001; Lin et al., 2001). Analysis in agrin, MuSK and Lrp4 mutant mice showed that these animals die after birth due a general defect in the assembly of the postsynaptic machinery when NMJ function is required for breathing (DeChiara et al., 1996; Gautam et al., 1996; Hesser et al., 2006; Weatherbee et al., 2006). On the other hand, in absence of ACh (or synaptic transmission) and agrin, the NMJs do form normally, but mice die at birth due to respiratory failure in the absence of neuromuscular transmission (Misgeld et al., 2005). Thus, agrin appears to be responsible for stabilizing the nascent postsynaptic apparatus formed through the action of MuSK/Lrp4.

The cell signaling downstream of MuSK requires the cytoplasmic adaptor protein (Dok-7), which is essential for both MuSK-mediated prepatterning of AChRs and agrin-stimulated AChR cluster stabilization upon innervation though the mechanism is not clear (Okada et al., 2006; Inoue et al., 2009); as might be expected, mice carrying loss of function mutations of Dok-7 die shortly after birth (Beeson et al., 2006; Okada et al., 2006).

In humans, 5–10% cases of myasthenia gravis are caused by autoantibodies against MuSK, which prevent binding between MuSK and Lrp4, and inhibit agrin-stimulated MuSK phosphorylation (Huijbers et al., 2013). LDL receptor-related protein 4 has also been shown to be a target of autoantibodies in some MG patients; being a new diagnostic marker for this disease (Pevzner et al., 2012).

ATP

ATP is recognized as an important signaling molecule that mediates diverse biological processes. In skeletal muscle, ATP is released at the NMJ by synaptic vesicles and by myofibers and has a postulated role in various regulatory processes including cell proliferation, differentiation, and muscle contraction.

Synaptic vesicles isolated from vertebrates contain ACh and ATP at a ratio of ~10:1 (Dowdall et al., 1974; Zimmermann and Denston, 1976; Volknandt and Zimmermann, 1986). An ADP/ATP translocase enables the synaptic vesicle to accumulate ATP (Luqmani, 1981; Lee and Witzemann, 1983; Stadler and Fenwick, 1983); inside the vesicle, ATP is not, however complexed with ACh (Kobos and Rechnitz, 1976). ATP is released with ACh at NMJ in a pulsatile way in response to nerve impulses (Silinsky, 1975; Redman and Silinsky, 1994; Silinsky and Redman, 1996; Vizi et al., 2000; Santos et al., 2003). The significance of this pulsatile release is not clear. In development, ATP bound to P2X receptors is equally effective with ACh acting through nicotinic receptors in calcium mobilization (Kolb and Wakelam, 1983; Häggblad and Heilbronn, 1988). In adults, the co-transmitter role of ATP is less prominent than during development. Adenosine generated by the hydrolysis of ATP was proposed as physiological mediator of prejunctional neuromuscular depression (Redman and Silinsky, 1994); at postjunctional sites, extracellular ATP facilitates the action of ACh (Ribeiro, 1977), increases AChR activity (Ewald, 1976; Akasu et al., 1981; Lu and Smith, 1991), and K^+ channel activation (Thomas and Hume, 1993), and inhibits Cl^- channels, by activating $P2Y_1$ receptors (Voss, 2009); overall, ATP enhances neuromuscular signaling in adult skeletal muscle.

During skeletal muscle contraction, ATP is released from muscle fibers (Cunha and Sebastião, 1993; Sandonà et al., 2005; Li et al., 2008). This ATP can be released through ATP permeable channels, including Cx HC and Panx channels (Bao et al., 2004; Kang et al., 2008; Buvinic et al., 2009; Riquelme et al., 2013). As was mentioned in Section Connexins and Skeletal Muscle, Cxs are not expressed in the adult skeletal muscle. However, the Panx1 is expressed and forms Panx1 HCs in T-tubules. Thus Panx1 channels could be responsible for the release of ATP in adult skeletal muscles. This ATP is necessary for the muscle potentiation that occurs during repetitive electrical stimulation (Riquelme et al., 2013).

The accumulation of ATP outside the sarcolemma has also been shown to be necessary for the increased membrane permeability observed in muscle in pathological conditions where it activates Cx HCs and Panx1 channels as well as $P2X_7Rs$ leading to membrane permeabilization to ions and small molecules (Cea et al., 2013); this accumulation of extracellular ATP is likely to be facilitated after denervation of skeletal muscles or spinal cord injury by the *de novo* expression of $P2X_7Rs$ and Cxs (39, 43 and 45) and up-regulation of Panx1 (Cea et al., 2013) as discussed above.

NEUROTROPHIC FACTORS

The neurotrophic factors are critical for the development of the nervous system (Skaper, 2012). In adulthood, there is well-established interdependence between glial cells and motor neurons (Michailov et al., 2004; Schulz et al., 2014). However, little is known about the relationship between neurons, neurotrophic factors, and trophic actions on myofibers.

Neuregulin 1 (NRG1) has been proposed as an extracellular signal that induces synapse-specific transcription, because NRG1 induces AChR transcription in cultured muscle cells (Falls, 2003).

However, mice lacking NRG1 in both motor neurons and skeletal muscles, or deficient for both the NRG-1 receptors ErbB2 and ErbB4 in skeletal muscles, have morphologically normal synapses, although the amounts of AChRs and AChR mRNA at synapses are modestly reduced (Escher et al., 2005; Jaworski and Burden, 2006). A recent work elucidates that NRG1/ErbB signaling maintains the efficacy of synaptic transmission by stabilizing the NMJs via phosphorylation of α-dystrobrevin (Schmidt et al., 2011).

Ciliary neurotrophic factor (CNTF), a member of the interleukin-6 (IL-6) superfamily, induces cachectic effects (Martin et al., 1996) and several inflammatory responses on innervated skeletal muscle, including induction of fever and a hepatic acute phase protein response (Espat et al., 1996). Also, it has been postulated to CNTF acts as a neurotrophic factor that regulates the expression of its receptor (CNTFR; Helgren et al., 1994; Ip et al., 1996) and acetylcholinesterase in adult rat skeletal muscle (Boudreau-Larivière et al., 1996).

Brain-derived neurotrophic factor (BDNF) is expressed at relatively high levels during muscle development and then down regulated postnatally (Griesbeck et al., 1995; Mousavi et al., 2004). In adult rat muscle the constitutive expression of BDNF is confined to the myofibers, satellite cells, Schwann cells and endothelial cells (Liem et al., 2001), and is up-regulated in muscles as response to acute or repeated exercise (Cuppini et al., 2007; Matthews et al., 2009), but its possible effects on myofibers are unknown.

In co-cultures of dissociated DRG neurons and skeletal myofibers nerve growth factor (NGF) and neurotrophin 3 (NT-3) increase the levels of messenger RNAs (mRNAs) of preprotachykinin (PPT), calcitonin-gene related peptide (CGRP), neurofilament 200 (NF-200), and microtubule associated protein 2 (MAP-2; Zhang et al., 2012), suggesting trophic effects although one could not assess how these growth factors altered neuromuscular function. Due to the limited information available, more studies are needed to elucidate the true importance of neurotrophic factors in the maintenance of muscle characteristics.

CONCLUDING REMARKS

Electrical and metabolic coupling mediated by Cx-based gap junctions are characteristics of smooth and cardiac muscles, which must achieve coordinated contraction of large groups of myocytes. The expression of Cx proteins is required for the formation of gap junction channels that play critical roles in coordinating diverse normal tissue functions in smooth and cardiac muscle, including propagation of cardiac action potentials and smooth muscle slow contraction waves. By contrast, skeletal muscles are characterized by precise and rapid contraction responses of single fibers or groups of fibers innervated by a common nerve fiber (motor units) that must be activated independently of muscle fibers in other motor units. This feature is accomplished by the direct command of the nervous system through nerve fibers with similar conduction velocity that innervate individual motor units with similar electrical threshold. Thus, electrical coupling of muscle fibers through gap junctions appears to be unnecessary for rapid and coordinated contraction of skeletal muscle fibers. Notably, neuromuscular activation represses the expression or translation of several non-selective ion channels during development.

Neuromuscular activation represses the expression or translation of several non-selective ion channels including HCs formed by Cxs 39, 43 or 45, $P2X_7R$, TRPV2 channel and alpha-7 nicotinic receptor in skeletal myofibers (Cea et al., 2013; Lee et al., 2014). However, disruption of neural continuity at any level between upper motor neuron and motor end plate elevates the membrane incorporation of those gene products, with major adverse effects on myofiber biology. Moreover, the membrane expression of Panx1 channels is up-regulated. The sarcolemmal incorporation of these protein subunits results in the cell surface expression of non-selective ion channels, all permeable to monovalent cations and Ca^{2+} and some of them are also permeable to small molecules (e.g., Cx HCs). Therefore, all of them could contribute to different extents to reducing the resting membrane potential of denervated myofibers as well as to the activation of intracellular metabolic response activated by free cytoplasmic Ca^{2+}, including protein degradation. The overlapping features of non-selective ion channels expressed in denervated muscle might be taken as evidence that their expression is controlled by a common mechanism (e.g., the same transcription factor). Thus, a critical issue to be unraveled in the future is the identification of the signal transduction mechanism activated at NMJs/motor end plates that repress the expression of all these non-selective ion channels. So far, it is known that early electrical stimulation of muscles under disuse due to denervation does not prevent the reduction in resting membrane potential and thus is insufficient to maintain the homeostasis of the sarcolemma. Likewise, the presence of an intact lower motor neuron and NMJ does not appear sufficient to prevent incorporation of these HCs and channels into the sarcolemma. Contact between the nerve terminal and motor end plate allows interaction of a series of molecules released by axon terminals and glial cells and their receptors present in myofibers. The main molecules are: ACh that is responsible for the end-plate potential and dispersal of Ach receptor clusters, ATP which is involved in muscle potentiation, agrin that acts as positive signal in the clustering of AChRs, and neurotrophins, whose effect on adult muscle fibers is poorly understood (**Figure 1**). Therefore, more studies are needed to elucidate the role of these substances in repressing incorporation of the above channels and HCs into the sarcolemma of skeletal muscle. miRNAs are recognized as regulators of diverse gene networks and pathways and bind to their target mRNAs, causing mRNA degradation or preventing protein translation. However, miRNA expression levels do not fully explain changes in myofiber expression levels of HCs and channels, and further studies are required to identify the role of miRNA, and to identify alternative mechanisms that determine sarcolemmal expression levels of them.

The discovery of the humoral factor that prevents the expression of protein subunits that form non-selective ion channels in denervated muscles might unveil a valuable molecular target to design a rational therapeutic to prevent degeneration of denervated myofibers that might also be useful to treat diverse myopathies with compromised NMJs.

ACKNOWLEDGMENTS

This work was partially funded by FONDECYT grants 1111033 (to Juan C. Sáez), millennium institute (to Juan C. Sáez), and Bruno A. Cisterna acknowledge the support of a CONICYT fellowship, and the Department of Veterans Affairs Rehabilitation Research and Development Service (B9212C) and the James J. Peters VA Medical Center.

REFERENCES

Adami, N., Kern, H., Mayr, W., Carraro, U., Biral, D., and Sandra, Z. (2007). Permanent denervation of rat Tibialis Anterior after bilateral sciatectomy: determination of chronaxie by surface electrode stimulation during progression of atrophy up to one year. *Basic Appl. Myol.* 17, 237–243.

Adams, L., and Goldman, D. (1998). Role for calcium from the sarcoplasmic reticulum in coupling muscle activity to nicotinic acetylcholine receptor gene expression in rat. *J. Neurobiol.* 35, 245–257. doi: 10.1002/(SICI)1097-4695(19980605)35:3%3C245::AID-NEU2%3E3.0.CO;2-Z

Akasu, T., Hirai, K., and Koketsu, K. (1981). Increase of acetylcholine-receptor sensitivity by adenosine triphosphate: a novel action of ATP on ACh-sensitivity. *Br. J. Pharmacol.* 74, 505–507. doi: 10.1111/j.1476-5381.1981.tb09997.x

Albuquerque, E. X., and Thesleff, S. (1968). A comparative study of membrane properties of innervated and chronically denervated fast and slow skeletal muscles of the rat. *Acta Physiol. Scand.* 73, 471–480. doi: 10.1111/j.1365-201x.1968.tb10886.x

Anderson, C., Catoe, H., and Werner, R. (2006). MIR-206 regulates connexin43 expression during skeletal muscle development. *Nucleic Acids Res.* 34, 5863–5871. doi: 10.1093/nar/gkl743

Araya, R., Eckardt, D., Riquelme, M. A., Willecke, K., and Sáez, J. C. (2003). Presence and importance of connexin43 during myogenesis. *Cell Commun. Adhes.* 10, 451–456. doi: 10.1080/cac.10.4-6.451.456

Araya, R., Riquelme, M. A., Brandan, E., and Sáez, J. C. (2004). The formation of skeletal muscle myotubes requires functional membrane receptors activated by extracellular ATP. *Brain Res. Brain Res. Rev.* 47, 174–188. doi: 10.1016/j.brainresrev.2004.06.003

Armstrong, D. L., Turin, L., and Warner, A. E. (1983). Muscle activity and the loss of electrical coupling between striated muscle cells in Xenopus embryos. *J. Neurosci.* 3, 1414–1421.

Award, E. A., Swaiman, K. F., and Kottke, F. J. (1965). Changes in the structure, innervation, electromyographic patterns and enzymes of skeletal muscle resulting from experimental treatment with triamcinolone. *Arch. Phys. Med. Rehabil.* 46, 297–306.

Bain, J. R., Veltri, K. L., Chamberlain, D., and Fahnestock, M. (2001). Improved functional recovery of denervated skeletal muscle after temporary sensory nerve innervation. *Neuroscience* 103, 503–510. doi: 10.1016/s0306-4522(00)00577-7

Baldi, J. C., Jackson, R. D., Moraille, R., and Mysiw, W. J. (1998). Muscle atrophy is prevented in patients with acute spinal cord injury using functional electrical stimulation. *Spinal Cord* 36, 463–469. doi: 10.1038/sj.sc.3100679

Baldwin, K. M., Haddad, F., Pandorf, C. E., Roy, R. R., and Edgerton, V. R. (2013). Alterations in muscle mass and contractile phenotype in response to unloading models: role of transcriptional/pretranslational mechanisms. *Front. Physiol.* 4:284. doi: 10.3389/fphys.2013.00284

Bao, L., Locovei, S., and Dahl, G. (2004). Pannexin membrane channels are mechanosensitive conduits for ATP. *FEBS Lett.* 572, 65–68. doi: 10.1016/j.febslet.2004.07.009

Beeson, D., Higuchi, O., Palace, J., Cossins, J., Spearman, H., Maxwell, S., et al. (2006). Dok-7 mutations underlie a neuromuscular junction synaptopathy. *Science* 313, 1975–1978. doi: 10.1126/science.1130837

Belluardo, N., Trovato-Salinaro, A., Mudò, G., and Condorelli, D. F. (2005). Expression of the rat connexin 39 (rCx39) gene in myoblasts and myotubes in developing and regenerating skeletal muscles: an in situ hybridization study. *Cell Tissue Res.* 320, 299–310. doi: 10.1007/s00441-005-1087-7

Berrih-Aknin, S., and Le Panse, R. (2014). Myasthenia gravis: a comprehensive review of immune dysregulation and etiological mechanisms. *J. Autoimmun.* 52, 90–100. doi: 10.1016/j.jaut.2013.12.011

Birks, R., Katz, B., and Miledi, R. (1960). Physiological and structural changes at the amphibian myoneural junction, in the course of nerve degeneration. *J. Physiol.* 150, 145–168.

Bjugn, R., Nyengaard, J. R., and Rosland, J. H. (1997). Spinal cord transection–no loss of distal ventral horn neurons. *Exp. Neurol.* 148, 179–186. doi: 10.1006/exnr.1997.6610

Boudreau-Larivière, C., Sveistrup, H., Parry, D. J., and Jasmin, B. J. (1996). Ciliary neurotrophic factor: regulation of acetylcholinesterase in skeletal muscle and distribution of messenger RNA encoding its receptor in synaptic versus extrasynaptic compartments. *Neuroscience* 73, 613–622. doi: 10.1016/0306-4522(96)00033-4

Bruusgaard, J. C., and Gundersen, K. (2008). In vivo time-lapse microscopy reveals no loss of murine myonuclei during weeks of muscle atrophy. *J. Clin. Invest.* 118, 1450–1457. doi: 10.1172/JCI34022

Bruzzone, S., Guida, L., Zocchi, E., Franco, L., and De Flora, A. (2001). Connexin 43 hemi channels mediate Ca2+-regulated transmembrane NAD+ fluxes in intact cells. *FASEB J.* 15, 10–12. doi: 10.1096/fj.00-0566fje

Burns, A. S., Jawaid, S., Zhong, H., Yoshihara, H., Bhagat, S., Murray, M., et al. (2007). Paralysis elicited by spinal cord injury evokes selective disassembly of neuromuscular synapses with and without terminal sprouting in ankle flexors of the adult rat. *J. Comp. Neurol.* 500, 116–133. doi: 10.1002/cne.21143

Buvinic, S., Almarza, G., Bustamante, M., Casas, M., López, J., Riquelme, M., et al. (2009). ATP released by electrical stimuli elicits calcium transients and gene expression in skeletal muscle. *J. Biol. Chem.* 284, 34490–34505. doi: 10.1074/jbc.M109.057315

Castro, M. J., Apple, D. F. Jr., Staron, R. S., Campos, G. E., and Dudley, G. A. (1999). Influence of complete spinal cord injury on skeletal muscle within 6 mo of injury. *J. Appl. Physiol. (1985)* 86, 350–358.

Cea, L. A., Cisterna, B. A., Puebla, C., Frank, M., Figueroa, X. F., Cardozo, C., et al. (2013). De novo expression of connexin hemichannels in denervated fast skeletal muscles leads to atrophy. *Proc. Natl. Acad. Sci. U S A* 110, 16229–16234. doi: 10.1073/pnas.1312331110

Cherian, P. P., Siller-Jackson, A. J., Gu, S., Wang, X., Bonewald, L. F., Sprague, E., et al. (2005). Mechanical strain opens connexin 43 hemichannels in osteocytes: a novel mechanism for the release of prostaglandin. *Mol. Biol Cell* 16, 3100–3106. doi: 10.1091/mbc.e04-10-0912

Choi, H. Y., Liu, Y., Tennert, C., Sugiura, Y., Karakatsani, A., Kroger, S., et al. (2013). APP interacts with LRP4 and agrin to coordinate the development of the neuromuscular junction in mice. *Elife* 2:e00220. doi: 10.7554/elife.00220

Constantin, B., and Cronier, L. (2000). Involvement of gap junctional communication in myogenesis. *Int. Rev. Cytol.* 196, 1–65. doi: 10.1016/S0074-7696(00)96001-7

Cunha, R. A., and Sebastião, A. M. (1993). Adenosine and adenine nucleotides are independently released from both the nerve terminals and the muscle fibres upon electrical stimulation of the innervated skeletal muscle of the frog. *Pflugers Arch.* 424, 503–510. doi: 10.1007/bf00374914

Cuppini, R., Sartini, S., Agostini, D., Guescini, M., Ambrogini, P., Betti, M., et al. (2007). Bdnf expression in rat skeletal muscle after acute or repeated exercise. *Arch. Ital. Biol.* 145, 99–110.

DeChiara, T. M., Bowen, D. C., Valenzuela, D. M., Simmons, M. V., Poueymirou, W. T., Thomas, S., et al. (1996). The receptor tyrosine kinase MuSK is required for neuromuscular junction formation in vivo. *Cell* 85, 501–512. doi: 10.1016/s0092-8674(00)81251-9

Dell, D. D. (2002). Cachexia in patients with advanced cancer. *Clin. J. Oncol. Nurs.* 6, 235–238. doi: 10.1188/02.CJON.235-238

Demontis, F., Piccirillo, R., Goldberg, A. L., and Perrimon, N. (2013). Mechanisms of skeletal muscle aging: insights from Drosophila and mammalian models. *Dis. Model. Mech.* 6, 1339–1352. doi: 10.1242/dmm.012559

Dennis, M. J., Ziskind-Conhaim, L., and Harris, A. J. (1981). Development of neuromuscular junctions in rat embryos. *Dev. Biol.* 81, 266–279. doi: 10.1016/0012-1606(81)90290-6

Dickinson, J. A., Hanrott, K. E., Mok, M. H., Kew, J. N., and Wonnacott, S. (2007). Differential coupling of alpha7 and non-alpha7 nicotinic acetylcholine receptors to calcium-induced calcium release and voltage-operated calcium channels in PC12 cells. *J. Neurochem.* 100, 1089–1096. doi: 10.1111/j.1471-4159.2006.04273.x

Ditunno, J. F., Little, J. W., Tessler, A., and Burns, A. S. (2004). Spinal shock revisited: a four-phase model. *Spinal Cord* 42, 383–395. doi: 10.1038/sj.sc.3101603

Dowdall, M. J., Boyne, A. F., and Whittaker, V. P. (1974). Adenosine triphosphate. A constituent of cholinergic synaptic vesicles. *Biochem. J.* 140, 1–12.

Dudley-Javoroski, S., and Shields, R. K. (2008). Muscle and bone plasticity after spinal cord injury: review of adaptations to disuse and to electrical muscle stimulation. *J. Rehabil. Res. Dev.* 45, 283–296. doi: 10.1682/jrrd.2007.02.0031

Dupont-Versteegden, E. E., Houlé, J. D., Gurley, C. M., and Peterson, C. A. (1998). Early changes in muscle fiber size and gene expression in response to spinal cord transection and exercise. *Am. J. Physiol.* 275, C1124–C1133.

Emmanouilidou, E., Melachroinou, K., Roumeliotis, T., Garbis, S. D., Ntzouni, M., Margaritis, L. H., et al. (2010). Cell-produced alpha-synuclein is secreted in a calcium-dependent manner by exosomes and impacts neuronal survival. *J. Neurosci.* 30, 6838–6851. doi: 10.1523/JNEUROSCI.5699-09.2010

Escher, P., Lacazette, E., Courtet, M., Blindenbacher, A., Landmann, L., Bezakova, G., et al. (2005). Synapses form in skeletal muscles lacking neuregulin receptors. *Science* 308, 1920–1923. doi: 10.1126/science.1108258

Espat, N. J., Auffenberg, T., Rosenberg, J. J., Rogy, M., Martin, D., Fang, C. H., et al. (1996). Ciliary neurotrophic factor is catabolic and shares with IL-6 the capacity to induce an acute phase response. *Am. J. Physiol.* 271, R185–R190.

Ewald, D. A. (1976). Potentiation of postjunctional cholinergic sensitivity of rat diaphragm muscle by high-energy-phosphate adenine nucleotides. *J. Membr. Biol.* 29, 47–65. doi: 10.1007/bf01868951

Eyzaguirre, C., Espildora, J., and Luco, J. V. (1952). Alterations of neuromuscular synapsis during Wallerian degeneration. *Acta Physiol. Lat. Am.* 2, 213–227.

Falls, D. L. (2003). Neuregulins and the neuromuscular system: 10 years of answers and questions. *J. Neurocytol.* 32, 619–647. doi: 10.1023/b:neur.0000020614.83883.be

Ferns, M., Hoch, W., Campanelli, J. T., Rupp, F., Hall, Z. W., and Scheller, R. H. (1992). RNA splicing regulates agrin-mediated acetylcholine receptor clustering activity on cultured myotubes. *Neuron* 8, 1079–1086. doi: 10.1016/0896-6273(92)90129-2

Fiori, M. C., Figueroa, V., Zoghbi, M. E., Sáez, J. C., Reuss, L., and Altenberg, G. A. (2012). Permeation of calcium through purified connexin 26 hemichannels. *J. Biol. Chem.* 287, 40826–40834. doi: 10.1074/jbc.M112.383281

Fu, A. K., Ip, F. C., Fu, W. Y., Cheung, J., Wang, J. H., Yung, W. H., et al. (2005). Aberrant motor axon projection, acetylcholine receptor clustering and neurotransmission in cyclin-dependent kinase 5 null mice. *Proc. Natl. Acad. Sci. U S A* 102, 15224–15229. doi: 10.1073/pnas.0507678102

Gage, P. W., Lamb, G. D., and Wakefield, B. T. (1989). Transient and persistent sodium currents in normal and denervated mammalian skeletal muscle. *J. Physiol.* 418, 427–439.

Gautam, M., Noakes, P. G., Moscoso, L., Rupp, F., Scheller, R. H., Merlie, J. P., et al. (1996). Defective neuromuscular synaptogenesis in agrin-deficient mutant mice. *Cell* 85, 525–535. doi: 10.1016/s0092-8674(00)81253-2

Glass, D. J. (2003). Molecular mechanisms modulating muscle mass. *Trends Mol. Med.* 9, 344–350. doi: 10.1016/s1471-4914(03)00138-2

Glass, D. J., Bowen, D. C., Stitt, T. N., Radziejewski, C., Bruno, J., Ryan, T. E., et al. (1996). Agrin acts via a MuSK receptor complex. *Cell* 85, 513–523. doi: 10.1016/s0092-8674(00)81252-0

Gossman, D. G., and Zhao, H. B. (2008). Hemichannel-mediated inositol 1,4,5-trisphosphate (IP3) release in the cochlea: a novel mechanism of IP3 intercellular signaling. *Cell Commun. Adhes.* 15, 305–315. doi: 10.1080/15419060802357217

Griesbeck, O., Parsadanian, A. S., Sendtner, M., and Thoenen, H. (1995). Expression of neurotrophins in skeletal muscle: quantitative comparison and significance for motoneuron survival and maintenance of function. *J. Neurosci. Res.* 42, 21–33. doi: 10.1002/jnr.490420104

Gutmann, E., Vodicka, Z., and Zelená, J. (1955). Veranderungen in quergestreiften Muskel bei Durchtrennung in Abhangigkeit von der Lange des peripheren Stumpfes. *Physiologia Bohemoslov.* 4, 200–204.

Häggblad, J., and Heilbronn, E. (1988). P2-purinoceptor-stimulated phosphoinositide turnover in chick myotubes. Calcium mobilization and the role of guanyl nucleotide-binding proteins. *FEBS Lett.* 235, 133–136. doi: 10.1016/0014-5793(88)81248-1

Harris, R. L. W., Bobet, J., Sanelli, L., and Bennett, D. J. (2006). Tail muscles become slow but fatigable in chronic sacral spinal rats with spasticity. *J. Neurophysiol.* 95, 1124–1133. doi: 10.1152/jn.00456.2005

Harris, J. B., and Thesleff, S. (1971). Studies on tetrodotoxin resistant action potentials in denervated skeletal muscle. *Acta Physiol. Scand.* 83, 382–388. doi: 10.1111/j.1748-1716.1971.tb05091.x

Helgren, M. E., Squinto, S. P., Davis, H. L., Parry, D. J., Boulton, T. G., Heck, C. S., et al. (1994). Trophic effect of ciliary neurotrophic factor on denervated skeletal muscle. *Cell* 76, 493–504. doi: 10.1016/0092-8674(94)90113-9

Hesser, B. A., Henschel, O., and Witzemann, V. (2006). Synapse disassembly and formation of new synapses in postnatal muscle upon conditional inactivation of MuSK. *Mol. Cell. Neurosci.* 31, 470–480. doi: 10.1016/j.mcn.2005.10.020

Hník, P., Kkorpil, V., and Vyklický, L. (1962). "Diagnosis and therapy of denervation muscle atrophy," in *The Denervated Muscle*, ed E. Gutmann (Prague: Publishing House of the Czechoslovak Academy of Science), 433–466.

Hník, P., and Skorpil, V. (1962). "Fibrillation activity in denervated muscle," in *The Denervated Muscle*, ed E. Gutmann (Prague: Publishing House of the Czechoslovak Academy of Sciences), 136–150.

Huijbers, M. G., Zhang, W., Klooster, R., Niks, E. H., Friese, M. B., Straasheijm, K. R., et al. (2013). MuSK IgG4 autoantibodies cause myasthenia gravis by inhibiting binding between MuSK and Lrp4. *Proc. Natl. Acad. Sci. U S A* 110, 20783–20788. doi: 10.1073/pnas.1313944110

Hyatt, J. P., Roy, R. R., Baldwin, K. M., and Edgerton, V. R. (2003). Nerve activity-independent regulation of skeletal muscle atrophy: role of MyoD and myogenin in satellite cells and myonuclei. *Am. J. Physiol. Cell Physiol.* 285, C1161–C1173. doi: 10.1152/ajpcell.00128.2003

Inoue, A., Setoguchi, K., Matsubara, Y., Okada, K., Sato, N., Iwakura, Y., et al. (2009). Dok-7 activates the muscle receptor kinase MuSK and shapes synapse formation. *Sci. Signal.* 2:ra7. doi: 10.1126/scisignal.2000113

Ip, F. C., Fu, A. K., Tsim, K. W., and Ip, N. Y. (1996). Differential expression of ciliary neurotrophic factor receptor in skeletal muscle of chick and rat after nerve injury. *J. Neurochem.* 67, 1607–1612. doi: 10.1046/j.1471-4159.1996.67041607.x

Ishii, W., Matsuda, M., Okamoto, N., Shimojima, Y., Yazaki, M., Motomura, M., et al. (2005). Myasthenia gravis with anti-MuSK antibody, showing progressive muscular atrophy without blepharoptosis. *Intern. Med.* 44, 671–672. doi: 10.2169/internalmedicine.44.671

Jackman, R. W., Cornwell, E. W., Wu, C. L., and Kandarian, S. C. (2013). Nuclear factor-kappaB signalling and transcriptional regulation in skeletal muscle atrophy. *Exp. Physiol.* 98, 19–24. doi: 10.1113/expphysiol.2011.063321

Jaworski, A., and Burden, S. J. (2006). Neuromuscular synapse formation in mice lacking motor neuron- and skeletal muscle-derived Neuregulin-1. *J. Neurosci.* 26, 655–661. doi: 10.1523/jneurosci.4506-05.2006

Jing, L., Lefebvre, J. L., Gordon, L. R., and Granato, M. (2009). Wnt signals organize synaptic prepattern and axon guidance through the zebrafish unplugged/MuSK receptor. *Neuron* 61, 721–733. doi: 10.1016/j.neuron.2008.12.025

Jirmanova, I., Sobotkova, M., Thesleff, S., and Zelena, J. (1964). Atrophy in skeletal muscles poisoned with botulinum toxin. *Physiol. Bohemoslov.* 13, 467–472.

Jones, R., and Vrbová, G. (1970). Effect of muscle activity on denervation hypersensitivity. *J. Physiol.* 210, 144P–145P.

Kaelan, C., Jacobsen, P. F., and Kakulas, B. A. (1988). An investigation of possible transynaptic neuronal degeneration in human spinal cord injury. *J. Neurol. Sci.* 86, 231–237. doi: 10.1016/0022-510x(88)90101-3

Kalderon, N., Epstein, M. L., and Gilula, N. B. (1977). Cell-to-cell communication and myogenesis. *J. Cell Biol.* 75, 788–806. doi: 10.1083/jcb.75.3.788

Kang, J., Kang, N., Lovatt, D., Torres, A., Zhao, Z., Lin, J., et al. (2008). Connexin 43 hemichannels are permeable to ATP. *J. Neurosci.* 28, 4702–4711. doi: 10.1523/JNEUROSCI.5048-07.2008

Keesey, J. C. (2004). Clinical evaluation and management of myasthenia gravis. *Muscle Nerve* 29, 484–505. doi: 10.1002/mus.20030

Kim, S. J., Roy, R. R., Kim, J. A., Zhong, H., Haddad, F., Baldwin, K. M., et al. (2008). Gene expression during inactivity-induced muscle atrophy: effects of brief bouts of a forceful contraction countermeasure. *J. Appl. Physiol. (1985)* 105, 1246–1254. doi: 10.1152/japplphysiol.90668.2008

Kinder, F. R. Jr., Versace, R. W., Bair, K. W., Bontempo, J. M., Cesarz, D., Chen, S., et al. (2001). Synthesis and antitumor activity of ester-modified analogues of bengamide B. *J. Med. Chem.* 44, 3692–3699. doi: 10.1021/jm010188c

Kobos, R. K., and Rechnitz, G. A. (1976). Acetylcholine-ATP binding by direct membrane electrode measurement. *Biochem. Biophys. Res. Commun.* 71, 762–767. doi: 10.1016/0006-291x(76)90896-2

Kolb, H. A., and Wakelam, M. J. (1983). Transmitter-like action of ATP on patched membranes of cultured myoblasts and myotubes. *Nature* 303, 621–623. doi: 10.1038/303621a0

Kotsias, B. A., and Venosa, R. A. (2001). Sodium influx during action potential in innervated and denervated rat skeletal muscles. *Muscle Nerve* 24, 1026–1033. doi: 10.1002/mus.1106

Lee, D. A., and Witzemann, V. (1983). Photoaffinity labeling of a synaptic vesicle specific nucleotide transport system from Torpedo marmorata. *Biochemistry* 22, 6123–6130. doi: 10.1021/bi00295a013

Lee, S., Yang, H. S., Sasakawa, T., Khan, M. A., Khatri, A., Kaneki, M., et al. (2014). Immobilization with atrophy induces de novo expression of neuronal nicotinic alpha7 acetylcholine receptors in muscle contributing to neurotransmission. *Anesthesiology* 120, 76–85. doi: 10.1097/ALN.0000000000000025

Li, C. L. (1960). Mechanism of fibrillation potentials in denervated mammalian skeletal muscle. *Science* 132, 1889–1890. doi: 10.1126/science.132.3443.1889

Li, J., Gao, Z., Kehoe, V., Xing, J., King, N., and Sinoway, L. (2008). Interstitial adenosine triphosphate modulates muscle afferent nerve-mediated pressor reflex. *Muscle Nerve* 38, 972–977. doi: 10.1002/mus.21014

Li, F., Sugishita, K., Su, Z., Ueda, I., and Barry, W. H. (2001). Activation of connexin-43 hemichannels can elevate [Ca(2+)]i and [Na(+)]i in rabbit ventricular myocytes during metabolic inhibition. *J. Mol. Cell. Cardiol.* 33, 2145–2155. doi: 10.1006/jmcc.2001.1477

Liem, R. S., Brouwer, N., and Copray, J. C. (2001). Ultrastructural localisation of intramuscular expression of BDNF mRNA by silver-gold intensified non-radioactive in situ hybridisation. *Histochem. Cell Biol.* 116, 545–551. doi: 10.1007/s00418-001-0349-z

Lin, W., Burgess, R. W., Dominguez, B., Pfaff, S. L., Sanes, J. R., and Lee, K. F. (2001). Distinct roles of nerve and muscle in postsynaptic differentiation of the neuromuscular synapse. *Nature* 410, 1057–1064. doi: 10.1038/35074025

Lin, W., Dominguez, B., Yang, J., Aryal, P., Brandon, E. P., Gage, F. H., et al. (2005). Neurotransmitter acetylcholine negatively regulates neuromuscular synapse formation by a Cdk5-dependent mechanism. *Neuron* 46, 569–579. doi: 10.1016/j.neuron.2005.04.002

Ling, Y., Appelt, D., Kelly, A. M., and Franzini-Armstrong, C. (1992). Differences in the histogenesis of EDL and diaphragm in rat. *Dev. Dyn.* 193, 359–369. doi: 10.1002/aja.1001930409

Lomo, T., and Rosenthal, J. (1972). Control of ACh sensitivity by muscle activity in the rat. *J. Physiol.* 221, 493–513.

Lomo, T., and Westgaard, R. H. (1975). Further studies on the control of ACh sensitivity by muscle activity in the rat. *J. Physiol.* 252, 603–626.

Lu, Z., and Smith, D. O. (1991). Adenosine 5′-triphosphate increases acetylcholine channel opening frequency in rat skeletal muscle. *J. Physiol.* 436, 45–56.

Luco, J. V., and Eyzaguirre, C. (1955). Fibrillation and hypersensitivity to ACh in denervated muscle: effect of length of degenerating nerve fibers. *J. Neurophysiol.* 18, 65–73.

Luqmani, Y. A. (1981). Nucleotide uptake by isolated cholinergic synaptic vesicles: evidence for a carrier of adenosine 5/-triphosphate. *Neuroscience* 6, 1011–1021. doi: 10.1016/0306-4522(81)90067-1

Ma, J., Shen, J., Garrett, J. P., Lee, C. A., Li, Z., Elsaidi, G. A., et al. (2007). Gene expression of myogenic regulatory factors, nicotinic acetylcholine receptor subunits and GAP-43 in skeletal muscle following denervation in a rat model. *J. Orthop. Res.* 25, 1498–1505. doi: 10.1002/jor.20414

Maeda, K., Ueda, M., Ohtaka, H., Koyama, Y., Ohgami, M., and Miyazaki, H. (1987). A massive dose of vincristine. *Jpn. J. Clin. Oncol.* 17, 247–253.

Martin, D., Merkel, E., Tucker, K. K., McManaman, J. L., Albert, D., Relton, J., et al. (1996). Cachectic effect of ciliary neurotrophic factor on innervated skeletal muscle. *Am. J. Physiol.* 271, R1422–R1428.

Matthews, V. B., Aström, M. B., Chan, M. H., Bruce, C. R., Krabbe, K. S., Prelovsek, O., et al. (2009). Brain-derived neurotrophic factor is produced by skeletal muscle cells in response to contraction and enhances fat oxidation via activation of AMP-activated protein kinase. *Diabetologia* 52, 1409–1418. doi: 10.1007/s00125-009-1364-1

Maynard, F. M., Karunas, R. S., Waring, W. P. 3rd (1990). Epidemiology of spasticity following traumatic spinal cord injury. *Arch. Phys. Med. Rehabil.* 71, 566–569.

McMahan, U. J. (1990). The agrin hypothesis. *Cold Spring Harb. Symp. Quant. Biol.* 55, 407–418. doi: 10.1101/sqb.1990.055.01.041

Michailov, G. V., Sereda, M. W., Brinkmann, B. G., Fischer, T. M., Haug, B., Birchmeier, C., et al. (2004). Axonal neuregulin-1 regulates myelin sheath thickness. *Science* 304, 700–703. doi: 10.1126/science.1095862

Miledi, R., and Slater, C. R. (1970). On the degeneration of rat neuromuscular junctions after nerve section. *J. Physiol.* 207, 507–528.

Misgeld, T., Kummer, T. T., Lichtman, J. W., and Sanes, J. R. (2005). Agrin promotes synaptic differentiation by counteracting an inhibitory effect of neurotransmitter. *Proc. Natl. Acad. Sci. U S A* 102, 11088–11093. doi: 10.1073/pnas.0504806102

Morley, J. E. (2012). Undernutrition in older adults. *Fam. Pract.* 29(Suppl. 1), i89–i93. doi: 10.1093/fampra/cmr054

Mousavi, K., Parry, D. J., and Jasmin, B. J. (2004). BDNF rescues myosin heavy chain IIB muscle fibers after neonatal nerve injury. *Am. J. Physiol. Cell Physiol.* 287, C22–C29. doi: 10.1152/ajpcell.00583.2003

Nikolic, M., Bajek, S., Bobinac, D., Soic Vranic, T., Starcevoc Klasan, G., Arbanas, J., et al. (2010). Expression of myogenic regulatory factors in rat skeletal muscles after denervation. *Periodicum Biologorum* 112, 83–88.

Nitkin, R. M., Smith, M. A., Magill, C., Fallon, J. R., Yao, Y. M., Wallace, B. G., et al. (1987). Identification of agrin, a synaptic organizing protein from Torpedo electric organ. *J. Cell Biol.* 105, 2471–2478. doi: 10.1083/jcb.105.6.2471

Okada, K., Inoue, A., Okada, M., Murata, Y., Kakuta, S., Jigami, T., et al. (2006). The muscle protein Dok-7 is essential for neuromuscular synaptogenesis. *Science* 312, 1802–1805. doi: 10.1126/science.1127142

Ollivier-Lanvin, K., Lemay, M. A., Tessler, A., and Burns, A. S. (2009). Neuromuscular transmission failure and muscle fatigue in ankle muscles of the adult rat after spinal cord injury. *J. Appl. Physiol. (1985)* 107, 1190–1194. doi: 10.1152/japplphysiol.00282.2009

Oyamada, M., Oyamada, Y., and Takamatsu, T. (2005). Regulation of connexin expression. *Biochim. Biophys. Acta* 1719, 6–23. doi: 10.1016/j.bbamem.2005.11.002

Pellegrino, C., and Franzini, C. (1963). An electron microscope study of denervation atrophy in red and white skeletal muscle fibers. *J. Cell Biol.* 17, 327–349. doi: 10.1083/jcb.17.2.327

Péréon, Y., Sorrentino, V., Dettbarn, C., Noireaud, J., and Palade, P. (1997). Dihydropyridine receptor and ryanodine receptor gene expression in long-term denervated rat muscles. *Biochem. Biophys. Res. Commun.* 240, 612–617. doi: 10.1006/bbrc.1997.7712

Pevzner, A., Schoser, B., Peters, K., Cosma, N. C., Karakatsani, A., Schalke, B., et al. (2012). Anti-LRP4 autoantibodies in AChR- and MuSK-antibody-negative myasthenia gravis. *J. Neurol.* 259, 427–435. doi: 10.1007/s00415-011-6194-7

Picken, J. R., and Kirby, A. C. (1976). Denervated frog skeletal muscle: calcium content and kinetics of exchange. *Exp. Neurol.* 53, 64–70. doi: 10.1016/0014-4886(76)90281-8

Proulx, A., Merrifield, P. A., and Naus, C. C. (1997). Blocking gap junctional intercellular communication in myoblasts inhibits myogenin and MRF4 expression. *Dev. Genet.* 20, 133–144. doi: 10.1002/(sici)1520-6408(1997)20:2<133::aid-dvg6>3.0.co;2-8

Purves, D., and Sakmann, B. (1974a). The effect of contractile activity on fibrillation and extrajunctional acetylcholine-sensitivity in rat muscle maintained in organ culture. *J. Physiol.* 237, 157–182.

Purves, D., and Sakmann, B. (1974b). Membrane properties underlying spontaneous activity of denervated muscle fibre. *J. Physiol.* 239, 125–153.

Qin, W., Bauman, W. A., and Cardozo, C. (2010). Bone and muscle loss after spinal cord injury: organ interactions. *Ann. N Y Acad. Sci.* 1211, 66–84. doi: 10.1111/j.1749-6632.2010.05806.x

Rahimov, F., and Kunkel, L. M. (2013). The cell biology of disease: cellular and molecular mechanisms underlying muscular dystrophy. *J. Cell Biol.* 201, 499–510. doi: 10.1083/jcb.201212142

Rana, S., and Dringen, R. (2007). Gap junction hemichannel-mediated release of glutathione from cultured rat astrocytes. *Neurosci. Lett.* 415, 45–48. doi: 10.1016/j.neulet.2006.12.043

Rao, P. K., Kumar, R. M., Farkhondeh, M., Baskerville, S., and Lodish, H. F. (2006). Myogenic factors that regulate expression of muscle-specific microRNAs. *Proc. Natl. Acad. Sci. U S A* 103, 8721–8726. doi: 10.1073/pnas.0602831103

Redman, R. S., and Silinsky, E. M. (1994). ATP released together with acetylcholine as the mediator of neuromuscular depression at frog motor nerve endings. *J. Physiol.* 477, 117–127.

Retamal, M. A., Froger, N., Palacios-Prado, N., Ezan, P., Sáez, P. J., Sáez, J. C., et al. (2007). Cx43 hemichannels and gap junction channels in astrocytes are regulated oppositely by proinflammatory cytokines released from activated microglia. *J. Neurosci.* 27, 13781–13792. doi: 10.1523/jneurosci.2042-07.2007

Ribeiro, J. A. (1977). Potentiation of postjunctional cholinergic sensitivity of rat diaphragm muscle by high-energy-phosphate adenine nucleotides. *J. Membr. Biol.* 33, 401–402. doi: 10.1007/bf01869526

Ringel, S. P., Bender, A. N., Engel, W. K., Daniels, M. P., and Vogel, Z. (1975). A sequential study of denervation - ultrastructural immunoperoxidase localization of alpha-bungarotoxin. *Trans. Am. Neurol. Assoc.* 100, 52–56.

Riquelme, M. A., Cea, L. A., Vega, J. L., Boric, M. P., Monyer, H., Bennett, M. V., et al. (2013). The ATP required for potentiation of skeletal muscle contraction is released via pannexin hemichannels. *Neuropharmacology* 75, 594–603. doi: 10.1016/j.neuropharm.2013.03.022

Rochester, L., Barron, M. J., Chandler, C. S., Sutton, R. A., Miller, S., and Johnson, M. A. (1995). Influence of electrical stimulation of the tibialis anterior muscle in paraplegic subjects. 2. Morphological and histochemical properties. *Paraplegia* 33, 514–522. doi: 10.1038/sc.1995.112

Rosenblueth, A., and Luco, J. V. (1937). A study of denervated mammalian skeletal muscle. *Am. J. Physiol.* 120, 781–797.

Rüegg, M. A., and Glass, D. J. (2011). Molecular mechanisms and treatment options for muscle wasting diseases. *Annu. Rev. Pharmacol. Toxicol.* 51, 373–395. doi: 10.1146/annurev-pharmtox-010510-100537

Ruegg, M. A., Tsim, K. W., Horton, S. E., Kröger, S., Escher, G., Gensch, E. M., et al. (1992). The agrin gene codes for a family of basal lamina proteins that differ in function and distribution. *Neuron* 8, 691–699. doi: 10.1016/0896-6273(92)90090-z

Sáez, J. C., Retamal, M. A., Basilio, D., Bukauskas, F. F., and Bennett, M. V. (2005). Connexin-based gap junction hemichannels: gating mechanisms. *Biochim. Biophys. Acta* 1711, 215–224. doi: 10.1016/j.bbamem.2005.01.014

Salmons, S., Ashley, Z., Sutherland, H., Russold, M. F., Li, F., and Jarvis, J. C. (2005). Functional electrical stimulation of denervated muscles: basic issues. *Artif. Organs* 29, 199–202. doi: 10.1111/j.1525-1594.2005.29034.x

Salzer, J. L., and Bunge, R. P. (1980). Studies of Schwann cell proliferation. I. An analysis in tissue culture of proliferation during development, Wallerian degeneration and direct injury. *J. Cell Biol.* 84, 739–752. doi: 10.1083/jcb.84.3.739

Sánchez, H. A., Orellana, J. A., Verselis, V. K., and Sáez, J. C. (2009). Metabolic inhibition increases activity of connexin-32 hemichannels permeable to Ca2+ in transfected HeLa cells. *Am. J. Physiol. Cell Physiol.* 297, C665–C678. doi: 10.1152/ajpcell.00200.2009

Sandonà, D., Danieli-Betto, D., Germinario, E., Biral, D., Martinello, T., Lioy, A., et al. (2005). The T-tubule membrane ATP-operated P2X4 receptor influences contractility of skeletal muscle. *FASEB J.* 19, 1184–1186. doi: 10.1096/fj.04-3333fje

Santos, D. A., Salgado, A. I., and Cunha, R. A. (2003). ATP is released from nerve terminals and from activated muscle fibres on stimulation of the rat phrenic nerve. *Neurosci. Lett.* 338, 225–228. doi: 10.1016/s0304-3940(02)01419-2

Schalper, K. A., Sánchez, H. A., Lee, S. C., Altenberg, G. A., Nathanson, M. H., and Sáez, J. C. (2010). Connexin 43 hemichannels mediate the Ca2+ influx induced by extracellular alkalinization. *Am. J. Physiol. Cell Physiol.* 299, C1504–C1515. doi: 10.1152/ajpcell.00015.2010

Schmidt, N., Akaaboune, M., Gajendran, N., Martinez-Pena y Valenzuela, I., Wakefield, S., Thurnheer, R., et al. (2011). Neuregulin/ErbB regulate neuromuscular junction development by phosphorylation of alpha-dystrobrevin. *J. Cell Biol.* 195, 1171–1184. doi: 10.1083/jcb.201107083

Schulz, A., Kyselyova, A., Baader, S. L., Jung, M. J., Zoch, A., Mautner, V. F., et al. (2014). Neuronal merlin influences ERBB2 receptor expression on Schwann cells through neuregulin 1 type III signalling. *Brain* 137, 420–432. doi: 10.1093/brain/awt327

Scremin, A. M., Kurta, L., Gentili, A., Wiseman, B., Perell, K., Kunkel, C., et al. (1999). Increasing muscle mass in spinal cord injured persons with a functional electrical stimulation exercise program. *Arch. Phys. Med. Rehabil.* 80, 1531–1536. doi: 10.1016/s0003-9993(99)90326-x

Sekiguchi, K., Kanda, F., Mitsui, S., Kohara, N., and Chihara, K. (2012). Fibrillation potentials of denervated rat skeletal muscle are associated with expression of cardiac-type voltage-gated sodium channel isoform Nav1.5. *Clin. Neurophysiol.* 123, 1650–1655. doi: 10.1016/j.clinph.2012.01.002

Shen, J., Ma, J., Lee, C., Smith, B. P., Smith, T. L., Tan, K. H., et al. (2006). How muscles recover from paresis and atrophy after intramuscular injection of botulinum toxin A: study in juvenile rats. *J. Orthop. Res.* 24, 1128–1135. doi: 10.1002/jor.20131

Shields, R. K., and Dudley-Javoroski, S. (2007). Musculoskeletal adaptations in chronic spinal cord injury: effects of long-term soleus electrical stimulation training. *Neurorehabil. Neural Repair* 21, 169–179. doi: 10.1177/1545968306293447

Silinsky, E. M. (1975). On the association between transmitter secretion and the release of adenine nucleotides from mammalian motor nerve terminals. *J. Physiol.* 247, 145–162.

Silinsky, E. M., and Redman, R. S. (1996). Synchronous release of ATP and neurotransmitter within milliseconds of a motor nerve impulse in the frog. *J. Physiol.* 492, 815–822.

Siu, P. M., and Alway, S. E. (2005). Mitochondria-associated apoptotic signalling in denervated rat skeletal muscle. *J. Physiol.* 565, 309–323. doi: 10.1113/jphysiol.2004.081083

Skaper, S. D. (2012). The neurotrophin family of neurotrophic factors: an overview. *Methods Mol. Biol.* 846, 1–12. doi: 10.1007/978-1-61779-536-7_1

Sköld, C., Levi, R., and Seiger, A. (1999). Spasticity after traumatic spinal cord injury: nature, severity and location. *Arch. Phys. Med. Rehabil.* 80, 1548–1557. doi: 10.1016/s0003-9993(99)90329-5

Smith, J. W., and Thesleff, S. (1976). Spontaneous activity in denervated mouse diaphragm muscle. *J. Physiol.* 257, 171–186.

Song, E. K., Rah, S. Y., Lee, Y. R., Yoo, C. H., Kim, Y. R., Yeom, J. H., et al. (2011). Connexin-43 hemichannels mediate cyclic ADP-ribose generation and its Ca2+-mobilizing activity by NAD+/cyclic ADP-ribose transport. *J. Biol. Chem.* 286, 44480–44490. doi: 10.1074/jbc.m111.307645

Squecco, R., Carraro, U., Kern, H., Pond, A., Adami, N., Biral, D., et al. (2009). A subpopulation of rat muscle fibers maintains an assessable excitation-contraction coupling mechanism after long-standing denervation despite lost contractility. *J. Neuropathol. Exp. Neurol.* 68, 1256–1268. doi: 10.1097/nen.0b013e3181c18416

Stadler, H., and Fenwick, E. M. (1983). Cholinergic synaptic vesicles from Torpedo marmorata contain an atractyloside-binding protein related to the mitochondrial ADP/ATP carrier. *Eur. J. Biochem.* 136, 377–382. doi: 10.1111/j.1432-1033.1983.tb07752.x

Stout, C. E., Costantin, J. L., Naus, C. C., and Charles, A. C. (2002). Intercellular calcium signaling in astrocytes via ATP release through connexin hemichannels. *J. Biol. Chem.* 277, 10482–10488. doi: 10.1074/jbc.m109902200

Tews, D. S. (2002). Apoptosis and muscle fibre loss in neuromuscular disorders. *Neuromuscul. Disord.* 12, 613–622. doi: 10.1016/s0960-8966(02)00030-5

Thesleff, S. (1963). "Spontaneous electrical activity in denervated rat skeletal muscle," in *In The Effect of Use and Disse on Neuromuwcular Functions*, eds E. Gutmann and P. Hnik (Prague: Czechoslovak Academy of Sciences), 41–62.

Thomas, S. A., and Hume, R. I. (1993). Single potassium channel currents activated by extracellular ATP in developing chick skeletal muscle: a role for second messengers. *J. Neurophysiol.* 69, 1556–1566.

Thomas, C. K., Zaidner, E. Y., Calancie, B., Broton, J. G., and Bigland-Ritchie, B. R. (1997). Muscle weakness, paralysis and atrophy after human cervical spinal cord injury. *Exp. Neurol.* 148, 414–423. doi: 10.1006/exnr.1997.6690

Tisdale, M. J. (2002). Cachexia in cancer patients. *Nat. Rev. Cancer* 2, 862–871. doi: 10.1038/nrc927

Tomanek, R. J., and Lund, D. D. (1973). Degeneration of different types of skeletal muscle fibres. I. Denervation. *J. Anat.* 116, 395–407.

Tomanek, R. J., and Lund, D. D. (1974). Degeneration of different types of skeletal muscle fibres. II. Immobilization. *J. Anat.* 118, 531–541.

Tsuneki, H., Salas, R., and Dani, J. A. (2003). Mouse muscle denervation increases expression of an alpha7 nicotinic receptor with unusual pharmacology. *J. Physiol.* 547, 169–179. doi: 10.1111/j..2002.00169.x

Ung, R. V., Rouleau, P., and Guertin, P. A. (2010). Effects of co-administration of clenbuterol and testosterone propionate on skeletal muscle in paraplegic mice. *J. Neurotrauma* 27, 1129–1142. doi: 10.1089/neu.2009.1211

Vizi, E. S., Nitahara, K., Sato, K., and Sperlágh, B. (2000). Stimulation-dependent release, breakdown and action of endogenous ATP in mouse hemidiaphragm preparation: the possible role of ATP in neuromuscular transmission. *J. Auton. Nerv. Syst.* 81, 278–284. doi: 10.1016/s0165-1838(00)00129-6

Volknandt, W., and Zimmermann, H. (1986). Acetylcholine, ATP and proteoglycan are common to synaptic vesicles isolated from the electric organs of electric eel and electric catfish as well as from rat diaphragm. *J. Neurochem.* 47, 1449–1462. doi: 10.1111/j.1471-4159.1986.tb00778.x von Maltzahn, J., Euwens, C., Willecke, K., and Söhl, G. (2004). The novel mouse connexin39 gene is expressed in developing striated muscle fibers. *J. Cell Sci.* 117, 5381–5392. doi: 10.1242/jcs.01413

Voss, A. A. (2009). Extracellular ATP inhibits chloride channels in mature mammalian skeletal muscle by activating P2Y1 receptors. *J. Physiol.* 587, 5739–5752. doi: 10.1113/jphysiol.2009.179275

Wallraff, A., Köhling, R., Heinemann, U., Theis, M., Willecke, K., and Steinhäuser, C. (2006). The impact of astrocytic gap junctional coupling on potassium buffering in the hippocampus. *J. Neurosci.* 26, 5438–5447. doi: 10.1523/jneurosci.0037-06.2006

Ware, F. Jr., Bennett, A. L., and Mc, I. A. R. (1954). Membrane resting potential of denervated mammalian skeletal muscle measured in vivo. *Am. J. Physiol.* 177, 115–118.

Weatherbee, S. D., Anderson, K. V., and Niswander, L. A. (2006). LDL-receptor-related protein 4 is crucial for formation of the neuromuscular junction. *Development* 133, 4993–5000. doi: 10.1242/dev.02696

Williams, A. H., Valdez, G., Moresi, V., Qi, X., McAnally, J., Elliott, J. L., et al. (2009). MicroRNA-206 delays ALS progression and promotes regeneration of neuromuscular synapses in mice. *Science* 326, 1549–1554. doi: 10.1126/science.1181046

Wu, X., Baer, L. A., Wolf, S. E., Wade, C. E., and Walters, T. J. (2010). The impact of muscle disuse on muscle atrophy in severely burned rats. *J. Surg. Res.* 164, e243–e251. doi: 10.1016/j.jss.2010.08.032

Wu, Y., Collier, L., Qin, W., Creasey, G., Bauman, W. A., Jarvis, J., et al. (2013). Electrical stimulation modulates Wnt signaling and regulates genes for the motor endplate and calcium binding in muscle of rats with spinal cord transection. *BMC Neurosci.* 14:81. doi: 10.1186/1471-2202-14-81

Wu, Y., Zhao, J., Zhao, W., Pan, J., Bauman, W. A., and Cardozo, C. P. (2012). Nandrolone normalizes determinants of muscle mass and fiber type after spinal cord injury. *J. Neurotrauma* 29, 1663–1675. doi: 10.1089/neu.2011.2203

Yang, X., Arber, S., William, C., Li, L., Tanabe, Y., Jessell, T. M., et al. (2001). Patterning of muscle acetylcholine receptor gene expression in the absence of motor innervation. *Neuron* 30, 399–410. doi: 10.1016/s0896-6273(01)00287-2

Yang, J., Dominguez, B., de Winter, F., Gould, T. W., Eriksson, J. E., and Lee, K. F. (2011). Nestin negatively regulates postsynaptic differentiation of the neuromuscular synapse. *Nat. Neurosci.* 14, 324–330. doi: 10.1038/nn.2747

Yang, X., Li, W., Prescott, E. D., Burden, S. J., and Wang, J. C. (2000). DNA topoisomerase IIbeta and neural development. *Science* 287, 131–134. doi: 10.1126/science.287.5450.131

Ye, Z. C., Wyeth, M. S., Baltan-Tekkok, S., and Ransom, B. R. (2003). Functional hemichannels in astrocytes: a novel mechanism of glutamate release. *J. Neurosci.* 23, 3588–3596.

Zeman, R. J., Zhao, J., Zhang, Y., Zhao, W., Wen, X., Wu, Y., et al. (2009). Differential skeletal muscle gene expression after upper or lower motor neuron transection. *Pflugers Arch.* 458, 525–535. doi: 10.1007/s00424-009-0643-5

Zhang, W., Li, H., Xing, Z., Yuan, H., Kindy, M. S., and Li, Z. (2012). Expression of mRNAs for PPT, CGRP, NF-200 and MAP-2 in cocultures of dissociated DRG neurons and skeletal muscle cells in administration of NGF or NT-3. *Folia Histochem. Cytobiol.* 50, 312–318. doi: 10.5603/fhc.2012.0041

Zhao, C., Veltri, K., Li, S., Bain, J. R., and Fahnestock, M. (2004). NGF, BDNF, NT-3 and GDNF mRNA expression in rat skeletal muscle following denervation and sensory protection. *J. Neurotrauma* 21, 1468–1478. doi: 10.1089/neu.2004.21.1468

Zimmermann, H., and Denston, C. R. (1976). Adenosine triphosphate in cholinergic vesicles isolated from the electric organ of Electrophorus electricus. *Brain Res.* 111, 365–376. doi: 10.1016/0006-8993(76)90780-0

8

Emerging functions of pannexin 1 in the eye

Sarah Kurtenbach[1], Stefan Kurtenbach[1] and Georg Zoidl[1,2]*

[1] Department of Psychology, Faculty of Health, York University, Toronto, ON, Canada
[2] Department of Biology, Faculty of Science, York University, Toronto, ON, Canada

Correspondence:
Georg Zoidl, Department of Biology, Faculty of Science, York University, Life Science Building 323A, 4700 Keele Street, Toronto, ON M3J 1P3, Canada
e-mail: gzoidl@yorku.ca

Pannexin 1 (Panx1) is a high-conductance, voltage-gated channel protein found in vertebrates. Panx1 is widely expressed in many organs and tissues, including sensory systems. In the eye, Panx1 is expressed in major divisions including the retina, lens and cornea. Panx1 is found in different neuronal and non-neuronal cell types. The channel is mechanosensitive and responds to changes in extracellular ATP, intracellular calcium, pH, or ROS/nitric oxide. Since Panx1 channels operate at the crossroad of major signaling pathways, physiological functions in important autocrine and paracrine feedback signaling mechanisms were hypothesized. This review starts with describing in depth the initial Panx1 expression and localization studies fostering functional studies that uncovered distinct roles in processing visual information in subsets of neurons in the rodent and fish retina. Panx1 is expressed along the entire anatomical axis from optical nerve to retina and cornea in glia, epithelial and endothelial cells as well as in neurons. The expression and diverse localizations throughout the eye points towards versatile functions of Panx1 in neuronal and non-neuronal cells, implicating Panx1 in the crosstalk between immune and neural cells, pressure related pathological conditions like glaucoma, wound repair or neuronal cell death caused by ischemia. Summarizing the literature on Panx1 in the eye highlights the diversity of emerging Panx1 channel functions in health and disease.

Keywords: pannexin 1 (Panx1), eye, retina, lens, cornea, purinergic receptor signaling, ATP, feedback mechanism

INTRODUCTION

In the year 2000, pannexin (Panx) genes (Panx, Latin: *pan* = complete, everywhere and *nexus* = junction) were described as putative gap junction proteins based on a distant sequence homology to innexins, the gap junction proteins of invertebrates. In higher vertebrates, the Panx gene family consists of the three glycosylated integral membrane proteins Panx1, Panx2 and Panx3 (Panchin et al., 2000; Bruzzone et al., 2003; Baranova et al., 2004). Although, initial studies suggested that under certain conditions pannexins might be able to form gap junction channels *in vitro* (Bruzzone et al., 2003, 2005; Vanden Abeele et al., 2006; Lai et al., 2007), experimental evidence suggests that pannexins function as unopposed single membrane channels *in vivo* (Sosinsky et al., 2011).

Today, Panx1 is the best-characterized family member. Panx1 is almost ubiquitously expressed and found in many organs as well as in several cell types of the blood and immune system (Dvoriantchikova et al., 2006a,b; Locovei et al., 2006a; Penuela et al., 2007; Schenk et al., 2008; Seminario-Vidal et al., 2009, 2011; Celetti et al., 2010; Sridharan et al., 2010; Woehrle et al., 2010; Kienitz et al., 2011; Hanner et al., 2012). In the CNS, Panx1 shows a widespread distribution and largely co-localizes with Panx2 (Bruzzone et al., 2003; Ray et al., 2005; Vogt et al., 2005; Dvoriantchikova et al., 2006a,b; Ray et al., 2006; Zappalà et al., 2006, 2007), where the expression is mainly neuronal (Ray et al., 2005, 2006; Zoidl et al., 2007). Evidence for glial expression was found in cultured astrocytes and oligodendrocytes (Boassa et al., 2007; Huang et al., 2007a; Iglesias et al., 2009a; Suadicani et al., 2012). Further, Panx1 expression has been described in major sensory systems including the eye, inner ear, taste buds, and the olfactory epithelium (Bruzzone et al., 2003; Huang et al., 2007b; Romanov et al., 2007; Tang et al., 2008; Dando and Roper, 2009; Wang et al., 2009; Zhang et al., 2012). Panx2 is expressed in the eye, thyroid, kidney and liver, with highest expression levels in the brain and spinal cord (Bruzzone et al., 2003; Baranova et al., 2004; Vogt et al., 2005; Dvoriantchikova et al., 2006a,b; Ray et al., 2006). In humans, Panx2 expression is presumably brain specific. Panx3 is mainly expressed in the skin and cartilage, but can also be found in the heart ventricle, cochlea as well as in lung, kidney, thymus, liver and spleen and possibly astrocytes (Bruzzone et al., 2003; Penuela et al., 2007; Wang et al., 2009; Celetti et al., 2010).

Panx1 forms large-conductance channels, activated by changes in membrane potential, ATP, intracellular calcium, stretch, pH, elevated extracellular potassium, and following purinergic receptor activation (Bao et al., 2004a; Locovei et al., 2006a,b; Qiu and Dahl, 2009; Kawamura et al., 2010; Kienitz et al., 2011; Qiu et al., 2011; Kurtenbach et al., 2013). Channel opening and closing are subject to a plethora of molecular mechanisms including protein interactions, and post-translational modifications like glycosylation and S-nitrosylation (Johnstone et al., 2012; Lohman et al., 2012b; D'hondt et al., 2013; Penuela et al., 2014b; Retamal, 2014). Since Panx1 channels operate at the crossroad of major signaling pathways, foremost those involving intracellular calcium, extracellular ATP or ROS/nitric oxide, physiological functions in important autocrine and paracrine feedback signaling mechanisms were hypothesized (Kawamura et al., 2010; Kronlage et al., 2010; Lohman et al., 2012a; Bao et al., 2013). Further interest in Panx1 derives from independent lines of evidence supporting a critical role of Panx1 in pathological conditions. There, opening of Panx1 channels, frequently together with activation of purinergic receptors by extracellular ATP, has been implicated in diverse conditions including, but not limited to, ischemia (Madry et al., 2010; Monaco and Friedlander, 2012), trauma (Minkiewicz et al., 2013), seizures (Zappalà et al., 2006; Thompson et al., 2008; Mylvaganam et al., 2010; Kim and Kang, 2011; Santiago et al., 2011), HIV infection (Séror et al., 2011; Orellana et al., 2013), inflammation (Kanneganti et al., 2007; Schenk et al., 2008; de Rivero Vaccari et al., 2009; Lamkanfi et al., 2009; Schalper et al., 2009; Silverman et al., 2009), migraine (Karatas et al., 2013), and tumor cell growth (Lai et al., 2007; Penuela et al., 2012).

Important developments in Panx research and emerging roles in health and disease have been covered recently in excellent reviews (MacVicar and Thompson, 2010; Dahl and Keane, 2012; Bond and Naus, 2014; Penuela et al., 2014b). This review will address emerging roles of Panx1 in the eye. The anatomical axis of the eye is used to guide the reader through each chapter. Original studies demonstrating the expression and localization of Panx1 in neuronal and non-neuronal cell types throughout the eye are summarized (overview see **Table 1**). Furthermore, known functions are described, otherwise emerging roles of Panx1 in physiological and pathological conditions are highlighted and challenges outlined. In concluding remarks, functional diversity is discussed as the overarching theme. Readers interested in the roles of connexins in the eye can consult recently published reviews (Volgyi et al., 2013; Vroman et al., 2013; Berthoud et al., 2014; Beyer and Berthoud, 2014).

PANX1 IN THE RODENT RETINA: PROCESSING VISUAL INFORMATION

After the initial discovery of Panx expression in the eye (Bruzzone et al., 2003), refined characterizations of the rodent retina using *in situ hybridization* analysis and RT-PCR analysis of tissue captured by laser microdissection revealed Panx1 and Panx2 expression in the ganglion cell layer (GCL), inner nuclear layer (INL), the outer nuclear layer (ONL) and to a lesser extent in the inner plexiform layer (IPL; Ray et al., 2005; Dvoriantchikova et al., 2006a). In general, two Panx1 protein isoforms (43 kDa, 58 kDa) were found in the retina and brain, of which the 58 kDa variant is post-translationally modified (Kranz et al., 2013). RT-PCR and immunohistochemistry (IHC) revealed mRNA and protein expression regulation during development with a transient expression peak around birth, declining with age (Ray et al., 2005; Dvoriantchikova et al., 2006a). In neonatal and P20 animals, Panx1 labeling was prominent in retinal ganglion cells (RGC), amacrine cells and horizontal cells (HC) in the outer plexiform layer (OPL), whereas in adult animals the labeling mainly occurred in RGCs. Panx1 proteins were localized at the cell surface and neuronal processes.

More detailed analyses of mice retinae revealed Panx1 proteins in a puncta-like pattern on the soma and dendritic branches, sparsely along the axon, and more intensively at the axon terminal of type 3a OFF bipolar cells (CB3a cells) and on non-invaginating cellular processes at the base of rod spherules, resembling the flat contact between rods and cone bipolar cells. The authors also confirmed the expression of Panx1 in HC using single cell RT-PCR, IHC and ultrastructural analysis and found Panx1 localized at the tips of HC dendrites invaginating in rod spherules. It is worth noting that only one HC type is present in the murine retina receiving cone input at the dendrites and rod input at the axon terminals (Peichl and Gonzalez-Soriano, 1993) in contrast to the zebrafish retina possessing two types of HC that either make contacts to rods or cones, with the zebrafish drPanx1a (see below) found only at the HC cone synaptic complexes (Prochnow et al., 2009). In mice, the P2X7 receptor (P2X7-R), like Panx1, is localized at the tip of HC dendrites (Puthussery et al., 2006), suggesting that P2X7-R and Panx1 functions could be linked and involve the local release of ATP.

Physiological function(s) of Panx1 in the processing of visual information were investigated using Panx1$^{-/-}$ mice, using *in vivo* and *in vitro* electroretinography (ERG). Kranz et al. (2013) demonstrated that under scotopic, but not photopic light conditions, the a- and b-wave amplitude of *in vivo* ERGs was increased in knockout animals at high light intensity. *In vitro* studies confirmed this result, and it was concluded that the ablation of Panx1 interferes with the activity of the dark-adapted retina without altering the temporal properties of signal transmission within the retina, but does not affect the cone pathway under light conditions. The authors proposed that Panx1 closes under conditions of prolonged light adaptation as previously observed for retinal gap junctions (Bloomfield et al., 1995; Xin and Bloomfield, 1999). In scotopic light conditions, the changes to the b-wave were explained by the Panx1 channels on the CB3a cells providing a competing current to the light-evoked current of the ON bipolar cells. This might lead to a Panx1-mediated reduction of radial current flow, resulting in an enhanced b-wave in the Panx1$^{-/-}$ mice. The a-wave reflects the reduction of the dark current during light stimulation and is made up of an extracellular radial current from the photoreceptor inner segment towards the outer segments (Penn and Hagins, 1969) and from currents arising from the photoreceptor synapse (Hagins et al., 1970). In this context, Panx1 channels on HC dendrites may take part in negative feedback mechanisms leading to a decreased current flow at the photoreceptor synapse, causing an enlarged a-wave if Panx1 channels are not present. A mechanism addressing this feedback has been previously put forward for connexin hemichannels in the fish and turtle retina and new insight will follow below (Kamermans et al., 2001; Pottek et al., 2003; Kamermans and Fahrenfort, 2004).

PANX1 ORTHOLOGS IN THE FISH RETINA: FEEDBACK REGULATION IN THE OUTER RETINA

The retina of zebrafish shares many properties with those of higher vertebrates, but color vision is more similar to the human trichromatic vision when compared to the murine dichromatic vision (Goldsmith and Harris, 2003; Conway, 2007). Series of genome duplication events in teleost evolution caused partial gene duplications approximately 320–350 million years ago (Jaillon et al., 2004; Postlethwait, 2007; Ravi and Venkatesh, 2008). As a

Table 1 | Summary of known and emerging functions of pannexins in the eye.

Organism	Protein	Localization	Cell type	Localization	Interaction/ Pathway	Function
mouse, rat	Panx1	retina: GCL, IPL, INL, ONL, OPL	RGC	neuronal processes	n.d.	neuroprotection, ischemia induced cell death (Dvoriantchikova et al., 2012)
					P2X7-R, purinergic signaling pathway	activation of P2X7-R by hypotonicity-incuded ATP release, glaucoma *
			HC	dendrites, invaginating rod spherules	n.d.	n.d.
			CB3a bipolar cells	soma, dendritic branches, axons	P2X7-R, purinergic signaling pathway	ATP release, photoreceptor-HC feedback regulation, neuronal processing (Kranz et al., 2013)
			amacrine cells	n.d.	n.d.	n.d.
			microglia	n.d.	AMPA/kainate receptor	ATP release
		lens	epithelial cells	predominantly in cytoplasm	n.d.	n.d.
			young elongated fiber cells	predominantly in cytoplasm	n.d.	n.d.
			mature fiber cells	plasma membrane	n.d.	n.d.
		tunica vasculosa lentis	endothelial cells	n.d.	n.d.	retinopathy of prematurity *
		cornea	epithelial cells		P2X7-R, purinergic signaling pathway	wound repair *
	Panx2	retina: all layers	n.d.	n.d.	n.d.	n.d.
		lens	epithelium and fiber cells	n.d.	n.d.	n.d.
zebrafish	drPanx1a	retina: OPL	HC	mainly cone synaptic complex, soma	n.d.	pH dependent photoreceptor-HC feedback mechanism, ATP release, neuronal processing (Vroman et al., 2014)
	drPanx1b	retina: GCL, INL	amacrine cells*	n.d.	n.d.	n.d.
	drPanx2	Retina	n.d.	n.d.	n.d.	n.d.
porcine	Panx1	lens	cuboidal epithelial cells	n.d.	Na^+/K^+-ATPase, P2Y-R, TRPV4	pressure regulation 1. Hyposmotic stress mediated ATP release 2. Hyposmotic stress mediated Ca^{2+} influx activates Panx1 mediating
bovine	Panx1	trabecular meshwork cells	n.d.	n.d.	P2X7-R, connexins	pressure-activated ATP release, glaucoma*
		ciliary epithelial cells	n.d	n.d.	connexins, P2X7-R and independent vesicular ATP release	hypotonicity-triggered ATP release

n.d. indicates category not determined.

** indicates emerging functions or indirect association with disease/disorders.*

consequence, the zebrafish genome encodes for two Panx1 genes, drPanx1a and drPanx1b (Bond et al., 2012; Kurtenbach et al., 2013), and single drPanx2 and drPanx3 genes. In the zebrafish retina, drPanx1a, drPanx1b and three differently spliced mRNA transcripts of drPanx2 are expressed (Zoidl et al., 2008; Prochnow et al., 2009; Kurtenbach et al., 2013). The retinal expression of drPanx1a and drPanx1b has been further characterized using IHC. drPanx1a was found in a band-like, horseshoe shape pattern in the OPL on HC dendrites, but also on HC somata (Prochnow et al., 2009; Kurtenbach et al., 2013). No labeling could be found on other cell types like ON bipolar cells or interplexiform cells. Ultrastructural analysis revealed the expression of drPanx1a on HC dendrites inserted deeply in the cone synaptic complex, where they are located more distal from the cones' glutamate release sites compared to Cx55.5 hemichannels. No expression in rod synaptic terminals was found. IHC located drPanx1b in the GCL and in the INL suggestive for a potential localization in amacrine cells (Kurtenbach et al., 2013). Like mammalian Panx1, drPanx1a and drPanx1b are glycoproteins, with drPanx1b having three possible N-glycosylation sites and drPanx1a one. Fish pannexins, like mammalian Panx channels, form voltage-gated single membrane channels with shared and unique properties, including activation under physiological conditions, modulation by extracellular ATP, intracellular Ca^{2+} and pH changes (Kurtenbach et al., 2013).

Exciting novel results highlighted the role of drPanx1a in neuronal processing at the first retinal synapse, where HCs inhibit photoreceptors. This interaction generates the center/surround organization of bipolar cell receptive fields and is crucial for contrast enhancement (Burkhardt and Fahey, 1998; Jackman et al., 2011). Two competing hypotheses considered an ephaptic or a proton-mediated mechanism to explain this fundamental process (Kamermans et al., 2001; Kamermans and Fahrenfort, 2004; Fahrenfort et al., 2009; Hirasawa et al., 2012; Klaassen et al., 2012; Vroman et al., 2013). Since zebrafish with a functional knock out of the HC specific Cx55.5 retained 40% hemichannel conductance, it was tempting to speculate that other connexins or a Panx could account for the residual 40% conductance (Klaassen et al., 2011; Sun et al., 2012). The Kamermans group was able to answer this long-standing, fundamental question demonstrating that HC feedback to photoreceptors via an unexpected synthesis of both mechanisms (Vroman et al., 2014). The first one is a very fast connexin driven ephaptic mechanism, which has no synaptic delay making it one of the fastest inhibitory synapses known. The second one is a relatively slow mechanism, depending on ATP released via Panx1 channels located on horizontal cell dendrites deeply invaginating the cone synaptic terminal.

The unique finding was that the extracellular ATP hydrolysis to AMP, phosphate groups and protons, formed a buffer with a pKa of 7.2. This inhibited Ca^{2+}-channels in cones, and consequently reduced the cones' glutamate release. Since cone photoreceptors and HCs form a reciprocal synapse with cones transmitting to HCs through an excitatory synapse and HCs feeding back to cones through an inhibitory synapse, the reduction of glutamate release caused hyperpolarization in HCs, a condition that evoked a decrease of Panx1 channels conductance. Decreasing ATP release caused alkalization in the synaptic cleft and consequently increased cone glutamate release. Surprisingly, the hydrolysis of ATP instead of ATP itself mediated the synaptic modulation, revealing a novel form of synaptic modulation. Since Panx1 channels and ecto-ATPases are strongly expressed in the nervous system and Panx1 function was previously implicated in synaptic plasticity, learning and behavior (Prochnow et al., 2012), the authors hypothesized that this novel form of synaptic modulation might be a wide spread phenomenon.

PANX1 IN RETINAL MICROGLIA: IMPLICATIONS FOR CROSSTALK OF IMMUNE AND NEURAL CELLS

Panx1-mediated ATP release has been associated with the regulation of the morphology and behavior of "resting" microglia in the retina, the primary resident immune cells in the CNS, because of endogenous, ionotropic glutamatergic neurotransmission (Fontainhas et al., 2011). In their "resting" state, microglia have a ramified morphology with fine, extended processes, which exhibit rapid dynamics to make repeated contacts with neurons, glia and blood vessels (Davalos et al., 2005; Nimmerjahn et al., 2005; Lee et al., 2008). The factors that regulate "resting" microglia and induce their transformation to the "activated" state are not fully understood, but a credible hypothesis is that gradients of "on" and "off" signals activate/repress microglial activation involving extracellular signals like ATP, purinergic receptors, glutamate receptors, chemokines and neurotransmission (Mertsch et al., 2001; Xiang et al., 2006; Liang et al., 2010; Wong et al., 2011; Domercq et al., 2013). Experimental evidence suggesting Panx1 involvement derived from *ex vivo* retinal mouse explants, pharmacological intervention of retinal neurotransmission and analysis of microglia morphology (Fontainhas et al., 2011). Microglial morphology and dynamic behavior was modulated by retinal neurotransmission. Endogenous ionotropic glutamate through AMPA/kainate receptors maintained and increased dendritic morphology and process motility, whereas ionotropic GABAergic neurotransmission negatively regulated the morphology and motility. Since the application of probenecid decreased the AMPA mediated effects on microglia, the authors concluded that the response to neurotransmitter was mediated indirectly via secondary ATP, released in response to glutamatergic neurotransmission through probenecid-sensitive Panx channels. Interestingly, the effects of fast AMPA/kainate receptors on microglia were larger compared to those of slow NMDA receptors, although NMDA receptor stimulation has been shown triggering Panx1 opening in pyramidal neurons leading to epileptiform seizure activity (Thompson et al., 2008). This mode of constitutive signaling between neural and immune cells of the CNS illustrates the versatility of Panx1 channels for distinct physiological processes in the retina and beyond.

PANX1 IN RETINAL GANGLION CELLS: IMPLICATIONS FOR ISCHEMIC DISORDERS OF THE RETINA

A role of Panx1 channels in the pathology of the retina was investigated using conditional $Panx1^{-/-}$ mice and oxygen-glucose deprivation (OGD) to demonstrate whether Panx1 is part of a pathological cascade leading to ischemia-induced RGC loss (Dvoriantchikova et al., 2012). They found that Panx1 deficiency protects RGCs from death induced by ischemia, most likely by preventing rapid membrane permeation and Panx1 activation as part of the activation of neuronal inflammasome. It was concluded that Panx1 is activated at a convergence point for several neurotoxic pathways. Further, it was suggested that the Panx1 channel is a potential target for therapeutic intervention in retinal ischemic disorders.

PANX1 IN THE LENS: IMPLICATIONS FOR A ROLE IN HYPOOSMOTIC STRESS

The vertebrate lens is an avascular tissue with extensive gap junctional and hemichannel coupling supporting growth, differentiation and homeostasis. Using qRT-PCR and *in situ* hybridization, Dvoriantchikova et al. (2006b) demonstrated Panx1 and Panx2 mRNA expression in the lens epithelium and fiber cells. Relative to Cx50, the most abundant connexin expressed in the mouse lens, Panx1 and Panx2 expression is significantly lower in the epithelium and the fiber cells. Western blot analyses revealed four Panx1 protein isoforms: two major protein bands (43 and 120 kDa) and two minor bands (58 and 62 kDa). Using differential centrifugation, the monomeric 43 kDa protein was found in the soluble protein fraction, whereas the 62 kDa variant was present in the microsomal, organelle-enriched fraction, likely associated with the ER and Golgi apparatus. Further detergent extraction with Triton X-100 and methyl-β-cyclodextrin suggested association of the 58 kDa and 120 kDa isoforms with lipid raft membrane microdomains. The water-insoluble Panx1 62 and 120 kDa isoforms were lens and retina specific, whereas 58 kDa species is found in non-CNS tissues. Expression was age and cell type dependent, with the 120 kDa protein predominantly expressed in mature and young fibers, whereas the 58 kDa species was mainly present in the lens epithelium and the young elongating fibers. IHC analysis revealed that Panx1 proteins were predominantly located in the lens cortex and that expression declined towards the lens nucleus. The subcellular distribution was consistent with the expected live cycle of the Panx1 channel protein. During development, redistribution occurred shifting from a predominant presence in the cytoplasm in epithelial and young elongated fiber cells to a more pronounced localization in the plasma membrane of mature fiber cells.

Potential function(s) of Panx1 in the lens epithelium were investigated using the porcine lens model (Shahidullah et al., 2012). In the lens the cuboidal lens epithelium is located in the anterior portion between the lens capsule and the lens fibers regulating important homeostatic functions and contains a variety of transporters as well as, amongst others, purinergic receptors (Delamere and Tamiya, 2009). Panx1 was correlated with Na^+/K^+ATPase activity, an antiporter enzyme with important roles in maintaining resting potential, availing transport, regulating cellular volume, and as a signal transducer/integrator to regulate MAPK pathway, ROS, as well as intracellular calcium. As ions, nutrients, and liquid enter the lens from the aqueous humor, Na^+/K^+-ATPases in the lens epithelial cells pump ions out of the lens to maintain appropriate lens osmolarity and volume, with equatorially positioned lens epithelium cells contributing most to this current. The activity of the Na^+/K^+-ATPases keeps water and current flowing through the lens from the poles and exiting through the equatorial regions. Similar to mice, 58 kDa and 120 kDa isoforms of Panx1 were detected, further Cx53 and Cx50, but not Panx2, Panx3 or Cx40. Hypoosmotic, but not a hyperosmotic solution triggered pannexin- and connexins-mediated ATP release increasing Na^+/K^+ATPase activity. This activity was pharmacologically reduced by the connexin inhibitor 18a-glycyrrhetinic acid (AGA), and abolished when AGA was combined with the Panx1 specific inhibitor probenecid, or when apyrase was added to catalyze the hydrolysis of ATP. Using the exocytosis inhibitor N-ethylmaleimide (NEM), the authors ruled that ATP is released via exocytosis. Using the purinergic P2 receptor antagonist reactive blue-2 and pertussis toxin (PTX), the authors showed that the increased Na^+/K^+-ATPase activity under hyposmotic stimulation is most probably mediated by G-protein coupled P2Y receptor activation. Thus, increased Na^+/K^+ATPase activity of the epithelium during a hyposmotic challenge was prevented when either ATP release or extracellular ATP accumulation was suppressed, or P2Y receptors were inhibited. This result correlated Panx and connexin activities with ATP release, most likely due to the channels' mechanosensitivity (Bao et al., 2004a,b), in response to swelling of the lens cells during the hyposmotic stimulation.

Pannexin/connexin-mediated ATP release during hyposmotic stress was linked to TRPV4 channel activation by Shahidullah et al. (2012). The TRPV4 encoded protein is a Ca^{2+}-permeable, nonselective cation channel that is involved in the regulation of systemic osmotic pressure. Like other members of the TRP superfamily, TRPV channels can be activated through seemingly disparate mechanisms. In the lens epithelium, TRPV4 inhibitors prevented ATP release and propidium iodide uptake of the lens epithelium and an agonist elicited ATP release and propidium iodide uptake even under isosmotic conditions. Similarly, in esophageal epithelial cells it was shown that TRPV4 activation triggers Cx43- and Panx1-mediated ATP release (Ueda et al., 2011), and in the airway epithelium TRPV4 transduced the membrane stretch signal upon leading to hypotonicity-induced, Panx1-mediated ATP release (Seminario-Vidal et al., 2009). Since the TRPV4 channel is permeable for Ca^{2+} ions, the activation of TRPV4 caused an increase of intracellular Ca^{2+} in response to the hyposmotic challenge, which was eliminated by the two TRPV4 antagonists and elicited by the agonist under isosmotic conditions. The intracellular $Ca2^+$ rise was also prevented when the bathing solution was calcium free. Thus, the Ca^{2+}, which is entering the cells through activated TRPV4 channels during the hyposmotic challenge, might in turn activate Panx1 channels (Locovei et al., 2006b) or connexin hemichannels like Cx32 (De Vuyst et al., 2006) or Cx43 (Retamal et al., 2007). Since inhibition of TRPV4 prevented ATP release during hyposmotic stress, no increased Na^+/K^+ATPase activity was evoked. Summing up, the results reported suggest that Panx1 together with connexins, ATP and purinergic receptors participates in osmotic regulation, through modulation of Na^+/K^+ATPase activity.

Panx1 was also identified in blood endothelial cells of the *tunica vasculosa lentis* (Dvoriantchikova et al., 2006b), an extensive capillary network deriving from the *vasa hyaloidea*, spreading over the posterior and lateral surfaces of the lens, which disappears after birth (Skapinker and Rothberg, 1987; Barishak, 1992). There, Panx1 was mostly found at the luminal side of the blood capillaries. No co-localization with Cx50 was detected.

Panx1 channels, alongside with connexins, are localized in a strategic position alongside connexins with potential to participate in differentiation and lens homeostasis. With Panx1 channels known to play a significant role in cell death, it is tempting to speculate about a role in the retinopathy of prematurity (Pau, 2008). It remains to be demonstrated whether pharmacological blocking of Panx1 in patients with this condition can modulate the disproportional regression of the *vasa hyaloidea* by cell death and the proliferation of retinal vessels, causing fibrovascular tissue to overgrow the remnants of retina-attaching hyaloidal structures through the vitreous cavity up to the *tunica vasculosa lentis*.

PANX1 AND AQUEOUS HUMOR OUTFLOW: IMPLICATIONS IN GLAUCOMA

A potential role of Panx1 channels in pressure regulation of the aqueous humor outflow pathway has emerged recently (Li et al., 2010, 2012). Two parallel pathways mediate the outflow of the aqueous humor. In the uveoscleral pathway, aqueous humor is flowing through ciliary muscle bundles and subsequent passage to pressure-insensitive routes (Toris et al., 2008). The pressure-sensitive trabecular meshwork pathway comprises the trabecular meshwork, juxtacanalicular tissue, the inner wall of the Schlemm's canal, collector channels and aqueous veins (Gong et al., 1996; Goel et al., 2010). Cells of the outflow pathway are important for the regulation of the outflow, with trabecular meshwork cells playing a critical role (Francis and Alvarado, 1997). Among the substances released by trabecular meshwork cells are components of the extracellular matrix, like metalloproteinases (MMPs) and their inhibitors modeling the extracellular matrix and therewith the outflow resistance (Aga et al., 2008). The release of MMP-2 is regulated by A1 adenosine receptors on trabecular meshwork cells (Shearer and Crosson, 2002), leading to reduced outflow resistance and reduced intraocular pressure (IOP). Inflow and outflow pressure regulation involves adenosine receptor pathways. The A3 receptor activation enhances the intraocular pressure by activating chloride channels on nonpigmented ciliary epithelial cell that secrete chloride and enhance aqueous humor formation (Mitchell et al., 1999; Avila et al., 2001). Both act in concert and are potential targets to reduce IOP in glaucoma patients. It is tempting to speculate whether a feedback control mechanism exists involving Panx1 released ATP that is metabolized by ecto-ATPases to adenosine (Shearer and Crosson, 2002; Husain et al., 2007). Results that provided first evidence were quantitative RT-PCR experiments demonstrating that Cx26, Cx31 and Cx43 were expressed in trabecular meshwork cells, with Panx1 and Cx43 being equally expressed in the human trabecular meshwork cell line TM5. In human explant-derived primary trabecular meshwork cells Cx43 levels were 10-fold higher compared to Panx1, whereas the expression of the P2X7-R is about a 100-fold lower. Applying 21 different inhibitors, Li et al. (2010) showed that Panx1 channels, connexin hemichannels, and P2X7-R are important in swelling-activated ATP release from trabecular meshwork cells, with Panx1 and connexins accounting for about 40% each of the released ATP, and P2X7-R for about 16% or 30% when Panx1 and connexin hemichannels were not blocked simultaneously. This supports a role of Panx1, most likely in cooperation with connexins and P2X7-R in modulating the pressure of the outflow pathway. Further, in the bovine ciliary epithelium, quantitative RT-PCR experiments revealed the expression of Panx1, Cx40, Cx43 and P2X7-R in mixed primary cultures of nonpigmented and pigmented ciliary epithelial cells and transformed nonpigmented and pigmented epithelial cells. Using 11 different inhibitors, Li et al. (2010) showed that in these cells during hypotonicity-triggered ATP release Panx1 and connexins contribute to the ATP release, but that P2X7-R-mediated ATP release was insignificant, whereas 20% of the ATP release was vesicular.

A role of Panx1 in modulating pressure is potentially of significant clinical relevance. Glaucoma represents a group of ocular disorders with multi-factorial etiology united by a clinically characteristic IOP associated optic neuropathy and is a leading cause for blindness. The treatment that delays the onset and slows the progression of glaucomatous blindness is reducing IOP. Evidence is accruing that Panx1 might be directly involved in the pathology of glaucoma. It has been shown that ATP levels are elevated with increased IOP (Resta et al., 2007; Li et al., 2011), even in correlation with the magnitude of increased pressure in the vitreous of patients with acute angle closure glaucoma (Zhang et al., 2007). It has been hypothesized that excess extracellular ATP might lead to neuronal death, since stimulation of P2X7-R on RGCs leads to an increase of the cytosolic Ca^{2+} concentration with subsequent excitotoxic cell death (Zhang et al., 2005). The P2X7-R is of special importance, as the receptor can also initiate inflammatory responses, which play a role in glaucoma (reviewed in Krizaj et al., 2014). Also, RGCs were rescued during rapidly increased ocular pressure by apyrase-mediated dephosphorylation of ATP and by inhibiting purinergic receptors (Resta et al., 2007). In accordance, mechanical perturbations, e.g., due to increased hydrostatic pressure, cell swelling or stretching, shear stress, are most efficiently triggering ATP release (discussed in Reigada et al., 2008). One study linked a possible involvement of Panx1 in glaucoma pathogenesis (Reigada et al., 2008) to Panx1 as a mechanosensitive ATP-release channel (Bao et al., 2004a,b). Using *ex vivo* bovine eyecup preparations, the authors showed that 10 min of increased pressure of 20 mmHg above the atmospheric pressure led to physiological ATP release in the vitreal chamber that declined back to the baseline within 30 min. In contrast, applying 70 mmHg led to a constant increase of the ATP levels throughout a 60 min experiment. Furthermore, the authors demonstrated that the amount of ATP strongly correlated with the applied pressure. Carbenoxolone, 5-nitro-2-(3-phenylpropylamino) benzoic acid (NPPB; Silverman et al., 2008), but not mefloquine, reduced the ATP release by about 90%, leading to the author's conclusion of a Panx1-mediated ATP release. Since mefloquine efficiently inhibits Panx1 channel in the nM concentration range (Iglesias et al., 2009b), but failed to reduce the pressure-induced ATP-increase in this study, the results raise the question whether connexin hemichannels or other ATP release mechanisms, like vesicular release were involved. Further, NPPB can inhibit other channels that have been associated with ATP release (discussed in Reigada et al., 2008), like volume sensitive channels (Furukawa et al., 1998), voltage-dependent anion channels (VDAC; Sabirov et al., 2001) and the cystic fibrosis transmembrane conductance regulator (CFTR; Cuthbert, 2001). Using isolated rat RGCs and applying blocker pharmacology, a subsequent study proved that

RGCs themselves mediate mechanosensitive ATP release, as they respond upon bathing in hypotonic solutions to swelling-induced mechanical stress with rapid and sustained Panx1-mediated ATP release, which in turn autostimulates P2X7-R on RGCs (Xia et al., 2012). In mixed retinal cultures, the hypotonicity-induced ATP levels rapidly rose to a transient peak and afterwards declined to a steady state, indicating a higher rate of ATP breakdown than ATP release. In contrast, in isolated RGCs the ATP levels constantly rose during the recording time, indicating less ecto-ATPase activity compared to the mixed cultures. Also, stretching mixed retinal cells or isolated RGCs lead to ATP release in a Panx1-dependent manner. Subsequent whole-cell patch clamp recordings from RGCs in mixed retinal cultures and isolated RGCs revealed that swelling induced ATP release activates P2X7-R on RGCs. In turn, the ecto-ATPase apyrase, probenecide or carbenoxolone or the P2X7-R antagonists A438079, AZ 10606120 and zinc reduced the swelling-activated currents. Since Panx1 and P2X7-R are expressed on RGC neurites, Xia et al. (2012) reasoned that the ATP release and P2X7-R autostimulation take place in the RGC axons. These results are unexpected, since other studies have shown that increased IOP first affects the dendritic field size, number of synapses, light-evoked responses and leads to abnormalities in dendritic arbors prior to a reduction of axon thickness and deformation of the optic nerve head (reviewed in Krizaj et al., 2014). Interestingly, Panx1-mediated ATP release seems to be important for RGCs cells for their regulatory volume decrease. Under control conditions during bathing in hypotonic solution, the RGCs initially swell to a peak size, but after 30 min decrease their volume again due to regulatory volume decrease, which was reduced by about 60% when probenecide was applied, indicating a beneficial contribution of Panx1 channels in this context. Still, in an event of chronic increased IOP, Panx1 may mediate the release of excessive ATP amounts leading to harmful autocrine stimulation of P2X7-R that drives the whole system in a pathological state, leading to the P2X7-R/Panx1 "death-complex". Thus, in accordance with being localized on neurites, this putative damaging signaling pathway can cause RGC death by damaging axonal processes, a possible mechanism in glaucoma pathology. A novel finding is that astrocytes isolated from optic nerve heads upregulated Panx1 expression and increased ATP release with sustained stretch (Beckel et al., 2014). This suggested a glia cell based mechanism for maintaining elevated extracellular ATP upon sustained pressure induced mechanical strain. Further, Beckel et al. concluded that physiological consequences could be either beneficial or detrimental for RGCs, depending on the relative levels of adenosine or P2X7-R/Panx1. Other mechanosensitive (like TRPV4) or inflammatory factors in glaucoma exist and have been recently reviewed (Krizaj et al., 2014).

PANX1 IN THE CORNEA: IMPLICATIONS FOR WOUND HEALING

IHC experiments revealed that Panx1 is also expressed in the cornea of mice (Mayo et al., 2008). In this study, the primary purpose was to analyze the role of P2X7-R in the repair of *in vivo* corneal epithelial debridement wounds and in the structural organization on the corneal stroma. After epithelial debridement was performed on P2X7-R$^{-/-}$ and wild-type mice, light microscopic, immunohistochemical, and electron microscopic analysis showed that Panx1 was localized between cells with a certain distance distal to the leading edge of epithelial debridement of mice corneas. In contrast, Panx1 was detected throughout the epithelium and at the leading edge of the injury in WT animals. This indicated that the absence of P2X7-R alters the localization of Panx1 in the cornea. Further, in P2X7$^{-/-}$ mice the rate of wound repair was negatively affected. The absence of Panx1 at the wound edge is in line with another study about Panx1 in skin development and wound healing (Penuela et al., 2014a).

CHALLENGES IN PANNEXIN RESEARCH

Research addressing the (patho)physiological roles of Panx1 needs to overcome biological, technical and logistical challenges. A significant biological challenge derives from the complex interactions of Panx1 with major signaling pathways involving ionotropic and metabotropic receptors (Isakson and Thompson, 2014), the inflammasome (Adamson and Leitinger, 2014) cell death pathway (Jackson et al., 2014), or the cytoskeleton (Boyce et al., 2014). The capacity of Panx1 to undergo multiple interactions is exciting, suggesting clinically relevant roles in many physiological and pathophysiological settings. Dissecting complex networks of interactions has high priority and will guide our understanding of the (patho)physiology of Panx1 in the eye and beyond.

Pharmacological manipulation of Panx1 is a second challenge despite the significant advances made in recent years. Blocking peptides like ^{10}Panx1 (Pelegrin and Surprenant, 2006), peptides interfering with Panx1 in a specific context like anoxia (Weilinger et al., 2012), or pharmacological blockers like probenecid (Silverman et al., 2008), mefloquine (Iglesias et al., 2009b), or more recently the food dye FD&C Blue No. 1 (Wang et al., 2013) have greatly improved our understanding of Panx1. These studies were mostly performed in the context of single cell types or oocytes and all compounds used acted by closing the Panx1 channel. However, it might be beneficial in certain conditions to open Panx1 channels when ATP efflux is needed. Further, there is a demand for selective agonists/antagonists to discriminate between the three Panx channels and connexin hemichannels frequently co-expressed in complex tissues like the eye.

The availability of Panx1 KO mouse models has greatly contributed to the understanding of Panx1 functions *in vivo*. A challenge emerging from mouse models and KO cell lines is the question of compensatory upregulation of other Panxs or connexin hemichannels. Recently, Panx3 has been shown to be upregulated in the arterial walls and skin of Panx1 knock out mice (Lohman and Isakson, 2014; Penuela et al., 2014a). In case compensatory regulation is more common, proving functional compensation will critically depend on specific blockers not available at present. Rational designed drugs to selectively target Panx proteins could derive from mimetic peptides with less cross-inhibition or pharmacological screens of natural compounds (Grek et al., 2014; Saez and Leybaert, 2014).

Dissecting complex interactions and solving pharmacological shortcomings are of general concern. A specific concern for vision and eye research is choosing appropriate animal models. While vertebrate eyes share a general anatomy and physiology, visual capabilities are distinct, frequently reflecting functional adaptations to habitats and/or behaviors. Complementary models are needed to weight advantages over disadvantages of each species when investigating the roles of Panxs in the eye. Already distinct functional roles of Panx1 in the mouse and zebrafish retina have emerged, reflecting unique properties of the two species (Kranz et al., 2013; Vroman et al., 2014). Beyond doubt, studies in mouse and fish models have and will aid our understanding of fundamental roles of Panx1 in the visual system, but to build a more generalized view of Panx functions complementary studies in other higher vertebrates like rats and rabbits, or in color vision competent non-human primates will be needed. Recent advances in genome engineering technologies based on transcription activator-like effector nucleases (TALEN) or the RNA-guided Cas9 nuclease (Cas9/CRISPR) have been successfully used to generate first functional knock downs or knock outs in a wide range of species including non-human primates (Mali et al., 2013a,b; Liu et al., 2014a,b; Niu et al., 2014; Yang et al., 2014) making targeting Panx1 in higher vertebrates an experimental option.

CONCLUDING REMARKS

Panx1 research in the eye has moved from early expression and localization studies to functional studies. This research is currently driven by the availability of knock out animal models, sophisticated *ex vivo* and *in vitro* preparations suitable to address the emerging roles of Panx1 in a physiological or pathological context, and the potential implication of Panx1 in leading causes of blindness. There is no confirmed unifying theme to be highlighted yet. Instead, we wish to communicate that Panx1 channels are highly versatile channels. Functions are mediated by the distinct localization of Panx1 proteins in different cell types, multiple protein interactions, as well as post-translational modifications. Since Panx1 channels operate at the crossroad of major signaling pathways, it is tempting to speculate that long lasting changes to this channel have the potential to impact physiological functions by altering major signaling pathways. ATP release, as found in the humor outflow system or as a mediator of cross talk between neurons and microglia, has the potential to be the unifying molecular function, connecting biophysical properties of Panx1 channels to ATP mediated signaling. However, this generalization might not be fully true. Bipolar cells like CB3a cells do not release ATP in any known physiological or pathological context. Further, HC release ATP, but it is the pH shift evoked by the breakdown of ATP to adenosine that alters neuronal communication. Since insight into the relevant molecular, cellular and physiological roles of Panx1 in conditions as diverse as primary processing of visual information in the outer retina or IOP regulation is emerging, we expect soon advances in conceptual insight related to the different roles of Panx1 in this physiologically and clinically important sensory organ.

ACKNOWLEDGMENTS

This work was supported by CRC/CIHR and NSERC-DG.

REFERENCES

Adamson, S. E., and Leitinger, N. (2014). The role of pannexin1 in the induction and resolution of inflammation. *FEBS Lett.* 588, 1416–1422. doi: 10.1016/j.febslet.2014.03.009

Aga, M., Bradley, J. M., Keller, K. E., Kelley, M. J., and Acott, T. S. (2008). Specialized podosome- or invadopodia-like structures (PILS) for focal trabecular meshwork extracellular matrix turnover. *Invest. Ophthalmol. Vis. Sci.* 49, 5353–5365. doi: 10.1167/iovs.07-1666

Avila, M. Y., Stone, R. A., and Civan, M. M. (2001). A(1)-, A(2A)- and A(3)-subtype adenosine receptors modulate intraocular pressure in the mouse. *Br. J. Pharmacol.* 134, 241–245. doi: 10.1038/sj.bjp.0704267

Bao, Y., Chen, Y., Ledderose, C., Li, L., and Junger, W. G. (2013). Pannexin 1 channels link chemoattractant receptor signaling to local excitation and global inhibition responses at the front and back of polarized neutrophils. *J. Biol. Chem.* 288, 22650–22657. doi: 10.1074/jbc.m113.476283

Bao, L., Locovei, S., and Dahl, G. (2004a). Pannexin membrane channels are mechanosensitive conduits for ATP. *FEBS Lett.* 572, 65–68. doi: 10.1016/j.febslet.2004.07.009

Bao, L., Sachs, F., and Dahl, G. (2004b). Connexins are mechanosensitive. *Am. J. Physiol. Cell Physiol.* 287, C1389–C1395. doi: 10.1152/ajpcell.00220.2004

Baranova, A., Ivanov, D., Petrash, N., Pestova, A., Skoblov, M., Kelmanson, I., et al. (2004). The mammalian pannexin family is homologous to the invertebrate innexin gap junction proteins. *Genomics* 83, 706–716. doi: 10.1016/j.ygeno.2003.09.025

Barishak, Y. R. (1992). Embryology of the eye and its adnexae. *Dev. Ophthalmol.* 24, 1–142.

Beckel, J. M., Argall, A. J., Lim, J. C., Xia, J., Lu, W., Coffey, E. E., et al. (2014). Mechanosensitive release of adenosine 5′-triphosphate through pannexin channels and mechanosensitive upregulation of pannexin channels in optic nerve head astrocytes: a mechanism for purinergic involvement in chronic strain. *Glia* 62, 1486–1501. doi: 10.1002/glia.22695

Berthoud, V. M., Minogue, P. J., Osmolak, P., Snabb, J. I., and Beyer, E. C. (2014). Roles and regulation of lens epithelial cell connexins. *FEBS Lett.* 588, 1297–1303. doi: 10.1016/j.febslet.2013.12.024

Beyer, E. C., and Berthoud, V. M. (2014). Connexin hemichannels in the lens. *Front. Physiol.* 5:20. doi: 10.3389/fphys.2014.00020

Bloomfield, S. A., Xin, D., and Persky, S. E. (1995). A comparison of receptive field and tracer coupling size of horizontal cells in the rabbit retina. *Vis. Neurosci.* 12, 985–999. doi: 10.1017/s0952523800009524

Boassa, D., Ambrosi, C., Qiu, F., Dahl, G., Gaietta, G., and Sosinsky, G. (2007). Pannexin1 channels contain a glycosylation site that targets the hexamer to the plasma membrane. *J. Biol. Chem.* 282, 31733–31743. doi: 10.1074/jbc.m702422200

Bond, S. R., and Naus, C. C. (2014). The pannexins: past and present. *Front. Physiol.* 5:58. doi: 10.3389/fphys.2014.00058

Bond, S. R., Wang, N., Leybaert, L., and Naus, C. C. (2012). Pannexin 1 ohnologs in the teleost lineage. *J. Membr. Biol.* 245, 483–493. doi: 10.1007/s00232-012-9497-4

Boyce, A. K., Wicki-Stordeur, L. E., and Swayne, L. A. (2014). Powerful partnership: crosstalk between pannexin 1 and the cytoskeleton. *Front. Physiol.* 5:27. doi: 10.3389/fphys.2014.00027

Bruzzone, R., Barbe, M. T., Jakob, N. J., and Monyer, H. (2005). Pharmacological properties of homomeric and heteromeric pannexin hemichannels expressed in Xenopus oocytes. *J. Neurochem.* 92, 1033–1043. doi: 10.1111/j.1471-4159.2004.02947.x

Bruzzone, R., Hormuzdi, S. G., Barbe, M. T., Herb, A., and Monyer, H. (2003). Pannexins, a family of gap junction proteins expressed in brain. *Proc. Natl. Acad. Sci. U S A* 100, 13644–13649. doi: 10.1073/pnas.2233464100

Burkhardt, D. A., and Fahey, P. K. (1998). Contrast enhancement and distributed encoding by bipolar cells in the retina. *J. Neurophysiol.* 80, 1070–1081.

Celetti, S. J., Cowan, K. N., Penuela, S., Shao, Q., Churko, J., and Laird, D. W. (2010). Implications of pannexin 1 and pannexin 3 for keratinocyte differentiation. *J. Cell Sci.* 123, 1363–1372. doi: 10.1242/jcs.056093

Conway, B. R. (2007). Color vision: mice see hue too. *Curr. Biol.* 17, R457–R460. doi: 10.1016/j.cub.2007.04.017

Cuthbert, A. W. (2001). Assessment of CFTR chloride channel openers in intact normal and cystic fibrosis murine epithelia. *Br. J. Pharmacol.* 132, 659–668. doi: 10.1038/sj.bjp.0703859

Dahl, G., and Keane, R. W. (2012). Pannexin: from discovery to bedside in 11+/−4 years? *Brain Res.* 1487, 150–159. doi: 10.1016/j.brainres.2012.04.058

Dando, R., and Roper, S. D. (2009). Cell-to-cell communication in intact taste buds through ATP signalling from pannexin 1 gap junction hemichannels. *J. Physiol.* 587, 5899–5906. doi: 10.1113/jphysiol.2009.180083

Davalos, D., Grutzendler, J., Yang, G., Kim, J. V., Zuo, Y., Jung, S., et al. (2005). ATP mediates rapid microglial response to local brain injury in vivo. *Nat. Neurosci.* 8, 752–758. doi: 10.1038/nn1472

Delamere, N. A., and Tamiya, S. (2009). Lens ion transport: from basic concepts to regulation of Na,K-ATPase activity. *Exp. Eye Res.* 88, 140–143. doi: 10.1016/j.exer.2008.05.005

de Rivero Vaccari, J. P., Lotocki, G., Alonso, O. F., Bramlett, H. M., Dietrich, W. D., Keane, R. W., et al. (2009). Therapeutic neutralization of the NLRP1 inflammasome reduces the innate immune response and improves histopathology after traumatic brain injury. *J. Cereb. Blood Flow Metab.* 29, 1251–1261. doi: 10.1038/jcbfm.2009.46

De Vuyst, E., Decrock, E., Cabooter, L., Dubyak, G. R., Naus, C. C., Evans, W. H., et al. (2006). Intracellular calcium changes trigger connexin 32 hemichannel opening. *EMBO J.* 25, 34–44. doi: 10.1038/sj.emboj.7600908

D'hondt, C., Iyyathurai, J., Vinken, M., Rogiers, V., Leybaert, L., Himpens, B., et al. (2013). Regulation of connexin- and pannexin-based channels by post-translational modifications. *Biol. Cell* 105, 373–398. doi: 10.1111/boc.201200096

Domercq, M., Vazquez-Villoldo, N., and Matute, C. (2013). Neurotransmitter signaling in the pathophysiology of microglia. *Front. Cell. Neurosci.* 7:49. doi: 10.3389/fncel.2013.00049

Dvoriantchikova, G., Ivanov, D., Barakat, D., Grinberg, A., Wen, R., Slepak, V. Z., et al. (2012). Genetic ablation of Pannexin1 protects retinal neurons from ischemic injury. *PLoS One* 7:e31991. doi: 10.1371/journal.pone.0031991

Dvoriantchikova, G., Ivanov, D., Panchin, Y., and Shestopalov, V. I. (2006a). Expression of pannexin family of proteins in the retina. *FEBS Lett.* 580, 2178–2182. doi: 10.1016/j.febslet.2006.03.026

Dvoriantchikova, G., Ivanov, D., Pestova, A., and Shestopalov, V. (2006b). Molecular characterization of pannexins in the lens. *Mol. Vis.* 12, 1417–1426.

Fahrenfort, I., Steijaert, M., Sjoerdsma, T., Vickers, E., Ripps, H., van Asselt, J., et al. (2009). Hemichannel-mediated and pH-based feedback from horizontal cells to cones in the vertebrate retina. *PLoS One* 4:e6090. doi: 10.1371/journal.pone.0006090

Fontainhas, A. M., Wang, M., Liang, K. J., Chen, S., Mettu, P., Damani, M., et al. (2011). Microglial morphology and dynamic behavior is regulated by ionotropic glutamatergic and GABAergic neurotransmission. *PLoS One* 6:e15973. doi: 10.1371/journal.pone.0015973

Francis, B. A., and Alvarado, J. (1997). The cellular basis of aqueous outflow regulation. *Curr. Opin. Ophthalmol.* 8, 19–27. doi: 10.1097/00055735-199704000-00005

Furukawa, T., Ogura, T., Katayama, Y., and Hiraoka, M. (1998). Characteristics of rabbit ClC-2 current expressed in Xenopus oocytes and its contribution to volume regulation. *Am. J. Physiol.* 274, C500–C512.

Goel, M., Picciani, R. G., Lee, R. K., and Bhattacharya, S. K. (2010). Aqueous humor dynamics: a review. *Open Ophthalmol. J.* 4, 52–59. doi: 10.2174/1874364101004010052

Goldsmith, P., and Harris, W. A. (2003). The zebrafish as a tool for understanding the biology of visual disorders. *Semin. Cell Dev. Biol.* 14, 11–18. doi: 10.1016/s1084-9521(02)00167-2

Gong, H., Tripathi, R. C., and Tripathi, B. J. (1996). Morphology of the aqueous outflow pathway. *Microsc. Res. Tech.* 33, 336–367. doi: 10.1002/(sici)1097-0029(19960301)33:4<336::aid-jemt4>3.0.co;2-n

Grek, C. L., Rhett, J. M., and Ghatnekar, G. S. (2014). Cardiac to cancer: connecting connexins to clinical opportunity. *FEBS Lett.* 588, 1349–1364. doi: 10.1016/j.febslet.2014.02.047

Hagins, W. A., Penn, R. D., and Yoshikami, S. (1970). Dark current and photocurrent in retinal rods. *Biophys. J.* 10, 380–412. doi: 10.1016/s0006-3495(70)86308-1

Hanner, F., Lam, L., Nguyen, M. T., Yu, A., and Peti-Peterdi, J. (2012). Intrarenal localization of the plasma membrane ATP channel pannexin1. *Am. J. Physiol. Renal Physiol.* 303, F1454–F1459. doi: 10.1152/ajprenal.00206.2011

Hirasawa, H., Yamada, M., and Kaneko, A. (2012). Acidification of the synaptic cleft of cone photoreceptor terminal controls the amount of transmitter release, thereby forming the receptive field surround in the vertebrate retina. *J. Physiol. Sci.* 62, 359–375. doi: 10.1007/s12576-012-0220-0

Huang, Y., Grinspan, J. B., Abrams, C. K., and Scherer, S. S. (2007a). Pannexin1 is expressed by neurons and glia but does not form functional gap junctions. *Glia* 55, 46–56. doi: 10.1002/glia.20435

Huang, Y. J., Maruyama, Y., Dvoryanchikov, G., Pereira, E., Chaudhari, N., and Roper, S. D. (2007b). The role of pannexin 1 hemichannels in ATP release and cell-cell communication in mouse taste buds. *Proc. Natl. Acad. Sci. U S A* 104, 6436–6441. doi: 10.1073/pnas.0611280104

Husain, S., Shearer, T. W., and Crosson, C. E. (2007). Mechanisms linking adenosine A1 receptors and extracellular signal-regulated kinase 1/2 activation in human trabecular meshwork cells. *J. Pharmacol. Exp. Ther.* 320, 258–265. doi: 10.1124/jpet.106.110981

Iglesias, R., Dahl, G., Qiu, F., Spray, D. C., and Scemes, E. (2009a). Pannexin 1: the molecular substrate of astrocyte "hemichannels". *J. Neurosci.* 29, 7092–7097. doi: 10.1523/jneurosci.6062-08.2009

Iglesias, R., Spray, D. C., and Scemes, E. (2009b). Mefloquine blockade of Pannexin1 currents: resolution of a conflict. *Cell Commun. Adhes.* 16, 131–137. doi: 10.3109/15419061003642618

Isakson, B. E., and Thompson, R. J. (2014). Pannexin-1 as a potentiator of ligand-gated receptor signaling. *Channels (Austin)* 8, 118–123. doi: 10.4161/chan.27978

Jackman, S. L., Babai, N., Chambers, J. J., Thoreson, W. B., and Kramer, R. H. (2011). A positive feedback synapse from retinal horizontal cells to cone photoreceptors. *PLoS Biol.* 9:e1001057. doi: 10.1371/journal.pbio.1001057

Jackson, D. G., Wang, J., Keane, R. W., Scemes, E., and Dahl, G. (2014). ATP and potassium ions: a deadly combination for astrocytes. *Sci. Rep.* 4:4576. doi: 10.1038/srep04576

Jaillon, O., Aury, J. M., Brunet, F., Petit, J. L., Stange-Thomann, N., Mauceli, E., et al. (2004). Genome duplication in the teleost fish Tetraodon nigroviridis reveals the early vertebrate proto-karyotype. *Nature* 431, 946–957. doi: 10.1038/nature03025

Johnstone, S. R., Billaud, M., Lohman, A. W., Taddeo, E. P., and Isakson, B. E. (2012). Posttranslational modifications in connexins and pannexins. *J. Membr. Biol.* 245, 319–332. doi: 10.1007/s00232-012-9453-3

Kamermans, M., and Fahrenfort, I. (2004). Ephaptic interactions within a chemical synapse: hemichannel-mediated ephaptic inhibition in the retina. *Curr. Opin. Neurobiol.* 14, 531–541. doi: 10.1016/j.conb.2004.08.016

Kamermans, M., Fahrenfort, I., Schultz, K., Janssen-Bienhold, U., Sjoerdsma, T., and Weiler, R. (2001). Hemichannel-mediated inhibition in the outer retina. *Science* 292, 1178–1180. doi: 10.1126/science.1060101

Kanneganti, T. D., Lamkanfi, M., Kim, Y. G., Chen, G., Park, J. H., Franchi, L., et al. (2007). Pannexin-1-mediated recognition of bacterial molecules activates the cryopyrin inflammasome independent of toll-like receptor signaling. *Immunity* 26, 433–443. doi: 10.1016/j.immuni.2007.03.008

Karatas, H., Erdener, S. E., Gursoy-Ozdemir, Y., Lule, S., Eren-Kocak, E., Sen, Z. D., et al. (2013). Spreading depression triggers headache by activating neuronal Panx1 channels. *Science* 339, 1092–1095. doi: 10.1126/science.1231897

Kawamura, M. Jr., Ruskin, D. N., and Masino, S. A. (2010). Metabolic autocrine regulation of neurons involves cooperation among pannexin hemichannels, adenosine receptors and KATP channels. *J. Neurosci.* 30, 3886–3895. doi: 10.1523/JNEUROSCI.0055-10.2010

Kienitz, M. C., Bender, K., Dermietzel, R., Pott, L., and Zoidl, G. (2011). Pannexin 1 constitutes the large conductance cation channel of cardiac myocytes. *J. Biol. Chem.* 286, 290–298. doi: 10.1074/jbc.M110.163477

Kim, J. E., and Kang, T. C. (2011). The P2X7 receptor-pannexin-1 complex decreases muscarinic acetylcholine receptor-mediated seizure susceptibility in mice. *J. Clin. Invest.* 121, 2037–2047. doi: 10.1172/JCI44818

Klaassen, L. J., Fahrenfort, I., and Kamermans, M. (2012). Connexin hemichannel mediated ephaptic inhibition in the retina. *Brain Res.* 1487, 25–38. doi: 10.1016/j.brainres.2012.04.059

Klaassen, L. J., Sun, Z., Steijaert, M. N., Bolte, P., Fahrenfort, I., Sjoerdsma, T., et al. (2011). Synaptic transmission from horizontal cells to cones is impaired by loss of connexin hemichannels. *PLoS Biol.* 9:e1001107. doi: 10.1371/journal.pbio.1001107

Kranz, K., Dorgau, B., Pottek, M., Herrling, R., Schultz, K., Bolte, P., et al. (2013). Expression of Pannexin1 in the outer plexiform layer of the mouse retina and physiological impact of its knockout. *J. Comp. Neurol.* 521, 1119–1135. doi: 10.1002/cne.23223

Krizaj, D., Ryskamp, D. A., Tian, N., Tezel, G., Mitchell, C. H., Slepak, V. Z., et al. (2014). From mechanosensitivity to inflammatory responses: new players in the pathology of glaucoma. *Curr. Eye Res.* 39, 105–119. doi: 10.3109/02713683.2013.836541

Kronlage, M., Song, J., Sorokin, L., Isfort, K., Schwerdtle, T., Leipziger, J., et al. (2010). Autocrine purinergic receptor signaling is essential for macrophage chemotaxis. *Sci. Signal.* 3:ra55. doi: 10.1126/scisignal.2000588

Kurtenbach, S., Prochnow, N., Kurtenbach, S., Klooster, J., Zoidl, C., Dermietzel, R., et al. (2013). Pannexin1 channel proteins in the zebrafish retina have shared and unique properties. *PLoS One* 8:e77722. doi: 10.1371/journal.pone.0077722

Lai, C. P., Bechberger, J. F., Thompson, R. J., Macvicar, B. A., Bruzzone, R., and Naus, C. C. (2007). Tumor-suppressive effects of pannexin 1 in C6 glioma cells. *Cancer Res.* 67, 1545–1554. doi: 10.1158/0008-5472.can-06-1396

Lamkanfi, M., Malireddi, R. K., and Kanneganti, T. D. (2009). Fungal zymosan and mannan activate the cryopyrin inflammasome. *J. Biol. Chem.* 284, 20574–20581. doi: 10.1074/jbc.m109.023689

Lee, J. E., Liang, K. J., Fariss, R. N., and Wong, W. T. (2008). Ex vivo dynamic imaging of retinal microglia using time-lapse confocal microscopy. *Invest. Ophthalmol. Vis. Sci.* 49, 4169–4176. doi: 10.1167/iovs.08-2076

Li, A., Leung, C. T., Peterson-Yantorno, K., Mitchell, C. H., and Civan, M. M. (2010). Pathways for ATP release by bovine ciliary epithelial cells, the initial step in purinergic regulation of aqueous humor inflow. *Am. J. Physiol. Cell Physiol.* 299, C1308–C1317. doi: 10.1152/ajpcell.00333.2010

Li, A., Leung, C. T., Peterson-Yantorno, K., Stamer, W. D., Mitchell, C. H., and Civan, M. M. (2012). Mechanisms of ATP release by human trabecular meshwork cells, the enabling step in purinergic regulation of aqueous humor outflow. *J. Cell. Physiol.* 227, 172–182. doi: 10.1002/jcp.22715

Li, A., Zhang, X., Zheng, D., Ge, J., Laties, A. M., and Mitchell, C. H. (2011). Sustained elevation of extracellular ATP in aqueous humor from humans with primary chronic angle-closure glaucoma. *Exp. Eye Res.* 93, 528–533. doi: 10.1016/j.exer.2011.06.020

Liang, J., Takeuchi, H., Jin, S., Noda, M., Li, H., Doi, Y., et al. (2010). Glutamate induces neurotrophic factor production from microglia via protein kinase C pathway. *Brain Res.* 1322, 8–23. doi: 10.1016/j.brainres.2010.01.083

Liu, H., Chen, Y., Niu, Y., Zhang, K., Kang, Y., Ge, W., et al. (2014a). TALEN-mediated gene mutagenesis in rhesus and cynomolgus monkeys. *Cell Stem Cell* 14, 323–328. doi: 10.1016/j.stem.2014.01.018

Liu, Z., Zhou, X., Zhu, Y., Chen, Z. F., Yu, B., Wang, Y., et al. (2014b). Generation of a monkey with MECP2 mutations by TALEN-based gene targeting. *Neurosci. Bull.* 30, 381–386. doi: 10.1007/s12264-014-1434-8

Locovei, S., Bao, L., and Dahl, G. (2006a). Pannexin 1 in erythrocytes: function without a gap. *Proc. Natl. Acad. Sci. U S A* 103, 7655–7659. doi: 10.1073/pnas.0601037103

Locovei, S., Wang, J., and Dahl, G. (2006b). Activation of pannexin 1 channels by ATP through P2Y receptors and by cytoplasmic calcium. *FEBS Lett.* 580, 239–244. doi: 10.1016/j.febslet.2005.12.004

Lohman, A. W., Billaud, M., and Isakson, B. E. (2012a). Mechanisms of ATP release and signalling in the blood vessel wall. *Cardiovasc. Res.* 95, 269–280. doi: 10.1093/cvr/cvs187

Lohman, A. W., and Isakson, B. E. (2014). Differentiating connexin hemichannels and pannexin channels in cellular ATP release. *FEBS Lett.* 588, 1379–1388. doi: 10.1016/j.febslet.2014.02.004

Lohman, A. W., Weaver, J. L., Billaud, M., Sandilos, J. K., Griffiths, R., Straub, A. C., et al. (2012b). S-nitrosylation inhibits pannexin 1 channel function. *J. Biol. Chem.* 287, 39602–39612. doi: 10.1074/jbc.m112.397976

MacVicar, B. A., and Thompson, R. J. (2010). Non-junction functions of pannexin-1 channels. *Trends Neurosci.* 33, 93–102. doi: 10.1016/j.tins.2009.11.007

Madry, C., Haglerod, C., and Attwell, D. (2010). The role of pannexin hemichannels in the anoxic depolarization of hippocampal pyramidal cells. *Brain* 133, 3755–3763. doi: 10.1093/brain/awq284

Mali, P., Esvelt, K. M., and Church, G. M. (2013a). Cas9 as a versatile tool for engineering biology. *Nat. Methods* 10, 957–963. doi: 10.1038/nmeth.2649

Mali, P., Yang, L., Esvelt, K. M., Aach, J., Guell, M., Dicarlo, J. E., et al. (2013b). RNA-guided human genome engineering via Cas9. *Science* 339, 823–826. doi: 10.1126/science.1232033

Mayo, C., Ren, R., Rich, C., Stepp, M. A., and Trinkaus-Randall, V. (2008). Regulation by P2X7: epithelial migration and stromal organization in the cornea. *Invest. Ophthalmol. Vis. Sci.* 49, 4384–4391. doi: 10.1167/iovs.08-1688

Mertsch, K., Hanisch, U. K., Kettenmann, H., and Schnitzer, J. (2001). Characterization of microglial cells and their response to stimulation in an organotypic retinal culture system. *J. Comp. Neurol.* 431, 217–227. doi: 10.1002/1096-9861(20010305)431:2<217::aid-cne1066>3.3.co;2-k

Minkiewicz, J., De Rivero Vaccari, J. P., and Keane, R. W. (2013). Human astrocytes express a novel NLRP2 inflammasome. *Glia* 61, 1113–1121. doi: 10.1002/glia.22499

Mitchell, C. H., Peterson-Yantorno, K., Carre, D. A., Mcglinn, A. M., Coca-Prados, M., Stone, R. A., et al. (1999). A3 adenosine receptors regulate Cl- channels of nonpigmented ciliary epithelial cells. *Am. J. Physiol.* 276, C659–C666.

Monaco, E. A. 3rd, and Friedlander, R. M. (2012). A novel target for ischemic stroke therapy: pannexins. *Neurosurgery* 70, N13–N14. doi: 10.1227/01.neu.0000413220.21264.df

Mylvaganam, S., Zhang, L., Wu, C., Zhang, Z. J., Samoilova, M., Eubanks, J., et al. (2010). Hippocampal seizures alter the expression of the pannexin and connexin transcriptome. *J. Neurochem.* 112, 92–102. doi: 10.1111/j.1471-4159.2009.06431.x

Nimmerjahn, A., Kirchhoff, F., and Helmchen, F. (2005). Resting microglial cells are highly dynamic surveillants of brain parenchyma in vivo. *Science* 308, 1314–1318. doi: 10.1126/science.1110647

Niu, Y., Shen, B., Cui, Y., Chen, Y., Wang, J., Wang, L., et al. (2014). Generation of gene-modified cynomolgus monkey via Cas9/RNA-mediated gene targeting in one-cell embryos. *Cell* 156, 836–843. doi: 10.1016/j.cell.2014.01.027

Orellana, J. A., Velasquez, S., Williams, D. W., Saez, J. C., Berman, J. W., and Eugenin, E. A. (2013). Pannexin1 hemichannels are critical for HIV infection of human primary CD4+ T lymphocytes. *J. Leukoc. Biol.* 94, 399–407. doi: 10.1189/jlb.0512249

Panchin, Y., Kelmanson, I., Matz, M., Lukyanov, K., Usman, N., and Lukyanov, S. (2000). A ubiquitous family of putative gap junction molecules. *Curr. Biol.* 10, R473–R474. doi: 10.1016/s0960-9822(00)00576-5

Pau, H. (2008). Retinopathy of prematurity: clinic and pathogenesis. Disproportion between apoptosis of vitreal and proliferation of retinal vascularization. *Ophthalmologica* 222, 220–224. doi: 10.1159/000130069

Peichl, L., and Gonzalez-Soriano, J. (1993). Unexpected presence of neurofilaments in axon-bearing horizontal cells of the mammalian retina. *J. Neurosci.* 13, 4091–4100. doi: 10.1098/rspb.1993.0051

Pelegrin, P., and Surprenant, A. (2006). Pannexin-1 mediates large pore formation and interleukin-1beta release by the ATP-gated P2X7 receptor. *EMBO J.* 25, 5071–5082. doi: 10.1038/sj.emboj.7601378

Penn, R. D., and Hagins, W. A. (1969). Signal transmission along retinal rods and the origin of the electroretinographic a-wave. *Nature* 223, 201–204. doi: 10.1038/223201a0

Penuela, S., Bhalla, R., Gong, X. Q., Cowan, K. N., Celetti, S. J., Cowan, B. J., et al. (2007). Pannexin 1 and pannexin 3 are glycoproteins that exhibit many distinct characteristics from the connexin family of gap junction proteins. *J. Cell Sci.* 120, 3772–3783. doi: 10.1242/jcs.009514

Penuela, S., Gyenis, L., Ablack, A., Churko, J. M., Berger, A. C., Litchfield, D. W., et al. (2012). Loss of pannexin 1 attenuates melanoma progression by reversion to a melanocytic phenotype. *J. Biol. Chem.* 287, 29184–29193. doi: 10.1074/jbc.m112.377176

Penuela, S., Kelly, J. J., Churko, J. M., Barr, K. J., Berger, A. C., and Laird, D. W. (2014a). Panx1 regulates cellular properties of keratinocytes and dermal fibroblasts in skin development and wound healing. *J. Invest. Dermatol.* 134, 2026–2035. doi: 10.1038/jid.2014.86

Penuela, S., Simek, J., and Thompson, R. J. (2014b). Regulation of pannexin channels by post-translational modifications. *FEBS Lett.* 588, 1411–1415. doi: 10.1016/j.febslet.2014.01.028

Postlethwait, J. H. (2007). The zebrafish genome in context: ohnologs gone missing. *J. Exp. Zool. B Mol. Dev. Evol.* 308, 563–577. doi: 10.1002/jez.b.21137

Pottek, M., Hoppenstedt, W., Janssen-Bienhold, U., Schultz, K., Perlman, I., and Weiler, R. (2003). Contribution of connexin26 to electrical feedback inhibition in the turtle retina. *J. Comp. Neurol.* 466, 468–477. doi: 10.1002/cne.10897

Prochnow, N., Abdulazim, A., Kurtenbach, S., Wildforster, V., Dvoriantchikova, G., Hanske, J., et al. (2012). Pannexin1 stabilizes synaptic plasticity and is needed for learning. *PLoS One* 7:e51767. doi: 10.1371/journal.pone.0051767

Prochnow, N., Hoffmann, S., Vroman, R., Klooster, J., Bunse, S., Kamermans, M., et al. (2009). Pannexin1 in the outer retina of the zebrafish, Danio rerio. *Neuroscience* 162, 1039–1054. doi: 10.1016/j.neuroscience.2009.04.064

Puthussery, T., Yee, P., Vingrys, A. J., and Fletcher, E. L. (2006). Evidence for the involvement of purinergic P2X receptors in outer retinal processing. *Eur. J. Neurosci.* 24, 7–19. doi: 10.1111/j.1460-9568.2006.04895.x

Qiu, F., and Dahl, G. (2009). A permeant regulating its permeation pore: inhibition of pannexin 1 channels by ATP. *Am. J. Physiol. Cell Physiol.* 296, C250–C255. doi: 10.1152/ajpcell.00433.2008

Qiu, F., Wang, J., Spray, D. C., Scemes, E., and Dahl, G. (2011). Two non-vesicular ATP release pathways in the mouse erythrocyte membrane. *FEBS Lett.* 585, 3430–3435. doi: 10.1016/j.febslet.2011.09.033

Ravi, V., and Venkatesh, B. (2008). Rapidly evolving fish genomes and teleost diversity. *Curr. Opin. Genet. Dev.* 18, 544–550. doi: 10.1016/j.gde.2008.11.001

Ray, A., Zoidl, G., Wahle, P., and Dermietzel, R. (2006). Pannexin expression in the cerebellum. *Cerebellum* 5, 189–192. doi: 10.1080/14734220500530082

Ray, A., Zoidl, G., Weickert, S., Wahle, P., and Dermietzel, R. (2005). Site-specific and developmental expression of pannexin1 in the mouse nervous system. *Eur. J. Neurosci.* 21, 3277–3290. doi: 10.1111/j.1460-9568.2005.04139.x

Reigada, D., Lu, W., Zhang, M., and Mitchell, C. H. (2008). Elevated pressure triggers a physiological release of ATP from the retina: possible role for pannexin hemichannels. *Neuroscience* 157, 396–404. doi: 10.1016/j.neuroscience.2008.08.036

Resta, V., Novelli, E., Vozzi, G., Scarpa, C., Caleo, M., Ahluwalia, A., et al. (2007). Acute retinal ganglion cell injury caused by intraocular pressure spikes is mediated by endogenous extracellular ATP. *Eur. J. Neurosci.* 25, 2741–2754. doi: 10.1111/j.1460-9568.2007.05528.x

Retamal, M. A. (2014). Connexin and Pannexin hemichannels are regulated by redox potential. *Front. Physiol.* 5:80. doi: 10.3389/fphys.2014.00080

Retamal, M. A., Schalper, K. A., Shoji, K. F., Bennett, M. V., and Saez, J. C. (2007). Opening of connexin 43 hemichannels is increased by lowering intracellular redox potential. *Proc. Natl. Acad. Sci. U S A* 104, 8322–8327. doi: 10.1073/pnas.0702456104

Romanov, R. A., Rogachevskaja, O. A., Bystrova, M. F., Jiang, P., Margolskee, R. F., and Kolesnikov, S. S. (2007). Afferent neurotransmission mediated by hemichannels in mammalian taste cells. *EMBO J.* 26, 657–667. doi: 10.1038/sj.emboj.7601526

Sabirov, R. Z., Dutta, A. K., and Okada, Y. (2001). Volume-dependent ATP-conductive large-conductance anion channel as a pathway for swelling-induced ATP release. *J. Gen. Physiol.* 118, 251–266. doi: 10.1085/jgp.118.3.251

Saez, J. C., and Leybaert, L. (2014). Hunting for connexin hemichannels. *FEBS Lett.* 588, 1205–1211. doi: 10.1016/j.febslet.2014.03.004

Santiago, M. F., Veliskova, J., Patel, N. K., Lutz, S. E., Caille, D., Charollais, A., et al. (2011). Targeting pannexin1 improves seizure outcome. *PLoS One* 6:e25178. doi: 10.1371/journal.pone.0025178

Schalper, K. A., Orellana, J. A., Berthoud, V. M., and Saez, J. C. (2009). Dysfunctions of the diffusional membrane pathways mediated by hemichannels in inherited and acquired human diseases. *Curr. Vasc. Pharmacol.* 7, 486–505. doi: 10.2174/157016109789043937

Schenk, U., Westendorf, A. M., Radaelli, E., Casati, A., Ferro, M., Fumagalli, M., et al. (2008). Purinergic control of T cell activation by ATP released through pannexin-1 hemichannels. *Sci. Signal.* 1:ra6. doi: 10.3410/f.1123131.580258

Seminario-Vidal, L., Kreda, S., Jones, L., O'Neal, W., Trejo, J., Boucher, R. C., et al. (2009). Thrombin promotes release of ATP from lung epithelial cells through coordinated activation of rho- and Ca2+-dependent signaling pathways. *J. Biol. Chem.* 284, 20638–20648. doi: 10.1074/jbc.M109.004762

Seminario-Vidal, L., Okada, S. F., Sesma, J. I., Kreda, S. M., van Heusden, C. A., Zhu, Y., et al. (2011). Rho signaling regulates pannexin 1-mediated ATP release from airway epithelia. *J. Biol. Chem.* 286, 26277–26286. doi: 10.1074/jbc.M111.260562

Séror, C., Melki, M. T., Subra, F., Raza, S. Q., Bras, M., Saidi, H., et al. (2011). Extracellular ATP acts on P2Y2 purinergic receptors to facilitate HIV-1 infection. *J. Exp. Med.* 208, 1823–1834. doi: 10.1084/jem.20101805

Shahidullah, M., Mandal, A., and Delamere, N. A. (2012). TRPV4 in porcine lens epithelium regulates hemichannel-mediated ATP release and Na-K-ATPase activity. *Am. J. Physiol. Cell Physiol.* 302, C1751–C1761. doi: 10.1152/ajpcell.00010.2012

Shearer, T. W., and Crosson, C. E. (2002). Adenosine A1 receptor modulation of MMP-2 secretion by trabecular meshwork cells. *Invest. Ophthalmol. Vis. Sci.* 43, 3016–3020.

Silverman, W. R., De Rivero Vaccari, J. P., Locovei, S., Qiu, F., Carlsson, S. K., Scemes, E., et al. (2009). The pannexin 1 channel activates the inflammasome in neurons and astrocytes. *J. Biol. Chem.* 284, 18143–18151. doi: 10.1074/jbc.M109.004804

Silverman, W., Locovei, S., and Dahl, G. (2008). Probenecid, a gout remedy, inhibits pannexin 1 channels. *Am. J. Physiol. Cell Physiol.* 295, C761–C767. doi: 10.1152/ajpcell.00227.2008

Skapinker, R., and Rothberg, A. D. (1987). Postnatal regression of the tunica vasculosa lentis. *J. Perinatol.* 7, 279–281.

Sosinsky, G. E., Boassa, D., Dermietzel, R., Duffy, H. S., Laird, D. W., MacVicar, B., et al. (2011). Pannexin channels are not gap junction hemichannels. *Channels (Austin)* 5, 193–197. doi: 10.4161/chan.5.3.15765

Sridharan, M., Adderley, S. P., Bowles, E. A., Egan, T. M., Stephenson, A. H., Ellsworth, M. L., et al. (2010). Pannexin 1 is the conduit for low oxygen tension-induced ATP release from human erythrocytes. *Am. J. Physiol. Heart Circ. Physiol.* 299, H1146–H1152. doi: 10.1152/ajpheart.00301.2010

Suadicani, S. O., Iglesias, R., Wang, J., Dahl, G., Spray, D. C., and Scemes, E. (2012). ATP signaling is deficient in cultured Pannexin1-null mouse astrocytes. *Glia* 60, 1106–1116. doi: 10.1002/glia.22338

Sun, Z., Risner, M. L., van Asselt, J. B., Zhang, D. Q., Kamermans, M., and McMahon, D. G. (2012). Physiological and molecular characterization of connexin hemichannels in zebrafish retinal horizontal cells. *J. Neurophysiol.* 107, 2624–2632. doi: 10.1152/jn.01126.2011

Tang, W., Ahmad, S., Shestopalov, V. I., and Lin, X. (2008). Pannexins are new molecular candidates for assembling gap junctions in the cochlea. *Neuroreport* 19, 1253–1257. doi: 10.1097/WNR.0b013e32830891f5

Thompson, R. J., Jackson, M. F., Olah, M. E., Rungta, R. L., Hines, D. J., Beazely, M. A., et al. (2008). Activation of pannexin-1 hemichannels augments aberrant bursting in the hippocampus. *Science* 322, 1555–1559. doi: 10.1126/science.1165209

Toris, C. B., Gabelt, B. T., and Kaufman, P. L. (2008). Update on the mechanism of action of topical prostaglandins for intraocular pressure reduction. *Surv. Ophthalmol.* 53(Suppl. 1), S107–S120. doi: 10.1016/j.survophthal.2008.08.010

Ueda, T., Shikano, M., Kamiya, T., Joh, T., and Ugawa, S. (2011). The TRPV4 channel is a novel regulator of intracellular Ca2+ in human esophageal epithelial cells. *Am. J. Physiol. Gastrointest. Liver Physiol.* 301, G138–G147. doi: 10.1152/ajpgi.00511.2010

Vanden Abeele, F., Bidaux, G., Gordienko, D., Beck, B., Panchin, Y. V., Baranova, A. V., et al. (2006). Functional implications of calcium permeability of the channel formed by pannexin 1. *J. Cell Biol.* 174, 535–546. doi: 10.1083/jcb.200601115

Vogt, A., Hormuzdi, S. G., and Monyer, H. (2005). Pannexin1 and Pannexin2 expression in the developing and mature rat brain. *Brain Res. Mol. Brain Res.* 141, 113–120. doi: 10.1016/j.molbrainres.2005.08.002

Volgyi, B., Kovacs-Oller, T., Atlasz, T., Wilhelm, M., and Gabriel, R. (2013). Gap junctional coupling in the vertebrate retina: variations on one theme? *Prog. Retin. Eye Res.* 34, 1–18. doi: 10.1016/j.preteyeres.2012.12.002

Vroman, R., Klaassen, L. J., Howlett, M. H., Cenedese, V., Klooster, J., Sjoerdsma, T., et al. (2014). Extracellular ATP hydrolysis inhibits synaptic transmission by increasing ph buffering in the synaptic cleft. *PLoS Biol.* 12:e1001864. doi: 10.1371/journal.pbio.1001864

Vroman, R., Klaassen, L. J., and Kamermans, M. (2013). Ephaptic communication in the vertebrate retina. *Front. Hum. Neurosci.* 7:612. doi: 10.3389/fnhum.2013.00612

Wang, J., Jackson, D. G., and Dahl, G. (2013). The food dye FD&C Blue No. 1 is a selective inhibitor of the ATP release channel Panx1. *J. Gen. Physiol.* 141, 649–656. doi: 10.1085/jgp.201310966

Wang, X. H., Streeter, M., Liu, Y. P., and Zhao, H. B. (2009). Identification and characterization of pannexin expression in the mammalian cochlea. *J. Comp. Neurol.* 512, 336–346. doi: 10.1002/cne.21898

Weilinger, N. L., Tang, P. L., and Thompson, R. J. (2012). Anoxia-induced NMDA receptor activation opens pannexin channels via Src family kinases. *J. Neurosci.* 32, 12579–12588. doi: 10.1523/JNEUROSCI.1267-12.2012

Woehrle, T., Yip, L., Elkhal, A., Sumi, Y., Chen, Y., Yao, Y., et al. (2010). Pannexin-1 hemichannel-mediated ATP release together with P2X1 and P2X4 receptors regulate T-cell activation at the immune synapse. *Blood* 116, 3475–3484. doi: 10.1182/blood-2010-04-277707

Wong, W. T., Wang, M., and Li, W. (2011). Regulation of microglia by ionotropic glutamatergic and GABAergic neurotransmission. *Neuron Glia Biol.* 7, 41–46. doi: 10.1017/S1740925x11000123

Xia, J., Lim, J. C., Lu, W., Beckel, J. M., Macarak, E. J., Laties, A. M., et al. (2012). Neurons respond directly to mechanical deformation with pannexin-mediated ATP release and autostimulation of P2X7 receptors. *J. Physiol.* 590, 2285–2304. doi: 10.1113/jphysiol.2012.227983

Xiang, Z., Chen, M., Ping, J., Dunn, P., Lv, J., Jiao, B., et al. (2006). Microglial morphology and its transformation after challenge by extracellular ATP in vitro. *J. Neurosci. Res.* 83, 91–101. doi: 10.1002/jnr.20709

Xin, D., and Bloomfield, S. A. (1999). Dark- and light-induced changes in coupling between horizontal cells in mammalian retina. *J. Comp. Neurol.* 405, 75–87. doi: 10.1002/(sici)1096-9861(19990301)405:1<75::aid-cne6>3.0.co;2-d

Yang, L., Mali, P., Kim-Kiselak, C., and Church, G. (2014). CRISPR-Cas-mediated targeted genome editing in human cells. *Methods Mol. Biol.* 1114, 245–267. doi: 10.1007/978-1-62703-761-7_16

Zappalà, A., Cicero, D., Serapide, M. F., Paz, C., Catania, M. V., Falchi, M., et al. (2006). Expression of pannexin1 in the CNS of adult mouse: cellular localization and effect of 4-aminopyridine-induced seizures. *Neuroscience* 141, 167–178. doi: 10.1016/j.neuroscience.2006.03.053

Zappalà, A., Li Volti, G., Serapide, M. F., Pellitteri, R., Falchi, M., La Delia, F., et al. (2007). Expression of pannexin2 protein in healthy and ischemized brain of adult rats. *Neuroscience* 148, 653–667. doi: 10.1016/j.neuroscience.2007.06.028

Zhang, H., Chen, Y., and Zhang, C. (2012). Patterns of heterogeneous expression of pannexin 1 and pannexin 2 transcripts in the olfactory epithelium and olfactory bulb. *J. Mol. Histol.* 43, 651–660. doi: 10.1007/s10735-012-9443-x

Zhang, X., Li, A., Ge, J., Reigada, D., Laties, A. M., and Mitchell, C. H. (2007). Acute increase of intraocular pressure releases ATP into the anterior chamber. *Exp. Eye Res.* 85, 637–643. doi: 10.1016/j.exer.2007.07.016

Zhang, J., Peng, S., Lin, M., and Du, C. (2005). [Clinical study of twenty-four-hour pattern of intraocular pressure in normal person and patient with primary open-angle glaucoma]. *Yan Ke Xue Bao* 21, 127–130.

Zoidl, G., Kremer, M., Zoidl, C., Bunse, S., and Dermietzel, R. (2008). Molecular diversity of connexin and pannexin genes in the retina of the zebrafish Danio rerio. *Cell Commun. Adhes.* 15, 169–183. doi: 10.1080/15419060802014081

Zoidl, G., Petrasch-Parwez, E., Ray, A., Meier, C., Bunse, S., Habbes, H. W., et al. (2007). Localization of the pannexin1 protein at postsynaptic sites in the cerebral cortex and hippocampus. *Neuroscience* 146, 9–16. doi: 10.1016/j.neuroscience.2007.01.061

9

The connexin43 mimetic peptide Gap19 inhibits hemichannels without altering gap junctional communication in astrocytes

Verónica Abudara[1†], John Bechberger[2], Moises Freitas-Andrade[2], Marijke De Bock[3], Nan Wang[3], Geert Bultynck[4], Christian C. Naus[2], Luc Leybaert[3‡] and Christian Giaume[1*‡]

[1] Center for Interdisciplinary Research in Biology, Centre National de la Recherche Scientifique, Collège de France, Paris, France
[2] Department of Cellular and Physiological Sciences, Faculty of Medicine, Life Sciences Institute, University of British Columbia, Vancouver, BC, Canada
[3] Department of Basic Medical Sciences - Physiology Group, Faculty of Medicine and Health Sciences, Ghent University, Ghent, Belgium
[4] Laboratory of Molecular and Cellular Signaling, Department of Cellular and Molecular Medicine, KU Leuven, Leuven, Belgium

***Correspondence:**
Christian Giaume, Center for Interdisciplinary Research in Biology, Centre National de la Recherche Scientifique UMR7241/INSERM U1050, Collège de France, 11 Place Marcelin Berthelot, 75005 Paris, France
e-mail: christian.giaume@college-de-france.fr

†Present address:
Verónica Abudara, Departamento de Fisiología, Facultad de Medicina, Universidad de la República, Montevideo, Uruguay

‡Shared senior authors.

In the brain, astrocytes represent the cellular population that expresses the highest amount of connexins (Cxs). This family of membrane proteins is the molecular constituent of gap junction channels and hemichannels that provide pathways for direct cytoplasm-to-cytoplasm and inside-out exchange, respectively. Both types of Cx channels are permeable to ions and small signaling molecules allowing astrocytes to establish dynamic interactions with neurons. So far, most pharmacological approaches currently available do not distinguish between these two channel functions, stressing the need to develop new specific molecular tools. In astrocytes two major Cxs are expressed, Cx43 and Cx30, and there is now evidence indicating that at least Cx43 operates as a gap junction channel as well as a hemichannel in these cells. Based on studies in primary cultures as well as in acute hippocampal slices, we report here that Gap19, a nonapeptide derived from the cytoplasmic loop of Cx43, inhibits astroglial Cx43 hemichannels in a dose-dependent manner, without affecting gap junction channels. This peptide, which not only selectively inhibits hemichannels but is also specific for Cx43, can be delivered *in vivo* in mice as TAT-Gap19, and displays penetration into the brain parenchyma. As a result, Gap19 combined with other tools opens up new avenues to decipher the role of Cx43 hemichannels in interactions between astrocytes and neurons in physiological as well as pathological situations.

Keywords: connexins, glial cells, gap junctions, astroglia, mimetic peptide

INTRODUCTION

Compared to neurons, astrocytes make up the brain cell population that expresses the highest amount of the gap junction proteins, named connexins (Cxs) (Ransom and Giaume, 2013). Connexin-mediated channel functions are essential for the dynamic and metabolic interactions that astrocytes establish with each other and at their interfaces with neurons and the vasculature (Giaume et al., 2010). Indeed, transgenic animals in which the two major astroglial Cxs, i.e., Cx43 and Cx30, have been deleted, exhibit impaired potassium clearance, synaptic transmission and plasticity (Wallraff et al., 2006; Pannasch et al., 2011), a dysmyelinating phenotype (Lutz et al., 2009) and a loss in blood-brain barrier integrity (Ezan et al., 2012). However, the exclusive use of such animals does not distinguish between the contributions of the two types of astroglial Cxs as well as between the channel and hemichannel functions that they support (Giaume and Theis, 2010). So far, the use of single Cx knock-out mice has provided key data demonstrating a role of Cx43 in neuronal migration (Elias et al., 2007; Cina et al., 2009), a synaptic activity-dependent modulation of Cx30 gap junctions in astrocytes in the olfactory bulb (Roux et al., 2011) and recently, it was reported that the lack of Cx30 impacts synaptic transmission through the modulation of astroglial glutamate transport (Pannasch et al., 2014). However, there is still a need to develop new pharmacological tools to design *in vitro* and *in vivo* experiments studying the role of Cxs in astrocytes.

Gap junction channels form junctional plaques that are composed of two docked hemichannels oligomerized from six Cx protein subunits. Usually, most of the unapposed/non-junctional hemichannels are closed but a fraction of Cx43 HCs can be open under resting conditions and have physiological roles (Stehberg et al., 2012; Chever et al., 2014) while they become more active in pathological situations (Giaume et al., 2013). Their activation results in gliotransmitter (ATP, glutamate) release, the entry of calcium ions (Ca^{2+}) and glucose, ionic imbalance, cellular volume overload and, in certain cases, cell death (Decrock et al., 2009; De Bock et al., 2013; Giaume et al., 2013). Currently, there are no tools available that allow selective targeting of hemichannels since all known pharmacological blockers, including glycyrrhetinic acid-derivatives such as carbenoxolone or related molecules with improved blood-brain barrier permeability (Takeuchi et al., 2011), poorly discriminate between gap junctions and hemichannels. Additionally, they mostly affect Cx channels composed of various distinct Cx types (Harris, 2001; Evans et al., 2006; Spray et al., 2006; Saez and Leybaert, 2014). Beside these derivatives of glycyrrhetinic acid, other compounds such as gadolinium (Gd^{3+}) and lanthanum (La^{3+}) are supposed to affect only hemichannels but, especially in the nervous system where neurons are present, they have side effects that limit the interpretation of their use. Connexins are tetraspan membrane proteins that have two extracellular (EL) loops and one intracellular cytoplasmic loop (CL). Synthetic peptides like Gap26 and Gap27 that mimic a short stretch of amino acids (AAs) on the extracellular loops have been developed more than two decades ago to inhibit gap junctional communication (Warner et al., 1995) (for Gap26 and Gap27 sequences see **Figure 1**). These peptides are thought to interact with the extracellular loops and block hemichannel activity within minutes (Wang et al., 2012; Giaume et al., 2013). They also prevent the docking of two facing hemichannels and thus affect gap junctional communication when applied for periods of several hours (Evans and Boitano, 2001; Decrock et al., 2009). Similarly, antibodies directed against the EL domains of the Cx protein rapidly inhibit hemichannels but they also display delayed inhibition of gap junction channels by preventing the processes of hemichannel docking and *de novo* gap junction channel formation (Orellana et al., 2011; Riquelme et al., 2013). In some cases, distinctive effects on hemichannels and gap junctions depend on the concentration at which they are applied: peptide5, which contains a sequence that comprises part of the Gap27 domain (SRPTEKT), inhibits hemichannels at low (5 μM) concentration while combined gap junction/hemichannel block is only observed at high (500 μM) concentration (O'carroll et al., 2008).

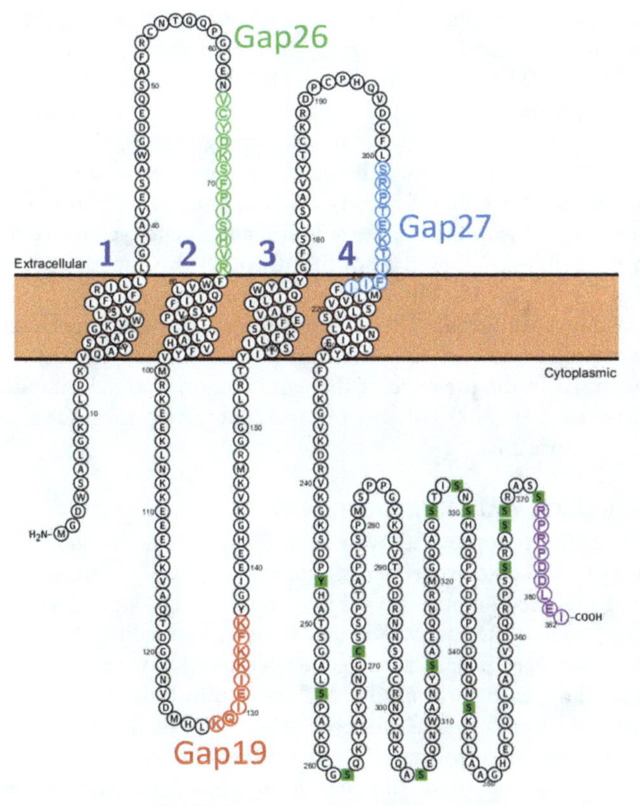

FIGURE 1 | Position of the Gap19 sequence in the intracellular cytoplasmic loop domain of human Cx43. One identified interaction site is located in the last 9 AAs of the CT-tail marked in purple (Wang et al., 2013a). The sequences of Gap19 (red) on the intracellular loop, Gap26 (green) and Gap27 (blue) on the extracellular loops are indicated on the drawing. The CT residues marked as green squares are sites of posttranslational modifications and have been added for illustration purposes [Illustration generated with the Protter tool (Omasits et al., 2014)].

Here, we describe the effect on astroglial hemichannels of a peptide, named Gap19, that is identical to a short sequence present on the intracellular CL domain of Cx43 (**Figure 1**). Peptides mimicking CL sequences have been frequently used as control peptides for gap junction work, since it was shown that these do not inhibit gap junctional coupling (Evans and Leybaert, 2007). Although this specificity has already been reported for the cardiac system (Wang et al., 2013b) we think that it is important to validate this property for the nervous system. Indeed, this is particular relevant for the astrocytes because Cx43 hemichannels contribute to "gliotransmission" and can thus be involved in neuroglial interactions that participate to the control of neuronal activity and survival (see Giaume et al., 2013). In line with this, we found that Gap19 did not inhibit gap junction coupling in astrocytes as measured with dye transfer assays. At the same time, however, Gap19 strongly inhibited Cx43 hemichannels as exemplified by ATP release and dye uptake assays. Finally, we provide evidence that the TAT version of Gap19 is able to cross the intact blood-brain barrier indicating that this peptide can be used to block astroglial Cx43 hemichannel activity when applied through a vascular route.

MATERIALS AND METHODS

ANIMALS

Cultures and slices were obtained from OF1, C57Bl/6 and GFAP-eGFP mice. All experiments were performed according to the European Community Council Directives of November 24, 1986 (86/609/EEC) and all efforts were made to minimize the number of animals used and their suffering. For the *in vivo* experiments with TAT-Gap19 all breeding and animal procedures were approved by The University of British Columbia Animal Care Committee or Ghent University Animal Experiment Ethical Committee and performed in accordance with the guidelines established by the Canadian Council on Animal Care or European Ethics Committee guidelines.

ASTROCYTE CULTURES

Primary astrocyte cultures were prepared from the cortex of newborn (1–2 days) OF1 mice as described previously (Meme et al., 2006). Briefly, cells were seeded into 100-mm diameter plastic dishes (Nunc, Roskilde, Denmark) at the density of 3 × 10^6 cells/dish in DMEM (Sigma-Aldrich, St-Louis MO, USA), supplemented with penicillin (5 U/ml), streptomycin (5 μg/ml) (Invitrogen, Carlsbad, CA, USA) and 10% FCS (Hyclone, Logan, UT, USA). The medium was changed twice a week. When cells reached confluence, around 10 days *in vitro* (DIV), they were harvested with trypsin-EDTA (Invitrogen). Then, cells were replated (2×10^5 cells per well), as secondary cultures, on glass coverslips (Gassalem, Limeil-Brévannes, France) placed inside 16-mm diameter 4 well plastic plates for hemichannel assay and on 35 mm diameter petri dishes for scrape loading dye transfer technique. Finally, they were grown to confluence (about 1 week) and the medium was changed twice a week until the experiments were carried out.

ACUTE HIPPOCAMPAL SLICES

GFAP-eGFP mice (Nolte et al., 2001) were decapitated and their brains were rapidly removed. Hippocampi were dissected and placed in ice-cold artificial cerebrospinal fluid (ACSF) equilibrated with 95% O_2–5% CO_2. Transverse hippocampal slices (300–400 μm thick) were cut on a vibroslicer (Leica VT 1000S, Wetzlar, Germany) and transferred to a holding chamber where they rested on a nylon mesh, submerged in oxygenated ACSF at room temperature for a stabilization period of 45 min before recodings. The ACSF solution contained in mM: NaCl 134; KCl 2.8; $NaHCO_3$ 29; NaH_2PO_4 1.1; glucose 12; $MgSO_4$ 1.5; $CaCl_2$ 2.5.

DYE UPTAKE EXPERIMENTS

Hemichannel activity in cultured astrocytes was induced either by treatment with two pro-inflammatory cytokines, TNF-α and IL-1β (Retamal et al., 2007) or by exposing the cells to a Ca^{2+}-free solution (Stout et al., 2002). The hemichannel-permeable fluorescent tracer ethidium bromide (Etd^+, 314 Da) was applied for 10 min at 4 μM final concentration and at 37°C. Then, cells were washed with Hank's balanced salt solution (HBSS) in mM: NaCl: 137; KCl: 5.4; Na_2HPO_4: 0.34; KH_2PO_4: 0.44, at pH 7.4 and supplemented with 1.2 mM $CaCl_2$ (HBSS-Ca^{2+}). Finally, astrocytes were mounted in Fluoromount and examined by epifluorescence (518 nm excitation and 605 nm emission) using an inverted microscope (Diaphot-Nikon, Nikon France S.A, Champigny sur Marne) equipped with a CCD camera (Nikon) associated with image analyzer software (Lucia-Nikon). Captured images of Etd^+ uptake were analyzed with the Image J program (NIH software).

In slices derived from GFAP-eGFP mice, astrocytes were identified by GFP fluorescence and hemichannel activity was investigated. In this mouse, not all astrocytes are GFP-positive (Nolte et al., 2001; Houades et al., 2006) and Etd^+ uptake was also observed in cells that were not positive for GFP. These cells could be GFP-negative astrocytes, neurons or microglial cells, in which hemichannel activity has been reported (see for instance Orellana et al., 2011). In order to be sure that we were quantifying Etd^+ uptake in astrocytes, in this study hemichannel activity was only considered in GFP-positive cells. As previously described (Giaume et al., 2012), living slices were incubated with Etd^+ for 10 min at room temperature and at 4 μM final concentration. Slices were also treated for 3 h with lipopolysaccharide (LPS, 1–100 ng/ml), a procedure know to activate microglial cells and induce Cx43 hemichannel activity in astrocytes (Retamal et al., 2007). Then, following the 10 min incubation with Etd^+, slices were rinsed 15 min in ACSF to stop dye uptake and reduce background labeling before submerging them for 2 h in fixing solution (4% paraformaldehyde in 0.12 M buffer phosphate). Fixed slices were then rinsed in PBS and mounted in Fluoromount-G medium until photomicrographs were taken. Labeled cells were visualized with a 40x objective in a microscope equipped with epifluorescence illumination and appropriate filters for Etd^+ (excitation wavelength, 528 nm; emission wavelength, 598 nm) and GFAP-eGFP (excitation wavelength, 488 nm; emission wavelength, 507 nm). Alternatively, immunolabeled astrocytes were examined at 63x and 20x with a confocal laser-scanning microscope (Leica TBCS SP2). Stacks of consecutive confocal images taken at 1 μm intervals were acquired sequentially with two lasers (argon 488 nm for GFAP-eGFP and 561 nm for Etd^+). Fluorescence was quantified in arbitrary units (AU), Image J program (NIH software). Dye uptake intensity was evaluated as the difference ($F—F_0$) between the fluorescence (F) from astrocytes (20—30 cells per slice) and the background fluorescence (F_0) measured in the same field where no labeled cells were detected. At least three fields were selected in every slice for background evaluation.

DETERMINATION OF GAP JUNCTIONAL COMMUNICATION

Experiments were performed by using the scrape loading dye transfer technique, as previously described (Meme et al., 2006). Cells were incubated at room temperature for 10 min in HEPES buffered salt solution containing (in mM): NaCl, 140; KCl, 5.5; $CaCl_2$, 1.8; $MgCl_2$, 1; glucose, 10; HEPES, 10 at pH 7.35. Cells were then washed in calcium-free HEPES solution for 1 min and the scrape loading and dye transfer assay (see Giaume et al., 2012) was carried out in the same calcium-free solution containing Lucifer yellow CH (427 Da, 1 mg/ml). After 1 min, cells were washed with the HEPES solution and Lucifer yellow loaded in the cells was allowed to diffuse through gap junction channels for 8 min. Photomicrographs were taken and data were quantified using NIS Nikon software. In all experiments, the fluorescence area of the first row of cells initially loaded, as measured in the presence of the gap junction channel inhibitor carbenoxolone (100 μM, 10 min), was subtracted from the total fluorescence area.

TAT-Gap19 *IN VIVO* ADMINISTRATION

Two different approaches were used for *in vivo* delivery of TAT-Gap19. For acute experiments, 4 months old C57Bl6 male mice were subjected to intra-carotid injection of 45 mg/kg TAT-Gap19 (GenScript, Pistcataway, NJ, US) dissolved in saline. After 1 h the mice were deeply anesthetized with sodium pentobarbital (120 mg/kg intraperitoneally) and were transcardially perfused with cold ice phosphate-buffered saline (PBS) followed by perfusion with 10% formalin (Sigma-Aldrich, Oakville, Canada). Brains were removed and stored in 10% formalin and the next day were cryoprotected in 30% sucrose in PBS solution. Cortical brain sections, 10 μm thick, were collected and mounted sequentially on glass microscope slides. Immunohistochemistry was performed on the sections as previously described (Ozog et al., 2002; Nakase et al., 2004; Kozoriz et al., 2010). Ten μm thick glass mounted sections were rehydrated in PBS and then blocked in PBS pH 7.4 containing 0.3% Triton-X-100 (TX100, Fisher Scientific) and 2% bovine serum albumin (BSA;

Invitrogen,Canada) for 1 h, and incubated overnight in PBS containing 1.0% BSA, 0.3% TX100 and anti-TAT primary antibody (1/100 dilution; Cat # CB0888, Cell Applications, Inc., USA). The following day, the slices were washed with PBS (3 × 10 min) and subsequently incubated for 1 h with secondary antibody (anti-mouse IgG tagged with Alexa 488; Molecular Probes, USA) in PBS containing 1% BSA, 0.3% TX100. Slices were then rinsed 3 times for 10 min with PBS and mounted with ProLong Gold antifade reagent with DAPI. Images were obtained using the same exposure times on a Ziess Axiophot Epifluorescent microscope, Zeiss Canada, Toronto).

For experiments with i.v. administration, 55 mg/kg TAT-Gap19 was injected *via* the tail vein to obtain an estimated 250 μM concentration in the blood (assuming a blood volume of 8% of the body weight). Twenty-four hours after injection of the peptide, mice were deeply anesthetized by i.p. injection of ketamine (240 mg/kg) and xylazine (12 mg/kg) and were transcardially perfused with PBS. Brains were removed, snap-frozen in liquid nitrogen-cooled isopentane and mounted in cryoprotectant (Klinipath) before storage at −80°C. For staining, brains were sliced with a cryostat into 25 μm thick coronal sections, mounted on superfrost plus microscope slides (Thermo Scientific) and fixed in 4% paraformaldehyde (25 min, RT). Slices were subsequently treated with PBS containing 0.2% TX100 (2 h, RT), blocked with PBS containing 0.05% TX100 and 10% normal goat serum (NGS, 2 h, RT), and incubated overnight (4°C) with anti-TAT antibody (1/50 dilution, Cell Signaling Technology, #2547S) in PBS containing 5% NGS and 0.05% TX100. The following day, slices were washed with PBS and incubated for 2 h at room temperature with secondary antibody (1/200 anti-mouse IgG linked to Alexa 488, Invitrogen Life Sciences) in PBS containing 5% NGS and 0.05% TX100. Slides were then rinsed with PBS and nuclei were counterstained with DAPI (1 μg/mL). Finally, slides were mounted in Vectashield mounting medium (Vector Labs). Images were acquired using a BD Pathway BioImaging System (Becton Dickinson) that includes stitching software, obtaining a montage image of the entire brain section. Intensity of Alexa 488 fluorescence in 10 analysis zones in the left and right cortex respectively was analyzed using ImageJ software.

TAT-Gap19 PEPTIDE

TAT-Gap19 (YGRKKRRQRRR-KQIEIKKFK) was synthesized by Pepnome Inc. (Hongkong, China) at a purity of 95%. It was dissolved in PBS and aliquoted/stored as a stock solution of 10 mM.

DATA ANALYSIS AND STATISTICS

The data are expressed as mean ± s.e.m., with "n" denoting the number of independent experiments. Two groups were compared by student's t-test and two-tail p-value. In experiments where the effects of different treatments were assessed on normalized data, non-parametric ANOVA Kruskal-Wallis tests were performed. Tests with significance of $p < 0.05$ were followed by a Dunn's multiple comparison *post-hoc* test using the GraphPad Prism version 5.00 (San Diego, CA, USA). Unless stated otherwise, significance as compared to control condition based on the raw data (before normalization) was assessed by two-tailed Wilcoxon signed-ranked tests. The level of significance was set at $p < 0.05$. Graphics were prepared using Microcal Origin 6.0 (Northampton, MA, USA) and Adobe Illustrator 10 (San Jose, CA, USA).

RESULTS

Recently, it was reported that Gap19 (KQIEIKKFK; see **Figure 1**) inhibits Cx43 hemichannel activity but not gap junctional communication in the heart (Wang et al., 2013b). To determine whether these findings also apply for brain cells, especially astrocytes which predominantly express Cx43 (Ransom and Giaume, 2013), we tested the effect of this peptide on hemichannel activity (Etd^+ uptake and ATP release assays) and gap junctional communication (scrape loading and dye transfer) in two *in vitro* preparations: primary cultures of astrocytes and acute hippocampal slices.

While *in vivo* astrocytes express two major Cxs, Cx43 and Cx30, astrocytes in primary culture offer the advantage that only one of them is expressed, namely Cx43 (Giaume et al., 1991a). Indeed, Cx30 is detected in astrocytes only either after 10–11 weeks of solo-culture (Kunzelmann et al., 1999) or in 3 week-old astrocytes after 1 week of co-culture with neurons (Koulakoff et al., 2008). Firstly, we tested whether Gap19 influences hemichannel activity in primary cultures of astrocytes stimulated by glutamate (100 μM, 15 min), a treatment that has been reported to trigger hemichannel-mediated ATP release in primary astrocytes (De Vuyst et al., 2009). As illustrated in **Figure 2**, we found that Gap19 (30 min treatment) inhibited glutamate-triggered ATP release (**Figure 2A**). Alternatively, Cx43 hemichannel activity was also monitored by an Etd^+ uptake assay in astrocytes treated with a combination of two pro-inflammatory cytokines, TNF-α and IL-1β (10 ng/ml for each, 3 h), a procedure that activates Cx43 hemichannels in cultured astrocytes (Retamal et al., 2007). Under those conditions, we observed that Etd^+ uptake was inhibited in the presence of Gap19 in a dose-dependent manner, with the peptide applied prior to (30 min) and during Etd^+ uptake (**Figures 2B,C**). Finally, since in confluent cultures of astrocytes intercellular communication through Cx43 gap junction channels is high (Giaume et al., 1991b), we tested whether Gap19 (344 μM and 688 μM 30 min) had any effect on the level of gap junctional coupling. Experiments performed with the scrape loading and dye transfer technique demonstrated that Gap19 was without any effect on gap junctional communication (**Figures 1D$_1$–D$_3$** and summary bar chart) in astrocytes which in culture express only Cx43 (see Koulakoff et al., 2008).

The inhibitory effect of Gap19 *in situ* was then tested by performing the Etd^+ uptake assay in acute hippocampal brain slices from GFAP-eGFP transgenic mice. While in normal saline solution Etd^+ uptake was very weak, as reported previously (Orellana et al., 2011), treatment of the slices with a Ca^{2+}-free solution (no added Ca^{2+} and 5 mM of the Ca^{2+}-buffer EGTA), a condition known to activate Cx43 hemichannels in astrocytes (Ye et al., 2003) and that can be inhibited by carbenoxolone in hippocampal slices (Rouach et al., 2008), induced Etd^+ uptake in GFP-positive cells. This increase in Etd^+ uptake was not inhibited by Gap19 (applied 30 min prior to and during Etd^+ uptake) used at 172 μM

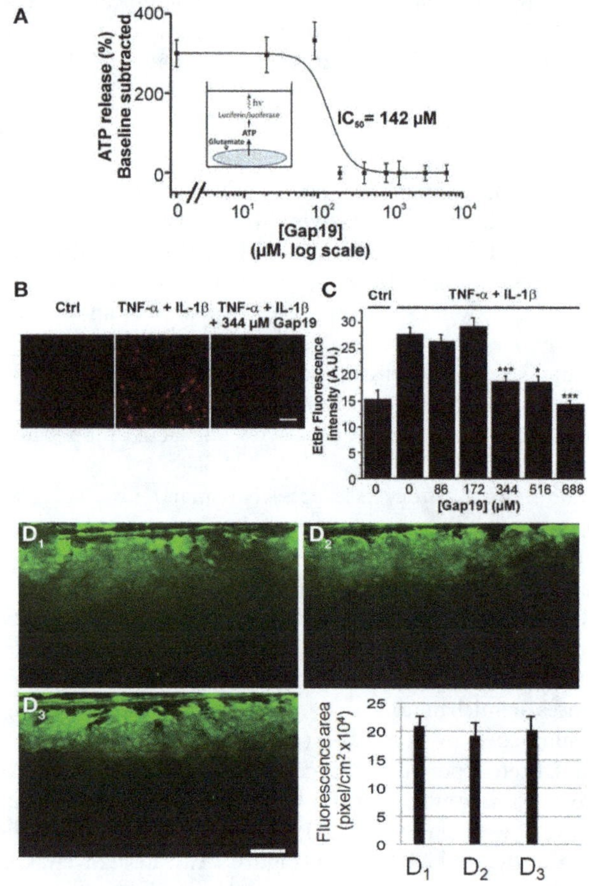

FIGURE 2 | Dose-dependent inhibition of ATP release and Etd$^+$ uptake through Cx43 hemichannels by Gap19 with lack of effect on gap junctional communication in cultured astrocytes. (A) Concentration-dependent inhibition by Gap19 (30 min pre-incubation) of ATP release in cultured cortical astrocytes triggered by glutamate (100 μM, 15 min application) ($n = 6$ independent experiments). **(B)** Representative images showing Etd$^+$ uptake (red) in cultured astrocytes under control conditions (Ctrl) and after TNF-α/IL-1β or TNF-α/IL-1β+ Gap19 treatment. Scale bar: 20 μm. ($^*p < 0.05$; $^{***}p < 0.001$). **(C)** Summary data of Etd$^+$ uptake studies in astrocytes, demonstrating inhibition by Gap19 ($n =$ 5–8 independent experiments). Statistical comparisons refer to the stimulus condition without Gap19 (zero Gap19 concentration). **(D_1–D_3)** Representative images of scrape-loading dye transfer experiment in confluent cultures of astrocytes. Compared to control condition **(D_1)**, with Gap19 344 μM (**D_2**, 30 min pre-incubation) or 688 μM (**D_3**, 30 min pre-incubation). Lower graph: Quantification of scrape-loading data indicating that Gap19 did not influence gap junctional coupling as measured in confluent cultures of astrocytes (from left to right, bars are from control and the two tested concentrations of Gap19, respectively). ($n =$ 3–5 independent experiments).

while concentrations of 344 and 688 μM inhibited the uptake of Etd$^+$ by GFP-positive cells (**Figures 3A,B**).

The possibility of *in vivo* application of hemichannel blockers was explored through peripheral administration of Gap19 linked to the TAT membrane translocation motif at its N-terminal end (TAT-Gap19) in adult mice. One hour after carotid injection (45 mg/kg), animals were sacrificed and the brain tissue prepared for immunohistochemistry to localize TAT-Gap19 making use of an antibody directed against the TAT sequence. **Figure 4A** shows that following this treatment, TAT immunoreactivity was readily detected throughout the brain as compared to the brain tissue of mice that received the vehicle (PBS) only. Gap19 has an intracellular target located on the C-terminal tail of Cx43 to which it binds with a K_d of ~2.5 μM (Wang et al., 2013b), making it possible that the peptide can be trapped and retained intracellulary. We therefore tested whether administration of a single dose of TAT-Gap19 i.v. (55 mg/kg, giving an estimated ~250 μM concentration assuming distribution in the blood volume) resulted in detectable peptide signal in the brain 24 h after administration, again based on TAT immunostaining. **Figure 4B** illustrates the average data of such experiments, demonstrating that the TAT-Gap19 signal was significantly above the background level, indicating that TAT-Gap19 was retained in the brain tissue.

DISCUSSION

Gap19 is a peptide that corresponds to a sequence on the cytoplasmic loop (CL) of Cx43. Its sequence is part of the L2 domain that is involved in interactions of the CL with the CT-tail of Cx43. Interestingly, CT-CL interaction has a differential effect on hemichannels and gap junction channels: it closes gap junction channels while it is necessary for hemichannel opening (Ponsaerts et al., 2010; D'hondt et al., 2013; Iyyathurai et al., 2013; Wang et al., 2013a,b). This differential effect is most probably the consequence of distinct channel configurations of hemichannels as compared to gap junction channels (Wang et al., 2013b). Gap19 peptide binds to the CT and thereby prevents interaction of the CT with the CL, making hemichannels unavailable for opening (Wang et al., 2013a). Gap19 has most extensively been characterized in C6 cells overexpressing Cx43 and in cardiomyocytes (Wang et al., 2013a). We here demonstrate that Gap19 also inhibits hemichannel activity in astrocytes while not affecting gap junctional coupling. The concentrations necessary to achieve hemichannel block in astrocytes appeared to be somewhat higher than in Cx43-expressing C6 cells: in the latter, half-maximal inhibition of hemichannel opening triggered by exposure to the Ca^{2+}-ionophore A23187 occurred at ~47 μM (Wang et al., 2013b) while the half-maximal effect concentration was ~142 μM for glutamate-triggered ATP release in astrocyte cultures (**Figure 2A**) and in the 250 μM range when hemichannels were opened with TNF-α/IL-1β in astrocyte cultures or with zero extracellular Ca^{2+} in brain slices (**Figure 3B**). These differences may relate to differences in the triggers used to induce hemichannel opening (increase of intracellular Ca^{2+} vs. decrease of extracellular Ca^{2+} or cytokine exposure) or differences in the hemichannel assays used (patch clamp, ATP release, dye uptake). Gap19 needs to enter the cell in order to reach its target located on the CT-tail of Cx43; it is endowed with some intrinsic plasma membrane-permeability that is related to the fact that 4 AAs of the nonapeptide are positively charged Lys residues. However, linking Gap19 to the HIV-derived TAT internalization sequence further promotes its membrane permeability and reduces the concentration for half-maximal inhibition of Cx43 hemichannels in C6 cells from ~47 μM to ~7 μM (Wang et al., 2013b), Supplemental Data. This 7 μM concentration is in good agreement with the

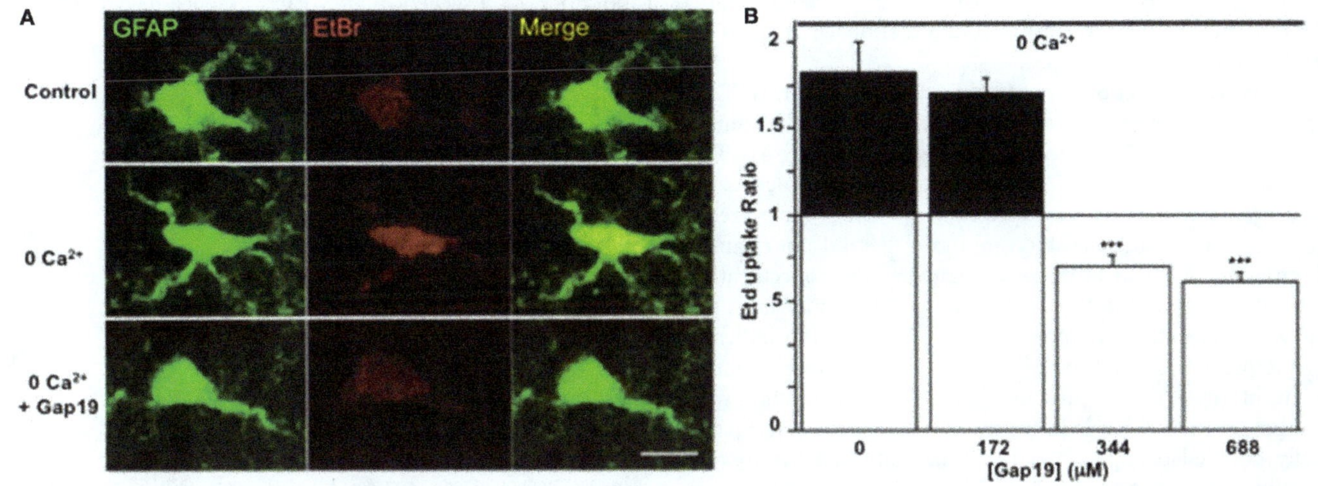

FIGURE 3 | Gap19 inhibits hemichannel activity in astrocytes studied in acute hippocampal slices. (A) Representative images of Etd$^+$ uptake (red) in astrocytes (green) in hippocampal slices from GFAP-eGFP transgenic mice, under control conditions (upper row) and after 10 min exposure to a Ca^{2+}-free solution without (middle row) or with (lower row) 344 μM Gap19 treatment. Scale bar: 10 μm. **(B)** Summary graph demonstrating significant Etd$^+$ uptake in astrocytes that was inhibited by Gap19 used at concentration of 344 and 688 μM. Statistical comparisons in **(B)** were done with the stimulus condition without Gap19 ($n = 3$ independent experiments; $^{***}p < 0.001$).

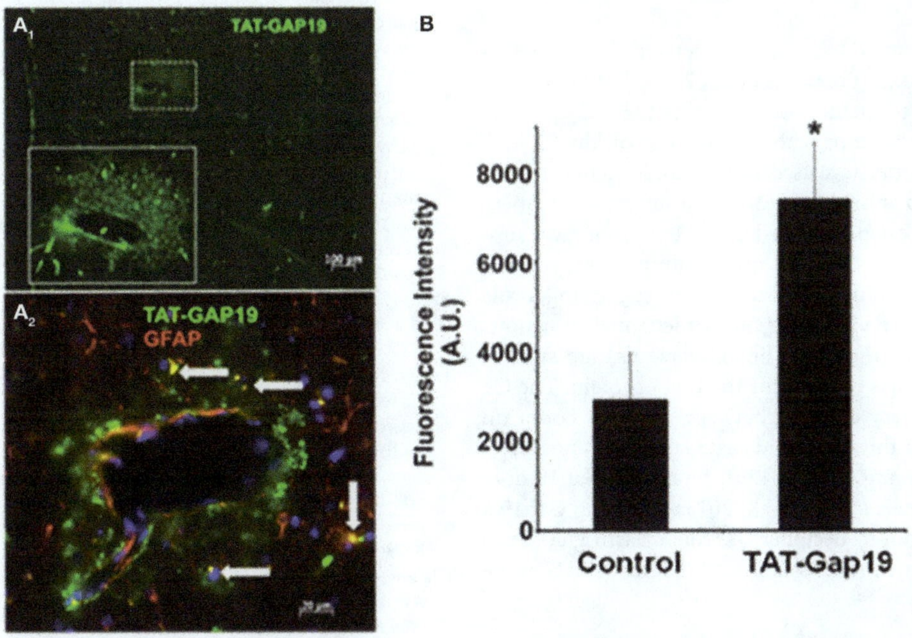

FIGURE 4 | Detection of TAT-Gap19 in the cortex of the mouse. (A_1–A_2) One hour after carotid injection of TAT-Gap19 the brain displayed clear TAT immunoreactivity compared to mice that received vehicle (PBS) only. **(A_1)**, Area in small box is enlarged in the lower right box of the panel; **(A_2)**, Double labeling with anti-TAT and anti-GFAP antibodies indicates that some GFAP-positive astrocytes (white arrows) have taken up the TAT peptide. Note that the green fluorescence is concentrated in the white box area probably because this represents a large vessel that would have more passage of TAT-Gap19 into the brain parenchyma surrounding the vessel. **(B)** A single i.v. injection of TAT-Gap19 gave significant immune signal in the brain 24 h later. Taken together, these experiments indicate that TAT-Gap19 traverses the blood-brain barrier and is retained in the cells. Fluorescence intensities of Alexa 488 signal of the secondary antibody determined in slices immunostained with a primary antibody directed against the TAT sequence. Control represents experiments in mice injected with PBS vehicle ($n = 5$ for control and TAT-Gap19; $^{*}p < 0.05$).

~6.5 μM half-maximal inhibition of unitary Cx43 hemichannel currents when Gap19 is dialyzed in the cell via the patch pipette. Finally, the effects of Gap19 on hemichannel activity reported here are similar to those previously described for Gap26 and Gap27 that also blocks hemichannel activity in astrocytes induced by LPS or pro-inflammatory treatments in culture astrocytes (see Retamal et al., 2007; Froger et al., 2010) and in acute brain slices treated with the amyloid peptide (Orellana et al., 2011).

We have also published that carbenoxolone blocks hemichannel activity induced by a calcium free solution (Rouach et al., 2008, Supplementary Data Figure S5).

The *in vivo* injection experiments demonstrate that TAT-Gap19 can be detected in the brain tissue, based on immunostaining for the TAT moiety of the peptide (**Figure 4**). This observation indicates that TAT-Gap19 can successfully cross the blood-brain barrier, which is expected as a result of the presence of the TAT membrane translocation motif. We did not quantify the immune signal in terms of estimates of the concentration attained. However, the fact that there is still significant TAT-Gap19 immune signal 24 h after a single i.v. injection indicates that the peptide is retained for several hours in the cells, which is plausible given the fact that the K_d for Gap19 interaction with its target on the CT-tail is in the micromolar range ($\sim$2.5 μM). Finally, double labeling with anti-TAT and anti-GFAP antibodies indicates that the peptide reaches GFAP-positive astrocytes, however, several TAT-positive cells are apparently not GFAP-positive. This is not surprising for experiments performed in the cortex of adult mice where it is well known that only a few astrocytes are GFAP-positive at this age (see for instance Nolte et al., 2001; Houades et al., 2006). Consequently, we cannot exclude that the TAT-peptide staining could also come from endothelial cells or microglia that have both been reported to express Cx43 (Orellana et al., 2011).

The L2 peptide, from which Gap19 is derived, is also a specific hemichannel blocker (Ponsaerts et al., 2010) but Gap19 has the advantage that it is smaller and contains the most crucial motif to engage in interactions with the target(s) on the CT-tail. Stehberg and collaborators have used TAT-L2 for local injection in the amygdala in *in vivo* animal studies on fear memory (Stehberg et al., 2012). Their work demonstrated that fear memory was suppressed by TAT-L2 and was rescued upon addition of a cocktail of several proposed gliotransmitter substances, suggesting a role of gliotransmitter release via astrocytic hemichannels. Of note, both L2 and Gap19 (or their TAT-linked versions) are specific not only for hemichannels but also for the Cx43 protein. The CL and CT domains are very different between different connexin species and in line with this, we found that Gap19 does not affect Cx40 hemichannels (Wang et al., 2013a). Moreover, Gap19 does not block Panx1 channels (Wang et al., 2013a), making Gap19 a potentially powerful tool to decipher the role of astroglial Cx43 hemichannels in brain functions and pathologies (see Giaume et al., 2013).

ACKNOWLEDGMENTS

This work was supported by the Fund for Scientific Research Flanders (FWO-Vlaanderen), Belgium (Grant N° G.0134.09, G.0298.11, G.0571.12 and G.0A54.13, to Luc Leybaert) and the InterUniversity Attraction Poles Program (Belgian Scoience Policy, Project P7/10, to Luc Leybaert); the Heart and Stroke Foundation of Canada (to Christian C. Naus and Moises Freitas-Andrade); the Collège de France (Paris, France) (ATER position to Verónica Abudara) and Comisión Sectorial de Investigación Científica de la Universidad de la República Oriental del Uruguay (to Verónica Abudara); the ANR funding grant "AstroSleep" N°ANR-12-BSV4-0013-01 and MEMOLIFE Laboratory of Excellence and Paris Science Lettre Research University, Paris, France (Christian Giaume).

REFERENCES

Chever, O., Lee, C. Y., and Rouach, N. (2014). Astroglial connexin43 hemichannels tune Basal excitatory synaptic transmission. *J. Neurosci.* 34, 11228–11232. doi: 10.1523/JNEUROSCI.0015-14.2014

Cina, C., Maass, K., Theis, M., Willecke, K., Bechberger, J. F., and Naus, C. C. (2009). Involvement of the cytoplasmic C-terminal domain of connexin43 in neuronal migration. *J. Neurosci.* 29, 2009–2021. doi: 10.1523/JNEUROSCI.5025-08.2009

D'hondt, C., Iyyathurai, J., Wang, N., Gourdie, R. G., Himpens, B., Leybaert, L., et al. (2013). Negatively charged residues (Asp378 and Asp379) in the last ten amino acids of the C-terminal tail of Cx43 hemichannels are essential for loop/tail interactions. *Biochem. Biophys. Res. Commun.* 432, 707–712. doi: 10.1016/j.bbrc.2013.01.066

De Bock, M., Wang, N., Decrock, E., Bol, M., Gadicherla, A. K., Culot, M., et al. (2013). Endothelial calcium dynamics, connexin channels and blood-brain barrier function. *Prog. Neurobiol.* 108, 1–20. doi: 10.1016/j.pneurobio.2013.06.001

De Vuyst, E., Wang, N., Decrock, E., De Bock, M., Vinken, M., Van Moorhem, M., et al. (2009). Ca(2+) regulation of connexin 43 hemichannels in C6 glioma and glial cells. *Cell Calcium* 46, 176–187. doi: 10.1016/j.ceca.2009.07.002

Decrock, E., De Vuyst, E., Vinken, M., Van Moorhem, M., Vranckx, K., Wang, N., et al. (2009). Connexin 43 hemichannels contribute to the propagation of apoptotic cell death in a rat C6 glioma cell model. *Cell Death Differ.* 16, 151–163. doi: 10.1038/cdd.2008.138

Elias, L. A., Wang, D. D., and Kriegstein, A. R. (2007). Gap junction adhesion is necessary for radial migration in the neocortex. *Nature* 448, 901–907. doi: 10.1038/nature06063

Evans, W. H., and Boitano, S. (2001). Connexin mimetic peptides: specific inhibitors of gap-junctional intercellular communication. *Biochem. Soc. Trans.* 29, 606–612. doi: 10.1042/BST0290606

Evans, W. H., De Vuyst, E., and Leybaert, L. (2006). The gap junction cellular internet: connexin hemichannels enter the signalling limelight. *Biochem. J.* 397, 1–14. doi: 10.1042/BJ20060175

Evans, W. H., and Leybaert, L. (2007). Mimetic peptides as blockers of connexin channel-facilitated intercellular communication. *Cell Commun. Adhes.* 14, 265–273. doi: 10.1080/15419060801891034

Ezan, P., Andre, P., Cisternino, S., Saubamea, B., Boulay, A. C., Doutremer, S., et al. (2012). Deletion of astroglial connexins weakens the blood-brain barrier. *J. Cereb. Blood Flow Metab.* 32, 1457–1467. doi: 10.1038/jcbfm.2012.45

Froger, N., Orellana, J. A., Calvo, C. F., Amigou, E., Kozoriz, M. G., Naus, C. C., et al. (2010). Inhibition of cytokine-induced connexin43 hemichannel activity in astrocytes is neuroprotective. *Mol. Cell. Neurosci.* 45, 37–46. doi: 10.1016/j.mcn.2010.05.007

Giaume, C., Fromaget, C., El Aoumari, A., Cordier, J., Glowinski, J., and Gros, D. (1991a). Gap junctions in cultured astrocytes: single-channel currents and characterization of channel-forming protein. *Neuron* 6, 133–143. doi: 10.1016/0896-6273(91)90128-M

Giaume, C., Koulakoff, A., Roux, L., Holcman, D., and Rouach, N. (2010). Astroglial networks: a step further in neuroglial and gliovascular interactions. *Nat. Rev. Neurosci.* 11, 87–99. doi: 10.1038/nrn2757

Giaume, C., Leybaert, L., Naus, C. C., and Saez, J. C. (2013). Connexin and pannexin hemichannels in brain glial cells: properties, pharmacology, and roles. *Front. Pharmacol.* 4:88. doi: 10.3389/fphar.2013.00088

Giaume, C., Marin, P., Cordier, J., Glowinski, J., and Premont, J. (1991b). Adrenergic regulation of intercellular communications between cultured striatal astrocytes from the mouse. *Proc. Natl. Acad. Sci. U.S.A.* 88, 5577–5581. doi: 10.1073/pnas.88.13.5577

Giaume, C., Orellana, J. A., Abudara, V., and Saez, J. C. (2012). Connexin-based channels in astrocytes: how to study their properties. *Methods Mol. Biol.* 814, 283–303. doi: 10.1007/978-1-61779-452-0_19

Giaume, C., and Theis, M. (2010). Pharmacological and genetic approaches to study connexin-mediated channels in glial cells of the central nervous system. *Brain Res. Rev.* 63, 160–176. doi: 10.1016/j.brainresrev.2009.11.005

Harris, A. L. (2001). Emerging issues of connexin channels: biophysics fills the gap. *Q. Rev. Biophys.* 34, 325–472. doi: 10.1017/S0033583501003705

Houades, V., Rouach, N., Ezan, P., Kirchhoff, F., Koulakoff, A., and Giaume, C. (2006). Shapes of astrocyte networks in the juvenile brain. *Neuron Glia Biol.* 2, 3–14. doi: 10.1017/S1740925X06000081

Iyyathurai, J., D'hondt, C., Wang, N., De Bock, M., Himpens, B., Retamal, M. A., et al. (2013). Peptides and peptide-derived molecules targeting the intracellular domains of Cx43: gap junctions versus hemichannels. *Neuropharmacology* 75, 491–505. doi: 10.1016/j.neuropharm.2013.04.050

Kozoriz, M. G., Bechberger, J. F., Bechberger, G. R., Suen, M. W., Moreno, A. P., Maass, K., et al. (2010). The connexin43 C-terminal region mediates neuroprotection during stroke. *J. Neuropathol. Exp. Neurol.* 69, 196–206. doi: 10.1097/NEN.0b013e3181cd44df

Koulakoff, A., Ezan, P., and Giaume, C. (2008). Neurons control the expression of connexin 30 and connexin 43 in mouse cortical astrocytes. *Glia* 56, 1299–1311. doi: 10.1002/glia.20698

Kunzelmann, P., Schroder, W., Traub, O., Steinhauser, C., Dermietzel, R., and Willecke, K. (1999). Late onset and increasing expression of the gap junction protein connexin30 in adult murine brain and long-term cultured astrocytes. *Glia* 25, 111–119.

Lutz, S. E., Zhao, Y., Gulinello, M., Lee, S. C., Raine, C. S., and Brosnan, C. F. (2009). Deletion of astrocyte connexins 43 and 30 leads to a dysmyelinating phenotype and hippocampal CA1 vacuolation. *J. Neurosci.* 29, 7743–7752. doi: 10.1523/JNEUROSCI.0341-09.2009

Meme, W., Calvo, C. F., Froger, N., Ezan, P., Amigou, E., Koulakoff, A., et al. (2006). Proinflammatory cytokines released from microglia inhibit gap junctions in astrocytes: potentiation by beta-amyloid. *FASEB J.* 20, 494–496. doi: 10.1096/fj.05-4297fje

Nakase, T., Söhl, G., Theis, M., Willecke, K., and Naus, C. C. (2004). Increased apoptosis and inflammation after focal brain ischemia in mice lacking connexin43 in astrocytes. *Am. J. Pathol.* 164, 2067–2075. doi: 10.1016/S0002-9440(10)63765-0

Nolte, C., Matyash, M., Pivneva, T., Schipke, C. G., Ohlemeyer, C., Hanisch, U. K., et al. (2001). GFAP promoter-controlled EGFP-expressing transgenic mice: a tool to visualize astrocytes and astrogliosis in living brain tissue. *Glia* 33, 72–86. doi: 10.1002/1098-1136(20010101)33:1<72::AID-GLIA1007>3.0.CO;2-A

O'carroll, S. J., Alkadhi, M., Nicholson, L. F., and Green, C. R. (2008). Connexin 43 mimetic peptides reduce swelling, astrogliosis, and neuronal cell death after spinal cord injury. *Cell Commun. Adhes.* 15, 27–42. doi: 10.1080/15419060802014164

Omasits, U., Ahrens, C. H., Muller, S., and Wollscheid, B. (2014). Protter: interactive protein feature visualization and integration with experimental proteomic data. *Bioinformatics* 30, 884–886. doi: 10.1093/bioinformatics/btt607

Orellana, J. A., Shoji, K. F., Abudara, V., Ezan, P., Amigou, E., Saez, P. J., et al. (2011). Amyloid beta-induced death in neurons involves glial and neuronal hemichannels. *J. Neurosci.* 31, 4962–4977. doi: 10.1523/JNEUROSCI.6417-10.2011

Ozog, M. A., Siushansian, R., and Naus, C. C. (2002). Blocked gap junctional coupling increases glutamate-induced neurotoxicity in neuron-astrocyte cocultures. *J. Neuropathol. Exp. Neurol.* 61, 132–141.

Pannasch, U., Freche, D., Dallerac, G., Ghezali, G., Escartin, C., Ezan, P., et al. (2014). Connexin 30 sets synaptic strength by controlling astroglial synapse invasion. *Nat. Neurosci.* 17, 549–558. doi: 10.1038/nn.3662

Pannasch, U., Vargova, L., Reingruber, J., Ezan, P., Holcman, D., Giaume, C., et al. (2011). Astroglial networks scale synaptic activity and plasticity. *Proc. Natl. Acad. Sci. U.S.A.* 108, 8467–8472. doi: 10.1073/pnas.1016650108

Ponsaerts, R., De Vuyst, E., Retamal, M., D'hondt, C., Vermeire, D., Wang, N., et al. (2010). Intramolecular loop/tail interactions are essential for connexin 43-hemichannel activity. *FASEB J.* 24, 4378–4395. doi: 10.1096/fj.09-153007

Ransom, B., and Giaume, C. (2013). "Gap junctions, hemichannels," in *Neuroglia*, eds H. Kettenmann and B. R. Ransom (New York, NY: Oxford University press), 292–305.

Retamal, M. A., Froger, N., Palacios-Prado, N., Ezan, P., Saez, P. J., Saez, J. C., et al. (2007). Cx43 hemichannels and gap junction channels in astrocytes are regulated oppositely by proinflammatory cytokines released from activated microglia. *J. Neurosci.* 27, 13781–13792. doi: 10.1523/JNEUROSCI.2042-07.2007

Riquelme, M. A., Kar, R., Gu, S., and Jiang, J. X. (2013). Antibodies targeting extracellular domain of connexins for studies of hemichannels. *Neuropharmacology* 75, 594–603. doi: 10.1016/j.neuropharm.2013.03.022

Rouach, N., Koulakoff, A., Abudara, V., Willecke, K., and Giaume, C. (2008). Astroglial metabolic networks sustain hippocampal synaptic transmission. *Science* 322, 1551–1555. doi: 10.1126/science.1164022

Roux, L., Benchenane, K., Rothstein, J. D., Bonvento, G., and Giaume, C. (2011). Plasticity of astroglial networks in olfactory glomeruli. *Proc. Natl. Acad. Sci. U.S.A.* 108, 18442–18446. doi: 10.1073/pnas.1107386108

Saez, J. C., and Leybaert, L. (2014). Hunting for connexin hemichannels. *FEBS Lett.* 588, 1205–1211. doi: 10.1016/j.febslet.2014.03.004

Spray, D. C., Ye, Z. C., and Ransom, B. R. (2006). Functional connexin "hemichannels": a critical appraisal. *Glia* 54, 758–773. doi: 10.1002/glia.20429

Stehberg, J., Moraga-Amaro, R., Salazar, C., Becerra, A., Echeverria, C., Orellana, J. A., et al. (2012). Release of gliotransmitters through astroglial connexin 43 hemichannels is necessary for fear memory consolidation in the basolateral amygdala. *FASEB J.* 26, 3649–3657. doi: 10.1096/fj.11-198416

Stout, C. E., Costantin, J. L., Naus, C. C., and Charles, A. C. (2002). Intercellular calcium signaling in astrocytes via ATP release through connexin hemichannels. *J. Biol. Chem.* 277, 10482–10488. doi: 10.1074/jbc.M109902200

Takeuchi, H., Mizoguchi, H., Doi, Y., Jin, S., Noda, M., Liang, J., et al. (2011). Blockade of gap junction hemichannel suppresses disease progression in mouse models of amyotrophic lateral sclerosis and Alzheimer's disease. *PLoS ONE* 6:e21108. doi: 10.1371/journal.pone.0021108

Wallraff, A., Kohling, R., Heinemann, U., Theis, M., Willecke, K., and Steinhauser, C. (2006). The impact of astrocytic gap junctional coupling on potassium buffering in the hippocampus. *J. Neurosci.* 26, 5438–5447. doi: 10.1523/JNEUROSCI.0037-06.2006

Wang, N., De Bock, M., Antoons, G., Gadicherla, A. K., Bol, M., Decrock, E., et al. (2012). Connexin mimetic peptides inhibit Cx43 hemichannel opening triggered by voltage and intracellular $Ca2^{+}$ elevation. *Basic Res. Cardiol.* 107, 304. doi: 10.1007/s00395-012-0304-2

Wang, N., De Bock, M., Decrock, E., Bol, M., Gadicherla, A., Bultynck, G., et al. (2013a). Connexin targeting peptides as inhibitors of voltage- and intracellular $Ca2^{+}$-triggered Cx43 hemichannel opening. *Neuropharmacology* 75, 506–516. doi: 10.1016/j.neuropharm.2013.08.021

Wang, N., De Vuyst, E., Ponsaerts, R., Boengler, K., Palacios-Prado, N., Wauman, J., et al. (2013b). Selective inhibition of Cx43 hemichannels by Gap19 and its impact on myocardial ischemia/reperfusion injury. *Basic Res. Cardiol.* 108, 309. doi: 10.1007/s00395-012-0309-x

Warner, A., Clements, D. K., Parikh, S., Evans, W. H., and Dehaan, R. L. (1995). Specific motifs in the external loops of connexin proteins can determine gap junction formation between chick heart myocytes. *J. Physiol.* 488 (Pt 3), 721–728.

Ye, Z. C., Wyeth, M. S., Baltan-Tekkok, S., and Ransom, B. R. (2003). Functional hemichannels in astrocytes: a novel mechanism of glutamate release. *J. Neurosci.* 23, 3588–3596.

10

Hemichannel composition and electrical synaptic transmission: Molecular diversity and its implications for electrical rectification

Nicolás Palacios-Prado[1,2], Wolf Huetteroth[2,3] and Alberto E. Pereda[1,2]*

[1] Dominick P. Purpura Department of Neuroscience, Albert Einstein College of Medicine, Bronx, NY, USA
[2] Marine Biological Laboratory, Woods Hole, Massachusetts, MA, USA
[3] Department of Neurobiology, University of Konstanz, Konstanz, Germany

***Correspondence:**
Alberto E. Pereda, Dominick P. Purpura Department of Neuroscience, Albert Einstein College of Medicine, 1300 Morris Park Ave, Bronx, New York, NY 10461, USA
e-mail: alberto.pereda@einstein.yu.edu

Unapposed hemichannels (HCs) formed by hexamers of gap junction proteins are now known to be involved in various cellular processes under both physiological and pathological conditions. On the other hand, less is known regarding how differences in the molecular composition of HCs impact electrical synaptic transmission between neurons when they form intercellular heterotypic gap junctions (GJs). Here we review data indicating that molecular differences between apposed HCs at electrical synapses are generally associated with rectification of electrical transmission. Furthermore, this association has been observed at both innexin and connexin (Cx) based electrical synapses. We discuss the possible molecular mechanisms underlying electrical rectification, as well as the potential contribution of intracellular soluble factors to this phenomenon. We conclude that asymmetries in molecular composition and sensitivity to cellular factors of each contributing hemichannel can profoundly influence the transmission of electrical signals, endowing electrical synapses with more complex functional properties.

Keywords: gap junction, connexin, innexin, electrical synapse, asymmetry, rectification

INTRODUCTION

Channels formed by connexin (Cx) or pannexin proteins (connexon and pannexon, respectively) were shown to impact cellular properties and underlie various pathological processes by serving as conduits for ions and various autocrine and paracrine signaling molecules (Contreras et al., 2002; Bennett et al., 2003; Scemes et al., 2007; Iglesias et al., 2009; **Figure 1A**). Some of these channels can assemble into intercellular structures. That is, docking of two connexons or "hemichannels" (HC) from two adjacent cells form intercellular channels that cluster into structures called "gap junctions" (GJs; Goodenough and Paul, 2009; **Figures 1B,C**), which mediate intercellular communication between neighboring cells in virtually all tissues of deuterostomes (Hervé et al., 2005; Abascal and Zardoya, 2013). Invertebrate GJ proteins, however, are part of an unrelated gene family called innexins (Inxs; Starich et al., 1996; Ganfornina et al., 1999). Three Inx-like genes were subsequently found in the genome of vertebrates, which were named pannexins (Panxs; Panchin et al., 2000; Bruzzone et al., 2003). Interestingly, while Panxs were shown to form intercellular channels when overexpressed in oocytes (Bruzzone et al., 2003), there is little evidence so far supporting that they form GJ channels under physiological conditions (Dahl and Locovei, 2006; Sahu et al., 2014). Since Inxs form GJ channels in invertebrates (Hervé et al., 2005; Phelan, 2005), it is speculated that Panxs might have evolved to function mainly as HCs in vertebrates (Dahl and Locovei, 2006). On the other hand, recent evidence suggests that Inxs can also function as HCs or "innexons" (Dahl and Muller, 2014). Due to the current uncertainty of Panx-based GJ channels, the distinction between Inx and Panx has remained in the literature to distinguish GJ forming proteins in protostomes and cnidarians (Inx) vs. HC forming proteins in deuterostomes (Panxs) (Abascal and Zardoya, 2013).

Currently, we know that the family of Cx proteins in humans is composed by 21 genes (Söhl and Willecke, 2004) whereas Inxs represent 25 genes in *C. elegans* or eight genes in *D. melanogaster* (Adams et al., 2000; Altun et al., 2009). Gap junction channels formed by different Cxs and Inxs were shown to exhibit differential permeability and distinct electrophysiological properties providing diversity to gap junctional communication (Goodenough and Paul, 2009; Samuels et al., 2010). Notably, GJ channels can be either formed by the docking of identical (homotypic configuration; **Figure 1B**) or different (heterotypic configuration) HCs (**Figure 1C**), further enhancing the functional diversity of GJs. That is, at heterotypic channels, the molecular and functional singularities of each of the apposed/contributing HC (aHC) influence the properties of the intercellular channel and, furthermore, can potentially endow heterotypic channels with properties which could not be predicted from those displayed in homotypic configuration (Verselis et al., 1994; Oh et al., 1999). In other words, asymmetries in molecular composition of each aHC could profoundly influence intercellular communication,

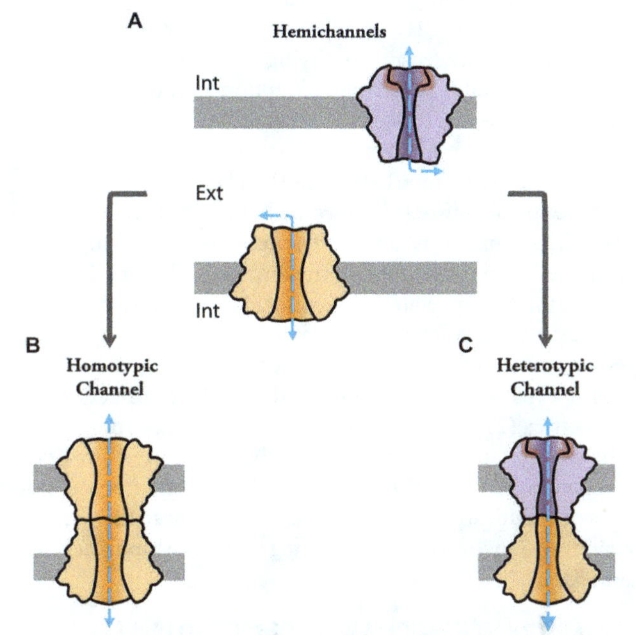

FIGURE 1 | Channels formed by connexin and innexin proteins can play functional roles as single "hemichannels" (HCs) or assemble into intercellular channels at gap junctions to provide cell-cell communication. **(A)** Hemichannels can be molecularly different (note difference in color and shape) and act as conduits for the release of messenger molecules. **(B)** Identical HCs can assemble into intercellular "homotypic" gap junction channels. **(C)** Hemichannels of different molecular composition can assemble into intercellular "heterotypic" gap junction channels.

providing GJs with complex functional properties. In addition, many types of cells co-express several Cx or Inx isoforms with the potential to form heteromeric HCs. In the same scenario, different homomeric HCs can cluster into the same junctional plaque forming bi-homotypic GJs (Li et al., 2008), in which HCs are docked with the same kind of HC, i.e., channels are homomeric homotypic. Heterotypic/heteromeric GJ channels are likely to be present in the retina, brain and peripheral system (Söhl et al., 2000; Vaney and Weiler, 2000). Twenty-one different Cx isoforms can potentially form 210 different heterotypic GJs, however not all different Cx isoforms are compatible with each other. More than forty functional heterotypic pairs have been found and analyzed so far (Palacios-Prado and Bukauskas, 2012). Although the expression of *Drosophila* Inxs was shown to overlap in some tissues (Stebbings et al., 2000) their functional compatibility remains, in contrast to Cxs, largely unexplored.

Gap junctions constitute the basis for electrical synaptic transmission in both vertebrate and invertebrate nervous systems (Bennett and Zukin, 2004; Pereda et al., 2013). Beyond their ability to allow the passage of small messenger molecules, neuronal GJs (or "electrical synapses") serve as low resistance pathways for the spread of electrical currents between coupled neurons, a key property for a cellular type that critically relies on electrical signaling (Bennett and Zukin, 2004; Connors and Long, 2004; Pereda et al., 2013). This article reviews data indicating that molecular differences between aHC at electrical synapses are generally associated with rectification of electrical transmission (differential resistance to current flow in one vs. the other direction between two coupled cells), a voltage-dependent property of GJ channels that has been observed at both Cx- and Inx-based electrical synapses (Furshpan and Potter, 1959; Auerbach and Bennett, 1969; Edwards et al., 1998; Rela and Szczupak, 2007; Rash et al., 2013). Here we discuss some molecular mechanisms underlying rectification of electrical transmission. We conclude that asymmetries in the molecular composition of individual HCs forming electrical synapses can strongly influence transmission of electrical signals between neurons coupled by GJs.

BI-DIRECTIONALITY AND SYMMETRY OF ELECTRICAL TRANSMISSION

Neurons operate by computing variations of the membrane potential evoked by synaptic currents and active processes, which are usually translated into trains of action potentials. The change in the membrane potential observed by the spread of presynaptic currents through GJs to a postsynaptic neuron is usually referred to as a "coupling potential". The amplitude of this "coupling potential" does not solely depend on the conductance of GJ channels but also on the passive properties determined by the capacitance and the input resistance (R), which is directly proportional to the membrane resistance (R_m), and indirectly proportional to the area of the membrane (size and geometry) of the coupled neurons (**Figure 2A**). The strength of an electrical synapse is generally expressed as the "coupling coefficient", a ratio that expresses the amplitude of the coupling potential normalized to the amplitude of the signal that originated it in a neighboring coupled cell (see equation in **Figure 2A**; this value is obtained once the capacitance of the membrane is charged, which is generally referred to as "steady state"). Electrical synaptic transmission is bidirectional and symmetric when the Rs of coupled cells are similar (**Figure 2B**). Gap junction channels at electrical synapses were shown in some cases to behave as electrical rectifiers, that is, to offer differential resistance to the flow of currents in one vs. the other direction across the junction between two coupled neurons. As a matter of fact, the first characterization of an unequivocally electrically mediated synapse came together with the description of electrical rectification (Furshpan and Potter, 1957, 1959). Rather than a simple bidirectional spread of electrotonic potential, the crayfish giant motor synapse transmitted depolarization signals from the giant axon to the motor fiber, but not in the opposite direction. Similarly, hyperpolarization signals were transmitted only from the motor fiber to the giant axon. Although the transmission of relative positive or negative potentials is unidirectional (only in one direction), rectifying junctions allow the spread of signals in either direction. In addition, postsynaptic signals reproduced the time course of presynaptic signals, and transmission was, surprisingly, voltage-dependent, thus challenging all the criteria established for chemical transmission. The rectification properties discovered in this preparation also helped to exclude the prevailing hypothesis of gross protoplasmic connections suggested earlier to explain symmetric electrotonic spread of current between

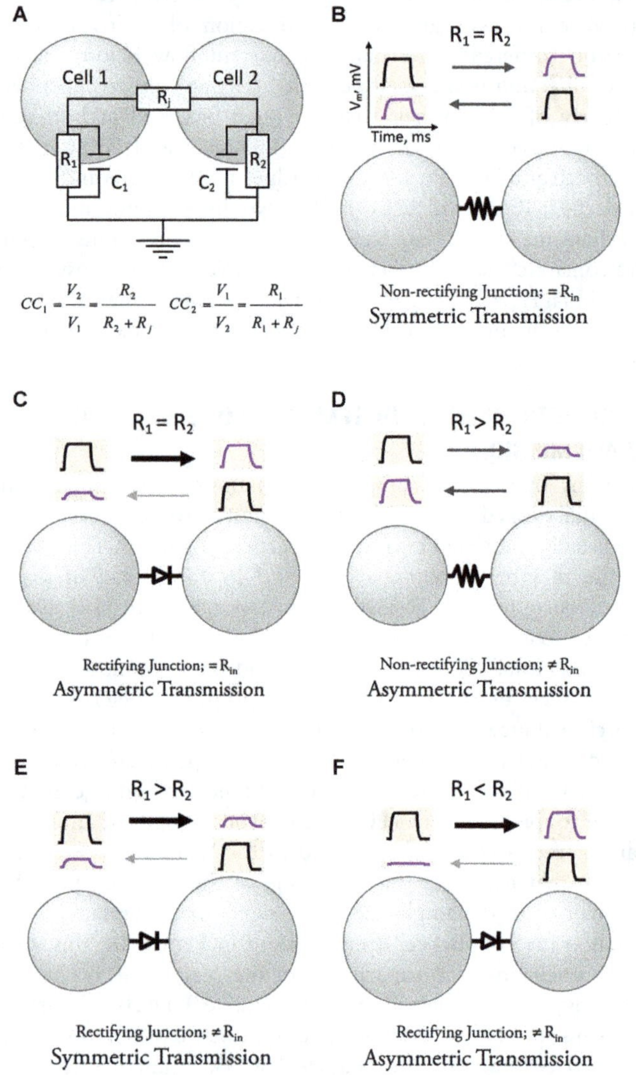

FIGURE 2 | Directionality and symmetry in electrical transmission. (A) Determinants of the strength of electrical transmission. The amplitude of the coupling potential, defined by the coupling coefficients in each direction (CC_1 & CC_2), once the capacitance of the membrane has been charged, is determined by both the resistance of the junction (R_j) and the resistances of the coupled cells (R_1 & R_2). **(B)** Electrical transmission is symmetric in cells with similar R and non-rectifying junctions (constant R_j). **(C)** Rectifying junctions make electrical transmission between neurons with similar R asymmetric. **(D)** Cells with different R create asymmetry of electrical transmission when junctions are non-rectifying. **(E)** Electrical transmission could be symmetric in cells with different R and rectifying junctions if the effects cancel each other. **(F)** The combination of differences in R and rectifying junctions can create strong asymmetry of electrical transmission for some polarities.

cardiac Purkinje cells (Weidmann, 1952) or among neurons of the lobster cardiac ganglion (Watanabe, 1958). Rectification was subsequently found in several electrical synapses in *in vivo* preparations exhibiting direction asymmetry of signal transfer (Smith et al., 1965; Auerbach and Bennett, 1969; Baylor and Nicholls, 1969; Ringham, 1975; Muller and Scott, 1981; Roberts et al., 1982; Margiotta and Walcott, 1983; Rash et al., 2013). Because of their properties, rectifying GJs can underlie asymmetric electrical transmission (**Figure 2C**). Asymmetry of electrical transmission does not necessarily require rectifying GJ channels, as differences in R of the coupled cells can make coupling coefficients stronger in the direction towards the cell with higher R (**Figure 2D**, Trenholm et al., 2013). Moreover, rectifying junctions can make a synapse bi-directional by counterbalancing the effect of differences in R of the coupled neurons (**Figure 2E**). Finally, the combination of rectifying junctions and differences in R of coupled cells can create strong asymmetric transmission (**Figure 2F**). Thus, although intimately related, directionality, rectification, and symmetry express different properties of electrical synaptic transmission and should not be considered interchangeable. In other words, electrical transmission could be: (1) bi-directional and asymmetric; (2) non-rectifying and markedly asymmetric; and (3) bidirectional and rectifying. Finally, while directionality and symmetry refer to electrical transmission (coupling potentials), the term rectification should be reserved to describe the asymmetric transjunctional current-voltage relationship of certain GJ channels.

ASYMMETRY IN HEMICHANNEL COMPOSITION IS ASSOCIATED WITH RECTIFICATION OF ELECTRICAL TRANSMISSION

The association of asymmetry of hemichannel composition and electrical rectification has been observed at both Inx- and Cx-based electrical synapses.

INNEXIN-BASED ELECTRICAL SYNAPSES

It was actually proposed that the formation of homomeric, homotypic channels is generally rare among fly Inxs (Phelan and Starich, 2001). Of the eight Inxs in *Drosophila*, a few were shown to form heterotypic GJs (reviewed in Hasegawa and Turnbull, 2014). Most information about electrical properties of heterotypic GJs in the fly brain exists for splice variants of the gene *shakingB*. Here we focus on two different circuits that both include shakB-based heterotypic electrical synapses in their architecture, the giant fiber system (GFS) and the antennal lobe. One forms a rather rigid reflex network, the other is a heavily modulated sensory information processor dealing with odors. In addition, we discuss the contribution of heterotypic GJs to memory formation in the fly brain and to *C. elegans* nervous system signaling.

The most complete picture of GJ function within a behavioral circuit is based on the GFS in *Drosophila*, an efficient escape reflex circuit that initiates a "jump and flight" response (reviewed in Allen et al., 2006). Threatening sudden stimuli can evoke a response in the giant fiber (GF) via giant commissural interneurons in the brain. The GF provides a fast connection of the brain with the ventral nerve cord in the thorax, where it forms mixed electrical and chemical synapses onto two cell types, a motor neuron that innervates a contralateral leg muscle, and an interneuron, which activates ipsilateral flight muscle motor neurons (Blagburn et al., 1999). The first evidence of the presence of GJs in this network stems from intracellular recordings on the GF with simultaneous brain stimulation and flight muscle recordings (Tanouye and Wyman, 1980). Subsequent work on the mutants *shaking-B*^{2} (*shakB*^{2}) and *passover* characterized the involved gene (Thomas and Wyman, 1984; Phelan et al., 1996; Sun and Wyman, 1996). It was later discovered that this *shakB* (or *inx8*) gene gives rise to five different splice variants, resulting in three confirmed protein isoforms: ShakB(Lethal), ShakB(Neural), and ShakB(Neural+16), which has 16 more amino acids on the amino-terminus (Zhang et al., 1999). Interestingly, while ShakB(Lethal) is capable of functional homotypic channel formation in *Xenopus* oocytes, the ShakB(Neural) variant is not (Phelan et al., 1998). Ten years later the same lab used the GFS to provide first evidence that electrical rectification emerges from differential aHC composition and formation of heterotypic GJs (Phelan et al., 2008). They identified two ShakB variants being responsible for heterotypic GJs in the GFS: ShakB(Neural+16) in the presynaptic GF, and ShakB(Lethal) in the postsynaptic motor neuron and interneuron. A more recent study using oocytic expression of chimeric ShakB proteins suggested a role for the amino-terminal end of ShakB in voltage gating and electrical rectification (Marks and Skerrett, 2014).

Heterotypic GJs involving ShakB are also found in the antennal lobe. The antennal lobe is the first integration center of olfactory information in insects and shows structural similarity to the vertebrate olfactory bulb; within neuropilar substructures called glomeruli, the olfactory sensory neurons converge onto projection neurons (PNs), which in turn relay the olfactory information into higher brain regions (Wilson, 2013). The glomeruli are interconnected by a third class of antennal lobe neurons, the amacrine local interneurons (LNs). The majority are GABAergic and mainly provide lateral inhibition on the presynaptic terminal of the sensory neurons (iLNs), but some are excitatory (eLNs) and either cholinergic (Shang et al., 2007) or glutamatergic (Chou et al., 2010; Das et al., 2011). The cholinergic eLNs form chemical synapses onto iLNs and electrical synapses with PNs (Huang et al., 2010; Yaksi and Wilson, 2010). The mutant *shakB*^{2} abolishes electrical transmission between eLNs and PNs, and RT-PCR identified shakB transcripts in PNs (Yaksi and Wilson, 2010). *ShakB*^{2} (which affects both Neural variants) was successfully rescued by ectopic expression of ShakB(Neural) in adult flies. Since homomeric ShakB(Neural) HCs fail to form functional GJs (Phelan et al., 1998; Curtin et al., 2002), heterotypic interaction with another Inx, probably ShakB(Lethal) like in the GFS circuit, seems likely. A more targeted rescue in either PNs or eLNs will resolve on which side the ShakB(Neural) HC is essential; based on the electrical properties of the junction it would be expected on the eLN side.

There is also evidence for the presence of heterotypic GJs in the mushroom body (MB). The MB is regarded as the homologous structure of the vertebrate pallium in the brain of protostomes (Tomer et al., 2010) and is crucial for associative memory processes (Perisse et al., 2013). This paired neuropil consists mainly of about 2000 Kenyon cells on each side, which can roughly be subdivided in a dendritic calyx region, two orthogonal, elongated lobes, which contain the majority of presynaptic sites, and a peduncle that connects calyx with lobes. Prevailing sensory input to the *Drosophila* calyx is of olfactory nature, coming from the antennal lobe. Certain Kenyon cell subdivisions exhibit preferential roles in memory acquisition and in memory retrieval. The MB is innervated by two large amacrine cells per hemisphere; the GABAergic anterior paired lateral cell (APL), which innervates all MB regions (Liu and Davis, 2009), and the serotonergic dorsal paired medial cell (DPM) which innervates peduncle and lobes only (Lee et al., 2011). A prominent role of the APL is to maintain signal sparseness by feedback inhibition (Lin et al., 2014), but it also shows involvement in labile appetitive memory (Pitman et al., 2011). The DPM is crucial for long-term memory consolidation, so its role can be separated from APL (Pitman et al., 2011). Both neurons seem to be electrically connected by heterotypic channels formed by Inx6 (DPM) and Inx7 (APL), especially in a subregion of the MB. This was inferred from contact marker expression and a combination of immunostainings, targeted RNAi expression against various Inxs in DPM and APL, dye coupling backfills and behavioral experiments (Wu et al., 2011; Pitman et al., 2011). Taken together, it is tempting to speculate that a rectifying electrical synapse between APL and DPM could contribute to generate a reverberant circuit, thus providing the ongoing cellular activity to consolidate a memory trace. The presence of this putative heterotypic channel is interesting for several reasons: it involves a novel pair of interacting Inxs, despite bigger spatial overlap between both contributing cells it seems to be segregated to a specific subcellular region, and, importantly, because of its potential contribution to a memory consolidation process.

The existence of numerous heterotypic GJs was suggested also to be the case for *C. elegans* Inxs, with the notable exception of UNC-7 and UNC-9, and possibly Inx14 with Inx8 or Inx9 (Simonsen et al., 2014). The *unc-7* gene gives rise to three protein isoforms, and the homomeric heterotypic channel formed by the UNC-7S (or UNC-7b) isoform and UNC-9 was shown to be rectifying (Starich et al., 2009). Since both Inxs are widely expressed in nerve cells (Altun et al., 2009), and ~10% of all synapses in *C. elegans* are electrical, this rectifying synapse might contribute significantly to direct signal transduction in the nematode nervous system. This is supported by the locomotion phenotype in mutants of both *unc-7* and *unc-9* (Starich et al., 1993; Barnes and Hekimi, 1997).

Gregarious behavior in *C. elegans* is determined by sensory integration in a hub-and-spoke circuit where the RMG neuron forms electrical synapses with many sensory neurons (Macosko et al., 2009). Activation or inhibition of this sensory integration induces gathering or solitary behaviors, respectively. Interestingly, RMG neurons are only labeled with the *unc-7a* promoter fragment while its sensory partners express different Inxs: IL2, ADL and AWB express UNC-9; IL2, ADL and ASK express Inx-18; and IL2, ADL and ASH express Inx-19 (Altun et al., 2009). Since UNC-7S and UNC-9 are known to form rectifying heterotypic junctions (Starich et al., 2009), it is possible that electrical rectification is involved in sensory integration in the RMG hub-and-spoke circuit, and therefore heterotypic GJ might be involved in gregarious behavior of *C. elegans*. This type of circuit motif—one integrating hub neuron connected to many sensory neurons by electrical synapses—are present in large numbers (more than 15 different hubs) in the nematode nervous system and may be a conserved functional unit for coincidence detection (Rabinowitch et al., 2013). Finally, although the relationship with heterotypic Inx-based GJs still needs to be established, multiple rectifying electrical synapses have been described in various invertebrates, such as crayfish (Furshpan and Potter, 1959), horseshoe crab (Smith et al., 1965) and leech (Baylor and Nicholls, 1969; Muller and Scott, 1981).

CONNEXIN-BASED ELECTRICAL SYNAPSES

Although the presence of asymmetric transmission has been reported at electrical synapses between several vertebrate cell types, such as the inferior olive (Devor and Yarom, 2002), striatum (Venance et al., 2004), cochlear nucleus (Apostolides and Trussell, 2013) and thalamus (Haas et al., 2011) and reportedly involving in some of these cases asymmetry of GJ conductance (Devor and Yarom, 2002; Venance et al., 2004; Haas et al., 2011), electrical rectification was demonstrated in only a few cases (Auerbach and Bennett, 1969; Ringham, 1975; Rash et al., 2013). Recent evidence suggests that, as observed in invertebrates, electrical rectification is also associated with asymmetry in the molecular composition of aHCs. That is, electrical synapses at auditory afferents and the teleost Mauthner cell known as "Club endings" are formed by two homologs of mammalian Cx36 (considered the main synaptic Cx in mammals due to its widespread expression in neurons (Condorelli et al., 2000)), Cx35 and Cx34.7 (Rash et al., 2013). As a result of additional genome duplication (Volff, 2005), teleost fish have more than one homologous gene for most mammalian Cxs (Eastman et al., 2006). Remarkably, while Cx35 is restricted to presynaptic GJ hemiplaques (the portion of the GJ plaque contributed by each cell), Cx34.7 is restricted to postsynaptic hemiplaques, forming heterotypic junctions (Rash et al., 2013). In contrast to many different Cxs that are compatible to form heterotypic GJs, Cx36 is known so far to form only "homotypic" GJs (Teubner et al., 2000; Li et al., 2004). From an evolutionary point of view, the existence of compatible Cx36 teleost homologs that form heterotypic channels provided neurons with the ability to connect through GJs with more complex properties. Estimates of junctional conductance (g_j) between Club endings and the Mauthner cell revealed a four-fold difference between the antidromic (from the postsynaptic Mauthner cell to the presynaptic Club ending) and orthodromic (from the Club ending to the Mauthner cell) directions (Rash et al., 2013). This rectifying property is thought to play an important functional role by promoting cooperativity between different auditory afferents (see below).

MECHANISMS UNDERLYING RECTIFICATION OF ELECTRICAL TRANSMISSION

GAP JUNCTIONS AS DIODES

The original mechanism proposed for rectification of electrical transmission was represented as a simple analogy to an electric rectifier or diode (Furshpan and Potter, 1959), in which separation of negative and positive permanent charges results in an asymmetric energy barrier. This barrier generates instantaneous transjunctional current (I_j) rectification with characteristics of a p-n junction in semiconductors. At that time GJ channels had not yet been discovered and thus the properties of the rectifier and the electrostatic effect were assigned to the "synaptic membrane". Nonetheless, the novel hypothesis of p-n junctions in biological membranes was examined (Mauro, 1962; Coster, 1965), and provided a theoretical framework for considering fixed charges in junctional membranes (Brink and Dewey, 1980) that could explain the steep rectification of the junctional conductance-voltage relation (g_j-V_j) in some electrical synapses.

The hypothesis that electrical rectification could arise from an asymmetry in aHC composition came in the late 70's (Bennett, 1977; Loewenstein, 1981). With the exogenous expression of different Cx isoforms, it was possible to examine this hypothesis. Indeed, electrical rectification of heterotypic GJ channels (originally called heteromolecular or hybrid cell-cell channels) was first studied in pair of oocytes overexpressing Cx32/Cx26 or Cx32/Cx43 GJs (Swenson et al., 1989; Werner et al., 1989; Barrio et al., 1991). In the case of heterotypic Cx32/Cx26 GJ channels, asymmetries in the instantaneous and steady-state g_j-V_j relationships were observed (Barrio et al., 1991). To make a clear distinction between instantaneous and steady-state asymmetries in the g_j-V_j relationship, we refer to instantaneous and steady-state asymmetries as "electrical rectification" and "asymmetric gating", respectively.

Based on single GJ channel and HC recordings showing multiple I_j substates, we know that the steady-state I_j-V_j relationship of GJ channels is the product of two V_j-sensitive gating mechanisms present in each aHC, the *fast* or "V_j" gate and the *slow* or "loop" gate (Bukauskas and Verselis, 2004). The probability of each V_j-sensitive gate to dwell in a closed state is a function of the intensity and relative polarity of V_j (gating polarity). The instantaneous I_j-V_j relationship is mostly determined by the electrical properties of the unitary conductance of the fully open state (γ_o) of GJ channels, which can rectify by allowing larger I_js in one direction than in the other. The unitary conductance of the residual state (γ_{res}; one or two *fast* gates in closed position) may also rectify (Bukauskas et al., 1995; Oh et al., 1999), and therefore can contribute to electrical rectification. In general, homotypic GJs show symmetric g_j-V_j relationships for either polarity of V_j (**Figure 3A**). However, asymmetry in the composition of aHCs (or transjunctional asymmetry in cytosolic factors; see below) may result in electrical rectification (**Figure 3B**) and asymmetric gating.

The mechanism for asymmetric gating observed in Cx32/Cx26 GJ channels was explained by a difference in gating polarity of voltage-sensitive gates present in Cx26 and Cx32 aHCs (Verselis et al., 1994). Heterotypic GJs that possess aHC with opposite gating polarity exhibit marked asymmetric gating since one polarity of V_j simultaneously opens the gates in both aHCs, and the opposite polarity closes them. In addition to opposite gating polarity, asymmetric gating can also be produced by differences in unitary conductances of aHCs ($\gamma_{o,H}$), or simply by differences in intrinsic sensitivity to V_j (Bukauskas et al., 1995; Rackauskas et al., 2007). When $\gamma_{o,H}$s are considerably dissimilar, like in the case of Cx43/Cx45 heterotypic GJs ($\gamma_{o,H}$ of Cx43 is ~4 times higher than that of Cx45), a bigger fraction of V_j drops across the aHC with higher resistance (Cx45), thus enhancing its sensitivity to V_j compared to the aHC with smaller

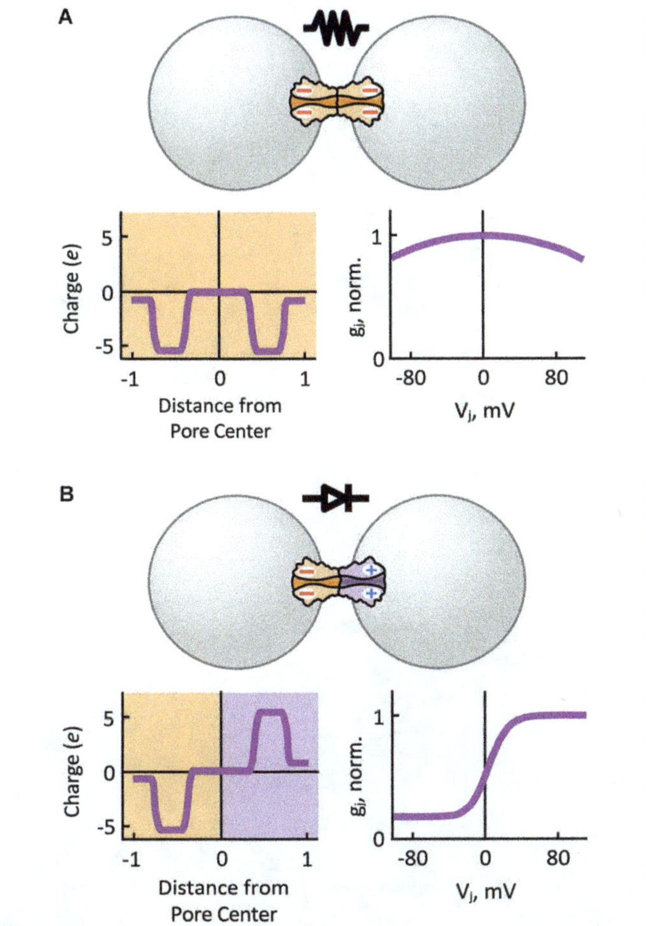

FIGURE 3 | Hemichannel composition determines the symmetry of electrical transmission. (A) Homotypic gap junction channels that comprise symmetric charge distribution with respect to the pore center of the channel, behave as passive resistors with symmetric junctional conductance (g_j) over transjunctional voltage (V_j) dependence (normalized to g_j value at V_j equal zero). **(B)** Heterotypic gap junction channels that comprise an asymmetry in positive and negative charge distribution with respect to the pore center of the channel behave as electrical rectifiers (p-n junction) with steep asymmetric g_j-V_j dependence. Normally, depolarizing (positive) potentials are more easily transmitted from the cell with negatively charged HCs to the cell with positively charged HCs.

resistance (Cx43). Therefore, differential drop of V_j in aHCs of heterotypic GJs may also result in asymmetric gating (Bukauskas et al., 2002). Although asymmetric gating may be important to determine I_j and transjunctional flux directionalities under long-lasting asymmetries in V_j (Palacios-Prado and Bukauskas, 2009), electrical rectification (instantaneous asymmetry) determines the directionality of synaptic electrical transmission between neurons with brief (millisecond) oscillatory changes in V_j during action potentials.

Analysis of heterotypic Cx32/Cx26 GJs at the single channel level revealed that γ_o rectifies depending on V_j (Bukauskas et al., 1995). Using this premise, an electrodiffusive model that solves the Poisson-Nernst-Planck (PNP) equations in one dimension (Chen and Eisenberg, 1993) was used to describe the asymmetric single channel fluxes and currents observed in heterotypic Cx32/Cx26 GJs (Oh et al., 1999). The PNP model in combination with site-directed mutagenesis successfully predicted that electrical rectification was produced by an asymmetric position of fixed charged amino acid residues present in the heterotypic Cx32/Cx26 channel pore. These findings demonstrated that the original diode hypothesis of p-n junctions could indeed generate electrical rectification of synaptic transmission based on the asymmetric position of charges near the channel-pore surface (**Figure 3B**) that, in turn, produce differences in ionic conductance and selectivity of HCs (Suchyna et al., 1999). Thus, heterotypic GJ channels that form rectifying junctions with steep asymmetric g_j-V_j relationship (**Figure 4A**) can make an electrical synapse nearly unidirectional by allowing the transmission of depolarizing and hyperpolarizing potentials in only one direction (opposite to each other), and restricting the transmission of depolarizing and hyperpolarizing potentials in the opposite direction (**Figure 4B**). The mechanism for electrical rectification and asymmetric gating observed in the *Drosophila* GFS is indeed associated with molecular asymmetry of HCs (Phelan et al., 2008). Since ShakB(Neural+16) and ShakB(Lethal) variants exhibit significant differences in amino acid sequence and sensitivity to V_j, it is likely that electrical rectification arises from asymmetry in position of charges (p-n junction), and asymmetric gating arises from differences in intrinsic V_j-sensitivity of aHCs rather than opposite gating polarities or differences in $\gamma_{o,H}$.

In addition to the p-n junction hypothesis for electrical rectification, a unique voltage-dependent gating mechanism was proposed after a detailed characterization of the rectifying crayfish giant motor synapse using high-quality voltage clamp at low temperatures (Jaslove and Brink, 1986). These studies suggested that, rather than an instantaneous electrostatic effect, the rectification profile of the I_j-V_j relationship contained a voltage-dependent kinetic component with a time constant in the order of milliseconds, which was attributed to a rapid gating mechanism present in one of the aHCs. The authors proposed that this gate was set to a low open probability at resting conditions and that changing the polarity of V_j would rapidly open the gates. This "millisecond timescale" gating mechanism has not been reported in any other Cx- or Inx-based rectifying electrical synapse; hence it is unclear whether rapid gating mechanisms may contribute to the observed electrical rectification in other invertebrate and vertebrate electrical synapses.

Gap junctions can occur in homocellular or heterocellular junctions; that is, coupled cells can be from the same or different cell types and perform similar or different functions, respectively. One remarkable similarity among electrical synapses showing steep rectification is that they occur mostly in heterocellular junctions and very often there is a difference in the resting potential of coupled neurons that give rise to a relatively constant V_j (Giaume and Korn, 1983; Ramón and Rivera, 1986). Regardless of the mechanism of rectification, the resting V_j derived from the difference in the resting potential of neurons forming rectifying electrical synapses in crayfish and leech is essential to produce steep rectification, since bidirectional transmission of depolarization pulses could be achieved by reversing the resting

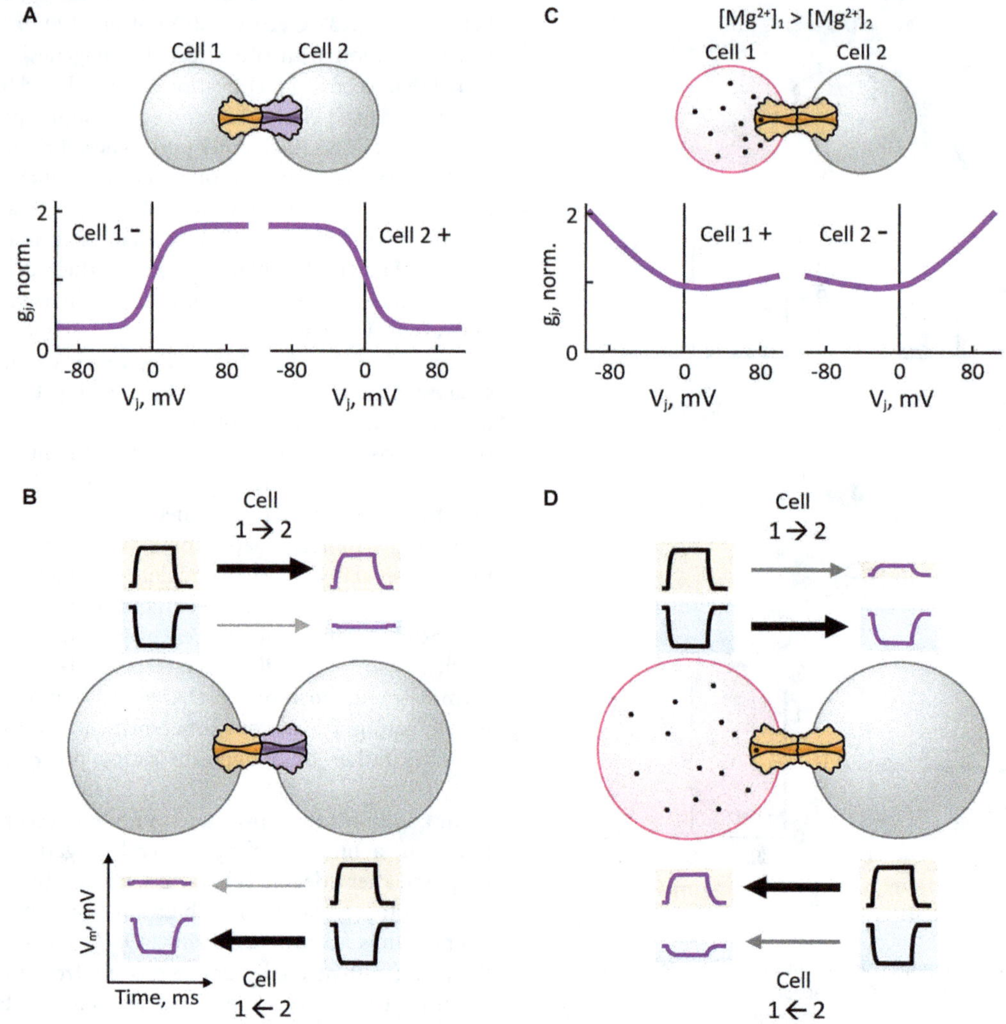

FIGURE 4 | Hemichannel composition and intercellular gradient of charged cytosolic factors can lead to rectification of electrical transmission. (A,B) Heterotypic gap junction channels with steep asymmetric g_j-V_j dependence **(A)** facilitate or attenuate the electrical transmission of depolarizing (positive) potentials from Cell 1 to Cell 2 (1→2) or 1←2, respectively **(B)**. The same junctions facilitate or attenuate the electrical transmission of hyperpolarizing (negative) potentials from 1←2 or 1→2, respectively **(B)**. **(C,D)** Transjunctional gradient of free magnesium ion concentration ($[Mg^{2+}]_i$) induces asymmetric g_j-V_j dependence in homotypic gap junction channels **(C)** that are hypersensitive to $[Mg^{2+}]_i$, such as Cx36 gap junction channels. Electrical transmission of depolarizing potentials is facilitated from 2→1 **(D)**, which is the opposite direction of the transjunctional $[Mg^{2+}]_i$ gradient (1→2). The same transjunctional $[Mg^{2+}]_i$ gradient facilitates the electrical transmission of hyperpolarizing potentials from 1→2 **(D)**.

V_j polarity (Giaume et al., 1987; Rela and Szczupak, 2007). Both p-n junction and rapid gating mechanisms imply a molecular asymmetry in aHC composition, and both require a resting V_j (difference in membrane potentials between the coupled cells) to exhibit significant electrical rectification or asymmetric gating, respectively. As an analogy to p-n junctions in semiconductors and silica nanochannels (Cheng and Guo, 2007), the resting V_j would normally set GJs to a low conductive state by producing a "reversed bias" effect (expression used when the flow of current is obstructed by increasing the resistance). Only action potentials that lower this resting V_j would produce a "forward bias" effect in the junction to allow the spread of electrotonic potentials.

CONTRIBUTION OF INTRACELLULAR SOLUBLE FACTORS

Gap junction channels and HCs are highly regulated according to cellular requirements and respond to various changes in the extracellular and intracellular environments. Besides their sensitivity to V_j, GJ channels and HCs are sensitive to phosphorylation, lipophilic molecules and other chemical factors (Baldridge et al., 1987; Bennett et al., 1991; Harris, 2001; Bukauskas and Verselis, 2004; Jouhou et al., 2007; Márquez-Rosado et al., 2012). Furthermore, GJ channels are sensitive to changes in intracellular ionic composition, such as intracellular pH, Mg^{2+} and Ca^{2+} that may vary under physiological conditions (Noma and Tsuboi, 1987; Cheng and Reynolds, 2000; Chesler, 2003; Matsuda et al., 2010; Shindo et al., 2010; Yamanaka et al., 2013). This suggests that modulation of electrical and metabolic gap junctional intercellular communication by these factors may be important for normal cell function.

It has been recently reported that Cx36-containing electrical synapses expressed in the mesencephalic nucleus of the trigeminal nerve (MesV) and the thalamic reticular nucleus (TRN) as well as heterologous expression systems transfected with Cx36 are bi-directionally modulated by changes in intracellular concentration of free Mg^{2+} ($[Mg^{2+}]_i$) (Palacios-Prado et al., 2013, 2014). This is a novel Mg^{2+}-dependent form of electrical synaptic plasticity where g_j can be augmented or reduced by lowering or increasing $[Mg^{2+}]_i$, respectively. These studies support the notion that $[Mg^{2+}]_i$ controls neuronal coupling via modulation of gating mechanisms of Cx36 GJs by interacting with a Mg^{2+}-sensitive domain located in the lumen of the GJ channel. Since intracellular levels of ATP determines $[Mg^{2+}]_i$ (Lüthi et al., 1999), Mg^{2+}-dependent plasticity of electrical synapses could be under control of neuronal metabolism and circadian rhythms (Dworak et al., 2010). In addition, electrical synaptic transmission could potentially decrease after neuronal depolarization and glutamate exposure, due to an increment in $[Mg^{2+}]_i$ (Kato et al., 1998; Shindo et al., 2010).

Electrical synapses formed by Cx36 show a unique Mg^{2+}-dependent instantaneous g_j-V_j relationship, in which instantaneous g_j increases over V_j under high $[Mg^{2+}]_i$, or remain constant over V_j under low $[Mg^{2+}]_i$. Interestingly, an intercellular gradient of Mg^{2+} (asymmetric transjunctional $[Mg^{2+}]_i$) produces electrical rectification (**Figure 4C**) and asymmetric gating in homotypic GJs by affecting the instantaneous and steady-state g_j-V_j relationship of Cx36, respectively (Palacios-Prado et al., 2013, 2014). Asymmetric transjunctional $[Mg^{2+}]_i$ produces greater transmission of depolarizing or hyperpolarizing potentials from the cell with lower or higher $[Mg^{2+}]_i$, respectively, compared to the opposite directions (**Figure 4D**). To explain this unique electrical rectification of Cx36, the authors proposed that a combination of two or more mechanisms are necessary: asymmetric fixed charges inside the Cx36 aHC pore that produce a p-n junction type of rectification; and a V_j-dependent modulation of Mg^{2+} interaction with its binding sites inside the pore. Mg^{2+}-dependent plasticity of Cx36 GJ channel properties is the only described mechanism so far by which transjunctional asymmetry is derived from a diffusible cytosolic factor that produces electrical rectification in homotypic GJs; all other examples arise from a molecular asymmetry in aHC composition. In principle, transjunctional asymmetry in ATP concentration may also induce electrical rectification by producing a transjunctional asymmetry in $[Mg^{2+}]_i$. It is noteworthy that other intracellular diffusible cations such as H^+, Ca^{2+} and spermine have been shown to affect cell-cell coupling via gating mechanisms in a Cx-specific manner (White et al., 1990; Musa et al., 2004; Harris and Contreras, 2014), but their effect on electrical rectification is yet to be demonstrated.

Asymmetry in the molecular composition of aHCs can also play a role in the effects of cytosolic factors. Heterotypic channels formed by expression of Cx35 and Cx34.7 in cell lines (the Cxs present at Club ending-Mauthner cell synapses) exhibited differential sensitivity to changes in $[Mg^{2+}]_i$, suggesting that molecular differences in heterotypic junctions might also contribute to generate electrical rectification by expressing a differential sensitivity to cytosolic factors (Rash et al., 2013).

FUNCTIONAL PROPERTIES OF RECTIFYING ELECTRICAL SYNAPSES

Rectifying electrical synapses have been proposed to play important functional roles within various neuronal networks (Furshpan and Potter, 1957; Edwards et al., 1999; Allen et al., 2006; Gutierrez and Marder, 2013). Providing directionality to electrical transmission between pre- and postsynaptic neurons, rectifying electrical synapses can significantly contribute to general signal transduction as in *C. elegans* (Starich et al., 2009) and are a feature in many escape networks (Furshpan and Potter, 1959; Edwards et al., 1999; Allen et al., 2006; Phelan et al., 2008). Rectifying electrical synapses were initially described at the giant motor synapses of the abdominal nerve cord of the crayfish between GFs and giant motor axons that innervate the flexor musculature of the tail (Furshpan and Potter, 1959). They also mediate directional communication in the *Drosophila* GFS (Allen et al., 2006) and between mechanoreceptor afferents and interneurons synapsing on the lateral giant neurons in crayfish (Edwards et al., 1999). Their ability to generate voltage-dependent directional transmission was also reported to be advantageous for certain motor behaviors in leech (Rela and Szczupak, 2003) and in fish spinal cord (Auerbach and Bennett, 1969), providing fast directional communication between identifiable interneurons and motor neurons.

Interestingly, rectifying electrical synapses can also underlie bidirectional communication between neuronal processes of dissimilar size, compensating for unfavorable electrical and geometrical conditions for the symmetrical spread of currents through the junctions. This is the case of a group of identifiable auditory synapses on the Mauthner cell known as Club endings (Pereda et al., 2004); the Mauthner cell network mediates auditory-evoked escape responses in fish (Faber and Pereda, 2011). Because electrical synapses at Club endings are bidirectional, the signals produced by a population of active Club endings in the Mauthner cell dendrite can influence the excitability of non-active neighboring Club endings, thus serving as a mechanism for "lateral excitation" (Pereda et al., 1995). Lateral excitation increases the sensitivity of sensory inputs (Herberholz et al., 2002). Electrical rectification favors this mechanism of lateral excitation by promoting the spread of currents originated in the dendrite to the presynaptic afferents, which otherwise would passively spread towards the lower input resistance soma of the Mauthner cell (Rash et al., 2013). Thus, by favoring the spread of currents to the presynaptic afferent, the rectification properties of electrical synapses between Club endings and the Mauthner cell enhance bi-directionality of electrical communication between these two cells of dissimilar size and geometry. From the functional point of view, lateral excitation promotes the coordinated activity of a population of Club endings, thus increasing the efficacy of the auditory input for the initiation of an escape response.

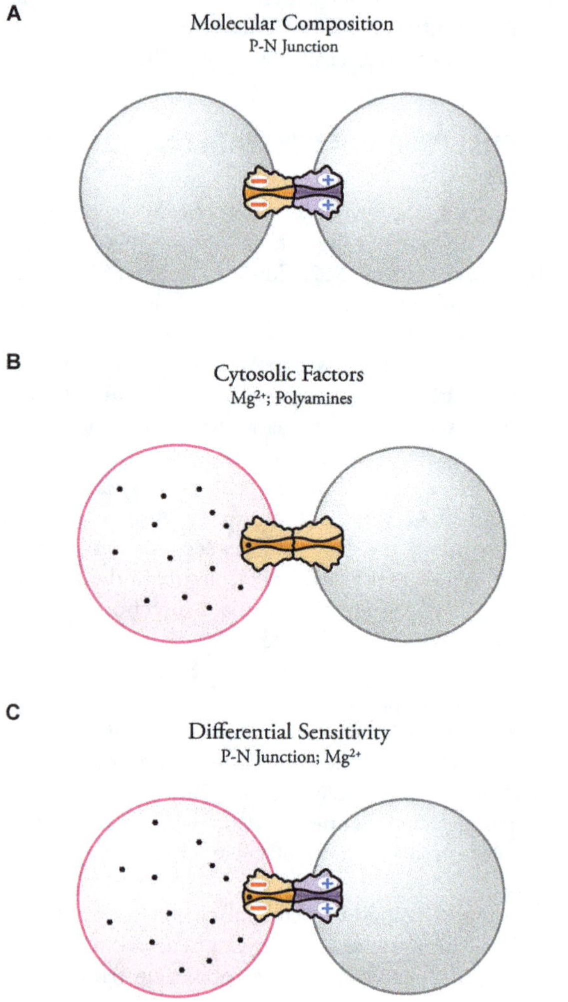

FIGURE 5 | Possible mechanisms underlying electrical rectification at electrical synapses. **(A)** Steep electrical rectification can be attainable by the separation of fixed positive and negative charges at opposite ends of heterotypic gap junction channels (p-n junction) resulting from asymmetries in the molecular composition of the HCs. **(B)** Electrical rectification can also result from the presence of an intercellular gradient of charged cytosolic factors, such as Mg^{2+} and spermine, which alter channel conductance. Molecular diversity can make some channels more susceptible of interacting with certain cytosolic factors. **(C)** Complex electrical rectification can arise from the combination of both mechanisms.

Recent studies suggest that rectifying electrical synapses are capable on endowing networks with more complex behaviors. Modeling studies explored the impact of rectifying electrical synapses in a pattern-generating neuronal network containing both chemical and electrical synapses (Gutierrez and Marder, 2013). The presence of rectifying electrical synapses was observed to have profound functional consequences, altering the sensitivity of the network dynamics to variations in the strength of chemical synapses (Gutierrez and Marder, 2013). Remarkably, the addition of rectifying electrical synapses to certain network configurations yielded robust circuit dynamics that were insensitive to variations in the strength of chemical synapses (Gutierrez and Marder, 2013), suggesting that the presence of rectifying electrical synapses is likely to play important roles in the stability and function of neural networks.

Finally, as a result on their voltage-dependent properties, rectifying electrical synapses were proposed to act as coincidence detectors (Edwards et al., 1998; Marder, 2009). Coincidence detection is an essential property of all nervous systems and is sustained by a variety of molecular, cellular and network properties. This phenomenon has been implicated in visual perception (Veruki and Hartveit, 2002), sound source localization (Joris et al., 1998), memory formation (Tsien, 2000), and motor control (Hjorth et al., 2009), amongst others. While non-rectifying electrical synapses are considered coincidence detectors of inputs arriving simultaneously at two different coupled neurons (Galarreta and Hestrin, 2001; Veruki and Hartveit, 2002), electrical rectification underlies the ability of the lateral giant neurons of crayfish to sum inputs that arrive synchronously (Edwards et al., 1998). Remarkably, this mechanism provides a significant temporal fidelity and it does not operate for inputs that are separated by only 100 ms or more. Because rectifying synapses in this system only allow bidirectional current flow when the presynaptic afferents are depolarized relative to the postsynaptic compartment (the lateral giant neuron), current flows increase during the presynaptic spike and remain electrically coupled after its completion (Edwards et al., 1998). Taking advantage of this property, synchronous inputs from mechanoreceptor afferents and interneurons integrate effectively and produce large excitatory responses. Asynchronous inputs, on the other hand, are much less efficient in activating the mechanism because: (1) the early arriving postsynaptic potential retards the opening of voltage-sensitive channels at additional synapses; and (2) the late arriving synaptic currents are shunted by the increase in g_j. Given the involvement of these neurons in escape responses, the coincidence detection mediated by the voltage-dependent properties of rectifying electrical synapses allows crayfish to elicit reflex escape responses only to particularly abrupt mechanical stimuli (Edwards et al., 1998).

CONCLUSIONS

Electrical transmission has become a topic of high interest in neuroscience. Together with the already established role of electrical synapses in invertebrates and cold-blooded vertebrates, evidence for the presence and importance of electrical synapses in the diverse areas of the mammalian brain continues to increase. Despite their wide distribution and functional relevance, the molecular complexity of electrical synapses and how this complexity affects synaptic function is still poorly understood. The evidence reviewed here indicates that the molecular composition of each aHC can endow neuronal GJs with important functional properties. More specifically, asymmetry in the molecular composition of aHCs has been associated with rectification of electrical transmission. The fact that such association was found at both Inx- and Cx-based electrical synapses emphasizes the contribution of the molecular asymmetry in underlying this voltage-dependent phenomenon. It has been shown that heterotypic channels with asymmetric position of charges near the channel-pore surface act as p-n junctions (diode hypothesis) with asymmetric transjunctional current flow (**Figure 5A**). Electrical rectification can also be observed at homotypic channels, arising from transjunctional asymmetries in the concentration of cytosolic factors that are capable of interacting with the channel pore (**Figure 5B**). Finally, cytosolic factors can contribute to electrical rectification at heterotypic junctions if one of the aHCs is more susceptible to interact with them, further enhancing the rectifying properties of the junction (**Figure 5C**).

Despite the presence of molecularly distinct pre- and postsynaptic sites, chemical synapses are considered indivisible functional units at which both sites are required to generate synaptic function. Electrical synapses can be viewed in a similar way. The fact that the docking of two HC is required does not necessarily imply that their molecular composition and that of the hemiplaques are the same. Hemiplaques should be different, suggesting that electrical synapses in analogy to chemical synapses can have distinct pre- and postsynaptic sites, endowing electrical synapses with more complex functional properties. While we emphasize in this article asymmetries in the composition of aHCs by GJ-forming proteins, asymmetries might also include the presence of associated scaffolding and regulatory proteins. Finally, an interesting scenario would be if asymmetries could be dynamically created by posttranslational modifications of Cxs in only one of the aHCs (asymmetric phosphorylation), or by differences in the intracellular concentration of soluble factors that affect channels properties as a result of metabolic changes in one of the coupled cells, providing electrical synapses with plastic rectifying properties.

ACKNOWLEDGMENTS

Portions of this work have been presented as part of a thesis dissertation (Nicolás Palacios-Prado). The authors are indebted to the Rainbow Kittens. Supported by the Grass Foundation, a Howard Hughes Medical Institute International Student Research Fellowship to Nicolás Palacios-Prado, a Marie-Curie Zukunftskolleg Incoming Fellowship to Wolf Huetteroth, and National Institutes of Health grants NIH DC03186, DC011099, NS055726, NS085772 and NS0552827 to Alberto E. Pereda.

REFERENCES

Abascal, F., and Zardoya, R. (2013). Evolutionary analyses of gap junction protein families. *Biochim. Biophys. Acta* 1828, 4–14. doi: 10.1016/j.bbamem.2012.02.007

Adams, M. D., Celniker, S. E., Holt, R. A., Evans, C. A., Gocayne, J. D., Amanatides, P. G., et al. (2000). The genome sequence of *Drosophila melanogaster*. *Science* 287, 2185–2195. doi: 10.1126/science.287.5461.2185

Allen, M. J., Godenschwege, T. A., Tanouye, M. A., and Phelan, P. (2006). Making an escape: development and function of the *Drosophila* giant fibre system. *Semin. Cell Dev. Biol.* 17, 31–41. doi: 10.1016/j.semcdb.2005.11.011

Altun, Z. F., Chen, B., Wang, Z. W., and Hall, D. H. (2009). High resolution map of *Caenorhabditis elegans* gap junction proteins. *Dev. Dyn.* 238, 1936–1950. doi: 10.1002/dvdy.22025

Apostolides, P. F., and Trussell, L. O. (2013). Regulation of interneuron excitability by gap junction coupling with principal cells. *Nat. Neurosci.* 16, 1764–1772. doi: 10.1038/nn.3569

Auerbach, A. A., and Bennett, M. V. L. (1969). A rectifying electrotonic synapse in the central nervous system of a vertebrate. *J. Gen. Physiol.* 53, 211–237. doi: 10.1085/jgp.53.2.211

Baldridge, W. H., Ball, A. K., and Miller, R. G. (1987). Dopaminergic regulation of horizontal cell gap junction particle density in goldfish retina. *J. Comp. Neurol.* 265, 428–436. doi: 10.1002/cne.902650310

Barnes, T. M., and Hekimi, S. (1997). The *Caenorhabditis elegans* avermectin resistance and anesthetic response gene unc-9 encodes a member of a protein family implicated in electrical coupling of excitable cells. *J. Neurochem.* 69, 2251–2260. doi: 10.1046/j.1471-4159.1997.69062251.x

Barrio, L. C., Suchyna, T., Bargiello, T., Xu, L. X., Roginski, R. S., Bennett, M. V., et al. (1991). Gap junctions formed by connexins 26 and 32 alone and in combination are differently affected by applied voltage. *Proc. Natl. Acad. Sci. U S A* 88, 8410–8414. doi: 10.1073/pnas.88.19.8410

Baylor, D. A., and Nicholls, J. G. (1969). Chemical and electrical synaptic connexions between cutaneous mechanoreceptor neurones in the central nervous system of the leech. *J. Physiol.* 203, 591–609.

Bennett, M. V. L. (1977). "Electrical transmission: a functional analysis and comparison to chemical transmission," in *The Handbook of Physiology*, ed E. Kandel (Washington: American Physiological Society), 357–416.

Bennett, M. V., Barrio, L. C., Bargiello, T. A., Spray, D. C., Hertzberg, E., and Sáez, J. C. (1991). Gap junctions: new tools, new answers, new questions. *Neuron* 6, 305–320. doi: 10.1016/0896-6273(91)90241-q

Bennett, M. V. L., Contreras, J. E., Bukauskas, F. F., and Sáez, J. C. (2003). New roles for astrocytes: gap junction hemichannels have something to communicate. *Trends Neurosci.* 26, 610–617. doi: 10.1016/j.tins.2003.09.008

Bennett, M. V. L., and Zukin, R. S. (2004). Electrical coupling and neuronal synchronization in the Mammalian brain. *Neuron* 41, 495–511. doi: 10.1016/s0896-6273(04)00043-1

Blagburn, J. M., Alexopoulos, H., Davies, J. A., and Bacon, J. P. (1999). Null mutation in shaking-B eliminates electrical, but not chemical, synapses in the *Drosophila* giant fiber system: a structural study. *J. Comp. Neurol.* 404, 449–458. doi: 10.1002/(sici)1096-9861(19990222)404:4<449::aid-cne3>3.3.co;2-4

Brink, P. R., and Dewey, M. M. (1980). Evidence for fixed charge in the nexus. *Nature* 285, 101–102. doi: 10.1038/285101a0

Bruzzone, R., Hormuzdi, S. G., Barbe, M. T., Herb, A., and Monyer, H. (2003). Pannexins, a family of gap junction proteins expressed in brain. *Proc. Natl. Acad. Sci. U S A* 11, 13644–13649. doi: 10.1073/pnas.2233464100

Bukauskas, F. F., Bukauskiene, A., Verselis, V. K., and Bennett, M. V. L. (2002). Coupling asymmetry of heterotypic connexin 45/connexin 43-EGFP gap junctions: properties of fast and slow gating mechanisms. *Proc. Natl. Acad. Sci. U S A* 99, 7113–7118. doi: 10.1073/pnas.032062099

Bukauskas, F. F., Elfgang, C., Willecke, K., and Weingart, R. (1995). Heterotypic gap junction channels (connexin26-connexin32) violate the paradigm of unitary conductance. *Pflugers Arch.* 429, 870–872. doi: 10.1007/bf00374812

Bukauskas, F. F., and Verselis, V. K. (2004). Gap junction channel gating. *Biochim. Biophys. Acta* 1662, 42–60. doi: 10.1016/j.bbamem.2004.01.008

Chen, D. P., and Eisenberg, R. S. (1993). Flux, coupling and selectivity in ionic channels of one conformation. *Biophys. J.* 65, 727–746. doi: 10.1016/s0006-3495(93)81099-3

Cheng, L.-J., and Guo, L. J. (2007). Rectified ion transport through concentration gradient in homogeneous silica nanochannels. *Nano Lett.* 7, 3165–3171. doi: 10.1021/nl071770c

Cheng, C., and Reynolds, I. J. (2000). Subcellular localization of glutamate-stimulated intracellular magnesium concentration changes in cultured rat forebrain neurons using confocal microscopy. *Neuroscience* 95, 973–979. doi: 10.1016/s0306-4522(99)00471-6

Chesler, M. (2003). Regulation and modulation of pH in the brain. *Physiol. Rev.* 83, 1183–1221. doi: 10.1152/physrev.00010.2003

Chou, Y.-H., Spletter, M. L., Yaksi, E., Leong, J. C. S., Wilson, R. I., and Luo, L. (2010). Diversity and wiring variability of olfactory local interneurons in the *Drosophila* antennal lobe. *Nat. Neurosci.* 13, 439–449. doi: 10.1038/nn.2489

Condorelli, D. F., Belluardo, N., Trovato-Salinaro, A., and Mudò, G. (2000). Expression of Cx36 in mammalian neurons. *Brain Res. Brain Res. Rev.* 32, 72–85. doi: 10.1016/s0165-0173(99)00068-5

Connors, B. W., and Long, M. A. (2004). Electrical synapses in the mammalian brain. *Annu. Rev. Neurosci.* 27, 393–418. doi: 10.1146/annurev.neuro.26.041002.131128

Contreras, J. E., Sánchez, H. A., Eugenín, E. A., Speidel, D., Theis, M., Willecke, K., et al. (2002). Metabolic inhibition induces opening of unapposed connexin 43 gap junction hemichannels and reduces gap junctional communication in cortical astrocytes in culture. *Proc. Natl. Acad. Sci. U S A* 99, 495–500. doi: 10.1073/pnas.012589799

Coster, H. G. (1965). A quantitative analysis of the voltage-current relationships of fixed charge membranes and the associated property of "punch-through". *Biophys. J.* 5, 669–686. doi: 10.1016/s0006-3495(65)86745-5

Curtin, K. D., Zhang, Z., and Wyman, R. J. (2002). Gap junction proteins expressed during development are required for adult neural function in the *Drosophila* optic lamina. *J. Neurosci.* 22, 7088–7096.

Dahl, G., and Locovei, S. (2006). Pannexin: to gap or not to gap, is that a question? *IUBMB Life* 58, 409–419. doi: 10.1080/15216540600794526

Dahl, G., and Muller, K. J. (2014). Innexin and pannexin channels and their signaling. *FEBS Lett.* 588, 1396–1402. doi: 10.1016/j.febslet.2014.03.007

Das, A., Chiang, A., Davla, S., Priya, R., Reichert, H., Vijayraghavan, K., et al. (2011). Identification and analysis of a glutamatergic local interneuron lineage in the adult *Drosophila* olfactory system. *Neural Syst. Circuits* 1:4. doi: 10.1186/2042-1001-1-4

Devor, A., and Yarom, Y. (2002). Electrotonic coupling in the inferior olivary nucleus revealed by simultaneous double patch recordings. *J. Neurophysiol.* 87, 3048–3058.

Dworak, M., McCarley, R. W., Kim, T., Kalinchuk, A. V., and Basheer, R. (2010). Sleep and brain energy levels: ATP changes during sleep. *J. Neurosci.* 30, 9007–9016. doi: 10.1523/JNEUROSCI.1423-10.2010

Eastman, S. D., Chen, T. H.-P., Falk, M. M., Mendelson, T. C., and Iovine, M. K. (2006). Phylogenetic analysis of three complete gap junction gene families reveals lineage-specific duplications and highly supported gene classes. *Genomics* 87, 265–274. doi: 10.1016/j.ygeno.2005.10.005

Edwards, D. H., Heitler, W. J., and Krasne, F. B. (1999). Fifty years of a command neuron: the neurobiology of escape behavior in the crayfish. *Trends Neurosci.* 22, 153–161. doi: 10.1016/s0166-2236(98)01340-x

Edwards, D. H., Yeh, S.-R., and Krasne, F. B. (1998). Neuronal coincidence detection by voltage-sensitive electrical synapses. *Proc. Natl. Acad. Sci. U S A* 95, 7145–7150. doi: 10.1073/pnas.95.12.7145

Faber, D. S., and Pereda, A. E. (2011). "Physiology of the mauthner cell: function", in *Encyclopedia of Fish Physiology: From Genome to Environment, vol. 1*, ed A. P. Farrell (San Diego: Academic Press), 73–79.

Furshpan, E. J., and Potter, D. D. (1957). Mechanism of nerve-impulse transmission at a crayfish synapse. *Nature* 180, 342–343. doi: 10.1038/180342a0

Furshpan, E. J., and Potter, D. D. (1959). Transmission at the giant motor synapses of the crayfish. *J. Physiol.* 145, 289–325.

Galarreta, M., and Hestrin, S. (2001). Spike transmission and synchrony detection in networks of GABAergic interneurons. *Science* 292, 2295–2299. doi: 10.1126/science.1061395

Ganfornina, M. D., Sánchez, D., Herrera, M., and Bastiani, M. J. (1999). Developmental expression and molecular characterization of two gap junction channel proteins expressed during embryogenesis in the grasshopper *Schistocerca americana. Dev. Genet.* 24, 137–150. doi: 10.1002/(sici)1520-6408(1999)24:1/2<137::aid-dvg13>3.3.co;2-z

Giaume, C., Kado, R. T., and Korn, H. (1987). Voltage-clamp analysis of a crayfish rectifying synapse. *J. Physiol.* 386, 91–112.

Giaume, C., and Korn, H. (1983). Bidirectional transmission at the rectifying electrotonic synapse: a voltage-dependent process. *Science* 220, 84–87. doi: 10.1126/science.6298940

Goodenough, D. A., and Paul, D. L. (2009). Gap junctions. *Cold Spring Harb. Perspect. Biol.* 1:a002576. doi: 10.1101/cshperspect.a002576

Gutierrez, G. J., and Marder, E. (2013). Rectifying electrical synapses can affect the influence of synaptic modulation on output pattern robustness. *J. Neurosci.* 33, 13238–13248. doi: 10.1523/JNEUROSCI.0937-13.2013

Haas, J. S., Zavala, B., and Landisman, C. E. (2011). Activity-dependent long-term depression of electrical synapses. *Science* 334, 389–393. doi: 10.1126/science.1207502

Harris, A. L. (2001). Emerging issues of connexin channels: biophysics fills the gap. *Q. Rev. Biophys.* 34, 325–427. doi: 10.1017/s0033583501003705

Harris, A. L., and Contreras, J. E. (2014). Motifs in the permeation pathway of connexin channels mediate voltage and Ca (2+) sensing. *Front. Physiol.* 5:113. doi: 10.3389/fphys.2014.00113

Hasegawa, D. K., and Turnbull, M. W. (2014). Recent findings in evolution and function of insect innexins. *FEBS Lett.* 588, 1403–1410. doi: 10.1016/j.febslet.2014.03.006

Herberholz, J., Antonsen, B. L., and Edwards, D. H. (2002). A lateral excitatory network in the escape circuit of crayfish. *J. Neurosci.* 22, 9078–9085.

Hervé, J. C., Phelan, P., Bruzzone, R., and White, T. W. (2005). Connexins, innexins and pannexins: bridging the communication gap. *Biochim. Biophys. Acta* 1719, 3–5. doi: 10.1016/j.bbamem.2005.11.013

Hjorth, J., Blackwell, K. T., and Kotaleski, J. H. (2009). Gap junctions between striatal fast-spiking interneurons regulate spiking activity and synchronization as a function of cortical activity. *J. Neurosci.* 29, 5276–5286. doi: 10.1523/JNEUROSCI.6031-08.2009

Huang, J., Zhang, W., Qiao, W., Hu, A., and Wang, Z. (2010). Functional connectivity and selective odor responses of excitatory local interneurons in *Drosophila* antennal lobe. *Neuron* 67, 1021–1033. doi: 10.1016/j.neuron.2010.08.025

Iglesias, R., Dahl, G., Qiu, F., Spray, D. C., and Scemes, E. (2009). Pannexin 1: the molecular substrate of astrocyte "hemichannels". *J. Neurosci.* 29, 7092–7097. doi: 10.1523/JNEUROSCI.6062-08.2009

Jaslove, S. W., and Brink, P. R. (1986). The mechanism of rectification at the electrotonic motor giant synapse of the crayfish. *Nature* 323, 63–65. doi: 10.1038/323063a0

Joris, P. X., Smith, P. H., and Yin, T. C. (1998). Coincidence detection in the auditory system. *Neuron* 21, 1235–1238. doi: 10.1016/s0896-6273(00)80643-1

Jouhou, H., Yamamoto, K., Iwasaki, M., and Yamada, M. (2007). Acidification decouples gap junctions but enlarges the receptive field size of horizontal cells in carp retina. *Neurosci. Res.* 57, 203–209. doi: 10.1016/j.neures.2006.10.006

Kato, H., Gotoh, H., Kajikawa, M., and Suto, K. (1998). Depolarization triggers intracellular magnesium surge in cultured dorsal root ganglion neurons. *Brain Res.* 779, 329–333. doi: 10.1016/s0006-8993(97)01232-8

Lee, P.-T., Lin, H.-W., Chang, Y.-H., Fu, T.-F., Dubnau, J., Hirsh, J., et al. (2011). Serotonin-mushroom body circuit modulating the formation of anesthesia-resistant memory in *Drosophila. Proc. Natl. Acad. Sci. U S A* 108, 13794–13799. doi: 10.1073/pnas.1019483108

Li, X., Kamasawa, N., Ciolofan, C., Olson, C. O., Lu, S., Davidson, K. G., et al. (2008). Connexin45-containing neuronal gap junctions in rodent retina also contain connexin36 in both apposing hemiplaques, forming bihomotypic gap junctions, with scaffolding contributed by zonula occludens-1. *J. Neurosci.* 28, 9769–9789. doi: 10.1523/JNEUROSCI.2137-08.2008

Li, X., Olson, C., Lu, S., Kamasawa, N., Yasumura, T., Rash, J. E., et al. (2004). Neuronal connexin36 association with zonula occludens-1 protein (ZO-1) in mouse brain and interaction with the first PDZ domain of ZO-1. *Eur. J. Neurosci.* 19, 2132–2146. doi: 10.1111/j.0953-816x.2004.03283.x

Lin, A. C., Bygrave, A. M., de Calignon, A., Lee, T., and Miesenböck, G. (2014). Sparse, decorrelated odor coding in the mushroom body enhances learned odor discrimination. *Nat. Neurosci.* 17, 559–568. doi: 10.1038/nn.3660

Liu, X., and Davis, R. L. (2009). The GABAergic anterior paired lateral neuron suppresses and is suppressed by olfactory learning. *Nat. Neurosci.* 12, 53–59. doi: 10.1038/nn.2235

Loewenstein, W. R. (1981). Junctional intercellular communication: the cell-to-cell membrane channel. *Physiol. Rev.* 61, 829–913.

Lüthi, D., Günzel, D., and McGuigan, J. A. (1999). Mg-ATP binding: its modification by spermine, the relevance to cytosolic Mg2+ buffering, changes in the intracellular ionized Mg2+ concentration and the estimation of Mg2+ by 31P-NMR. *Exp. Physiol.* 84, 231–252. doi: 10.1111/j.1469-445x.1999.01799.x

Macosko, E. Z., Pokala, N., Feinberg, E. H., Chalasani, S. H., Butcher, R. A., Clardy, J., et al. (2009). A hub-and-spoke circuit drives pheromone attraction and social behaviour in *C. elegans. Nature* 458, 1171–1175. doi: 10.1038/nature07886

Marder, E. (2009). Electrical synapses: rectification demystified. *Curr. Biol.* 19, R34–R35. doi: 10.1016/j.cub.2008.11.008

Margiotta, J. F., and Walcott, B. (1983). Conductance and dye permeability of a rectifying electrical synapse. *Nature* 305, 52–55. doi: 10.1038/305052a0

Marks, W. D., and Skerrett, I. M. (2014). Role of the amino terminus in voltage gating and junctional rectification of Shaking B Innexins. *J. Neurophysiol.* 111, 1383–1395. doi: 10.1152/jn.00385.2013

Márquez-Rosado, L., Solan, J. L., Dunn, C. A., Norris, R. P., and Lampe, P. D. (2012). Connexin43 phosphorylation in brain, cardiac, endothelial and epithelial tissues. *Biochim. Biophys. Acta* 1818, 1985–1992. doi: 10.1016/j.bbamem.2011.07.028

Matsuda, H., Kurata, Y., Oka, C., Matsuoka, S., and Noma, A. (2010). Magnesium gating of cardiac gap junction channels. *Prog. Biophys. Mol. Biol.* 103, 102–110. doi: 10.1016/j.pbiomolbio.2010.05.009

Mauro, A. (1962). Space charge regions in fixed charge membranes and the associated property of capacitance. *Biophys. J.* 2, 179–198. doi: 10.1016/s0006-3495(62)86848-9

Muller, K. J., and Scott, S. A. (1981). Transmission at a "direct" electrical connexion mediated by an interneurone in the leech. *J. Physiol.* 311, 565–583.

Musa, H., Fenn, E., Crye, M., Gemel, J., Beyer, E. C., and Veenstra, R. D. (2004). Amino terminal glutamate residues confer spermine sensitivity and affect voltage gating and channel conductance of rat connexin40 gap junctions. *J. Physiol.* 557, 863–878. doi: 10.1113/jphysiol.2003.059386

Noma, A., and Tsuboi, N. (1987). Dependence of junctional conductance on proton, calcium and magnesium ions in cardiac paired cells of guinea-pig. *J. Physiol.* 382, 193–211.

Oh, S., Rubin, J. B., Bennett, M. V., Verselis, V. K., and Bargiello, T. A. (1999). Molecular determinants of electrical rectification of single channel conductance in gap junctions formed by connexins 26 and 32. *J. Gen. Physiol.* 114, 339–364. doi: 10.1085/jgp.114.3.339

Palacios-Prado, N., and Bukauskas, F. F. (2009). Heterotypic gap junction channels as voltage-sensitive valves for intercellular signaling. *Proc. Natl. Acad. Sci. U S A* 106, 14855–14860. doi: 10.1073/pnas.0901923106

Palacios-Prado, N., and Bukauskas, F. F. (2012). Modulation of metabolic communication through gap junction channels by transjunctional voltage; synergistic and antagonistic effects of gating and ionophoresis. *Biochim. Biophys. Acta* 1818, 1884–1894. doi: 10.1016/j.bbamem.2011.09.001

Palacios-Prado, N., Chapuis, S., Panjkovich, A., Fregeac, J., Nagy, J. I., and Bukauskas, F. F. (2014). Molecular determinants of magnesium-dependent synaptic plasticity at electrical synapses formed by connexin36. *Nat. Commun.* 5:4667. doi: 10.1038/ncomms5667

Palacios-Prado, N., Hoge, G., Marandykina, A., Rimkute, L., Chapuis, S., Paulauskas, N., et al. (2013). Intracellular magnesium-dependent modulation of gap junction channels formed by neuronal connexin36. *J. Neurosci.* 33, 4741–4753. doi: 10.1523/JNEUROSCI.2825-12.2013

Panchin, Y., Kelmanson, I., Matz, M., Lukyanov, K., Usman, N., and Lukyanov, S. (2000). A ubiquitous family of putative gap junction molecules. *Curr. Biol.* 10, R473–R474. doi: 10.1016/s0960-9822(00)00576-5

Pereda, A. E., Bell, T. D., and Faber, D. S. (1995). Retrograde synaptic communication via gap junctions coupling auditory afferents to the Mauthner cell. *J. Neurosci.* 15, 5943–5955.

Pereda, A. E., Curti, S., Hoge, G., Cachope, R., Flores, C. E., and Rash, J. E. (2013). Gap junction-mediated electrical transmission: regulatory mechanisms and plasticity. *Biochim. Biophys. Acta* 1828, 134–146. doi: 10.1016/j.bbamem.2012.05.026

Pereda, A. E., Rash, J. E., Nagy, J. I., and Bennett, M. V. L. (2004). Dynamics of electrical transmission at club endings on the Mauthner cells. *Brain Res. Brain Res. Rev.* 47, 227–244. doi: 10.1016/j.brainresrev.2004.06.010

Perisse, E., Burke, C., Huetteroth, W., and Waddell, S. (2013). Shocking revelations and saccharin sweetness in the study of *Drosophila* olfactory memory. *Curr. Biol.* 23, R752–R763. doi: 10.1016/j.cub.2013.07.060

Phelan, P. (2005). Innexins: members of an evolutionarily conserved family of gap-junction proteins. *Biochim. Biophys. Acta* 1711, 225–245. doi: 10.1016/j.bbamem.2004.10.004

Phelan, P., Goulding, L. A., Tam, J. L., Allen, M. J., Dawber, R. J., Davies, J. A., et al. (2008). Molecular mechanism of rectification at identified electrical synapses in the *Drosophila* giant fiber system. *Curr. Biol.* 18, 1955–1960. doi: 10.1016/j.cub.2008.10.067

Phelan, P., Nakagawa, M., Wilkin, M. B., Moffat, K. G., O'Kane, C. J., Davies, J. A., et al. (1996). Mutations in shaking-B prevent electrical synapse formation in the *Drosophila* giant fiber system. *J. Neurosci.* 16, 1101–1113.

Phelan, P., and Starich, T. A. (2001). Innexins get into the gap. *Bioessays* 23, 388–396. doi: 10.1002/bies.1057

Phelan, P., Stebbings, L. A., Baines, R. A., Bacon, J. P., Davies, J. A., and Ford, C. (1998). *Drosophila* Shaking-B protein forms gap junctions in paired *Xenopus* oocytes. *Nature* 391, 181–184. doi: 10.1038/34426

Pitman, J. L., Huetteroth, W., Burke, C. J., Krashes, M. J., Lai, S. L., Lee, T., et al. (2011). A pair of inhibitory neurons are required to sustain labile memory in the *Drosophila* mushroom body. *Curr. Biol.* 21, 855–861. doi: 10.1016/j.cub.2011.03.069

Rabinowitch, I., Chatzigeorgiou, M., and Schafer, W. R. (2013). A gap junction circuit enhances processing of coincident mechanosensory inputs. *Curr. Biol.* 23, 963–967. doi: 10.1016/j.cub.2013.04.030

Rackauskas, M., Kreuzberg, M. M., Pranevicius, M., Willecke, K., Verselis, V. K., and Bukauskas, F. F. (2007). Gating properties of heterotypic gap junction channels formed of connexins 40, 43 and 45. *Biophys. J.* 92, 1952–1965. doi: 10.1529/biophysj.106.099358

Ramón, F., and Rivera, A. (1986). Gap junction channel modulation—a physiological viewpoint. *Prog. Biophys. Mol. Biol.* 48, 127–153. doi: 10.1016/0079-6107(86)90010-6

Rash, J., Curti, S., Vanderpool, K., Kamasawa, N., Nannapaneni, S., Palacios-Prado, N., et al. (2013). Molecular and functional asymmetry at a vertebrate electrical synapse. *Neuron* 79, 957–969. doi: 10.1016/j.neuron.2013.06.037

Rela, L., and Szczupak, L. (2003). Coactivation of motoneurons regulated by a network combining electrical and chemical synapses. *J. Neurosci.* 23, 682–692.

Rela, L., and Szczupak, L. (2007). In situ characterization of a rectifying electrical junction. *J. Neurophysiol.* 97, 1405–1412. doi: 10.1152/jn.00973.2006

Ringham, G. L. (1975). Localization and electrical characteristics of a giant synapse in the spinal cord of the lamprey. *J. Physiol.* 251, 395–407.

Roberts, A., Krasne, F. B., Hagiwara, G., Wine, J. J., and Kramer, A. P. (1982). Segmental giant: evidence for a driver neuron interposed between command and motor neurons in the crayfish escape system. *J. Neurophysiol.* 47, 761–781.

Sahu, G., Sukumaran, S., and Bera, A. K. (2014). Pannexins form gap junctions with electrophysiological and pharmacological properties distinct from connexins. *Sci. Rep.* 4:4955. doi: 10.1038/srep04955

Samuels, S. E., Lipitz, J. B., Dahl, G., and Muller, K. J. (2010). Neuroglial ATP release through innexin channels controls microglial cell movement to a nerve injury. *J. Gen. Physiol.* 136, 425–442. doi: 10.1085/jgp.201010476

Scemes, E., Suadicani, S. O., Dahl, G., and Spray, D. C. (2007). Connexin and pannexin mediated cell-cell communication. *Neuron Glia Biol.* 3, 199–208. doi: 10.1017/S1740925X08000069

Shang, Y., Claridge-Chang, A., Sjulson, L., Pypaert, M., and Miesenböck, G. (2007). Excitatory local circuits and their implications for olfactory processing in the fly antennal lobe. *Cell* 128, 601–612. doi: 10.1016/j.cell.2006.12.034

Shindo, Y., Fujimoto, A., Hotta, K., Suzuki, K., and Oka, K. (2010). Glutamate-induced calcium increase mediates magnesium release from mitochondria in rat hippocampal neurons. *J. Neurosci. Res.* 88, 3125–3132. doi: 10.1002/jnr.22467

Simonsen, K. T., Moerman, D. G., and Naus, C. C. (2014). Gap junctions in *C. elegans*. *Front. Physiol.* 5:40. doi: 10.3389/fphys.2014.00040

Smith, T. G., Baumann, F., and Fuortes, M. G. (1965). Electrical connections between visual cells in the ommatidium of limulus. *Science* 147, 1446–1448. doi: 10.1126/science.147.3664.1446

Söhl, G., Güldenagel, M., Traub, O., and Willecke, K. (2000). Connexin expression in the retina. *Brain Res. Brain Res. Rev.* 32, 138–145. doi: 10.1016/s0165-0173(99)00074-0

Söhl, G., and Willecke, K. (2004). Gap junctions and the connexin protein family. *Cardiovasc. Res.* 62, 228–232. doi: 10.1016/j.cardiores.2003.11.013

Starich, T. A., Herman, R. K., and Shaw, J. E. (1993). Molecular and genetic analysis of unc-7, a *Caenorhabditis elegans* gene required for coordinated locomotion. *Genetics* 133, 527–541.

Starich, T. A., Lee, R. Y., Panzarella, C., Avery, L., and Shaw, J. E. (1996). eat-5 and unc-7 represent a multigene family in *Caenorhabditis elegans* involved in cell-cell coupling. *J. Cell Biol.* 134, 537–548. doi: 10.1083/jcb.134.2.537

Starich, T. A., Xu, J., Skerrett, I. M., Nicholson, B. J., and Shaw, J. E. (2009). Interactions between innexins UNC-7 and UNC-9 mediate electrical synapse specificity in the *Caenorhabditis elegans* locomotory nervous system. *Neural Dev.* 4:16. doi: 10.1186/1749-8104-4-16

Stebbings, L. A., Todman, M. G., Phelan, P., Bacon, J. P., and Davies, J. A. (2000). Two *Drosophila* innexins are expressed in overlapping domains and cooperate to form gap-junction channels. *Mol. Biol. Cell* 11, 2459–2470. doi: 10.1091/mbc.11.7.2459

Suchyna, T. M., Nitsche, J. M., Chilton, M., Harris, A. L., Veenstra, R. D., and Nicholson, B. J. (1999). Different ionic selectivities for connexins 26 and 32 produce rectifying gap junction channels. *Biophys. J.* 77, 2968–2987. doi: 10.1016/s0006-3495(99)77129-8

Sun, Y.-A., and Wyman, R. J. (1996). Passover eliminates gap junctional communication between neurons of the giant fiber system in *Drosophila*. *J. Neurobiol.* 30, 340–348. doi: 10.1002/(sici)1097-4695(199607)30:3<340::aid-neu3>3.3.co;2-b

Swenson, K. I., Jordan, J. R., Beyer, E. C., and Paul, D. L. (1989). Formation of gap junctions by expression of connexins in *Xenopus* oocyte pairs. *Cell* 57, 145–155. doi: 10.1016/0092-8674(89)90180-3

Tanouye, M. A., and Wyman, R. J. (1980). Motor outputs of giant nerve fiber in *Drosophila. J. Neurophysiol.* 44, 405–421.

Teubner, B., Degen, J., Sohl, G., Guldenagel, M., Bukauskas, F. F., Trexler, E. B., et al. (2000). Functional expression of the murine connexin 36 gene coding for a neuron-specific gap junctional protein. *J. Membr. Biol.* 176, 249–262. doi: 10.1007/s002320001094

Thomas, J. B., and Wyman, R. J. (1984). Mutations altering synaptic connectivity between identified neurons in *Drosophila. J. Neurosci.* 4, 530–538.

Tomer, R., Denes, A. S., Tessmar-Raible, K., and Arendt, D. (2010). Profiling by image registration reveals common origin of annelid mushroom bodies and vertebrate pallium. *Cell* 142, 800–809. doi: 10.1016/j.cell.2010.07.043

Trenholm, S., McLaughlin, A. J., Schwab, D. J., and Awatramani, G. B. (2013). Dynamic tuning of electrical and chemical synaptic transmission in a network of motion coding retinal neurons. *J. Neurosci.* 33, 14927–14938. doi: 10.1523/jneurosci.0808-13.2013

Tsien, J. Z. (2000). Linking Hebb's coincidence-detection to memory formation. *Curr. Opin. Neurobiol.* 10, 266–273. doi: 10.1016/S0959-4388(00)00070-2

Vaney, D. I., and Weiler, R. (2000). Gap junctions in the eye: evidence for heteromeric, heterotypic and mixed-homotypic interactions. *Brain Res. Brain Res. Rev.* 32, 115–120. doi: 10.1016/s0165-0173(99)00070-3

Venance, L., Glowinski, J., and Giaume, C. (2004). Electrical and chemical transmission between striatal GABAergic output neurones in rat brain slices. *J. Physiol.* 559, 215–230. doi: 10.1113/jphysiol.2004.065672

Verselis, V. K., Ginter, C. S., and Bargiello, T. A. (1994). Opposite voltage gating polarities of two closely related connexins. *Nature* 368, 348–351. doi: 10.1038/368348a0

Veruki, M. L., and Hartveit, E. (2002). AII (Rod) amacrine cells form a network of electrically coupled interneurons in the mammalian retina. *Neuron* 33, 935–946. doi: 10.1016/s0896-6273(02)00609-8

Volff, J.-N. (2005). Genome evolution and biodiversity in teleost fish. *Heredity (Edinb).* 94, 280–294. doi: 10.1038/sj.hdy.6800635

Watanabe, A. (1958). The interaction of electrical activity among neurons of lobster cardiac ganglion. *Jpn. J. Physiol.* 8, 305–318. doi: 10.2170/jjphysiol.8.305

Weidmann, S. (1952). The electrical constants of Purkinje fibres. *J. Physiol.* 118, 348–360.

Werner, R., Levine, E., Rabadan Diehl, C., and Dahl, G. (1989). Formation of hybrid cell-cell channels. *Proc. Natl. Acad. Sci. U S A* 86, 5380–5384. doi: 10.1073/pnas.86.14.5380

White, R. L., Doeller, J. E., Verselis, V. K., and Wittenberg, B. A. (1990). Gap junctional conductance between pairs of ventricular myocytes is modulated synergistically by H+ and Ca++. *J. Gen. Physiol.* 95, 1061–1075. doi: 10.1085/jgp.95.6.1061

Wilson, R. I. (2013). Early olfactory processing in *Drosophila*: mechanisms and principles. *Annu. Rev. Neurosci.* 36, 217–241. doi: 10.1146/annurev-neuro-062111-150533

Wu, C.-L., Shih, M.-F. M., Lai, J. S.-Y., Yang, H.-T., Turner, G. C., Chen, L., et al. (2011). Heterotypic gap junctions between two neurons in the *Drosophila* brain are critical for memory. *Curr. Biol.* 21, 848–854. doi: 10.1016/j.cub.2011.02.041

Yaksi, E., and Wilson, R. I. (2010). Electrical coupling between olfactory glomeruli. *Neuron* 67, 1034–1047. doi: 10.1016/j.neuron.2010.08.041

Yamanaka, R., Shindo, Y., Hotta, K., Suzuki, K., and Oka, K. (2013). NO/cGMP/PKG signaling pathway induces magnesium release mediated by mitoKATP channel opening in rat hippocampal neurons. *FEBS Lett.* 587, 2643–2648. doi: 10.1016/j.febslet.2013.06.049

Zhang, Z., Curtin, K. D., Sun, Y.-A., and Wyman, R. J. (1999). Nested transcripts of gap junction gene have distinct expression patterns. *J. Neurobiol.* 40, 288–301. doi: 10.1002/(sici)1097-4695(19990905)40:3<288::aid-neu2>3.0.co;2-o

11

Opening of pannexin- and connexin-based channels increases the excitability of nodose ganglion sensory neurons

Mauricio A. Retamal[1,2*†], Julio Alcayaga[3*†], Christian A. Verdugo[1], Geert Bultynck[4], Luc Leybaert[5], Pablo J. Sáez[2], Ricardo Fernández[6], Luis E. León[7] and Juan C. Sáez[2,8]

[1] *Facultad de Medicina, Centro de Fisiología Celular e Integrativa, Clínica Alemana Universidad del Desarrollo, Santiago, Chile*
[2] *Departamento de Fisiología, Facultad de Ciencias Biológicas, Pontificia Universidad Católica de Chile, Santiago, Chile*
[3] *Laboratorio de Fisiología Celular, Departamento de Biología, Facultad de Ciencias, Universidad de Chile, Santiago, Chile*
[4] *KU Leuven, Laboratory of Molecular and Cellular Signaling, Department of Cellular and Molecular Medicine, Leuven, Belgium*
[5] *Physiology Group, Department of Basic Medical Sciences, Faculty of Medicine and Health Sciences, Ghent University, Ghent, Belgium*
[6] *Facultad de Ciencias Biológicas y Facultad de Medicina, Universidad Andrés Bello, Santiago, Chile*
[7] *Facultad de Medicina, Centro de Genética Humana, Clínica Alemana Universidad del Desarrollo, Santiago, Chile*
[8] *Centro Interdisciplinario de Neurociencias de Valparaíso, Instituto Milenio, Valparaíso, Chile*

***Correspondence:**
Mauricio A. Retamal, Facultad de Medicina, Centro de Fisiología Celular e Integrativa, Clínica Alemana Universidad del Desarrollo, Av. Las Condes 12438, Lo Barnechea, Santiago 7710162, Chile e-mail: mretamal@udd.cl;
Julio Alcayaga, Laboratorio de Fisiología Celular, Departamento de Biología, Facultad de Ciencias, Universidad de Chile, Las Palmeras 3425, Santiago 7800003, Chile e-mail: jalcayag@uchile.cl

† *These authors have contributed equally to this work.*

Satellite glial cells (SGCs) are the main glia in sensory ganglia. They surround neuronal bodies and form a cap that prevents the formation of chemical or electrical synapses between neighboring neurons. SGCs have been suggested to establish bidirectional paracrine communication with sensory neurons. However, the molecular mechanism involved in this cellular communication is unknown. In the central nervous system (CNS), astrocytes present connexin43 (Cx43) hemichannels and pannexin1 (Panx1) channels, and the opening of these channels allows the release of signal molecules, such as ATP and glutamate. We propose that these channels could play a role in glia-neuron communication in sensory ganglia. Therefore, we studied the expression and function of Cx43 and Panx1 in rat and mouse nodose-petrosal-jugular complexes (NPJcs) using confocal immunofluorescence, molecular and electrophysiological techniques. Cx43 and Panx1 were detected in SGCs and in sensory neurons, respectively. In the rat and mouse, the electrical activity of vagal nerve increased significantly after nodose neurons were exposed to a Ca^{2+}/Mg^{2+}-free solution, a condition that increases the open probability of Cx hemichannels. This response was partially mimicked by a cell-permeable peptide corresponding to the last 10 amino acids of Cx43 (TAT-Cx43CT). Enhanced neuronal activity was reduced by Cx hemichannel, Panx1 channel and $P2X_7$ receptor blockers. Moreover, the role of Panx1 was confirmed in NPJc, because in those from Panx1 knockout mice showed a reduced increase of neuronal activity induced by Ca^{2+}/Mg^{2+}-free extracellular conditions. The data suggest that Cx hemichannels and Panx channels serve as paracrine communication pathways between SGCs and neurons by modulating the excitability of sensory neurons.

Keywords: glial satellite cells, connexon, peripheral glial cells, sensory ganglia, nodose ganglia

INTRODUCTION

Plasma membrane hemichannels are composed of six protein subunits named connexins (Cxs). When two hemichannels from apposing cells contact, a gap junction channel (GJC) can be formed. GJCs communicate the cytoplasm of neighboring cells, whereas hemichannels allow the exchange of ions and small molecules between intra- and extracellular media. In addition, hemichannel opening is associated with the release of signaling molecules such as ATP, NAD^+, glutamate and prostaglandin-E_2 (Bruzzone et al., 2001; Stout et al., 2002; Ye et al., 2003; Cherian et al., 2005). On the other hand, pannexins (Panxs) are membrane proteins that also form channels at the plasma membrane (Bruzzone et al., 2003) and that share some properties with Cx hemichannels. For example, Panx1 channels are permeable to large molecules such as calcein and HEPES (Thompson et al., 2006; Romanov et al., 2012a,b). However, the permeability of Panx1 channels to ATP remains a matter of controversy, mainly because (i) Panx1 channels seem to be impermeable to ATP (Romanov et al., 2012a,b) and (ii) a high extracellular ATP concentration acts as blocker of Panx channels (Qiu and Dahl, 2009).

In the central nervous system (CNS), the expression and function of channels formed by Cxs or by Panxs have received

Abbreviations: NPJcs, nodose-petrosal-jugular complexes; Cx, connexin (e.g., Cx43); Panx, pannexin (e.g., Panx1); SGCs, satellite glial cells; GJC, gap junction cannel; βGA, 18β-glycyrrhetinic acid; HBSS, Hanks' balanced salt solution; GS, glutamine synthetase; BzATP, 2′(3′)-O-(4-benzoylbenzoyl) adenosine 5′-triphosphate triethylammonium; oATP, periodate oxidized adenosine 5′-triphosphate; OCT, Optimal cutting temperature compound.

particular attention (Sáez et al., 2003; Nagy et al., 2004; Theis et al., 2005; Thompson and Macvicar, 2008). However, in the peripheral nervous system (PNS), these channels have been less studied (Pannese, 2010). In the PNS, cell bodies of sensory neurons are located in sensory ganglia. These neurons have a pseudomonopolar shape and each perikaryon is fully surrounded by a cellular layer of glia, called satellite glial cells (SGCs). It has been proposed that these SGCs separate each neuronal body from one another, precluding the establishment of any type of synapse (Pannese, 1981, 2010). Consistent with this proposal, numerous ultrastructural and dye coupling studies have failed to find morphological or functional evidence of GJCs between sensory neurons or between sensory neurons and SGCs (Stensaas and Fidone, 1977; Shinder et al., 1998; Sakuma et al., 2001; Zuriel and Devor, 2001; Chen et al., 2002; Hanani et al., 2002; Pannese et al., 2003; Huang et al., 2006). In contrast, SGCs are coupled electrically and metabolically through GJCs (Sakuma et al., 2001; Huang et al., 2006; Pannese, 2010). It is known that SGCs influence neuronal excitability (Hanani, 2005; Huang and Hanani, 2005) by modifying the extracellular K^+ concentration and the intracellular free Ca^{2+} concentration (Pannese, 1981; Suadicani et al., 2009). Additionally, SGCs can modify neuronal responses to neurotransmitters (Pannese, 1981; Mandelzys and Cooper, 1992; Heblich et al., 2001).

The molecular mechanism of the autocrine/paracrine communication between SGCs and neurons is not fully understood. Hence, in the present work we tested whether cells from visceral sensory ganglia present functional Cx- and/or Panx-based channels in their plasma membranes. The mRNAs and proteins of Cx43 and Panx1 were detected in the nodosal petrosal jugular complex (NPJc) by RT-PCR, Western blot and confocal immunofluorescence. Electrophysiological studies *in vitro* showed hemichannels opening in response to a Ca^{2+}/Mg^{2+}-free solution (mHBSS), which is associated with increased electrical activity of nodose neurons. Compared with NPJc of wild type mice, ganglia from Panx1 knockout mice exposed to Ca^{2+}/Mg^{2+}-free solution showed a decreased response. Similar results were obtained when the $P2X_7$ receptors were pharmacologically inhibited. Thus, we postulate that Cx hemichannels and Panx channels serve as paracrine communication pathways in sensory ganglia, determining the electrical excitability of these PNS neurons.

MATERIALS AND METHODS

CHEMICALS

Fluoromount-G was purchased from Electron Microscopy Science (Ft. Washington, PA, USA). Distilled water, collagenase type A, deoxyribonuclease I, poly-D-lysine, 18β-glycyrrhetinic acid (βGA), 2′(3′)-O-(4-benzoylbenzoyl) adenosine 5′-triphosphate triethylammonium salt (BzATP), periodate oxidized adenosine 5′-triphosphate (oATP), acetyl choline and Probenecid were obtained from Sigma-Aldrich (St. Louis, MO, USA). Mouse nerve growth factor (NGF 7S) was obtained from Invitrogen (Carlsbad, CA, USA). Gap27 peptide was obtained from AnaSpec (Fremont, CA, USA). Mouse monoclonal glial glutamine synthetase (GS) antibody was obtained from Santa Cruz Biotechnology. Previously described rabbit polyclonal anti-Cx43 (see Brañes et al., 2002) and rabbit polyclonal anti-Panx1 (see Riquelme et al., 2013) sera were used.

ANIMALS

Male Sprague-Dawley rats and male and female C57BL/6 mice were obtained from the animal research facilities of the Faculty of Biological Sciences of the Pontificia Universidad Católica de Chile. Panx1 knock-out (KO) C57BL/6 mice previously described by Bargiotas et al. (2011) were kindly provided by Dr. Hannah Monyer, University Heidelberg, Germany. These animals were bred in the animal research facilities of the Pontifícia Universidad Católica de Chile. Wild type C57BL/6 mice were used as controls. The use of KO mice was limited to crucial experiments to reduce the number of animals sacrificed.

The Commission of Bioethics and Biosafety from our respective universities approved all experimental protocols, which were performed according to the "Guide for the Care and Use of Laboratory Animals," Institute of Laboratory Animal Research Commission on Life Sciences, National Research Council (National Academy Press, Washington, DC 1996).

GANGLION EXTRACTION

NPJc were obtained from 6-8-week-old Sprague-Dawley rats and from C57BL/6 mice (wild type and Panx1 knock out). Rats and mice of both sexes were anesthetized with sodium pentobarbitone 60 mg/kg which was administered intraperitoneally (i.p.) and supplemented with additional doses when necessary to maintain a light level of surgical anesthesia (Stage 3, plane 2). The neck was opened through a midline incision. Then, the vagus nerve was dissected, and its peripheral processes were cut ~1 centimeter distal to the ganglion. Next, each NPJc was exposed and its central process was cut approximately 1 mm from its apparent central border. After both NPJc were removed, the animals were euthanized by an overdose (180 mg/kg) of pentobarbitone.

IMMUNOBLOT

Ganglia were dissected as indicated above and then placed in ice cold phosphate buffered saline solution (PBS) containing 200 μg/mL soybean trypsin inhibitor, 1 mg/mL benzamidine, 1 mg/mL ε-aminocaproic acid and 2 mM phenylmethylsulfonyl fluoride and phosphatase inhibitors (20 mM $Na_4P_2O_7$ and 100 mM NaF). Then, ganglia were cut in small pieces with thin scissors and lysed by sonication. Samples were resuspended in Laemmli buffer and stored at −80°C, or proteins were resolved immediately in 8% SDS-PAGE. After electrophoresis, proteins were electrotransferred to VDF membranes incubated in PBS-BLOTTO (5% non-fat milk in PBS) for 45 min to block non-specific binding sites. Then, blots were incubated with primary antibodies for 1 h at room temperature, followed by several washes in PBS, and then incubated with HRP-conjugated goat anti-rabbit IgGs (secondary antibodies) for 1 h at room temperature. An ECL SuperSignal kit was used according to the manufacturer's instructions to detect immunoreactivity.

RT-PCR PROCEDURE

NPJc from male Sprague-Dawley rats anesthetized with pentobarbitone (60 mg/kg i.p.) were excised as described above and immediately transferred into cold modified (Ca^{2+}/Mg^{2+}-free)

Table 1 | Sequences of RT-PCR primers used to detect Cxs and Panxs mRNAs in rat's NPJcs.

Connexin	Sequence 5′–3′
Cx26F	ACGTTGGCCTTTTGGTTATG
Cx26R	TGTTGCGGGCTGTACTCAG
Cx37F	TTCTGGCCACCCTGGGGGGC
Cx37R	GGCTGGACCATGGAGCCGGT
*Cx43F	TACCACGCCACCACTGGCCCA
*Cx43R	ATTCTGGTTGTCGTCGGGGAAATC
Cx45F	GCAGAACAAAGCCAATATCGCCCA
Cx45R	TTCTGGTGATGGTAGGCCTGGATT
Pannexin	**Sequence 5′–3′**
*Panx1F	AGAGCTAGCTTTGTTCCCGG
*Panx1R	AGCTTATCTGGGTACCGATGG A
Panx2F	TGGACATCGTATTGCTCTGC
Panx2R	CCACGTTGTCGTACATGAGG

Hanks' balanced salt solution (mHBSS). Due to the small size of NPJc, they were pooled from 4 rats (8 NPJc) and stored in TRIzol-reagent for RNA extraction (For details, see Fernández et al., 2011). Briefly, RNA was prepared by the acid guanidinium-phenol-chloroform method, using TRIzol reagent (Invitrogen, Carlsbad, CA, USA) according to the manufacturer's instructions. Tissues were homogenized on ice. Then, quantification and purity checks of total RNA were performed spectrophotometrically and electrophoretically, respectively. After DNase treatment, total RNA (2 μg) was reverse transcribed into single strand cDNA using random primers and Moloney murine leukemia virus reverse transcriptase (M-MLV RT, 200 U/μL, Invitrogen, Carlsbad, CA. USA). Primer pairs used to amplify the coding regions for Cxs and for Panxs are listed in **Table 1**. Total RNA from rat brains was processed as mentioned above and used as positive control. As negative controls, reactions were also performed using samples without RNA or with cDNA prepared in the absence of RT enzyme. The amplified products were separated on a 2% agarose gel, which was subsequently stained with ethidium bromide (Sigma-Aldrich) and photographed under UV illumination. Images were taken with a digital camera.

ELECTROPHYSIOLOGY

NPJc were placed in ice-chilled Hanks' balanced salt solution (HBSS), and the connective tissue over the ganglia was carefully removed. The NPJc were transferred into a two-compartment chamber kept at 38.0 ± 0.5°C and superfused with HBSS supplemented with 5 mM HEPES buffer, pH 7.43, which was equlibrated with air and flowed at approximately 1.2 mL/min. Ganglia were placed in the 0.2 mL capacity lower compartment, over a pair of platinum electrodes, and pinned to the bottom of the chamber. The electrodes were connected to a stimulator, and a thermistor was in the superfusion channel near the ganglion surface. The vagus nerve (VN) was placed on paired Pt recording electrodes and lifted into the upper compartment of the chamber which was filled with mineral oil. Recording electrodes were connected to an AC-preamplifier (Model 1800; A-M Systems, USA) and the resulting electroneurogram was amplified, displayed on an oscilloscope, and recorded on video cassette tapes. The electroneurogram was also fed to a spike amplitude discriminator whose standarized pulses were counted at 1-s intervals to assess the frequency of discharge (f_{vn}), which was also digitized online through an AD board displayed on a computer using custom software and saved as ASCII encoded text files for later analysis. Drugs were applied in 10–50 μL boluses by micropipettes whose tips were placed approximately 1 mm distance from the exposed surface of the NPJc. All peptides were applied in a 50 μl bolus at a final concentration of 100 μM in a stop flow configuration for a maximum of 15 min. The mean basal activity (bas f_{VN}) was computed in the 30-s period prior before any experimental procedure. The steady-state frequency was computed in a 30-s interval at the end of an experimental superfusion period. The initial increase in frequency was computed in a 2 min interval after 1 min of an experimental superfusion period. The discharge frequency rate of change ($\Delta f/\Delta t$) was computed in 3-s intervals for each frequency point (1-s). The maximal values for the onset and end of the responses were used to compute the mean absolute value of $\Delta f/\Delta t$ ($|\Delta f/\Delta t|$).

CONFOCAL MICROSCOPY ANALYSIS

Rat NPJc were embedded in a resin (OCT) to allow sectioning of frozen tissue, frozen in liquid nitrogen and stored at −80°C. Sagittal cryostat sections (10 μm thick) were prepared fixed with 4% paraformaldehyde for 20 min at room temperature, washed three times with PBS and stored at 4°C. A blocking solution, which contained 1% IgG-free BSA, 50 mM NH_4Cl and 0.05% Triton X-100 in PBS was used to permeabilize and to block unspecific reactive sites. Panx1 and Cx43 were detected with rabbit polyclonal anti-Panx1 and anti-Cx43 sera, respectively. These antibodies were properly diluted in blocking solution and incubated overnight. After washing with PBS, Cy2 conjugated goat anti-rabbit (1:300) IgGF(ab′) fragments were incubated for 45 min at room temperature to detect bound primary antibody. After Cy2 washing, a monoclonal anti-glutamine synthetase (GS) antibody was added to samples for 3 h at proper dilution. Then, samples were washed and incubated with a Cy3 conjugated anti-mouse IgGF(ab′) and fragments were incubated as described previously for Cy2. DAPI Fluoromount-G (Electron Microscopy Sciences, Washington, PA) was used as an anti-fade solution to mount samples. Images were examined using a spectral two-photon confocal laser scanning microscope (Zeiss, Spectral Confocal Microscope, LSM780). Images with an optical thickness of 0.3 μm were obtained using a Plan-Apochromat 63x/1.40 Oil DIC M27 objective and then analyzed using Carl Zeiss image analysis software (Zen 2011).

STATISTICAL ANALYSES

Results are presented as the mean ± standard error (SE). Two populations were compared using Student's t-test or a nonparametric test, according to the data structure. Multiple populations were analyzed using repeated measures two-way ANOVA with multiple comparisons post-hoc tests. Analyses were performed using GraphPad Prism 6.03 (GraphPad Software, La Jolla, CA, USA) or Microsoft Excel programs. $P < 0.05$ was considered statistically significant. All comparisons of experimental data were performed with two-tailed tests, whereas the comparison of indexes was performed with one-tail statistics.

RESULTS

EXPRESSION AND CELLULAR DISTRIBUTION OF Cx43 AND PANX1 IN NPJc

The presence and distribution of Cx43 and Panx1 were studied in NPJc by RT-PCR, Western blot and confocal immunofluorescence analyses. In NPJc extracts, Cx43 mRNA was detected in each preparation analyzed (**Figure 1A**). We used extracts of lung and heart as positive controls (**Figure 1A**) and an enzyme mix without RNA as the negative control [**Figure 1A**, (-)]. In Western blot analyses, Cx43 was detected in NPJc samples (**Figure 1B**, line NPJc) and in rat brain extracts which were used as positive control. In both cases, several bands with electrophoretic mobility of approximately 43 kDa were observed. We suggest that in both samples, Cx43 present different degrees of phosphorylation, which is reflected as bands with different electrophoretic mobility (Kadle et al., 1991).

In rat NPJc sections, Cx43 reactivity was extremely weak and distributed as small dots all over the ganglion cross section (**Figure 1C**, nuclei were stained with DAPI). By zooming in an area containing a single neuron, Cx43 immunoreactivity was clearly observed not in the neuronal body but surrounding the neural body (**Figure 1D**, white arrows). To confirm that Cx43 was localized in SGCs, we performed double immunostaining with glutamine synthetase (GS), which is a SGC marker (Jasmin et al., 2010). Higher reactivity of Cx43 was usually detected in small zones of the ganglion periphery. In this case, a clear co-immunolabeling between GS and Cx43 was observed (**Figure 1E**). The digital analysis of the fluorescence signals obtained in **Figure 1E** (Red arrow) confirmed that Cx43 expression (**Figure 1F**, green line) was strongly associated with presence of GS (**Figure 1F**, red line). To clarify these findings, we used these data to generate a Z-plane rendered reconstruction (**Figure 1F**, lower panel). When Cx43 primary antibody was omitted, no immunolabeling was observed (**Figures 1G,H**).

Panx1 mRNA was also detected in NPJc extracts (**Figure 2A**). Samples from rat brain were used as positive control. In Western blot analyses, a band with electrophoretic mobility of approximately 50 kDa was evident in NPJc extracts. Other bands with higher and lower electrophoretic mobility were also detected (**Figure 2B**, NPJc). In rat brain samples, an approximately 50 kDa band was also detected; however, bands with less electrophoretic mobility were also evident (**Figure 2B**, Brain). These bands could be the result of post-translational modification, such as glycosylation (Boassa et al., 2007). Similar results have been obtained in samples from rat brain using different Panx1 antibodies (Cone et al., 2013). We analyzed the localization of Panx1 in NPJc and found that Panx1 reactivity was localized primarily as dotted marks in the cytoplasm of sensory neurons (**Figure 2C**). Panx1 was located mostly in neurons with variable degrees of reactivity (**Figure 2D**, upper panel). Analyses of DAPI and Panx1 localization (**Figure 2D**, lower panel) revealed that Panx1 (green line) is excluded from zones with strong

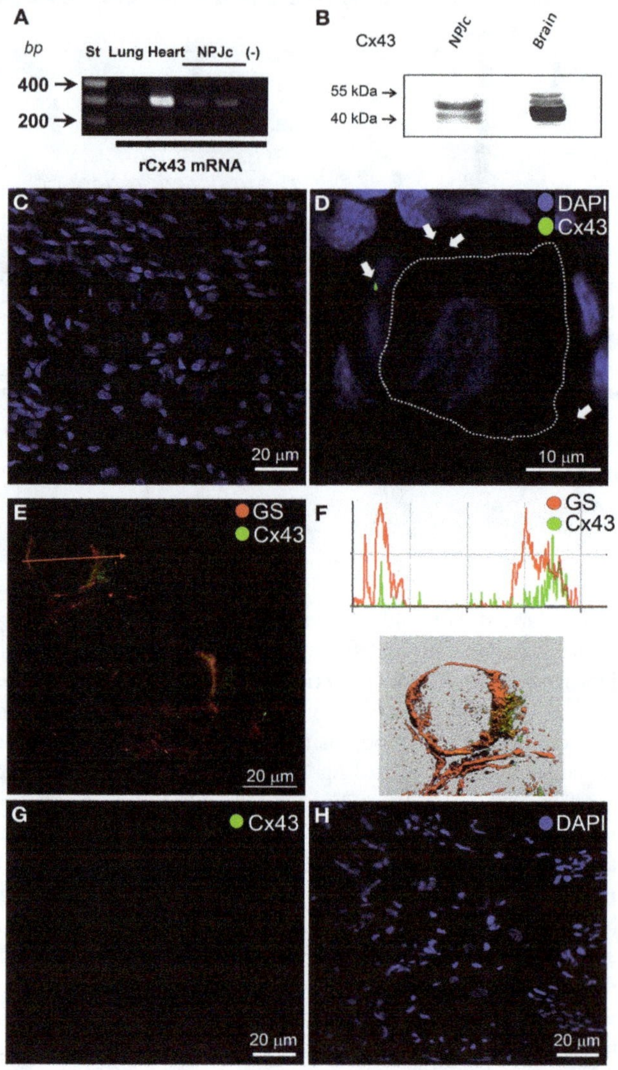

FIGURE 1 | Cx43 is expressed in NPJcs. (A) RT-PCR detection of Cx43 mRNA in total extracts of rat NPJcs. Each lane shows Etd$^+$-stained amplicons: lanes 1 and 2, lung and heart used as positive controls, lanes 3 and 4 two independent NPJc samples and lane 5 a reaction performed without RNA (negative control) ($n = 4$). **(B)** Levels of Cx43 in total NPJc homogenates were analyzed by Western blot. Cx43 was detected in total homogenates of NPJc and brain (50 µg of protein). **(C)** Indirect immunofluorescence of Cx43 in a slice of NPJc. **(D)** Digital zoom of a single neuron soma, where white arrows indicate the presence of Cx43 and the neuronal soma was drawn with a white line. **(E)** Co-localization of Cx43 (green) and GS (red) was evident in a zone (NPJc edge) where both neuronal bodies and SGC are present. **(F)** Digital analyses of the fluorescence observed in **(E)** (Red arrow), and the Z-reconstruction of this fluorescent signal. **(G)** Indirect immunofluorescence in a slice of NPJc when Cx43 primary antibody was omitted. **(H)** DAPI staining shows cell nucleus of same slice as in panel **(G)**.

DAPI fluorescence, corresponding mostly to SGC nuclei (blue line). This finding suggests that Panx1 is expressed, mainly, if not exclusively, in sensory neurons. To confirm this finding, double immunostaining of Panx1 and GS was performed, which showed Panx1 localization in the soma of sensory neurons (**Figure 2E**), and not in GS reactive zones. Rendering of a single neuron confirmed that Panx1 is localized in neurons (green) and not in surrounding SGCs (red) (**Figure 2F**). When Panx1 primary antibody was omitted, no immunolabeling was observed (**Figures 2G,H**).

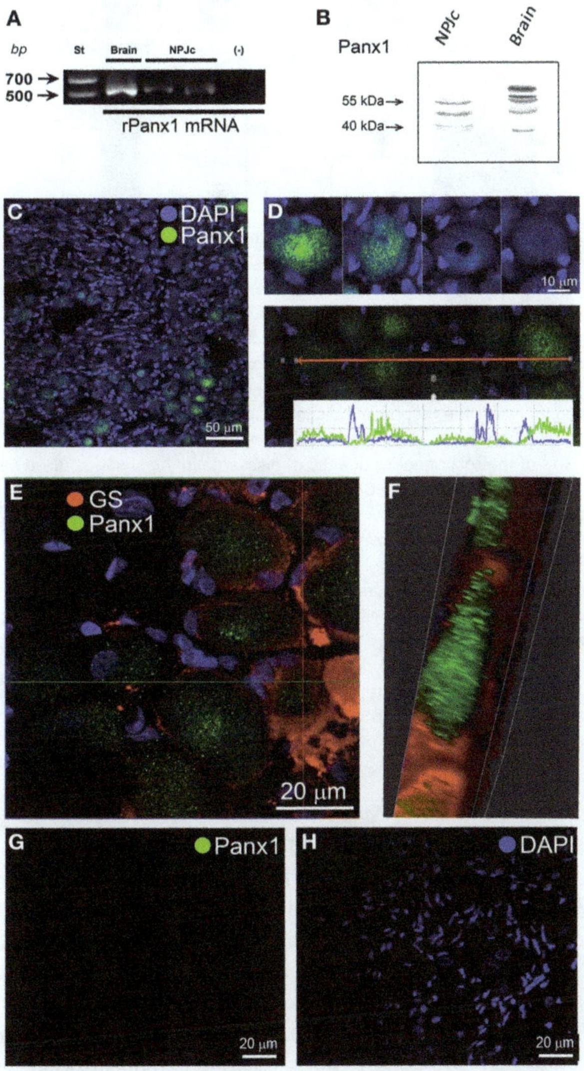

FIGURE 2 | Panx1 is expressed in NPJcs. (A) RT-PCR detection of Panx1 mRNA in total extracts of NPJcs. Each lane shows Etd^{+}-stained amplicons: lane 1 correspond to brain extracts used as positive control; lanes 2 and 3 correspond to two independent NPJc samples and lane 4 correspond to a reaction that was performed without RNA (negative control). **(B)** Panx1 protein was detected by Western blot analyses. In NPJc homogenates, several reactive bands with electrophoretic mobility were found. However, a band of approximately 50 kDa was clearly detected in rat NPJc and brain samples. **(C)** Confocal immunofluorescence detection of Panx1 in NPJc slices. **(D)** Four examples of neurons expressing Panx1 are presented (upper panel). Digital analysis of a zone of the NPJc expressing Panx1 (lower panel, green line) and DAPI staining (lower panel, blue line) were performed, showing that Panx1 does not co-localize with DAPI, which mainly marks the SGC nuclei. **(E)** Co-immunolabeling of Panx1 (Green) and GS (Red) **(F)** 3D module of ZEN 2011 used to visualize the data by transparent rendering Z-reconstruction. **(G)** Indirect immunofluorescence in a slice of NPJc when Panx1 primary antibody was omitted. **(H)** DAPI staining showing cell nuclei of the same field presented in panel **(G)**.

Cx43 HEMICHANNEL OPENING INCREASES NODOSAL NEURONAL ACTIVITY

We demonstrated that NPJc cells express at least Cx43. The opening of channels formed by these proteins is known to permit the release of several neurotransmitters (Sáez et al., 2003). TATCx43CT is a peptide corresponding to the last 10 amino acids of the C-terminal tail of Cx43 which favors the opening of Cx43 hemichannels by preventing their closure at high cytoplasmic Ca^{2+} concentration (Ponsaerts et al., 2010; De Bock et al., 2012; Iyyathurai et al., 2013). We explored whether this peptide could increase the electrical activity of sensory neurons projecting through the vagus nerve in the stop-flow mode. Stop-flow *per se* (for a maximum of 15 min) did not affect neuronal basal discharges (**Figure 3A**, blue line). The application of 100 μM TATCx43CT to rat NPJc *in vitro* induced a fast and sustained increase in the nodose neuronal discharge frequency recorded in the vagus nerve (f_{VN}) (**Figure 3A**, black line). This response was maintained for at least 15 min. The application of 100 μM TATCx43Rev [a peptide that has the same sequence as TATCx43CT but in a reversed order and that does not interact with Cx43 hemichannels (Ponsaerts et al., 2010)] only induced a small and transient (30–90 s) increase in neuronal activity (**Figure 3A**, red line). This finding suggests that Cx43 hemichannel opening increases the neuronal discharge of action potentials.

A method to induce Cx43 hemichannel opening is to expose cells to an extracellular solution without divalent cations (Sáez et al., 2010). Our previous results supported the idea that Cx43

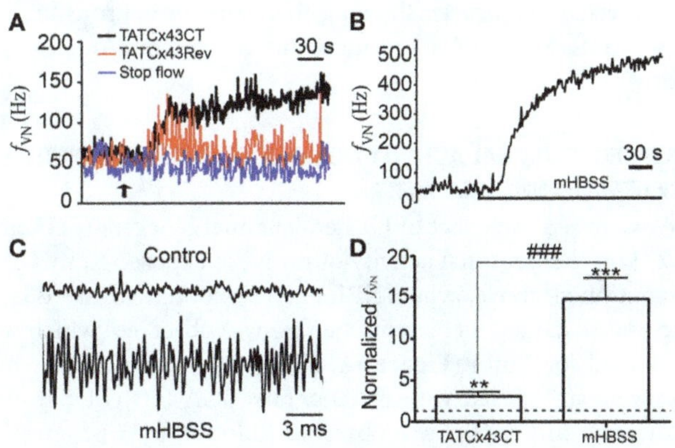

FIGURE 3 | TATCx43CT and mHBSS enhance the neuronal activity of rat NPJcs. Rat NPJc were placed in a recoding chamber superfused with Hanks' solution (HBSS) and the frequency of discharge was measured in the vagus nerve (f_{VN}). **(A)** Effect of (10 μL) (black arrow) 100 μM TATCx43CT (black trace) or 100 μM TATCx43Rev (red trace) applied to NPJcs *ex vivo* on f_{VN}. Stop-flow technique did not modify the basal frequency of discharge (blue trace). **(B)** f_{VN} recorded before and during (indicated by continuous line) superfusion with mHBSS. **(C)** Spontaneous activity recorded under control conditions (upper trace) and during superfusion with Ca^{2+}/Mg^{2+}-free HBSS (mHBSS; lower trace). **(D)** Graph showing the normalized effect of TATCx43CT ($n = 9$) and mHBSS ($n = 21$) over the electrical activity of NPJcs compared with control conditions (dashed line). Statistical significances; $^{**}P < 0.01$ and $^{***}P < 0.001$ different compared with the basal (pre stimulus) condition, $^{###}P < 0.001$ between stimuli. Wilcoxon matched-pairs signed rank test.

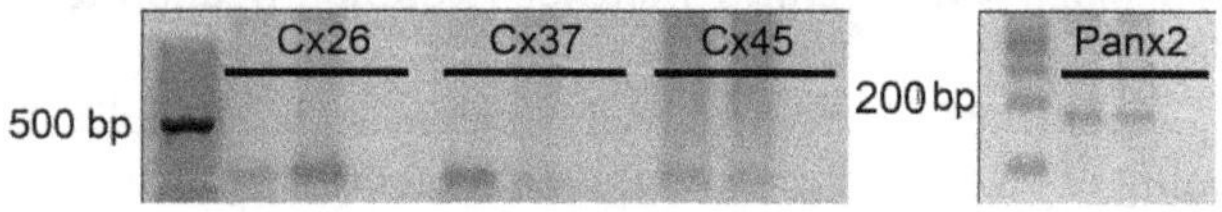

FIGURE 4 | Presence of Cxs 26, 37, 45, and Panx2 in NPJcs. mRNAs of Cxs 26, 37 and 45 and Panx2 were detected in total RNA extracted from rat NPJcs and in RT-PCR analyses. The primers used are listed in **Table 1**.

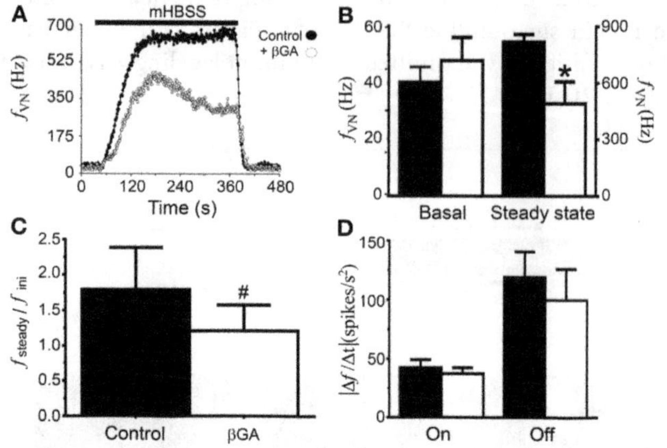

FIGURE 5 | The enhanced neuronal activity induced by mHBSS is sensitive to βGA. (A) Rat NPJcs were placed in a recording chamber superfused with Hanks' solution (HBSS), followed by superfusion with mHBSS alone (filled circles) or supplemented with 70 μM βGA (empty circles). Bar over the response indicates superfusion with mHBSS. **(B)** Graph showing the average of the basal (pre-stimulus) and steady-state (last 30 s of the response) frequency of discharge in the absence (filled bar) or presence (empty bar) of βGA in the superfusion medium ($n = 6$). **(C)** Graph showing the ratio between the mean steady-state frequency of discharge and mean initial frequency (2 min average after 1 min of the beginning of the stimulation) for each condition. **(D)** Calculated absolute values for the increase and decrease rates of the discharge frequency in the absence (filled bar) or presence (empty bar) of βGA in the superfusion solution. ON represents the rise in frequency when the NPJc is superfused with mHBSS and OFF is the decrease in the frequency of discharge when the ganglion is re-superfused with HBSS. Statistical significances; $P < 0.05$ with respect to control (mHBSS) condition. #Wilcoxon matched-pairs signed rank test; *Student's paired *t*-test. Deviation marks: SE.

hemichannels are present in NPJc cells. Thus, we tested the effect of Hanks' solution with nominal zero Ca^{2+} and Mg^{2+} (mHBSS) on neuronal activity. In rat NPJc superfused with mHBSS, an increase of the basal frequency of discharge was recorded (**Figure 3B**). This increase in neuronal activity began 0.5–3 min after changing regular Hanks' solution (HBSS) to mHBSS and persisted for at least 30 min. According to these results, few spontaneous action potentials were detected in control conditions (**Figure 3C**, upper recording), however, during mHBSS or after TATCx43CT addition, spontaneous action potentials were frequently recorded (**Figure 3C**, lower recording). The quantification of these results shows that mHBSS increased the neuronal activity 14.8 ± 2.4 times ($n = 21$), which was 4.2 times larger than that of TATCx43CT (3.1 ± 0.5 times; $n = 7$) (**Figure 3D**, $P < 0.001$). This result can be explained by the incomplete stimulation of Cx43 hemichannel opening by TATCx43CT, and by the presence of hemichannels composed of other Cxs. We analyzed the presence of mRNAs of other Cxs and Panxs present in the nervous system (Nagy et al., 2004; Thompson and Macvicar, 2008). The mRNAs for Cx26, Cx37, Cx45, and Panx2 were detected (**Figure 4**), showing that sensory neurons and/or SGCs express several Cx isoforms and at least one other Panx isoform.

INCREASED NEURONAL ACTIVITY INDUCED BY mHBSS IS SENSITIVE TO Cx HEMICHANNEL BLOCKERS

Here, we tested the effect of Cx hemichannel blockers (βGA and Gap27) on the neuronal activity of rat NPJc exposed to mHBSS. As mentioned above, when NPJcs were exposed to mHBSS, a rapid and sustained increase in the f_{VN} was observed, which was maintained for 5 min (**Figure 5A**, filled circles). However, when NPJc were superfused with mHBSS plus βGA (70 μM), an initial increase in the f_{VN} was observed followed by a progressive reduction in the maximal frequency until reaching a steady-state. Both, the maximal f_{VN} and f_{VN} in steady-state were lower compared with values under control conditions (**Figure 5A**). Analyses of these results showed no differences between the basal f_{VN} under control conditions compared with those values before βGA (**Figure 5B**; $P > 0.05$). However, the steady-state frequency of discharge was reduced from 815.5 ± 42.6 Hz under control conditions to 491.6 ± 116.7 Hz during βGA (**Figure 5B**, $P < 0.05$). Additionally, the ratio between the steady-state and the maximal initial f_{VN} was significantly higher in NPJcs under control conditions compared with those values with βGA (**Figure 5C**; 1.79 ± 0.59 before vs. 1.22 ± 0.37 during βGA; $P < 0.05$). Then, we analyzed the absolute increase rate of the f_{VN} when HBSS was changed to mHBSS and vice versa. In this case, no difference was found before (42.5 ± 6.7 spikes/s^2) and during superfusion with βGA (37.6 ± 5.0 spikes/s^2) (**Figure 5D**). Similar results were observed in the decrease rate of the f_{VN} when HBSS superfusion was restored (before 118.8 ± 21.9, v/s during βGA 99.7 ± 26.2 spikes/s^2) (**Figure 5D**). These results reveal that Cx hemichannels are partly responsible for the increase in the f_{VN} when superfused with mHBSS. To test the role of Cx43 hemichannels in this system, we used a Cx43 mimetic peptide (Gap27) (Evans and Leybaert, 2007; Retamal et al., 2007). In mHBSS stop-flow experiments, a final concentration of 50 μM Gap27 was used, and this application reduced neuronal activity by 86.9 ± 6.6%. Similar to that observed with mHBSS, the TATCx43CT effect was also sensitive to Cx- hemichannel blockers. For instance, βGA (70 μM) reduced the TATCx43CT effect by 75.9 ± 14.8% ($n = 6$) and the Gap27 effect by 70.9 ± 7.4% ($n = 4$). Therefore, at least Cxs form functional hemichannels in sensory ganglia cells, and their opening modulates the electrical properties of sensory neurons.

NPJcs FROM PANX1 KNOCK-OUT MICE ARE LESS SENSITIVE TO mHBSS

Our data suggest that increased neuronal activity induced by mHBSS may also depend on Panx1 channel opening. Therefore, we tested whether sensory neurons from Panx1 Knock out (Panx1-KO) mice could be less sensitive to mHBSS. We found

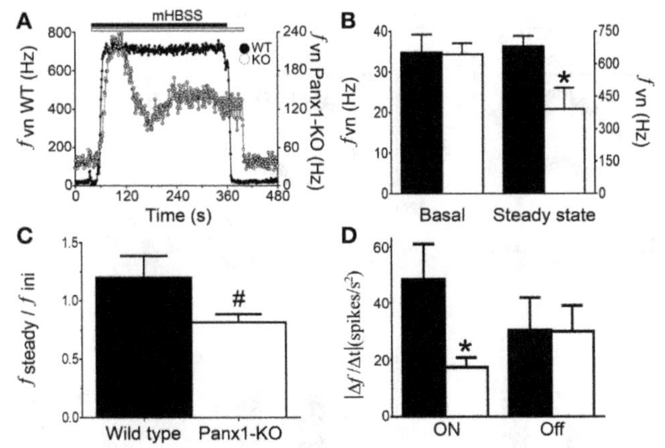

FIGURE 6 | NPJcs from Panx1 knock-out mice are less responsive to mHBSS. NPJcs were obtained from wild type (WT) and Panx1 knock-out (Panx1-KO) mice and the mHBSS effect on the vagus nerve activity was tested. **(A)** Representative records showing changes in f_{VN} induced by mHBSS in NPJcs of WT (filled circles; left Y-axis) and Panx1-KO (empty circles; right Y-axis) mice. Bars on top of the recordings indicate the duration of each stimulus. **(B)** f_{VN} of NPJcs of WT (filled bar, $n = 8$) and Panx1-KO mice (empty bar, $n = 9$) recorded when the NPJcs were superfused with HBSS (Basal; left Y-axis) or mHBSS (Steady state; right Y-axis). **(C)** Graph showing the ratio between the f_{VN} in steady state divided by the maximal f_{VN} reached after mHBSS superfusion (see **Figure 3** for definition). **(D)** Graph showing the rate (in absolute value) of the initial changes induced by mHBSS (ON) or the rate of changes when mHBSS was changed to HBSS (OFF) in NPJcs of WT (filled bar, $n = 8$) and Panx1-KO mice (empty bar, $n = 9$). Statistical significance; $P < 0.05$ with respect to WT. #Because these data are unrelated and not normally distributed, we used the Mann-Whitney test; *Student's t-test.

that WT and Panx1-KO mice of similar age and weight presented no obvious morphological differences in their NPJcs (not shown). When HBSS was changed to mHBSS, an abrupt increase in the f_{VN} was observed in both WT and Panx1-KO NPJcs (**Figure 6A**). However, in NPJcs of Panx1-KO mice the f_{VN} subsequently decreased until reaching a new steady-state (**Figure 6A**, empty circles). Analyses of these data revealed that the basal f_{VN} values from WT (34.8 ± 4.4 Hz) and Panx1-KO (34.4 ± 2.7 Hz) mice under control conditions did not show significant differences (**Figure 6B**; $P > 0.05$). However, the steady-state of the f_{VN} reached under constant mHBSS superfusion showed that the WT activity was ~60% larger than that recorded in NPJcs from Panx1-KO mice (680.6 ± 47.8 Hz, WT vs. 391.6 ± 98.1 Hz, Panx1-KO; $P < 0.05$, Student's t-test; **Figure 6B**). The ratio between the steady state and maximal initial frequency of discharge was significantly lower in NPJcs from Panx1-KO mice, compared with that of WT mice (**Figure 6C**; 1.20 ± 0.19 WT vs. 0.81 ± 0.07 KO; $P < 0.05$, Mann-Whitney test). Interestingly, the rate of the initial f_{VN} increase in NPJcs of Panx1-KO animals (17.4 ± 3.5 spikes/s^2) was slower than that observed in NPJcs of WT mice (48.6 ± 12.5 Hz/s; $P < 0.05$, Student's t-test; **Figure 6D**). Nevertheless, the rate of decrease (when mHBSS was changed to HBSS) of f_{VN} was similar, as demonstrated when comparing recordings obtained in NPJc of WT and Panx1-KO mice (**Figure 6D**; $P > 0.05$, Student's t-test).

THE ACTIVATION OF THE P2X$_7$ RECEPTOR-PANX1 CHANNEL PATHWAY MODULATES THE SENSORY NEURON RESPONSE TO mHBSS

It is known that Panx1 channels can be open after P2X$_7$ receptor activation (Iglesias et al., 2008; Gulbransen et al., 2012). We used BzATP, which is a P2X$_7$ receptor agonist to determine whether the activation of P2X$_7$ receptors could mimic the mHBSS effect. We found that a single application of BzATP (200 μM) induced a large (~200 Hz), but brief, increase in f_{VN} (**Figure 7A**). However, we noted that a second application of BzATP induced a small response compared with the first response (**Figure 7A**). This finding indicates that the P2X$_7$ receptor in the rat NPJc is desensitized for at least this period of time. Because P2X$_7$ receptors can be desensitized by repetitive applications of BzATP, we blocked these receptors. After P2X$_7$ receptors were desensitized by repetitive applications of BzATP (200 μM), NPJc were immediately superfused with mHBSS. An increase in f_{VN} was observed; however, this increase was always smaller than that observed under control conditions. After the initial f_{VN} increase, the neuronal activity reached a lower steady-state (**Figure 7B**, $n = 4$). Notably, these responses were similar to those observed in NPJcs of Panx1-KO mice superfused with mHBSS (**Figure 6A**, empty circles). This finding suggests that the coupled P2X$_7$ receptor-Panx1 channel is responsible for maintaining the high level of neuronal activity in the absence of divalent cations. To confirm that P2X$_7$ receptors are involved in this phenomenon, we superfused NPJcs with mHBSS supplemented with oATP (100 μM), which is a P2X$_7$ receptor antagonist. We found that oATP drastically reduced the maximal frequency of discharge of the vagus nerve in response to mHBSS compared with to control conditions (**Figure 7C**, $n = 3$). This result is similar to that obtained when P2X$_7$ receptors were desensitized. Finally, we tested the effect of probenecid (1 mM), a Panx1 channel blocker. When NPJcs were superfused with mHBSS with probenecid, a reduction in neuronal activity (**Figure 7D**, empty circles, $n = 4$) compared with control conditions was evident (**Figure 7D**, filled circles). All of these data support the idea that the Cx hemichannel-P2X$_7$ receptor-Panx1 channel pathway is present in the NPJc and modulates the activity of sensory neurons.

NEUROTRANSMITTERS MODULATE SENSORY NEURON RESPONSES TO mHBSS

Sensory neurons express ATP and dopamine receptors (Burnstock, 2009b; Peiser et al., 2009) that upon activation modulate the electrical activity of visceral sensory neurons (Alcayaga et al., 2003). Additionally, these two neurotransmitters are known to modulate channels formed by Panxs and Cxs (Kothmann et al., 2009; Qiu and Dahl, 2009). Therefore, we studied whether ATP and dopamine mediate the enhanced neuronal activity induced by mHBSS. In NPJcs under constant superfusion with mHBSS, boluses of either ATP (10 μl of 500 μg) or dopamine (10 μl of 500 μg) were applied. ATP induces fast and transient decreases in maximal f_{VN} (**Figure 8A**, $n = 5$), however, such inhibition was variable between preparations. Similarly, dopamine decreased the maximal f_{VN}, abolishing the increase in f_{VN} in 3 of 5 tests. (**Figure 8B**, $n = 5$).

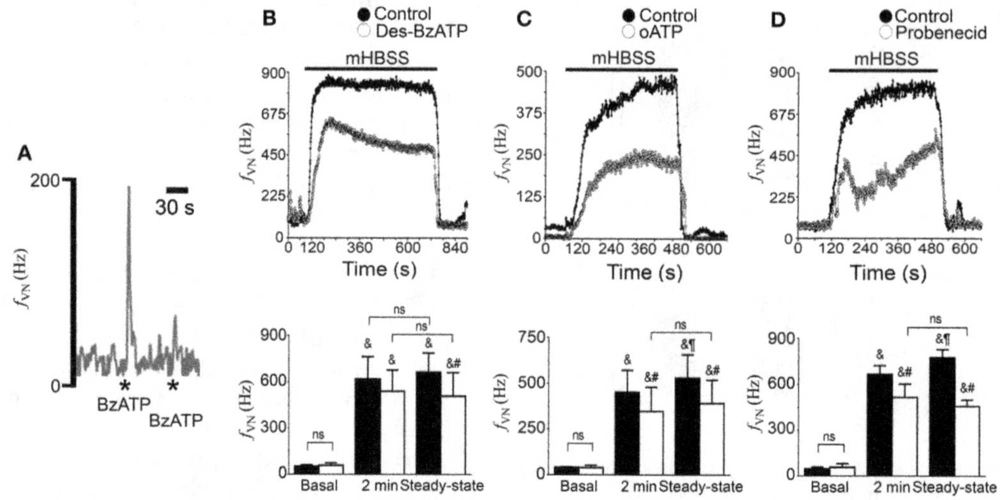

FIGURE 7 | Neuronal activation induced by mHBSS is sensitive to $P2X_7$ receptor-Panx1 channel blockers and neurotransmitters. Representative recordings of rat NPJcs superfused with mHBSS in the absence (control) or presence of P2X7/Panx blockers. **(A)** 200 μM of BzATP induced a rapid and transient increase in f_{VN}; however, a second application of BzATP induced only a small response (desensitization). Asterisks show the moment of application of a BzATP bolus to the bath solution. Neuronal frequency of discharge recorded from NPJcs superfused with mHBSS before (Control, filled circles) and after (empty circles) $P2X_7$ desensitization (**B**; $n = 4$). Neuronal frequency of discharge recorded from NPJcs superfused with mHBSS alone (Control, filled circles) or supplemented with 100 μM oATP (**C**, empty circles; $n = 4$) with 1 mM probenecid (**D**, empty circles; $n = 5$). **(B–D**; Lower panels). Quantification of the f_{VN} before superfusion with mHBSS (basal), during 2 min after 1 min in mHBSS and when the f_{VN} reached a steady state (30 s before the end of mHBSS superfusion) for each corresponding treatment. [&]Significantly different from basal in the same condition; Newman-Keuls multiple comparisons test. [#]Significantly different from control conditions (mHBSS), Bonferroni's multiple comparisons test. [¶]Significantly different from 1 to 3 min (2 min) activity in the same condition. Newman-Keuls multiple comparisons test. Repeated measures 2-Way ANOVA.

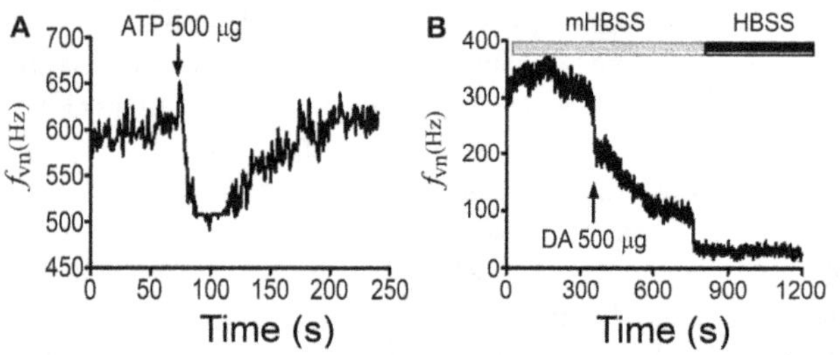

FIGURE 8 | Neuronal activation induced by mHBSS is sensitive to neurotransmitters. Representative recordings of electrical activity of rat NPJcs superfused with mHBSS and during steady state activity reached after application of 10 μL of **(A)** 500 μg ATP ($n = 5$) or **(B)** 500 μg dopamine (DA) ($n = 5$).

DISCUSSION

In this work, we demonstrated that neurons from rat and mouse nodose ganglia present a dramatic increase in activity upon exposure to a Ca^{2+}/Mg^{2+}-free solution (mHBSS). This enhanced neuronal activity was mimicked to some extent by a peptide that favors Cx43 hemichannel opening and was partially inhibited by Cx hemichannel and Panx channel blockers. According to these results, NPJcs from Panx1 KO mice showed a reduced response to mHBSS, which was similar to the response observed when $P2X_7$ receptors were pharmacologically blocked. Therefore, $P2X_7$ receptors, Panx1 channels, and Cx43 hemichannels play a role in the enhanced neuronal activity induced by mHBSS, and this activity can be modulated by neurotransmitters, such as ATP and dopamine.

It is accepted that low extracellular Ca^{2+} concentration increases neuronal excitability by direct changes in electrical properties of the neuronal plasma membrane (Frankenhaeuser and Hodgkin, 1957; Hille, 1968). However, in this study, we show that the mHBSS-triggered increase in sensory neuronal activity is partially explained by the activation of Cx hemichannels, Panx channels and $P2X_7$ receptors. In support of this notion we found that: (i) this response was sensitive to Cx hemichannel and Panx channel blockers, (ii) the activity of sensory neurons is increased by mHBSS, (iii) $P2X_7$ receptor desensitization and oATP partially reduced the increased neuronal activity induced by mHBSS, and (iv) NPJcs of Panx1-KO mice showed a reduced response to mHBSS. Therefore, these results strongly support the notion that the activity of channels formed by Cxs and Panxs does modulate the neuronal activity of the nodose ganglion. Interestingly, the exogenous application of ATP decreased neuronal activity in NPJcs superfused with mHBSS. Accordingly, ATP has been shown to exert a negative feedback control over Panx1 channels, inducing their closure (Qiu and Dahl, 2009). An alternative explanation can be that ATP activates P2X receptors, which in turn overexcites neurons inducing a transient inactivation of voltage-gated Na^+ channels, which will decrease the number of action potentials, decreasing the f_{VN}. The inhibitory effect of ATP under mHBSS can be favored by the fact that Mg^{2+} decreases ATP-mediated currents in nodose neurons (Li et al., 1997). Therefore, when Mg^{2+} is removed, ATP can have a greater chance to overexcite sensory neurons, decreasing their activity.

It has been shown that TATCx43CT can restore Cx43 hemichannel activity blocked by high intracellular Ca^{2+} concentrations, thrombin or TATCx43L2 in mammalian cells (Ponsaerts et al., 2010, 2012). Cx43 hemichannels present in sensory ganglia cells appear to be preferentially closed under resting conditions, whereas TATCx43CT releases these hemichannels from this closed state, at physiological Ca^{2+} concentration. In our experiments, the effect of mHBSS on increasing neuronal activity was much more potent than that of TATCx43CT. This result can be due to a combination of direct effects of low extracellular Ca^{2+} concentration on membrane properties and the contribution of other types of Cxs in addition to Cx43. In support of the above hypothesis, we detected Cx26, Cx37 and Cx45 mRNAs in NPJcs. Moreover, it has been reported that injections of capsaicin into a temporomandibular joint transiently increase Cx36 and Cx40 levels in sensory neurons, and Cx26 levels in SGCs of the trigeminal ganglion (Garrett and Durham, 2008). Therefore, the expression of more than one type of Cxs in SGCs and/or sensory neurons of NPJc cannot be ruled out. Thus, we strongly suggest that more than one type of Cx hemichannel is involved in the mHBSS-induced enhanced neuronal activity. Nevertheless, Cx43 appears to have a prominent role because the mHBSS-induced enhanced neuronal activity was drastically inhibited by Gap27, which is a well-known Cx43 hemichannel blocker (Piao et al., 2011; Wang et al., 2012).

Western blot analyses showed several bands of Panx1 that would most likely correspond to glycosylated and non-glycosylated forms because N-glycosylation of Panx1 can change its apparent molecular weight (Penuela et al., 2009). We believe that the non-glycosylated Panx1 is in the cytoplasm and that the glycosylated form forms channels at the plasma membrane of sensory neurons. Consistent with this interpretation, it has been shown that non-glycosylated Panx1 is not delivered to cell membrane and glycosylated Panx1 is observed in the cell surface (Boassa et al., 2007). However, a biochemical study is required to test this hypothesis.

It has been suggested that SGCs and neurons of sensory ganglia present bidirectional communication (Suadicani et al., 2009; Pannese, 2010). Chemical "crosstalk" of sensory neurons has been shown in dorsal root ganglia (Amir and Devor, 1996) and in nodose ganglia (Oh and Weinreich, 2002). However, as mentioned earlier, the somata of sensory neurons are capped by SGCs. Thus, when the somata of sensory neurons release neurotransmitters, such as serotonin, ATP, substance P or acetylcholine (Fueri et al., 1984; Palouzier-Paulignan et al., 1992; Matsuka et al., 2001; Zhang et al., 2007), we propose that these transmitters can reach neighboring SGCs, generating neuronal-glia communication (Suadicani et al., 2009). It is also possible that SGCs release gliotransmitters, such as ATP, that may reach sensory neurons. SGC activation increases intracellular free Ca^{2+} concentration in neighboring neurons via a suramin-sensitive pathway (Suadicani et al., 2009). Therefore, Cx hemichannels and Panx channels could serve as pathways for releasing neuroactive molecules (e.g., ATP and/or glutamate) directly involved in communication between SGCs and neurons. An alternative hypothesis is that sensory neurons and SGCs are connected by gap junction. However, there is no evidence of gap junction plaques between sensory neurons and SGCs and also there is no dye transfer between these cells *in vivo*. However, the above evidence does not exclude the possibility of electrical coupling mediated by small gap junctions between neurons and glial cells that are difficult to see by electron microscopy or to detect with a negatively charged permeability probe such as Lucifer yellow.

The initial increase in f_{VN} induced by mHBSS was only partially affected by Cx hemichannels and Panx channel inhibitors. However, the maximal f_{VN} and its maintenance were drastically modified. This result could suggest that these processes are dependent of neurotransmitters released through Cx- and/or Panx-based channels. Because the $P2X_7$ receptor is involved, we speculate that at least ATP is important in this process. This hypothesis is consistent with the knowledge that ATP-mediated communication between SGCs and neurons is extremely important (Gu et al., 2010). However, $P2X_7$ receptors have been found only in SGCs (Gu et al., 2010). We also observed that sensory neurons express Panx1. Thus, the couple $P2X_7$ receptor-Panx1 channel can only be possible if neurons express low levels of $P2X_7$ receptor not detected by Gu et al. (2010) and if SGCs present low levels of Panx1 that were not detected in the present work. Single cell mRNA analyses will be required to clarify this point.

In the CNS, when a spinal cord injury occurs, it may induce chronic neuropathic pain. Cx43 expressed in astrocytes play a relevant role in this phenomenon. It has been shown that carbenoxolone (an inhibitor of Cx- and Panx1-based channels) significantly reduces oxaliplatin-induced astrogliosis and mechanical hypersensitivity (Yoon et al., 2013). Additionally, neuropathic pain evoked by weight-drop induces astrogliosis, heat hyperalgesia and mechanical allodynia in WT mice after 4-week post-treatment (Chen et al., 2012). However, these responses almost completely disappeared when Cx43 was deleted (Chen et al., 2012), suggesting that hemichannels and/or gap junction channels formed by Cx43 are critical factors for the development of neuropathic pain. Similarly, SGCs in the PNS may play an important role in the genesis and maintenance of chronic pain (Jasmin et al., 2010; Villa et al., 2010; Hanani, 2012). Additionally, the cascade leading to chronic pain seems to be Cx-dependent (Vit et al., 2006; Ohara et al., 2008; Hanani, 2012). For example, the injection of a GJC and hemichannel blocker prevents the inflammation-induced decrease in the pain threshold in rats injected with Freund's adjuvant (Dublin and Hanani, 2007). Interestingly, SGCs show a reduction in membrane resistance after axotomy of trigeminal neurons (Cherkas et al., 2004), suggesting that some ion channels (e.g., hemichannels) are activated. In contrast, Cx43 hemichannels have been proposed as molecular targets to reduce the inflammatory response of glial and immune cells (Eltzschig et al., 2006; Retamal et al., 2007). This finding raises a question concerning how hemichannels could play a role in pain processing in the PNS. A possible answer is that astrocytes of the CNS control several neuronal functions through ATP, glutamate and NAD^+ released via Cx43 and Panx1 based-channels (Parpura et al., 1994; Parpura and Haydon, 2000; Higashida et al., 2001; Newman, 2003; Bao et al., 2004). However, the permeability of Panx1 channels to ATP remains a matter of controversy, mainly because Panx1 channels seem to be impermeable to ATP and are blocked by high extracellular ATP concentrations (Qiu and Dahl,

2009; Romanov et al., 2012a,b). Similar to astrocytes of the CNS, SGCs could modulate the activity of sensory neurons through the release of neuroactive substances that may modify the activity of neighboring neurons through Cx43 hemichannels that are permeable to ATP (Stout et al., 2002; Kang et al., 2008), which is an important neurotransmitter related to pain (Burnstock, 2009a). In this study, we observed that mHBSS increased neuronal activity *in vitro* in a Cx43- and Panx1-dependent manner, suggesting that SGCs may release neuroactive molecules that increase the neuronal activity through these channels.

To our knowledge, this work is the first to suggest that $P2X_7$ receptors, Cx43 hemichannels, and Panx1 channels may serve as communication pathways between SGCs and sensory neurons somata projecting their axons through the vagus nerve. However, further studies concerning the localization and function of Cx43 and Panx1 in the SGCs are required to understand the physiological and/or pathophysiological role of these pathways in sensory ganglia.

AUTHOR CONTRIBUTIONS

Mauricio A. Retamal, Julio Alcayaga, and Juan C. Sáez designed research; Mauricio A. Retamal, Julio Alcayaga, Pablo J. Sáez, Christian A. Verdugo, Ricardo Fernández, and Luis E. León performed research; Mauricio A. Retamal, Julio Alcayaga, Christian A. Verdugo, and Juan C. Sáez analyzed data; and Mauricio A. Retamal, Julio Alcayaga, Geert Bultynck, Luc Leybaert, and Juan C. Sáez wrote the paper.

ACKNOWLEDGMENTS

We would like to thank Ms. Paola Fernández and Ms. Paulina Arias for their technical support. We also thank to Mrs Carolina Larrain for her assistance in the preparation of the manuscript and to the Centro de Microscopia Avanzada (CMA, BIO BIO), Universidad de Concepción for their help with the imaging capture and analyses. This work was funded by grants Fondecyt #1120214 and Anillo ACT 1104 (to Mauricio A. Retamal), 1090157 and 1130177 (to Julio Alcayaga), 1120976 (to Ricardo Fernández), the Scientific Research Foundation—Flanders FWO grant # G.0298.11 (to Geert Bultynck and Luc Leybaert), Fondecyt #1111033, Anillo ATC-71 and the Chilean Science Millennium Institute grant # P09-022-F (to Juan C. Sáez).

REFERENCES

Alcayaga, J., Retamal, M., Cerpa, V., Arroyo, J., and Zapata, P. (2003). Dopamine inhibits ATP-induced responses in the cat petrosal ganglion *in vitro*. *Brain Res.* 966, 283–287. doi: 10.1016/S0006-8993(02)04215-4

Amir, R., and Devor, M. (1996). Chemically mediated cross-excitation in rat dorsal root ganglia. *J. Neurosci.* 16, 4733–4741.

Bao, L., Locovei, S., and Dahl, G. (2004). Pannexin membrane channels are mechanosensitive conduits for ATP. *FEBS Lett.* 572, 65–68. doi: 10.1016/j.febslet.2004.07.009

Bargiotas, P., Krenz, A., Hormuzdi, S. G., Ridder, D. A., Herb, A., Barakat, et al. (2011). Pannexins in ischemia-induced neurodegeneration. *Proc. Natl. Acad. Sci. U.S.A.* 108, 20772–20777. doi: 10.1073/pnas.1018262108

Boassa, D., Ambrosi, C., Qiu, F., Dahl, G., Gaietta, G., and Sosinsky, G. (2007). Pannexin1 channels contain a glycosylation site that targets the hexamer to the plasma membrane. *J. Biol. Chem.* 282, 31733–31743. doi: 10.1074/jbc.M702422200

Brañes, M. C., Contreras, J. E., and Sáez, J. C. (2002). Activation of human polymorphonuclear cells induces formation of functional gap junctions and expression of connexins. *Med. Sci. Monit.* 8, BR313–BR323.

Bruzzone, R., Hormuzdi, S. G., Barbe, M. T., Herb, A., and Monyer, H. (2003). Pannexins, a family of gap junction proteins expressed in brain. *Proc. Natl. Acad. Sci. U.S.A.* 100, 13644–13649. doi: 10.1073/pnas.2233464100

Bruzzone, S., Guida, L., Zocchi, E., Franco, L., and De Flora, A. (2001). Connexin 43 hemi channels mediate Ca^{2+}-regulated transmembrane NAD^+ fluxes in intact cells. *FASEB J.* 15, 10–12. doi: 10.1096/fj.00-0566fje

Burnstock, G. (2009a). Purinergic receptors and pain. *Curr. Pharm. Des.* 15, 1717–1735. doi: 10.2174/138161209788186335

Burnstock, G. (2009b). Purines and sensory nerves. *Handb. Exp. Pharmacol.* 194, 333–392. doi: 10.1007/978-3-540-79090-7_10

Chen, J., He, L., Dinger, B., Stensaas, L., and Fidone, S. (2002). Chronic hypoxia upregulates connexin43 expression in rat carotid body and petrosal ganglion. *J. Appl. Physiol.* 92, 1480–1486. doi: 10.1152/japplphysiol.00077.2001

Chen, M. J., Kress, B., Han, X., Moll, K., Peng, W., Ji, R. R. et al. (2012). Astrocytic Cx43 hemichannels and gap junctions play a crucial role in development of chronic neuropathic pain following spinal cord injury. *Glia* 60, 1660–1670. doi: 10.1002/glia.22384

Cherian, P. P., Siller-Jackson, A. J., Gu, S., Wang, X., Bonewald, L. F., Sprague, E., et al. (2005). Mechanical strain opens connexin 43 hemichannels in osteocytes: a novel mechanism for the release of prostaglandin. *Mol. Biol. Cell.* 16, 3100–3106. doi: 10.1091/mbc.E04-10-0912

Cherkas, P. S., Huang, T. Y., Pannicke, T., Tal, M., Reichenbach, A., and Hanani, M. (2004). The effects of axotomy on neurons and satellite glial cells in mouse trigeminal ganglion. *Pain* 110, 290–298. doi: 10.1016/j.pain.2004.04.007

Cone, A. C., Ambrosi, C., Scemes, E., Martone, M. E., and Sosinsky, G. E. (2013). A comparative antibody analysis of pannexin1 expression in four rat brain regions reveals varying subcellular localizations. *Front. Pharmacol.* 6:6. doi: 10.3389/fphar.2013.00006

De Bock, M., Wang, N., Bol, M., Decrock, E., Ponsaerts, R., Bultynck, G., et al. (2012). Connexin 43 hemichannels contribute to cytoplasmic Ca^{2+} oscillations by providing a bimodal Ca^{2+}-dependent Ca^{2+} entry pathway. *J. Biol. Chem.* 287, 12250–12266. doi: 10.1074/jbc.M111.299610

Dublin, P., and Hanani, M. (2007). Satellite glial cells in sensory ganglia: their possible contribution to inflammatory pain. *Brain Behav. Immun.* 21, 592–598. doi: 10.1016/j.bbi.2006.11.011

Eltzschig, H. K., Eckle, T., Mager, A., Kuper, N., Karcher, C., Weissmuller, T., et al. (2006). ATP release from activated neutrophils occurs via connexin 43 and modulates adenosine-dependent endothelial cell function. *Circ. Res.* 99, 1100–1108. doi: 10.1161/01.RES.0000250174.31269.70

Evans, W. H., and Leybaert, L. (2007). Mimetic peptides as blockers of connexin channel-facilitated intercellular communication. *Cell Commun. Adhes.* 14, 265–273. doi: 10.1080/15419060801891034

Fernández, R., Nardocci, G., Simon, F., Martin, A., Becerra, A., Rodríguez-Tirado, C., et al. (2011). Lipopolysaccharide signaling in the carotid chemoreceptor pathway of rats with sepsis syndrome. *Respir. Physiol. Neurobiol.* 175, 336–348. doi: 10.1016/j.resp.2010.12.014

Frankenhaeuser, B., and Hodgkin, A. L. (1957). The action of calcium on the electrical properties of squid axons. *J. Physiol.* 137, 218–244.

Fueri, C., Faudon, M., Hery, M., and Hery, F. (1984). Release of serotonin from perikarya in cat nodose ganglia. *Brain Res.* 304, 173–177. doi: 10.1016/0006-8993(84)90877-1

Garrett, F. G., and Durham, P. L. (2008). Differential expression of connexins in trigeminal ganglion neurons and satellite glial cells in response to chronic or acute joint inflammation. *Neuron Glia Biol.* 4, 295–306. doi: 10.1017/S1740925X09990093

Gu, Y., Chen, Y., Zhang, X., Li, G. W., Wang, C., and Huang, L. Y. (2010). Neuronal soma-satellite glial cell interactions in sensory ganglia and the participation of purinergic receptors. *Neuron Glia Biol.* 6, 53–62. doi: 10.1017/S1740925X10000116

Gulbransen, B. D., Bashashati, M., Hirota, S. A., Gui, X., Roberts, J. A., MacDonald, J. A., et al. (2012). Activation of neuronal $P2X_7$ receptor-pannexin-1 mediates death of enteric neurons during colitis. *Nat. Med.* 18, 600–604. doi: 10.1038/nm.2679

Hanani, M. (2005). Satellite glial cells in sensory ganglia: from form to function. *Brain Res. Rev.* 48, 457–476. doi: 10.1016/j.brainresrev.2004.09.001

Hanani, M. (2012). Intercellular communication in sensory ganglia by purinergic receptors and gap junctions: implications for chronic pain. *Brain Res.* 1487, 183–191. doi: 10.1016/j.brainres.2012.03.070

Hanani, M., Huang, T. Y., Cherkas, P. S., Ledda, M., and Pannese, E. (2002). Glial cell plasticity in sensory ganglia induced by nerve damage. *Neuroscience* 114, 279–283. doi: 10.1016/S0306-4522(02)00279-8

Heblich, F., England, S., and Docherty, R. J. (2001). Indirect actions of bradykinin on neonatal rat dorsal root ganglion neurones: a role for non-neuronal cells as nociceptors. *J. Physiol.* 536, 111–121. doi: 10.1111/j.1469-7793.2001.00111.x

Higashida, H., Hashii, M., Yokoyama, S., Hoshi, N., Asai, K., and Kato, T. (2001). Cyclic ADP-ribose as a potential second messenger for neuronal Ca^{2+} signaling. *J. Neurochem.* 76, 321–331. doi: 10.1046/j.1471-4159.2001.00082.x

Hille, B. (1968). Charges and potentials at the nerve surface. Divalent ions and pH. *J. Gen. Physyol.* 51, 221–236. doi: 10.1085/jgp.51.2.221

Huang, T. Y., and Hanani, M. (2005). Morphological and electrophysiological changes in mouse dorsal root ganglia after partial colonic obstruction. *Am. J. Physiol. Gastrointest. Liver Physiol.* 289, G670–G678. doi: 10.1152/ajpgi.00028.2005

Huang, T. Y., Hanani, M., Ledda, M., De Palo, S., and Pannese, E. (2006). Aging is associated with an increase in dye coupling and in gap junction number in satellite glial cells of murine dorsal root ganglia. *Neuroscience* 137, 1185–1192. doi: 10.1016/j.neuroscience.2005.10.020

Iglesias, R., Locovei, S., Roque, A., Alberto, A. P., Dahl, G., Spray, D. C., et al. (2008). P2X7 receptor-Pannexin1 complex: pharmacology and signaling. *Am. J. Physiol. Cell Physiol.* 295, C752–C760. doi: 10.1152/ajpcell.00228.2008

Iyyathurai, J., D'Hondt, C., Wang, N., De Bock, M., Himpens, B., Retamal, M. A., et al. (2013). Peptides and peptide-derived molecules targeting the intracellular domains of Cx43: gap junctions versus hemichannels. *Neuropharmacology* 75, 491–505. doi: 10.1016/j.neuropharm

Jasmin, L., Vit, J. P., Bhargava, A., and Ohara, P. T. (2010). Can satellite glial cells be therapeutic targets for pain control? *Neuron Glia Biol.* 6, 63–71. doi: 10.1017/S1740925X10000098

Kadle, R., Zhang, J. T., and Nicholson, B. J. (1991). Tissue-specific distribution of differentially phosphorylated forms of Cx43. *Mol Cell Biol.* 11, 363–369.

Kang, J., Kang, N., Lovatt, D., Torres, A., Zhao, Z., Lin, J., et al. (2008). Connexin 43 hemichannels are permeable to ATP. *J. Neurosci.* 28, 4702–4711. doi: 10.1523/JNEUROSCI.5048-07.2008

Kothmann, W. W., Massey, S. C., and O'Brien, J. (2009). Dopamine-stimulated dephosphorylation of connexin 36 mediates AII amacrine cell uncoupling. *J. Neurosci.* 29, 14903–14911. doi: 10.1523/JNEUROSCI.3436-09.2009

Li, C., Peoples, R. W., and Weight, F. F. (1997). Mg2+ inhibition of ATP-activated current in rat nodose ganglion neurons: evidence that Mg2+ decreases the agonist affinity of the receptor. *J. Neurophysiol.* 77, 3391–3395.

Mandelzys, A., and Cooper, E. (1992). Effects of ganglionic satellite cells and NGF on the expression of nicotinic acetylcholine currents by rat sensory neurons. *J. Neurophysiol.* 67, 1213–1221.

Matsuka, Y., Neubert, J. K., Maidment, N. T., and Spigelman, I. (2001). Concurrent release of ATP and substance P within guinea pig trigeminal ganglia *in vivo*. *Brain Res.* 915, 248–255. doi: 10.1016/S0006-8993(01)02888-8

Nagy, J. I., Dudek, F. E., and Rash, J. E. (2004). Update on connexins and gap junctions in neurons and glia in the mammalian nervous system. *Brain Res. Rev.* 47, 191–215. doi: 10.1016/j.brainresrev.2004.05.005

Newman, E. A. (2003). Glial cell inhibition of neurons by release of ATP. *J. Neurosci.* 23, 1659–1666.

Oh, E. J., and Weinreich, D. (2002). Chemical communication between vagal afferent somata in nodose Ganglia of the rat and the Guinea pig *in vitro*. *J. Neurophysiol.* 87, 2801–2807. doi: 10.1152/jn.00768.2001

Ohara, P. T., Vit, J. P., Bhargava, A., and Jasmin, L. (2008). Evidence for a role of connexin 43 in trigeminal pain using RNA interference *in vivo*. *J. Neurophysiol.* 100, 3064–3073. doi: 10.1152/jn.90722.2008

Palouzier-Paulignan, B., Chamoin, M. C., and Temaux, J. P. (1992). Somatic acetylcholine release in rabbit nodose ganglion. *Eur. J. Neurosci.* 4, 1123–1129. doi: 10.1111/j.1460-9568.1992.tb00139.x

Pannese, E. (1981). The satellite cells of the sensory ganglia. *Adv. Anat. Embryol. Cell. Biol.* 65, 1–111. doi: 10.1007/978-3-642-67750-2_1

Pannese, E. (2010). The structure of the perineuronal sheath of satellite glial cells (SGCs) in sensory ganglia. *Neuron Glia Biol.* 6, 3–10. doi: 10.1017/S1740925X10000037

Pannese, E., Ledda, M., Cherkas, P. S., Huang, T. Y., and Hanani, M. (2003). Satellite cell reactions to axon injury of sensory ganglion neurons: increase in number of gap junctions and formation of bridges connecting previously separate perineuronal sheaths. *Anat Embryol. (Berl).* 206, 337–347.

Parpura, V., Basarsky, T. A., Liu, F., Jeftinija, K., Jeftinija, S., and Haydon, P. G. (1994). Glutamate-mediated astrocyte-neuron signalling. *Nature* 369, 744–747. doi: 10.1038/369744a0

Parpura, V., and Haydon, P. G. (2000). Physiological astrocytic calcium levels stimulate glutamate release to modulate adjacent neurons. *Proc. Natl. Acad. Sci. U.S.A.* 97, 8629–8634. doi: 10.1073/pnas.97.15.8629

Peiser, C., Andreae, D. A., Groneberg, D. A., Dinh, Q. T., Müller, B., Wahn, U., et al. (2009). Dopamine D2 receptor mRNA expression is increased in the jugular-nodose ganglia of rats with nitrogen dioxide-induced chronic bronchitis. *Neurosci. Lett.* 465, 143–146. doi: 10.1016/j.neulet.2009.09.004

Penuela, S., Bhalla, R., Nag, K., and Laird, D. W. (2009). Glycosylation regulates pannexin intermixing and cellular localization. *Mol. Biol. Cell.* 20, 4313–4323. doi: 10.1091/mbc.E09-01-0067

Piao, H., Sato, A., Nozawa, Y., Sun, W., Morioka, T., and Oite, T. (2011). Effects of connexin-mimetic peptides on perfusion pressure in response to phenylephrine in isolated, perfused rat kidneys. *Clin. Exp. Nephrol.* 15, 203–211. doi: 10.1007/s10157-010-0382-0

Ponsaerts, R., De Vuyst, E., Retamal, M., D'Hondt, C., Vermeire, D., Wang, N., et al. (2010). Intramolecular loop/tail interactions are essential for connexin 43-hemichannel activity. *FASEB J.* 24, 4378–4395. doi: 10.1096/fj.09-153007

Ponsaerts, R., Wang, N., Himpens, B., Leybaert, L., and Bultynck, G. (2012). The contractile system as a negative regulator of the connexin 43 hemichannel. Biol Cell. 104, 367–377. doi: 10.1111/boc.201100079

Qiu, F., and Dahl, G. A. (2009). permeant regulating its permeation pore: inhibition of pannexin 1 channels by ATP. *Am. J. Physiol. Cell. Physiol.* 296, C250–C255. doi: 10.1152/ajpcell.00433.2008

Retamal, M. A., Froger, N., Palacios-Prado, N., Ezan, P., Sáez, P. J., Sáez, J. C. et al. (2007). Cx43 hemichannels and gap junction channels in astrocytes are regulated oppositely by proinflammatory cytokines released from activated microglia. *J. Neurosci.* 27, 13781–13792. doi: 10.1523/JNEUROSCI.2042-07.2007

Riquelme, M. A., Cea, L. A., Vega, J. L., Boric, M. P., Monyer, H., Bennett, M. V., et al. (2013). The ATP required for potentiation of skeletal muscle contraction is released via pannexin hemichannels. *Neuropharmacology* 75, 594–603. doi: 10.1016/j.neuropharm.2013.03.022

Romanov, R. A., Bystrova, M. F., Rogachevskaya, O. A., Sadovnikov, V. B., Shestopalov, V. I., and Kolesnikov, S. S. (2012a). The ATP permeability of pannexin 1 channels in a heterologous system and in mammalian taste cells is dispensable. *J Cell Sci.* 125(Pt 22), 5514–523. doi: 10.1242/jcs.111062

Romanov, R. A., Rogachevskaia, O. A., Kolesnikova, A. S., Khokhlov, A. A., and Kolesnikov, S. S. (2012b). Permeability of pannexin 1 channels to large anions. *Ross Fiziol Zh Im I M Sechenova.* 98, 1578–1586.

Sáez, J. C., Contreras, J. E., Bukauskas, F. F., Retamal, M. A., and Bennett, M. V. (2003). Gap junction hemichannels in astrocytes of the CNS. *Acta Physiol. Scand.* 179, 9–22. doi: 10.1046/j.1365-201X.2003.01196.x

Sáez, J. C., Schalper, K. A., Retamal, M. A., Orellana, J. A., Shoji, K. F., and Bennett, M. V. (2010). Cell membrane permeabilization via connexin hemichannels in living and dying cells. *Exp. Cell. Res.* 316, 2377–23789. doi: 10.1016/j.yexcr.2010.05.026

Sakuma, E., Wang, H. J., Asai, Y., Tamaki, D., Amano, K., Mabuchi, Y., et al. (2001). Gap junctional communication between the satellite cells of rat dorsal root ganglia. *Kaibogaku Zasshi.* 76, 297–302.

Shinder, V., Amir, R., and Devor, M. (1998). Cross-excitation in dorsal root ganglia does not depend on close cell-to-cell apposition. *Neuroreport* 9, 3997–4000. doi: 10.1097/00001756-199812210-00002

Stensaas, L. J., and Fidone, S. J. (1977). An ultrastructural study of cat petrosal ganglia: a search for autonomic ganglion cells. *Brain Res.* 124, 29–39. doi: 10.1016/0006-8993(77)90861-7

Stout, C. E., Costantin, J. L., Naus, C. C., and Charles, A. C. (2002). Intercellular calcium signaling in astrocytes via ATP release through connexin hemichannels. *J. Biol. Chem.* 277, 10482–10488. doi: 10.1074/jbc.M109902200

Suadicani, S. O., Cherkas, P. S., Zuckerman, J., Smith, D. N., Spray, D. C., and Hanani, M. (2009). Bidirectional calcium signaling between satellite glial cells and neurons in cultured mouse trigeminal ganglia. *Neuron Glia Biol.* 6, 43–51. doi: 10.1017/S1740925X09990408

Theis, M., Sohl, G., Eiberger, J., and Willecke, K. (2005). Emerging complexities in identity and function of glial connexins. *Trends Neurosci.* 28, 188–195. doi: 10.1016/j.tins.2005.02.006

Thompson, R. J., and Macvicar, B. A. (2008). Connexin and pannexin hemichannels of neurons and astrocytes. *Channels (Austin)* 2, 81–86. doi: 10.4161/chan.2.2.6003

Thompson, R. J., Zhou, N., and MacVicar, B. A. (2006). Ischemia opens neuronal gap junction hemichannels. *Science* 312, 924–927. doi: 10.1126/science.1126241

Villa, G., Fumagalli, M., Verderio, C., Abbracchio, M. P., and Ceruti, S. (2010). Expression and contribution of satellite glial cells purinoceptors to pain transmission in sensory ganglia: an update. *Neuron Glia Biol.* 6, 31–42. doi: 10.1017/S1740925X10000086

Vit, J. P., Jasmin, L., Bhargava, A., and Ohara, P. T. (2006). Satellite glial cells in the trigeminal ganglion as a determinant of orofacial neuropathic pain. *Neuron Glia Biol.* 2, 247–257. doi: 10.1017/S1740925X07000427

Wang, N., De Bock, M., Antoons, G., Gadicherla, A. K., Bol, M., Decrock, E., et al. (2012). Connexin mimetic peptides inhibit Cx43 hemichannel opening triggered by voltage and intracellular Ca^{2+} elevation. *Basic Res. Cardiol.* 107:304. doi: 10.1007/s00395-012-0304-2.

Ye, Z. C., Wyeth, M. S., Baltan-Tekkok, S., and Ransom, B. R. (2003). Functional hemichannels in astrocytes: a novel mechanism of glutamate release. *J. Neurosci.* 23, 3588–35896.

Yoon, S. Y., Robinson, C. R., Zhang, H., and Dougherty, P. M. (2013). Spinal astrocyte gap junctions contribute to oxaliplatin-induced mechanical hypersensitivity. *J. Pain.* 14, 205–214. doi: 10.1016/j.jpain.2012.11.002

Zhang, X., Chen, Y., Wang, C., and Huang, L. Y. (2007). Neuronal somatic ATP release triggers neuron-satellite glial cell communication in dorsal root ganglia. *Proc. Natl. Acad. Sci. U.S.A.* 104, 9864–9869. doi: 10.1073/pnas.0611048104

Zuriel, E., and Devor, M. (2001). Dye coupling does not explain functional crosstalk within dorsal root ganglia. *J. Peripher. Nerv. Syst.* 6, 227–231 doi: 10.1046/j.1529-8027.2001.01024.x

12

Prenatal nicotine exposure enhances Cx43 and Panx1 unopposed channel activity in brain cells of adult offspring mice fed a high-fat/cholesterol diet

Juan A. Orellana[1]*, Dolores Busso[2,3], Gigliola Ramírez[1], Marlys Campos[4], Attilio Rigotti[2,3], Jaime Eugenín[4] and Rommy von Bernhardi[1]

[1] *Departamento de Neurología, Escuela de Medicina, Pontificia Universidad Católica de Chile, Santiago, Chile*
[2] *Departamento de Nutrición, Diabetes y Metabolismo, Escuela de Medicina, Pontificia Universidad Católica de Chile, Santiago, Chile*
[3] *Centro de Nutrición Molecular y Enfermedades Crónicas, Escuela de Medicina, Pontificia Universidad Católica de Chile, Santiago, Chile*
[4] *Laboratorio de Sistemas Neurales, Departamento de Biología, Facultad de Química y Biología, Universidad de Santiago de Chile, Santiago, Chile*

***Correspondence:**
Juan A. Orellana, Departamento de Neurología, Escuela de Medicina, Pontificia Universidad Católica de Chile, Marcoleta 391, Santiago, 8330024, Chile
e-mail: jaorella@uc.cl

Nicotine, the most important neuroteratogen of tobacco smoke, can reproduce brain and cognitive disturbances *per se* when administered prenatally. However, it is still unknown if paracrine signaling among brain cells participates in prenatal nicotine-induced brain impairment of adult offspring. Paracrine signaling is partly mediated by unopposed channels formed by connexins hemichannels (HCs) and pannexins serving as aqueous pores permeable to ions and small signaling molecules, allowing exchange between the intra- and extracellular milieus. Our aim was to address whether prenatal nicotine exposure changes the activity of those channels in adult mice offspring under control conditions or subjected to a second challenge during young ages: high-fat/cholesterol (HFC) diet. To induce prenatal exposure to nicotine, osmotic minipumps were implanted in CF1 pregnant mice at gestational day 5 to deliver nicotine bitartrate or saline (control) solutions. After weaning, offspring of nicotine-treated or untreated pregnant mice were fed ad libitum with chow or HFC diets for 8 weeks. The functional state of connexin 43 (Cx43) and pannexin 1 (Panx1) unopposed channels was evaluated by dye uptake experiments in hippocampal slices from 11-week-old mice. We found that prenatal nicotine increased the opening of Cx43 HCs in astrocytes, and Panx1 channels in microglia and neurons only if offspring mice were fed with HFC diet. Blockade of inducible nitric oxide synthase (iNOS), cyclooxygenase 2 (COX_2) and prostaglandin E receptor 1 (EP_1), ionotropic ATP receptor type 7 ($P2X_7$) and NMDA receptors, showed differential inhibition of prenatal nicotine-induced channel opening in glial cells and neurons. Importantly, inhibition of the above mentioned enzymes and receptors, or blockade of Cx43 and Panx1 unopposed channels greatly reduced adenosine triphosphate (ATP) and glutamate release from hippocampal slices of prenatally nicotine-exposed offspring. We propose that unregulated gliotransmitter release through Cx43 and Panx1 unopposed channels may participate in brain alterations observed in offspring of mothers exposed to tobacco smoke during pregnancy.

Keywords: hemichannels, connexins, pannexins, nicotine, brain, glia, fat diet

INTRODUCTION

A growing body of evidence indicates that the risk of developing chronic diseases throughout life is related to environmental factors acting on tissue plasticity at specific windows during fetal development. Maternal cigarette smoking is a well established environmental risk factor associated with adverse effects on fetal outcome, increasing perinatal morbidity and mortality, and evoking long-term behavioral, learning, and memory impairment in the offspring (Naeye and Peters, 1984; Rantakallio and Koiranen, 1987; DiFranza and Lew, 1995; Jacobsen et al., 2006; Julvez et al., 2007). Nicotine is the most important neuroteratogen component of tobacco smoke and, given that it easily crosses the placental and blood–brain barriers (Luck et al., 1985), it is believed to have a dramatic influence on healthy brain development via activation of fetal nicotinic receptors (Dwyer et al., 2008). Indeed, nicotine delivery during pregnancy, eliciting plasma levels comparable to those found in heavy smokers, induces activation of apoptosis-associated genes, long-lasting morphological alterations of neurons, reduced neuronal cell layer thickness, increased number of glial cells, and behavioral impairment (Navarro et al., 1989; Roy and Sabherwal, 1998; Roy et al., 2002; Eugenín et al., 2008). Several studies have attempted to understand the mechanism underlying these nicotine-induced consequences by examining alterations in neurotransmitters (Navarro et al., 1989; Muneoka et al., 1997), changes in nicotinic receptor availability (van de Kamp and Collins, 1994; Coddou et al., 2009), modifications in gene expression (Toledo-Rodriguez et al., 2010; Schneider et al., 2011), and early adaptations of dendritic morphology (Roy and Sabherwal, 1994, 1998). However, the impact of maternal nicotine exposure on the communication of neurons with their partnership cells, the glia, has received little attention.

It is nowadays well established that glial cells express virtually all known neurotransmitter receptors types, allowing them to sense neuronal activity and microenvironmental changes by responding locally through the Ca^{2+}-dependent release of bioactive molecules termed "gliotransmitters" (e.g., glutamate, adenosine triphosphate (ATP), adenosine, GABA, and D-serine; Perea et al., 2009; Perea and Araque, 2010). In the central nervous system (CNS), gliotransmitter release is in part mediated by the opening of unopposed membrane channels formed by connexins hemichannels (HCs) or pannexins (Wang et al., 2013b). These channels serve as aqueous pores permeable to ions and small molecules, allowing for diffusional exchange between the intra- and extracellular compartments. In glial cells, HCs and pannexin channels (PCHs) grant the release of gliotransmitters that are necessary for different brain functions including glucosensing (Orellana et al., 2012), ischemic tolerance (Lin et al., 2008), fear memory consolidation (Stehberg et al., 2012), neuron-glia crosstalk (Torres et al., 2012), and chemoreception (Huckstepp et al., 2010). However, several independent studies have pointed out that onset and progression of neurodegenerative homeostatic imbalances may be associated to impairment in permeability properties of these channels in the CNS (Takeuchi et al., 2006; Thompson et al., 2006; Karpuk et al., 2011; Orellana et al., 2011a,b; Gulbransen et al., 2012; Burkovetskaya et al., 2014). Therefore, we decided to investigate whether prenatal nicotine exposure could affect the functional activity of HCs and PCHs in glial cells and neurons in the offspring. Different maternal conditions during pregnancy, including prenatal nicotine, have been shown to sensitize the brain of the adult offspring on its response to a subsequent environmental challenge (Slotkin et al., 1991; Bilbo et al., 2005). Given that fat and cholesterol-enriched diets impair synaptic transmission and glial cell function (Dufour et al., 2006; Triviño et al., 2006; Ya et al., 2013), we studied if dyslipidemia induced by feeding a high-fat/cholesterol (HFC) diet in combination with prenatal nicotine exposure could enhance the opening of connexin and pannexin unopposed channels in the offspring brain. We chose dyslipidemia as a second environmental hit due to the high prevalence of this metabolic condition as a consequence of sedentary lifestyle and overnutrition in Western populations in the last decades.

In this work, we show that prenatal nicotine can increase the opening of unopposed channels formed by connexin 43 (Cx43) in astrocytes and pannexin 1 (Panx1) in microglia and neurons. Interestingly, these responses were only detected when offspring mice were subjected to dyslipidemia induced by feeding them a HFC diet.

MATERIALS AND METHODS

REAGENTS AND ANTIBODIES

Gap26, Gap19; YGRKKRRQRRRDGANVDMHLKQIEIKKFKY GIEEHGK (TAT-L2) and ^{10}panx1 peptides were obtained from Genscript (New Jersey, USA). HEPES, DMEM, DNAse I, poly-L-lysine, LN-6, ns-398, sc-19220, polyclonal anti-Cx43 antibody, 3-(2-carboxypiperazin-4-yl)propyl-1-phosphonic acid (CPP), Brilliant blue G (BBG), oATP, ethidium (Etd) bromide, and probenecid (Prob) were purchased from Sigma-Aldrich (St. Louis, MO, USA). Fetal calf serum (FCS) was obtained from Hyclone (Logan, UT, USA). Penicillin, streptomycin, polyclonal anti-Panx1 antibody (PI488000), goat anti-mouse Alexa Fluor 488 and goat anti-mouse Alexa Fluor 555 were obtained from Invitrogen (Carlsbad, CA, USA). Anti-NeuN monoclonal antibody was obtained from Chemicon (Martinsried/Munich, Germany). Normal goat serum (NGS) was purchased from Zymed (San Francisco, CA, USA). Anti-GFAP monoclonal antibody was purchased from ICN Chemicals, (Irvine, CA). Anti-Cx43 monoclonal antibody was obtained from BD Biosciences (Franklin Lakes, NJ, USA).

ANIMAL CARE AND USE

All animal experimentation was conducted in accordance with the guidelines for care and use of experimental animals of the National Institute of Health (NIH) and local guidance documents generated by the *ad hoc* committee of the Chilean National Commission of Scientific and Technological Research (CONICYT). The procedures and research plan were approved by the Universidad de Santiago Bioethics Committee. CF-1 mice were obtained from the Public Health Institute and housed at the animal facility of the Laboratory of Neural Systems, Universidad de Santiago de Chile. Mice were housed under a 12-h light-darkness condition, with access to *ad libitum* fresh water and food in a temperature (18–26°C)- and humidity (40–70%)-controlled and well ventilated environment.

PRENATAL NICOTINE EXPOSURE AND POSTNATAL FEEDING WITH HFC DIET

Subcutaneous implantation of osmotic minipumps (model 2004, Alzet) was performed in CF1 pregnant mice at gestational day 5 as previously described (Eugenín et al., 2008). In brief, pumps were implanted through an incision made between scapulae, using strict aseptic conditions, under anesthesia with 60–80/20 mg/kg ketamine/xylazine by intraperitoneal (i.p.) injection. Pumps delivered saline (controls) or nicotine bitartrate (60 mg kg^{-1} day^{-1}) at a rate of 0.25 $\mu l\ h^{-1}$. Recovery from anesthesia was performed under controlled temperature. Dams were maintained in separate cages and daily supervision was done based on the protocol by Morton and Griffiths (Morton and Griffiths, 1985). After weaning, offspring of nicotine-treated or vehicle-treated pregnant mice were fed *ad libitum* with chow or HFC diet (1.25% cholesterol, 15% total fat, and 0.5% cholic acid; Harlan Teklad, USA) for 8 weeks. This dietary condition induced a ~2–3 fold increase in total plasma cholesterol with no significant effect in body weight (not shown).

ACUTE HIPPOCAMPAL SLICES

Eleven-week-old offspring mice were decapitated, and their brains were dissected and placed in ice-cold artificial cerebrospinal fluid (ACSF) containing (in mM): 125 NaCl, 2.5 KCl, 25 glucose, 25 $NaHCO_3$, 1.25 NaH_2PO_4, 2 $CaCl_2$, and 1 $MgCl_2$, bubbled with 95% O_2/5% CO_2, pH 7.4. Hippocampal coronal brain slices (400 μm) were cut using a vibratome (Leica, VT 1000GS; Leica, Wetzlar, Germany) filled with ice-cold ACSF. The slices were transferred at room temperature (20–22°C) to a holding chamber and immersed in oxygenated ACSF, pH 7.4, for a stabilization period of 30 min before being used.

DYE UPTAKE AND CONFOCAL MICROSCOPY

For "snapshot" experiments, acute slices were incubated with 100 μM Etd for 15 min in a chamber with oxygenated (95% O_2 and 5% CO_2) ACSF, pH 7.4. Then, they were washed five times with ACSF, fixed at room temperature with 4% paraformaldehyde for 30 min, rinsed extensively in phosphate buffered saline (PBS) and stored overnight at 4°C with cryoprotectant (30% sucrose). Next day, slices were frozen and cut into 12–16 μm-thick cryosections with a cryostat. Cryosections were incubated in 0.1% PBS-Triton X-100 containing 10% NGS for 30 min. Afterwards, they were incubated overnight at 4°C with anti-Iba-1 polyclonal antibody (1:300, Wako), anti-GFAP monoclonal antibody (1:300, Sigma), anti-NeuN monoclonal antibody (1:400, Chemicon), polyclonal anti-Cx43 (1:600, Sigma) or polyclonal anti-Panx1 (1:600, Invitrogen) diluted in 0.1% PBS-Triton X-100 with 2% NGS. After five rinses in 0.1% PBS-Triton X-100, cryosections were incubated with goat anti-rabbit Alexa Fluor 488 (1:1500), goat anti-mouse Alexa Fluor 488 (1:1500) or goat anti-mouse Alexa Fluor 647 (1:1500) at room temperature for 1 h. After several washes, coverslips were mounted in Fluoromount and examined in a confocal laser-scanning microscope (Olympus Fluoview FV1000, Tokyo, Japan). Stacks of consecutive confocal images taken with a 63 X objective at 500 nm intervals were sequentially acquired with two lasers (argon 488 nm and helium/neon 543 nm), and Z projections were reconstructed using Fluoview software. The dye uptake ratio was calculated as the subtraction (F-F0) between the fluorescence (F) from respective cell and the background fluorescence (F0). At least six cells by field were selected from at least three fields in each hippocampal slice. Gap26, Gap19; TAT-L2, ^{10}panx1, probenecid, L-N6, ns-398, sc-19220, CPP, BBG and oATP were pre-incubated 15 min and then coapplied with Etd for "snapshot" experiments.

MEASUREMENT OF EXTRACELLULAR ATP AND GLUTAMATE CONCENTRATION

Acute hippocampal slices were immersed in oxygenated ACSF, pH 7.4, at room temperature (20–22°C) for 30 min. Then, ATP and glutamate concentration in the extracellular solution were measured using a luciferin/luciferase bioluminescence and glutamate assay kit (Sigma-Aldrich), respectively. The amount of ATP and glutamate in each sample were calculated from standard curves and normalized by the protein concentration. Briefly, after the experiments, slices were washed twice with ACSF solution and sonicated in ice-cold PBS containing 5 μM ethylenediaminetetraacetic acid (EDTA), Halt (78440) and T-PER protein extraction cocktail (78510), according to manufacturer instructions (Pierce, Rockford, IL). Proteins were measured using the Bio-Rad protein assay. Gap 26, Gap 19; TAT-L2, ^{10}panx1, probenecid, LN-6, ns-398, sc-19220, CPP, BBG and oATP were pre-incubated 30 min before the measurements.

IL-1β AND TNF-α ASSAY

IL-1β and TNF-α were determined in 100 μL of sera. Samples were centrifuged at 14,000 g for 40 min. Supernatants were collected and protein content assayed by the BCA method. IL-1β and TNF-α levels were determined by sandwich ELISA, according to the manufacturer's protocol (eBioscience, San Diego, CA, USA). For the assay, 100 μl of samples were added per ELISA plate well and incubated at 4°C overnight. A calibration curve with recombinant cytokine was included. Detection antibody was incubated at room temperature for 1 h and the reaction developed with avidin–HRP and substrate solution. Absorbance was measured at 450 nm with reference to 570 nm with the microplate reader Synergy HT (Biotek Instruments).

DATA ANALYSIS AND STATISTICS

For each data group, results were expressed as mean ± standard error (SEM); n refers to the number of independent experiments. For statistical analysis, each treatment was compared with its corresponding control, and significance was determined using a one-way ANOVA followed, in case of significance, by a Tukey *post hoc* test.

RESULTS

PRENATAL NICOTINE ENHANCES CX43 AND PANX1 UNOPPOSED CHANNEL ACTIVITY IN BRAIN CELLS OF OFFSPRING MICE FED A HIGH-FAT/CHOLESTEROL DIET

Nicotine delivery during pregnancy induces neuronal alterations and behavioral impairment (Navarro et al., 1989; Roy and Sabherwal, 1998; Roy et al., 2002). On the other hand, the opening of HCs and PCHs has been linked to glial cell dysfunction and neuronal impairment (Takeuchi et al., 2006; Orellana et al., 2011a,b, 2013; Shestopalov and Slepak, 2014). Therefore, we investigated whether prenatal nicotine exposure could affect the functional activity of these channels in microglia, astrocytes and neurons of the offspring. To address this, we examined HC and PCH activity by measuring Etd uptake in acute hippocampal slices from 11-week-old offspring. Etd is a molecule that crosses the plasma membrane in healthy cells by passing through specific large channels, including connexin and pannexin unopposed channels. Upon binding to intracellular nucleic acids, Etd becomes fluorescent, indicating channel opening when appropriate blockers are employed (Schalper et al., 2008; Sáez and Leybaert, 2014). Etd uptake by Iba-1-positive microglia, GFAP-positive astrocytes and NeuN-positive neurons on acute hippocampal slices was evaluated in "snapshot" experiments. Microglia, astrocytes and neurons from offspring of control dams showed a low Etd uptake ratio (**Figures 1A–H, 2A–H, 3A–H**) as previously reported (Orellana et al., 2010; Karpuk et al., 2011). Interestingly, prenatal exposure to nicotine did not change Etd uptake in brain cells in the offspring compared to control conditions (**Figures 1Q, 2Q, 3Q**).

Further, we evaluated whether prenatal nicotine sensitized the brain of adult offspring on its response to a HFC diet. Similarly to what was observed in the offspring from control dams fed a chow diet, Etd uptake remained low when the offspring from control dams was fed a HFC diet (**Figures 1Q, 2Q, 3Q**). However, HFC diet increased Etd uptake in microglia, astrocytes and pyramidal

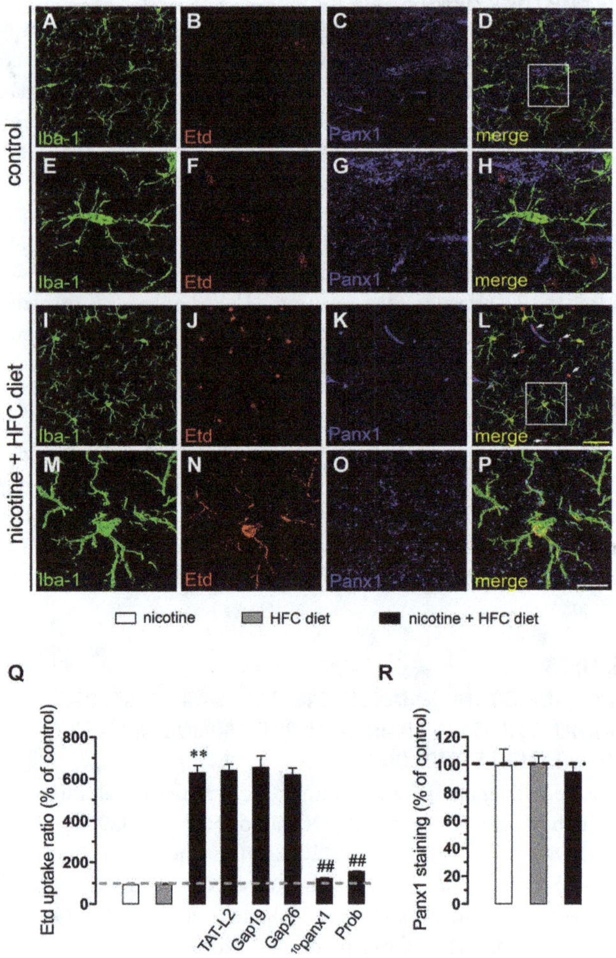

FIGURE 1 | Exposure to prenatal nicotine and postnatal HFC diet increased Panx1 channel activity in microglia. Representative images showing Iba-1 (green), Etd (red) uptake and Panx1 (blue) staining of acute hippocampal brain slices made from control mice **(A–H)** or subjected to prenatal nicotine and postnatal HFC diet **(I–P)**. Images of microglia displayed on the bottom insets **E–H** and **M–P** were taken from the zone depicted by the white square in respective panels **D** and **L**. Calibration bars: yellow = 30 μm and white = 10 μm. **(Q)** Averaged data of Etd uptake ratio normalized by the control condition (dashed line) of microglia from mice subjected to prenatal nicotine (white bars), postnatal cholesterol-enriched diet (gray bars), or a combination of both (black bars). Also shown are the effects of the following blockers applied during Etd uptake recordings: TAT-L2 (100 μM), Gap19 (100 μM), Gap26 (100 μM), ^{10}panx1 (100 μM) or probenecid (Prob, 500 uM). **(R)** Averaged data of Panx1 staining normalized to control conditions (dashed line) of microglia from mice exposed to prenatal nicotine (white bars), postnatal HFC diet (gray bars), or a combination of both (black bars). $^{**}p < 0.005$, effect of prenatal nicotine and postnatal HFC diet compared with control conditions. $^{\#\#}p < 0.005$, effect of each blocker compared with the respective effect induced by prenatal nicotine and postnatal HFC diet. Averaged data were obtained from at least three independent experiments.

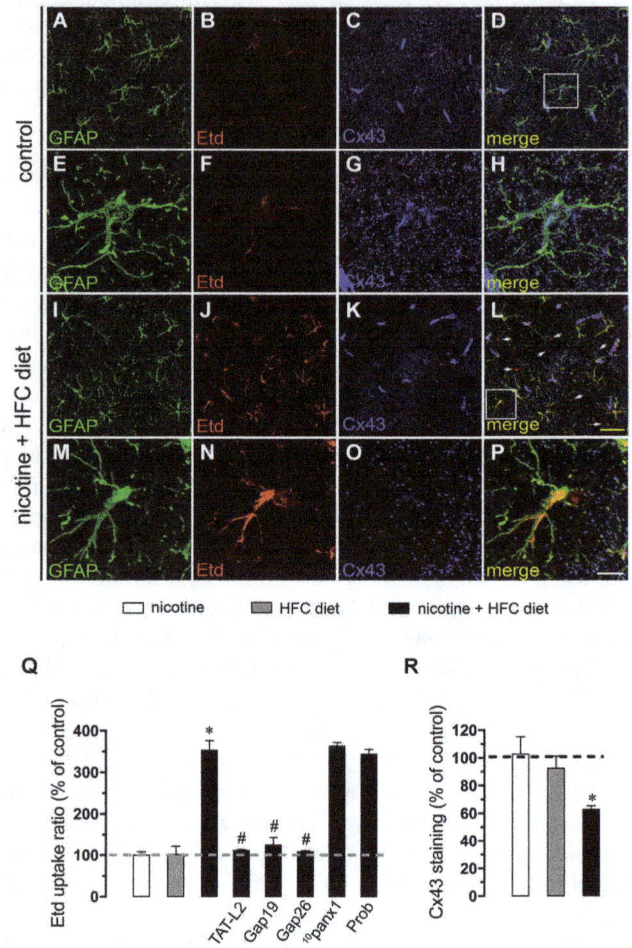

FIGURE 2 | Exposure to prenatal nicotine and postnatal HFC diet increased Cx43 HC activity in astrocytes. Representative images showing GFAP (green), Etd (red) uptake and Cx43 (blue) staining of acute hippocampal brain slices made from control mice **(A–H)** or subjected to prenatal nicotine and postnatal HFC diet **(I–P)**. Images of hippocampal astrocytes displayed on the bottom insets **E–H** and **M–P** were taken from the zone depicted by the white square in respective panels **D** and **L**. Calibration bars: yellow = 70 μm and white = 20 μm **(Q)** Averaged data of Etd uptake ratio normalized to control conditions (dashed line) of astrocytes from mice exposed to prenatal nicotine (white bars), postnatal HFC diet (gray bars), or a combination of both (black bars). Also shown are the effects of the following blockers applied during Etd uptake recordings: TAT-L2 (100 μM), Gap19 (100 μM), Gap26 (100 μM), ^{10}panx1 (100 μM) or probenecid (Prob, 500 uM). **(R)** Averaged data of Cx43 staining normalized to control conditions (dashed line) of astrocytes from mice exposed to prenatal nicotine (white bars), postnatal HFC diet (gray bars), or a combination of both (black bars). $^{*}p < 0.05$, effect of prenatal nicotine and postnatal HFC diet compared with control conditions. $^{\#}p < 0.05$, effect of each blocker compared with the effect induced by prenatal nicotine plus a postnatal HFC diet. Averaged data were obtained from at least three independent experiments.

neurons when the offspring came from nicotine-treated dams (**Figures 1I–Q, 2I–Q, 3I–Q**). Microglia express functional unopposed channels formed by Panx1 and Cx43 (Orellana et al., 2011b). The possible role of Panx1 channels in nicotine-evoked Etd uptake was studied using probenecid and the mimetic peptide ^{10}panx1 with an amino acid sequence homologous to the second loop of Panx1 (Pelegrin and Surprenant, 2006; Silverman et al., 2008). Probenecid (500 μM) and ^{10}panx1 (200 μM) nearly abolished the increased microglial cell Etd uptake triggered by prenatal nicotine and postnatal HFC diet (**Figures 1I–Q**). In contrast, mimetic peptides homologous to the cytoplasmic (TAT-L2 and Gap19; 100 μM) or first extracellular (Gap26; 100 μM) loop of Cx43 (Wang et al., 2013a), did not reduce nicotine-induced Etd uptake by microglia (**Figure 1Q**).

Astrocytes express functional unopposed channels formed by Cx43 (Contreras et al., 2002) and Panx1 (Iglesias et al., 2009). Thereby, we used TAT-L2, Gap19, Gap26, probenecid and ^{10}panx1 to determine the contribution of both channels in the nicotine-induced Etd uptake by astrocytes. TAT-L2 (100 μM), Gap19 (100 μM) and Gap26 (100 μM) fully reduced astroglial cell Etd uptake evoked by prenatal nicotine and postnatal HFC diet (**Figure 2Q**). In contrast, ^{10}panx1 (100 μM) and probenecid (500 μM) failed to induce the same inhibition (**Figure 2Q**).

For neurons, most evidence support that they express unopposed channels formed by Panx1 (Thompson et al., 2006). In agreement with that evidence, ^{10}panx1 and probenecid strongly reduced the nicotine and postnatal HFC diet-induced Etd uptake observed in pyramidal neurons (**Figure 3Q**), whereas TAT-L2, Gap19 and Gap26 failed to cause a similar response (**Figure 3Q**). Overall, these data support the idea that prenatal nicotine plus postnatal HFC diet increases the opening of unopposed channels formed by Cx43 in astrocytes and Panx1 in microglia and neurons.

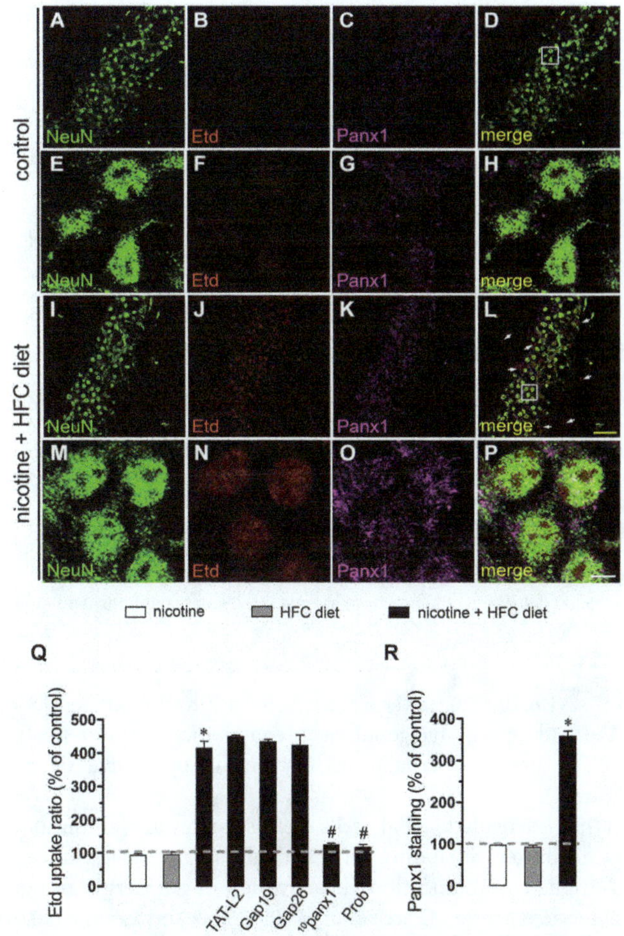

FIGURE 3 | Exposure to prenatal nicotine and postnatal HFC diet increased Panx1 channel activity in neurons. Representative images showing NeuN (green), Etd (red) uptake and Panx1 (magenta) staining of acute hippocampal brain slices made from control mice **(A–H)** or subjected to prenatal nicotine and postnatal HFC diet **(I–P)**. Images of hippocampal neurons displayed on the bottom insets **E–H** and **M–P** were taken from the zone depicted by the white square in respective panels **C** and **I**. Calibration bars: yellow = 90 μm and white = 10 μm. **(Q)** Averaged data of Etd uptake ratio normalized with control conditions (dashed line) of neurons from mice subjected to prenatal nicotine (white bars), postnatal HFC diet (gray bars), or a combination of both (black bars). Also shown are the effect of the following blockers applied during Etd uptake recordings: TAT-L2 (100 μM), Gap19 (100 μM), Gap26 (100 μM), ^{10}panx1 (100 μM) or probenecid (Prob, 500 uM). **(R)** Averaged data of Panx1 staining normalized with control conditions (dashed line) of neurons from mice subjected to prenatal nicotine (white bars), postnatal HFC diet (gray bars), or a combination of both (black bars). $^{*}p < 0.05$, effect of prenatal nicotine and postnatal HFC diet compared with control conditions. $^{\#}p < 0.05$, effect of each blocker compared with the effect induced by prenatal nicotine plus a postnatal HFC diet. Averaged data were obtained from at least three independent experiments.

PRENATAL NICOTINE AFFECTS LEVELS OF CX43 AND PANX1 IN BRAIN CELLS OF OFFSPRING MICE FED A HIGH-FAT/CHOLESTEROL DIET

Given that pathological conditions affect the expression of connexins and pannexins in the CNS (Rouach et al., 2002; Orellana et al., 2009), we examined whether prenatal nicotine plus postnatal HFC diet could modulate Cx43 and Panx1 levels in brain cells by confocal analysis. As expected, neither prenatal nicotine nor postnatal HFC diet alone affected Cx43 and Panx1 levels in astrocytes and neurons, respectively (**Figures 2R, 3R**). However, combination of prenatal nicotine plus feeding a HFC diet during adulthood reduced Cx43 levels in astrocytes (**Figure 2R**), whereas in pyramidal neurons, immunodetection of Panx1 was increased (**Figure 3R**). For all tested conditions, Panx1 remained unchanged in microglia (**Figure 1R**).

INCREASED OPENING OF CX43 AND PANX1 UNOPPOSED CHANNELS IN NICOTINE AND HIGH-FAT/CHOLESTEROL EXPOSED OFFSPRING BRAIN DEPENDS ON INOS/COX2/EP1 RECEPTOR PATHWAY AND PURINERGIC/GLUTAMATERGIC SIGNALING

Under neuroinflammatory conditions, glial cells exhibit a prominent activation of inducible nitric oxide (NO) synthase (iNOS) and cyclooxygenase 2 (COX_2; Tzeng et al., 2005; Amitai, 2010), two enzymes that produce mediators (NO and prostaglandins, respectively) linked to the opening of Cx43 and Panx1 unopposed channels (Retamal et al., 2007; Orellana et al., 2013). Accordingly, we investigated the contribution of iNOS and COX_2 activation on Etd uptake induced by prenatal nicotine and postnatal HFC diet. Notably, iNOS and COX_2 inhibition by L-N6 (1 μM) and ns-398 (5 μM), respectively, greatly reduced Etd uptake induced by prenatal nicotine and postnatal HFC diet in microglia, astrocytes and pyramidal neurons (**Figure 4**). It has been previously shown that iNOS-dependent release of NO increases COX_2 activity and prostaglandin E_2 (PEG_2) production in macrophages (Salvemini et al., 1993). Importantly, PEG_2 activates G protein-coupled receptor 1 (EP_1), increasing the intracellular free Ca^{2+} concentration ($[Ca^{2+}]_i$; Woodward et al., 2011), causing opening of Panx1 unopposed channels (Orellana et al., 2013). Thus, we assessed whether the EP_1 receptor participated in the above mentioned responses. Inhibition of the EP_1 receptor

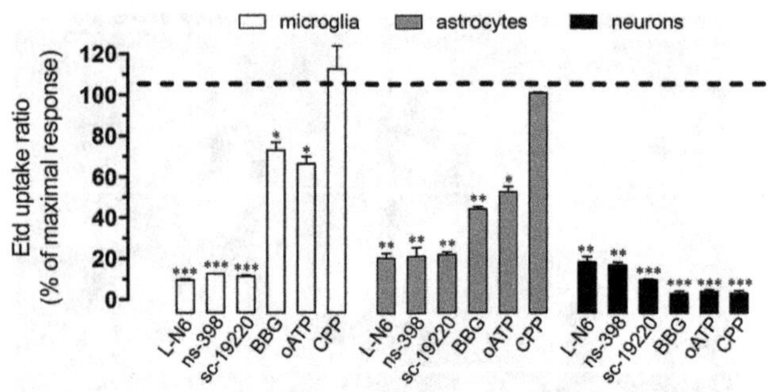

FIGURE 4 | Cx43 and Panx1 unopposed channel activity evoked by prenatal nicotine and postnatal HFC diet depended on the iNOS/COX_2/EP_1 receptor pathway and NMDA/$P2X_7$ receptor signaling. Averaged data normalized to maximal Etd uptake ratio (dashed line) achieved by prenatal nicotine and postnatal HFC diet in microglia (white bars), astrocytes (gray bars) and neurons (black bars) treated with the following blockers during Etd uptake recordings: L-N6 (1 μM), ns-398 (5 μM), sc-19220 (20 μM), BBG (200 μM), oATP (10 μM) and CPP (20 μM). * $p < 0.05$, ** $p < 0.005$; *** $p < 0.001$, effect of each blocker compared with the effect induced by prenatal nicotine plus a postnatal HFC diet. Averaged data were obtained from at least three independent experiments.

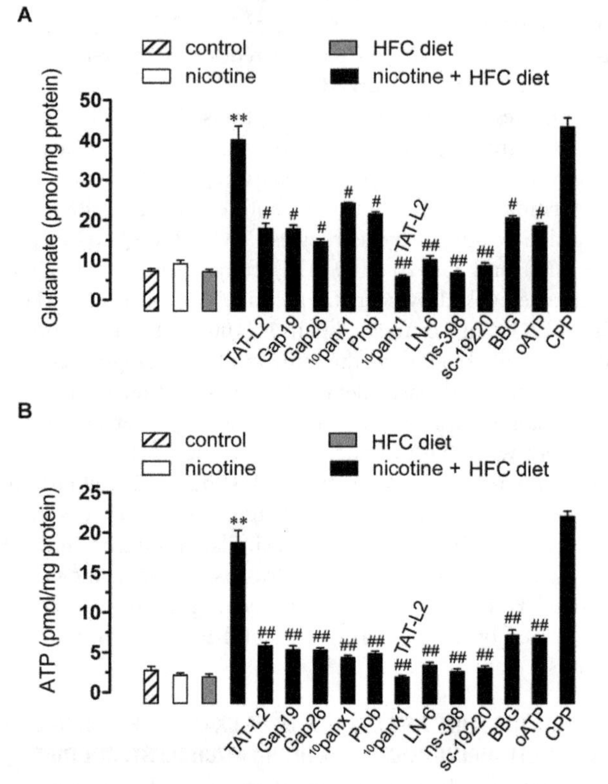

FIGURE 5 | Opening of Cx43 and Panx1 unopposed channels allowed for the release of glutamate and ATP via the iNOS/COX_2/EP_1 receptor pathway and NMDA/$P2X_7$ receptor signaling. Averaged data of glutamate **(A)** or ATP **(B)** release by hippocampal slices made from mice maintained under control conditions (dashed bar), or exposed to prenatal nicotine (white bar), postnatal HFC diet (gray bars) or a combination of both (black bars). Also shown are the effect of the following blockers applied during the experiments: TAT-L2 (100 μM), Gap19 (100 μM), Gap26 (100 μM), ^{10}panx1 (100 μM), probenecid (Prob, 500 uM), L-N6, ns-398, sc-19220, BBG, oATP or CPP. ** $p < 0.005$, effect of prenatal nicotine plus postnatal HFC diet compared with control conditions. $^{\#}$ $p < 0.05$, $^{\#\#}$ $p < 0.005$; effect of each blocker compared with the effect induced by prenatal nicotine plus a postnatal HFC diet. Averaged data were obtained from at least three independent experiments.

with sc-19220 (20 μM) resulted in a prominent reduction of the Etd uptake triggered by prenatal nicotine and postnatal HFC diet in microglia, astrocytes and pyramidal neurons (**Figure 4**).

When activated, glial cells release relevant amounts of gliotransmitters including ATP and glutamate, which underlie glia-to-glia communication via activation of purinergic and glutamatergic receptors (Perea et al., 2009; Perea and Araque, 2010). Because opening of HCs and PCHs has been associated with purinergic and glutamatergic signaling (Locovei et al., 2006; Thompson et al., 2008; Orellana et al., 2011a,b), we examined if NDMA and ionotropic ATP receptor type 7 ($P2X_7$) receptors were involved in the Etd uptake induced by prenatal nicotine and postnatal HFC diet. Remarkably, BBG (10 μM) and oATP (200 μM), two blockers of $P2X_7$ receptors, partially reduced the nicotine-induced Etd uptake in microglia and astrocytes, whereas it achieved an almost complete inhibition on pyramidal neurons (**Figure 4**). In addition, the NMDA receptor blocker CPP (20 μM), completely abolished Etd uptake evoked by prenatal nicotine and postnatal HFC diet in pyramidal neurons, whereas it failed to show the same inhibitory effect in glial cells (**Figure 4**). Taken together, these data indicate that the increase in Etd uptake induced by prenatal nicotine and postnatal HFC diet depended on activation of the iNOS/COX_2/EP_1 receptor pathway and signaling via $P2X_7$/NMDA receptors.

PRENATAL NICOTINE INDUCES CX43 AND PANX1-DEPENDENT RELEASE OF ATP AND GLUTAMATE IN BRAIN CELLS OF OFFSPRING MICE FED A HIGH-FAT/CHOLESTEROL DIET

Recently, it has been demonstrated that gliotransmitters elicit their own release in an autocrine manner via Cx43 and Panx1 unopposed channels (Orellana et al., 2012, 2013). Given that NMDA/$P2X_7$ receptors were involved in the Etd uptake observed in the offspring exposed to prenatal nicotine and fed a HFC diet during adulthood, we next evaluated whether glutamate and ATP release from hippocampal slices via Cx43 and/or Panx1 unopposed channels were also affected in this condition. Similarly to that observed in Etd uptake experiments, neither prenatal nicotine nor postnatal HFC diet by themselves affected the release of both gliotransmitters compared with control conditions (**Figures 5A,B**). However, the exposure to nicotine prenatally combined with a HFC diet during adult life strongly increased the release of glutamate and ATP (**Figures 5A,B**). Interestingly, TAT-L2 (100 μM), Gap19 (100 μM) and Gap26 (100 μM) prominently reduced the release of glutamate and ATP induced by prenatal nicotine and postnatal HFC diet (**Figures 5A,B**). Similar effects were observed on the release of glutamate and ATP upon treatment with ^{10}panx1 and probenecid (**Figures 5A,B**). Taken together, these results indicate that exposure to prenatal nicotine plus postnatal HFC diet increased the release of glutamate and

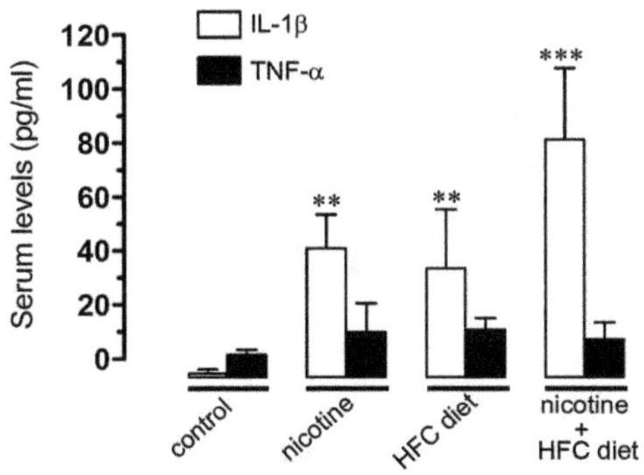

FIGURE 6 | Prenatal nicotine and postnatal HFC diet increased serum levels of IL-1β. Averaged data of serum levels of IL-1β (white bars) and TNF-α (black bars) from mice under control conditions, or subjected to prenatal nicotine, postnatal HFC diet, or a combination of both conditions. ** $p < 0.005$, *** $p < 0.001$; effect of treatments compared with control conditions. Averaged data were obtained from at least five animals.

ATP by the opening of unopposed channels formed by Cx43 and Panx1.

As expected, L-N6 (1 μM), ns-398 (5μM) and sc-19220 (20 μM) fully inhibited the release of glutamate and ATP triggered by prenatal nicotine exposure and HFC diet (**Figures 5A,B**). Furthermore, supporting the idea that gliotransmitters can elicit their own release, we found that BBG (10 μM) and oATP (200 μM) almost completely abolished the release of glutamate and ATP induced by prenatal nicotine and postnatal HFC diet. However, blockade of NMDA receptors with CPP did not show the same effect (**Figures 5A,B**). The evidence suggest that ATP but not glutamate, could partially evoke its own release by an autocrine pathway possibly mediated by Cx43 and Panx1 unopposed channels.

PRENATAL NICOTINE AND POSTNATAL HIGH-FAT/CHOLESTEROL DIET INCREASED SERUM LEVELS OF IL-1β

Given that previous studies have shown that IL-1β and TNF-α increase the opening of HCs and PCHs in glial cells (Retamal et al., 2007; Froger et al., 2009, 2010; Sáez et al., 2013), we evaluated serum levels of IL-1β and TNF-α in the offspring. TNF-α levels remained unchanged at the various conditions. However, IL-1β was notably increased by prenatal nicotine or postnatal HFC diet alone, and by the combination of prenatal nicotine plus postnatal HFC diet, being the latest the condition achieving the most robust increase (**Figure 6**). This evidence indicates that opening of Cx43 and Panx1 unopposed channels evoked by prenatal nicotine exposure and postnatal HFC diet occurred concomitantly with an increased pro-inflammatory state of the offspring.

DISCUSSION

In this study, we showed that prenatal nicotine and postnatal HFC diet for 8 weeks after weaning increased the opening of unopposed channels formed by Cx43 in astrocytes and Panx1 in microglia and neurons. This enhanced opening occurred by a mechanism depending on iNOS/COX_2/EP_1 receptor pathway activation and signaling via $P2X_7$/NMDA receptors. In addition, unopposed channel opening resulted in the release of two major gliotransmitters: glutamate and ATP.

Previous studies have demonstrated that nicotine delivery during pregnancy induces neuronal defects, increased number of glial cells, and behavioral impairment in the offspring (Navarro et al., 1989; Roy and Sabherwal, 1998; Roy et al., 2002). Our results suggest that the effect induced by prenatal nicotine could be mediated in part by enhanced release of gliotransmitters. It has been shown that gliotransmitter release through HCs and PCHs underlies crucial functions in the physiology of the CNS (Lin et al., 2008; Huckstepp et al., 2010; Orellana et al., 2012; Stehberg et al., 2012; Torres et al., 2012). Moreover, several studies indicate that uncontrolled opening of these channels results in exacerbated release of gliotransmitters, which in high concentrations can be toxic for neighboring cells (Takeuchi et al., 2006; Orellana et al., 2011a,b). Here, we found that prenatal nicotine in combination with a postnatal HFC diet increased the opening of HCs and PHCs in brain cells. In agreement with their sensor role in the CNS (Block et al., 2007), microglia showed the highest Etd uptake compared with astrocytes and neurons in mice exposed to both environmental stressors. As this response was nearly abolished by Panx1 but not Cx43 unopposed channel blockers, the former protein appeared to be the major responsible for the increased permeability. In agreement with our data, recent studies have shown that pro-inflammatory conditions increase the opening of Panx1 channels in microglia (Orellana et al., 2013; Sáez et al., 2013). Both astrocytes and neurons exhibited similar increases on Etd uptake after prenatal nicotine and postnatal HFC diet. However, this effect was due to Cx43 HCs in the former, as mimetic peptides known to block these channels (Wang et al., 2013a), completely inhibited astroglial cell Etd uptake. In contrast, ^{10}panx1 and probenecid did not affect astroglial cell Etd uptake. On the other hand, neuronal Etd uptake induced by prenatal nicotine and postnatal HFC diet was drastically blocked by ^{10}panx1 and probenecid but not by TAT-L2, Gap19 or Gap26, indicating that Panx1 channels were the main contributors for this response.

Glutamate and ATP are considered key mediators on neuron-glia crosstalk. Thereby, their release through membrane proteins and vesicles is tightly regulated (Fields and Burnstock, 2006; Perea and Araque, 2010). In fact, high concentrations of glutamate and ATP at the synaptic cleft under pathological conditions could result in neurotoxicity (Lau and Tymianski, 2010; Arbeloa et al., 2012; Ashpole et al., 2013). As mentioned before, part of neuronal damage could depend on the release of glutamate and ATP via HCs and PCHs (Takeuchi et al., 2006; Garré et al., 2010; Orellana et al., 2011a,b). Both glutamate and ATP released by glial cells trigger the activation of neuronal NMDA and P2X7 receptors, which result in the opening of neuronal Panx1 channels and further cell death (Orellana et al., 2011a,b). Our results indicate that release of glutamate and ATP evoked by prenatal nicotine and postnatal HFC diet occurred via Cx43 and Panx1 unopposed channels, as it was inhibited by TAT-L2, Gap19, Gap26, ^{10}panx1 and probenecid. Nevertheless, given that neuronal Etd uptake still

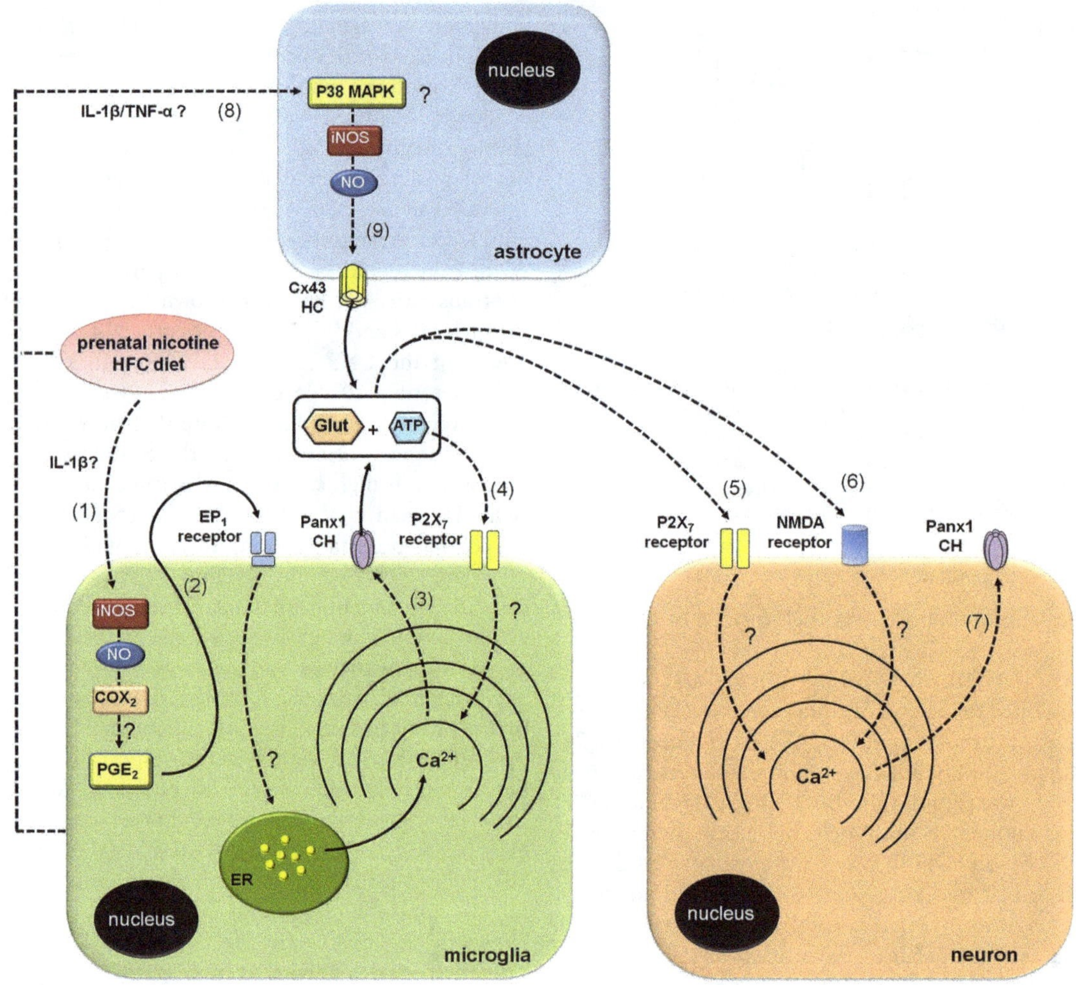

FIGURE 7 | Modulation of brain cell hemichannels by prenatal nicotine and postnatal high-fat/cholesterol diet. Prenatal nicotine and postnatal high-fat/cholesterol diet increase serum levels of IL-1β, leading to iNOS/COX_2 activation in microglia (1) and likely in astrocytes and neurons (not depicted). Possibly, PGE_2 released from microglia binds its EP_1 metabotropic receptors and produces the release of Ca^{2+} from intracellular stores (2). The increases in $[Ca^{2+}]_i$ is a known condition that evokes opening of Panx1 channels (Panx1 CHs), allowing the release of glutamate and ATP through them (3). ATP released via Panx1 CHs activates $P2X_7$ receptors (4), triggering a self-perpetuating mechanism, in which high levels of $[Ca^{2+}]_i$ could reactivate Panx1 CHs in microglia. In addition, paracrine release of glutamate and ATP from microglia could act on neighboring or distant neurons, resulting in the activation of $P2X_7$ (5) and NMDA (6) receptors, respectively. The latter enhances the activity of neuronal Panx1 CHs (7), allowing more release of ATP/glutamate and altering physiological functions of neurons (not depicted). Prenatal nicotine and postnatal high-fat/cholesterol diet could increase the release of glutamate and ATP from astrocytes via opening of Cx43 hemichannels (Cx43 HCs; 9). It is possible that microglia through the release of pro-inflammatory molecules (e.g., IL-1β and TNF-α) could contribute to the opening of astroglial Cx43 HCs via the activation of a p38MAPK/iNOS-dependent pathway.

persist under Cx43 but not Panx1 channel blockade, it seems that glutamate and ATP released from microglia rather than astrocytes are the major contributors to the opening of Panx1 unopposed channels in neurons (**Figure 7**).

How does the exposure to prenatal nicotine and postnatal HFC diet induce the opening of Cx43 and Panx1 unopposed channels? Previous studies have demonstrated that opening of these channels in microglia and astrocytes results on the activation of an iNOS/COX_2/EP_1 receptor- and p38MAPK/iNOS-dependent pathway, respectively (Retamal et al., 2007; Orellana et al., 2013). In agreement with that mechanism, Etd uptake and gliotransmitter release were nearly abolished by blockers of iNOS, COX_2 and EP_1 receptor, suggesting that activation of Cx43 and Panx1 unopposed channels likely occurred downstream of this pathway. Given that activation of EP_1 receptors raises $[Ca^{2+}]_i$ (Woodward et al., 2011), opening of these channels possibly occurred by this mechanism, which is coherent with previous studies showing that increased levels of $[Ca^{2+}]_i$ are necessary for gliotransmitter release via HCs and PCHs (Locovei et al., 2006; Torres et al., 2012). This is also in agreement with the fact that $P2X_7$ receptor activation, a well known mechanism that increases $[Ca^{2+}]_i$, was required to induce the release of glutamate and ATP we observed. By contrast, blockade of NMDA receptors, whose activation also enhances $[Ca^{2+}]_i$ levels, did not induce the same response. These data support the idea that ATP, but not glutamate, evokes its own release via Panx1 unopposed channels, and subsequent activation of purinergic receptors in microglia, as has been previously observed (Orellana et al., 2013; **Figure 7**). Previous studies have described that astrocytes exposed to activated microglia exhibit an increased Cx43 hemichannel opening sensitive to L-NAME (a broad range NOS inhibitor) and p38 MAPK inhibitors (Retamal et al., 2007). Therefore, it is conceivable to speculate that along with direct effect of nicotine and HFC diet on astrocytes, microglia might also contribute to the opening of astroglial HCs by releasing pro-inflammatory cytokines (see below; **Figure 7**). Whether specific crosstalk (e.g., through $P2X_7$ receptors, HCs and PCHs) between astrocytes and microglia could explain their different contribution to neuronal Panx1 channel opening will be a matter of future investigation.

It has been described that nicotine exposure increases peripheral and brain levels of inflammatory cytokines, including IL-1β (Lau et al., 2006; Bradford et al., 2011). We speculate that prenatal nicotine exposure could affect the inflammatory state of dams, resulting in epigenetic modification of brain genes, leading to permanent changes in gene expression and long-term changes in structure and function (Boksa, 2010). Here, we found that the brain of adult offspring from nicotine-treated dams are sensitized to postnatal HFC diet, as has been previously described to occur with others environmental challenges (Slotkin et al., 1991; Bilbo et al., 2005). Given that feeding mice a cholesterol-enriched diet during adulthood results in a general inflammatory state (Thirumangalakudi et al., 2008; Lewis et al., 2010), this condition could act as a second inflammatory challenge affecting the CNS. Accordingly, we found that prenatal nicotine and postnatal cholesterol-enriched diet induced higher serum levels of IL-1β compared to control conditions.

Elevated blood levels of cytokines correlate with increased brain levels of cytokines (Erickson and Banks, 2011), being the latter closely linked to activation of iNOS, COX_2 and EP_1 receptors (Vinukonda et al., 2010; Sheng et al., 2011; Samy and Igwe, 2012). Therefore, it is plausible to speculate that increased brain levels of IL-1β and activation of iNOS/COX_2/EP_1 receptor pathway could be involved in the increased opening of Cx43 and Panx1 unopposed channels observed in our model (**Figure 7**). Supporting this idea, IL-1β causes opening of HCs and changes connexin expression in brain cells (Retamal et al., 2007; Froger et al., 2009, 2010; Orellana et al., 2011a; Xiong et al., 2012). Here, we observed that prenatal nicotine and postnatal HFC diet reduced Cx43 expression in astrocytes. Given that surface HCs account for ~11% of total Cx43 under resting conditions (Schalper et al., 2008), making them less detectable by immunofluorescence than gap junctions plaques, a reduction on Cx43 immunodetection not necessarily implicates a decrease on surface HCs or in their activity. On the other hand, Panx1 expression was increased in pyramidal neurons. It is possible that part of Etd uptake observed in pyramidal neurons could rely on this phenomenon. Further studies are required to elucidate whether changes in protein expression could contribute as well to the Cx43 and Panx1 unopposed channel activity triggered by prenatal nicotine and postnatal HFC diet.

Diverse studies have shown that cell and tissue responses to injuries depend on properties of the cells (e.g., age, hormonal exposure, and stage of cell cycle) and insult (e.g., duration, intensity, and quality). Moreover, CNS responses depend on interactions between their constituent cells, including chemical and electrical transmission as well as paracrine and autocrine signaling (e.g., by cytokines and ROS), possibly mediated by HCs and PCHs. In most chronic diseases, additional mechanisms are progressively added to the primary cause and thus, complicating the assignment of contribution of each factor to the final condition. Under this view, we speculate that the combined effect of two stressors (exposure to prenatal nicotine and postnatal HFC diet) resulted in our system in an synergic outcome on the increased activity of HCs and PCHs in brain cells, as has been previously observed in other inflammatory models (Orellana et al., 2010). Despite the difficulty of assigning contributions to connexin and pannexin unopposed channels in the pathogenesis of neurodegenerative diseases, recent studies using homo and/or heterocellular cultures have provided clues to elucidate this matter (Sáez and Leybaert, 2014). Although our model does not recapitulate completely mechanisms underlying brain abnormalities induced by maternal cigarette smoking, it allows us to dissect the specific contribution of HCs and PCHs expressed by individual brain cell types. Our findings bring new light on how gliotransmitters and the unbalance on paracrine signaling mediated by HCs and PCHs could contribute to developing brain abnormalities induced by different stressors during pregnancy.

ACKNOWLEDGMENTS

This work was partially supported by CONICYT collaborative grant 79090028 to Juan A. Orellana and Dolores Busso and FONDECYT grants 11121133 to Juan A. Orellana, 11090064 and 1141236 to Dolores Busso, 1110712 to Attilio Rigotti, 1130874 to Jaime Eugenín, 1131025 to Rommy von Bernhardi.

REFERENCES

Amitai, Y. (2010). Physiologic role for "inducible" nitric oxide synthase: a new form of astrocytic-neuronal interface. *Glia* 58, 1775–1781. doi: 10.1002/glia.21057

Arbeloa, J., Pérez-Samartin, A., Gottlieb, M., and Matute, C. (2012). P2X7 receptor blockade prevents ATP excitotoxicity in neurons and reduces brain damage after ischemia. *Neurobiol. Dis.* 45, 954–961. doi: 10.1016/j.nbd.2011.12.014

Ashpole, N. M., Chawla, A. R., Martin, M. P., Brustovetsky, T., Brustovetsky, N., and Hudmon, A. (2013). Loss of calcium/calmodulin-dependent protein kinase II activity in cortical astrocytes decreases glutamate uptake and induces neurotoxic release of ATP. *J. Biol. Chem.* 288, 14599–14611. doi: 10.1074/jbc.M113.466235

Bilbo, S. D., Levkoff, L. H., Mahoney, J. H., Watkins, L. R., Rudy, J. W., and Maier, S. F. (2005). Neonatal infection induces memory impairments following an immune challenge in adulthood. *Behav. Neurosci.* 119, 293–301. doi: 10.1037/0735-7044.119.1.293

Block, M. L., Zecca, L., and Hong, J. S. (2007). Microglia-mediated neurotoxicity: uncovering the molecular mechanisms. *Nat. Rev. Neurosci.* 8, 57–69. doi: 10.1038/nrn2038

Boksa, P. (2010). Effects of prenatal infection on brain development and behavior: a review of findings from animal models. *Brain Behav. Immun.* 24, 881–897. doi: 10.1016/j.bbi.2010.03.005

Bradford, S. T., Stamatovic, S. M., Dondeti, R. S., Keep, R. F., and Andjelkovic, A. V. (2011). Nicotine aggravates the brain postischemic inflammatory response. *Am.*

J. Physiol. Heart Circ. Physiol. 300, H1518–H1529. doi: 10.1152/ajpheart.00928.2010

Burkovetskaya, M., Karpuk, N., Xiong, J., Bosch, M., Boska, M. D., Takeuchi, H., et al. (2014). Evidence for Aberrant Astrocyte Hemichannel activity in Juvenile Neuronal Ceroid Lipofuscinosis (JNCL). *PLoS One* 9:e95023. doi: 10.1371/journal.pone.0095023

Coddou, C., Bravo, E., and Eugenín, J. (2009). Alterations in cholinergic sensitivity of respiratory neurons induced by pre-natal nicotine: a mechanism for respiratory dysfunction in neonatal mice. *Philos. Trans. R. Soc. Lond. B Biol. Sci.* 364, 2527–2535. doi: 10.1098/rstb.2009.0078

Contreras, J. E., Sánchez, H. A., Eugenin, E. A., Speidel, D., Theis, M., Willecke, K., et al. (2002). Metabolic inhibition induces opening of unapposed connexin 43 gap junction hemichannels and reduces gap junctional communication in cortical astrocytes in culture. *Proc. Natl. Acad. Sci. U S A* 99, 495–500. doi: 10.1073/pnas.012589799

DiFranza, J. R., and Lew, R. A. (1995). Effect of maternal cigarette smoking on pregnancy complications and sudden infant death syndrome. *J. Fam. Pract.* 40, 385–394.

Dufour, F., Liu, Q. Y., Gusev, P., Alkon, D., and Atzori, M. (2006). Cholesterol-enriched diet affects spatial learning and synaptic function in hippocampal synapses. *Brain Res.* 1103, 88–98. doi: 10.1016/j.brainres.2006.05.086

Dwyer, J. B., Broide, R. S., and Leslie, F. M. (2008). Nicotine and brain development. *Birth Defects Res. C Embryo Today* 84, 30–44. doi: 10.1002/bdrc.20118

Erickson, M. A., and Banks, W. A. (2011). Cytokine and chemokine responses in serum and brain after single and repeated injections of lipopolysaccharide: multiplex quantification with path analysis. *Brain Behav. Immun.* 25, 1637–1648. doi: 10.1016/j.bbi.2011.06.006

Eugenín, J., Otárola, M., Bravo, E., Coddou, C., Cerpa, V., Reyes-Parada, M., et al. (2008). Prenatal to early postnatal nicotine exposure impairs central chemoreception and modifies breathing pattern in mouse neonates: a probable link to sudden infant death syndrome. *J. Neurosci.* 28, 13907–13917. doi: 10.1523/JNEUROSCI.4441-08.2008

Fields, R. D., and Burnstock, G. (2006). Purinergic signalling in neuron-glia interactions. *Nat. Rev. Neurosci.* 7, 423–436. doi: 10.1038/nrn1928

Froger, N., Orellana, J. A., Calvo, C. F., Amigou, E., Kozoriz, M. G., Naus, C. C., et al. (2010). Inhibition of cytokine-induced connexin43 hemichannel activity in astrocytes is neuroprotective. *Mol. Cell. Neurosci.* 45, 37–46. doi: 10.1016/j.mcn.2010.05.007

Froger, N., Orellana, J. A., Cohen-Salmon, M., Ezan, P., Amigou, E., Sáez, J. C., et al. (2009). Cannabinoids prevent the opposite regulation of astroglial connexin43 hemichannels and gap junction channels induced by pro-inflammatory treatments. *J. Neurochem.* 111, 1383–1397. doi: 10.1111/j.1471-4159.2009.06407.x

Garré, J. M., Retamal, M. A., Cassina, P., Barbeito, L., Bukauskas, F. F., Saez, J. C., et al. (2010). FGF-1 induces ATP release from spinal astrocytes in culture and opens pannexin and connexin hemichannels. *Proc. Natl. Acad. Sci. U S A* 107, 22659–22664. doi: 10.1073/pnas.1013793107

Gulbransen, B. D., Bashashati, M., Hirota, S. A., Gui, X., Roberts, J. A., Macdonald, J. A., et al. (2012). Activation of neuronal P2X7 receptor-pannexin-1 mediates death of enteric neurons during colitis. *Nat. Med.* 18, 600–604. doi: 10.1038/nm.2679

Huckstepp, R. T. R., id Bihi, R., Eason, R., Spyer, K. M., Dicke, N., Willecke, K., et al. (2010). Connexin hemichannel-mediated CO2-dependent release of ATP in the medulla oblongata contributes to central respiratory chemosensitivity. *J. Physiol.* 588, 3901–3920. doi: 10.1113/jphysiol.2010.192088

Iglesias, R., Dahl, G., Qiu, F., Spray, D. C., and Scemes, E. (2009). Pannexin 1: the molecular substrate of astrocyte "hemichannels". *J. Neurosci.* 29, 7092–7097. doi: 10.1523/JNEUROSCI.6062-08.2009

Jacobsen, L. K., Slotkin, T. A., Westerveld, M., Mencl, W. E., and Pugh, K. R. (2006). Visuospatial memory deficits emerging during nicotine withdrawal in adolescents with prenatal exposure to active maternal smoking. *Neuropsychopharmacology* 31, 1550–1561. doi: 10.1038/sj.npp.1300981

Julvez, J., Ribas-Fitó, N., Torrent, M., Forns, M., Garcia-Esteban, R., and Sunyer, J. (2007). Maternal smoking habits and cognitive development of children at age 4 years in a population-based birth cohort. *Int. J. Epidemiol.* 36, 825–832. doi: 10.1093/ije/dym107

Karpuk, N., Burkovetskaya, M., Fritz, T., Angle, A., and Kielian, T. (2011). Neuroinflammation leads to region-dependent alterations in astrocyte gap junction communication and hemichannel activity. *J. Neurosci.* 31, 414–425. doi: 10.1523/JNEUROSCI.5247-10.2011

Lau, P. P., Li, L., Merched, A. J., Zhang, A. L., Ko, K. W., and Chan, L. (2006). Nicotine induces proinflammatory responses in macrophages and the aorta leading to acceleration of atherosclerosis in low-density lipoprotein receptor(-/-) mice. *Arterioscler. Thromb. Vasc. Biol.* 26, 143–149. doi: 10.1161/01.ATV.0000193510.19000.10

Lau, A., and Tymianski, M. (2010). Glutamate receptors, neurotoxicity and neurodegeneration. *Pflugers Arch.* 460, 525–542. doi: 10.1007/s00424-010-0809-1

Lewis, D. K., Bake, S., Thomas, K., Jezierski, M. K., and Sohrabji, F. (2010). A high cholesterol diet elevates hippocampal cytokine expression in an age and estrogen-dependent manner in female rats. *J. Neuroimmunol.* 223, 31–38. doi: 10.1016/j.jneuroim.2010.03.024

Lin, J. H., Lou, N., Kang, N., Takano, T., Hu, F., Han, X., et al. (2008). A central role of connexin 43 in hypoxic preconditioning. *J. Neurosci.* 28, 681–695. doi: 10.1523/JNEUROSCI.3827-07.2008

Locovei, S., Wang, J., and Dahl, G. (2006). Activation of pannexin 1 channels by ATP through P2Y receptors and by cytoplasmic calcium. *FEBS Lett.* 580, 239–244. doi: 10.1016/j.febslet.2005.12.004

Luck, W., Nau, H., Hansen, R., and Steldinger, R. (1985). Extent of nicotine and cotinine transfer to the human fetus, placenta and amniotic fluid of smoking mothers. *Dev. Pharmacol. Ther.* 8, 384–395.

Morton, D. B., and Griffiths, P. H. (1985). Guidelines on the recognition of pain, distress and discomfort in experimental animals and an hypothesis for assessment. *Vet. Rec.* 116, 431–436. doi: 10.1136/vr.116.16.431

Muneoka, K., Ogawa, T., Kamei, K., Muraoka, S., Tomiyoshi, R., Mimura, Y., et al. (1997). Prenatal nicotine exposure affects the development of the central serotonergic system as well as the dopaminergic system in rat offspring: involvement of route of drug administrations. *Brain Res. Dev. Brain Res.* 102, 117–126. doi: 10.1016/S0165-3806(97)00092-8

Naeye, R. L., and Peters, E. C. (1984). Mental development of children whose mothers smoked during pregnancy. *Obstet. Gynecol.* 64, 601–607.

Navarro, H. A., Seidler, F. J., Schwartz, R. D., Baker, F. E., Dobbins, S. S., and Slotkin, T. A. (1989). Prenatal exposure to nicotine impairs nervous system development at a dose which does not affect viability or growth. *Brain Res. Bull.* 23, 187–192. doi: 10.1016/0361-9230(89)90146-9

Orellana, J. A., Froger, N., Ezan, P., Jiang, J. X., Bennett, M. V., Naus, C. C., et al. (2011a). ATP and glutamate released via astroglial connexin 43 hemichannels mediate neuronal death through activation of pannexin 1 hemichannels. *J. Neurochem.* 118, 826–840. doi: 10.1111/j.1471-4159.2011.07210.x

Orellana, J. A., Hernández, D. E., Ezan, P., Velarde, V., Bennett, M. V., Giaume, C., et al. (2010). Hypoxia in high glucose followed by reoxygenation in normal glucose reduces the viability of cortical astrocytes through increased permeability of connexin 43 hemichannels. *Glia* 58, 329–343. doi: 10.1002/glia.20926

Orellana, J. A., Montero, T. D., and von Bernhardi, R. (2013). Astrocytes inhibit nitric oxide-dependent Ca(2+) dynamics in activated microglia: involvement of ATP released via pannexin 1 channels. *Glia* 61, 2023–2037. doi: 10.1002/glia.22573

Orellana, J. A., Sáez, P. J., Cortés-Campos, C., Elizondo, R. J., Shoji, K. F., Contreras-Duarte, S., et al. (2012). Glucose increases intracellular free Ca(2+) in tanycytes via ATP released through connexin 43 hemichannels. *Glia* 60, 53–68. doi: 10.1002/glia.21246

Orellana, J. A., Sáez, P. J., Shoji, K. F., Schalper, K. A., Palacios-Prado, N., Velarde, V., et al. (2009). Modulation of brain hemichannels and gap junction channels by pro-inflammatory agents and their possible role in neurodegeneration. *Antioxid. Redox Signal.* 11, 369–399. doi: 10.1089/ars.2008.2130

Orellana, J. A., Shoji, K. F., Abudara, V., Ezan, P., Amigou, E., Sáez, P. J., et al. (2011b). Amyloid β-induced death in neurons involves glial and neuronal hemichannels. *J. Neurosci.* 31, 4962–4977. doi: 10.1523/JNEUROSCI.6417-10.2011

Pelegrin, P., and Surprenant, A. (2006). Pannexin-1 mediates large pore formation and interleukin-1beta release by the ATP-gated P2X7 receptor. *EMBO J.* 25, 5071–5082. doi: 10.1038/sj.emboj.7601378

Perea, G., and Araque, A. (2010). Glia modulates synaptic transmission. *Brain Res. Rev.* 63, 93–102. doi: 10.1016/j.brainresrev.2009.10.005

Perea, G., Navarrete, M., and Araque, A. (2009). Tripartite synapses: astrocytes process and control synaptic information. *Trends Neurosci.* 32, 421–431. doi: 10.1016/j.tins.2009.05.001

Rantakallio, P., and Koiranen, M. (1987). Neurological handicaps among children whose mothers smoked during pregnancy. *Prev. Med.* 16, 597–606. doi: 10.1016/0091-7435(87)90043-0

Retamal, M. A., Froger, N., Palacios-Prado, N., Ezan, P., Sáez, P. J., Sáez, J. C., et al. (2007). Cx43 hemichannels and gap junction channels in astrocytes are regulated oppositely by proinflammatory cytokines released from activated microglia. *J. Neurosci.* 27, 13781–13792. doi: 10.1523/JNEUROSCI.2042-07.2007

Rouach, N., Avignone, E., Même, W., Koulakoff, A., Venance, L., Blomstrand, F., et al. (2002). Gap junctions and connexin expression in the normal and pathological central nervous system. *Biol. Cell* 94, 457–475. doi: 10.1016/S0248-4900(02)00016-3

Roy, T. S., and Sabherwal, U. (1994). Effects of prenatal nicotine exposure on the morphogenesis of somatosensory cortex. *Neurotoxicol. Teratol.* 16, 411–421. doi: 10.1016/0892-0362(94)90030-2

Roy, T. S., and Sabherwal, U. (1998). Effects of gestational nicotine exposure on hippocampal morphology. *Neurotoxicol. Teratol.* 20, 465–473. doi: 10.1016/S0892-0362(97)00137-2

Roy, T. S., Seidler, F. J., and Slotkin, T. A. (2002). Prenatal nicotine exposure evokes alterations of cell structure in hippocampus and somatosensory cortex. *J. Pharmacol. Exp. Ther.* 300, 124–133. doi: 10.1124/jpet.300.1.124

Sáez, J. C., and Leybaert, L. (2014). Hunting for connexin hemichannels. *FEBS Lett.* 588, 1205–1211. doi: 10.1016/j.febslet.2014.03.004

Sáez, P. J., Shoji, K. F., Retamal, M. A., Harcha, P. A., Ramirez, G., Jiang, J. X., et al. (2013). ATP is required and advances cytokine-induced gap junction formation in microglia in vitro. *Mediators Inflamm.* 2013:216402. doi: 10.1155/2013/216402

Salvemini, D., Misko, T. P., Masferrer, J. L., Seibert, K., Currie, M. G., and Needleman, P. (1993). Nitric oxide activates cyclooxygenase enzymes. *Proc. Natl. Acad. Sci. U S A* 90, 7240–7244. doi: 10.1073/pnas.90.15.7240

Samy, A. S., and Igwe, O. J. (2012). Regulation of IL-1beta-induced cyclooxygenase-2 expression by interactions of Abeta peptide, apolipoprotein E and nitric oxide in human neuroglioma. *J. Mol. Neurosci.* 47, 533–545. doi: 10.1007/s12031-011-9670-8

Schalper, K. A., Palacios-Prado, N., Orellana, J. A., and Sáez, J. C. (2008). Currently used methods for identification and characterization of hemichannels. *Cell Commun. Adhes.* 15, 207–218. doi: 10.1080/15419060802014198

Schneider, T., Ilott, N., Brolese, G., Bizarro, L., Asherson, P. J., and Stolerman, I. P. (2011). Prenatal exposure to nicotine impairs performance of the 5-choice serial reaction time task in adult rats. *Neuropsychopharmacology* 36, 1114–1125. doi: 10.1038/npp.2010.249

Sheng, W., Zong, Y., Mohammad, A., Ajit, D., Cui, J., Han, D., et al. (2011). Pro-inflammatory cytokines and lipopolysaccharide induce changes in cell morphology and upregulation of ERK1/2, iNOS and sPLA(2)-IIA expression in astrocytes and microglia. *J. Neuroinflammation* 8:121. doi: 10.1186/1742-2094-8-121

Shestopalov, V. I., and Slepak, V. Z. (2014). Molecular pathways of pannexin1-mediated neurotoxicity. *Front. Physiol.* 5:23. doi: 10.3389/fphys.2014.00023

Silverman, W., Locovei, S., and Dahl, G. (2008). Probenecid, a gout remedy, inhibits pannexin 1 channels. *Am. J. Physiol. Cell Physiol.* 295, C761–C767. doi: 10.1152/ajpcell.00227.2008

Slotkin, T. A., Lappi, S. E., Tayyeb, M. I., and Seidler, F. J. (1991). Chronic prenatal nicotine exposure sensitizes rat brain to acute postnatal nicotine challenge as assessed with ornithine decarboxylase. *Life Sci.* 49, 665–670. doi: 10.1016/0024-3205(91)90113-P

Stehberg, J., Moraga-Amaro, R., Salazar, C., Becerra, A., Echeverria, C., Orellana, J. A., et al. (2012). Release of gliotransmitters through astroglial connexin 43 hemichannels is necessary for fear memory consolidation in the basolateral amygdala. *FASEB J.* 26, 3649–3657. doi: 10.1096/fj.11-198416

Takeuchi, H., Jin, S., Wang, J., Zhang, G., Kawanokuchi, J., Kuno, R., et al. (2006). Tumor necrosis factor-alpha induces neurotoxicity via glutamate release from hemichannels of activated microglia in an autocrine manner. *J. Biol. Chem.* 281, 21362–21368. doi: 10.1074/jbc.M600504200

Thirumangalakudi, L., Prakasam, A., Zhang, R., Bimonte-Nelson, H., Sambamurti, K., Kindy, M. S., et al. (2008). High cholesterol-induced neuroinflammation and amyloid precursor protein processing correlate with loss of working memory in mice. *J. Neurochem.* 106, 475–485. doi: 10.1111/j.1471-4159.2008.05415.x

Thompson, R. J., Jackson, M. F., Olah, M. E., Rungta, R. L., Hines, D. J., Beazely, M. A., et al. (2008). Activation of pannexin-1 hemichannels augments aberrant bursting in the hippocampus. *Science* 322, 1555–1559. doi: 10.1126/science.1165209

Thompson, R. J. R., Zhou, N. N., and Macvicar, B. A. (2006). Ischemia opens neuronal gap junction hemichannels. *Science* 312, 924–927. doi: 10.1126/science.1126241

Toledo-Rodriguez, M., Lotfipour, S., Leonard, G., Perron, M., Richer, L., Veillette, S., et al. (2010). Maternal smoking during pregnancy is associated with epigenetic modifications of the brain-derived neurotrophic factor-6 exon in adolescent offspring. *Am. J. Med. Genet. B Neuropsychiatr. Genet.* 153B, 1350–1354. doi: 10.1002/ajmg.b.31109

Torres, A., Wang, F., Xu, Q., Fujita, T., Dobrowolski, R., Willecke, K., et al. (2012). Extracellular Ca(2)(+) acts as a mediator of communication from neurons to glia. *Sci. Signal.* 5:ra8. doi: 10.1126/scisignal.2002160

Triviño, A., Ramírez, A. I., Salazar, J. J., De Hoz, R., Rojas, B., Padilla, E., et al. (2006). A cholesterol-enriched diet induces ultrastructural changes in retinal and macroglial rabbit cells. *Exp. Eye Res.* 83, 357–366. doi: 10.1016/j.exer.2005.12.020

Tzeng, S. F., Hsiao, H. Y., and Mak, O. T. (2005). Prostaglandins and cyclooxygenases in glial cells during brain inflammation. *Curr. Drug Targets Inflamm. Allergy* 4, 335–340. doi: 10.2174/1568010054022051

van de Kamp, J. L., and Collins, A. C. (1994). Prenatal nicotine alters nicotinic receptor development in the mouse brain. *Pharmacol. Biochem. Behav.* 47, 889–900. doi: 10.1016/0091-3057(94)90293-3

Vinukonda, G., Csiszar, A., Hu, F., Dummula, K., Pandey, N. K., Zia, M. T., et al. (2010). Neuroprotection in a rabbit model of intraventricular haemorrhage by cyclooxygenase-2, prostanoid receptor-1 or tumour necrosis factor-alpha inhibition. *Brain* 133, 2264–2280. doi: 10.1093/brain/awq107

Wang, N., De Bock, M., Decrock, E., Bol, M., Gadicherla, A., Bultynck, G., et al. (2013a). Connexin targeting peptides as inhibitors of voltage- and intracellular Ca2+-triggered Cx43 hemichannel opening. *Neuropharmacology* 75, 506–516. doi: 10.1016/j.neuropharm.2013.08.021

Wang, N., De Bock, M., Decrock, E., Bol, M., Gadicherla, A., Vinken, M., et al. (2013b). Paracrine signaling through plasma membrane hemichannels. *Biochim. Biophys. Acta* 1828, 35–50. doi: 10.1016/j.bbamem.2012.07.002

Woodward, D. F., Jones, R. L., and Narumiya, S. (2011). International Union of Basic and Clinical Pharmacology. LXXXIII: classification of prostanoid receptors, updating 15 years of progress. *Pharmacol. Rev.* 63, 471–538. doi: 10.1124/pr.110.003517

Xiong, J., Burkovetskaya, M., Karpuk, N., and Kielian, T. (2012). IL-1RI (interleukin-1 receptor type I) signalling is essential for host defence and hemichannel activity during acute central nervous system bacterial infection. *ASN Neuro* 4:e00085. doi: 10.1042/AN20120008

Ya, B. L., Liu, W. Y., Ge, F., Zhang, Y. X., Zhu, B. L., and Bai, B. (2013). Dietary cholesterol alters memory and synaptic structural plasticity in young rat brain. *Neurol. Sci.* 34, 1355–1365. doi: 10.1007/s10072-012-1241-4

13

Aberrant Cx26 hemichannels and keratitis-ichthyosis-deafness syndrome: Insights into syndromic hearing loss

Helmuth A. Sanchez and Vytas K. Verselis *

Dominick P. Purpura Department of Neuroscience, Albert Einstein College of Medicine, Bronx, NY, USA

***Correspondence:**
Vytas K. Verselis, Dominick P. Purpura Department of Neuroscience, Albert Einstein College of Medicine, 1300 Morris Park Ave, Bronx, NY, 10461, USA
e-mail: vytas.verselis@einstein.yu.edu

Mutation of the *GJB2* gene, which encodes the connexin 26 (Cx26) gap junction (GJ) protein, is the most common cause of hereditary, sensorineural hearing loss. Cx26 is not expressed in hair cells, but is widely expressed throughout the non-sensory epithelial cells of the cochlea. Most *GJB2* mutations produce non-syndromic deafness, but a subset produces syndromic deafness in which profound hearing loss is accompanied by a diverse array of infectious and neoplastic cutaneous disorders that can be fatal. Although GJ channels, which are assembled by the docking of two, so-called hemichannels (HCs), have been the main focus of deafness-associated disease models, it is now evident that the HCs themselves can function in the absence of docking and contribute to signaling across the cell membrane as a novel class of ion channel. A notable feature of syndromic deafness mutants is that the HCs exhibit aberrant behaviors providing a plausible basis for disease that is associated with excessive or altered contributions of Cx26 HCs that, in turn, lead to compromised cell integrity. Here we discuss some of the aberrant Cx26 HC properties that have been described for mutants associated with keratitis-ichthyosis-deafness (KID) syndrome, a particularly severe Cx26-associated syndrome, which shed light on genotype-phenotype relationships and causes underlying cochlear dysfunction.

Keywords: connexin, hemichannel, deafness, cochlea, gating, permeability, calcium

Cx26 MUTATIONS AND DEAFNESS

It has been over 15 years since the *GJB2* gene encoding the Cx26 gap junction (GJ) protein was identified as a susceptibility gene for sensorineural deafness (Kelsell et al., 1997). In the original study, Cx26 mutations were shown to result in premature stop codons in several autosomal recessive non-syndromic deafness pedigrees. Cx26 mutations are now known to represent one of the most common causes of inherited, non-syndromic deafness in the human population (Apps et al., 2007; Chan et al., 2010; Duman and Tekin, 2012). More than 100 Cx26 mutations have been identified and, for the most part, they are recessive and produce loss of function as a result of deletions, insertions and frameshifts (Hoang Dinh et al., 2009; Lee and White, 2009). However, a subset of these mutations has been shown to cause sensorineural deafness that is accompanied by severe skin disorders. In contrast to non-syndromic deafness, syndromic deafness is characteristically cause by missense mutations that result in single amino acid substitutions and behave in an autosomal dominant manner.

To date, mutations at 18 positions in Cx26 have been identified in association with syndromic deafness (**Figure 1**). Based primarily on the nature of the cutaneous manifestations patients experience and the extent of ophthalmic involvement, mutations have been classified in one of several syndromes including keratitis-ichthyosis-deafness syndrome (KID), palmoplantar keratoderma (PPK), Bart-Pumphrey syndrome (BPS), Vohwinkel syndrome (VS) and hystrix-like ichthyosis deafness syndrome (HID; reviewed in Lee and White, 2009; Xu and Nicholson, 2013).

Cx channels are unique in that they can adopt two very different structural configurations, as intercellular GJ channels formed by the docking of two hemichannels (HCs), one contributed by each of two apposed cells, and as undocked HCs operating in the plasma membrane (**Figures 1A,B**). These two channel configurations perform very different functions, with GJs providing direct signaling between cells and HCs providing signaling across the plasma membrane. Given that loss of GJ function does not lead to skin disorders and that some syndromic deafness mutants function as GJs whereas others do not, the disease mechanisms in non-syndromic and syndromic cases likely differ. A property in common among syndromic deafness mutants is functional HCs that show aberrant properties, suggesting that aberrant signaling across the plasma membrane mediated by mutant HCs may be the principal cause of cellular dysfunction. For mutants that function as GJ channels and HCs, both channel configurations can contribute to disease mechanisms.

ROLES OF GJs AND HCs IN THE COCHLEA

Of the three fluid filled compartments in the cochlea, the scala media consisting of endolymph plays a special role in the

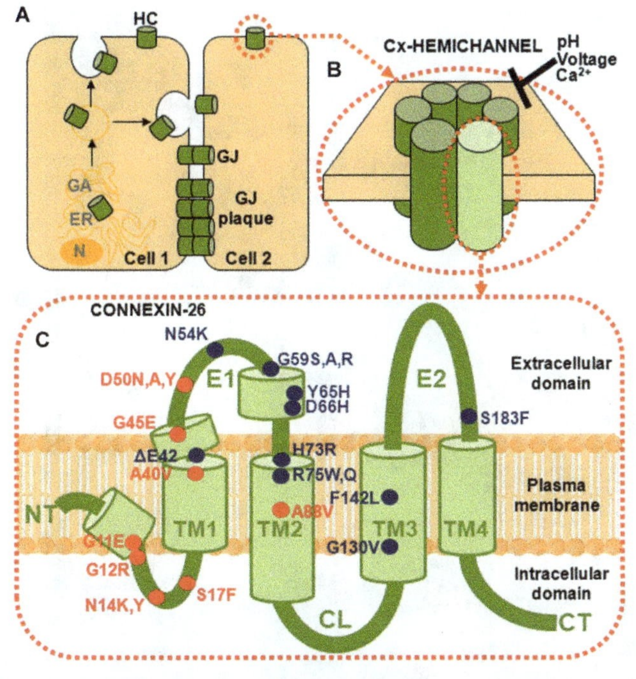

FIGURE 1 | Connexins (Cxs) form gap junction (GJ) channels and hemichannels (HCs). (A) GJ channels are formed by of docking of two HCs from neighboring cells. GJ channels cluster into plaques and provide a direct intercellular communication pathway between cells. **(B)** HCs, also called connexons, can function in the plasma membrane in the absence of docking. They are hexamers of six Cx subunits and are strongly regulated by pH, voltage and divalent cations, principally Ca^{2+}. **(C)** Topology of a Cx subunit. The amino (NT) and carboxy (CT) domains are oriented to the cytoplasm. The NT domain loops back into the membrane to contribute to the aqueous pore. There are four helical transmembrane domains (TM1–TM4) and two extracellular loop domains, E1 and E2. The short helical segments in NT and E1 identified in the crystal structure are also shown (Maeda et al., 2009). Locations of the 18 residues that have been associated with mutations causing syndromic deafness are indicated; those associated with KID syndrome are indicated in red.

transduction of electrical signals by sensory hair cells. The epithelium of the cochlea enclosing the endolymphatic compartment consists of sensory hair cells and surrounding support cells of the organ of Corti, Reissner's membrane and the cells of the stria vascularis; the latter produces endolymph, which is high in K^+ and low in Ca^{2+}, and generates the endocochlear potential, EP (**Figure 2**). Two Cxs, Cx26 and Cx30, are expressed in the cochlea and immunostaining has shown expression of both Cxs is widespread in the support cells of the organ of Corti and in the stria vascularis; Cx expression is absent in hair cells (Lautermann et al., 1998; Ahmad et al., 2003; Forge et al., 2003; Zhao and Yu, 2006; Liu and Zhao, 2008). **Figure 3** shows images of cultured cochlea explants from mice immunostained for Cx26 and Cx30. Regions of cell-cell contact show strong staining indicative of large clusters or plaques of GJ channels.

The expression patterns for Cx26 and Cx30 suggest that GJs contribute to the formation of extensive interconnected cellular networks and it has been proposed that these networks function as part of the recycling circuit for K^+ between perilymph and endolymph (Lang et al., 2007; Zdebik et al., 2009; Adachi et al., 2013). K^+ entering hair cells from the endolymph through mechano-sensitive channels in the apical surface exit at the basolateral membrane and are taken up by support cells via K^+/Cl^- co-transporters and diffuse via GJs towards the lateral wall. The cells of the stria vascularis, via a series of transporters and K^+ channels, produce movement of K^+ back into the endolymph and the generation of the EP (Adachi et al., 2013), which is an important driving force for the influx of K^+ that depolarizes hair cells during mechano-transduction. However, the availability of perilymphatic extracellular pathways for the movement of K^+ to the fibrocytes (FC) of the spiral ligament, which are electrically coupled to the basal cells (BC) of stria vascularis (Liu and Zhao, 2008), has questioned the need for GJs in the support cells of the sensory epithelium for K^+ recycling (Zdebik et al., 2009; Patuzzi, 2011). In addition, the establishment of the EP at P12 was unaffected in conditional Cx26 null mice, in which Cx26 was deleted from the support cells of the sensory epithelium, and also was unaffected in transgenic mice expressing the dominant-negative Cx26 mutant R75W (Cohen-Salmon et al., 2002; Kudo et al., 2003). In the latter study, the R75W mutant gene was introduced using Cre-loxP recombination under a CAG promoter, which has widespread activity and resulted in deafness associated with deformity of the supporting cells, loss of the tunnel of Corti and degeneration of sensory hair cells. EP does decline with Cx26 deficiency as postnatal development proceeds, likely due to loss of integrity of the epithelium. Interestingly, the deafness at birth caused by Cx26 deficiency is not associated with cell degeneration or a reduction in EP (Liang et al., 2012; Chen et al., 2014). Thus, alternatively, GJs in support cells likely play a developmental role that impacts on cochlear function early on (Wang et al., 2009) and by providing local K^+ buffering and/or metabolic coupling that ultimately impacts on cell integrity. Since the organ of Corti is avascular, it was suggested that GJs between the support cells may help deliver glucose from the highly-vascularized stria vascularis (Chang et al., 2008). GJs also were reported to mediate the spread of Ca^{2+} waves in support cells of the organ of Corti through the intercellular diffusion of IP_3 and subsequent Ca^{2+} release (Beltramello et al., 2005). Ca^{2+} acting on the Ca^{2+}-dependent Cl–K co-transport systems may function to maintain the ionic balance of cochlear fluids.

Although Cx30 is co-localized with Cx26 and deletion of Cx30 produces deafness in mice, it is now apparent that the loss of Cx30 itself is not the cause of deafness. Cx26 and Cx30 appear to be co-regulated and expression of Cx26 protein is reduced in the cochlea of Cx30 null mice (Ahmad et al., 2007; Ortolano et al., 2008; Kamiya et al., 2014). When Cx26 expression was increased in Cx30 null mice by introducing a bacterial artificial chromosome containing additional copies of the Cx26 gene, hearing was rescued (Ahmad et al., 2007). The reciprocal experiment of increasing Cx30 expression in Cx26 null mice did not restore hearing (Qu et al., 2012). Furthermore, deletion of Cx30 using a strategy in which Cx26 expression was preserved did not result in hearing loss (Boulay et al., 2013). These results indicate that deafness in original reports of Cx30 KO mice were likely due to reduced expression of Cx26. Thus, although the specific roles

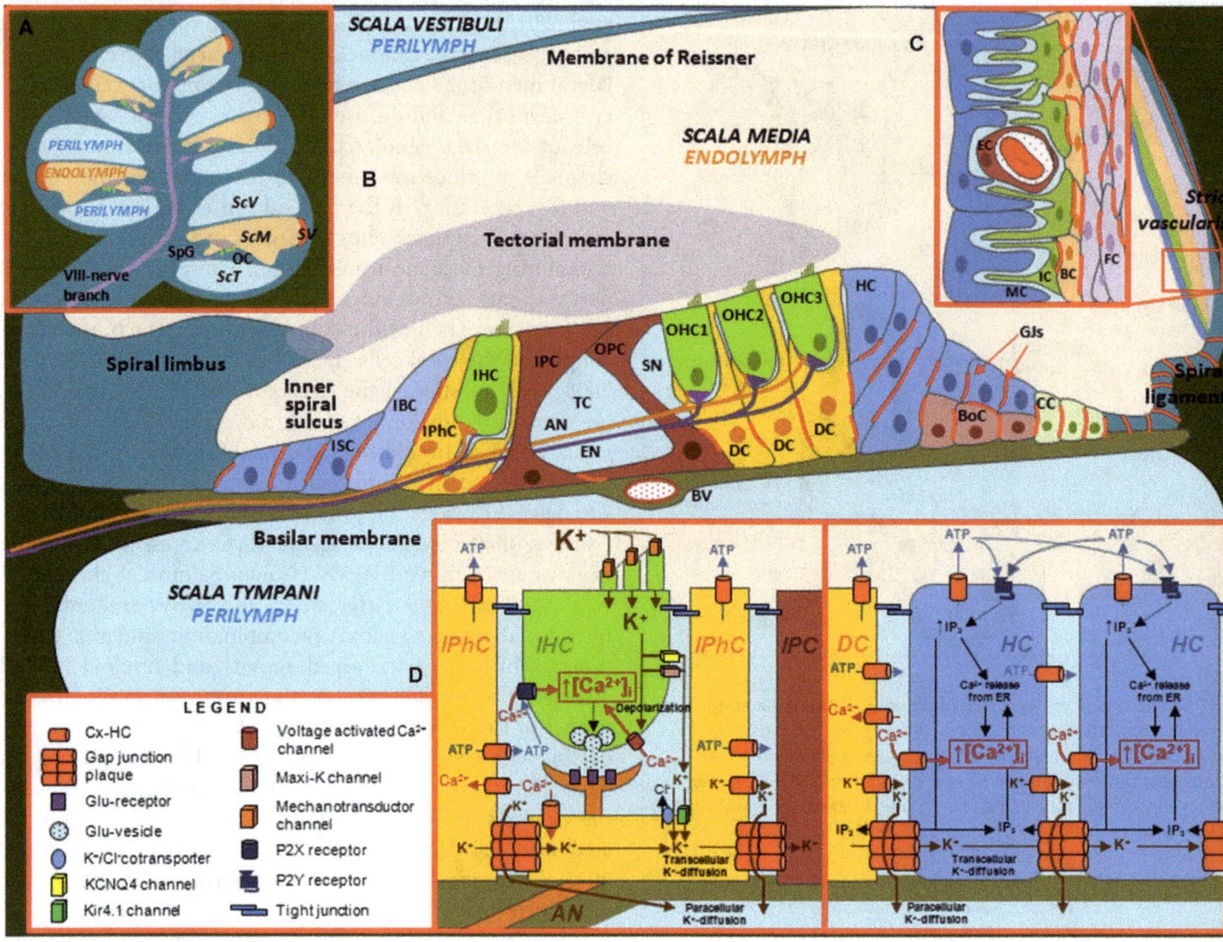

FIGURE 2 | Illustrations depicting the cellular organization of the mammalian cochlea and the putative roles of Cx channels. (A) Transverse section through the cochlea. OC: organ of Corti. SpG: Spiral ganglia. *SV: stria vascularis. ScV: scala vestibuli. ScM: scala media. ScT: scala tympani.* **(B)** Diagram of the organ of Corti. In this representation, the relative sizes of the different structures are altered to emphasize the organ of Corti, which contains inner hair cells (IHC), outer hair cells (OHC), different types of support cells and structures including afferent (AN) and efferent (EN) nerve fibers, blood vessel (BV), Boettcher cells (BoC), Claudius cells (CC), Deiter's cells (DC), Hensen cells (HC), Inner border cells (IBC), Inner pilar cells (IPC), Inner phalangeal cells (IPhC), Inner Sulcus cells (ISC), Outer pilar cells (OPC), space of Nuel (SN) and tunnel of Corti (TC). Inset **(C)** Structure of the *Stria vascularis* containing endothelial cells (EC), fibrocytes (FC), basal cells (BC), intermediate cells (IC) and marginal cells (MC). Inset **(D)** Putative roles of Cx channels in support cells. Cx HCs and GJ channels in support cells have been proposed to mediate the flux of K^+, ATP, IP_3 and Ca^{2+} across the membrane and between cells, respectively.

of GJs in the cochlea have yet to be defined, it is apparent that expression of Cx26, but not Cx30, is crucial for auditory function. Interestingly, replacement of Cx26 with Cx32 under the Cx26 promoter resulted in mice with normal hearing (Degen et al., 2011) suggesting Cx32 and Cx26 are interchangeable, at least in terms of Cx26 loss-of-function models for deafness.

What about the function of Cx HCs in the cochlea? Dye uptake studies using multiple fluorescent probes reported increased uptake only in cochlear support cells that was ascribed to functional HCs (Zhao, 2005). Using isolated cochleae or explant cultures, studies have reported that HCs act as conduits for the release of ATP and inositol 1,4,5 triphosphate, IP_3, from support cells (Zhao et al., 2005; Anselmi et al., 2008; Gossman and Zhao, 2008). There is Cx expression both in basolateral and apical regions of support cells, suggesting that signaling molecules could be released both into perilymphatic and endolymphatic compartments. Most studies have focused on ATP release. Using cochleae isolated form adult guinea pigs, ATP release into the endolymph was detected suggesting the possibility of direct modulation of outer hair cell (OHC) electromotility through activation of purinergic receptors (Zhao et al., 2005; Yu and Zhao, 2008). In mouse cochlea, ATP release through HCs was suggested to mediate Ca^{2+} wave propagation within the support cell network (Anselmi et al., 2008). The link between Ca^{2+} signaling, wave propagation and hearing may reside, in part, through effects on coordinated connexin expression in cochlear support cells (Rodriguez et al., 2012; Mammano, 2013). It is also possible that IP_3 signaling modulates other components, such as K^+ channels and co-transporters in the non-sensory epithelial cells and, through propagation, to the intermediate

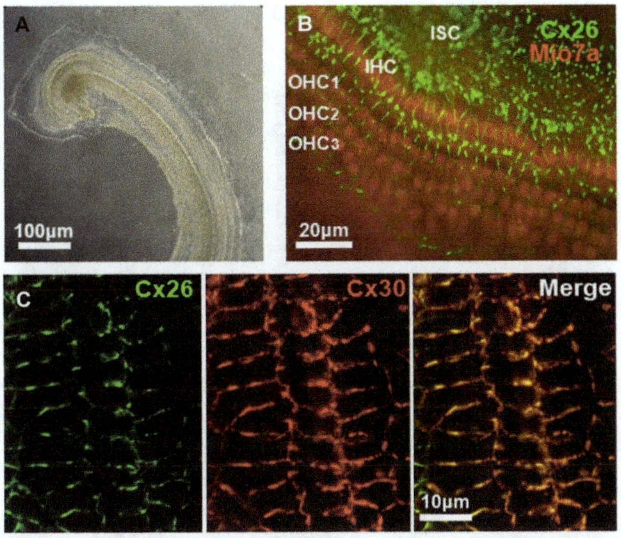

FIGURE 3 | Connexin expression in mouse cochlea. (A) Low-power transmitted-light image of a primary explant culture of cochlea (72 h *in vitro*) prepared from a P4 mouse. **(B)** Immunoreactivity for Cx26 (green), and MyosinVIIa (red), marker of hair cells, obtained from an *in vitro* primary culture of mouse cochlea after 4 days in culture. Cx26 staining is widespread in cochlea support cells in regions of cell-cell contact. Inner (IHC) and outer (OHC) hair cells are indicated. ISC: Inner support cells. **(C)** Expression of Cx26 and Cx30 in support cells of organ of Corti in an *in vitro* primary culture of mouse cochlea after 5 days in culture. Merged image of Cx26 and Cx30 immunostaining shows extensive co-localization of both connexins.

cells (IC) of the stria vascularis that regulates K^+ flux into the intrastrial space and ultimately into marginal cells (MC) and the endolymphatic compartment. Developmentally, spontaneous release of ATP through HCs in support cells in the developing rat cochlea was suggested as a mechanism by which hair cells depolarize and synchronize their outputs to promote firing of the auditory nerve, which could help refine the development of central auditory pathways prior to the onset of hearing (Tritsch et al., 2007).

ATP and IP_3 release from cochlear support cells was attributed to Cx HCs after ruling out other pathways such as P2X receptors, pannexin channels and anion channels through the use of blockers, although blocker specificity is lacking. In the case of ATP release via Cx HCs, both Cx26 and Cx30 presumably can serve this role. The presence of HC activity in support cells of the cochlea has been inferred from dye uptake, but has not been confirmed electrophysiologically. To date, no specific blockers have been identified for Cx HCs, although ATP release was blocked by La^{3+}, which generally has been shown to block dye uptake attributed to open Cx HCs in other preparations (Sáez et al., 2010).

SYNDROMIC DEAFNESS MUTATIONS CLUSTER IN NT AND E1 DOMAINS

The Cx subunits that constitute a HC have four transmembrane domains (TM1–TM4), with N-terminal (NT) and carboxy-terminal (CT) domains located on the cytoplasmic side (**Figure 1C**). The two extracellular loop domains, E1 and E2, each consisting of ~30–35 residues, mediate docking of the HCs to form GJ channels. An interesting feature that emerges from the mapping of the syndromic deafness mutations onto topology of a Cx subunit is that they largely cluster in the NT and E1 domains (**Figure 1C**). In contrast, the ~100 non-syndromic Cx26 deafness mutations (not shown) are scattered throughout the Cx protein (see Lee and White, 2009).

Although members of the Cx family can be classified into phylogenetic groups (Bennett et al., 1994; Cruciani and Mikalsen, 2007) there is notable sequence conservation in the transmembrane and extracellular domains and HCs and GJ channels formed of different Cxs share the same overall properties in that they permit the passage of hydrophilic dyes, they possess the same voltage gating mechanisms and respond to the same classes of inhibitors, e.g., acidification, long-chain alkanols, glycyrrhetinic acid and fenamates. Channel opening is controlled by voltage and there are two distinct gating mechanisms, both of which are properties of HCs, whether in docked or undocked configurations (Bukauskas et al., 1995; Trexler et al., 1996). One mechanism is characterized by gating transitions to a stable subconducting state, termed the residual conductance state, and is generally referred to as V_j gating because it was the first mechanism described that gated GJ channels in response to the voltage difference between two cells or the transjunctional voltage, V_j (Harris et al., 1981; Spray et al., 1981). The polarity of V_j gating is Cx-dependent and can be of either sign (Verselis et al., 1994). The second voltage gating mechanism closes channels fully, with gating transitions comprised a series of transient subconductance states, which gives them the appearance of being slow when recorded at typical filtering frequencies. This mechanism is termed "loop gating" as evidence suggests closure involves conformational changes in the extracellular loops (Trexler et al., 1996; Tang et al., 2009; Verselis et al., 2009). V_j and loop gating are also referred to as fast and slow gating, respectively, according to the kinetics of the gating transitions (Bukauskas and Verselis, 2004; Fasciani et al., 2013).

Biophysical and structural studies of HCs and GJ channels composed of several different connexins, e.g., Cx46, Cx50, Cx26 and the chimera Cx32*Cx43E1, have shown that the NT and E1 domains constitute the bulk of the aqueous pore and contain elements essential for voltage gating and regulation (Verselis et al., 1994; Zhou et al., 1997; Pfahnl and Dahl, 1998; Oh et al., 1999; Purnick et al., 2000; Kronengold et al., 2003; Maeda et al., 2009; Tang et al., 2009; Verselis et al., 2009; Sánchez et al., 2010). In fact, exchange of the NT halves (NT through CL) of Cx46 and Cx50 was shown to result in HCs in which the gating and unitary conductance properties remarkably correspond to the WT Cx constituting the N-terminal half of the protein (Kronengold et al., 2012). The co-segregation of biophysical properties with the NT half extends to Ca^{2+}- and K^+-dependent regulatory mechanisms in Cx46 and Cx50 HCs (Srinivas et al., 2006). Thus, syndromic deafness mutants map onto the domains that constitute the core elements that determine gating, permeability and regulatory characteristics of Cx HCs and GJ channels.

KID SYNDROME AND THE "LEAKY" HC HYPOTHESIS

Cx HCs have large aqueous pores, which not only permits inorganic cations and anions to pass, but also larger molecules such as metabolites, second messengers and other active biomolecules. In addition to ATP, signaling molecules reported to permeate through HCs include glutamate, IP_3 and Ca^{2+} (reviewed by Evans et al., 2006). Although the large pore of a Cx HC makes it a potentially important contributor to tissue function, it also makes it potentially harmful if activity is too high, which could run-down ionic gradients and allow entry/exit of molecules deleterious to cell function and survival.

Undocked Cx HCs are closed at negative potentials by the loop-gating mechanism and opening is promoted by membrane depolarization. KID syndrome mutant HCs have been described as behaving in a "leaky" manner (Stong et al., 2006; Gerido et al., 2007; Lee et al., 2009; Mese et al., 2011; Mhaske et al., 2013). The "leaky" behavior broadly refers to increased HC activity, i.e., opening, in the membrane and has been attributed, almost solely, to impaired regulation of HCs by extracellular Ca^{2+}. Extracellular Ca^{2+} shifts HC activation positive along the voltage axis (Ebihara and Steiner, 1993) and has been shown to act selectively on loop gating (Verselis and Srinivas, 2008). Thus, Ca^{2+} can act in conjunction with loop gating to effectively close Cx HCs over a wide range of voltages. In Cx26 HCs, V_j gating operates only at larger inside positive voltages and is unlikely to play much of a role in gating Cx26 HCs. For loop gating, the shift in activation is substantial (**Figure 4A**) with $V_{1/2}$, the voltage at which activation is half maximal, shifting ~50–60 mV with a 10-fold change in the extracellular Ca^{2+} concentration (Sánchez et al., 2010). Thus, normal plasma levels of Ca^{2+}, 1–2 mM, would tend to keep Cx26 HCs closed at resting potentials with opening requiring a large depolarizing stimulus. At lower extracellular Ca^{2+} concentrations, less depolarization becomes necessary to elicit Cx26 HC opening, which can lead to substantial opening even at resting membrane potentials. Loss or weakened regulation by Ca^{2+} caused by KID mutations could increase Cx26 HC opening without a need for a reduction in extracellular Ca^{2+} levels. Since Cx26 HCs are not selective for Na^+ and K^+, increased opening of the mutant HCs would tend to collapse the resting membrane potential leading to cell dysfunction. **Figure 4B** shows the effects of expressing KID mutant A40V, G45E and D50N HCs on cell membrane potentials. After 48 h of expression in *Xenopus* oocytes, resting membrane potentials in oocytes expressing A40V and D50N were lower than those expressing WT Cx26; interestingly, G45E showed similar resting potentials to WT Cx26. After 4 days, the mutants tended to accelerate rundown of the membrane potential and to lead to cell death, although the extent to which this occurred differed among the mutants (**Figure 4C**).

Thus far, impaired inhibition by extracellular Ca^{2+} has been reported for G11E, G12R, N14K, A40V, G45E, D50N/A mutant Cx26 HCs (Gerido et al., 2007; Lee et al., 2009; Sánchez et al., 2010; Terrinoni et al., 2010; Sanchez et al., 2013). However, the degree to which inhibition by Ca^{2+} is impaired does not correlate with the severity of the disease phenotype. Thus, as we discuss in the following sections, other aberrant HC properties, other than simply increased activity due to impaired inhibition by Ca^{2+}, are emerging as potential contributors to disease pathogenesis.

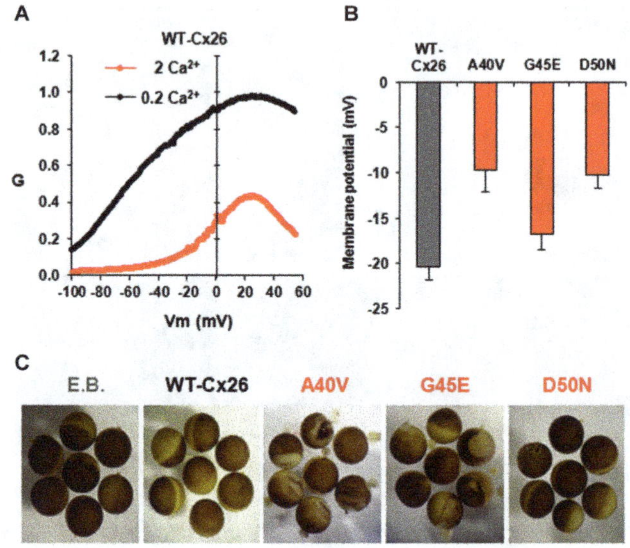

FIGURE 4 | HC activity in *Xenopus* oocytes injected with WT and mutant Cx26-mRNAs. (A) Normalized G-V relationships of WT Cx26 HCs at two different external Ca^{2+} concentrations, 0.2 and 2 mM. Data were obtained by applying slow (600 s) voltage ramps from +60 to −100 mV from a holding potential of −20 mV. Ramps were obtained for each oocyte in 0.2 and 2 mM Ca^{2+} and normalized to the maximum value measured in 0.2 mM Ca^{2+}. In WT Cx26, increasing Ca^{2+} shifted activation in the depolarizing direction and suppressed current magnitude. Adapted from (Sanchez et al., 2013). **(B)** Plot showing resting membrane potentials measured in oocytes 24–48 h after injection of RNA for WT-Cx26 and A40V, G45E and D50N mutants. Adapted from (Sánchez et al., 2010). **(C)** KID syndrome mutant HCs induce different degrees of cell damage in oocytes 4 days after RNA injection. As a control, oocytes were injected with elution buffer, E.B. (RNA vehicle). Oocytes expressing A40V, in particular, as well as G45E exhibit widespread disorganization of pigmentation and blebbing of the membrane. D50N exhibits delayed blebbing and cell death.

KID MUTANT HCs: ALTERED CA^{2+} PERMEABILITY

More extensive biophysical examination of three KID mutants, A40V, G45E and D50N, which are clustered near the TM1/E1 border and in the proximal segment of E1, revealed that two of the three residues, G45 and D50, are pore-lining (Sánchez et al., 2010; Sanchez et al., 2013) Unitary conductance of A40V HCs was indistinguishable from WT Cx26, whereas both G45E and D50N HCs showed substantially altered conductances (Sánchez et al., 2010). For G45E, unitary conductance was ~20% higher than WT Cx26 whereas for D50N it was ~50% lower and exhibited strong outward rectification of the open HC current. Furthermore, cysteine substitutions at these positions and subsequent application of thiol-modifying, methane-thiosulfonate reagents from either side of the membrane confirmed that G45 and D50, but not A40, are exposed to the aqueous pore in the open state of the Cx26 HC. Modification of G45C to a positively charged side chain with MTSET irreversibly reduced unitary conductance, whereas modification to a negatively charged side chain with MTSES had the opposite effect, increasing unitary conductance, much like the G45E KID mutation. Modification of D50C with oppositely charged reagents had the same qualitative effect, with MTSES essentially restoring WT

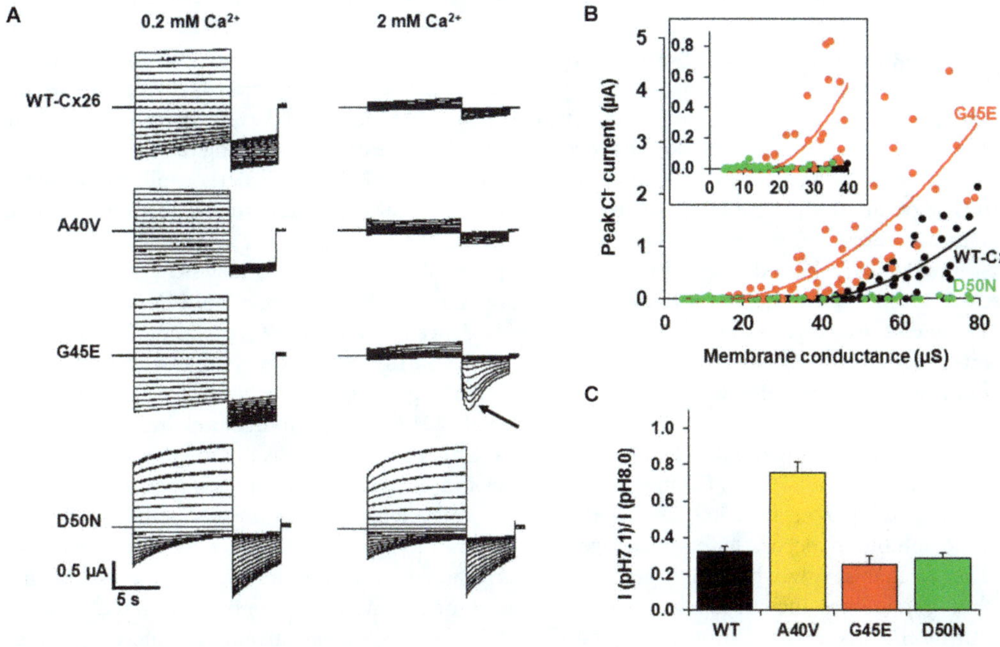

FIGURE 5 | KID syndrome mutations produce functional HCs, but with differential effects on regulation by Ca^{2+} and pH and permeability to Ca^{2+}. (A) Representative currents elicited by a series of voltage steps (10-s steps from +60 to −100 mV in intervals of 10 mV followed by a 5-s step to −110 mV) applied to oocytes expressing WT Cx26, A40V, G45E and D50N. Oocytes were voltage clamped to −20 mV. Currents shown in each case are from the same oocyte exposed to 0.2 and 2.0 mM Ca^{2+}. WT Cx26, A40V, and G45E all showed substantial reductions in current magnitude in 2 mM Ca^{2+}, whereas D50N was nearly insensitive to Ca^{2+}. A large transient inward Ca^{2+}-activated Cl current developed in oocytes expressing G45E when the voltage was stepped to −110 mV after depolarizing steps that activated these HCs (arrow). This was due to increased Ca^{2+} entry through G45E HCs. Data from (Sanchez et al., 2013). **(B)** Ca^{2+} activated Cl^- current plotted as a function of the macroscopic conductances of WT (black circles), G45E (red circles) and D50N (green circles) HCs in *Xenopus* oocytes. Solid lines correspond to regression lines fit to the data (best fit with a single exponential function). Inset: expanded view highlighting that only G45E induces Cl^- currents at considerably lower HC conductance values. Adapted from (Sánchez et al., 2010). **(C)** Bar graph showing the percent remaining HC current at pH 7.1 relative to a maximum at pH 8.0 ($I_{pH\ 7.1}/I_{pH\ 8.0}$). Currents were measured at a holding potential of −40 mV. Adapted from (Sanchez et al., 2014). Each bar represents the mean ratio ± SE. *n* = 5 for WT Cx26, 5 for A40V, 5 for G45E, and 5 for D50N.

conductance and rectification. Thus, charges at these positions strongly influence ion flux through the HC pore, explaining the effects of the KID mutations on HC conductance and rectification.

A potentially significant impact of the G45E mutation was uncovered by Sánchez et al. (2010). As shown in **Figure 5A**, steps to positive membrane potentials in the presence of 2 mM extracellular Ca^{2+} in oocytes expressing WT or G45E HCs produced relatively small outward currents due to the robust inhibitory effect of Ca^{2+} on HC activation. For D50N HCs, however, outward currents in 2 mM extracellular Ca^{2+} were large due to loss of the inhibitory effect of Ca^{2+}. However, if at the end of the activation step, the membrane was stepped to a large negative potential, the peak tail current for G45E was followed by activation of a large inward current. Using ion substitutions, blockers and intracellular chelation of Ca^{2+}, this current was established to be a Ca^{2+}-activated Cl current endogenous to *Xenopus* oocytes. Thus, it appeared that Ca^{2+} entry through G45E HCs was activating the Cl channel. Examination of the relationship between the Cl current and HC activation showed that substantially fewer G45E HCs were needed to activate Cl currents than WT Cx26 HC (**Figure 5B**). D50N HCs failed to activate Cl currents even with many HCs activated. These data demonstrate that the G45E KID mutation substantially increases permeability of Cx26 HCs to Ca^{2+}. Conversely, D50N appears to greatly diminish or abolish Ca^{2+} permeability.

OTHER POTENTIAL HC REGULATORS IN KID SYNDROME

Cx HCs are not only regulated by extracellular Ca^{2+}, but also by other divalent cations, including Zn^{2+}, as well as pH (Ebihara and Steiner, 1993; Trexler et al., 1999; Chappell et al., 2004; Ripps et al., 2004). For Cx26 HCs, sensitivity to pH falls in a physiological range such that substantial inhibition occurs at normal plasma pH levels of ~7.4. Examination of KID syndrome mutant HCs shows a reduced sensitivity to pH for A40V (Sanchez et al., 2014). **Figure 5C** shows that a change in extracellular pH from 8.0 to 7.1 produces substantial inhibition of WT Cx26 HCs, reducing current ~70%. However, this inhibition is substantially less for A40V, ~25%. For comparison, G45E and D50N mutant HCs show the same responses to pH as WT Cx26 HCs. Thus, among these three mutants, pH sensitivity is only affected in A40V. Interestingly,

A40V also selectively showed a reduction in sensitivity to Zn^{2+} (Sanchez et al., 2014). Thus, the A40V substitution appears to exert its effects mainly through increased opening resulting from alterations in a multiplicity of HC regulatory mechanisms.

A hallmark of patients afflicted with syndromic deafness due to mutations in Cx26 is susceptibility to skin infections and neoplasms that can lead to squamous cell carcinomas (Coggshall et al., 2013). Examination of the effects of peptidoglycans from *Staphylococcus aureus*, an opportunistic pathogen, on a keratinocyte cell line, HaCaT, showed induced ATP release attributed to increased Cx26 HC activity (Donnelly et al., 2012). Moreover, ATP release was significantly higher in cells expressing KID mutants suggesting these mutant HCs trigger pro-inflammatory events in response to peptidoglycans from opportunistic pathogens.

Several studies have indicated opening of HCs can be induced by metabolic inhibition (John et al., 1999; Contreras et al., 2002; Vergara et al., 2003; Sánchez et al., 2009). Induced HC opening with metabolic inhibition occurs without changes in extracellular Ca^{2+}. The mechanism by which HCs are opened is not understood, but possibilities include depletion of ATP leading to increased intracellular Ca^{2+} and possibly altered Cx phosphorylation. Interestingly, intracellular Ca^{2+} has been shown to modulate HC opening (De Vuyst et al., 2006, 2009; Schalper et al., 2008). The response of HCs to intracellular Ca^{2+} is bell-shaped, with opening increasing up to ~500 nM, and deceasing with larger Ca^{2+} concentrations (De Vuyst et al., 2009).

HC regulatory mechanisms, whether related to intracellular Ca^{2+}, peptidoglycans or to pathways involving ATP depletion, changes in kinase activity or free radical generation indicate that HC opening can be regulated in many ways and Cx mutations can affect any one of these processes. At this point, effects of syndromic deafness mutations on HC opening under metabolic stress or ischemia have not been evaluated.

GENOTYPE-PHENOTYPE RELATIONS

D50N HCs exhibit a near loss of inhibition by Ca^{2+}, but hearing loss in patients carrying this mutation can be moderate to severe with serious, although not fatal, cutaneous manifestations (Richard et al., 2002; van Steensel et al., 2002; Yotsumoto et al., 2003). On the other hand, G45E HCs exhibit near normal inhibition by Ca^{2+}, but this mutant leads to profound deafness and causes a lethal form of KID syndrome with fatality often occurring in the first year of life due to uncontrollable skin infections (Janecke et al., 2005; Jonard et al., 2008; Sbidian et al., 2010).

The finding that G45E HCs exhibit increased permeability to Ca^{2+} (Sánchez et al., 2010) provides a compelling argument for the severity of the phenotype in patients carrying this mutation. Increased Ca^{2+} entry, even with modest HC activity, could trigger Ca^{2+}-induced cascades that lead to broad dysfunction and/or cell death. An autopsy of a patient carrying the G45E mutation showed widespread vestibulo-cochlear dysplasia in the organ of Corti with no evidence of a developed sensory epithelium (Griffith et al., 2006). Thus, the G45E mutation, indeed, leads to devastating effects in the organ of Corti. In patients carrying the D50N mutation, the cochlea appears to develop normally, and hearing loss is less profound. Although D50N HCs are not appreciably inhibited by Ca^{2+}, which by inference should produce a severe phenotype, these HCs also show a gating phenotype characterized by a substantial shift in activation in the hyperpolarizing direction (Lopez et al., 2013; Sanchez et al., 2013). Thus, HC opening is reduced even when resting potentials are modest, thereby potentially mitigating the loss of inhibition by Ca^{2+}. This scenario would be particularly applicable to cochlear support cells that typically maintain robust resting potentials of ~−80 mV (Lang et al., 2007). D50N also shows no or greatly reduced Ca^{2+} permeability, which also may help mitigate the loss of inhibition by Ca^{2+}. A40V is not a pore-lining mutation and shows no evidence that it alters HC conductance and permeability (Sánchez et al., 2010). However, A40V exhibits a multiplicity of effects including impaired inhibition by Ca^{2+}, pH and Zn^{2+} (Sanchez et al., 2014). The combined impairment of these regulatory mechanisms may be responsible for the severe A40V phenotype that would not be predicted from impaired Ca^{2+} inhibition alone.

Although combinations of altered biophysical properties may explain the phenotypic diversity among KID mutants, increased HC activity due to more efficient expression and/or trafficking to the membrane must also be considered. Certainly, in exogenous expression systems, currents produced by some mutant HCs are larger even when protein expression levels are taken into account (Gerido et al., 2007). Whether this increased expression of KID mutants holds true in native tissue remains to be determined. In skin, Cx26 plays a key role in epidermal development and wound healing (Lucke et al., 1999; Coutinho et al., 2003; Richard, 2005; Djalilian et al., 2006). Persistent expression of Cx26 has been shown to maintain wounded epidermis in a hyperproliferative state and to block remodeling leading to infiltration of immune cells (Djalilian et al., 2006). Furthermore, ectopic Cx26 expression was shown to increase ATP release, resulting in delayed recovery of the epidermal barrier and promotion of an inflammatory response. However, overexpression of WT Cx26 leads to psoriatic skin and not the more severe phenotypes characteristic of KID syndrome. Thus, increased expression of Cx26 HCs alone does not appear to be sufficient to cause KID syndrome suggesting that aberrant biophysical properties are important. Even with a small sample of three KID mutants, genotype-phenotype relationships are emerging. Further biophysical examination of other KID mutants and syndromic should continue to shed light on the roles of HCs in cochlea in normal and pathologic conditions.

REFERENCES

Adachi, N., Yoshida, T., Nin, F., Ogata, G., Yamaguchi, S., Suzuki, T., et al. (2013). The mechanism underlying maintenance of the endocochlear potential by the K+ transport system in fibrocytes of the inner ear. *J. Physiol.* 591, 4459–4472. doi: 10.1113/jphysiol.2013.258046

Ahmad, S., Chen, S., Sun, J., and Lin, X. (2003). Connexins 26 and 30 are co-assembled to form gap junctions in the cochlea of mice. *Biochem. Biophys. Res. Commun.* 307, 362–368. doi: 10.1016/s0006-291x(03)01166-5

Ahmad, S., Tang, W., Chang, Q., Qu, Y., Hibshman, J., Li, Y., et al. (2007). Restoration of connexin26 protein level in the cochlea completely rescues hearing in a mouse model of human connexin30-linked deafness. *Proc. Natl. Acad. Sci. U S A* 104, 1337–1341. doi: 10.1073/pnas.0606855104

Anselmi, F., Hernandez, V. H., Crispino, G., Seydel, A., Ortolano, S., Roper, S. D., et al. (2008). ATP release through connexin hemichannels and gap junction transfer of second messengers propagate Ca2+ signals across the inner ear. *Proc. Natl. Acad. Sci. U S A* 105, 18770–18775. doi: 10.1073/pnas.0800793105

Apps, S. A., Rankin, W. A., and Kurmis, A. P. (2007). Connexin 26 mutations in autosomal recessive deafness disorders: a review. *Int. J. Audiol.* 46, 75–81. doi: 10.1080/14992020600582190

Beltramello, M., Piazza, V., Bukauskas, F. F., Pozzan, T., and Mammano, F. (2005). Impaired permeability to Ins(1,4,5)P3 in a mutant connexin underlies recessive hereditary deafness. *Nat. Cell Biol.* 7, 63–69. doi: 10.1038/ncb1205

Bennett, M. V., Zheng, X., and Sogin, M. L. (1994). The connexins and their family tree. *Soc. Gen. Physiol. Ser.* 49, 223–233.

Boulay, A. C., del Castillo, F. J., Giraudet, F., Hamard, G., Giaume, C., Petit, C., et al. (2013). Hearing is normal without connexin30. *J. Neurosci.* 33, 430–434. doi: 10.1523/JNEUROSCI.4240-12.2013

Bukauskas, F. F., Elfgang, C., Willecke, K., and Weingart, R. (1995). Biophysical properties of gap junction channels formed by mouse connexin40 in induced pairs of transfected human HeLa cells. *Biophys. J.* 68, 2289–2298. doi: 10.1016/s0006-3495(95)80411-x

Bukauskas, F. F., and Verselis, V. K. (2004). Gap junction channel gating. *Biochim. Biophys. Acta* 1662, 42–60. doi: 10.1016/j.bbamem.2004.01.008

Chan, D. K., Schrijver, I., and Chang, K. W. (2010). Connexin-26-associated deafness: phenotypic variability and progression of hearing loss. *Genet. Med.* 12, 174–181. doi: 10.1097/GIM.0b013e3181d0d42b

Chang, Q., Tang, W., Ahmad, S., Zhou, B., and Lin, X. (2008). Gap junction mediated intercellular metabolite transfer in the cochlea is compromised in connexin30 null mice. *PLoS One* 3:e4088. doi: 10.1371/journal.pone.0004088

Chappell, R. L., Qian, H., Zakevicius, J., and Ripps, H. (2004). Histidine suppresses zinc modulation of connexin hemichannels. *Biol. Bull.* 207, 188–190. doi: 10.2307/1543206

Chen, J., Chen, J., Zhu, Y., Liang, C., and Zhao, H. B. (2014). Deafness induced by Connexin 26 (GJB2) deficiency is not determined by endocochlear potential (EP) reduction but is associated with cochlear developmental disorders. *Biochem. Biophys. Res. Commun.* 448, 28–32. doi: 10.1016/j.bbrc.2014.04.016

Coggshall, K., Farsani, T., Ruben, B., McCalmont, T. H., Berger, T. G., Fox, L. P., et al. (2013). Keratitis, ichthyosis and deafness syndrome: a review of infectious and neoplastic complications. *J. Am. Acad. Dermatol.* 69, 127–134. doi: 10.1016/j.jaad.2012.12.965

Cohen-Salmon, M., Ott, T., Michel, V., Hardelin, J. P., Perfettini, I., Eybalin, M., et al. (2002). Targeted ablation of connexin26 in the inner ear epithelial gap junction network causes hearing impairment and cell death. *Curr. Biol.* 12, 1106–1111. doi: 10.1016/s0960-9822(02)00904-1

Contreras, J. E., Sánchez, H. A., Eugenin, E. A., Speidel, D., Theis, M., Willecke, K., et al. (2002). Metabolic inhibition induces opening of unapposed connexin 43 gap junction hemichannels and reduces gap junctional communication in cortical astrocytes in culture. *Proc. Natl. Acad. Sci. U S A* 99, 495–500. doi: 10.1073/pnas.012589799

Coutinho, P., Qiu, C., Frank, S., Tamber, K., and Becker, D. (2003). Dynamic changes in connexin expression correlate with key events in the wound healing process. *Cell Biol. Int.* 27, 525–541. doi: 10.1016/s1065-6995(03)00077-5

Cruciani, V., and Mikalsen, S. O. (2007). Evolutionary selection pressure and family relationships among connexin genes. *Biol. Chem.* 388, 253–264. doi: 10.1515/bc.2007.028

Degen, J., Schütz, M., Dicke, N., Strenzke, N., Jokwitz, M., Moser, T., et al. (2011). Connexin32 can restore hearing in connexin26 deficient mice. *Eur. J. Cell Biol.* 90, 817–824. doi: 10.1016/j.ejcb.2011.05.001

De Vuyst, E., Decrock, E., Cabooter, L., Dubyak, G. R., Naus, C. C., Evans, W. H., et al. (2006). Intracellular calcium changes trigger connexin 32 hemichannel opening. *EMBO J.* 25, 34–44. doi: 10.1038/sj.emboj.7600908

De Vuyst, E., Wang, N., Decrock, E., De Bock, M., Vinken, M., Van Moorhem, M., et al. (2009). Ca(2+) regulation of connexin 43 hemichannels in C6 glioma and glial cells. *Cell Calcium* 46, 176–187. doi: 10.1016/j.ceca.2009.07.002

Djalilian, A. R., McGaughey, D., Patel, S., Seo, E. Y., Yang, C., Cheng, J., et al. (2006). Connexin 26 regulates epidermal barrier and wound remodeling and promotes psoriasiform response. *J. Clin. Invest.* 116, 1243–1253. doi: 10.1172/jci27186

Donnelly, S., English, G., de Zwart-Storm, E. A., Lang, S., van Steensel, M. A., and Martin, P. E. (2012). Differential susceptibility of Cx26 mutations associated with epidermal dysplasias to peptidoglycan derived from Staphylococcus aureus and Staphylococcus epidermidis. *Exp. Dermatol.* 21, 592–598. doi: 10.1111/j.1600-0625.2012.01521.x

Duman, D., and Tekin, M. (2012). Autosomal recessive nonsyndromic deafness genes: a review. *Front. Biosci. (Landmark Ed.)* 17, 2213–2236. doi: 10.2741/4046

Ebihara, L., and Steiner, E. (1993). Properties of a nonjunctional current expressed from a rat connexin46 cDNA in Xenopus oocytes. *J. Gen. Physiol.* 102, 59–74. doi: 10.1085/jgp.102.1.59

Evans, W. H., De Vuyst, E., and Leybaert, L. (2006). The gap junction cellular internet: connexin hemichannels enter the signalling limelight. *Biochem. J.* 397, 1–14. doi: 10.1042/bj20060175

Fasciani, I., Temperán, A., Pérez-Atencio, L. F., Escudero, A., Martínez-Montero, P., Molano, J., et al. (2013). Regulation of connexin hemichannel activity by membrane potential and the extracellular calcium in health and disease. *Neuropharmacology* 75, 479–490. doi: 10.1016/j.neuropharm.2013.03.040

Forge, A., Becker, D., Casalotti, S., Edwards, J., Marziano, N., and Nevill, G. (2003). Gap junctions in the inner ear: comparison of distribution patterns in different vertebrates and assessement of connexin composition in mammals. *J. Comp. Neurol.* 467, 207–231. doi: 10.1002/cne.10916

Gerido, D. A., DeRosa, A. M., Richard, G., and White, T. W. (2007). Aberrant hemichannel properties of Cx26 mutations causing skin disease and deafness. *Am. J. Physiol. Cell Physiol.* 293, C337–C345. doi: 10.1152/ajpcell.00626.2006

Gossman, D. G., and Zhao, H. B. (2008). Hemichannel-mediated inositol 1,4,5-trisphosphate (IP3) release in the cochlea: a novel mechanism of IP3 intercellular signaling. *Cell Commun. Adhes.* 15, 305–315. doi: 10.1080/15419060802357217

Griffith, A. J., Yang, Y., Pryor, S. P., Park, H. J., Jabs, E. W., Nadol, J. B. Jr., et al. (2006). Cochleosaccular dysplasia associated with a connexin 26 mutation in keratitis-ichthyosis-deafness syndrome. *Laryngoscope* 116, 1404–1408. doi: 10.1097/01.mlg.0000224549.75161.ca

Harris, A. L., Spray, D. C., and Bennett, M. V. (1981). Kinetic properties of a voltage-dependent junctional conductance. *J. Gen. Physiol.* 77, 95–117. doi: 10.1085/jgp.77.1.95

Hoang Dinh, E., Ahmad, S., Chang, Q., Tang, W., Stong, B., and Lin, X. (2009). Diverse deafness mechanisms of connexin mutations revealed by studies using in vitro approaches and mouse models. *Brain Res.* 1277, 52–69. doi: 10.1016/j.brainres.2009.02.008

Janecke, A. R., Hennies, H. C., Günther, B., Gansl, G., Smolle, J., Messmer, E. M., et al. (2005). GJB2 mutations in keratitis-ichthyosis-deafness syndrome including its fatal form. *Am. J. Med. Genet. A* 133, 128–131. doi: 10.1002/ajmg.a.30515

John, S. A., Kondo, R., Wang, S. Y., Goldhaber, J. I., and Weiss, J. N. (1999). Connexin-43 hemichannels opened by metabolic inhibition. *J. Biol. Chem.* 274, 236–240. doi: 10.1074/jbc.274.1.236

Jonard, L., Feldmann, D., Parsy, C., Freitag, S., Sinico, M., Koval, C., et al. (2008). A familial case of Keratitis-Ichthyosis-Deafness (KID) syndrome with the GJB2 mutation G45E. *Eur. J. Med. Genet.* 51, 35–43. doi: 10.1016/j.ejmg.2007.09.005

Kamiya, K., Yum, S. W., Kurebayashi, N., Muraki, M., Ogawa, K., Karasawa, K., et al. (2014). Assembly of the cochlear gap junction macromolecular complex requires connexin 26. *J. Clin. Invest.* 124, 1598–1607. doi: 10.1172/JCI67621

Kelsell, D. P., Dunlop, J., Stevens, H. P., Lench, N. J., Liang, J. N., Parry, G., et al. (1997). Connexin 26 mutations in hereditary non-syndromic sensorineural deafness. *Nature* 387, 80–83. doi: 10.1038/387080a0

Kronengold, J., Srinivas, M., and Verselis, V. K. (2012). The N-terminal half of the connexin protein contains the core elements of the pore and voltage gates. *J. Membr. Biol.* 245, 453–463. doi: 10.1007/s00232-012-9457-z

Kronengold, J., Trexler, E. B., Bukauskas, F. F., Bargiello, T. A., and Verselis, V. K. (2003). Single-channel SCAM identifies pore-lining residues in the first extracellular loop and first transmembrane domains of Cx46 hemichannels. *J. Gen. Physiol.* 122, 389–405. doi: 10.1085/jgp.200308861

Kudo, T., Kure, S., Ikeda, K., Xia, A. P., Katori, Y., Suzuki, M., et al. (2003). Transgenic expression of a dominant-negative connexin26 causes degeneration of the organ of Corti and non-syndromic deafness. *Hum. Mol. Genet.* 12, 995–1004. doi: 10.1093/hmg/ddg116

Lang, F., Vallon, V., Knipper, M., and Wangemann, P. (2007). Functional significance of channels and transporters expressed in the inner ear and kidney. *Am. J. Physiol. Cell Physiol.* 293, C1187–C1208. doi: 10.1152/ajpcell.00024.2007

Lautermann, J., ten Cate, W. J., Altenhoff, P., Grümmer, R., Traub, O., Frank, H., et al. (1998). Expression of the gap-junction connexins 26 and 30 in the rat cochlea. *Cell Tissue Res.* 294, 415–420. doi: 10.1007/s004410051192

Lee, J. R., Derosa, A. M., and White, T. W. (2009). Connexin mutations causing skin disease and deafness increase hemichannel activity and cell death when expressed in Xenopus oocytes. *J. Invest. Dermatol.* 129, 870–878. doi: 10.1038/jid.2008.335

Lee, J. R., and White, T. W. (2009). Connexin-26 mutations in deafness and skin disease. *Expert Rev. Mol. Med.* 11:e35. doi: 10.1017/s1462399409001276

Liang, C., Zhu, Y., Zong, L., Lu, G. J., and Zhao, H. B. (2012). Cell degeneration is not a primary causer for Connexin26 (GJB2) deficiency associated hearing loss. *Neurosci. Lett.* 528, 36–41. doi: 10.1016/j.neulet.2012.08.085

Liu, Y. P., and Zhao, H. B. (2008). Cellular characterization of Connexin26 and Connnexin30 expression in the cochlear lateral wall. *Cell Tissue Res.* 333, 395–403. doi: 10.1007/s00441-008-0641-5

Lopez, W., Gonzalez, J., Liu, Y., Harris, A. L., and Contreras, J. E. (2013). Insights on the mechanisms of Ca (2+) regulation of connexin26 hemichannels revealed by human pathogenic mutations (D50N/Y). *J. Gen. Physiol.* 142, 23–35. doi: 10.1085/jgp.201210893

Lucke, T., Choudhry, R., Thom, R., Selmer, I. S., Burden, A. D., and Hodgins, M. B. (1999). Upregulation of connexin 26 is a feature of keratinocyte differentiation in hyperproliferative epidermis, vaginal epithelium and buccal epithelium. *J. Invest. Dermatol.* 112, 354–361. doi: 10.1046/j.1523-1747.1999.00512.x

Maeda, S., Nakagawa, S., Suga, M., Yamashita, E., Oshima, A., Fujiyoshi, Y., et al. (2009). Structure of the connexin 26 gap junction channel at 3.5 a resolution. *Nature* 458, 597–602. doi: 10.1038/nature07869

Mammano, F. (2013). ATP-dependent intercellular Ca2+ signaling in the developing cochlea: facts, fantasies and perspectives. *Semin. Cell Dev. Biol.* 24, 31–39. doi: 10.1016/j.semcdb.2012.09.004

Mese, G., Sellitto, C., Li, L., Wang, H. Z., Valiunas, V., Richard, G., et al. (2011). The Cx26–G45E mutation displays increased hemichannel activity in a mouse model of the lethal form of keratitis-ichthyosis-deafness syndrome. *Mol. Biol. Cell* 22, 4776–4786. doi: 10.1091/mbc.E11-09-0778

Mhaske, P. V., Levit, N. A., Li, L., Wang, H. Z., Lee, J. R., Shuja, Z., et al. (2013). The human Cx26–D50A and Cx26–A88V mutations causing keratitis-ichthyosis-deafness syndrome display increased hemichannel activity. *Am. J. Physiol. Cell Physiol.* 304, C1150–C1158. doi: 10.1152/ajpcell.00374.2012

Oh, S., Rubin, J. B., Bennett, M. V., Verselis, V. K., and Bargiello, T. A. (1999). Molecular determinants of electrical rectification of single channel conductance in gap junctions formed by connexins 26 and 32. *J. Gen. Physiol.* 114, 339–364. doi: 10.1085/jgp.114.3.339

Ortolano, S., Di Pasquale, G., Crispino, G., Anselmi, F., Mammano, F., and Chiorini, J. A. (2008). Coordinated control of connexin 26 and connexin 30 at the regulatory and functional level in the inner ear. *Proc. Natl. Acad. Sci. U S A* 105, 18776–18781. doi: 10.1073/pnas.0800831105

Patuzzi, R. (2011). Ion flow in stria vascularis and the production and regulation of cochlear endolymph and the endolymphatic potential. *Hear. Res.* 277, 4–19. doi: 10.1016/j.heares.2011.01.010

Pfahnl, A., and Dahl, G. (1998). Localization of a voltage gate in connexin46 gap junction hemichannels. *Biophys. J.* 75, 2323–2331. doi: 10.1016/s0006-3495(98)77676-3

Purnick, P. E., Oh, S., Abrams, C. K., Verselis, V. K., and Bargiello, T. A. (2000). Reversal of the gating polarity of gap junctions by negative charge substitutions in the N-terminus of connexin 32. *Biophys. J.* 79, 2403–2415. doi: 10.1016/s0006-3495(00)76485-x

Qu, Y., Tang, W., Zhou, B., Ahmad, S., Chang, Q., Li, X., et al. (2012). Early developmental expression of connexin26 in the cochlea contributes to its dominate functional role in the cochlear gap junctions. *Biochem. Biophys. Res. Commun.* 417, 245–250. doi: 10.1016/j.bbrc.2011.11.093

Richard, G. (2005). Connexin disorders of the skin. *Clin. Dermatol.* 23, 23–32. doi: 10.1016/j.clindermatol.2004.09.010

Richard, G., Rouan, F., Willoughby, C. E., Brown, N., Chung, P., Ryynänen, M., et al. (2002). Missense mutations in GJB2 encoding connexin-26 cause the ectodermal dysplasia keratitis-ichthyosis-deafness syndrome. *Am. J. Hum. Genet.* 70, 1341–1348. doi: 10.1086/339986

Ripps, H., Qian, H., and Zakevicius, J. (2004). Properties of connexin26 hemichannels expressed in Xenopus oocytes. *Cell. Mol. Neurobiol.* 24, 647–665. doi: 10.1023/b:cemn.0000036403.43484.3d

Rodriguez, L., Simeonato, E., Scimemi, P., Anselmi, F., Calì, B., Crispino, G., et al. (2012). Reduced phosphatidylinositol 4,5-bisphosphate synthesis impairs inner ear Ca2+ signaling and high-frequency hearing acquisition. *Proc. Natl. Acad. Sci. U S A* 109, 14013–14018. doi: 10.1073/pnas.1211869109

Sáez, J. C., Schalper, K. A., Retamal, M. A., Orellana, J. A., Shoji, K. F., and Bennett, M. V. (2010). Cell membrane permeabilization via connexin hemichannels in living and dying cells. *Exp. Cell Res.* 316, 2377–2389. doi: 10.1016/j.yexcr.2010.05.026

Sanchez, H. A., Bienkowski, R., Slavi, N., Srinivas, M., and Verselis, V. K. (2014). Altered inhibition of Cx26 hemichannels by pH and Zn2+ in the A40V mutation associated with keratitis-ichthyosis-deafness syndrome. *J. Biol. Chem.* 289, 21519–21532. doi: 10.1074/jbc.m114.578757

Sánchez, H. A., Mese, G., Srinivas, M., White, T. W., and Verselis, V. K. (2010). Differentially altered Ca2+ regulation and Ca2+ permeability in Cx26 hemichannels formed by the A40V and G45E mutations that cause keratitis ichthyosis deafness syndrome. *J. Gen. Physiol.* 136, 47–62. doi: 10.1085/jgp.2010 10433

Sánchez, H. A., Orellana, J. A., Verselis, V. K., and Saez, J. C. (2009). Metabolic inhibition increases activity of connexin-32 hemichannels permeable to Ca2+ in transfected HeLa cells. *Am. J. Physiol. Cell Physiol.* 297, C665–C678. doi: 10.1152/ajpcell.00200.2009

Sanchez, H. A., Villone, K., Srinivas, M., and Verselis, V. K. (2013). The D50N mutation and syndromic deafness: altered Cx26 hemichannel properties caused by effects on the pore and intersubunit interactions. *J. Gen. Physiol.* 142, 3–22. doi: 10.1085/jgp.201310962

Sbidian, E., Feldmann, D., Bengoa, J., Fraitag, S., Abadie, V., de Prost, Y., et al. (2010). Germline mosaicism in keratitis-ichthyosis-deafness syndrome: prenatal diagnosis in a familial lethal form. *Clin. Genet.* 77, 587–592. doi: 10.1111/j.1399-0004.2009.01339.x

Schalper, K. A., Palacios-Prado, N., Retamal, M. A., Shoji, K. F., Martínez, A. D., and Sáez, J. C. (2008). Connexin hemichannel composition determines the FGF-1-induced membrane permeability and free [Ca2+]i responses. *Mol. Biol. Cell* 19, 3501–3513. doi: 10.1091/mbc.e07-12-1240

Spray, D. C., Harris, A. L., and Bennett, M. V. (1981). Equilibrium properties of a voltage-dependent junctional conductance. *J. Gen. Physiol.* 77, 77–93. doi: 10.1085/jgp.77.1.77

Srinivas, M., Calderon, D. P., Kronengold, J., and Verselis, V. K. (2006). Regulation of connexin hemichannels by monovalent cations. *J. Gen. Physiol.* 127, 67–75. doi: 10.1085/jgp.200509397

Stong, B. C., Chang, Q., Ahmad, S., and Lin, X. (2006). A novel mechanism for connexin 26 mutation linked deafness: cell death caused by leaky gap junction hemichannels. *Laryngoscope* 116, 2205–2210. doi: 10.1097/01.mlg.0000241944.77192.d2

Tang, Q., Dowd, T. L., Verselis, V. K., and Bargiello, T. A. (2009). Conformational changes in a pore-forming region underlie voltage-dependent "loop gating" of an unapposed connexin hemichannel. *J. Gen. Physiol.* 133, 555–570. doi: 10.1085/jgp.200910207

Terrinoni, A., Codispoti, A., Serra, V., Didona, B., Bruno, E., Nisticò, R., et al. (2010). Connexin 26 (GJB2) mutations, causing KID Syndrome, are associated with cell death due to calcium gating deregulation. *Biochem. Biophys. Res. Commun.* 394, 909–914. doi: 10.1016/j.bbrc.2010.03.073

Trexler, E. B., Bennett, M. V., Bargiello, T. A., and Verselis, V. K. (1996). Voltage gating and permeation in a gap junction hemichannel. *Proc. Natl. Acad. Sci. U S A* 93, 5836–5841. doi: 10.1073/pnas.93.12.5836

Trexler, E. B., Bukauskas, F. F., Bennett, M. V., Bargiello, T. A., and Verselis, V. K. (1999). Rapid and direct effects of pH on connexins revealed by the connexin46 hemichannel preparation. *J. Gen. Physiol.* 113, 721–742. doi: 10.1085/jgp.113.5.721

Tritsch, N. X., Yi, E., Gale, J. E., Glowatzki, E., and Bergles, D. E. (2007). The origin of spontaneous activity in the developing auditory system. *Nature* 450, 50–55. doi: 10.1038/nature06233

van Steensel, M. A., van Geel, M., Nahuys, M., Smitt, J. H., and Steijlen, P. M. (2002). A novel connexin 26 mutation in a patient diagnosed with keratitis-ichthyosis-deafness syndrome. *J. Invest. Dermatol.* 118, 724–727. doi: 10.1046/j.1523-1747.2002.01735.x

Vergara, L., Bao, X., Cooper, M., Bello-Reuss, E., and Reuss, L. (2003). Gap-junctional hemichannels are activated by ATP depletion in human renal proximal tubule cells. *J. Membr. Biol.* 196, 173–184. doi: 10.1007/s00232-003-0636-9

Verselis, V. K., Ginter, C. S., and Bargiello, T. A. (1994). Opposite voltage gating polarities of two closely related connexins. *Nature* 368, 348–351. doi: 10.1038/368348a0

Verselis, V. K., and Srinivas, M. (2008). Divalent cations regulate connexin hemichannels by modulating intrinsic voltage-dependent gating. *J. Gen. Physiol.* 132, 315–327. doi: 10.1085/jgp.200810029

Verselis, V. K., Trelles, M. P., Rubinos, C., Bargiello, T. A., and Srinivas, M. (2009). Loop gating of connexin hemichannels involves movement of pore-lining residues in the first extracellular loop domain. *J. Biol. Chem.* 284, 4484–4493. doi: 10.1074/jbc.m807430200

Wang, Y., Chang, Q., Tang, W., Sun, Y., Zhou, B., Li, H., et al. (2009). Targeted connexin26 ablation arrests postnatal development of the organ of Corti. *Biochem. Biophys. Res. Commun.* 385, 33–37. doi: 10.1016/j.bbrc.2009.05.023

Xu, J., and Nicholson, B. J. (2013). The role of connexins in ear and skin physiology - functional insights from disease-associated mutations. *Biochim. Biophys. Acta* 1828, 167–178. doi: 10.1016/j.bbamem.2012.06.024

Yotsumoto, S., Hashiguchi, T., Chen, X., Ohtake, N., Tomitaka, A., Akamatsu, H., et al. (2003). Novel mutations in GJB2 encoding connexin-26 in Japanese patients with keratitis-ichthyosis-deafness syndrome. *Br. J. Dermatol.* 148, 649–653. doi: 10.1046/j.1365-2133.2003.05245.x

Yu, N., and Zhao, H. B. (2008). ATP activates P2x receptors and requires extracellular Ca(++) participation to modify outer hair cell nonlinear capacitance. *Pflugers Arch.* 457, 453–461. doi: 10.1007/s00424-008-0522-5

Zdebik, A. A., Wangemann, P., and Jentsch, T. J. (2009). Potassium ion movement in the inner ear: insights from genetic disease and mouse models. *Physiology (Bethesda)* 24, 307–316. doi: 10.1152/physiol.00018.2009

Zhao, H. B. (2005). Connexin26 is responsible for anionic molecule permeability in the cochlea for intercellular signalling and metabolic communications. *Eur. J. Neurosci.* 21, 1859–1868. doi: 10.1111/j.1460-9568.2005.04031.x

Zhao, H. B., and Yu, N. (2006). Distinct and gradient distributions of connexin26 and connexin30 in the cochlear sensory epithelium of guinea pigs. *J. Comp. Neurol.* 499, 506–518. doi: 10.1002/cne.21113

Zhao, H. B., Yu, N., and Fleming, C. R. (2005). Gap junctional hemichannel-mediated ATP release and hearing controls in the inner ear. *Proc. Natl. Acad. Sci. U S A* 102, 18724–18729. doi: 10.1073/pnas.0506481102

Zhou, X. W., Pfahnl, A., Werner, R., Hudder, A., Llanes, A., Luebke, A., et al. (1997). Identification of a pore lining segment in gap junction hemichannels. *Biophys. J.* 72, 1946–1953. doi: 10.1016/s0006-3495(97)78840-4

14

Connexons and pannexons: Newcomers in neurophysiology

Giselle Cheung, Oana Chever[†] **and Nathalie Rouach** *

Neuroglial Interactions in Cerebral Physiopathology, Center for Interdisciplinary Research in Biology, Collège de France, CNRS UMR 7241, INSERM U1050, Labex Memolife, PSL Research University, Paris, France

Correspondence:
Nathalie Rouach, Neuroglial Interactions in Cerebral Physiopathology, Center for Interdisciplinary Research in Biology, Collège de France, CNRS UMR 7241, INSERM U1050, 11 Place Marcelin Berthelot, Paris 75005, France
e-mail: nathalie.rouach@college-de-france.fr

[†]***Present address:***
Oana Chever, Institute of Molecular and Cellular Pharmacology, CNRS UMR 7275 and University of Nice-Sophia Antipolis, Valbonne, France

Connexin hemichannels are single membrane channels which have been traditionally thought to work in pairs to form gap junction channels across two opposing cells. In astrocytes, gap junction channels allow direct intercellular communication and greatly facilitate the transmission of signals. Recently, there has been growing evidence demonstrating that connexin hemichannels, as well as pannexin channels, on their own are open in various conditions. They allow bidirectional flow of ions and signaling molecules and act as release sites for transmitters like ATP and glutamate into the extracellular space. While much attention has focused on the function of connexin hemichannels and pannexons during pathological situations like epilepsy, inflammation, neurodegeneration or ischemia, their potential roles in physiology is often ignored. In order to fully understand the dynamic properties and roles of connexin hemichannels and pannexons in the brain, it is essential to decipher whether they also have some physiological functions and contribute to normal cerebral processes. Here, we present recent studies in the CNS suggesting emerging physiological functions of connexin hemichannels and pannexons in normal neuronal activity and behavior. We also discuss how these pioneer studies pave the way for future research to extend the physiological relevance of connexons and pannexons, and some fundamental issues yet to be addressed.

Keywords: neurons, astrocytes, synapses, plasticity, learning and memory, connexins, pannexins, hemichannels

INTRODUCTION

A typical feature of glia, in particular astrocytes, is to express high levels of connexins (Cxs), which have long been thought to only provide the molecular basis for the formation of gap junction (GJ) channels, mediating the extensive direct glial intercellular communication (Spray et al., 1999; Rozental et al., 2000; Theis et al., 2005; Pannasch and Rouach, 2013). Indeed Cxs, which represent a family of so far ~20 isoforms identified in mice and humans (Willecke et al., 2002), form homomeric or heteromeric hexamers on cell plasma membranes. These structures also called connexons, can align and dock with other connexons provided by neighboring cells to form GJ channels (Sáez et al., 2003). Such GJ channels thus connect the cytoplasm of adjacent cells. They allow direct exchange of a variety of small molecules <1.5 kDa (Loewenstein, 1981), including ions, energy metabolites, neurotransmitters and signaling molecules, coordinating electrical and metabolic activities of connected cells (Dermietzel and Spray, 1993; Pannasch and Rouach, 2013). Importantly, GJ functions have been attributed to various CNS pathologies as both protective and destructive (Rouach et al., 2002; Eugenin et al., 2012). However, over the last years, increasing evidence has emerged showing that astrocytic gap-junctional networks can also modulate physiological activities like synaptic transmission, plasticity and information processing (Lutz et al., 2009; Pannasch et al., 2011; Pannasch and Rouach, 2013; Han et al., 2014).

Nevertheless, besides the classical formation of GJ channels, connexons do exist on their own as single membrane channels, named hemichannels (HCs), which directly connect the cell cytoplasm to the extracellular space. Because HCs are thought to be poorly selective large pore channels permeable to numerous low molecular weight molecules, their opening is commonly viewed as deleterious due to potential loss of cytoplasmic integrity and neurotoxic damage that may be induced by the released factors (Giaume et al., 2013). Such membrane HCs were thus at first presumed to be closed and to serve as a reserve pool of connexons ready to be assembled into GJ channels at junctional plaques. However, this concept has been challenged in the nineties, when Cx HCs were for the first time reported to open in a number of conditions, albeit mostly related to pathological situations, as recently reviewed (Orellana et al., 2012b, 2013; Giaume et al., 2013).

Remarkably another family of proteins, the pannexins (Panxs), was discovered in the early 2000s to be homologous to the invertebrate GJ forming proteins innexins (Baranova et al., 2004). Although Panxs were initially suspected to form GJ channels due to their structural similarities to Cxs, thus far they have actually been shown in native systems to only form large pore membrane channels, similar to Cx HCs. Differential expression and properties of Cx HCs and Panx subtypes in glial cells (Cx43 and Panx1 in astrocytes; Cx43, Cx32, and Panx1 in microglia; Panx1 and possibly Cx29 or Cx32 in oligodendrocytes) and neurons (Cx36 and Panx1) have already been summarized in several comprehensive reviews (Sáez et al., 2003; Thompson and Macvicar, 2008; Orellana et al., 2011c; Giaume et al., 2013). In addition, like Cx HCs, the activation of Panx channels is generally also viewed as detrimental and has accordingly been mostly reported during pathological conditions, as recently extensively reviewed (Bennett et al., 2012; Giaume et al., 2013; Mylvaganam et al., 2014; Orellana et al., 2014; Velasquez and Eugenin, 2014). Interestingly, with these growing findings on HCs properties and roles inside and outside the brain, it was hypothesized that these channels may also have their importance in physiology. Indeed, a few pioneer studies have shown that HCs do open in physiological conditions in the retina (Pearson et al., 2005) or in the inner ear (Anselmi et al., 2008). This has led to the emergence of several recent studies over the last few years focusing on their physiological functions. In particular, Cx43 HCs, and Panx1 channels, among others, have been most commonly investigated in various physiological contexts. Although not much is known about Cx and Panx single membrane channels in the brain during physiological conditions and if they could modulate synaptic transmission and behavior, a few recent studies have explored this possibility. Here, we review these findings on connexons and pannexons focusing on their novel neurophysiological and behavioral roles in the CNS, as well as the experimental approaches used. In addition, future directions and prospects in unraveling the physiological relevance of these channels are also discussed.

UNIQUE FEATURES OF CONNEXINS AND PANNEXINS AS SINGLE MEMBRANE CHANNELS

Cx HCs and Panx channels are unique membrane channels due to their large pores with specific conductance properties (Thompson and Macvicar, 2008). This type of channels potentially represents a more dynamic aspect of such transmembrane pores, as their opening to the extracellular space may promote direct regulation of neurotransmission and paracrine signaling. While functions of Cx GJ channels have been extensively investigated and characterized, the properties of Cxs and Panxs as unapposed single membrane channels, or HCs, have only recently been given attention and explored. Over the last years, different experimental approaches have been developed to investigate the properties of HCs. As summarized in **Table 1**, these include the use of tracers permeable to HCs, quantification of released molecules and electrophysiological recordings of HC currents. Moreover, HC activities can also be manipulated by the addition of mimetic peptides, pharmacological blockers, antibodies or various transgenic animals. These techniques have greatly contributed to the identification of many unique features of Cx HCs and Panx channels.

CONNEXIN HEMICHANNELS

The first studies to show evidence for functional HCs was on rat lens fiber Cx46 expressed in xenopus oocytes (Paul et al., 1991; Ebihara and Steiner, 1993). In particular, the authors have identified non-junctional voltage dependent currents in Cx46 expressing oocytes, which they have later confirmed in lens fiber (Ebihara et al., 2011). Functional Cx30, Cx46 and Cx50 HCs have also been detected using HeLa cell expression system, but were found to be closed during physiological conditions (Valiunas and Weingart, 2000). Indeed, extracellular binding of Ca^{2+}, intracellular phosphorylation and strong voltage-dependence are hindering the opening of Cx HCs near resting membrane potentials. More subsequent studies have observed similar HC properties of Cx30.2 and Cx31.9 (Bukauskas et al., 2006), Cx43 (Contreras et al., 2003; Kang et al., 2008), and Cx26 (Gonzalez et al., 2006), although human Cx26 HCs have been shown in the inner ear to open in physiological conditions, where they sustain long range intercellular Ca^{2+} signals (Anselmi et al., 2008). Noteworthy, a physiologically low Ca^{2+} level inside cochlear endolymph (Bosher and Warren, 1978) is likely the reason why Cx26 HCs are open and participate in physiological signaling in the inner ear. Remarkably, HCs properties differ according to the Cx composition, which confers distinctive permeability and biophysical properties, including open probability, conductance and selectivity (Giaume et al., 2013). In addition, Cx HCs are strongly regulated, and some triggers for their opening include transmembrane voltage, changes in Ca^{2+} and K^{+} concentrations, ATP or post-transcriptional modifications of the channels, which have been extensively reviewed (Sáez et al., 2005; Macvicar and Thompson, 2010; Giaume et al., 2013; Orellana et al., 2013; Wang et al., 2013a). Finally, these HCs provide a direct means for signaling molecules like ATP and glutamate to travel between intra- and extracellular space (Giaume et al., 2013). In particular, the release of extracellular messengers like ATP via Cx HCs are of significant importance and has been shown to contribute to astroglial Ca^{2+} waves (Leybaert and Sanderson, 2012). Since their discoveries, many other signaling molecules like NAD^{+}, glutamate, glutathione and prostaglandin E2 have been found to be released via Cx HCs (Wang et al., 2013a), suggesting multiple dynamic functions of Cx HCs.

PANNEXIN CHANNELS

In 2004, a new type of GJ proteins homologous to the invertebrate innexins family was cloned (Baranova et al., 2004). Three members of this family, the Pannexins (Panxs) have been identified, with Panx1 expressed ubiquitously, Panx2 specifically in the brain, and Panx3 in osteoblasts and synovial fibroblasts (Baranova et al., 2004). In addition, Panx1 are found to be more abundant during early neuronal development and Panx2 later (Penuela et al., 2007). Interestingly, with little sequence homology, the Panxs share very similar topology and structure with the Cxs, displaying four transmembrane domains, two extracellular- and one intracellular- loop, as well as intracellular N- and C-termini (D'hondt et al., 2009). However, in contrast to Cxs, Panxs normally act in native systems as single membrane large pore channels rather than GJs (Penuela et al., 2007; Sosinsky et al., 2011). In fact, the only evidence of Panx GJ formation was demonstrated in overexpression systems in specific cell lines and showed distinct properties from Cx GJ channels (Bruzzone et al., 2003; Lai et al., 2007; Sahu et al., 2014).

Table 1 | Overview of available techniques and tools to study hemichannels.

How to study	Technique/Tool	Description	Useful for	Advantages	Disadvantages
HC properties	Dye uptake or efflux	Fluorescence quantification of the uptake or efflux of HC-permeable tracers (e.g., Ethidium bromide, Calcein)	Assessing HC activity based on their permeability to fluorescent tracers *In vitro* and *in situ*	Simple and fast assay Assesses HC activity of all cells with open HCs in a preparation Can reveal cell type specific uptake using immunostaining Potential use *in vivo*	Low temporal resolution Dye uptake: overestimation of the permeability of HCs due to diffusion of the tracers to neighboring cells via GJ upon uptake
	Efflux of biologically active molecules	Quantification of the release of active molecules (e.g., ATP, glutamate, NAD^+) via HCs when combined with molecular Cx/Panx knockdown/knockout or pharmacological blockers	Assessing HC activity based on relevant molecules released *In vitro* and *in situ*	Functionally relevant Quantitative	Low temporal resolution Not cell specific
	Electro-physiology	Patch-clamp recordings of HC-mediated ionic currents in whole cell, inside-out or cell-attached configurations	Measuring HC selectivity, kinetics and gating properties *In vitro*	Precise and dynamic measures of biophysical properties of HCs	Invasive Not appropriate for *in situ* recordings, due to low space-clamp of cells such as astrocytes
Impact of HCs	Pharmacological blockers	Carbenoxolone, lanthanum, divalent ions, flufenamic acid, alcohols, probenecid	Studying acutely the impact of HC activity *In vitro, in situ* and *in vivo*	Easy to use Fast action Local application and washout possible Antibodies: allow localization of HCs using immunostaining Gap19/TAT-Gap19 and TAT-L2: specific blockers for Cx43 HCs but not GJs	Not cell-specific Not selective for Cxs presenting high sequence homology Potential side effects Most blocking agents also inhibit GJ functions. This also applies to mimetic peptides when they are used over long periods (Samoilova et al., 2008)
	Antibodies	aEL2-186 (Hofer and Dermietzel, 1998), $Cx43^{E2}$ (Orellana et al., 2012b; Riquelme et al., 2013)			
	Mimetic peptides	Blockers: Gap26, Gap27, Gap19/TAT-Gap19, L2/TAT-L2, Panx1 and Panx2 mimetic peptides Enhancer: TAT-Cx43CT10 or TAT-Cx43CT9 (favors the opening of Cx43 HCs by preventing their closure at high cytoplasmic Ca^{2+} concentration) (Ponsaerts et al., 2010; De Bock et al., 2012)			
	Small interfering RNAs and Oligo-nucleotides	Genetic downregulation of the expression of Cx43 (Valiunas and Weingart, 2000; Figiel et al., 2007) and Panx1 (Wicki-Stordeur and Swayne, 2013)	Investigating the effect of lowering the expression of HC proteins *In vitro, in situ* and *in vivo*	High selectivity Spatial and temporal control of gene knockdown	Effect dependent on transfection/infection efficiency Off-target effects

(Continued)

Table 1 | Continued

How to study	Technique/Tool	Description	Useful for	Advantages	Disadvantages
	Transgenic mice	**Loss of function** Panx and Cx knockout mice Cx30$^{T5M/T5M}$ mice (defective Cx30 channel pore due to a single point mutation) (Grifa et al., 1999; Essenfelder et al., 2004; Schutz et al., 2010) **Gain of function** hGFAP-CreCx43$^{+/G138Rfl}$ mice (enhanced Cx43 HC activity in GFAP positive cells due to a single point mutation) (Dobrowolski et al., 2008; Torres et al., 2012)	Studying the impact of chronic alterations in HC activity *In vitro, in situ* and *in vivo*	Cell specificity using conditional knockout animals Investigation *in vivo* Can be used to test selectivity of Cx and Panx mimetic peptides	Possible developmental and compensatory effects Irreversible Gene alteration also affects GJ and non-channel function of Cx and Panxs

aEL2-186, antibody against position 186–206 on the extracellular loop of Cx43; Cx, Connexin; Cx43CT9 and 10, peptides targeting the last 9 or 10 amino acids of the Cx43 C terminus; Cx43^{E2}, antibody against the second extracellular loop (E2 domain) of Cx43; fl, floxed; G138R, substitution of a glycine by an arginine at position 138 of Cx43; GFAP, glial fibrillary acidic protein; GJ, gap junction; HC, hemichannel; Panx, Pannexin; TAT, peptide variant with enhanced plasma membrane permeability; T5M, substitution of a threonine by methionine at position 5 of Cx30.

Moreover, like Cx HCs, Panx channels can release numerous signaling molecules such as ATP or glutamate (Ye et al., 2003; Iglesias et al., 2009; Giaume et al., 2013). Remarkably, a unique property of Panxs is that they are glycosylated extracellularly at an arginine residue (Boassa et al., 2008). This glycosylation process has been proposed to be the limiting factor preventing Panx channels on adjacent cells from forming GJ channels. Thus, rare gap-junctional formation was only observed when Panx1 was overexpressed to a level exceeding glycosylation capacity (Bruzzone et al., 2003; Lai et al., 2007; Sahu et al., 2014). Concerning the biophysical properties of Panx channels, they are still elusive and controversial, most likely due to the recent discovery of Panxs compared to Cx HCs. Indeed, for instance unitary conductance of Panx1 channels has been reported to range from ~70 to 550 pS, which may reflect different recording conditions, cell type properties or sub-conductance states (Locovei et al., 2006; Kienitz et al., 2011; Ma et al., 2012). While the expression of Panx1 overlaps with that of Cxs mediating extensive GJ coupling in cellular networks (Ray et al., 2005), it was found to be distinctively regulated as compared to the Cxs (Penuela et al., 2007), implying unique functional significance of Panx channels. For instance, while many reports have demonstrated sensitivity of Cx HCs to extracellular Ca^{2+} ($[Ca^{2+}]_e$), which is commonly thought to keep HCs in a closed state at physiological $[Ca^{2+}]_e$ (Ebihara et al., 2003), Panxs appear to be insensitive to this (Bruzzone et al., 2005). In addition, Panx1 channels do not show the strong voltage-dependence typical of Cx HCs (Bruzzone et al., 2003). Thus, this type of insensitivity suggests that Panx channel activity does not depend on neuronal activity and that Panx channels might be open during basal physiological conditions.

EMERGING ROLES OF CONNEXON AND PANNEXON CHANNELS IN NEUROPHYSIOLOGY AND BEHAVIOR

Cx HCs and Panxs as single membrane channels have undoubtedly many roles in CNS pathologies involving inflammation (Orellana et al., 2011a; Bennett et al., 2012; Makarenkova and Shestopalov, 2014), ischemia (Contreras et al., 2004; Bargiotas et al., 2011), epilepsy (Thompson et al., 2008) and neurodegeneration (Orellana et al., 2011b, 2012b). For a long time, it was believed that Cx HCs and Panx channels with large channel conductance would remain closed in normal conditions in order to maintain cellular integrity and to prevent unregulated transmitter release into the extracellular space. It is indeed conceivable that, like many membrane channels, these Cx HCs and Panxs serve as a direct and effective way to relieve cells from any homeostatic imbalances found in many pathological situations (Bennett et al., 2012; Giaume et al., 2013). Given that specific HCs are found to be active outside the brain in physiological settings (Anselmi et al., 2008), and that Panx channels have unique properties favoring channel opening under the same conditions, a few pioneer groups began to postulate that these channels might also have some relevance in basal neuronal processes. Here, we review the few recent studies examining roles of Cx and Panx single membranes large pore channels on physiological functions ranging from cell division to learning and memory.

DEVELOPMENT

During CNS development including embryonic and adult neurogenesis, purinergic signaling is believed to have crucial contributions (Zimmermann, 2011). Indeed, ATP receptor activation was found to play a role in proliferation and DNA synthesis in astrocytes (Neary and Zhu, 1994) and neural stem cells (Ryu et al., 2003). The fact that both Cx43 HCs and Panx1 channels are permeable to ATP makes them candidates for purinergic signaling during development. The involvement of Cx43 HCs in these fundamental cellular processes was first explored in embryonic retina. As a result, they were found to play a role in cell division and proliferation (Pearson et al., 2005). The authors have demonstrated that upon spontaneous Ca^{2+} increase in a "trigger cell" in the retinal pigment epithelium, ATP is released via Cx43 HCs into the extracellular space adjacent to retinal progenitor cells during development, promoting mitosis and proliferation. In addition, the released ATP also triggers the spread of Ca^{2+} waves through neighboring retinal pigment epithelial cells. Both of these mechanisms involve P2 receptor activation. A short application (30 min) of the mimetic peptide Gap26, which specifically blocks Cx43 HCs without affecting GJ functions, was used to test the involvement of Cx43 HCs, whereas HC activity in retinal pigment epithelial cells was measured by Alexa 488 dye efflux assay. In a more recent study, functional Panx1 channels were identified in postnatal neural stem and progenitor cells both *in vitro* and *in vivo* (Wicki-Stordeur et al., 2012). Panx1 channel specific effect was investigated using Panx1 siRNA knockdown *in vitro* and/or probenecid, a blocker of Panx1 channels without affecting Cx HCs. These results were found to positively regulate proliferation of these cell types via ATP release and subsequent P2 receptor activation. The same group later demonstrated that Panx1 also plays a role in cell migration and neurite extension through its interaction with actin cytoskeleton (Wicki-Stordeur and Swayne, 2013). These results are interesting, since adult neurogenesis plays important roles in both physiological brain development as well as brain repair during pathology and diseases (Berg et al., 2013). Although astrocytic Cx43 and Cx30 have been shown to differentially regulate adult neurogenesis in knockout mice (Liebmann et al., 2013), it was not clear whether it was via their GJ or HC properties. The potential involvement of Cx43 HCs in embryonic retinal development and Panx1 channels in adult neurogenesis is encouraging and may offer important insights into other possible non-synaptic physiological functions of single membrane channels yet to be discovered.

GLUCOSE SENSING AND SIGNAL TRANSDUCTION

Apart from promoting cell division and proliferation, the activities of Cx43 HCs and Panx1 channels have also recently been proposed to be modulated by changes in extracellular glucose concentrations. Neuronal glucose-sensing is an important physiological process in the hypothalamus, in which action potentials are driven by changes in extracellular glucose concentration. This has been shown to largely contribute to feeding and satiety behavior, sleep-wake cycles, and energy expenditure (Levin et al., 2004; Burdakov et al., 2005). In astrocytes, glucose uptake via Cx43 HCs has already been demonstrated under inflammatory conditions (Retamal et al., 2007). However, whether this is also true in physiology remains unclear. Recently, the role of glial Cx43 HCs in glucose-sensing was explored using cultured tanycytes, which are specialized glial cells in the hypothalamus (Orellana et al., 2012a). The authors showed that, upon a rise in extracellular glucose concentration, glucose transporters (GLUTs) and to a lesser extent Cx43 HCs, allow diffusion of glucose into tanycytes, where it leads to increase in ATP. Subsequently, ATP is released via Cx43 HCs, which then stimulates P2Y receptors locally leading to the rise of $[Ca^{2+}]_i$. Both ATP release and $[Ca^{2+}]_i$ responses were inhibited by Cx43 HC blocking agents like La^{3+}, Gap26, and $Cx43^{E2}$ (an antibody against the second extracellular loop of Cx43). The opening of Cx43 HCs was found to be promoted by the closing of K_{ATP} channels. Interestingly, this serie of events does not require extracellular Ca^{2+}. Using ethidium bromide uptake assay, electrophysiology, and surface biotinylation, they have demonstrated that the open probability of Cx43 HCs, rather than their number, was enhanced by glucose. Although it was recently demonstrated that Cx30 but not Cx43 HCs expressed on oocytes are permeable to glucose (Hansen et al., 2014), these results have revealed a new role of Cx43 HCs in physiological situation, namely to sense and metabolize extracellular glucose. Further studies on how this directly affects glucose metabolism in the hypothalamus, as well as behavioral modifications *in vivo* would largely complement these data and draw physiological relevance. Since tanycytes express a majority of Cx43 rather than Panxs, it is not surprising that Panxs were found not to be involved in this process. However, it would be interesting to explore whether astrocytic Panxs participate in similar signaling pathways. In fact, the involvement of Panxs in metabolic autocrine regulation has been previously demonstrated (Kawamura et al., 2010). The authors showed that non-pathological changes in extracellular glucose concentration induce a purinergic autoregulation in hippocampal CA3 neurons. This was found to be mediated by the opening of Panx1 channels on neurons, but not astrocytes. Although this study was designed to determine the effect of the anti-epileptic ketogenic diet (high fat and low carbohydrates) on increasing seizure threshold, it gave important insights into how Panx1 channels could be involved in regulating physiological metabolic perturbations.

NEURONAL EXCITABILITY AND SYNAPTIC TRANSMISSION

Astrocytes express a large repertoire of ions channels, neurotransmitter receptors and transporters, allowing them to sense and modulate neuronal activity through multiple mechanisms. These include in part intracellular calcium signaling, morphological changes or gliotransmission, in which neuroactive substances such as glutamate or ATP are released from astrocytes and act on neurons. The opening of Cx HCs and Panx channels during physiological conditions suggests that they represent a release pathway of gliotransmitters and ions, which can regulate neuronal excitability and basal synaptic transmission. Here, we first present findings related to ATP release via astroglial Cx43 HCs and synaptic transmission. In particular, among several recent studies demonstrating physiological relevance of Cx43 HCs, we have very recently shown that astroglial Cx43 HCs are not only open in acute hippocampal slices under physiological conditions, but can also modulate basal excitatory synaptic transmission through ATP signaling (Chever et al., 2014).

Specifically, we reported that ethidium bromide uptake into astrocytes under basal conditions was decreased in the presence of the HC blocker carbenoxolone (CBX) or the Cx43 HC mimetic blocking peptide Gap26. These blocking agents had no effect on ethidium bromide uptake in brain slices prepared from $Cx43^{fl/fl}$ hGFAP-Cre mice ($Cx43^{-/-}$). These strongly suggested that functional astroglial Cx43 HCs are present under physiological conditions (**Figures 1A,B**). Interestingly, excitatory postsynaptic currents recorded in CA1 pyramidal neurons and ATP release from hippocampal slices detected using a luciferin-luciferase luminescence assay were also concurrently decreased in the presence of Gap26. Furthermore, pretreatment with ATP P2X and P2Y receptor antagonists abolished the inhibitory effect of Gap26 on excitatory synaptic activity, suggesting the involvement of P2 receptors (**Figure 1C**). Our results thus suggest a novel physiological pathway involving functional astroglial Cx43 HCs and ATP release in tuning excitatory synaptic transmission in the hippocampus.

The involvement of HCs in promoting astrocytic Ca^{2+} waves by ATP release (Cotrina et al., 1998) suggested another possible means for HCs to contribute to physiological cell-cell interaction and propagation of signals. In fact, this has already been demonstrated in the inner ear. Both Cx26 and Cx30 were found to play a role in propagating long-range intercellular Ca^{2+} signaling in cochlear organotypic cultures (Anselmi et al., 2008). These Cxs contributed by HC-mediated ATP release as well as gap-junctional diffusion of Ca^{2+}-mobilizing second messengers. A recent study (Torres et al., 2012) has taken a different approach to study HC function by directly decreasing $[Ca^{2+}]_e$ (by 0.5 mM), to a range thought to occur during neuronal network activity

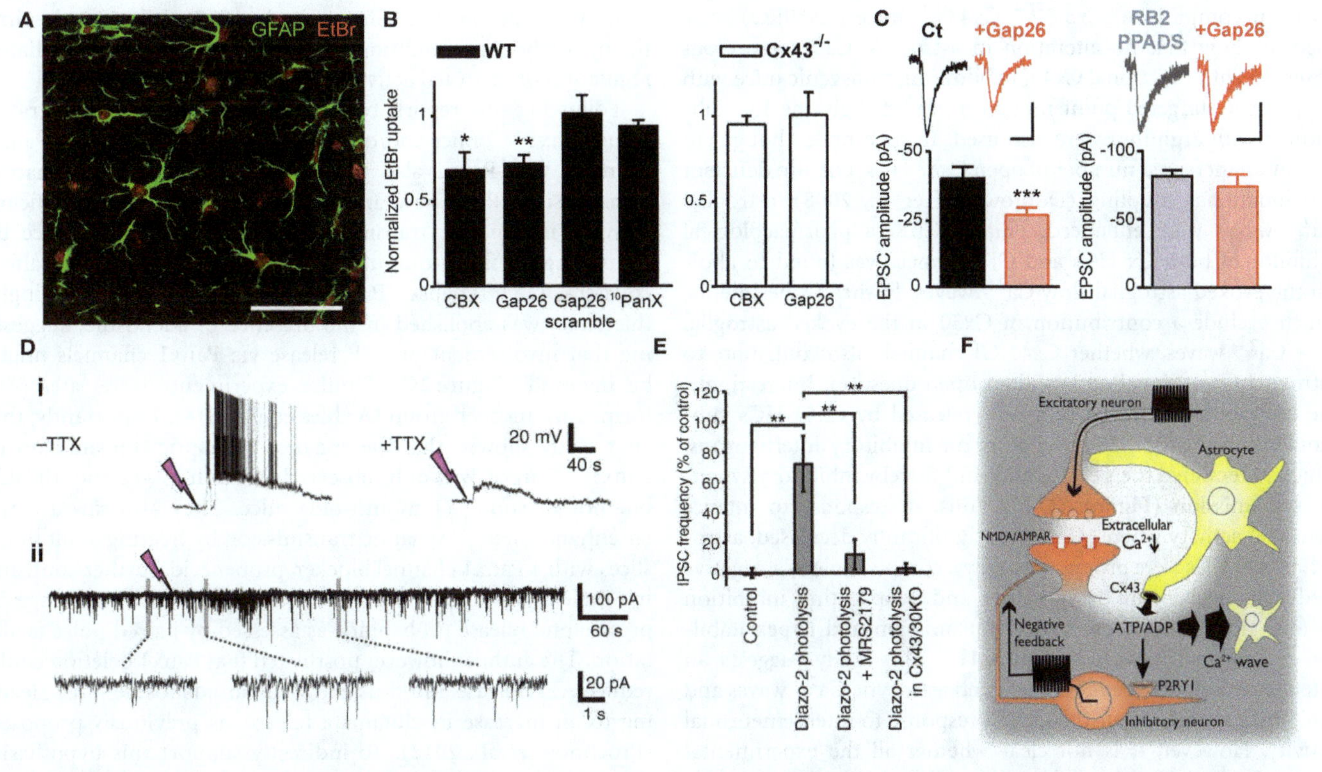

FIGURE 1 | Astrocytic Cx43 HCs modulate synaptic transmission in hippocampal slices. (A–C) Basal astroglial Cx43 HC activity enhances excitatory synaptic transmission via ATP signaling. **(A)** Representative image showing ethidium bromide uptake (EtBr; red) in astrocytes (immunostained for GFAP; green) of the stratum radiatum in an acute hippocampal slice. Scale bar, 50 μm. **(B)** Bar graphs showing astrocytic EtBr uptake in brain slices obtained from wild-type (WT) and astroglial conditional Cx43 KO ($Cx43^{-/-}$) mice normalized to control (untreated) conditions. Uptake was significantly deceased in WT slices treated with carbenoxolone (CBX, 200 μM) and Gap26 (100 μM), but not Gap26 scramble (100 μM) and ^{10}panx (400 μM) peptides. In $Cx43^{-/-}$ slices, however, both CBX and Gap26 had no significant effect. **(C)** Bar graph on the left showing a decrease in amplitude of evoked EPSC recorded in CA1 pyramidal neurons during Gap26 application (red) as compared to before (Ct, black). Bar graph on the right showing that pretreatment with ATP P2 receptor antagonists (RB2 + PPADS, gray) occludes the effect of Gap26 (red). Sample traces of corresponding evoked EPSCs are shown above. Scale bar: 20 pA, 20 ms (left); 40 pA, 40 ms (right). **(D–F)** Cx43 HCs in astrocytes promote feedback inhibitory transmission by releasing ATP. **(D)** Representative recordings showing that photolysis of diazo-2, represented by lightning bolts, (i) evokes depolarization and bursting in interneurons, with the depolarization persisting with 1 μM TTX, and (ii) transiently increases the frequency of spontaneous inhibitory postsynaptic currents (IPSCs) in CA1 pyramidal neurons. **(E)** Bar graph indicating increased IPSC frequency in pyramidal neurons after diazo-2 photolysis compared to control condition. This effect was blocked by the P2Y1 receptor antagonist MRS2179 (50 μM) or in brain slices prepared from Cx43/Cx30KO mice. **(F)** Schematic diagram illustrating a proposed negative feedback mechanism during excitatory transmission. During glutamatergic signaling, Ca^{2+} influx into neurons results in a localized decrease in $[Ca^{2+}]_e$, which in turn opens Cx43 HCs on astrocytes through which ATP is released. ATP can either trigger slowly propagating astrocytic Ca^{2+} waves or, when degraded to ADP, depolarize and increase firing in interneurons via P2Y1 receptors, thereby enhancing inhibitory transmission. $^{*}p < 0.05$; $^{**}p < 0.01$; $^{***}p < 0.001$. Adapted, with permission, from Torres et al. (2012) **(D–F)**.

(Massimini and Amzica, 2001; Amzica et al., 2002). Since low $[Ca^{2+}]_e$ has been found to open Cx HCs (Sáez et al., 2005), the effect of lowering $[Ca^{2+}]_e$ on synaptic transmission and the involvement of HCs in this process was investigated. Specifically, $[Ca^{2+}]_e$ in acute hippocampal slices was decreased using several methods including uncaging Ca^{2+} chelator (diazo2) or glutamate (MNI-glutamate) extracellularly. In an attempt to draw physiological relevance, high frequency stimulation-induced decrease in $[Ca^{2+}]_e$ was also assessed. With these manipulations to decrease $[Ca^{2+}]_e$, they detected an increase in extracellular ATP concentration, which triggered a slow astrocytic Ca^{2+} wave. Such ATP release was proposed to be mediated by astroglial Cx43 HCs, using a combination of molecular and pharmacological tools. $Cx30^{-/-}$ $Cx43^{fl/fl}$ hGFAP-Cre mice, with conditional deletion of Cx43 in astrocytes and total deletion of Cx30 (Wallraff et al., 2006), were used to avoid compensatory up-regulation of Cx30 (Theis et al., 2003). Astroglial Ca^{2+} waves were abolished in these mice. In comparison, $Cx30^{-/-}$ $Cx43^{fl/fl}$ mice (Cx30KO) were used to show that no alteration in astrocytic Ca^{2+} waves was observed with functional Cx43. In addition, transgenic mice with an astrocyte-targeted point mutation of Cx43 (glycine 138 substituted with arginine) were also used. In such mice, thought to display an increased number of open Cx43 HCs, but also deficient gap junctional coupling (Dobrowolski et al., 2008), astroglial Ca^{2+} waves were enhanced. Finally, CBX, a pharmacological inhibitor of both Cx HCs and GJ channels was found to abolish the evoked astroglial slow Ca^{2+} waves. In these experiments, which exclude a contribution of Cx30 in the evoked astroglial slow Ca^{2+} waves, whether Cx43 GJ channels also contribute to astroglial Ca^{2+} signaling is still an open question. Interestingly, the authors also showed that ATP released by Cx43 HCs subsequently activated P2Y1 receptors on inhibitory interneurons, which potentiated their excitability and thereby inhibitory synaptic transmission (**Figures 1D,E**). Thus, in response to intense neuronal activity where $[Ca^{2+}]_e$ is significantly decreased, astrocytic Cx43 HCs were proposed to play a complex role in a negative feedback mechanism by initiating and propagating inhibition to tone down strong excitatory transmission and hypexcitability of neuronal networks (**Figure 1F**). This study suggests an interesting link between HC-mediated astrocytic Ca^{2+} waves and inhibitory synaptic transmission in response to intense neuronal activity. However, it is not clear whether all the experimental protocols used in this study generated a decrease in $[Ca^{2+}]_e$ that is within a physiologically relevant range. Additional studies are therefore warranted to demonstrate that the regulation of neuronal activity by astroglial Cx43 HCs also applies to physiological conditions. In addition, it is unclear whether the decrease in $[Ca^{2+}]_e$ used in this study induces significant Cx HC opening in astrocytes. Indeed, although slight decrease in $[Ca^{2+}]_e$, from 1.8 to 1.6 mM, has been found to open Cx HCs in cell lines (Quist et al., 2000), in cultured astrocytes $[Ca^{2+}]_e$ below 1 mM is necessary to significantly open HCs, as assessed by HC-mediated glutamate release (Ye et al., 2003). Finally, it still remains unknown whether ATP activated P2Y1 receptors in other cells than interneurons, such as astrocytes, where they can contribute to glutamatergic gliotransmission (Jourdain et al., 2007; Pascual et al., 2012).

ATP-mediated modulations of neurotransmission are not only restricted to astroglial Cx43 HCs. Indeed, an interesting link between Panx channels, which also release ATP, and basal neuronal excitability was reported (Kawamura et al., 2010). Using whole-cell patch clamp recordings, it was demonstrated that changes in extracellular glucose concentration from 11 to 3 mM opened Panx1 channels on rat hippocampal CA3 pyramidal neurons through which ATP is released. Its metabolite adenosine then activates adenosine A1 receptors, leading to the opening of ATP-sensitive K^+ channels. As a result of this autocrine regulation, neuronal hyperpolarization occurs, downregulating neuronal activity. The involvement of Panx1 channels was confirmed using CBX, octanol and the selective peptide blocker ^{10}panx (**Figures 2A,B**). Of note, although decreasing glucose concentration to 3 mM was confirmed to be non-pathological using extracellular recordings, an initial concentration of 11 mM is almost four-fold higher than extracellular glucose concentration in brain tissues *in vivo* (Shram et al., 1997). Whether this initial hyperglycemic condition contributed to the Panx1-mediated regulation of neuronal activity thus remains to be clarified.

Following this report, two other studies have demonstrated using $Panx1^{-/-}$ mice and/or pharmacological blockade of Panx1 channels that Panx1 also plays an important role in synaptic transmission. Prochnow and colleagues observed a significant increase in synaptic transmission in adult $Panx1^{-/-}$ mice by measuring the input-output curves at the hippocampal Schaffer-collateral CA1 synapse (Prochnow et al., 2012). Interestingly, this effect was abolished in the presence of adenosine, suggesting that involvement of ATP release via Panx1 channels might be involved (**Figure 2C**). Similar experiments were later performed by another group (Ardiles et al., 2014). Importantly, this later study showed that the enhanced synaptic transmission in $Panx1^{-/-}$ mice was only observed in adult (9–12 month-old) but not in young (1 month-old) mice. They also saw a similar enhancement in synaptic transmission by treating adult brain slices with a Panx1 channel blocker, probenecid, further confirming their observations. This effect was not related to change in presynaptic release probability, as assessed by paired pulse facilitation. The authors however postulated that Panx1 deletion could reduce ATP release and thus extracellular adenosine levels, leading to an increase in glutamate release, as previously proposed (Prochnow et al., 2012). To indirectly support this hypothesis, they experimentally increased synaptic glutamate levels using a glutamate transporter blocker (TBOA), and observed a larger increase in fEPSP in adult $Panx1^{-/-}$ and probenecid-treated control than in untreated control. However, direct evidence of pre- and/or postsynaptic effects is still lacking. It would also be interesting to determine whether the effects observed are mediated by neuronal and/or glial Panx1 channels.

SYNAPTIC PLASTICITY, LEARNING, AND MEMORY

With these growing reports suggesting important roles of Cx HCs and Panx channels in neuronal excitability and synaptic transmission, some groups became interested in studying their functional significance in terms of synaptic plasticity as well as learning and memory. It is indeed believed that astrocytes have many essential roles in synaptic functions, including synaptic plasticity

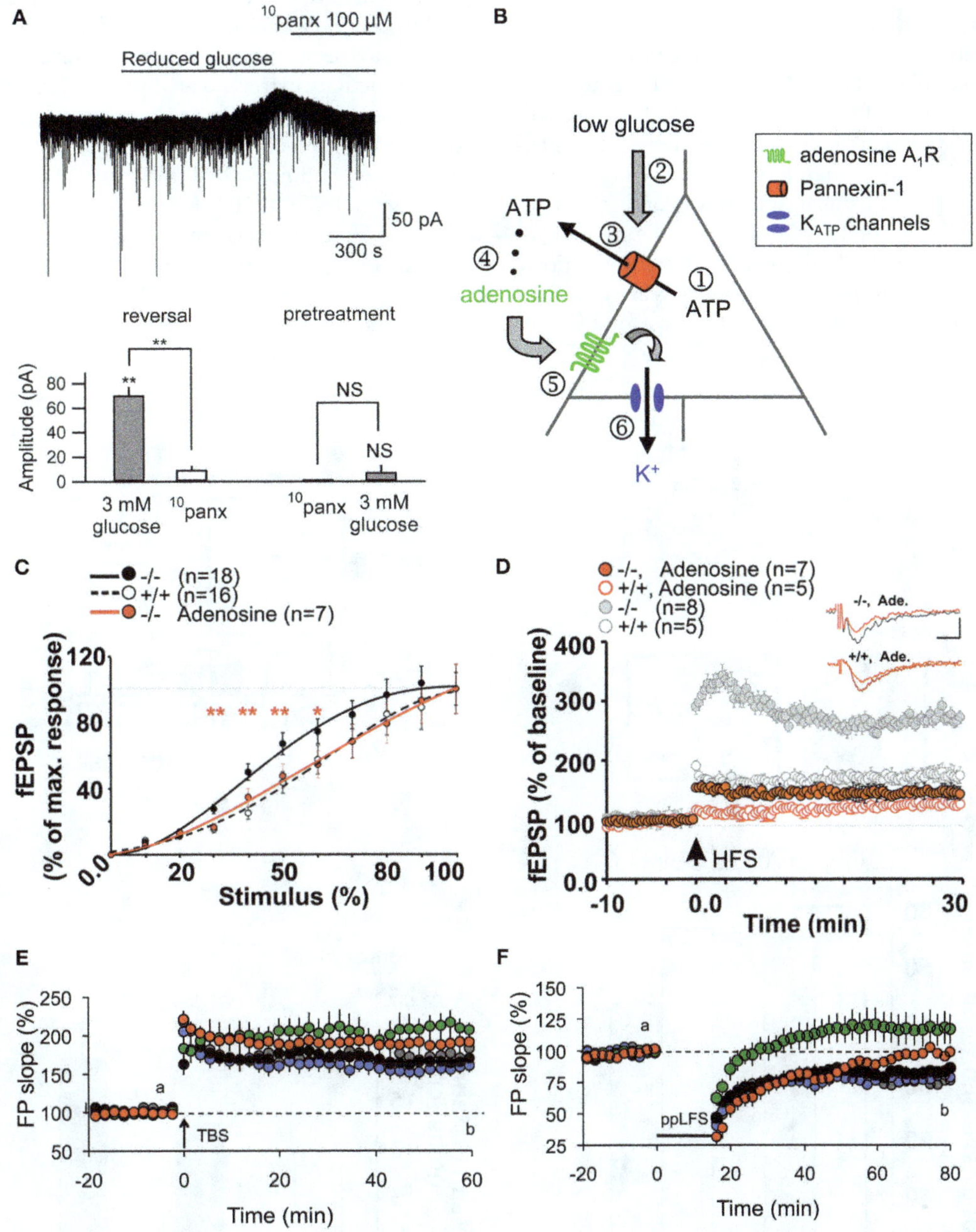

FIGURE 2 | Panx1 channels modulate neuronal excitability, synaptic transmission and plasticity in hippocampal slices. (A,B) Metabolic autocrine regulation of neuronal activity via Panx1 channels and adenosine. **(A)** Sample trace showing increased outward current upon reduced extracellular glucose (from 11 to 3 mM) and subsequent reversal to baseline with ^{10}panx application (100 μM) in CA3 pyramidal neurons. Bar graphs showing the reversal of reduced glucose-induced outward current amplitude with ^{10}panx (left), and that pretreatment with ^{10}panx prevented reduced glucose-induced outward current. $^{**}p < 0.01$. **(B)** Schematic showing a proposed model of purinergic autocrine regulation in CA3 pyramidal neurons. When $[ATP]_i$ is sufficient (1), low $[Glucose]_e$ (2) induces ATP release from Panx1 channels on neurons (3). ATP is then dephosphorylated to adenosine (4) which activates adenosice A_1R rceptors (5). K_{ATP} channels are then opened leading to a decrease in neuronal excitability. **(C–F)** Panx1 regulates synaptic transmission, LTP and LTD. **(C)** Input-Output curves showing increased synaptic transmission in Panx1$^{-/-}$ (black line) compared to control Panx1$^{+/+}$ (dashed line) mice. Such effect was abolished in Panx1$^{-/-}$ slices treated with 3 μM adenosine (red line). $^{*}p < 0.01$; $^{**}p < 0.001$. **(D)** LTP evoked by high frequency stimulation (four trains of 10 shocks at 100 Hz every 1 s; HFS) is enhanced in Panx1$^{-/-}$ (filled gray) compared to control Panx1$^{+/+}$ (open gray) mice. Adenosine treatment in Panx1$^{-/-}$ slices (filled red) restores LTP levels to that of untreated control mice. Figure insets illustrate responses before and 30 min post HFS. Scale bar: 0.5 mV, 10 ms. **(E)** LTP induced by the delivery of theta burst stimulation protocol (TBS) is increased in adult Panx1$^{-/-}$ (green) compared to Panx1$^{+/+}$ (black) mice, whereas no difference was observed between young mice (+/+, gray; −/−, blue). In the presence of 100 μM probenecid (Panx1 channel blocker; red), only transient LTP was enhanced. **(F)** Similarly, LTD induced by paired-pulse low frequency stimulation protocol (1 Hz for 15 min; PP-LFS) are impaired in adult Panx1$^{-/-}$ (green) compared to Panx1$^{+/+}$ (black) mice, whereas no difference was observed between young equivalent (+/+, gray; −/−, blue). In the presence of 100 μM probenecid, only a transient LTD was observed. Adapted, with permission, from Kawamura et al. (2010) **(A,B)**, Prochnow et al. (2012) **(C,D)** and Ardiles et al. (2014) **(E,F)**.

and information processing (Perea et al., 2009; Dallerac et al., 2013; Pannasch and Rouach, 2013). Whether this could be conducted via astroglial HC activity was first investigated in 2012. In particular, a possible role of HCs in learning and memory was reported (Stehberg et al., 2012). The authors selectively blocked Cx43 HCs using TAT-Cx43L2. This peptide was designed to interfere with the intracellular loop/tail interactions of Cx43 necessary for its HC but not gap-junctional activities (Ponsaerts et al., 2010). After microinfusion of TAT-Cx43L2 into the rat basolateral amygdala, an area of the brain associated with emotional memory (Ledoux, 2007), amnesia for auditory fear conditioning was observed (**Figures 3A,B**). This effect was found to be specific to memory consolidation, as the blocker applied 6 h after learning did not lead to memory deficits. Their results were corroborated using Gap27, a peptide commonly used to block Cx43 HC activity extracellularly, further suggesting specific contributions from Cx43 HCs. In order to investigate the molecular mechanisms involved, a cocktail of gliotransmitters containing glutamate, glutamine, lactate, D-Serine, glycine, and ATP was co-infused with TAT-Cx43L2. As a result, a rescue of the memory defect was observed (**Figure 3B**). Since they showed that TAT-Cx43L2 had no effect on ATP and glutamate release in neuronal cultures, and that Cx43 was found on astrocytes but not neurons, the authors concluded that the effect was specific for astroglial

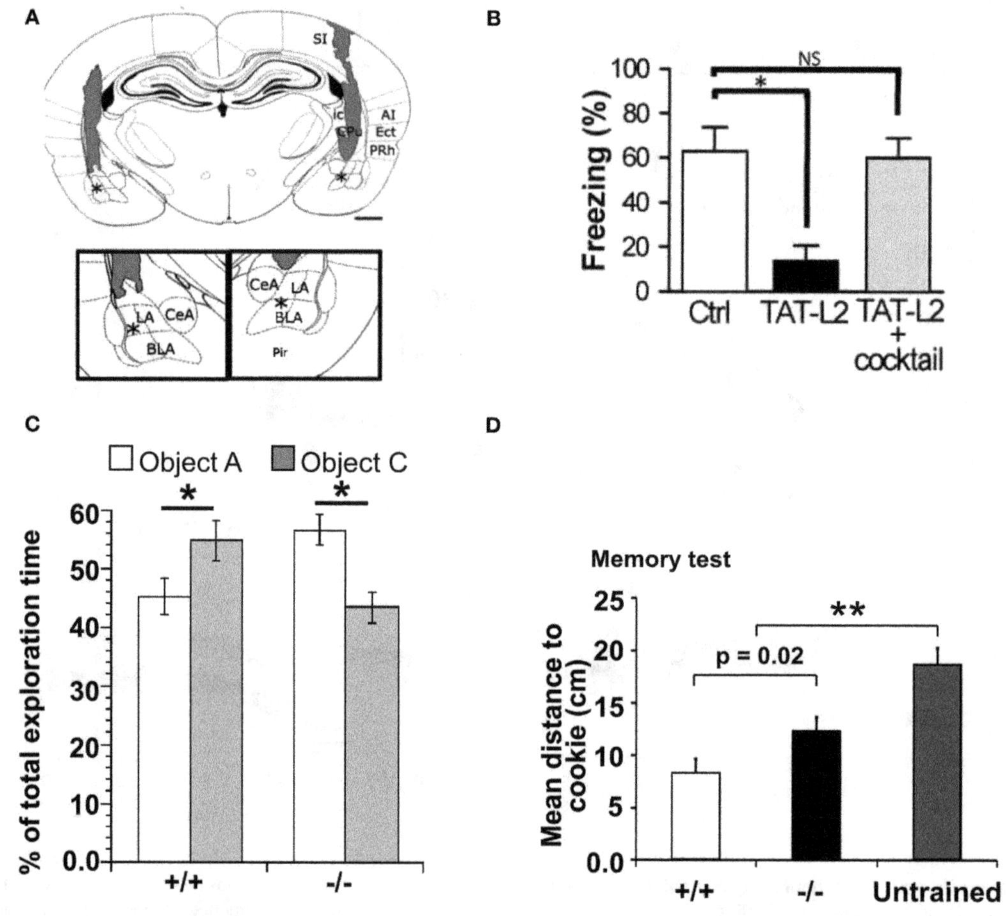

FIGURE 3 | Cx HCs and Panx1 channels have significant roles in learning and memory. (A,B) Cx43 HC function is required in fear conditioning memory consolidation. **(A)** Diagram showing site of microinfusions of the TAT-Cx43L2, a selective Cx43 HC blocker, into the basolateral amydala (BLA). Asterisks indicate the tips of injection cannula (shaded regions). Enlargement is shown in insets. LA, lateral amygdala; CeA, central amygdala; SI, somatosensory primary; PRh, perirhinal; Ect, ectorhinal; Pir, piriform; and AI, auditory primary cortices; CPu, caudoputamen; ic, internal capsula. Scale bar: 1 mm. **(B)** Fear conditioning memory was tested by first training rats to associate a tone with a foot shock. Their memory of this association was then assessed by how long they remain immobile (freeze) in response to the same tone alone 24 h later. TAT-Cx43L2 (10 nM) microinfusion prior to training impaired fear conditioning memory (decreased in freezing time) compared to control. Such effect was rescued by co-microinfusion of a mixture of gliotransmitters including D-serine, glutamate, glutamine, glycine, ATP, and lactate (cocktail) rescued such effect. **(C,D)** Panx1 deletion leads to dysfunctions in learning and memory. **(C)** To assess object recognition, mice were allowed to explore two novel objects (A and B) for 5 min. 1 h later, they were allowed to explore the familiar object A together with a novel object C. Panx1$^{-/-}$ mice spent more time on object A than C compared to Panx1$^{+/+}$ mice which did the opposite, indicating a deficit in object recognition. **(D)** Another memory test was carried out where mice were trained to remember locations of hidden cookies which were later removed. During the test, Panx1$^{-/-}$ mice explored locations further from the original locations of cookies compared to the Panx1$^{+/+}$ mice but not as far as the untrained mice, indicating an impairment but not abolishment in memory. $^{*}p < 0.05$; $^{**}p < 0.001$. Adapted, with permission, from Stehberg et al. (2012) **(A,B)** and Prochnow et al. (2012) **(C,D)**.

gliotransmitter release. This pioneer *in vivo* study has suggested an important contribution of gliotransmitter release via Cx43 HCs, presumed to be of astrocytic origin, in memory consolidation. However, it still remains elusive which gliotransmitters are involved and their downstream targets. Furthermore, direct release of such gliotransmitters via Cx43 HCs is yet to be demonstrated, as there are also other possible mechanisms like vesicular exocytosis (Parpura et al., 2004), Panxs (Iglesias et al., 2009), P2X7 channels (Suadicani et al., 2006), Bestrophin 1 (Han et al., 2013), and volume-regulated anion channels (Kimelberg, 2004), which could be downstream effects to Cx43 HC activation.

Shortly following the study by Stehberg and colleagues, another report linking HC activities to learning and memory was revealed. This time, the focus was on Panx channels and their well-characterized role in purinergic signaling. Since Panx1 is strongly expressed in postsynaptic terminals (Zoidl et al., 2007) and presumably also in astrocytes (Iglesias et al., 2009), this study explored the contributions of Panx1 to synaptic plasticity, which in turn may influence learning. Prochnow and colleagues have observed that in Panx1$^{-/-}$ hippocampus, neurons have increased synaptic transmission and enhanced long-term potentiation (LTP) triggered by high frequency stimulation (Prochnow et al., 2012), as shown in **Figure 2D**. Interestingly, these changes in neuronal properties could be rescued by supplying extracellular adenosine, a metabolite of ATP, as well as blockage of postsynaptic NMDARs, indicating that this process involves Panx1-mediated ATP release and an inhibition of glutamatergic excitatory responses. They have also found that only in adult mice, an upregulation of mGluR4 accompanies chronic deletion of Panx1, which accounted for the enhanced persistent LTP observed. The authors suggested that this might be an adaptive compensatory response to the enhanced neuronal transmission. Consequently, these modifications in Panx1$^{-/-}$ mice led to behavioral defects like altered sensory motor gating capabilities, assessed using pre-pulse inhibition of the acoustic startle response, which may reflect stress and anxiety. Cognitively, novel object recognition (**Figure 3C**) and spatial learning involving the location of a treat were also impaired (**Figure 3D**). These deficits were expected, since it has been shown that saturation of LTP is related to impaired learning (Moser et al., 1998). Indeed, this study demonstrated that ATP released through Panx1 channels participates in maintaining synaptic strength and plasticity in hippocampal CA1 neurons, and in turn plays a role in learning and memory. Furthermore, the authors suggested that Panx1 mediates a feedback response through presynaptic activation of adenosine A1 receptors and inhibition of glutamate release, which is essential in cognitive functioning. However, this study did not distinguish between neuronal and glial Panx1 channels. Given that astrocytes are thought to also express ATP-releasing Panx1 channels (Iglesias et al., 2009) and that they have been implicated in the modulation of neuronal activities (Dallerac et al., 2013), it is of interest to determine specific astrocytic contributions in these behavioral and cognitive processes, possibly with the use of conditional knockouts.

Interestingly, very recently, another study showed that not only LTP but also long-term depression (LTD) is altered in Panx1$^{-/-}$ mice (Ardiles et al., 2014). Specifically, they observed that LTD was impaired in either Panx1$^{-/-}$ hippocampus or during pharmacological blockage of Panx1 channels. Moreover, they have also demonstrated that both the enhancement in LTP and impairment in LTD were only observed in adult but not in young Panx1$^{-/-}$ mice (**Figures 2E,F**). This study revealed that Panx1, which may be of neuronal or glial origin, participates in bidirectional synaptic plasticity by modulating both potentiation and depression in an age-dependent manner. Thus, the authors suggested that Panx1 channels may influence learning and memory in adults by restraining the threshold for the induction of synaptic plasticity. The functional significance of this age-dependent bidirectional regulation, however, is yet to be explored.

VISION

In addition to their contributions to synaptic and higher level brain functions, some fundamental visual processes have also been shown to be mediated by Cx HCs and Panx channels. In particular, two groups have demonstrated the involvement of these channels in visual processes in zebrafish. Between the two channels, Cx HCs were first implicated in synaptic transmission essential for vision (Kamermans and Fahrenfort, 2004; Klaassen et al., 2011). In the retina, Cx52.6 and Cx55.5 form GJ channels between horizontal cells, while Cx55.5 also form HCs at the tips of the horizontal cell dendrites and can open at physiological membrane potentials (Shields et al., 2007). Klaassen et al. (2011) generated zebrafish Cx55.5 mutant, with a stop codon in the first extracellular loop of Cx55.5 (C54X), which led to decrease in both, Cx55.5 expression on the tips of the horizontal cell dendrites, and HC currents from horizontal cells. They reported that the total gap-junctional surface was preserved, although alterations in Cx52.6 expression and in the general organization of gap-junctional plaques were found. This study investigated the negative feedback from horizontal cells to cones, essential for contrast enhancement in vision. In these mutants, light-induced feedback from horizontal cells to cones was significantly reduced (**Figure 4A**). As a consequence, these mutant zebrafishes exhibited impairment in contrast sensitivity (**Figure 4B**). Supported by a mathematical model, the authors suggested that current flowing through Cx HCs at the tips of horizontal cell dendrites actually exert an ephaptic modulation of synaptic transmission through activation of adjacent voltage-dependent Ca^{2+} channels in presynaptic photoreceptor terminals. Interestingly, they also reported an upregulation of Panx1 expression at the tips of the horizontal cell dendrites in these mutants, suggesting that it could be partly accountable for the residual 40% HC activity in the mutant. The same group later demonstrated that Cx55.5 HC currents possess an inward component that is active at physiological membrane potentials and extracellular Ca^{2+} levels (Sun et al., 2012). These studies suggest an unconventional role of horizontal cell Cx HCs in synaptic transmission in the retina, where a carp ortholog of mammalian Cx43 is also expressed among several other Cxs (Dermietzel et al., 2000). While similar mechanisms involving other Cxs and other CNS regions are yet to be investigated, these research findings further confirmed the dynamic nature of physiological roles of HCs in synaptic transmission leading to functional outcome, which in this case is visual acuity.

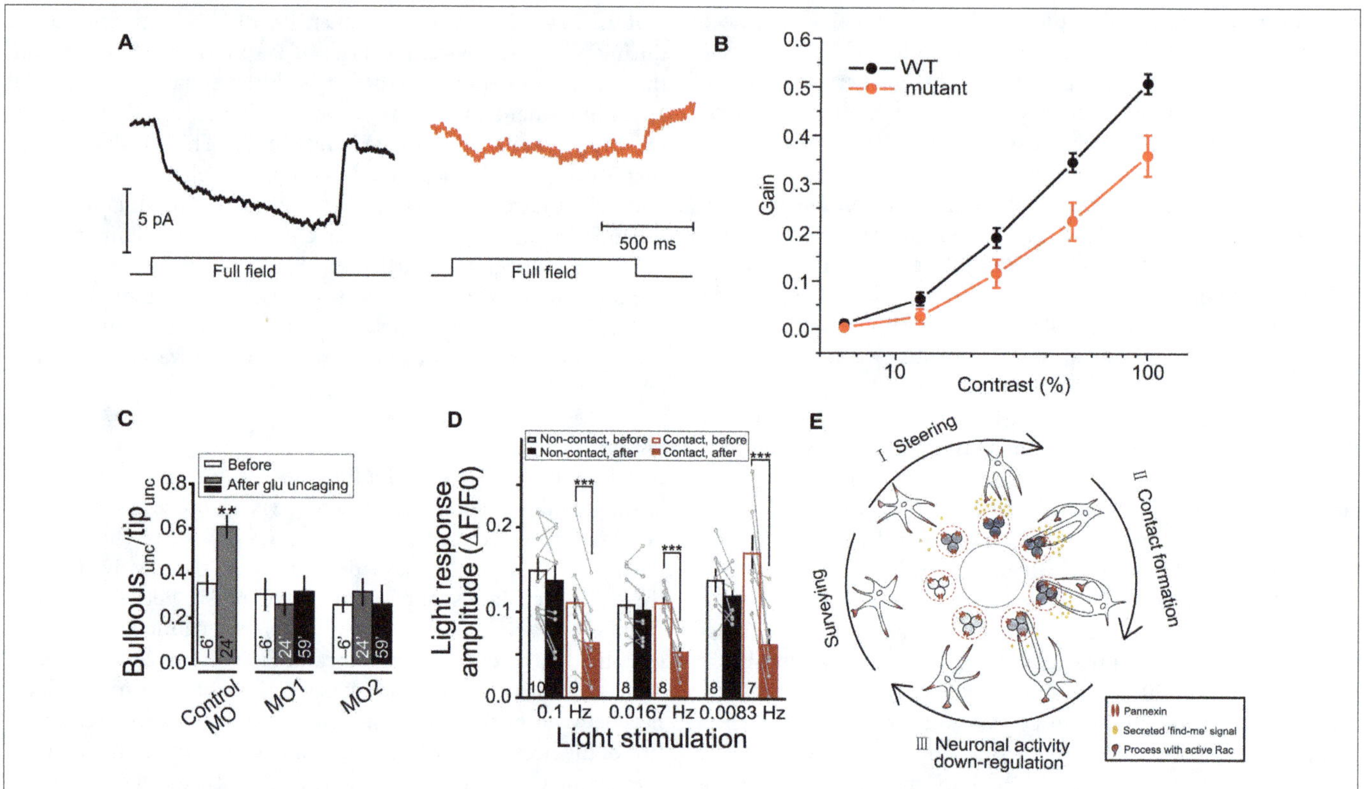

FIGURE 4 | Cx HCs and Panx1 channels have significant roles in synaptic transmission essential for vision. (A,B) Cx55.5 HCs are important for contrast sensitivity in zebrafish retina. **(A)** To measure light-induced feedback responses, cones were first saturated with a 20 μm spot of light. A full-field light flash induced an inward current in cones due to negative feedback from horizontal cells. Cx55.5 mutant (red) cones showed a decreased feedback response compared to wild-type (black), as shown in sample traces. **(B)** Optokinetic gain, as a measure of contrast sensitivity, was determined by dividing the eye movement velocity by the velocity of the stimulus over a range of contrast in zebrafish larvae. This was significantly decreased in mutant compared to wild-type zebrafish. **(C–E)** Reciprocal regulation between resting microglia and neuronal activity via Panx1 channels. **(C)** Glutamate uncaging was performed in the intact zebrafish larvae to evoke Ca^{2+} activities of tectal neurons within 20 μm around the uncaging point of 1 μm in the soma layer of the optic tectum. From the side of microglia facing the uncaging point ("unc"), the proportion of the number of bulbous normalized to all process tips ("Bulbous$_{unc}$/Tip$_{unc}$") is shown for larvae injected with splice morpholino oligonucleotides (MO) 6-min before (clear) and 24-min (gray) and 59-min (black) after uncaging. The increased in bulbous endings is shown in control MO, but abolished in Panx1 expression downregulation MO1 and MO2. **(D)** Normalized intensities of Ca^{2+} activities (light response amplitude) of tectal neurons *in vivo* evoked by moving bars at indicated frequencies are shown. Response is significantly reduced in neurons after microglial contact (red filled vs. clear bars) as compared to non-contact (black filled vs. clear bars). Numbers of neurons examined are shown on bars. **(E)** Schematic diagram showing a proposed model of microglial modulations of neuronal activity via Panx1 channels. During neuronal activity, neurons secrete "find me" signal locally (ATP being a candidate) via Panx1 channels, which steer microglial processes toward them (from "Surveying" to "I"). Bulbous endings are then formed on these processes promoting contact with neurons ("II"). Upon such contact, neuronal activity is downregulated ("III"). $^{**}p < 0.01$; $^{***}p < 0.001$. Adapted, with permission, from Klaassen et al. (2011) **(A,B)**, Li et al. (2012) **(C,D)**.

Another study revealed that neuronal activity in the optic tectum of larval zebrafish could modulate motility of resting microglia toward active neurons via Panx1 channels (Li et al., 2012). This in turn downregulates both spontaneous and visually-evoked neuronal activities under physiological condition. Specifically, using *in vivo* imaging techniques and local glutamate uncaging, the authors observed an enhanced formation of microglial processes with bulbous endings, which contact active neurons (**Figure 4C**). This was blocked by probenecid and CBX, as well as by injection of oligonucleotides which downregulate Panx1 expression, suggesting a potential contribution of Panx1 channel activity. As Panx1-mediated currents were recorded in tectal neurons but not in microglia, they concluded that this was due to the opening of neuronal Panx1 channels. The possible downstream involvement of ATP and P2 receptors was also tested. As a result, similar abolishment of activity-induced microglial bulbous ending formation was found in the presence of apyrase, an ATP-hydrolyzing enzyme, and suramin, a P2 receptor blocker. Remarkably, while microglia was found to preferentially contact neurons with initially high levels of spontaneous activity, the authors discovered that this spontaneous activity was reduced upon microglia-neuron contact. In addition, moving bar-evoked Ca^{2+} activity in these neurons was also decreased upon microglial contact (**Figure 4D**). These results suggest an interesting reciprocal modulation between neurons and microglia via neuronal Panx1 channels and ATP/P2 receptors (**Figure 4E**). The involvement of ATP as a "find-me" signal was expected, as activated microglia can be attracted toward sites of brain injury by ATP (Davalos et al., 2005). However, direct evidence of ATP release via Panx1 is yet to be shown. The authors postulated that since microglial processes mainly contact the soma, but not dendrites, of tectal neurons, such contact-induced downregulation is due to reduction of neuronal excitability rather than synaptic inputs.

CONCLUSIONS, REMAINING QUESTIONS AND FUTURE DIRECTIONS

Unlike in pathology, the functional roles of Cx HCs and Panxs in physiological situations have only recently been considered and explored. This review summarizes the latest studies supporting their neurophysiological relevance in the CNS. In particular, aspects concerning development, glucose sensing, synaptic transmission and plasticity, learning and memory, as well as vision are discussed (**Table 2**). One of the first studies showed that ATP is released via Cx43 HCs from retinal pigment epithelial cells, which promotes division and proliferation of progenitor cells in the developing retina (Pearson et al., 2005). In the hypothalamic tanycytes, the same HC subtype contributes to the uptake of extracellular glucose, triggering downstream Ca^{2+} response via ATP release (Orellana et al., 2012a). Moreover, ATP released via Cx43 HCs in astrocytes was found to increase hippocampal excitatory transmission via P2 receptors (Chever et al., 2014). Upon decrease in extracellular Ca^{2+} levels, however, ATP released from astrocytic Cx43 HCs may also participate in the generation of feedback inhibitory transmission in hippocampal interneurons (Torres et al., 2012). Functionally, these channels have also been shown to play a role in fear memory consolidation in adult rats (Stehberg et al., 2012). Apart from Cx43, Cx55.5 HCs located at the tips of retinal horizontal cell dendrites were also found to conduct light-induced feedback transmission between horizontal cells and cones, and as a functional consequence, enhance contrast sensitivity of the eye (Klaassen et al., 2011). Similar to Cx43 HCs, ATP release from Panx1 channels was found to promote neurogenesis in the postnatal mouse ventricular zone (Wicki-Stordeur et al., 2012), as well as trigger a dowregulation of neuronal excitability upon decrease in extracellular glucose in the hippocampus (Kawamura et al., 2010). Interestingly, also in the hippocampus, they reduce synaptic transmission and exert an age-dependent bidirectional control (decreasing LTP while increasing LTD) over synaptic plasticity (Prochnow et al., 2012; Ardiles et al., 2014). In the zebrafish, Panx1 channels were found to be involved in promoting microglial motility toward active neurons, leading to a decrease in neuronal activity related to vision (Li et al., 2012). Lastly, they have also been shown to be important for sensorimotor gating, as well as object recognition and spatial memory (Prochnow et al., 2012). Taken together, these important studies suggest that Cx HCs and Panxs are not only open during physiological conditions, but also play important and dynamic roles in a variety of neurophysiological processes and behavior. With these encouraging findings, it is expected that more future evidence will emerge to strengthen the physiological relevance of these large pore membrane channels in the CNS. However, in order to achieve this, some important issues need to be considered and addressed.

WHAT ARE THE COMPOSITION OF FUNCTIONAL HCs AND SUBSTANCES RELEASED?

Current understanding of the physiological roles of Cx HCs and Panx channels is largely focused on two specific subtypes, namely Cx43 and Panx1, both permeable to ATP. The lack of focus on other channel isoforms however does not rule out their potential contributions in physiology. Although Torres et al. (2012) have shown that Cx30 does not account for the hippocampal defects that they observed in Cx30/Cx43 double knockout mice, it is unclear whether Cx30 has any HC-specific effect on neurotransmission. Indeed, one major difficulty in the study of HCs is the lack of experimental tools with high isoform selectivity, as well as specificity for HCs over GJ channels. For instance, the pharmacological blocker CBX acts on Cx HCs and Panx channels, as well as on GJs (Schalper et al., 2008). Similarly, Gap26/27, which are mimetic peptides commonly used to block Cx43 HCs, can also affect GJ channels when applied for several hours (Samoilova et al., 2008). Nevertheless, a few inhibitors that block HCs but not GJs, such as Gap19/TAT-Gap19 and TAT-L2 (see **Table 1**), have been developed (Ponsaerts et al., 2010; Wang et al., 2013b). Therefore, advances in the generation of more specific mimetic peptides and even antibodies (Riquelme et al., 2013) should improve selectivity among HC isoforms. Transgenic organisms with impairment in either gap-junctional or HC functions would also be ideal. Furthermore, it is interesting that most of the studies described in this review have demonstrated effects mediated by ATP release or its metabolite, adenosine. As HCs are also permeable to signaling molecules like NAD^+, glutamate, glutathione and prostaglandin E2 (Wang et al., 2013a), which are known to be important for synaptic activity and cognitive functions, it is of interest to expand our search toward the effect of the release of these substances.

WHERE ARE FUNCTIONAL HCs LOCATED?

The fact that physiologically active HCs are detected in a number of cell types ranging from stem cells to neurons and glia implies their ubiquitous functions in the CNS (see **Table 2**). In fact, Cx43 proteins are largely expressed in vessels (Simard et al., 2003; Rouach et al., 2008) and supposedly near synapses (Ormel and Gundersen, 2011), suggesting their roles as regulators of synaptic activity, vasomotricity and hyperemia. In addition, their preferential expression on astrocytes facilitates their roles in astroglial network excitability, mediating paracrine (Torres et al., 2012) and autocrine regulations within astrocyte network (Wang et al., 2013a). Some neurons also express HC proteins like Cx36 in GABAergic interneurons or Panxs in pyramidal neurons, but their properties as HCs *in vivo* is still under investigation. Furthermore, both Cx HCs and Panx1 channels have also been found to be expressed on microglia and oligodendrocytes with activities associated with pathological ATP and glutamate release, as well as ion gradient imbalance (Orellana et al., 2011c). Their functions in physiology, however, are yet to be explored.

Other than cell-type specific expression, it is also unclear in which subcellular domain functional HCs are present. For instance, Cx43 HCs expressed on astroglial plasma membranes have been thought to serve as a reserve pool for GJ formation. Thus, it has been a challenge to differentiate between functional HCs and those presented on the membrane for GJ activities using only immunostaining. The development of new techniques and tools to not only study, but to also visualize the location of active HCs, is therefore warranted. We have recently shown that

Table 2 | Roles of connexin hemichannels and pannexin channels in neurophysiology.

Function	Channel subtype	Proposed mechanisms	Experimental models and tools	References
DEVELOPMENT AND GLUCOSE SENSING				
(1) Development				
↑ Cell division and proliferation	Cx43	ATP released from RPE cells via Cx43 HCs triggers Ca^{2+} waves through RPE cells and stimulates P2 receptors on retinal progenitor cells	Isolated neural retina from chick embryos *Pharmacological blocker (Carbenoxolone); mimetic peptide (Gap26)*	Pearson et al., 2005
	Panx1	ATP released via Panx1 channels stimulates NSC/NPC via P2 receptors	Neuro2a neuroblastoma cell line, cultures of postnatal mouse ventricular zone NSC/NPCs *Pharmacological blocker (Probenecid); Panx1 siRNA*	Wicki-Stordeur et al., 2012
(2) Glucose sensing and signal transduction				
↑ Hypothalamic glucose-sensing	Cx43	Glucose uptake and glycolysis opens Cx43 HCs allowing the release of ATP, which activates local P2Y1 receptors and $[Ca^{2+}]_i$ increase	Cultured rat hypothalamic tanycytes *Pharmacological blockers (La^{3+}, $Cx43^{E2}$, Probenecid); mimetic peptides (Gap26, ^{10}panx)*	Orellana et al., 2012a
↑ Metabolic autocrine regulation	Panx1	Decrease in extracellular glucose triggers the opening of Panx1 channels through which ATP is released	Acute hippocampal slices from juvenile rats *Pharmacological blocker (Carbenoxolone, Octanol); mimetic peptide (^{10}panx)*	Kawamura et al., 2010
SYNAPTIC TRANSMISSION AND PLASTICITY				
(1) Neuronal excitability and synaptic transmission				
↑ Synaptic transmission	Cx43	ATP released via Cx43 HCs in astrocytes promotes excitatory synaptic transmission via P2 receptors	Acute hippocampal slices from juvenile mice *Transgenic mice: $Cx43^{fl/fl}$:hGFAP-Cre = Astrocytic Cx43 conditional knockout; pharmacological blocker (Carbenoxolone), mimetic peptide (Gap26, ^{10}panx)*	Chever et al., 2014
↑ Feedback inhibitory synaptic transmission	Cx43	Decrease in $[Ca^{2+}]_e$ during excitatory transmission opens Cx43 HCs in astrocytes allowing the release of ATP, which triggers depolarization and firing of inhibitory interneurons and a slow Ca^{2+} wave in astrocytes both via P2Y receptors	Acute hippocampal slices from juvenile mice *Transgenic mice: (1) $Cx30^{-/-}Cx43^{fl/fl}$:hGFAP-Cre = Cx30 total- and Cx43 astrocytic conditional knockout; (2) $Cx30^{-/-}Cx43^{fl/fl}$ = Cx30 total knockout; and 3) $Cx43^{+/G138Rfl}$:hGFAP-Cre mice = enhanced astrocytic Cx43 HC activity; pharmacological blocker (Carbenoxolone)*	Torres et al., 2012
↓ Neuronal excitability	Panx1	Decrease in extracellular glucose triggers the opening of Panx1 channels through which ATP is released. Its metabolite adenosine then activates adenosine A1 receptors and opens K_{ATP} channels leading to decreased neuronal excitability	Acute hippocampal slices from juvenile rats *Pharmacological blocker (Carbenoxolone, Octanol); mimetic peptide (^{10}panx)*	Kawamura et al., 2010
↓ Synaptic transmission	Panx1	Adenosine metabolized from ATP released via Panx1 channels decreases neurotransmission	Acute hippocampal slices from adult mice *Transgenic mice: $Panx1^{-/-}$ (Panx1 knockout)*	Prochnow et al., 2012
	Panx1	Panx1 channels decrease neurotransmission in adult but not young mice	Acute hippocampal slices from young and adult mice *Transgenic mice: $Panx1^{-/-}$ (Panx1 knockout); Pharmacological blocker (Probenecid)*	Ardiles et al., 2014
↑ Light-induced feedback transmission	Cx55.5	Cx55.5 HC current at the tips of the horizontal cell dendrites induces local voltage drop near voltage dependent Ca^{2+} channels of cones mediating feedback ephaptic transmission	Isolated retina, dissociated horizontal cells and cones from zebrafish *Transgenic zebrafish: Cx55.5 mutant (no functional Cx55.5 protein formation)*	Klaassen et al., 2011

(Continued)

Table 2 | Continued

Function	Channel subtype	Proposed mechanisms	Experimental models and tools	References
↓ Spontaneous neuronal activity	Panx1	Panx1 channels on active neurons promote microglial motility via ATP/P2 receptors, which in turn downregulates spontaneous neuronal activity	Zebrafish larvae *Pharmacological blocker (Carbenoxolone, Probenecid); Panx1 downregulation oligonucleotides*	Li et al., 2012
(2) Synaptic plasticity				
↓ LTP	Panx1	Adenosine metabolized from ATP released via Panx1 channels suppresses the induction of LTP	Acute hippocampal slices from adult mice *Transgenic mice: Panx1$^{-/-}$ (Panx1 knockout); Pharmacological blocker (mefloquine)*	Prochnow et al., 2012
↓ LTP, ↑ LTD	Panx1	Panx1 channels control the threshold of the bidirectional induction of synaptic plasticity in adult but not in young mice	Acute hippocampal slices from young and adult mice *Transgenic mice: Panx1$^{-/-}$ (Panx1 knockout); Pharmacological blocker (Probenecid)*	Ardiles et al., 2014
INTEGRATIVE FUNCTIONS				
(1) Sensorimotor gating				
↓ Sensorimotor gating capabilities	Panx1	Panx1 channels inhibit pre-pulse inhibition of the acoustic startle response	Mice (adult) *Transgenic mice: Panx1$^{-/-}$ (Panx1 knockout)*	Prochnow et al., 2012
(2) Learning and memory				
↑ Fear memory consolidation	Cx43	Gliotransmitters release through astrocytic Cx43 HCs promotes the consolidation of fear memory	Rats (adult) *Mimetic peptides injection into basolateral amygdala (Gap26, TAT-Cx43L2)*	Stehberg et al., 2012
↑ Object recognition memory	Panx1	Panx1 channels improve the ability of a mouse to recognize an object as familiar	Mice (adult) *Transgenic mice: Panx1$^{-/-}$ (Panx1 knockout)*	Prochnow et al., 2012
↑ Spatial memory	Panx1	Panx1 channels enhance the ability of a mouse to remember the location of a treat	Mice (adult) *Transgenic mice: Panx1$^{-/-}$ (Panx1 knockout)*	Prochnow et al., 2012
(3) Vision				
↑ Contrast sensitivity	Cx55.5	Cx55.5 HC mediated feedback from horizontal cells to cones leads to contrast enhancement	Zebrafish larvae *Transgenic zebrafish: Cx55.5 mutant zebrafish (no functional Cx55.5 protein formation)*	Klaassen et al., 2011
↓ Visually-evoked neuronal activity	Panx1	Microglial contact promoted by Panx1 channels on active neurons via ATP/P2 receptors downregulates evoked neuronal activity	Zebrafish larvae *Pharmacological blocker (Carbenoxolone, Probenecid); Panx1 downregulation oligonucleotides*	Li et al., 2012

Cx, Connexin; fl, floxed; HC, hemichannel; G138R, single point mutation of glycine 138 to arginine of Cx43; LTD, long-term depression; LTP, long-term potentiation; NSC/NPS, neural stem cells and progenitor cells; Panx, Pannexin; RPE, retinal pigment epithelium.

astrocytes can extend their fine processes near or even within synaptic clefts (Pannasch et al., 2014), suggesting that the expression of functional HCs on these processes would allow them to more efficiently modulate neuronal transmission due to their proximity to targeted sites. Finally, if HCs are concentrated on subcellular domains, it would be interesting to investigate how they are targeted and trafficked to these locations.

WHEN ARE HCs OPEN?

While numerous *in vitro* studies have been performed to characterize the biophysical properties of these large membrane pores and to determine aspects that would trigger their opening, HC activation *in situ* or *in vivo*, particularly during physiological conditions, is less understood. Torres et al. (2012) proposed that the increase in glutamate and decrease in extracellular Ca^{2+} occurring during synaptic transmission favor astroglial Cx43 HC opening, suggesting that the conductance or open probability or number of active channels, or possibly recruitment of HCs near synaptic cleft, is dependent on neuronal activity. Although they also used high frequency stimulation to mimic neuronal activity, it is not clear as to whether these manipulations represent conditions *in vivo* and if the suggested mechanism is involved in the modulation of cognitive functions (Stehberg et al., 2012). Apart from extracellular Ca^{2+}, an increase in intracellular Ca^{2+} within the physiological range has also been found to open Cx43 and Cx32 HCs, as assessed by ATP release and dye uptake (De Vuyst et al., 2006, 2009). On the other hand, studies on Panx1 channels have reported cognitive and behavioral deficits in knockout mice (Prochnow et al., 2012). However, long-term effect due to chronic deletion of the gene cannot be ruled out. The development in the application of current techniques *in vivo*, especially those that allow acute manipulation or detection of channel opening and closing, would be useful for addressing these issues. More importantly, the use of such *in vivo* systems will offer the benefit of behavioral studies, cognitive tests as well as the characterization of developmental defects.

It is conceivable that, due to their large pore size and conductance, the opening of Cx HCs and Panx channels is different from other plasma membrane ion channels and likely requires tighter control to preserve cell integrity. This more stringent regulation could be related to minimizing channel opening duration. Perhaps various modulators could differentially open the same channel with different timescales. Transient channel opening to release substances in short pulses may also prevent a sudden loss of intracellular substances. In contrast, Cx HC and Panx channels opening might merely be a passive response toward transient changes in concentration gradients of molecules like glucose, lactate, or ATP in an activity-dependent manner contributing to the maintenance of basal homeostasis. Additionally, HCs are permeable to water and could play an important role in regulation of intracellular osmolarity and cell volume (Quist et al., 2000). To test these possibilities, it becomes essential to focus on the moment-to-moment action of HC activity on brain processes like neurotransmission.

HOW ARE HC OPENING AND SUBSTANCE RELEASE REGULATED FOR DOWNSTREAM EFFECTS?

HCs are rather non-selective large pore membrane channels which, if not regulated, could lead to loss of important intracellular molecules and more seriously cell death. Thus, the open probability of these channels, their membrane distribution, trafficking and proximity to target receptors are crucial aspects requiring tight controls. While the opening and closing of Cx GJ channels are tightly regulated (Dermietzel and Spray, 1993; Pannasch and Rouach, 2013), it is not well understood whether the same pathways also apply to HCs. More importantly, it is still unclear which molecular mechanisms underlie the formation of HCs over GJ channels and vice versa. It has been shown in pathological situations that GJ channels and HCs can display similar or opposite responses depending on differential activation of intracellular pathways. For example, lipopolysaccharide and basic fibroblast growth factor have been found to differentially influence Cx HCs and GJ channels (De Vuyst et al., 2007). In relation to this, aspects of how they are targeted to and maintained on plasma membranes as single channels are also fundamental to understand physiological HC properties.

Interestingly, Cx43 HCs and Panx1 channels seem to have opposite roles on neuronal activity and behavior. While Stehberg et al. (2012) showed that inhibiting Cx43 HCs with mimetic peptides *in vivo* strongly impaired fear memory consolidation, two other subsequent studies on Panx1$^{-/-}$ mice demonstrated a strong enhancement in LTP and an abolishment in LTD (Prochnow et al., 2012; Ardiles et al., 2014). This suggests that Cx43 and Panx1 HCs are involved in two distinct neuromodulatory pathways. One can also argue that such differential effect might rely on differences in experimental models using acute blockage of Cx43 HC as opposed to Panx1$^{-/-}$ mice with chronic deletion of channels. Indeed, the effects on LTP and LTD were not observed in young mice, but only in adults, in which they also observed a compensatory transcriptional up-regulation of metabotropic glutamate receptor 4, proposed to be due to chronic decrease in extracellular ATP. Furthermore, Stehberg et al. only showed that the release via Cx43 HCs involved one or more gliotransmitters, whereas Panx1$^{-/-}$ defects were specifically rescued by adenosine. This raises the questions of whether these HCs could exhibit cell-type specific regulations (in this case neurons vs. astrocytes) or even release different combinations of molecules under varying circumstances. In addition, the quantity of release and the proximity to targeted sites, as well as the nature and localization of receptors, also need to be taken into account. It is of great importance that such opposing effects between HC subtypes are clarified such that any differential modulations of downstream effects can be better understood. The use of global and conditional knockout systems will provide valuable insights into whether the observed effects are cell-type or developmentally specific. It would also be important to determine whether HC uptake and release of various transmitters, ions and metabolites are co-regulated and whether it is synergistic with other release pathways. This may enhance the efficiency of cell-to-cell communication and allow dynamic actions on downstream mechanisms.

On a more complex level, it would be of great interest to determine if HC activity could display plasticity similar to postsynaptic glutamate receptors. It is probable that in response to neuronal activity, HC conductance or phosphorylation is altered leading to differential expression/presentation and distribution accross plasma membrane. Indeed, several Cx HCs, like Cx43 and Cx46, have been shown to have increased open probability depending on their phosphorylation state (Li et al., 1996; Ngezahayo et al., 1998; Sáez et al., 2005). Lastly, while most physiological actions of HCs thus far involve ATP release, it is necessary to clarify whether different channels indeed have specific and/or exclusive roles in certain biophysical processes contributing to neurotransmission and behavior. If so, what underlying mechanisms would allow them to be differentially regulated. As ATP is an important substance regulating neurotransmission and many basal processes, similar roles of HCs on other functions like hyperemia (Pelligrino et al., 2011) and brain oscillations (Schulz et al., 2012) are likely and yet to be characterized.

While there has been decades of research to establish the pathological contributions of HCs in the CNS, our understanding of their potential roles in physiological processes has just begun. Indeed, as more encouraging findings emerge, many fundamental issues must be addressed. Nevertheless, it is hoped that by unraveling the properties of these membrane channels in physiology, we will also be able to gain insights into their functions in pathologies and to explore their potential roles as therapeutic targets.

ACKNOWLEDGMENTS

This work was supported by grants from the French Research Agency (ANR, Programme Blanc), City of Paris (Programme Emergence) INSERM, CNRS and Collège de France to Nathalie Rouach, Labex Memolife to Giselle Cheung and FRM (Fondation pour la Recherche Médicale) to Oana Chever.

REFERENCES

Amzica, F., Massimini, M., and Manfridi, A. (2002). Spatial buffering during slow and paroxysmal sleep oscillations in cortical networks of glial cells *in vivo*. *J. Neurosci.* 22, 1042–1053.

Anselmi, F., Hernandez, V. H., Crispino, G., Seydel, A., Ortolano, S., Roper, S. D., et al. (2008). ATP release through connexin hemichannels and gap junction transfer of second messengers propagate Ca^{2+} signals across the inner ear. *Proc. Natl. Acad. Sci. U.S.A.* 105, 18770–18775. doi: 10.1073/pnas.0800793105

Ardiles, A. O., Flores-Muñoz, C., Toro-Ayala, G., Cárdenas, A. M., Palacios, A. G., Muñoz, P., et al. (2014). Pannexin 1 regulates bidirectional hippocampal synaptic plasticity in adult mice. *Front. Cell Neurosci.* 8:326. doi: 10.3389/fncel.2014.00326

Baranova, A., Ivanov, D., Petrash, N., Pestova, A., Skoblov, M., Kelmanson, I., et al. (2004). The mammalian pannexin family is homologous to the invertebrate innexin gap junction proteins. *Genomics* 83, 706–716. doi: 10.1016/j.ygeno.2003.09.025

Bargiotas, P., Krenz, A., Hormuzdi, S. G., Ridder, D. A., Herb, A., Barakat, W., et al. (2011). Pannexins in ischemia-induced neurodegeneration. *Proc. Natl. Acad. Sci. U.S.A.* 108, 20772–20777. doi: 10.1073/pnas.1018262108

Bennett, M. V., Garre, J. M., Orellana, J. A., Bukauskas, F. F., Nedergaard, M., and Sáez, J. C. (2012). Connexin and pannexin hemichannels in inflammatory responses of glia and neurons. *Brain Res.* 1487, 3–15. doi: 10.1016/j.brainres.2012.08.042

Berg, D. A., Belnoue, L., Song, H., and Simon, A. (2013). Neurotransmitter-mediated control of neurogenesis in the adult vertebrate brain. *Development* 140, 2548–2561. doi: 10.1242/dev.088005

Boassa, D., Qiu, F., Dahl, G., and Sosinsky, G. (2008). Trafficking dynamics of glycosylated pannexin 1 proteins. *Cell Commun. Adhes.* 15, 119–132. doi: 10.1080/15419060802013885

Bosher, S. K., and Warren, R. L. (1978). Very low calcium content of cochlear endolymph, an extracellular fluid. *Nature* 273, 377–378. doi: 10.1038/273377a0

Bruzzone, R., Barbe, M. T., Jakob, N. J., and Monyer, H. (2005). Pharmacological properties of homomeric and heteromeric pannexin hemichannels expressed in Xenopus oocytes. *J. Neurochem.* 92, 1033–1043. doi: 10.1111/j.1471-4159.2004.02947.x

Bruzzone, R., Hormuzdi, S. G., Barbe, M. T., Herb, A., and Monyer, H. (2003). Pannexins, a family of gap junction proteins expressed in brain. *Proc. Natl. Acad. Sci. U.S.A.* 100, 13644–13649. doi: 10.1073/pnas.2233464100

Bukauskas, F. F., Kreuzberg, M. M., Rackauskas, M., Bukauskiene, A., Bennett, M. V., Verselis, V. K., et al. (2006). Properties of mouse connexin 30.2 and human connexin 31.9 hemichannels: implications for atrioventricular conduction in the heart. *Proc. Natl. Acad. Sci. U.S.A.* 103, 9726–9731. doi: 10.1073/pnas.0603372103

Burdakov, D., Luckman, S. M., and Verkhratsky, A. (2005). Glucose-sensing neurons of the hypothalamus. *Philos. Trans. R. Soc. Lond. B Biol. Sci.* 360, 2227–2235. doi: 10.1098/rstb.2005.1763

Chever, O., Lee, C. Y., and Rouach, N. (2014). Astroglial connexin43 hemichannels tune Basal excitatory synaptic transmission. *J. Neurosci.* 34, 11228–11232. doi: 10.1523/JNEUROSCI.0015-14.2014

Contreras, J. E., Sáez, J. C., Bukauskas, F. F., and Bennett, M. V. (2003). Gating and regulation of connexin 43 (Cx43) hemichannels. *Proc. Natl. Acad. Sci. U.S.A.* 100, 11388–11393. doi: 10.1073/pnas.1434298100

Contreras, J. E., Sanchez, H. A., Veliz, L. P., Bukauskas, F. F., Bennett, M. V., and Sáez, J. C. (2004). Role of connexin-based gap junction channels and hemichannels in ischemia-induced cell death in nervous tissue. *Brain Res. Brain Res. Rev.* 47, 290–303. doi: 10.1016/j.brainresrev.2004.08.002

Cotrina, M. L., Lin, J. H., Alves-Rodrigues, A., Liu, S., Li, J., Azmi-Ghadimi, H., et al. (1998). Connexins regulate calcium signaling by controlling ATP release. *Proc. Natl. Acad. Sci. U.S.A.* 95, 15735–15740. doi: 10.1073/pnas.95.26.15735

Dallerac, G., Chever, O., and Rouach, N. (2013). How do astrocytes shape synaptic transmission? Insights from electrophysiology. *Front. Cell. Neurosci.* 7:159. doi: 10.3389/fncel.2013.00159

Davalos, D., Grutzendler, J., Yang, G., Kim, J. V., Zuo, Y., Jung, S., et al. (2005). ATP mediates rapid microglial response to local brain injury *in vivo*. *Nat. Neurosci.* 8, 752–758. doi: 10.1038/nn1472

De Bock, M., Wang, N., Bol, M., Decrock, E., Ponsaerts, R., Bultynck, G., et al. (2012). Connexin 43 hemichannels contribute to cytoplasmic Ca^{2+} oscillations by providing a bimodal Ca^{2+}-dependent Ca^{2+} entry pathway. *J. Biol. Chem.* 287, 12250–12266. doi: 10.1074/jbc.M111.299610

Dermietzel, R., Kremer, M., Paputsoglu, G., Stang, A., Skerrett, I. M., Gomes, D., et al. (2000). Molecular and functional diversity of neural connexins in the retina. *J. Neurosci.* 20, 8331–8343.

Dermietzel, R., and Spray, D. C. (1993). Gap junctions in the brain: where, what type, how many and why? *Trends Neurosci.* 16, 186–192. doi: 10.1016/0166-2236(93)90151-B

De Vuyst, E., Decrock, E., Cabooter, L., Dubyak, G. R., Naus, C. C., Evans, W. H., et al. (2006). Intracellular calcium changes trigger connexin 32 hemichannel opening. *EMBO J.* 25, 34–44. doi: 10.1038/sj.emboj.7600908

De Vuyst, E., Decrock, E., De Bock, M., Yamasaki, H., Naus, C. C., Evans, W. H., et al. (2007). Connexin hemichannels and gap junction channels are differentially influenced by lipopolysaccharide and basic fibroblast growth factor. *Mol. Biol. Cell* 18, 34–46. doi: 10.1091/mbc.E06-03-0182

De Vuyst, E., Wang, N., Decrock, E., De Bock, M., Vinken, M., Van Moorhem, M., et al. (2009). $Ca^{(2+)}$ regulation of connexin 43 hemichannels in C6 glioma and glial cells. *Cell Calcium* 46, 176–187. doi: 10.1016/j.ceca.2009.07.002

D'hondt, C., Ponsaerts, R., De Smedt, H., Bultynck, G., and Himpens, B. (2009). Pannexins, distant relatives of the connexin family with specific cellular functions? *Bioessays* 31, 953–974. doi: 10.1002/bies.200800236

Dobrowolski, R., Sasse, P., Schrickel, J. W., Watkins, M., Kim, J. S., Rackauskas, M., et al. (2008). The conditional connexin43G138R mouse mutant represents a new model of hereditary oculodentodigital dysplasia in humans. *Hum. Mol. Genet.* 17, 539–554. doi: 10.1093/hmg/ddm329

Ebihara, L., Liu, X., and Pal, J. D. (2003). Effect of external magnesium and calcium on human connexin46 hemichannels. *Biophys. J.* 84, 277–286. doi: 10.1016/S0006-3495(03)74848-6

Ebihara, L., and Steiner, E. (1993). Properties of a nonjunctional current expressed from a rat connexin46 cDNA in Xenopus oocytes. *J. Gen. Physiol.* 102, 59–74. doi: 10.1085/jgp.102.1.59

Ebihara, L., Tong, J. J., Vertel, B., White, T. W., and Chen, T. L. (2011). Properties of connexin 46 hemichannels in dissociated lens fiber cells. *Invest. Ophthalmol. Vis. Sci.* 52, 882–889. doi: 10.1167/iovs.10-6200

Essenfelder, G. M., Bruzzone, R., Lamartine, J., Charollais, A., Blanchet-Bardon, C., Barbe, M. T., et al. (2004). Connexin30 mutations responsible for hidrotic ectodermal dysplasia cause abnormal hemichannel activity. *Hum. Mol. Genet.* 13, 1703–1714. doi: 10.1093/hmg/ddh191

Eugenin, E. A., Basilio, D., Sáez, J. C., Orellana, J. A., Raine, C. S., Bukauskas, F., et al. (2012). The role of gap junction channels during physiologic and pathologic conditions of the human central nervous system. *J. Neuroimmune Pharmacol.* 7, 499–518. doi: 10.1007/s11481-012-9352-5

Figiel, M., Allritz, C., Lehmann, C., and Engele, J. (2007). Gap junctional control of glial glutamate transporter expression. *Mol. Cell. Neurosci.* 35, 130–137. doi: 10.1016/j.mcn.2007.02.009

Giaume, C., Leybaert, L., Naus, C. C., and Sáez, J. C. (2013). Connexin and pannexin hemichannels in brain glial cells: properties, pharmacology, and roles. *Front. Pharmacol.* 4:88. doi: 10.3389/fphar.2013.00088

Gonzalez, D., Gomez-Hernandez, J. M., and Barrio, L. C. (2006). Species specificity of mammalian connexin-26 to form open voltage-gated hemichannels. *FASEB J.* 20, 2329–2338. doi: 10.1096/fj.06-5828com

Grifa, A., Wagner, C. A., D'ambrosio, L., Melchionda, S., Bernardi, F., Lopez-Bigas, N., et al. (1999). Mutations in GJB6 cause nonsyndromic autosomal dominant deafness at DFNA3 locus. *Nat. Genet.* 23, 16–18. doi: 10.1038/12612

Han, K. S., Woo, J., Park, H., Yoon, B. J., Choi, S., and Lee, C. J. (2013). Channel-mediated astrocytic glutamate release via Bestrophin-1 targets synaptic NMDARs. *Mol. Brain* 6:4. doi: 10.1186/1756-6606-6-4

Han, Y., Yu, H. X., Sun, M. L., Wang, Y., Xi, W., and Yu, Y. Q. (2014). Astrocyte-restricted disruption of connexin-43 impairs neuronal plasticity in mouse barrel cortex. *Eur. J. Neurosci.* 39, 35–45. doi: 10.1111/ejn.12394

Hansen, D. B., Braunstein, T. H., Nielsen, M. S., and Macaulay, N. (2014). Distinct permeation profiles of the connexin 30 and 43 hemichannels. *FEBS Lett.* 588, 1446–1457. doi: 10.1016/j.febslet.2014.01.036

Hofer, A., and Dermietzel, R. (1998). Visualization and functional blocking of gap junction hemichannels (connexons) with antibodies against external loop domains in astrocytes. *Glia* 24, 141–154.

Iglesias, R., Dahl, G., Qiu, F., Spray, D. C., and Scemes, E. (2009). Pannexin 1: the molecular substrate of astrocyte "hemichannels". *J. Neurosci.* 29, 7092–7097. doi: 10.1523/JNEUROSCI.6062-08.2009

Jourdain, P., Bergersen, L. H., Bhaukaurally, K., Bezzi, P., Santello, M., Domercq, M., et al. (2007). Glutamate exocytosis from astrocytes controls synaptic strength. *Nat. Neurosci.* 10, 331–339. doi: 10.1038/nn1849

Kamermans, M., and Fahrenfort, I. (2004). Ephaptic interactions within a chemical synapse: hemichannel-mediated ephaptic inhibition in the retina. *Curr. Opin. Neurobiol.* 14, 531–541. doi: 10.1016/j.conb.2004.08.016

Kang, J., Kang, N., Lovatt, D., Torres, A., Zhao, Z., Lin, J., et al. (2008). Connexin 43 hemichannels are permeable to ATP. *J. Neurosci.* 28, 4702–4711. doi: 10.1523/JNEUROSCI.5048-07.2008

Kawamura, M. Jr., Ruskin, D. N., and Masino, S. A. (2010). Metabolic autocrine regulation of neurons involves cooperation among pannexin hemichannels, adenosine receptors, and KATP channels. *J. Neurosci.* 30, 3886–3895. doi: 10.1523/JNEUROSCI.0055-10.2010

Kienitz, M. C., Bender, K., Dermietzel, R., Pott, L., and Zoidl, G. (2011). Pannexin 1 constitutes the large conductance cation channel of cardiac myocytes. *J. Biol. Chem.* 286, 290–298. doi: 10.1074/jbc.M110.163477

Kimelberg, H. K. (2004). Increased release of excitatory amino acids by the actions of ATP and peroxynitrite on volume-regulated anion channels (VRACs) in astrocytes. *Neurochem. Int.* 45, 511–519. doi: 10.1016/j.neuint.2003.11.002

Klaassen, L. J., Sun, Z., Steijaert, M. N., Bolte, P., Fahrenfort, I., Sjoerdsma, T., et al. (2011). Synaptic transmission from horizontal cells to cones is impaired by loss of connexin hemichannels. *PLoS Biol.* 9:e1001107. doi: 10.1371/journal.pbio.1001107

Lai, C. P., Bechberger, J. F., Thompson, R. J., Macvicar, B. A., Bruzzone, R., and Naus, C. C. (2007). Tumor-suppressive effects of pannexin 1 in C6 glioma cells. *Cancer Res.* 67, 1545–1554. doi: 10.1158/0008-5472.CAN-06-1396

Ledoux, J. (2007). The amygdala. *Curr. Biol.* 17, R868–R874. doi: 10.1016/j.cub.2007.08.005

Levin, B. E., Routh, V. H., Kang, L., Sanders, N. M., and Dunn-Meynell, A. A. (2004). Neuronal glucosensing: what do we know after 50 years? *Diabetes* 53, 2521–2528. doi: 10.2337/diabetes.53.10.2521

Leybaert, L., and Sanderson, M. J. (2012). Intercellular $Ca^{(2+)}$ waves: mechanisms and function. *Physiol. Rev.* 92, 1359–1392. doi: 10.1152/physrev.00029.2011

Li, H., Liu, T. F., Lazrak, A., Peracchia, C., Goldberg, G. S., Lampe, P. D., et al. (1996). Properties and regulation of gap junctional hemichannels in the plasma membranes of cultured cells. *J. Cell Biol.* 134, 1019–1030. doi: 10.1083/jcb.134.4.1019

Li, Y., Du, X. F., Liu, C. S., Wen, Z. L., and Du, J. L. (2012). Reciprocal regulation between resting microglial dynamics and neuronal activity *in vivo*. *Dev. Cell* 23, 1189–1202. doi: 10.1016/j.devcel.2012.10.027

Liebmann, M., Stahr, A., Guenther, M., Witte, O. W., and Frahm, C. (2013). Astrocytic Cx43 and Cx30 differentially modulate adult neurogenesis in mice. *Neurosci. Lett.* 545, 40–45. doi: 10.1016/j.neulet.2013.04.013

Locovei, S., Wang, J., and Dahl, G. (2006). Activation of pannexin 1 channels by ATP through P2Y receptors and by cytoplasmic calcium. *FEBS Lett.* 580, 239–244. doi: 10.1016/j.febslet.2005.12.004

Loewenstein, W. R. (1981). Junctional intercellular communication: the cell-to-cell membrane channel. *Physiol. Rev.* 61, 829–913.

Lutz, S. E., Zhao, Y., Gulinello, M., Lee, S. C., Raine, C. S., and Brosnan, C. F. (2009). Deletion of astrocyte connexins 43 and 30 leads to a dysmyelinating phenotype and hippocampal CA1 vacuolation. *J. Neurosci.* 29, 7743–7752. doi: 10.1523/JNEUROSCI.0341-09.2009

Ma, W., Compan, V., Zheng, W., Martin, E., North, R. A., Verkhratsky, A., et al. (2012). Pannexin 1 forms an anion-selective channel. *Pflugers Arch.* 463, 585–592. doi: 10.1007/s00424-012-1077-z

Macvicar, B. A., and Thompson, R. J. (2010). Non-junction functions of pannexin-1 channels. *Trends Neurosci.* 33, 93–102. doi: 10.1016/j.tins.2009.11.007

Makarenkova, H. P., and Shestopalov, V. I. (2014). The role of pannexin hemichannels in inflammation and regeneration. *Front. Physiol.* 5:63. doi: 10.3389/fphys.2014.00063

Massimini, M., and Amzica, F. (2001). Extracellular calcium fluctuations and intracellular potentials in the cortex during the slow sleep oscillation. *J. Neurophysiol.* 85, 1346–1350.

Moser, E. I., Krobert, K. A., Moser, M. B., and Morris, R. G. (1998). Impaired spatial learning after saturation of long-term potentiation. *Science* 281, 2038–2042. doi: 10.1126/science.281.5385.2038

Mylvaganam, S., Ramani, M., Krawczyk, M., and Carlen, P. L. (2014). Roles of gap junctions, connexins, and pannexins in epilepsy. *Front. Physiol.* 5:172. doi: 10.3389/fphys.2014.00172

Neary, J. T., and Zhu, Q. (1994). Signaling by ATP receptors in astrocytes. *Neuroreport* 5, 1617–1620. doi: 10.1097/00001756-199408150-00019

Ngezahayo, A., Zeilinger, C., Todt, I., Marten, I., and Kolb, H. A. (1998). Inactivation of expressed and conducting rCx46 hemichannels by phosphorylation. *Pflugers Arch.* 436, 627–629. doi: 10.1007/s004240050681

Orellana, J. A., Avendano, B. C., and Montero, T. D. (2014). Role of connexins and pannexins in ischemic stroke. *Curr. Med. Chem.* 21, 2165–2182. doi: 10.2174/0929867321666131228191714

Orellana, J. A., Figueroa, X. F., Sanchez, H. A., Contreras-Duarte, S., Velarde, V., and Sáez, J. C. (2011a). Hemichannels in the neurovascular unit and white matter under normal and inflamed conditions. *CNS Neurol. Disord. Drug Targets* 10, 404–414. doi: 10.2174/187152711794653869

Orellana, J. A., Froger, N., Ezan, P., Jiang, J. X., Bennett, M. V., Naus, C. C., et al. (2011b). ATP and glutamate released via astroglial connexin 43 hemichannels mediate neuronal death through activation of pannexin 1 hemichannels. *J. Neurochem.* 118, 826–840. doi: 10.1111/j.1471-4159.2011.07210.x

Orellana, J. A., Giaume, C., and Sáez, J. C. (2011c). "Role of connexin hemichannels in neurodegeneration," in *Neurodegenerative Diseases—Processes, Precvention, Protection and Monitoring,* ed R. Chuen-Chung Chang (Rijeka: In Tech.), 235–254.

Orellana, J. A., Martinez, A. D., and Retamal, M. A. (2013). Gap junction channels and hemichannels in the CNS: regulation by signaling molecules. *Neuropharmacology* 75, 567–582. doi: 10.1016/j.neuropharm.2013.02.020

Orellana, J. A., Saez, P. J., Cortes-Campos, C., Elizondo, R. J., Shoji, K. F., Contreras-Duarte, S., et al. (2012a). Glucose increases intracellular free $Ca^{(2+)}$ in tanycytes via ATP released through connexin 43 hemichannels. *Glia* 60, 53–68. doi: 10.1002/glia.21246

Orellana, J. A., Von Bernhardi, R., Giaume, C., and Sáez, J. C. (2012b). Glial hemichannels and their involvement in aging and neurodegenerative diseases. *Rev. Neurosci.* 23, 163–177. doi: 10.1515/revneuro-2011-0065

Ormel, L., and Gundersen, V. (2011). "Immunogold localization of connexin 43 in astrocytes: localization of hemichannels close to synapses," in *IBRO 8th World Congress Abstract C049* (Florence).

Pannasch, U., Freche, D., Dallerac, G., Ghezali, G., Escartin, C., Ezan, P., et al. (2014). Connexin 30 sets synaptic strength by controlling astroglial synapse invasion. *Nat. Neurosci.* 17, 549–558. doi: 10.1038/nn.3662

Pannasch, U., and Rouach, N. (2013). Emerging role for astroglial networks in information processing: from synapse to behavior. *Trends Neurosci.* 36, 405–417. doi: 10.1016/j.tins.2013.04.004

Pannasch, U., Vargova, L., Reingruber, J., Ezan, P., Holcman, D., Giaume, C., et al. (2011). Astroglial networks scale synaptic activity and plasticity. *Proc. Natl. Acad. Sci. U.S.A.* 108, 8467–8472. doi: 10.1073/pnas.1016650108

Parpura, V., Scemes, E., and Spray, D. C. (2004). Mechanisms of glutamate release from astrocytes: gap junction "hemichannels", purinergic receptors and exocytotic release. *Neurochem. Int.* 45, 259–264. doi: 10.1016/j.neuint.2003.12.011

Pascual, O., Ben Achour, S., Rostaing, P., Triller, A., and Bessis, A. (2012). Microglia activation triggers astrocyte-mediated modulation of excitatory neurotransmission. *Proc. Natl. Acad. Sci. U.S.A.* 109, E197–205. doi: 10.1073/pnas.1111098109

Paul, D. L., Ebihara, L., Takemoto, L. J., Swenson, K. I., and Goodenough, D. A. (1991). Connexin46, a novel lens gap junction protein, induces voltage-gated currents in nonjunctional plasma membrane of Xenopus oocytes. *J. Cell Biol.* 115, 1077–1089. doi: 10.1083/jcb.115.4.1077

Pearson, R. A., Dale, N., Llaudet, E., and Mobbs, P. (2005). ATP released via gap junction hemichannels from the pigment epithelium regulates neural retinal progenitor proliferation. *Neuron* 46, 731–744. doi: 10.1016/j.neuron.2005.04.024

Pelligrino, D. A., Vetri, F., and Xu, H. L. (2011). Purinergic mechanisms in gliovascular coupling. *Semin. Cell Dev. Biol.* 22, 229–236. doi: 10.1016/j.semcdb.2011.02.010

Penuela, S., Bhalla, R., Gong, X. Q., Cowan, K. N., Celetti, S. J., Cowan, B. J., et al. (2007). Pannexin 1 and pannexin 3 are glycoproteins that exhibit many distinct characteristics from the connexin family of gap junction proteins. *J. Cell Sci.* 120, 3772–3783. doi: 10.1242/jcs.009514

Perea, G., Navarrete, M., and Araque, A. (2009). Tripartite synapses: astrocytes process and control synaptic information. *Trends Neurosci.* 32, 421–431. doi: 10.1016/j.tins.2009.05.001

Ponsaerts, R., De Vuyst, E., Retamal, M., D'hondt, C., Vermeire, D., Wang, N., et al. (2010). Intramolecular loop/tail interactions are essential for connexin 43-hemichannel activity. *FASEB J.* 24, 4378–4395. doi: 10.1096/fj.09-153007

Prochnow, N., Abdulazim, A., Kurtenbach, S., Wildforster, V., Dvoriantchikova, G., Hanske, J., et al. (2012). Pannexin1 stabilizes synaptic plasticity and is needed for learning. *PLoS ONE* 7:e51767. doi: 10.1371/journal.pone.0051767

Quist, A. P., Rhee, S. K., Lin, H., and Lal, R. (2000). Physiological role of gap-junctional hemichannels. Extracellular calcium-dependent isosmotic volume regulation. *J. Cell Biol.* 148, 1063–1074. doi: 10.1083/jcb.148.5.1063

Ray, A., Zoidl, G., Weickert, S., Wahle, P., and Dermietzel, R. (2005). Site-specific and developmental expression of pannexin1 in the mouse nervous system. *Eur. J. Neurosci.* 21, 3277–3290. doi: 10.1111/j.1460-9568.2005.04139.x

Retamal, M. A., Froger, N., Palacios-Prado, N., Ezan, P., Saez, P. J., Sáez, J. C., et al. (2007). Cx43 hemichannels and gap junction channels in astrocytes are regulated oppositely by proinflammatory cytokines released from activated microglia. *J. Neurosci.* 27, 13781–13792. doi: 10.1523/JNEUROSCI.2042-07.2007

Riquelme, M. A., Kar, R., Gu, S., and Jiang, J. X. (2013). Antibodies targeting extracellular domain of connexins for studies of hemichannels. *Neuropharmacology* 75, 525–532. doi: 10.1016/j.neuropharm.2013.02.021

Rouach, N., Avignone, E., Meme, W., Koulakoff, A., Venance, L., Blomstrand, F., et al. (2002). Gap junctions and connexin expression in the normal and pathological central nervous system. *Biol. Cell* 94, 457–475. doi: 10.1016/S0248-4900(02)00016-3

Rouach, N., Koulakoff, A., Abudara, V., Willecke, K., and Giaume, C. (2008). Astroglial metabolic networks sustain hippocampal synaptic transmission. *Science* 322, 1551–1555. doi: 10.1126/science.1164022

Rozental, R., Giaume, C., and Spray, D. C. (2000). Gap junctions in the nervous system. *Brain Res. Brain Res. Rev.* 32, 11–15. doi: 10.1016/S0165-0173(99)00095-8

Ryu, J. K., Choi, H. B., Hatori, K., Heisel, R. L., Pelech, S. L., McLarnon, J. G., et al. (2003). Adenosine triphosphate induces proliferation of human neural stem cells: role of calcium and p70 ribosomal protein S6 kinase. *J. Neurosci. Res.* 72, 352–362. doi: 10.1002/jnr.10507

Sáez, J. C., Berthoud, V. M., Branes, M. C., Martinez, A. D., and Beyer, E. C. (2003). Plasma membrane channels formed by connexins: their regulation and functions. *Physiol. Rev.* 83, 1359–1400.

Sáez, J. C., Retamal, M. A., Basilio, D., Bukauskas, F. F., and Bennett, M. V. (2005). Connexin-based gap junction hemichannels: gating mechanisms. *Biochim. Biophys. Acta* 1711, 215–224. doi: 10.1016/j.bbamem.2005.01.014

Sahu, G., Sukumaran, S., and Bera, A. K. (2014). Pannexins form gap junctions with electrophysiological and pharmacological properties distinct from connexins. *Sci. Rep.* 4:4955. doi: 10.1038/srep04955

Samoilova, M., Wentlandt, K., Adamchik, Y., Velumian, A. A., and Carlen, P. L. (2008). Connexin 43 mimetic peptides inhibit spontaneous epileptiform activity in organotypic hippocampal slice cultures. *Exp. Neurol.* 210, 762–775. doi: 10.1016/j.expneurol.2008.01.005

Schalper, K. A., Palacios-Prado, N., Orellana, J. A., and Sáez, J. C. (2008). Currently used methods for identification and characterization of hemichannels. *Cell Commun. Adhes.* 15, 207–218. doi: 10.1080/15419060802014198

Schulz, S. B., Klaft, Z. J., Rosler, A. R., Heinemann, U., and Gerevich, Z. (2012). Purinergic P2X, P2Y and adenosine receptors differentially modulate hippocampal gamma oscillations. *Neuropharmacology* 62, 914–924. doi: 10.1016/j.neuropharm.2011.09.024

Schutz, M., Scimemi, P., Majumder, P., De Siati, R. D., Crispino, G., Rodriguez, L., et al. (2010). The human deafness-associated connexin 30 T5M mutation causes mild hearing loss and reduces biochemical coupling among cochlear non-sensory cells in knock-in mice. *Hum. Mol. Genet.* 19, 4759–4773. doi: 10.1093/hmg/ddq402

Shields, C. R., Klooster, J., Claassen, Y., Ul-Hussain, M., Zoidl, G., Dermietzel, R., et al. (2007). Retinal horizontal cell-specific promoter activity and protein expression of zebrafish connexin 52.6 and connexin 55.5. *J. Comp. Neurol.* 501, 765–779. doi: 10.1002/cne.21282

Shram, N. F., Netchiporouk, L. I., Martelet, C., Jaffrezic-Renault, N., and Cespuglio, R. (1997). Brain glucose: voltammetric determination in normal and hyperglycaemic rats using a glucose microsensor. *Neuroreport* 8, 1109–1112. doi: 10.1097/00001756-199703240-00009

Simard, M., Arcuino, G., Takano, T., Liu, Q. S., and Nedergaard, M. (2003). Signaling at the gliovascular interface. *J. Neurosci.* 23, 9254–9262.

Sosinsky, G. E., Boassa, D., Dermietzel, R., Duffy, H. S., Laird, D. W., Macvicar, B., et al. (2011). Pannexin channels are not gap junction hemichannels. *Channels (Austin).* 5, 193–197. doi: 10.4161/chan.5.3.15765

Spray, D. C., Duffy, H. S., and Scemes, E. (1999). Gap junctions in glia. Types, roles, and plasticity. *Adv. Exp. Med. Biol.* 468, 339–359. doi: 10.1007/978-1-4615-4685-6_27

Stehberg, J., Moraga-Amaro, R., Salazar, C., Becerra, A., Echeverria, C., Orellana, J. A., et al. (2012). Release of gliotransmitters through astroglial connexin 43 hemichannels is necessary for fear memory consolidation in the basolateral amygdala. *FASEB J.* 26, 3649–3657. doi: 10.1096/fj.11-198416

Suadicani, S. O., Brosnan, C. F., and Scemes, E. (2006). P2X7 receptors mediate ATP release and amplification of astrocytic intercellular Ca^{2+} signaling. *J. Neurosci.* 26, 1378–1385. doi: 10.1523/JNEUROSCI.3902-05.2006

Sun, Z., Risner, M. L., Van Asselt, J. B., Zhang, D. Q., Kamermans, M., and McMahon, D. G. (2012). Physiological and molecular characterization of connexin hemichannels in zebrafish retinal horizontal cells. *J. Neurophysiol.* 107, 2624–2632. doi: 10.1152/jn.01126.2011

Theis, M., Jauch, R., Zhuo, L., Speidel, D., Wallraff, A., Doring, B., et al. (2003). Accelerated hippocampal spreading depression and enhanced locomotory activity in mice with astrocyte-directed inactivation of connexin43. *J. Neurosci.* 23, 766–776.

Theis, M., Sohl, G., Eiberger, J., and Willecke, K. (2005). Emerging complexities in identity and function of glial connexins. *Trends Neurosci.* 28, 188–195. doi: 10.1016/j.tins.2005.02.006

Thompson, R. J., Jackson, M. F., Olah, M. E., Rungta, R. L., Hines, D. J., Beazely, M. A., et al. (2008). Activation of pannexin-1 hemichannels augments aberrant bursting in the hippocampus. *Science* 322, 1555–1559. doi: 10.1126/science.1165209

Thompson, R. J., and Macvicar, B. A. (2008). Connexin and pannexin hemichannels of neurons and astrocytes. *Channels (Austin).* 2, 81–86. doi: 10.4161/chan.2.2.6003

Torres, A., Wang, F., Xu, Q., Fujita, T., Dobrowolski, R., Willecke, K., et al. (2012). Extracellular $Ca^{(2+)}$ acts as a mediator of communication from neurons to glia. *Sci. Signal.* 5, ra8. doi: 10.1126/scisignal.2002160

Valiunas, V., and Weingart, R. (2000). Electrical properties of gap junction hemichannels identified in transfected HeLa cells. *Pflugers Arch.* 440, 366–379. doi: 10.1007/s004240000294

Velasquez, S., and Eugenin, E. A. (2014). Role of Pannexin-1 hemichannels and purinergic receptors in the pathogenesis of human diseases. *Front. Physiol.* 5:96. doi: 10.3389/fphys.2014.00096

Wallraff, A., Kohling, R., Heinemann, U., Theis, M., Willecke, K., and Steinhauser, C. (2006). The impact of astrocytic gap junctional coupling on potassium buffering in the hippocampus. *J. Neurosci.* 26, 5438–5447. doi: 10.1523/JNEUROSCI.0037-06.2006

Wang, N., De Bock, M., Decrock, E., Bol, M., Gadicherla, A., Vinken, M., et al. (2013a). Paracrine signaling through plasma membrane hemichannels. *Biochim. Biophys. Acta* 1828, 35–50. doi: 10.1016/j.bbamem.2012.07.002

Wang, N., De Vuyst, E., Ponsaerts, R., Boengler, K., Palacios-Prado, N., Wauman, J., et al. (2013b). Selective inhibition of Cx43 hemichannels by Gap19 and its impact on myocardial ischemia/reperfusion injury. *Basic Res. Cardiol.* 108, 309. doi: 10.1007/s00395-012-0309-x

Wicki-Stordeur, L. E., Dzugalo, A. D., Swansburg, R. M., Suits, J. M., and Swayne, L. A. (2012). Pannexin 1 regulates postnatal neural stem and progenitor cell proliferation. *Neural Dev.* 7, 11. doi: 10.1186/1749-8104-7-11

Wicki-Stordeur, L. E., and Swayne, L. A. (2013). Panx1 regulates neural stem and progenitor cell behaviours associated with cytoskeletal dynamics and interacts with multiple cytoskeletal elements. *Cell Commun. Signal.* 11:62. doi: 10.1186/1478-811X-11-62

Willecke, K., Eiberger, J., Degen, J., Eckardt, D., Romualdi, A., Guldenagel, M., et al. (2002). Structural and functional diversity of connexin genes in the mouse and human genome. *Biol. Chem.* 383, 725–737. doi: 10.1515/BC.2002.076

Ye, Z. C., Wyeth, M. S., Baltan-Tekkok, S., and Ransom, B. R. (2003). Functional hemichannels in astrocytes: a novel mechanism of glutamate release. *J. Neurosci.* 23, 3588–3596.

Zimmermann, H. (2011). Purinergic signaling in neural development. *Semin. Cell Dev. Biol.* 22, 194–204. doi: 10.1016/j.semcdb.2011.02.007

Zoidl, G., Petrasch-Parwez, E., Ray, A., Meier, C., Bunse, S., Habbes, H. W., et al. (2007). Localization of the pannexin1 protein at postsynaptic sites in the cerebral cortex and hippocampus. *Neuroscience* 146, 9–16. doi: 10.1016/j.neuroscience.2007.01.061

15

Investigation of olfactory function in a Panx1 knock out mouse model

Stefan Kurtenbach[1]*, Paige Whyte-Fagundes[1], Lian Gelis[2], Sarah Kurtenbach[1], Émerson Brazil[1], Christiane Zoidl[1], Hanns Hatt[2], Valery I. Shestopalov[3,4] and Georg Zoidl[1]

[1] Department of Psychology, Faculty of Health, York University, Toronto, ON, Canada
[2] Department of Cell Physiology, Ruhr University Bochum, Bochum, Germany
[3] Department of Ophthalmology, Bascom Palmer Eye Institute, Miller School of Medicine, University of Miami, Miami, FL, USA
[4] Vavilov Institute of General Genetics, Russian Academy of Sciences, Moscow, Russia

Correspondence:
Stefan Kurtenbach, Department of Physiology, Faculty of Health, York University, 4700 Keele Street, Toronto, ON M3J 1P3, Canada
e-mail: stefan.kurtenbach@me.com

Pannexin 1 (Panx1), the most extensively investigated member of a channel-forming protein family, is able to form pores conducting molecules up to 1.5 kDa, like ATP, upon activation. In the olfactory epithelium (OE), ATP modulates olfactory responsiveness and plays a role in proliferation and differentiation of olfactory sensory neurons (OSNs). This process continuously takes place in the OE, as neurons are replaced throughout the whole lifespan. The recent discovery of Panx1 expression in the OE raises the question whether Panx1 mediates ATP release responsible for modulating chemosensory function. In this study, we analyzed pannexin expression in the OE and a possible role of Panx1 in olfactory function using a Panx1$^{-/-}$ mouse line with a global ablation of Panx1. This mouse model has been previously used to investigate Panx1 functions in the retina and adult hippocampus. Here, qPCR, *in-situ* hybridization, and immunohistochemistry (IHC) demonstrated that Panx1 is expressed in axon bundles deriving from sensory neurons of the OE. The localization, distribution, and expression of major olfactory signal transduction proteins were not significantly altered in Panx1$^{-/-}$ mice. Further, functional analysis of Panx1$^{-/-}$ animals does not reveal any major impairment in odor perception, indicated by electroolfactogram (EOG) measurements and behavioral testing. However, ATP release evoked by potassium gluconate application was reduced in Panx1$^{-/-}$ mice. This result is consistent with previous reports on ATP release in isolated erythrocytes and spinal or lumbar cord preparations from Panx1$^{-/-}$ mice, suggesting that Panx1 is one of several alternative pathways to release ATP in the olfactory system.

Keywords: pannexin, Panx1, olfaction, knock out mouse, extracellular ATP, electroolfactogram, behavior

INTRODUCTION

Genomes of higher vertebrates contain three pannexin genes (Panx1, Panx2, and Panx3) that share homologies with invertebrate gap junction proteins named innexins (Baranova et al., 2004) and membrane topologies similar to vertebrate connexins, a novel class of integral membrane glycoproteins. Unlike connexins, pannexins function as unopposed channels (Sosinsky et al., 2011) that open through various stimuli like activation of purinergic receptors and high intracellular calcium (Locovei et al., 2006), cellular stretch (Bao et al., 2004), high extracellular potassium levels (Silverman et al., 2009), and depolarization (Bruzzone et al., 2003; Pelegrin and Surprenant, 2006). Pannexins are known to be expressed in sensory systems like the cochlea (Wang et al., 2009), retina (Dvoriantchikova et al., 2006; Zoidl et al., 2008), taste buds (Huang et al., 2007), and olfactory epithelium (OE) (Zhang et al., 2012).

Panx1 has moved into the focus of purinome research due to the ability to form large channel pores capable of conducting big molecules like ATP. The presence of pannexins in different chemosensory systems raises the general question of whether they may contribute to sensory perception by releasing ATP and modulating purinergic signaling. Perception of sensory stimuli also involves purinergic signaling at various levels: ATP can influence cochlear function through multiple mechanisms, including modulation of hearing sensitivity, sound transduction, neurotransmission, and even influencing gap-junctional coupling (Bobbin and Thompson, 1978; Muñoz et al., 1995; Zhu and Zhao, 2012). ATP release plays a primary role in signal transmission of taste cells to afferent nerve fibers (Finger et al., 2005) and a similar mechanism seems to hold true for keratinocytes (Azorin et al., 2011). In the retina, purinergic signals act as neuro- and gliotransmitters, most likely modulating retinal responses on various levels (Prochnow et al., 2009; Wurm et al., 2011; Vessey and Fletcher, 2012). In the OE, extracellular ATP elicits increases in $[Ca^{2+}]_i$ in both olfactory sensory neurons (OSNs) and sustentacular cells (SCs), leading to a suppression in odor responsiveness (Hegg et al., 2003). ATP also evokes inward currents and increased $[Ca^{2+}]_i$ in sensory neurons in the vomeronasal organ (VNO), by activation of P2X receptors (Vick and Delay, 2012).

The introduction of several Panx1 knock out animal models, with distinct genetic ablation strategies (Bargiotas et al., 2011, 2012; Dvoriantchikova et al., 2012; Hanstein et al., 2013), has initiated ample opportunities to investigate Panx1 functions from genes through to systems and behavioral outcomes. Since sensory inputs to the CNS can be altered in Panx1$^{-/-}$ mice, as shown recently for the retina (Kranz et al., 2013; Vroman et al., 2014), it was hypothesized that similar changes could be found in other sensory organs. In this study, we characterize the impact of genetic ablation of Panx1 in the olfactory system. Expression and localization studies demonstrated Panx1 expression in the OE. Surprisingly, Panx1 was mainly localized in OE axon bundles and not at the apical surface of the cilia, where the olfactory receptors are present and the olfactory signal transduction cascade is initiated. The OE of Panx1$^{+/+}$ and Panx1$^{-/-}$ mice showed no altered response to odors in electroolfactogram (EOG) measurements, suggesting that Panx1 is dispensable for acute olfactory function. However, a quantitative comparison of extracellular ATP concentration revealed significant differences between the two genotypes after exposure to potassium gluconate. Further, behavioral outcomes revealed small, but significant abnormalities in the processing of olfactory information, accompanied by a higher mobility of Panx1$^{-/-}$ mice. Our data do not support a prominent role of Panx1 channels in olfaction, and it was concluded that the behavioral abnormalities observed in Panx1$^{-/-}$ mice derive from alterations of integrating neuronal processes, as observed in the hippocampus. However, our data suggest that Panx1 is one of several alternative pathways to release ATP in the olfactory system and that dissecting these pathways will be a critical step to define the exact role(s) of Panx1 in sensory systems.

METHODS

Panx1 KNOCKOUT MICE

Panx1$^{+/+}$ mice (Panx1fl/fl) with three LoxP consensus sequences integrated into the Panx1 gene and knockout mice with global loss of Panx1 (Panx1$^{-/-}$, CMV-Cre/Panx1) were described previously (Gründken et al., 2011; Dvoriantchikova et al., 2012; Prochnow et al., 2012). Handling and housing of animals used in this study was performed in compliance with the German Animal Rights law and approved by the Landesamt für Natur, Umwelt und Verbraucherschutz Nordrhein-Westfalen, Germany (Permission No. 50.8735.1 Nr. 100/4) and formal approval by the Animal Care Committee (York University, Canada). Adult male mice (4–8 months of age) were housed individually 1 week prior to, and during, behavior testing.

IN-SITU HYBRIDIZATION (ISH)

Digoxigenin (dig)-labeled sense and antisense cRNA probes were prepared from a full length Panx1 cloned into the pcDNA3 plasmid as described previously (Ray et al., 2006). After linearization of the plasmid, sense and antisense cRNA probes were transcribed using T7 and SP6 RNA polymerase with dig-RNA labeling mix (Roche, Germany). The ISH was performed as described (Larsson et al., 2004) with minor modifications. OE from P7 mice were dissected and immediately embedded in tissue freezing medium (Leica, Germany) at −30°C and cryostat sections (12 μm) were cut immediately. Slides were subsequently fixed in 4% paraformaldehyde in PBS at 4°C for 20 min, washed in PBS and acetylated by a 15 min treatment in 0.1 M triethanolaminhydrochloride solution with 0.25% acetic anhydride on a stir plate. Sections were rinsed in 2× SSC (30 mM NaCl and 3 mM sodium citrate) and prehybridized in hybridization buffer (50% formamide, 5× SSC, 5× Denhardts' solution, 2.5 mM EDTA, 50 μg/ml heparin, 250 μg/ml tRNA, 500 μg/ml salmon sperm DNA, and 0.1% Tween-20) for 1 h at 55°C. Riboprobes were added to the hybridization buffer (50 ng in 200 μl hybridization buffer), denaturized at 80°C for 2 min and applied to sections. Sections were incubated over night at 55°C for hybridization. Post-hybridization, slides were washed with 0.2× SSC for 1 h and then with 0.1× SSC for 15 min, to remove non-specific binding. Sections were subsequently equilibrated for 10 min in PBS containing 0.1 % TritonX-100 (PBST), blocked with 10% goat serum in PBST buffer for 1 h, and then incubated with 1:1000 alkaline phosphatase (AP) conjugated anti-dig Fab fragment (Roche, Germany) in blocking solution overnight (ON) at 4°C. Finally, slides were washed in PBST, equilibrated in B3-Buffer (0.1 Tris-HCl, 0.1 M NaCl, 50 mM $MgCl_2$, 0.1% Tween-20), followed by treatment with NBT/BCIP (Roche, Germany) (20 μl/ml B3) to visualize the hybridized probes.

IMMUNOHISTOCHEMISTRY (IHC)

After the fur and palate were removed, heads from adult male mice were fixed in 4% PFA at 4°C ON, then immersed in 30% sucrose at 4°C ON. 12 μm cryosections were prepared, blocked with 5% cold-water fish skin gelatine for 1 h at RT, and primary antibodies (1:250, Santa Cruz, CA, $G_\alpha$olf sc-383; CNG sc-13700, ACIII sc-588, acetylated tubulin sc-23950) were applied in 1% cold-water fish skin gelatin in PBS containing 0.1% Triton X-100, at 4°C ON. After 30 min washing in PBS, secondary goat anti-rabbit antibodies Alexa Fluor 568 (Invitrogen, Germany) were applied for 30 min at RT in PBS. After 30 min washing in PBS, sections were embedded in ProlongGold Antifade (Invitrogen, Germany).

The Laird laboratory generously provided an antibody for Panx1 IHC (Penuela et al., 2007). For Panx1 detection the following modifications were introduced. For antigen retrieval, fixed cryostat sections were incubated for 5 min with 1% SDS, followed by three washes for 5 min with PBS. After blocking for 1 h at RT with 5% normal goat serum (NGS), 1% bovine serum albumin (BSA), and 0.1% Triton X100 in PBS, the primary antibody was diluted (1:100) in 1% BSA, 0.1% Triton X100 in PBS. After ON incubation, specimen were washed three times with PBS for 10 min each. The secondary goat anti-rabbit Alexa Fluor 488 antibody was diluted (1:1000) in PBS and applied for 30 min at RT, followed by three washes for 10 min with PBS.

Confocal microscopy was performed using a ZEISS LSM700 microscope. ZEISS ZEN software was used to control all parameters during imaging. Identical settings were used to allow a direct comparison of IHCs of Panx1$^{+/+}$ and Panx1$^{-/-}$ mice. LSM images were exported into tiff format and assembled using Photoshop CS.

qPCR

RNA was isolated from adult male mice using the RNAeasy Fibrous Tissue Mini Kit (Invitrogen, Canada) and cDNA was synthesized from 1 μg total RNA with the ReadyScript cDNA Synthesis Kit (Sigma-Aldrich, Canada), according to the manufacturer's instructions. qPCR was performed using the SsoFast EvaGreen Supremix (Bio-Rad, Canada) and the following oligonucleotide pairs: Panx1fw: CAGGCTGCCTTTGTGGATTC Panx1rev: CGGGCAGG TACAGGAGTATG Panx2fw: GGTACCAAGAAGGCCAAGA CT Panx2rev: GGGGTACGGGATTTCCTTCTC Panx3fw: CTTA CAACCGTTCCATCCGC Panx3rev: CAGGTACCGCTCTAGC AAGG 18Sfw: TGACTCTTTCGAGGCCCTGTA 18Srev: TGGA ATTACCGCGGCTGCTG. fw, forward; rev, reverse. Experiments were performed in triplicates, using three biological replicates. Relative gene expression was calculated using the REST software (2009) (Pfaffl et al., 2002). 18S served as the reference gene.

EOG RECORDINGS

Skulls from adult male mice were cut parasagittal to the septum to expose the nasal cavity. The turbinates were removed and EOGs were recorded from the OE on the septum. A constant, deodorized, humidified air stream was delivered to the OE and adjusted to 2.4 l/min. Henkel 100 (H100, Henkel, Germany), a mixture of 100 different odors (Wetzel et al., 1999), was used as a stimulus. Dilutions were made in distilled water and soaked into a piece of felt, which was placed into a custom-made injection device. Recording electrodes were made from pulled glass capillaries containing chloride silver wires and filled with Ringer's solution (140 mM NaCl, 5 mM KCl, 1 mM $CaCl_2$, 1 mM $MgCl_2$, 10 mM HEPES, pH 7.4). During recording, heads were placed on 2% agarose dissolved in Ringer's solution containing the reference electrode. Student's t-test was used for statistical analysis.

EX VIVO ATP ASSAY

The intact OE was exposed after skulls from adult male mice were cut parasagittal to the septum to expose the nasal cavity. Once the OE was entirely exposed, the tissue was positioned in upside down orientation. Since the OE rapidly degenerates once axons get transsected, this *ex vivo* preparation was used instead of more invasive dissection procedures to extract extracellular ATP from the cilia surface. To avoid mechanical stimulation small droplets (25 μl) of Ringers solution, H100 [Henkel 100, 1:10000 in Ringers solution, Henkel, Germany (Wetzel et al., 1999)], and potassium gluconate (25 mM, Sigma-Aldrich, diluted in Ringers solution) were subsequently and gently pipetted onto the cilia surface at the center of the OE. The tissue was incubated for the indicated timespans. Between steps droplets were fully removed. Solutions contained 100 μM ARL 67156 trisodium salt hydrate (Sigma-Aldrich) to inhibit ATPases. Samples were heated at 95°C for 1 min after extraction, flash frozen, and stored at −80°C until used. ATP assays were performed in 96 well format; using the Molecular Probes® ATP Determination Kit (Life Technologies, USA) and the Synergy H4 hybrid multiwell plate reader (Biotek, USA). ATP concentrations were determined from ATP standard curves included in each assay. To test for statistical significance the student's T-test was used.

BEHAVIORAL TESTS

Cookie-finding test

To familiarize the mice with the experimental setup, adult male mice were trained for 1 day to find a cookie (1 g) (Leibniz Butterkeks; Bahlsen, Germany) buried beneath 6 cm of woodchip bedding in their home cage. On the following days, smaller cookies of defined weight were hidden (see Results). The time to locate the cookie was determined from video recordings. We defined finding the cookie as the time point when mice held it in both fore paws. If a mouse did not find the cookie within 15 min, the test was aborted and most of the bedding above the cookie was removed, enabling the mouse to find the cookie very easily (usually within 1 min). Cookie-finding tests were analyzed using the U-test due to data truncation.

Mobility analysis

Animal mobility was analyzed with the EthoVision XT7 Software from Noldus (Wageningen, Netherland). Mobility was defined as distance per time. Tracking was stopped when the animal found the cookie. To test for statistical significance, Student's T-test was used.

RESULTS

PANNEXIN EXPRESSION AND LOCALIZATION IN THE OLFACTORY EPITHELIUM

Pannexin expression was determined in the mouse OE and whole brain lysate using qPCR and primer pairs specific for Panx1 (GI:86262133). Panx1 mRNA expression was found in the brain and OE adult mice (**Figure 1**), with mean cycle threshold (C_t) values of 29.8 ± 0.5 (brain) and 31.3 ± 0.9 (OE). Panx2 was expressed in the OE at significantly lower expression levels compared to the brain ($p < 0.001$). Mean cycle threshold (C_t) values were 27.2 ± 0.7 (brain) and 33.9 ± 0.7 (OE). No Panx3 expression was detected ($C_t < 35$). Neither in the brain, nor in the OE a compensatory upregulation of Panx2 or Panx3 mRNA was observed in Panx$^{-/-}$ mice. Consistent with the genetic phenotype, no Panx1 mRNA expression ($C_t < 35$) was found in Panx1$^{-/-}$ animals lacking exons 3 and 4 (Dvoriantchikova et al., 2012).

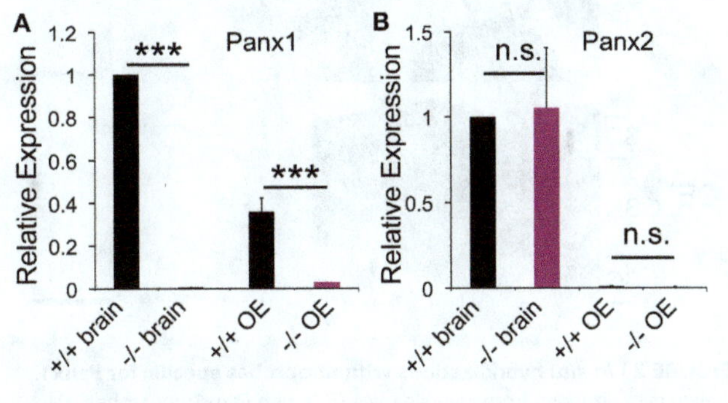

FIGURE 1 | Pannexin expression in the brain and OE. qPCR data for pannexin expression in the OE ($N = 3$) and brain ($N = 3$) of adult mice. **(A)** Panx1$^{-/-}$ mice lacked Panx1 expression. **(B)** No difference in Panx2 expression was detected in the OE. Expression of Panx3 was not detected. Primers specific for 18S were used as the reference. Experiments were performed in triplicates. $^{***}p < 0.001$. n.s., not significant. Error bars: s.e.m.

Localization of Panx1 expression sites in the OE were determined using *in-situ* hybridization technology and cRNA probes specific for mouse Panx1 (Ray et al., 2006). **Figure 2** shows a representative distribution of Panx1 mRNAs in the OE. The sense cRNA probes generated only a very weak background staining, demonstrating the antisense probes' specificity. In the OE, a strong staining of the OSN layer and a less intense staining of the sustentacular (SC, arrows) and basal cell (BC) layers were observed.

To validate mRNA localization data, immunohistochemistry (IHC) was performed using an antibody specific for mouse Panx1 (Penuela et al., 2007). In Panx1$^{+/+}$ animals (**Figure 3**), the IHC of the OE revealed a significant labeling of the OSN axon bundles, projecting to the olfactory bulb. Virtually no staining was found in OSN axon bundles of Panx1$^{-/-}$ mice. However, a diffuse signal was observed in the basal-, neuronal-, and SC layers. No pronounced staining was found in OSN cilia, which inherit olfactory receptor proteins and signaling proteins involved in OSN signaling/depolarization after odorant activation.

ODORANT DETECTION IS NOT ALTERED IN Panx1$^{-/-}$ MICE

When an olfactory receptor (OR) is activated, the signal transduction cascade elicited through the OR typically evokes a massive influx of calcium mediated by cyclic nucleotide gated (CNG) channels. Since Panx1 channels can be activated by raised intracellular calcium levels (Locovei et al., 2006), we investigated whether the lack of Panx1 alters the response of the OE upon odor application by performing electroolfactogram (EOG) measurements in adult male mice. A mixture of 100 different odorants [Henkel 100 (H100) (Wetzel et al., 1999)] was applied via a constant, humidified airstream to activate a broad range of ORs. **Figure 4A** shows a typical response amplitude after odorant application for 100 ms. Results were quantified and summarized in **Figure 4B**. No significant difference in the response amplitudes for Panx1$^{+/+}$ ($N = 7$) and Panx1$^{-/-}$ ($N = 8$) mice were detected (minimum 10 measurements on different locations per mouse, Panx1$^{+/+}$ = 4.2 ± 0.2 mV, Panx1$^{-/-}$ = 4.7 ± 0.3 mV, $p = 0.17$). Further, the response kinetics, namely the rise time (Panx1$^{+/+}$ = 86 ± 3 ms, Panx1$^{-/-}$ = 94 ± 4 ms, $p = 0.1$) and decay time (Panx1$^{+/+}$ = 989 ± 191 ms, Panx1$^{-/-}$ = 984 ± 128 ms, $p = 0.98$), did not differ between both animal groups.

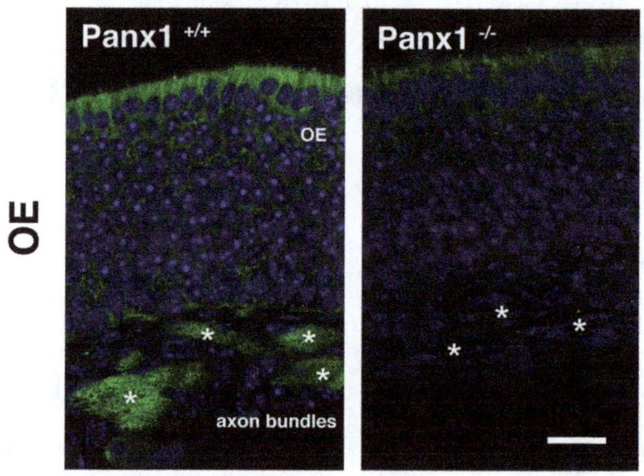

FIGURE 3 | Immunohistochemical detection of Panx1 in the OE. IHC of histological section deriving from OE of adult Panx1$^{+/+}$ and Panx$^{-/-}$ mice. In the OE, a prominent Panx1 staining (in green) was identified in OE axon bundles (asterisks), which was not detected in Panx1$^{-/-}$ mice. Nuclei were stained with DAPI (in blue). Images were recorded using identical conditions. Scale bars = 200 μm

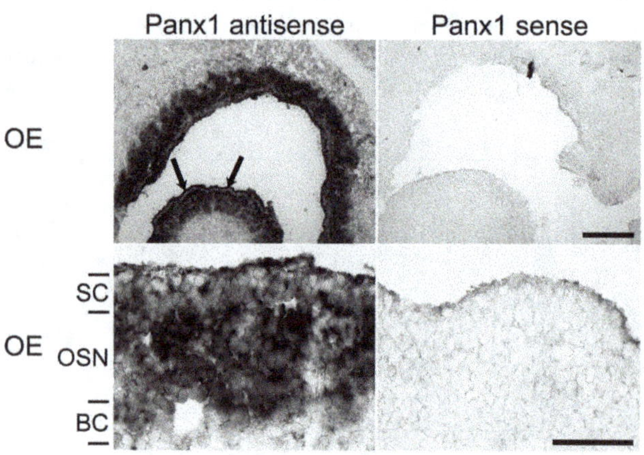

FIGURE 2 | *In-situ* hybridizations with riboprobes specific for Panx1. *In-situ* hybridizations from juvenile mice (P7) with Panx1 riboprobes. Panx1 antisense probes delivered strong labeling in the OE. Strong staining in the OE was present in the OSN layer. Weaker staining of SCs was more obvious in the overview (upper left, indicated by arrows). The BC layer showed weaker staining than the OSN layer. Scale bars = 200 μm, and 100 μm (close-up, middle) (OE, olfactory epithelium; SC, sustentacular cells; OSN, olfactory sensory neuron; BC, basal cell layer).

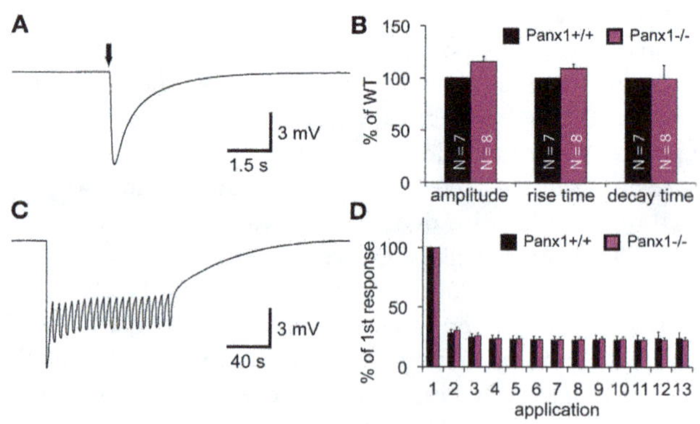

FIGURE 4 | EOG recordings of Panx1$^{+/+}$ and Panx1$^{-/-}$ olfactory epithelium. (A) Sample EOG response after application of a single odorant pulse (100 ms; 1:1000 H100). **(B)** Quantification of single odorant pulses. Amplitude and response kinetics [rise (10–90%) and decay (90–10%) time] are not significantly different in Panx1$^{-/-}$ mice ($p > 0.05$). Data were normalized to Panx1$^{+/+}$ values. $N_{Panx1+/+} = 7$; $N_{Panx1-/-} = 8$; minimum 10 measurements per mouse. **(C)** Sample recording of multiple odor applications (1 s; 4 s interstimulus interval; 1:1000 H100). **(D)** Quantification of 13 adaptation measurements (depicted in **C**). Responses (Δ of individual responses) were normalized to the first response. No significant differences were found ($N_{Panx1+/+} = 6$; $N_{Panx1-/-} = 6$; 1 measurement per mouse). Error bars indicate s.e.m. (EOG, electroolfactogram; H100, Henkel 100).

To test the ability of the OE to adapt to odorants, we measured odor responses after a repetitive stimulation paradigm (Brunert et al., 2009) ($N_{Panx1^{+/+}} = 6$, $N_{Panx1^{-/-}} = 6$, 1 s stimulus duration, 4 s interstimulus interval, 1 measurement per mouse), as depicted in **Figures 4C,D**. Using this paradigm, we achieved a robust adaptation to less than 30 % of the control level, but adaptation kinetics observed in Panx1$^{-/-}$ animals were statistically insignificant to those in wild type mice. Since Panx1$^{+/+}$ and Panx1$^{-/-}$ responded equally to both experimental conditions, we concluded that loss of Panx1 did not alter physiological odorant responses. Further confirming this finding, ablation of Panx1 did not cause a significant change in the localization and distribution of the four major olfactory signal transduction proteins detecting adenylyl cyclase 3 (ADCYIII), cyclic nucleotide gated channel alpha 2 (CNGA2), olfactory neuron specific-G protein (G_{olf}), and acetylated tubulin (AcTub) (Supplementary Figure 1). Similarly, quantification of the mRNA expression of adenylyl cyclase 3 (ADCYIII), cyclic nucleotide gated channel alpha 2 (CNGA2), and olfactory neuron specific-G protein (G_{olf})using qPCR revealed no difference (Supplementary Figure 2).

ATP RELEASE IN THE OE OF Panx1$^{-/-}$ MICE

Panx1 is considered to be a major ATP release channel in various tissues. We quantified extracellular ATP using *ex vivo* preparations of the OE (**Figure 5**). During this procedure the OE is placed upside down for extracellular ATP extraction. Due to the small size of the OE, ATP was extracted in a 25 μl droplet of Ringers solution gently placed at the center of the OE. After equilibration for 3 min, droplets recovered from the Panx1$^{+/+}$ and Panx1$^{-/-}$OEs had an ATP concentration in the pM range. Extracellular ATP concentrations in the 25 μl droplet increased more than 100fold when this extraction step was repeated with Ringers solution containing the ATPase inhibitor (ARL 67156). No significant difference was detected ($p = 0.66$), suggesting that ablation of Panx1 did not compromise efflux of extracellular ATP. After Ringers solution with ATPase inhibitor was completely removed, the odorant mixture Henkel 100 was applied for 90 s in a small droplet (25 μl) placed again at the center of the OE. No significant differences in extracellular ATP release in both mouse strains were detectable. The concentration of extractable ATP was slightly reduced to the previous step reflecting the reduced extraction time. It was concluded that alternative ATP release pathways primarily promoted the observed ATP efflux in response to the odorant. Subsequently 25 mM potassium gluconate in Ringers solution (25 μl) was applied for 90 s. Significant differences in extracellular ATP levels were detected (Panx1$^{+/+}$, 3.4 ± 0.4 pM; Panx1$^{-/-}$ 2.2 ± 0.3 pM; $p = 0.022$), suggesting that Panx1 channels contributed significantly to ATP release in this experimental condition.

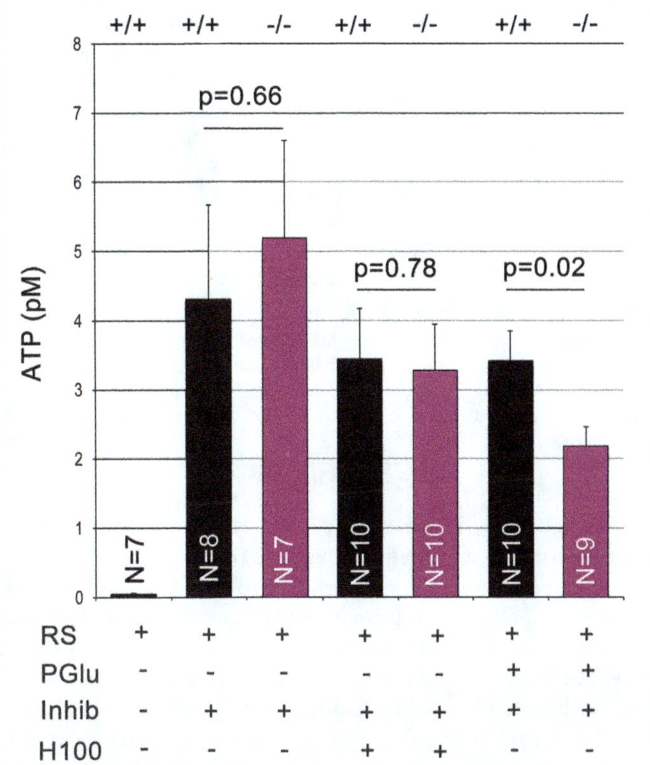

FIGURE 5 | Quantification of extracellular ATP. Extracellular ATP was extracted from the ciliary surface of the OE after skulls from adult male mice were cut parasagittal to the septum to expose the nasal cavity. ATP concentrations were determined using this *ex vivo* preparation after serial application and complete extraction in small droplets containing Ringers solution (equilibration time: 3 min), Ringers solution with the ATPase inhibitor ARL 67156 (100 μM, equilibration time: 3 min), Henkel100 (equilibration time: 90 s) and finally potassium gluconate (25 mM, equilibration time: 90 s). Low levels of ATP did not differ in wild type and Panx1$^{-/-}$ mice. Addition of the ATPase inhibitor increased ATP levels more than 100fold. Stimulation with Henkel100 caused significant ATP efflux in 90 s. Since no differences were detectable, we concluded that alternative ATP release pathways primarily promoted ATP efflux in response to the odorant. Subsequent stimulation with potassium gluconate for 90 s caused a reduction in ATP release in Panx1$^{-/-}$ mice, demonstrating a significant role of Panx1 channels. Error bars indicate s.e.m.

ODORANT PERCEPTION IN Panx1$^{-/-}$ MICE

The results shown suggested that loss of Panx1 does not impair detection of odorants in the OE. However, Panx1 is expressed in other cells of the olfactory system, including subsets of neurons in the olfactory bulb and the piriform cortex (Bruzzone et al., 2003; Ray et al., 2005). To test for odorant perception, a cookie-finding test was performed. In this behavioral test, an odorous cookie was hidden beneath the bedding in the home cage of each mouse and the latency until the mice found the treat was determined. Tests were conducted on four subsequent days (**Figure 6A**). On the training day (T), a large cookie (1 g) was hidden. The size (and odor) allowed the mice to familiarize with the test. Panx1$^{+/+}$ ($N = 15$) and Panx1$^{-/-}$ ($N = 14$) mice performed equally on test day 1, when test conditions were more difficult for the mice (50 mg cookie; Panx1$^{+/+}$ = 627 ± 83 s, Panx1$^{-/-}$ = 630 ± 82 s). We performed two more tests using a large [150 mg (day 2)] and a small [50 mg (day 3)] cookie. On day 2, Panx1$^{+/+}$ animals showed a tendency to find the cookie faster than Panx1$^{-/-}$ mice, however, this difference was not significant (Panx1$^{+/+}$ = 212 ± 83 s, Panx1$^{-/-}$ = 428 ± 82 s, $p = 0.1$). When the difficulty of the test was increased on day 3 (50 mg cookie), Panx1$^{-/-}$ mice took significantly longer to find the cookie (Panx1$^{+/+}$ = 149 ±

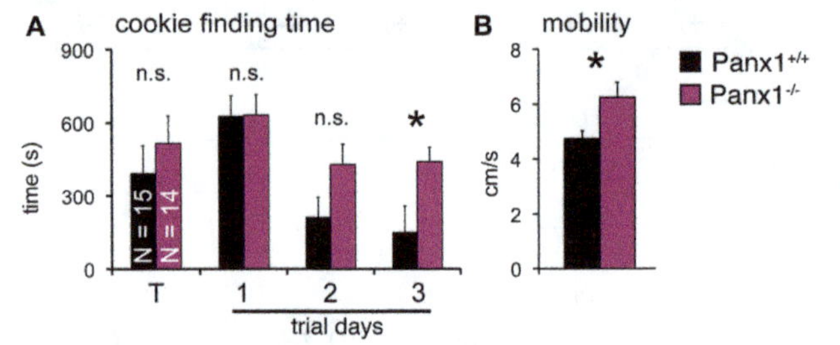

FIGURE 6 | Cookie-finding test. (A) Latency to find a hidden cookie on four trials. Cookie sizes used: Training (T) trial: 1 g; trial 1: 50 mg; trial 2: 150 mg; trial 3: 50 mg. Panx1$^{-/-}$ mice took significantly longer on trial 3 to retrieve the cookie ($p = 0.02$, $N_{Panx1^{-/-}} = 14$, $N_{Panx1^{+/+}} = 15$). **(B)** Velocity analysis showed a significantly higher mobility of Panx1$^{-/-}$ mice. *$p < 0.05$. Error bars indicate s.e.m.

58 s, Panx1$^{-/-}$ = 439 ± 107 s, $p = 0.02$). Further, they did not improve compared to the previous days (day 2) and training day (T) when conditions were easier (**Figure 6A**). This suggests that processing of olfactory information was not affected, as both animal cohorts did find the 50 mg cookie on test day one at the same time, but the ability of the animals to learn the test was significantly altered. We also tracked the velocity of the animals and found that Panx1$^{-/-}$ mice showed a significantly higher mobility (Panx1$^{+/+}$ = 4.7 ± 0.3 cm/s, Panx1$^{-/-}$ = 6.2 ± 0.5 cm/s, $p = 0.02$) compared to Panx1$^{+/+}$ mice (**Figure 6B**).

DISCUSSION

Our results show that Panx1 and Panx2 are expressed in the OE, with the expected lack of Panx1 expression in Panx1$^{-/-}$ mice. No compensatory regulation of Panx2 was observed. *In-situ* hybridizations of OE slices revealed that Panx1 mRNA is expressed in OSNs, and to a lower extent in SCs and basal cells, which is in accordance to a recent study (Zhang et al., 2012). We further showed that the Panx1 protein is localized in axon bundles of OSNs. The role of Panx1 channels in axon bundles is unclear. However, this localization and the apparent lack of Panx1 channels in the ciliary layer of the OE could explain why Panx1 is not directly involved in the olfactory signal transduction and activation of OSNs, in which case a ciliary localization would be most likely. Stimulation with short, single odorant pulses, or testing of adaptation, did evoke similar responses in Panx1$^{-/-}$ and Panx1$^{+/+}$ mice in EOG recordings. The kinetics of the EOG responses, namely the rise and decay times, were similar in both mouse groups, indicating no major difference in signal amplification and termination mechanisms, further substantiated by IHC staining and qPCR data for major signal transduction proteins, which showed no significant changes in localization and steady state mRNA expression (Supplementary Figures 1, 2), Behavioral testing was used to investigate the animals' capability to detect odorant cues. Since processing of olfactory information appeared unaffected, whereas the ability of the animals to learn the test was significantly altered, we conclude that learning and memory capabilities are impaired in Panx1$^{-/-}$ animals (Prochnow et al., 2012).

Extracellular ATP is known to modulate olfactory responsiveness and influence proliferation and neuronal differentiation (Hassenklöver et al., 2009; Jia et al., 2009, 2011; Gao et al., 2010). It was tempting to hypothesize that the ablation of Panx1, a major release channel for ATP (Locovei et al., 2006; Qiu and Dahl, 2009; Dahl and Keane, 2012), alters OE function, either by influencing neuronal turnover in the OE, or being directly involved in the olfactory signal transduction/adaptation. Therefore, we quantified extracellular ATP concentrations *ex vivo* in the OEs of unchallenged Panx1$^{+/+}$ and Panx1$^{-/-}$ animals, and after exposure to H100 and potassium gluconate. The latter has previously been shown to open Panx1 channels *in vitro* (Silverman et al., 2009). Application of potassium gluconate revealed significant changes between wild type and Panx1$^{-/-}$ animals, with ablation of Panx1 channels causing a significant reduction of potassium gluconate stimulated ATP release. The observed reduction is consistent with two previous studies. Similar to our results, extracellular ATP deriving from isolated erythrocytes (Qiu et al., 2011), or lumbar and sacral spinal cord slices (Lutz et al., 2013) were virtually identical in Panx1$^{+/+}$ and Panx1$^{-/-}$ mice. In contrast to human erythrocytes, where Panx1 channel inhibitors almost completely abolish ATP release, a substantial release remained in mouse erythrocytes even after probenecid treatment and the reduction in ATP release upon potassium stimulation was less than 50% (Qiu et al., 2011). Taken together, these findings are remarkable consistent in two independently generated Panx1KO models, supporting the role of Panx1 in ATP release and pointing at the existence of alternative ATP release pathways.

A challenge in future investigations will be to dissect the alternative ATP release pathways. Candidates are manifold and include vesicular release, connexin hemichannels, and ABC transporters, all present in the OE (Hayoz et al., 2012). In this study, we provide first evidence that LRRC8 channels, recently described as proteins with structural similarities to connexins and pannexins (Abascal and Zardoya, 2012), are expressed in the OE (Supplementary Figure 3). However, it remains to be demonstrated whether these channels are capable of releasing ATP. It is worth noting that expression of CALHM1, another channel with structural similarities to pannexins and ATP release capability (Siebert et al., 2013; Taruno et al., 2013) was not detected in the OE (unpublished data). Finally, compensatory regulation of connexins or pannexin proteins (Lohman and Isakson, 2014; Penuela et al., 2014) needs to be investigated to clarify synergistic or competing mechanisms of ATP efflux without Panx1 channels.

Arguably, the results presented in this study strongly advocate for a detailed analysis of ATP release mechanisms in the absence of Panx1. As shown, the characterization of Panx1$^{-/-}$ mice does not support a prominent role of Panx1 channels in olfaction, and it was concluded that the behavioral abnormalities observed in Panx1$^{-/-}$ mice derive from alterations of integrating neuronal processes, as observed in the hippocampus. In summary, our findings highlight the role of Panx1 as a major ATP release site operating in the OE alongside alternative pathways.

AUTHOR CONTRIBUTIONS

Stefan Kurtenbach and Georg Zoidl planned the project. Stefan Kurtenbach, Paige Whyte-Fagundes, Lian Gelis, Christiane Zoidl performed experiments. Stefan Kurtenbach, Georg Zoidl wrote the manuscript. Sarah Kurtenbach, Valery I. Shestopalov, and Hanns Hatt helped with data evaluation, interpretation, and manuscript preparation. Valery I. Shestopalov generated the mouse model.

ACKNOWLEDGMENTS

We thank Drs. Silvia Penuela and Dale Laird (Western University, ON, Canada) for kindly providing the anti-Panx1 antibody. NIH grant EY021517 and Russian Federal Special Program Grant 2012-1.5-12-000-1002-018 (Valery I. Shestopalov). The International Max Planck Research School (IMPRS). NSERC-DG (Georg Zoidl). The Research School and International Graduate School of Neuroscience (IGSN), Ruhr University Bochum, Germany.

REFERENCES

Abascal, F., and Zardoya, R. (2012). LRRC8 proteins share a common ancestor with pannexins, and may form hexameric channels involved in cell-cell communication. *Bioessays* 34, 551–560. doi: 10.1002/bies.201100173

Azorin, N., Raoux, M., Rodat-Despoix, L., Merrot, T., Delmas, P., and Crest, M. (2011). ATP signalling is crucial for the response of human keratinocytes to mechanical stimulation by hypo-osmotic shock. *Exp. Dermatol.* 20, 401–407. doi: 10.1111/j.1600-0625.2010.01219.x

Bao, L., Locovei, S., and Dahl, G. (2004). Pannexin membrane channels are mechanosensitive conduits for ATP. *FEBS Lett.* 572, 65–68. doi: 10.1016/j.febslet.2004.07.009

Baranova, A., Ivanov, D., Petrash, N., Pestova, A., Skoblov, M., Kelmanson, I., et al. (2004). The mammalian pannexin family is homologous to the invertebrate innexin gap junction proteins. *Genomics* 83, 706–716. doi: 10.1016/j.ygeno.2003.09.025

Bargiotas, P., Krenz, A., Hormuzdi, S. G., Ridder, D. A., Herb, A., Barakat, W., et al. (2011). Pannexins in ischemia-induced neurodegeneration. *Proc. Natl. Acad. Sci. U.S.A.* 108, 20772–20777. doi: 10.1073/pnas.1018262108

Bargiotas, P., Krenz, A., Monyer, H., and Schwaninger, M. (2012). Functional outcome of pannexin-deficient mice after cerebral ischemia. *Channels (Austin)* 6, 453–456. doi: 10.4161/chan.22315

Bobbin, R. P., and Thompson, M. H. (1978). Effects of putative transmitters on afferent cochlear transmission. *Ann. Otol. Rhinol. Laryngol.* 87, 185–190.

Brunert, D., Kurtenbach, S., Isik, S., Benecke, H., Hatt, H., and Wetzel, C. H. (2009). Odorant-dependent generation of nitric oxide in mammalian olfactory sensory neurons. *PLoS ONE* 4:e5499. doi: 10.1371/journal.pone.0005499

Bruzzone, R., Hormuzdi, S. G., Barbe, M. T., Herb, A., and Monyer, H. (2003). Pannexins, a family of gap junction proteins expressed in brain. *Proc. Natl. Acad. Sci. U.S.A.* 100, 13644–13649. doi: 10.1073/pnas.2233464100

Dahl, G., and Keane, R. W. (2012). Pannexin: from discovery to bedside in 11±4 years? *Brain Res.* 1487, 150–159. doi: 10.1016/j.brainres.2012.04.058

Dvoriantchikova, G., Ivanov, D., Barakat, D., Grinberg, A., Wen, R., Slepak, V. Z., et al. (2012). Genetic ablation of Pannexin1 protects retinal neurons from ischemic injury. *PLoS ONE* 7:e31991. doi: 10.1371/journal.pone.0031991

Dvoriantchikova, G., Ivanov, D., Panchin, Y., and Shestopalov, V. I. (2006). Expression of pannexin family of proteins in the retina. *FEBS Lett.* 580, 2178–2182. doi: 10.1016/j.febslet.2006.03.026

Finger, T. E., Danilova, V., Barrows, J., Bartel, D. L., Vigers, A. J., Stone, L., et al. (2005). ATP signaling is crucial for communication from taste buds to gustatory nerves. *Science* 310, 1495–1499. doi: 10.1126/science.1118435

Gao, L., Cao, L., Qiu, Y., Su, Z., Burnstock, G., Xiang, Z., et al. (2010). Blocking P2X receptors can inhibit the injury-induced proliferation of olfactory epithelium progenitor cells in adult mouse. *Int. J. Pediatr. Otorhinolaryngol.* 74, 747–751. doi: 10.1016/j.ijporl.2010.03.030

Gründken, C., Hanske, J., Wengel, S., Reuter, W., Abdulazim, A., Shestopalov, V. I., et al. (2011). Unified patch clamp protocol for the characterization of Pannexin 1 channels in isolated cells and acute brain slices. *J. Neurosci. Methods* 15, 15–25. doi: 10.1016/j.jneumeth.2011.04.016

Hanstein, R., Negoro, H., Patel, N. K., Charollais, A., Meda, P., Spray, D. C., et al. (2013). Promises and pitfalls of a Pannexin1 transgenic mouse line. *Front. Pharmacol.* 4:61. doi: 10.3389/fphar.2013.00061

Hassenklöver, T., Schwartz, P., Schild, D., and Manzini, I. (2009). Purinergic signaling regulates cell proliferation of olfactory epithelium progenitors. *Stem Cells* 27, 2022–2031. doi: 10.1002/stem.126

Hayoz, S., Jia, C., and Cosgrove Hegg, C. (2012). Mechanisms of constitutive and ATP-evoked ATP release in neonatal mouse olfactory epithelium. *BMC Neurosci.* 13:53. doi: 10.1186/1471-2202-13-53

Hegg, C. C., Greenwood, D., Huang, W., Han, P., and Lucero, M. T. (2003). Activation of purinergic receptor subtypes modulates odor sensitivity. *J. Neurosci.* 23, 8291–8301.

Huang, Y.-J., Maruyama, Y., Dvoryanchikov, G., Pereira, E., Chaudhari, N., and Roper, S. D. (2007). The role of pannexin 1 hemichannels in ATP release and cell-cell communication in mouse taste buds. *Proc. Natl. Acad. Sci. U.S.A.* 104, 6436–6441. doi: 10.1073/pnas.0611280104

Jia, C., Doherty, J. P., Crudgington, S., and Hegg, C. C. (2009). Activation of purinergic receptors induces proliferation and neuronal differentiation in Swiss Webster mouse olfactory epithelium. *Neuroscience* 163, 120–128. doi: 10.1016/j.neuroscience.2009.06.040

Jia, C., Sangsiri, S., Belock, B., Iqbal, T. R., Pestka, J. J., and Hegg, C. C. (2011). ATP mediates neuroprotective and neuroproliferative effects in mouse olfactory epithelium following exposure to satratoxin G *in vitro* and *in vivo*. *Toxicol. Sci.* 124, 169–178. doi: 10.1093/toxsci/kfr213

Kranz, K., Dorgau, B., Pottek, M., Herrling, R., Schultz, K., Bolte, P., et al. (2013). Expression of Pannexin1 in the outer plexiform layer of the mouse retina and physiological impact of its knockout. *J. Comp. Neurol.* 521, 1119–1135. doi: 10.1002/cne.23223

Larsson, M. C., Domingos, A. I., Jones, W. D., Chiappe, M. E., Amrein, H., and Vosshall, L. B. (2004). Or83b encodes a broadly expressed odorant receptor essential for Drosophila olfaction. *Neuron* 43, 703–714. doi: 10.1016/j.neuron.2004.08.019

Locovei, S., Wang, J., and Dahl, G. (2006). Activation of pannexin 1 channels by ATP through P2Y receptors and by cytoplasmic calcium. *FEBS Lett.* 580, 239–244. doi: 10.1016/j.febslet.2005.12.004

Lohman, A. W., and Isakson, B. E. (2014). Differentiating connexin hemichannels and pannexin channels in cellular ATP release. *FEBS Lett.* 588, 1379–1388. doi: 10.1016/j.febslet.2014.02.004

Lutz, S. E., Gonzalez-Fernandez, E., Ventura, J. C., Perez-Samartin, A., Tarassishin, L., Negoro, H., et al. (2013). Contribution of pannexin1 to experimental autoimmune encephalomyelitis. *PLoS ONE* 8:e66657. doi: 10.1371/journal.pone.0066657

Muñoz, D. J., Thorne, P. R., Housley, G. D., Billett, T. E., and Battersby, J. M. (1995). Extracellular adenosine 5′-triphosphate (ATP) in the endolymphatic compartment influences cochlear function. *Hear. Res.* 90, 106–118.

Pelegrin, P., and Surprenant, A. (2006). Pannexin-1 mediates large pore formation and interleukin-1beta release by the ATP-gated P2X7 receptor. *EMBO J.* 25, 5071–5082. doi: 10.1038/sj.emboj.7601378

Penuela, S., Bhalla, R., Gong, X.-Q., Cowan, K. N., Celetti, S. J., Cowan, B. J., et al. (2007). Pannexin 1 and pannexin 3 are glycoproteins that exhibit many distinct characteristics from the connexin family of gap junction proteins. *J. Cell Sci.* 120, 3772–3783. doi: 10.1242/jcs.009514

Penuela, S., Kelly, J. J., Churko, J. M., Barr, K. J., Berger, A. C., and Laird, D. W. (2014). Panx1 regulates cellular properties of keratinocytes and dermal fibroblasts in skin development and wound healing. *J. Invest. Dermatol.* 134, 2026–2035. doi: 10.1038/jid.2014.86

Pfaffl, M. W., Horgan, G. W., and Dempfle, L. (2002). Relative expression software tool (REST) for group-wise comparison and statistical analysis of relative expression results in real-time PCR. *Nucleic Acids Res.* 30:e36. doi: 10.1093/nar/30.9.e36

Prochnow, N., Abdulazim, A., Kurtenbach, S., Wildförster, V., Dvoriantchikova, G., Hanske, J., et al. (2012). Pannexin1 stabilizes synaptic plasticity and is needed for learning. *PLoS ONE* 7:e51767. doi: 10.1371/journal.pone.0051767

Prochnow, N., Hoffmann, S., Dermietzel, R., and Zoidl, G. (2009). Replacement of a single cysteine in the fourth transmembrane region of zebrafish pannexin 1 alters hemichannel gating behavior. *Exp. Brain Res.* 199, 255–264. doi: 10.1007/s00221-009-1957-4

Qiu, F., and Dahl, G. (2009). A permeant regulating its permeation pore: inhibition of pannexin 1 channels by ATP. *Am. J. Physiol. Cell Physiol.* 296, C250–C255. doi: 10.1152/ajpcell.00433.2008

Qiu, F., Wang, J., Spray, D. C., Scemes, E., and Dahl, G. (2011). Two non-vesicular ATP release pathways in the mouse erythrocyte membrane. *FEBS Lett.* 585, 3430–3435. doi: 10.1016/j.febslet.2011.09.033

Ray, A., Zoidl, G., Weickert, S., Wahle, P., and Dermietzel, R. (2005). Site-specific and developmental expression of pannexin1 in the mouse nervous system. *Eur. J. Neurosci.* 21, 3277–3290. doi: 10.1111/j.1460-9568.2005.04139.x

Ray, A., Zoidl, G., Wahle, P., and Dermietzel, R. (2006). Pannexin expression in the cerebellum. *Cerebellum* 5, 189–192. doi: 10.1080/147342205005 30082

Siebert, A. P., Ma, Z., Grevet, J. D., Demuro, A., Parker, I., and Foskett, J. K. (2013). Structural and functional similarities of calcium homeostasis modulator 1 (CALHM1) ion channel with connexins, pannexins, and innexins. *J. Biol. Chem.* 288, 6140–6153. doi: 10.1074/jbc.M112.409789

Silverman, W. R., de Rivero Vaccari, J. P., Locovei, S., Qiu, F., Carlsson, S. K., Scemes, E., et al. (2009). The pannexin 1 channel activates the inflammasome in neurons and astrocytes. *J. Biol. Chem.* 284, 18143–18151. doi: 10.1074/jbc.M109.004804

Sosinsky, G. E., Boassa, D., Dermietzel, R., Duffy, H. S., Laird, D. W., MacVicar, B., et al. (2011). Pannexin channels are not gap junction hemichannels. *Channels* 5, 193–197. doi: 10.4161/chan.5.3.15765

Taruno, A., Vingtdeux, V., Ohmoto, M., Ma, Z., Dvoryanchikov, G., Li, A., et al. (2013). CALHM1 ion channel mediates purinergic neurotransmission of sweet, bitter and umami tastes. *Nature* 495, 223–226. doi: 10.1038/nature11906

Vessey, K. A., and Fletcher, E. L. (2012). Rod and cone pathway signalling is altered in the P2X7 receptor knock out mouse. *PLoS ONE* 7:e29990. doi: 10.1371/journal.pone.0029990

Vick, J. S., and Delay, R. J. (2012). ATP excites mouse vomeronasal sensory neurons through activation of P2X receptors. *Neuroscience* 220, 341–350. doi: 10.1016/j.neuroscience.2012.06.004

Vroman, R., Klaassen, L. J., Howlett, M. H. C., Cenedese, V., Klooster, J., Sjoerdsma, T., et al. (2014). Extracellular ATP hydrolysis inhibits synaptic transmission by increasing ph buffering in the synaptic cleft. *PLoS Biol.* 12:e1001864. doi: 10.1371/journal.pbio.1001864

Wang, X.-H., Streeter, M., Liu, Y.-P., and Zhao, H.-B. (2009). Identification and characterization of pannexin expression in the mammalian cochlea. *J. Comp. Neurol.* 512, 336–346. doi: 10.1002/cne.21898

Wetzel, C. H., Oles, M., Wellerdieck, C., Kuczkowiak, M., Gisselmann, G., and Hatt, H. (1999). Specificity and sensitivity of a human olfactory receptor functionally expressed in human embryonic kidney 293 cells and *Xenopus laevis* oocytes. *J. Neurosci.* 19, 7426–7433.

Wurm, A., Pannicke, T., Iandiev, I., Francke, M., Hollborn, M., Wiedemann, P., et al. (2011). Purinergic signaling involved in Müller cell function in the mammalian retina. *Prog. Retin. Eye Res.* 30, 324–342. doi: 10.1016/j.preteyeres.2011.06.001

Zhang, H., Chen, Y., and Zhang, C. (2012). Patterns of heterogeneous expression of pannexin 1 and pannexin 2 transcripts in the olfactory epithelium and olfactory bulb. *J. Mol. Histol.* 43, 651–660. doi: 10.1007/s10735-012-9443-x

Zhu, Y., and Zhao, H.-B. (2012). ATP activates P2X receptors to mediate gap junctional coupling in the cochlea. *Biochem. Biophys. Res. Commun.* 426, 528–532. doi: 10.1016/j.bbrc.2012.08.119

Zoidl, G., Kremer, M., Zoidl, C., Bunse, S., and Dermietzel, R. (2008). Molecular diversity of connexin and pannexin genes in the retina of the zebrafish *Danio rerio*. *Cell Commun. Adhes.* 15, 169–183. doi: 10.1080/15419060802014081

16

Pannexin 2 protein expression is not restricted to the CNS

Maxence Le Vasseur, Jonathan Lelowski, John F. Bechberger, Wun-Chey Sin and Christian C. Naus *

Department of Cellular and Physiological Sciences, The Life Science Institute, University of British Columbia, Vancouver, BC, Canada

Correspondence:
Christian C. Naus, Department of Cellular and Physiological Sciences, The Life Science Institute, University of British Columbia, 2350 Health Science Mall, Vancouver, BC V6T 1Z3, Canada
e-mail: christian.naus@ubc.ca

Pannexins (Panx) are proteins homologous to the invertebrate gap junction proteins called innexins (Inx) and are traditionally described as transmembrane channels connecting the intracellular and extracellular compartments. Three distinct Panx paralogs (Panx1, Panx2 and Panx3) have been identified in vertebrates but previous reports on Panx expression and functionality focused primarily on Panx1 and Panx3 proteins. Several gene expression studies reported that Panx2 transcript is largely restricted to the central nervous system (CNS) hence suggesting that Panx2 might serve an important role in the CNS. However, the lack of suitable antibodies prevented the creation of a comprehensive map of Panx2 protein expression and Panx2 protein localization profile is currently mostly inferred from the distribution of its transcript. In this study, we characterized novel commercial monoclonal antibodies and surveyed Panx2 expression and distribution at the mRNA and protein level by real-time qPCR, Western blotting and immunofluorescence. Panx2 protein levels were readily detected in every tissue examined, even when transcriptional analysis predicted very low Panx2 protein expression. Furthermore, our results indicate that Panx2 transcriptional activity is a poor predictor of Panx2 protein abundance and does not correlate with Panx2 protein levels. Despite showing disproportionately high transcript levels, the CNS expressed less Panx2 protein than any other tissues analyzed. Additionally, we showed that Panx2 protein does not localize at the plasma membrane like other gap junction proteins but remains confined within cytoplasmic compartments. Overall, our results demonstrate that the endogenous expression of Panx2 protein is not restricted to the CNS and is more ubiquitous than initially predicted.

Keywords: pannexin 2, gap junction, gene transcription, protein expression, protein distribution, central nervous system (CNS), mouse, mRNA

INTRODUCTION

Gap junction proteins are traditionally described as aqueous plasma membrane channels which allow rapid cell-to-cell communication by directly connecting the cytoplasm of adjacent cells. In chordates, connexins (Cxs) are the canonical gap junction proteins while gap junctions in invertebrates are formed exclusively by the evolutionarily unrelated innexin (Inx) family. In 2000, another small gene family named pannexin (Panx) was identified based on sequence homology with the Inx family and was found to be expressed alongside Cxs in chordates (Panchin et al., 2000). Three distinct Panx paralogs (Panx1, Panx2 and Panx3) were initially identified in vertebrates (Panchin et al., 2000; Panchin, 2005; Barbe, 2006) but recent studies showed that Panx1 has been retained as two independent ohnologs in teleost as a result of an ancestral whole genome duplication (Bond et al., 2012; Kurtenbach et al., 2013). Despite the lack of sequence similarity between Inxs/Panxs and Cxs, both families share structural resemblance. Cxs and Panxs both have a predicted topology consisting of four membrane-spanning domains, two extracellular loops, a cytoplasmic loop, and cytoplasmic N- and C-termini (Panchin, 2005). Despite sharing structural resemblance with Cxs, the ability of Panx channels to form gap junctional coupling remains controversial. A few groups reported that Panx1 and Panx3 can form cell-cell junctional channels (Bruzzone et al., 2003; Vanden Abeele et al., 2006; Lai et al., 2007; Ishikawa et al., 2011; Sahu et al., 2014) but their observations were limited primarily to heterologous or over-expression systems and undisputable evidence supporting Panx-based coupling is still lacking. In contrast to Cxs, all three Panxs are glycosylated at their extracellular loops (Penuela et al., 2009) with carbohydrate moieties that sterically hinder the docking of channels from adjacent cells (Boassa et al., 2007). Therefore, it is largely accepted that under physiological conditions, Panx channels primarily form non-junctional membrane channels controlling the exchange of ions and small molecules between the cytoplasm and extracellular space and do not significantly contribute to direct cell-to-cell gap junctional communication (Sosinsky et al., 2011).

Abbreviations: Cx, connexin; Inx, innexin; Panx, pannexin; qPCR, quantitative real-time PCR; GFAP, glial fibrillary acidic protein; HRPO, horseradish peroxidase; DEPC, diethylpyrocarbonate; Polr2a, DNA-directed RNA polymerase II subunit RPB1; SDS-PAGE, sodium dodecyl sulfate polyacrylamide gel electrophoresis; BCA, bicinchoninic acid; TCE, 2,2,2-trichloroethanol; NCL, nitrocellulose; PB, phosphate buffer.

Several gene expression profiling studies reported that Panx2 transcriptional activity is largely restricted to the central nervous system in human (Baranova et al., 2004), rat (Bruzzone et al., 2003) and zebrafish (Zoidl et al., 2008; Bond et al., 2012). Minimal Panx2 mRNA levels have also been detected in some non-neural tissues such as the eye, thyroid, prostate, kidney, liver, heart and olfactory epithelium (Bruzzone et al., 2003; Dvoriantchikova et al., 2006; Bond et al., 2012; Zhang et al., 2012) but given the much larger Panx2 mRNA levels found in the CNS, Panx2 transcript and corresponding protein are largely assumed to be primarily expressed in the CNS. In the healthy brain, Panx2 protein was shown to have a complex distribution pattern and is expressed in pyramidal cells and interneurons alike (Zappalà et al., 2007). Interestingly, Panx2 protein was also detected in astrocytes following ischemia in the rat but not in the healthy brain (Zappalà et al., 2007). Panx2 protein is also present in hippocampal neural progenitors and mature neurons both *in vitro* and *in vivo* (Swayne et al., 2010). However, because Panx2 is believed to be primarily CNS-specific, the mapping of Panx2 protein distribution in other tissues has not been undertaken.

In this study, we compared Panx2 gene transcription and protein expression profiles in mouse tissues using a combination of real-time qPCR, Western blot and immunofluorescence. Our results reveal that Panx2 mRNA and protein levels are not correlated and demonstrate that Panx2 protein expression is more ubiquitous than initially predicted.

MATERIALS AND METHODS

ANIMAL CARE

All experiments were performed in accordance with the guidelines established by the Canadian Council on Animal Care and were approved by the University of British Columbia Animal Care Committee (protocol number A11-0169).

ANTIBODIES

The two Panx2 mouse monoclonal antibodies (clones N121A/1 and N121A/31) were generated by UC Davis/NIH NeuroMab Facility (Davis, CA, USA) using an immunogen made of the entire rat Panx2 protein sequence (accession number P60571) minus the first 10 amino acids. Both clones were used at 20 μg/mL for immunofluorescence and 5 μg/mL for Western immunoblotting or dot blotting. The rabbit anti-Panx1 polyclonal antibody was generously provided by Dr. Dale Laird from the University of Western Ontario (London, ON, Canada) and was used at 2 μg/mL for immunofluorescence and 0.4 μg/mL for Western immunoblotting. The rabbit anti-GFAP (Sigma, St. Louis, MO, USA) was used at 1:500. Purified immunoglobulin from non-immunized mouse was obtained from Jackson Immunoresearch (cat# 015-000-003; West Grove, PA, USA) and was used at the same concentration as the anti-Panx2 antibodies. AlexaFluor- and HRPO-conjugated goat secondary antibodies were obtained from Invitrogen (Carlsbad, CA, USA) and Sigma (St. Louis, MO, USA) respectively.

CELL CULTURE

Wild-type C6 glioma cells as well as C6-Panx1GFP, C6-Panx2 and C6-Panx2GFP stable transfectants were cultured as previously described (Lai et al., 2007, 2009). Briefly, cells were grown in low glucose DMEM (Sigma-Aldrich, St. Louis, MO, USA) containing 10% fetal bovine serum, 10 units/mL penicillin, and 10 μg/mL streptomycin at 37°C and 5% CO_2. Primary cultures of astrocytes were prepared as previously described (Le et al., 2014). Briefly, cortices were dissected from early postnatal (P0–P1) mouse pups, freed of meninges, minced and mechanically triturated in DMEM. The cell suspension was then strained through a 70 μm filter and seeded into T75 flasks (2 cortices/flask). Cells were cultured in DMEM (Sigma-Aldrich, St. Louis, MO, USA) containing 10% fetal bovine serum, 10 units/mL penicillin, and 10 μg/mL streptomycin at 37°C and 5% CO_2 and the medium was initially replaced 3 days after plating and every other day subsequently. After 7–8 days, the flasks were vigorously shaken to remove loosely attached cells and primary astrocytes were harvested with trypsin-EDTA (Invitrogen, Carlsbad, CA, USA) and frozen in DMEM, 10% FBS, and 8% DMSO. Frozen astrocytes were thawed and plated on glass coverslips coated with poly-L-ornithine (0.01% solution, Sigma-Aldrich, St. Louis, MO, USA). Cultures were maintained for 5, 10 or 15 days prior to staining. The percentage of Panx2-positive astrocytes was defined as the number of cells that stained positively for Panx2 divided by the number of nuclei. For each time point, that ratio was calculated by averaging the values obtained from three coverslips with 10 field of views (168 × 225 μm) per coverslip.

EPITOPE MAPPING

A library of 70 overlapping peptides covering the entire sequence of the rat Panx2 (Uniprot accession number P60571) minus the four transmembrane domains was obtained from Genscript (Piscataway, NJ, USA). Peptides were 15 amino acids in length with a 7 amino acids overlap. A total of 100 μg of peptide was pre-incubated overnight at 4°C in 100 μL of dot-blot buffer (8 M Urea, 100 mM NaH_2PO_4, 10 mM Tris, pH 8.0) containing 10 μg bovine serum albumin. Peptides were then dot-blotted on nitrocellulose membrane (Bio-Rad, Hercules, CA, USA), washed with sodium phosphate buffered saline (PBS, pH 7.4), dried at 37°C and blocked for 1 h in milk solution (4% nonfat milk, 20 mM Tris, 150 mM NaCl pH 7.4). The membrane was then immunoprobed for 2 h at room temperature with the primary antibody followed by HRPO-conjugated secondary antibodies (Sigma, St. Louis, MO, USA) for 1 h at room temperature.

RNA ISOLATION AND REAL-TIME QUANTITATIVE PCR

Three 3–5 month old mice were deeply anesthetized by intraperitoneal injection of sodium pentobarbital and perfused transcardially with 10–15 mL of PBS (pH 7.4) followed by 10–15 ml of aqueous ammonium sulfate solution (5.3 M ammonium sulfate, 25 mM sodium citrate, 10 mM EDTA, pH5.2) to precipitate degenerative RNases. Organs were rapidly harvested and stored at −80°C in the same solution. Total RNA was harvested from 50 mg of tissue using Trizol Reagent (Invitrogen, Carlsbad, CA, USA) according to the manufacturer's directions. Air-dried RNA samples were re-solubilized in DEPC-treated double distilled H_2O and RNA quantity and purity was assessed using a NanoDrop 1000 Spectrophotometer (Thermo Scientific, Waltham, MA, USA). All samples had A260/280 and A260/230 ratios above 1.9 and 2.3 respectively. A total of 500 ng per sample was reverse transcribed into cDNA in a 10 μL reaction volume using qScript (Quanta Biosciences, Gaithersburg, MD, USA) according to the manufacturer's instructions. Real-time qPCR was performed in 18 μL reaction volume containing 45 ng of cDNA and 0.4 μM primers diluted in 2× Fast Plus EvaGreen® qPCR Master Mix (Biotium Inc., Hayward, CA, USA). The following primer pairs were used to amplify Panx2 cDNA (forward 5′-AGAAGGCCAAGACTGAGGCG-3′ and reverse 5′- GGAGCATCTTTGGTGGGTGC-3′) and the reference gene DNA-directed RNA polymerase II subunit (RPB1) (Polr2a) cDNA (forward 5′- AGCTGGTCCTTCGAATCCGC-3′ and reverse 5′- TGGACTCAATGCATCGCAGGA-3′). Primers were designed to span an exon junction to prevent amplification of genomic DNA. Samples were amplified in duplicate using the CFX96 Real-Time PCR Detection System (Bio-Rad, Hercules, CA, USA) with the following cycling scheme: 2 min at 95°C followed by 50 cycles consisting of 5 s denaturation at 95°C, 5 s annealing at 60°C and 25 s elongation at 72°C. Raw data were exported as text files and analyzed with the qPCR package for R (Ritz and Spiess, 2008). Amplification efficiencies and Cy0 crossing points (Guescini et al., 2008) were calculated from a 5-parameter log-logistic model fitted to the raw fluorescence data to accommodate asymmetrical amplification curves (Spiess et al., 2008). Values from duplicated qPCR runs were averaged. Expression ratios were calculated from three biological replicates by the Pfaffl method to correct for variation in PCR efficiency (Pfaffl, 2001) and normalized against the reference gene Polr2a. Spinal cord mRNA levels were used as baseline to compare Panx2 expression across tissues. Propagation of error was estimated by a Monte Carlo simulation with 10,000 iterations as described in the qPCR package documentation.

PROTEIN ISOLATION AND WESTERN BLOTTING

Organs were quickly collected after transcardial perfusion with PBS, flash frozen in liquid nitrogen and stored at −80°C until needed. Tissues or cells were homogenized in RIPA buffer (150 mM NaCl, 50 mM Tris-HCl pH 8.0, 0.5% Sarkosyl, 1% IGEPAL, 0.1% SDS) containing protease inhibitors (Pierce, Rockford, IL, USA) and phosphatase inhibitors (Sigma, St. Louis, MO, USA). Protein concentration was determined using a bicinchoninic acid (BCA) assay kit (Pierce, Rockford, IL, USA) and 50 μg was separated on 10% Tris-glycine SDS-PAGE gels containing 0.5% 2,2,2-trichloroethanol (TCE; Sigma, St. Louis, MO, USA). Upon electrophoresis completion, protein bands were visualized at 300 nm on an AlphaImager 3400 transilluminator (AlphaInnotech, San Leandro, CA, USA) as previously described (Ladner et al., 2004) and electroblotted on nitrocellulose (NCL) membrane (Bio-Rad, Hercules, CA, USA). Protein bands on NCL were re-visualized under UV for quantification and total protein normalization (Gürtler et al., 2013). For analysis which did not require quantification, TCE was omitted and PVDF membranes were used (Bio-Rad, Hercules, CA, USA). Membranes were blocked at room temperature for 1 h in milk solution (4% nonfat milk, 20 mM Tris, 150 mM NaCl, pH 7.4) and probed with primary antibodies at 4°C overnight followed by HRPO-conjugated secondary antibodies (Sigma, St. Louis, MO, USA) for 1 h at room temperature. All antibodies were diluted in blocking solution. HRPO activity was visualized with Amersham ECL Prime Western Blotting Detection Reagent (GE Healthcare Life Sciences, Pittsburgh, PA, USA) or SuperSignal West Femto Chemiluminescent Substrate (Thermo Scientific, Waltham, MA, USA) and exposed on Bioflex Econo films (Clonex, Markham, ON, Canada). Image acquisition for Western blot quantification was done as previously described (Gassmann et al., 2009). Briefly, film images were acquired on an AlphaImager 3400 (AlphaInnotech, San Leandro, CA, USA) under stable transillumination and fitted with CCD camera lacking automatic gain control. Final 16-bit 1392×1040 pixel images were corrected for shading to compensate for non-homogenous illumination and densitometry analysis was performed using the Image Studio Lite software (LI-COR, Lincoln, NE, USA). Panx2 protein ratios were calculated by dividing the band density of each tissue by the band density of the spinal cord. A tissue lysate from spinal cord was resolved on each gel to permit between gel comparisons. Panx2 protein ratios were calculated from three independent biological replicates.

IMMUNOFLUORESCENCE

Three 3–5 month old mice were transcardially perfused with PBS followed by 10–15 mL 10% formalin (Fisher) or 4% paraformaldehyde (PFA) in PBS. Tissues were rapidly harvested and postfixed overnight at 4°C in the same fixative. Tissues were equilibrated at 4°C in 30% sucrose in PBS containing 0.05% sodium azide for cryoprotection, embedded in Tissue-Tek O.C.T. compound (Sakura Finetek, Torrance, CA, USA), frozen, cryosectioned at 10 μm thickness and air-dried. Tissue sections were rehydrated in 0.1 M phosphate buffer (PB, pH 7.4), post-fixed with 4% PFA for 10 min and washed twice with PB. Antigen retrieval was performed by incubating the sections in 10 mM sodium citrate (pH 8.5) at 80°C for 30 min. Sections were cooled to room temperature, washed with PB and treated with 1% sodium borohydride in PB for 30 min. After several washes in PB, samples were blocked for 1 h at room temperature with 5% goat serum in PB containing 0.2% Triton X-100 and incubated overnight at 4°C in primary antibody diluted in 1% goat serum in PB. Sections were then washed in PB and incubated for 1 h at room temperature with the appropriate AlexaFluor 488 or 568 secondary antibodies (Invitrogen, Carlsbad, CA, USA), followed by mounting in ProLong Gold antifade reagent with DAPI (Invitrogen). Cells grown on coverslips were simply fixed with 4% PFA for 15–20 min prior blocking and immunostaining. Imaging was performed on a Leica TCS SP5 confocal microscope (Leica, Mannheim, Germany). When indicated, images were further processed with an iterative Lucy-Richardson deconvolution algorithm (Vonesch and Unser, 2008). All images were displayed as individual optical sections taken from a z-stack and not as maximal projection. All image acquisition and post-acquisition processing steps were kept constant when comparing sections stained with anti-Panx2 antibodies and immunoglobulins from non-immunized animal.

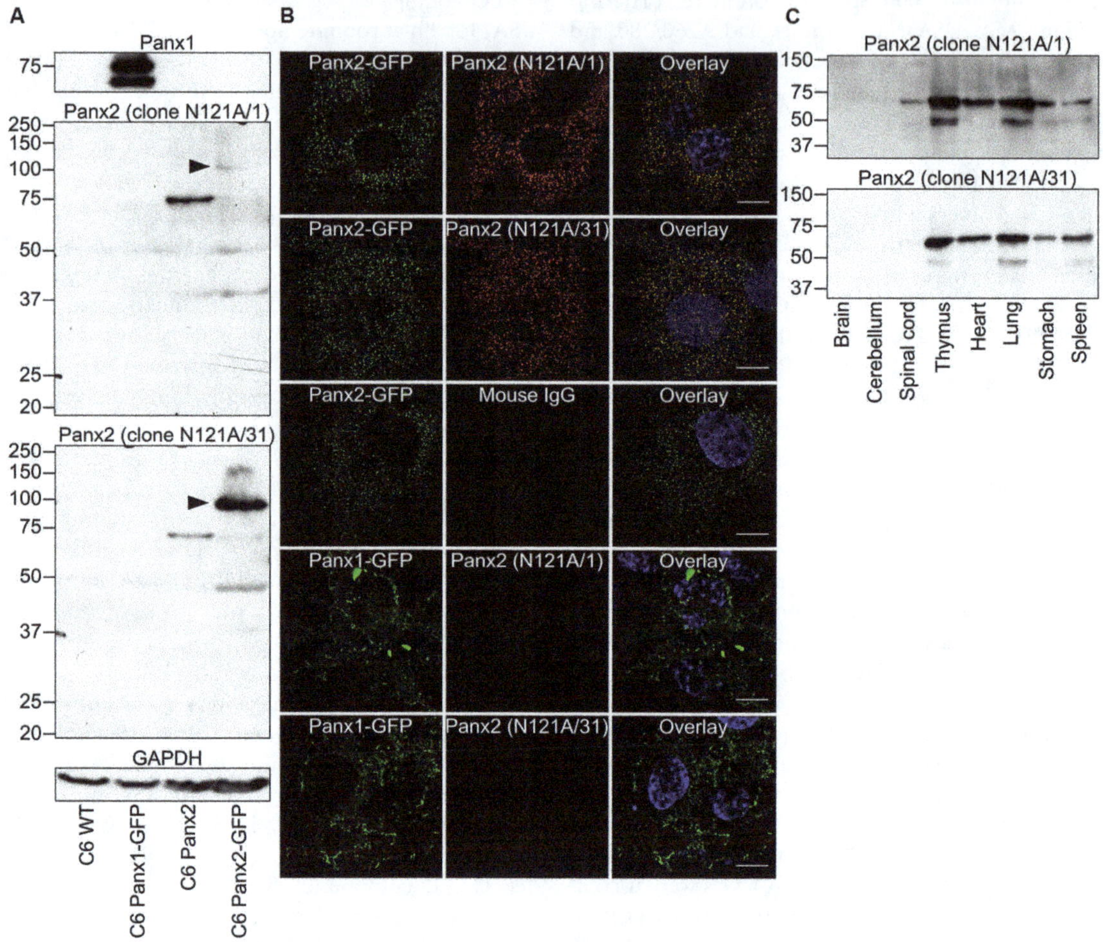

FIGURE 1 | Specificity of two novel anti-Panx2 monoclonal antibodies. Two anti-Panx2 monoclonal antibodies (clones N121A/1 and N121A/31) showed high specificity both in Western blot **(A,C)** and immunolabeling **(B)**. **(A)** Both clones detected a single band of the proper size in C6 cells overexpressing Panx2 but did not cross-react with Panx1 or identify unspecific protein bands in wild-type C6 cells. The molecular weight of the protein identified by both clones shifted by ~27 kDa when the Panx2 C-terminal was tagged with GFP (black arrowheads). Interestingly, the avidity of clone N121A/1 was drastically reduced when the Panx2 was tagged with GFP. This is explained by the fact that clone N121A/1 recognizes an epitope immediately adjacent to the GFP tag (**Figure 2**). **(B)** Immunolabeling signal from both clones co-localized with GFP fluorescence emitted from Panx2GFP but not Panx1GFP when expressed in C6 cells. Immunoglobulin from non-immunized mouse did not give any signal thereby ruling out the possibility of unspecific binding of mouse IgG. Scale bars = 20 μm. **(C)** Both clones were equally specific in Western blot performed with various tissue lysates and identified a protein of ~70 kDa. The occasional detection of a ~50 kDa band in some tissues suggest that the antibodies might recognize Panx2 degradation products or alternatively that the anti-mouse secondary antibody detects endogenous immunoglobulin heavy chains.

RESULTS

CHARACTERIZATION OF TWO NOVEL MONOCLONAL ANTIBODIES SPECIFIC FOR Panx2

We initially characterized the selectivity of two novel anti-Panx2 monoclonal antibodies (N121A/1 and N121A/31) by showing that both clones identified a single band of the proper size in C6 cells overexpressing rat Panx2 (**Figure 1A**). Importantly, none of the antibody cross-reacted with Panx1 (**Figure 1A**) which has been shown to be co-expressed with Panx2 in the CNS (Vogt et al., 2005). Selectivity was further supported by an electrophoretic mobility assay showing that the band identified by both clones shifted by ~27 kDa when the Panx2 C-terminal tail was tagged with GFP (**Figure 1A**). Intriguingly, tagging Panx2 with GFP decreased the avidity of clone N121A/1. Using a library of overlapping peptides covering the amino acid sequence of Panx2, we mapped the epitope of clone N121A/1 to the last 15 amino acids of the Panx2 C-terminal tail (**Figure 2**). It is therefore likely that the addition of a GFP tag adjacent to the epitope caused steric hindrance and reduced the avidity of clone N121A/1. We also tested the antibodies by immunofluorescence and showed that the labelling from both clones overlapped with the GFP fluorescence signal emitted by Panx2-GFP but not by Panx1-GFP (**Figure 1B**).

To test whether the novel antibodies could also be used on samples expressing endogenous Panx2 protein we immunoprobed protein lysates prepared from eight tissues and separated by SDS-PAGE (**Figure 1C**). Our results show that both clones

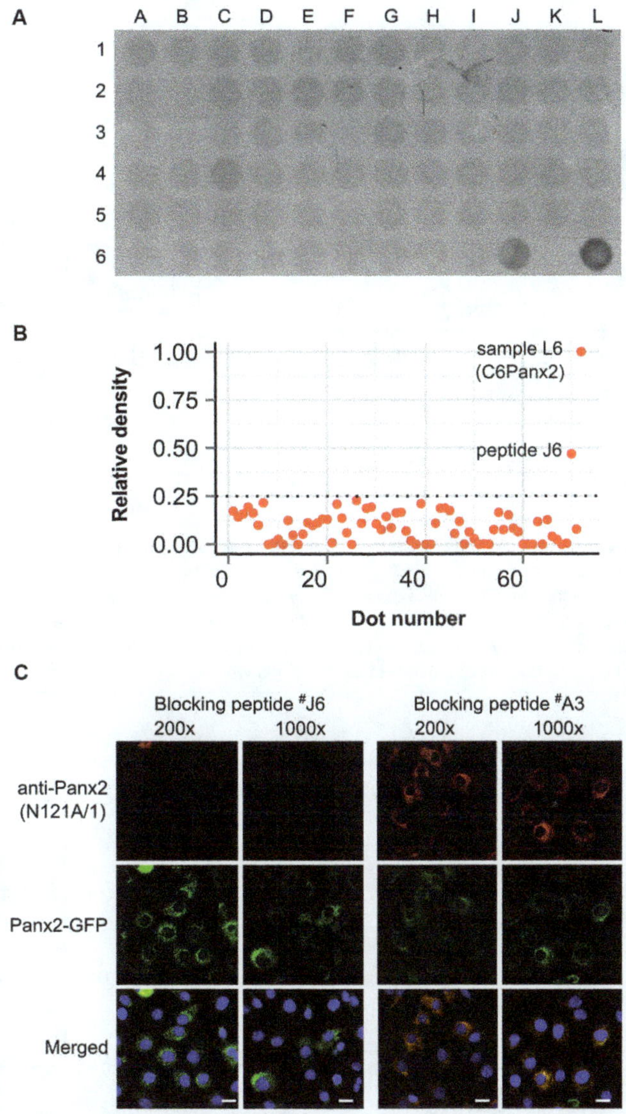

FIGURE 2 | Clone N121A/1 recognizes an epitope located within the last 15 amino acids of the Panx2 C-terminal. (A) A library of 70 overlapping peptides spanning the entire Panx2 amino acid sequence minus the four transmembrane domains was dot blotted (A1 to J6) and immunoprobed with the N121A/1 anti-Panx2 antibody. A total of 25 μg of protein lysate from wild-type C6 and C6Panx2 glioma cells (K6 and L6 respectively) was also dot blotted alongside peptides to provide negative and positive control respectively. The clone N121A/1 specifically recognized C6Panx2 protein lysate and peptide J6 corresponding to the last 15 amino acids of the Panx2 C-terminal (TFEEPRTVVSTVEF). **(B)** Densitometry analysis of the dot blot staining shown in **(A)**. Only peptide J6 showed signal above threshold. Values were normalized against the density of the C6Panx2 sample. **(C)** The epitope identity was further confirmed by a blocking peptide assay. Pre-absorbing the N121A/1 anti-Panx2 antibody with a 200- or 1000-fold molar excess of peptide J6 dramatically reduced immunolabeling while pre-absorption with a peptide randomly selected along the Panx2 amino acid sequence (peptide A3) did not alter the labeling intensity. Scale bars = 20 μm.

recognized a band of approximately 70 kDa corresponding to the expected size of endogenous Panx2 as previously reported (Zappalà et al., 2007; **Figure 1C**). Overall our results identified two different commercial monoclonal antibodies specific for Panx2. Both clones recognize rat and mouse Panx2 and we also successfully used clone N121A/1 to detect human Panx2 (data not shown). To avoid redundancy and because both clones were equally specific we selected clone N121A/1 for subsequent analysis.

Panx2 HAS A UBIQUITOUS PROTEIN EXPRESSION PROFILE

Interestingly our initial results showed substantial Panx2 protein amount outside the nervous system (**Figure 1C**). To further compare the Panx2 protein expression profile of different tissues, we carried out semi-quantitative densitometry analysis on protein lysates obtained from 16 tissues, separated by SDS-PAGE and immunoprobed for Panx2 (**Figure 3A**). Because the expression of common reference proteins was subjected to important fluctuations across tissues (data not shown), a stain-free total protein

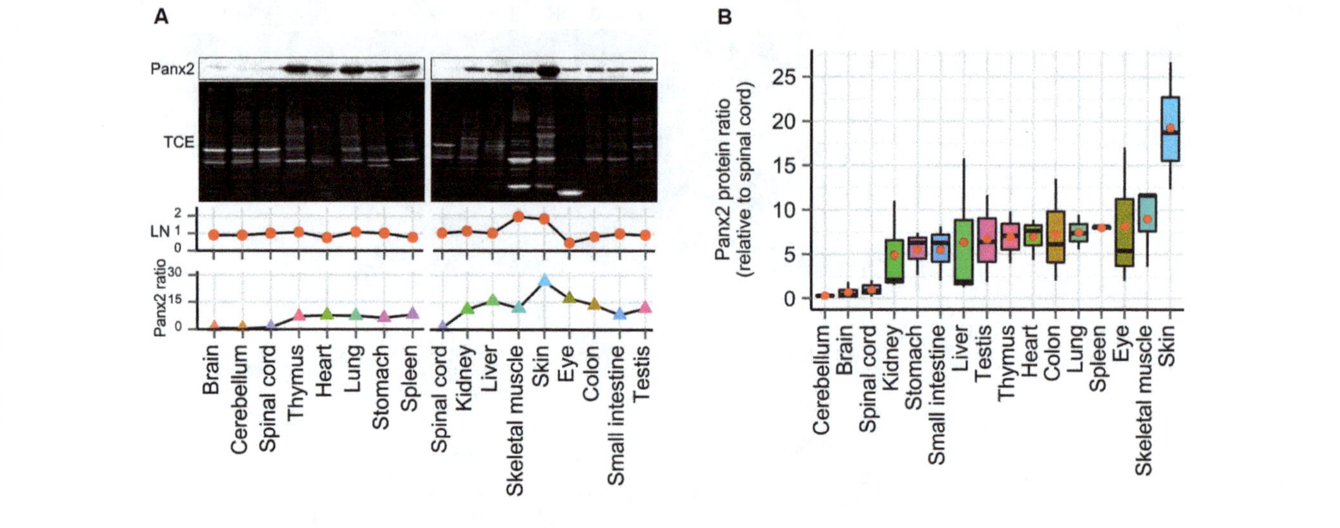

FIGURE 3 | Panx2 protein is ubiquitously expressed. (A) Panx2 protein levels were semi-quantified in 16 tissues using stain-free total protein quantification to normalize protein levels across samples. Following exposure to TCE and UV, protein bands electroblotted on nitrocellulose (NTC) were visualized by fluorescence (TCE, second panel) and loading normalization (LN) was performed by dividing the fluorescence density from an entire individual lane by the total fluorescent density measured from the spinal cord lane on the corresponding NTC membrane (LN, third panel). Normalized Panx2 protein ratios were expressed relative to Panx2 levels found in the spinal cord (fourth panel). **(B)** Panx2 protein ratios were calculated from three different mice (relative to spinal cord, red dots represent mean values). Panx2 protein levels were lower in the CNS than any other tissues.

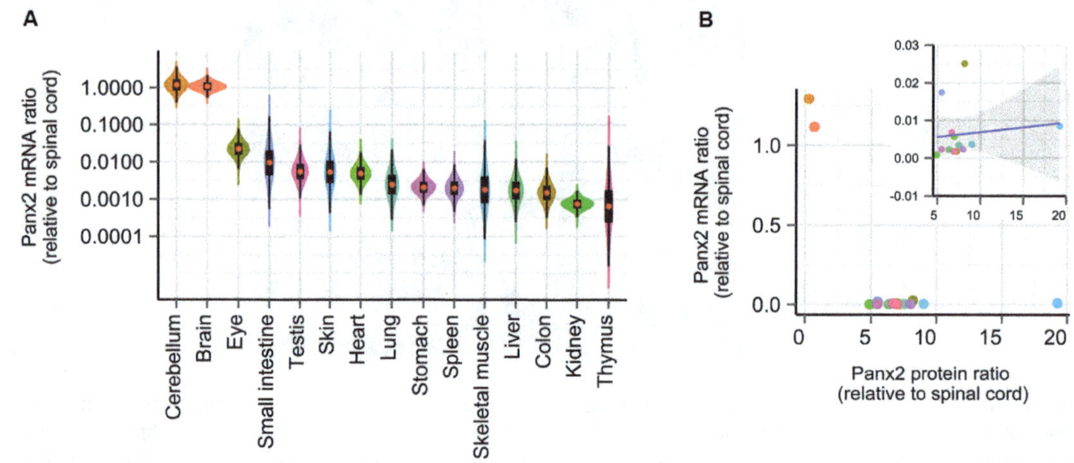

FIGURE 4 | Panx2 transcriptional activity does not correlate with protein levels. (A) Panx2 mRNA levels were measured by real-time qPCR in 16 tissues and expressed relative to Panx2 mRNA levels found in the spinal cord. Error propagation was estimated by a Monte Carlo simulation and data distribution represented using a combination of violin and box plots (red dots represent mean values). Data were plotted on a logarithmic scale. Panx2 mRNA levels were several orders of magnitude higher in the CNS than in any other tissues. **(B)** Panx2 protein and mRNA levels were not correlated. No significant correlation was observed even after values from the CNS were eliminated (inset).

normalization strategy was employed to control for even loading as previously described (Ladner et al., 2004; Gürtler et al., 2013). Briefly, we incorporated TCE in the gel formulation which, upon ultraviolet (UV) irradiation, catalyzes a covalent reaction with tryptophan residues. This reaction emits fluorescence that can be imaged and documented in gel and following protein transfer on membranes (Ladner et al., 2004; Gürtler et al., 2013). The Panx2 staining density for each tissue was then normalized against the intensity of the TCE-fluorescence measured for the entire lane after protein transfer on nitrocellulose membrane (**Figure 3A**). Our data indicate that Panx2 protein is present at substantial levels in every tissue studied (**Figures 3A,B**). More surprisingly, in contradiction with current predictions, Panx2 protein was lower in the nervous system than in any other tissues (**Figure 3B**).

TRANSCRIPTIONAL ACTIVITY DOES NOT PREDICT Panx2 PROTEIN LEVELS

Gene profiling studies have reported that Panx2 mRNA expression is largely restricted to the CNS in human (Baranova et al.,

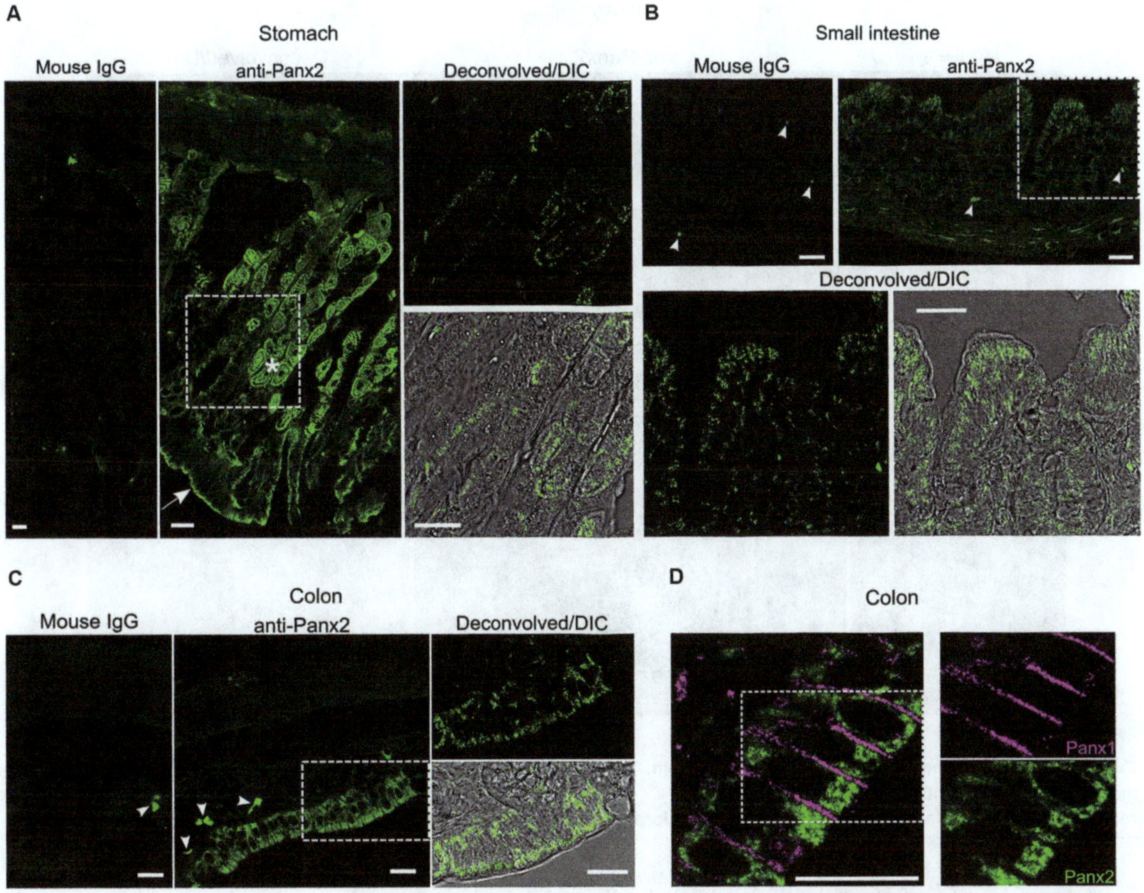

FIGURE 5 | Panx2 protein is expressed in the gastrointestinal tract. Sections of mouse stomach **(A)** small intestine **(B)** and colon **(C)** were stained with an immunoglobulin from non-immunized mouse (Mouse IgG panels) or the N121A/1 monoclonal anti-Panx2 antibody (anti-Panx2 and Deconvolved/DIC panels). Panx2 protein was heavily expressed in the parietal (asterisk) and epithelial (arrow) cells of the stomach **(A)** and the epithelial cells of the small **(B)** and large **(C)** intestine. Panx2 distribution was primarily perinuclear and cytoplasmic. No specific staining was observed with the IgG from non-immunized mouse. Putative lymphocytes expressing endogenous immunoglobulins were occasionally labeled by the anti-mouse secondary antibody (arrowheads). **(D)** Panx1 and Panx2 showed distinct subcellular distribution in the colon. Scale bars: 20 μm.

2004), rat and mouse (Bruzzone et al., 2003; Dvoriantchikova et al., 2006) and zebrafish (Zoidl et al., 2008; Bond et al., 2012) but a similar profiling study has not been completed in mouse. To determine whether our observations on Panx2 protein expression could be explained by species-specific Panx2 transcriptional activity, we compared the Panx2 transcription profile using RNA isolated from 16 mouse tissues, reverse-transcribed into cDNA and analyzed by real-time qPCR. Although primers were designed to span an exon junction, end-point PCR was initially performed using non-transcribed RNA as template to confirm the absence of genomic DNA amplification (data not shown). We also tested the specificity of our primer pair by visualizing the amplification product by gel electrophoresis and by analyzing the amplicon's melting curve (data not shown). Our results are in accordance with previous studies (Bruzzone et al., 2003; Baranova et al., 2004; Dvoriantchikova et al., 2006; Zoidl et al., 2008; Bond et al., 2012) and showed that Panx2 transcriptional activity largely predominates in the CNS (**Figure 4A**). Panx2 transcript levels detected in non-neural tissues were several orders of magnitude lower than in the CNS (**Figure 4A**). Furthermore, we demonstrated that there is no significant correlation between Panx2 transcript and corresponding protein levels (**Figure 4B**).

Our results showed that Panx2 mRNA and protein levels are not correlated when compared across different tissues but does not exclude a possible correlation within a specific tissue as opposed to between different tissues. This scenario appears unlikely however as we have shown that Panx2 protein levels remain surprisingly constant in the brain over a developmental period during which Panx2 mRNA levels have been shown to be temporally up-regulated (data not shown) (Vogt et al., 2005). Overall, these results suggest that regulatory mechanisms unrelated to transcriptional activity must also control Panx2 protein levels and indicate that Panx2 protein levels cannot be directly inferred from the quantification of its transcript levels.

Panx2 PROTEIN IS LOCALIZED TO CYTOPLASMIC COMPARTMENTS

We next characterized the expression and distribution of Panx2 in different tissues by immunofluorescence. In the gastrointestinal

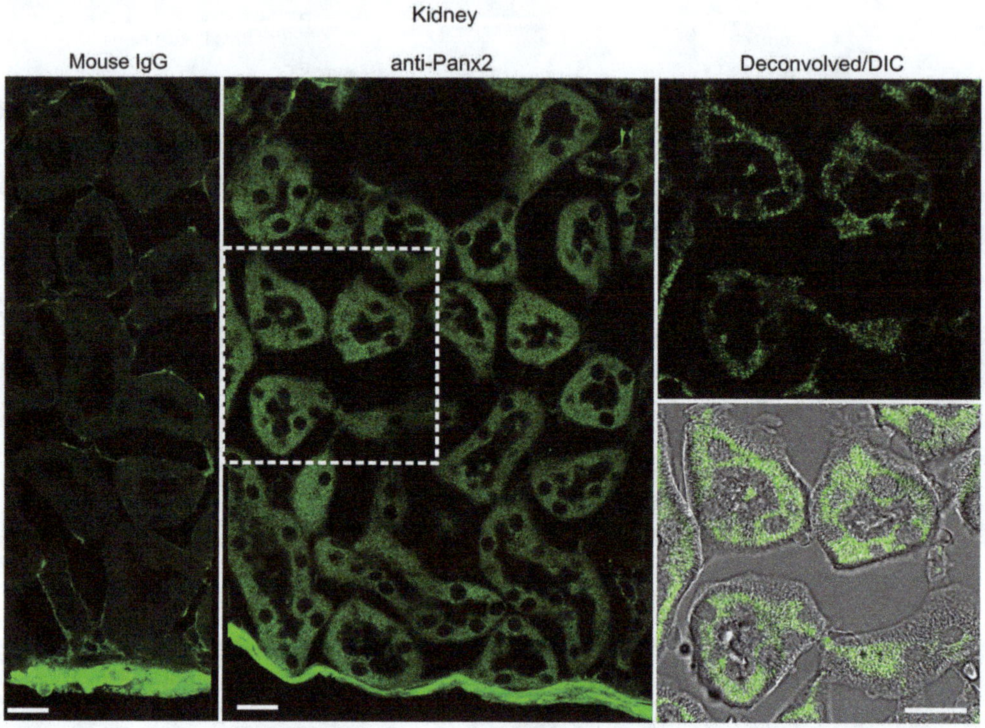

FIGURE 6 | Cuboidal cell from kidney tubule express Panx2 protein. Sections of mouse kidney labeled with an immunoglobulin from non-immunized mouse (Mouse IgG left panel) or the N121A/1 monoclonal anti-Panx2 antibody (middle and right panels). Panx2 protein was distributed in the cytoplasm of the epithelial cells lining the lumen of renal tubule but was not discernible at the plasma membrane. Scale bars: 20 μm.

tract, an important population of glandular and epithelial cells displayed strong Panx2 immunoreactivity (**Figure 5**). Parietal cells, which secrete gastric acid, and the apical surface of epithelial cells of the stomach were strongly reactive for Panx2 (**Figure 5A**). In the small and large intestine, a population of columnar epithelial cells were also strongly reactive for Panx2 (**Figures 5B,C**). As Panx1 has also been shown to be expressed in the columnar epithelial cells of the human colon (Diezmos et al., 2013), we tested whether Panx1 and Panx2 could co-localize in these cell types. Interestingly, Panx1 and Panx2 did not co-localize but showed quite different subcellular distribution patterns (**Figure 5D**). Panx1 expression was largely restricted to the plasma membrane between the epithelial cells. In contrast, Panx2 was not discernible at the plasma membrane but remained largely confined to the cytoplasmic area (**Figure 5D**).

In the kidney, cuboidal cells forming the single layered epithelium of tubules were strongly labeled (**Figure 6**) whereas glomeruli cells were not (data not shown). Panx2 staining was predominantly cytoplasmic and could not be detected at the plasma membrane. Similarly, germ cells from testis seminiferous tubules showed abundant perinuclear and cytoplasmic but no plasma membrane staining (**Figure 7**).

Panx2 immunoreactivity displayed a distinct pattern in the mouse retina (**Figure 8**). Photoreceptor inner segments protruding into the subretinal space were densely decorated with Panx2-labeled aggregates (**Figure 8**). Only sparse immunoreactivity was observed in the outer nuclear layer which forms the compact layer containing photoreceptor cell bodies (**Figure 8**). Substantial staining was also observed in the outer plexiform layer which comprises a dense network of neuronal synapses formed between photoreceptors and bipolar and horizontal cell dendrites (**Figure 8**).

Despite showing lower Panx2 protein levels than any other tissues (**Figure 3**), Panx2 immunoreactivity was easily distinguishable by immunofluorescence in the CNS (**Figure 9**) and showed a complex expression pattern as previously reported (Zappalà et al., 2007). Panx2 was widely distributed in the cytoplasm of neurons throughout the CNS but was not readily detected in astrocytes *in vivo* (**Figures 9A–C**). Interestingly, we showed that the majority of primary astrocytes (63.8 ± 0.9%) expressed cytoplasmic Panx2 at 5 days *in vitro* (**Figure 9D**). However, the percentage of Panx2-positive astrocytes rapidly declined after 10 and 15 days *in vitro* (6.9 ± 1.2% and 7.3 ± 1.1%) hereby suggesting that Panx2 is expressed by immature but not mature astrocytes. That observation could explain the up-regulation of Panx2 expression seen in astrocytes following ischemia (Zappalà et al., 2007) as ischemia is characterized by astrocyte proliferation.

DISCUSSION

This study reveals that Panx2 protein is more ubiquitous than initially predicted. By performing real-time qPCR and

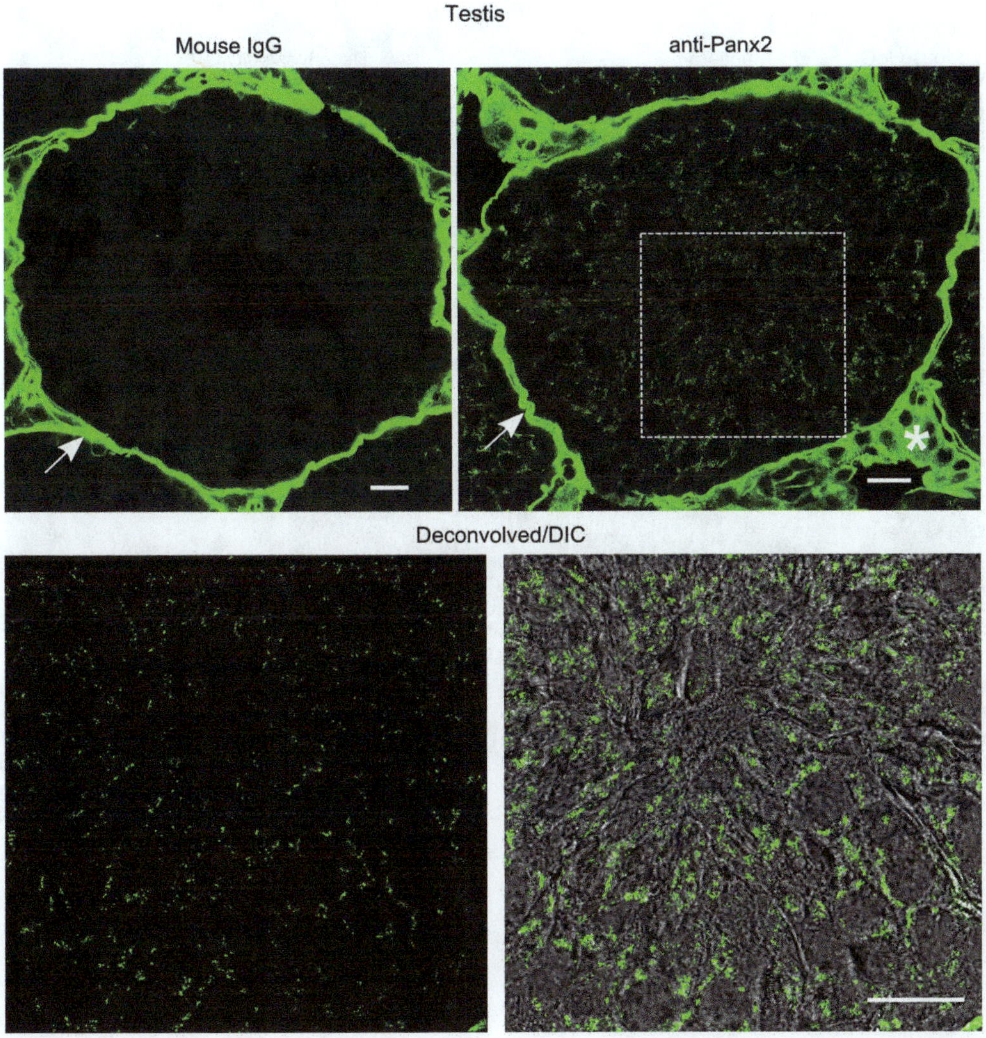

FIGURE 7 | Germ cells from mouse testis express Panx2 protein. Sections of mouse testis were labeled with an immunoglobulin from non-immunized mouse (Mouse IgG panel) or the N121A/1 monoclonal anti-Panx2 antibody (anti-Panx2 and Deconvolved/DIC panels). Panx2 protein was localized in the cytoplasm of germ cells in the seminiferous epithelium. Smooth muscle cells (arrow) and interstitial tissue (asterisk) showed high autofluorescence levels. Scale bars: 20 μm.

semi-quantitative Western blot analysis on a panel of mouse tissues, we showed that fluctuations in Panx2 mRNA abundance do not predict changes in Panx2 protein levels. We showed that Panx2 protein levels are surprisingly more abundant in non-neural tissues than in the CNS; an observation opposite to Panx2 transcriptional activity which is weak in non-neural tissues and largely predominant in the CNS. The ubiquitous expression of Panx2 protein suggests a more fundamental function than the CNS-specific role which was originally proposed. Although the exact function of Panx2 remains elusive, based on our *in vivo* immunofluorescence results we hypothesize that Panx2 channels do not significantly contribute to communication exchange between the intracellular and extracellular spaces but rather control intracellular signaling through cytoplasmic compartments.

ABSENCE OF CORRELATION BETWEEN Panx2 mRNA AND PROTEIN LEVELS

The initial Panx2 gene expression profiles were obtained from Northern blots using commercial rat (Bruzzone et al., 2003) and human (Baranova et al., 2004) mRNA. Although notable differences exist between the two studies, both groups reported that Panx2 mRNA is largely predominant in the CNS; an observation that was subsequently confirmed in zebrafish using real-time qPCR (Zoidl et al., 2008; Bond et al., 2012). Our results show that Panx2 mRNA follows a similar expression profile in the mouse since Panx2 transcript levels are 40 to over 1600 times higher in the CNS than in other tissues. Because of this dramatic disparity in Panx2 mRNA expression it has long been assumed that Panx2 protein was preferentially, if not exclusively, expressed in the CNS.

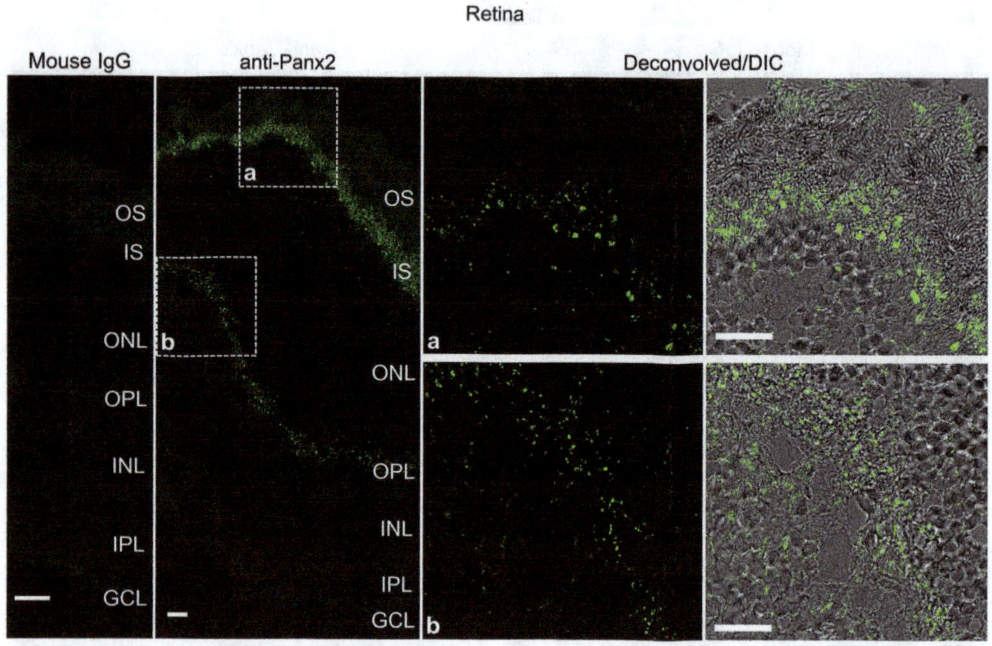

FIGURE 8 | Panx2 protein is expressed in the photoreceptor inner segment and outer plexiform layer of the mouse retina. Sections from mouse retina were labeled with an immunoglobulin from non-immunized mouse (Mouse IgG panel) or the N121A/1 monoclonal anti-Panx2 antibody (anti-Panx2 and Deconvolved/DIC panels). Panx2 protein was primarily expressed in the photoreceptor inner segment (IS) (inset a) or the outer plexiform layer (OPL) (inset b). Panx2 clustered in small to large aggregates which appeared to be primarily cytoplasmic. However, because of the high degree of cellular compaction we could not reach definitive conclusions regarding Panx2 subcellular distribution in the retina. OS: photoreceptor outer segment, IS: photoreceptor inner segment, ONL: outer nuclear layer, OPL: outer plexiform layer, INL: inner nuclear layer, IPL: inner plexiform layer, GCL: ganglion cell layer. Scale bars: 20 μm.

However, in almost every organism steady-state transcript concentrations only partially correlate with protein expression levels (de Sousa Abreu et al., 2009) and the assumption that transcripts can predict protein abundances has been heavily challenged. Post-transcriptional regulatory mechanisms have overwhelming influence on changes observed at the proteome level (Foss et al., 2011) and protein levels cannot be accurately extrapolated from transcript levels because several factors unrelated to transcriptional control also directly influence protein levels. For example, protein degradation rate has been shown to influence the correlation between transcripts and corresponding protein levels as stable proteins are less affected by perturbations in mRNA levels than proteins with high turnover rates (Raj et al., 2006). Hence, the long half-life of Panx proteins (Penuela et al., 2007) could efficiently buffer important fluctuations in mRNA levels and decrease the impact of Panx2 transcripts on Panx2 protein levels. Bearing this information in mind, it is safe to affirm that variations of Panx2 transcript levels should be interpreted restrictively, without assuming equivalent changes at the protein level.

Mass-spectrometry-based proteomics can perform large-scale unbiased analyses of biological systems and examine which genes are translated into proteins in specific tissues. Recently, two groups assembled and published mass-spectrometry-based drafts of the human proteome into databases available online for real-time analysis (Kim et al., 2014; Wilhelm et al., 2014). Interestingly, unique Panx2 peptides were identified in the ileum, colon and ovary (Wilhelm et al., 2014) as well as the gut, spinal cord, urinary bladder, liver, ovary, testis and prostate (Kim et al., 2014). Although substantial improvements are still needed to achieve a complete and quantitative proteome coverage, these independent studies nonetheless corroborate our results and demonstrate that Panx2 protein expression is not restricted to the CNS.

It is important to note that Panx2 protein ratios showed high variability in some tissues (**Figure 3B**). This is more likely attributable to the limited dynamic range of the chemiluminescence technique that was used for the quantification of Panx2 protein expression. An alternative would have been to use a ratiometric analysis based on infrared detection of protein bands to increase the linear detection range and increased reproducibility (Zellner et al., 2008).

Panx2: A CYTOPLASMIC UNUSUAL SUSPECT

Technical reasons such as prevalent autofluorescence prevented the analysis of certain tissues by immunofluorescence. Nonetheless, our study shows that Panx2 protein was heavily distributed in the cytoplasmic compartment and could not be readily detected at the plasma membrane in all tissues analyzed by immunofluorescence (9 out of 16) or in cultured primary astrocytes expressing endogenous Panx2 protein. Previous studies had reported cytoplasmic Panx2 in transfected overexpression

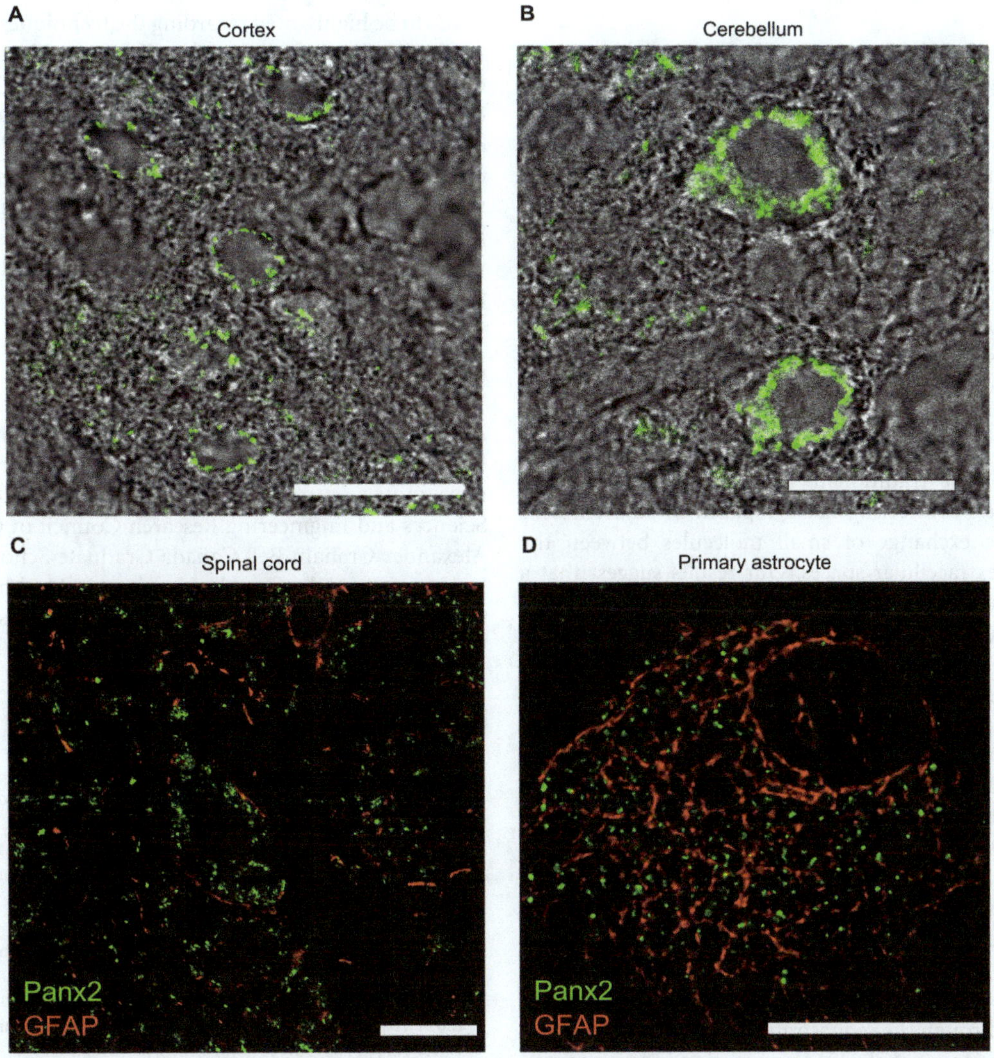

FIGURE 9 | Panx2 protein localizes in the cytoplasm of CNS neurons. Panx2 heavily labeled the cytoplasm, but not the plasma membrane, of cortical neurons **(A)** Purkinje cells **(B)** and spinal cord motoneurons **(C)**. Panx2 was not detected in GFAP-positive astrocytes in the brain (data not shown) or the spinal cord **(C)** but was detected in over 60% of primary astrocytes cultured for 5 days **(D)**. Scale bars: 20 μm.

systems (Lai et al., 2009; Penuela et al., 2009; Bhalla-Gehi et al., 2010) or in neurons and neural progenitor cells (Zappalà et al., 2007; Swayne et al., 2010) but we are the first group to identify endogenous cytoplasmic Panx2 in such a large variety of tissues.

The unique intracellular distribution of Panx2 protein is in striking contrast with Panx1 and Panx3 proteins which are primarily localized at the plasma membrane (Penuela et al., 2007, 2009; Bhalla-Gehi et al., 2010). The cellular localization of Panx proteins is influenced by glycosylation (Penuela et al., 2009). All three Panx paralogs are glycosylated to a high mannose form in the endoplasmic reticulum (ER; Penuela et al., 2009; Bhalla-Gehi et al., 2010) but interestingly only Panx1 and Panx3 proteins form complex glycoprotein species requiring post-translational modifications occurring in the Golgi (Penuela et al., 2009; Bhalla-Gehi et al., 2010). Panx1 and Panx3 proteins follow a COPII-dependent ER to Golgi secretory pathway prior to being trafficked to the plasma membrane (Bhalla-Gehi et al., 2010). In contrast, the absence of complex glycosylated Panx2 suggests Panx2 protein follows a different trafficking pathway which might not involve transition through the Golgi and subsequent trafficking to the plasma membrane. Intriguingly, Panx2 has been shown to colocalize with the endolysosomal enriched mannose-6-phosphate receptor in N2a neuroblastoma cells expressing Panx2 tagged with GFP (Wicki-Stordeur et al., 2013). However, the study used transient overexpression of tagged Panx2 protein which might have resulted in the accumulation of misfolded or misassembled proteins and increased the likelihood of artifactual missorting in endomembrane compartments. Therefore, additional localization studies detecting endogenous Panx2 protein using a combination of different approaches are still needed for the accurate identification of Panx2-positive cytoplasmic compartments.

Others have detected putative Panx2 at the plasma membrane of mature primary hippocampal neurons (Swayne et al., 2010) or in cell types overexpressing Panx2 (Ambrosi et al., 2010). The ectopic expression of Panx1 and Panx2 in NRK cells has also been shown to increase Panx2 trafficking to the plasma membrane (Penuela et al., 2009). However, the physiological relevance of this increase in Panx2 at the plasma membrane is unclear because Panx1/Panx2 heteromeric channels are unstable (Ambrosi et al., 2010). Although we cannot exclude the possibility that undetectable levels of Panx2 are distributed at the plasma membrane in some cell types, we conclude that under physiological conditions Panx2 protein is primarily localized in the cytoplasmic compartment in most, if not all, tissues.

Gap junctions have traditionally been described as plasma membrane channels connecting the cytoplasm of adjacent cells or controlling the exchange of small molecules between the intracellular and extracellular spaces. Our results suggest that a different model must apply to Panx2 because its range of action seems to be restricted to the cytoplasmic milieu. Consequently, we hypothesize that Panx2 can modulate cell activity through nonconventional routes and novel intracellular signaling pathways. In that aspect, it is interesting to note that over-expression of Panx1 and Panx3 can form calcium permeable channels in the ER (Vanden Abeele et al., 2006; Ishikawa et al., 2011). As over-expression of Panx2 in C6 cells showed a prominent signal overlap with the ER (Lai et al., 2009) it is possible that Panx2 can also modulate ER calcium signaling. Moreover, although an essential property of gap junction proteins is their ability to oligomerize to form transmembrane channels, it should be emphasized that gap junctions also have channel-independent functions. For example, connexin 43 (Cx43) has recently been shown to control the biogenesis of autophagosomes through the sequestration of several autophagy-related proteins (Bejarano et al., 2014); a function independent of Cx43 channel activity. Since Panx2 has a long C-terminal tail (301 a.a) it is reasonable to suggest that protein-protein interactions involving its C-terminus are likely to play an important role in the function of Panx2. However, until the exact nature of Panx2 subcellular compartment remains unknown, formulating hypothesis regarding the function of Panx2 remains rather difficult.

Our understanding of Panx2 protein currently assumes that Panx2 function can be extrapolated from our knowledge of the other Panx proteins. More precisely, Panx2 is often perceived as a CNS-specific protein assuming a role complementary, if not redundant, to the function of Panx1 channel. However, our study shows that this assumption is misleading and unlikely to increase our knowledge on any of the Panx channels. Prior to our work, several studies investigating the role of Panx channels outside of the CNS focused exclusively on Panx1 and Panx3 but completely neglected the potential implication of Panx2. As our study shows that Panx2 protein expression is more ubiquitous than initially predicted it would be interesting to revisit these original studies while taking into account the presence of Panx2. This is especially important in the context of Panx1 knockout mice since the deletion of Panx1 could have compensatory effects by altering the expression level of Panx2. Another cautionary note needs to be highlighted regarding the techniques that are currently used to assay Panx2 functionality. Several studies use patch-clamping of the plasma membrane to address the functionality of Panx2 channels (for example see Bargiotas et al., 2011; Poon et al., 2014). Although we cannot totally exclude the presence of Panx2 at the plasma membrane, our study nonetheless shows that Panx2 protein is predominantly in cytoplasmic compartments. Consequently, Panx2 channel properties cannot be solely investigated through electrophysiological recordings at the plasma membrane.

ACKNOWLEDGMENTS

We would like to thank Drs. Stephen Bond and Moises Freitas-Andrade for constructive comments during the preparation of this manuscript and Dr. Hoa T. Le for the preparation of primary mouse astrocytes. This work was supported by the Natural Sciences and Engineering Research Council of Canada (NSERC) Alexander Graham Bell Canada Graduate Scholarship (Maxence Le Vasseur) and the Canadian Institute of Health Research (Christian C. Naus, Wun-Chey Sin). Christian Naus holds a Canada Research Chair.

REFERENCES

Ambrosi, C., Gassmann, O., Pranskevich, J. N., Boassa, D., Smock, A., Wang, J., et al. (2010). Pannexin1 and pannexin2 channels show quaternary similarities to connexons and different oligomerization numbers from each other. *J. Biol. Chem.* 285, 24420–24431. doi: 10.1074/jbc.m110.115444

Baranova, A., Ivanov, D., Petrash, N., Pestova, A., Skoblov, M., Kelmanson, I., et al. (2004). The mammalian pannexin family is homologous to the invertebrate innexin gap junction proteins. *Genomics* 83, 706–716. doi: 10.1016/j.ygeno.2003.09.025

Barbe, M. T. (2006). Cell-cell communication beyond connexins: the pannexin channels. *Physiology* 21, 103–114. doi: 10.1152/physiol.00048.2005

Bargiotas, P., Krenz, A., Hormuzdi, S. G., Ridder, D. A., Herb, A., Barakat, W., et al. (2011). Pannexins in ischemia-induced neurodegeneration. *Proc. Natl. Acad. Sci. U S A* 108, 20772–20777. doi: 10.1073/pnas.1018262108

Bejarano, E., Yuste, A., Patel, B., Stout, R. F. Jr., Spray, D. C., and Cuervo, A. M. (2014). Connexins modulate autophagosome biogenesis. *Nat. Cell Biol.* 16, 401–414. doi: 10.1038/ncb2934

Bhalla-Gehi, R., Penuela, S., Churko, J. M., Shao, Q., and Laird, D. W. (2010). Pannexin1 and pannexin3 delivery, cell surface dynamics and cytoskeletal interactions. *J. Biol. Chem.* 285, 9147–9160. doi: 10.1074/jbc.m109.082008

Boassa, D., Ambrosi, C., Qiu, F., Dahl, G., Gaietta, G., and Sosinsky, G. (2007). Pannexin1 channels contain a glycosylation site that targets the hexamer to the plasma membrane. *J. Biol. Chem.* 282, 31733–31743. doi: 10.1074/jbc.m702422200

Bond, S. R., Wang, N., Leybaert, L., and Naus, C. C. (2012). Pannexin 1 ohnologs in the teleost lineage. *J. Membr. Biol.* 245, 483–493. doi: 10.1007/s00232-012-9497-4

Bruzzone, R., Hormuzdi, S. G., Barbe, M. T., Herb, A., and Monyer, H. (2003). Pannexins, a family of gap junction proteins expressed in brain. *Proc. Natl. Acad. Sci. U S A* 100, 13644–13649. doi: 10.1073/pnas.2233464100

de Sousa Abreu, R., Penalva, L. O., Marcotte, E. M., and Vogel, C. (2009). Global signatures of protein and mRNA expression levels. *Mol. Biosyst.* 5, 1512–1526. doi: 10.1039/b908315d

Diezmos, E. F., Sandow, S. L., Markus, I., Shevy Perera, D., Lubowski, D. Z., King, D. W., et al. (2013). Expression and localization of pannexin-1 hemichannels in human colon in health and disease. *Neurogastroenterol. Motil.* 25, e395–e405. doi: 10.1111/nmo.12130

Dvoriantchikova, G., Ivanov, D., Panchin, Y., and Shestopalov, V. I. (2006). Expression of pannexin family of proteins in the retina. *FEBS Lett.* 580, 2178–2182. doi: 10.1016/j.febslet.2006.03.026

Foss, E. J., Radulovic, D., Shaffer, S. A., Goodlett, D. R., Kruglyak, L., and Bedalov, A. (2011). Genetic variation shapes protein networks mainly through non-transcriptional mechanisms. *PLoS Biol.* 9:e1001144. doi: 10.1371/journal.pbio.1001144

Gassmann, M., Grenacher, B., Rohde, B., and Vogel, J. (2009). Quantifying western blots: pitfalls of densitometry. *Electrophoresis* 30, 1845–1855. doi: 10.1002/elps.200800720

Guescini, M., Sisti, D., Rocchi, M. B. L., Stocchi, L., and Stocchi, V. (2008). A new real-time PCR method to overcome significant quantitative inaccuracy due to slight amplification inhibition. *BMC Bioinformatics* 9:326. doi: 10.1186/1471-2105-9-326

Gürtler, A., Kunz, N., Gomolka, M., Hornhardt, S., Friedl, A. A., McDonald, K., et al. (2013). Stain-free technology as a normalization tool in western blot analysis. *Anal. Biochem.* 433, 105–111. doi: 10.1016/j.ab.2012.10.010

Ishikawa, M., Iwamoto, T., Nakamura, T., Doyle, A., Fukumoto, S., and Yamada, Y. (2011). Pannexin 3 functions as an ER Ca(2+) channel, hemichannel and gap junction to promote osteoblast differentiation. *J. Cell Biol.* 193, 1257–1274. doi: 10.1083/jcb.201101050

Kim, M.-S., Pinto, S. M., Getnet, D., Nirujogi, R. S., Manda, S. S., Chaerkady, R., et al. (2014). A draft map of the human proteome. *Nature* 509, 575–581. doi: 10.1038/nature13302

Kurtenbach, S., Prochnow, N., Kurtenbach, S., Klooster, J., Zoidl, C., Dermietzel, R., et al. (2013). Pannexin1 channel proteins in the zebrafish retina have shared and unique properties. *PloS One* 8:e77722. doi: 10.1371/journal.pone.0077722

Ladner, C. L., Yang, J., Turner, R. J., and Edwards, R. A. (2004). Visible fluorescent detection of proteins in polyacrylamide gels without staining. *Anal. Biochem.* 326, 13–20. doi: 10.1016/j.ab.2003.10.047

Lai, C., Bechberger, J., and Naus, C. (2009). Pannexin2 as a novel growth regulator in C6 glioma cells. *Oncogene* 28, 4402–4408. doi: 10.1038/onc.2009.283

Lai, C. P. K., Bechberger, J. F., Thompson, R. J., MacVicar, B. A., Bruzzone, R., and Naus, C. C. (2007). Tumor-suppressive effects of pannexin 1 in C6 glioma cells. *Cancer Res.* 67, 1545–1554. doi: 10.1158/0008-5472.can-06-1396

Le, H. T., Sin, W. C., Lozinsky, S., Bechberger, J., Vega, J. L., Guo, X. Q., et al. (2014). Gap junction intercellular communication mediated by connexin43 in astrocytes is essential for their resistance to oxidative stress. *J. Biol. Chem.* 289, 1345–1354. doi: 10.1074/jbc.m113.508390

Panchin, Y. V. (2005). Evolution of gap junction proteins–the pannexin alternative. *J. Exp. Biol.* 208, 1415–1419. doi: 10.1242/jeb.01547

Panchin, Y., Kelmanson, I., Matz, M., Lukyanov, K., Usman, N., and Lukyanov, S. (2000). A ubiquitous family of putative gap junction molecules. *Curr. Biol.* 10, R473–R474. doi: 10.1016/s0960-9822(00)00576-5

Penuela, S., Bhalla, R., Gong, X. Q., Cowan, K. N., Celetti, S. J., Cowan, B. J., et al. (2007). Pannexin 1 and pannexin 3 are glycoproteins that exhibit many distinct characteristics from the connexin family of gap junction proteins. *J. Cell Sci.* 120, 3772–3783. doi: 10.1242/jcs.009514

Penuela, S., Bhalla, R., Nag, K., and Laird, D. W. (2009). Glycosylation regulates pannexin intermixing and cellular localization. *Mol. Biol. Cell* 20, 4313–4323. doi: 10.1091/mbc.e09-01-0067

Pfaffl, M. W. (2001). A new mathematical model for relative quantification in real-time RT-PCR. *Nucleic Acids Res.* 29:e45. doi: 10.1093/nar/29.9.e45

Poon, I. K. H., Chiu, Y.-H., Armstrong, A. J., Kinchen, J. M., Juncadella, I. J., Bayliss, D. A., et al. (2014). Unexpected link between an antibiotic, pannexin channels and apoptosis. *Nature* 507, 329–334. doi: 10.1038/nature13147

Raj, A., Peskin, C. S., Tranchina, D., Vargas, D. Y., and Tyagi, S. (2006). Stochastic mRNA synthesis in mammalian cells. *PLoS Biol.* 4:e309. doi: 10.1371/journal.pbio.0040309

Ritz, C., and Spiess, A.-N. (2008). qpcR: an R package for sigmoidal model selection in quantitative real-time polymerase chain reaction analysis. *Bioinformatics* 24, 1549–1551. doi: 10.1093/bioinformatics/btn227

Sahu, G., Sukumaran, S., and Bera, A. K. (2014). Pannexins form gap junctions with electrophysiological and pharmacological properties distinct from connexins. *Sci. Rep.* 4:4955. doi: 10.1038/srep04955

Sosinsky, G. E., Boassa, D., Dermietzel, R., Duffy, H. S., Laird, D. W., MacVicar, B., et al. (2011). Pannexin channels are not gap junction hemichannels. *Channels (Austin)* 5, 193–197. doi: 10.4161/chan.5.3.15765

Spiess, A.-N., Feig, C., and Ritz, C. (2008). Highly accurate sigmoidal fitting of real-time PCR data by introducing a parameter for asymmetry. *BMC Bioinformatics* 9:221. doi: 10.1186/1471-2105-9-221

Swayne, L. A., Sorbara, C. D., and Bennett, S. A. L. (2010). Pannexin 2 is expressed by postnatal hippocampal neural progenitors and modulates neuronal commitment. *J. Biol. Chem.* 285, 24977–24986. doi: 10.1074/jbc.m110.130054

Vanden Abeele, F., Bidaux, G., Gordienko, D., Beck, B., Panchin, Y. V., Baranova, A. V., et al. (2006). Functional implications of calcium permeability of the channel formed by pannexin 1. *J. Cell Biol.* 174, 535–546. doi: 10.1083/jcb.200601115

Vogt, A., Hormuzdi, S. G., and Monyer, H. (2005). Pannexin1 and pannexin2 expression in the developing and mature rat brain. *Brain Res. Mol. Brain Res.* 141, 113–120. doi: 10.1016/j.molbrainres.2005.08.002

Vonesch, C., and Unser, M. (2008). A fast thresholded landweber algorithm for wavelet-regularized multidimensional deconvolution. *IEEE Trans. Image Process.* 17, 539–549. doi: 10.1109/tip.2008.917103

Wicki-Stordeur, L. E., Boyce, A. K. J., and Swayne, L. A. (2013). Analysis of a pannexin 2-pannexin 1 chimeric protein supports divergent roles for pannexin C-termini in cellular localization. *Cell Commun. Adhes.* 20, 73–79. doi: 10.3109/15419061.2013.791681

Wilhelm, M., Schlegl, J., Hahne, H., Gholami, A. M., Lieberenz, M., Savitski, M. M., et al. (2014). Mass-spectrometry-based draft of the human proteome. *Nature* 509, 582–587. doi: 10.1038/nature13319

Zappalà, A., Li Volti, G., Serapide, M. F., Pellitteri, R., Falchi, M., La Delia, F., et al. (2007). Expression of pannexin2 protein in healthy and ischemized brain of adult rats. *Neuroscience* 148, 653–667. doi: 10.1016/j.neuroscience.2007.06.028

Zellner, M., Babeluk, R., Diestinger, M., Pirchegger, P., Skeledzic, S., and Oehler, R. (2008). Fluorescence-based western blotting for quantitation of protein biomarkers in clinical samples. *Electrophoresis* 29, 3621–3627. doi: 10.1002/elps.200700935

Zhang, H., Chen, Y., and Zhang, C. (2012). Patterns of heterogeneous expression of pannexin 1 and pannexin 2 transcripts in the olfactory epithelium and olfactory bulb. *J. Mol. Histol.* 43, 651–660. doi: 10.1007/s10735-012-9443-x

Zoidl, G., Kremer, M., Zoidl, C., Bunse, S., and Dermietzel, R. (2008). Molecular diversity of connexin and pannexin genes in the retina of the zebrafish danio rerio. *Cell Commun. Adhes.* 15, 169–183. doi: 10.1080/15419060802014081

17

Cxs and Panx hemichannels in peripheral and central chemosensing in mammals

Edison Pablo Reyes[1,2], Verónica Cerpa[1], Liliana Corvalán[1] and Mauricio Antonio Retamal[1]*

[1] Centro de Fisiología Celular e Integrativa, Facultad de Medicina, Clínica Alemana Universidad del Desarrollo, Santiago, Chile
[2] Dirección de Investigación, Universidad Autónoma de Chile, Santiago, Chile

***Correspondence:**
Mauricio Antonio Retamal, Centro de Fisiología Celular e Integrativa, Facultad de Medicina, Clínica Alemana Universidad del Desarrollo, Av. Las Condes, 12438 Santiago, Chile
e-mail: mretamal@udd.cl

Connexins (Cxs) and Pannexins (Panx) form hemichannels at the plasma membrane of animals. Despite their low open probability under physiological conditions, these hemichannels release signaling molecules (i.e., ATP, Glutamate, PGE_2) to the extracellular space, thus subserving several important physiological processes. Oxygen and CO_2 sensing are fundamental to the normal functioning of vertebrate organisms. Fluctuations in blood PO_2, PCO_2 and pH are sensed at the carotid bifurcations of adult mammals by glomus cells of the carotid bodies. Likewise, changes in pH and/or PCO_2 of cerebrospinal fluid are sensed by central chemoreceptors, a group of specialized neurones distributed in the ventrolateral medulla (VLM), raphe nuclei, and some other brainstem areas. After many years of research, the molecular mechanisms involved in chemosensing process are not completely understood. This manuscript will review data regarding relationships between chemosensitive cells and the expression of channels formed by Cxs and Panx, with special emphasis on hemichannels.

Keywords: gap junctions, carotid body, glomus cells, connexins, astrocytes, hypoxia, hypercapnia

INTRODUCTION

In vertebrates, the family of membrane proteins that forms gap junction channels (GJCs) is called connexin. To date, 21 connexin isoforms in the human genome and 20 in the mouse genome (Willecke et al., 2002) have been described in almost all cell types, except for in mature sperm cells, differentiated skeletal muscle (in physiological conditions) and erythrocytes. The topological organization of the connexin-protein family is highly preserved and consist of four transmembrane regions linked by one intercellular cytoplasmic loop (CL), two extracellular loops (E1 and E2) (Hertzbergs et al., 1988; Milks et al., 1988; Falk et al., 1994; reviewed by Bruzzone et al., 1996; Kumar and Gilula, 1996) and both protein termini are located at the cytoplasmic side. The C-terminus length is variable and it is subjected to post-translational modifications related to intracellular signal cascades (Sáez et al., 1998; **Figure 1**). Connexins are abbreviated "Cx" followed by the molecular mass in kDa, e.g., Cx43 (Beyer et al., 1987). However, an alternative nomenclature categorized Cxs into five different groups: α, β, γ, δ and ϵ based on homology (specifically "the extent of sequence identity") and length of their CL (Bennett et al., 1991; Nielsen et al., 2012).

The expression of Cxs has a distinctive spatial, temporal and overlapping pattern (reviewed by Oyamada et al., 2005; Rackauskas et al., 2010) and the physiological relevance of different Cxs has been studied using several approaches (Cx knock-out animals, specific mutations in Cx genes and down regulating or changing the expression pattern of GJC). These studies showed that, in different organs, the disruption of GJCs can lead to pathological conditions such as, cataract formation, epidermal disease, hearing loss, apoptosis or cancer (Baruch et al., 2001; Saito et al., 2001; Common et al., 2005; Aishah et al., 2008; Kameritsch et al., 2013).

Cxs oligomerize forming aqueous hexameric hemichannels called connexons. Oligomerization occurs in intracellular compartments depending on the connexin type, e.g., Cx43 assemble in the trans-Golgi network (Musil and Goodenough, 1993; George, 1999) and Cx32 in the endoplasmic reticulum (Das Sarma et al., 2002). Connexons can be built of one or different Cxs isoforms assembling homomeric or heteromeric hemichannels, respectively. GJCs result from the association of two hemichannels, each provided by one of the two participating cells (Perkins et al., 1997; Unger et al., 1999). A connexon may dock with either, an identical or a different hemichannel forming homotypic or heterotypic channels, respectively (Kumar and Gilula, 1996). Henceforth, four arrangements of channels are possible (**Figure 1**). As the majority of cells expresses more than one Cx isoform, channels formed by heteromeric connexons should be the obvious study matter. However, since Cxs have different intracellular pathways to oligomerize, and at least two different pathways to be transported to the plasma membrane (George, 1999; Martin et al., 2001), the research focused on homomeric heterotypic channels (Werner et al., 1989; Elfgang et al., 1995; Falk et al., 1997; Gemel et al., 2004).

Gap junction (GJs) are clusters of intercellular channels (GJCs) present in almost all cell types. Initially, GJCs were described as nonspecific passive pores permeable to all soluble second messengers (e.g., amino acids, nucleotides, Ca^{2+}, glucose and metabolites smaller than 1.2 kDa) (reviewed by Bruzzone et al., 1996). These

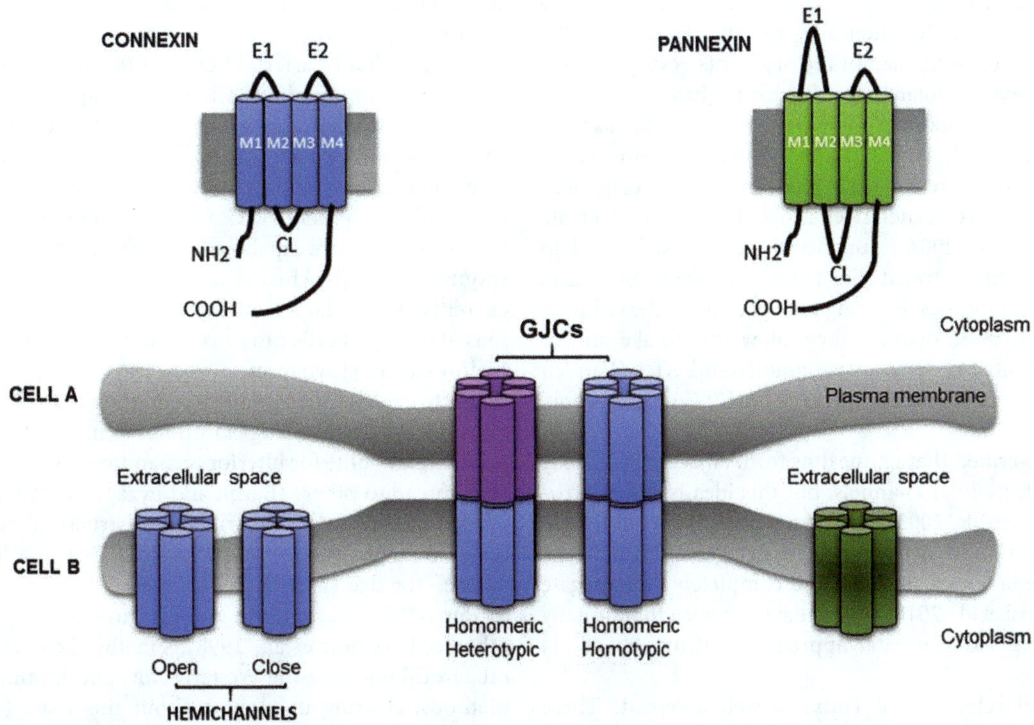

FIGURE 1 | Schematic illustration showing the topological structure of connexin and pannexin. Six connexins (Cxs) (a protein with four membrane domains (M1–M4), two extracellular loops (E1, E2), one cytoplasmic loop (CL) and the N and C termini exposed to the cytoplasm) subtype oligomerize into homomeric or heteromeric hemichannel. Under physiological circumstances, hemichannel remain closed. Gap junction channels (GJCs) (homotypic or heterotypic) connect the cytoplasm of two adyacent cells (Cell A and Cell B), allowing the passage of a variety of small molecules. Pannexin showed a similar topological structure.

channels provide cytoplasmic connections between two adjacent cells allowing the exchange of signaling molecules (ions, second messengers and small metabolites) (reviewed by Bennett et al., 1991; Bruzzone et al., 1996; Goodenough et al., 1996). This direct cell to cell electric and metabolic communication is essential in many physiological processes (e.g., embryonic development, propagation of action potential, cell growth and differentiation), synchronizing the function of organs including heart, liver, testis, skin and brain.

Hemichannels were considered a non-functional part of an intercellular communication pore. The rationale was: since GJCs are nonspecific if hemichannels were open at the plasma membrane, important components of the cytoplasm may "leak" to the extracellular medium and the cell would have to spent enormous amounts of energy to maintain its homeostasis. Today it is well known that "functional hemichannels" expressed in non-junctional plasma membrane of several cell types providing direct communication between intra- and extra-cellular environments (reviewed in Sáez et al., 2003, 2005). Under normal circumstances, hemichannels are closed and maintains cells isolated from external conditions. They become open after membrane depolarization, extracellular alkalization, metabolic inhibition, mechanical stimulation or in low extracellular calcium (DeVries and Schwartz, 1992; Ebihara et al., 1995; John, 1999; Contreras et al., 2002; Retamal et al., 2006; Schalper et al., 2010). Interestingly, removal of extracellular calcium in isosmotic condition also induce reversible changes in the cellular volume of different cells that normally express Cxs (e.g., fibroblast, endothelial and epithelial cells) (Quist et al., 2000). Even more, studies performed in *Xenopus* oocytes showed that hemichannels act as cationic channels, having distinctive voltage-dependent properties (reviewed by Bukauskas and Verselis, 2004).

Once functional hemichannels opened, they release NAD+, ATP, glutamate and prostaglandin E_2 to the extracellular space (Bruzzone et al., 2001; Stout et al., 2002; Ye et al., 2003; Cherian et al., 2005). These molecules play a critical role in central nervous system (CNS) physiology, hepatic homeostasis, and several paracrine/autocrine signaling (Corriden and Insel, 2010; Vinken, 2011; Orellana et al., 2013; Wang et al., 2013a). Pathological situations, such as oxidative stress or metabolism inhibition, may also open hemichannels allowing the movement of above molecules, which contribute to cell damage activating apoptotic mechanisms or altering cell physiology (Lin et al., 2003; Retamal et al., 2006; Ramachandran et al., 2007; Schalper et al., 2008).

Recently, it has been identified a novel family of integral membrane proteins, which share some structural and functional characteristics with Cx: pannexins (Panxs; Panchin et al., 2000; Yen and Saier, 2007; Sosinsky et al., 2011). Panxs are encoded by three genes: pannexin 1 (Panx1); pannexin 2 (Panx2) and pannexin 3 (Panx3), showing a 50–60% of sequence similarity (Sosinsky et al., 2011). Topologically, Cx and Panx have the same structure (four transmembrane segments, cytoplasmatic termini and two extracellular loops; **Figure 1**). Panx1 is ubiquitously expressed (e.g., brain, kidney, liver, retina, testis, skeletal and heart muscle, etc.), Panx2 is predominantly expressed in CNS and Panx3 is expressed in embryonic tissue, osteoblast and synovial fibroblast (Panchin et al., 2000; Bruzzone et al., 2003; Baranova et al., 2004). Functional studies performed in *Xenopus* oocytes demonstrated that Panx could be expressed in non-junctional membranes, forming hemichannels referred to as pannexons. When these opened, they allow the uptake and/or release of metabolites such as Ca^{2+}, anions and ATP (Vanden et al., 2006; Ambrosi et al., 2010; Ma et al., 2012; Romanov et al., 2012).

It has been described that pannexons from adjacent cells, may dock forming intercellular channels, but this idea is still controversial (Bruzzone et al., 2003). Just like Cxs, Panxs could form homotypic (Panx1) and heterotypic (Panx1/Panx2) functional channels, but the latter are unstable and completely disaggregate after 24 h (Ambrosi et al., 2010). Panx3 have not been functionally expressed in these experimental approaches (Bruzzone et al., 2003).

The biological relevance of Panxs is well accepted. There are some studies revealing their participation in specific processes, e.g., skeletal muscle release ATP through Panx1 after repetitive stimulation (Riquelme et al., 2013; Valladares et al., 2013). Likewise, neurons—and possibly also astrocytes—release arachidonic acid derivatives through Panx1. Those derivatives may be involved in a novel way of calcium wave propagation (MacVicar and Thompson, 2010). Also Panx1 may be involved in the regulation of the vascular tone regulating the release of ATP throughout the arterial network (Billaud et al., 2011; Lohman et al., 2012). Interestingly, Panx1 appears to be involved in a novel tri or quadripartite synapse at the carotid body (CB) chemoreceptors, amplifying the ATP signaling (Zhang et al., 2012; Piskuric and Nurse, 2013). The biological function of Panx3 is not clear and remains to be studied, but it has been related with osteoblast differentiation by functioning as calcium channels at the endoplasmic reticulum (Ishikawa et al., 2011) and to the differentiation of keratinocytes of the epidermis (Celetti et al., 2010).

Considering all physiological processes in which Cx and Panx are involved and the fact that both proteins are expressed in chemosensory systems, this review will outline the key features related to their biological relevance in the homeostasis of PO_2, PCO_2 and pH.

Finally, the evidence suggests that GJCs in non-excitable tissue may contribute to the spread of calcium waves (Gomes et al., 2006).

Cx AND Panx CHANNELS IN ARTERIAL CHEMORECEPTION

The main peripheral chemoreceptor is the CB. It is located at the carotid bifurcation and innervated by the carotid nerve (CN), a branch of the glossopharyngeal nerve (IXth pair). The CB is a compound receptor, where clusters of chemosensory units—glomus (Type I) cells—are surrounded by sustentacular (Type II) cells, and are found in close contact with *en calyce* endings of CN (**Figure 2**).

Information of arterial PO_2 and PCO_2 is conveyed from glomus cells to nerve endings. Chemical synapses in the CB have been extensively studied and many transmitters have been described so far (for a review see Zapata, 1997a,b). However, the electrical transmission only appears as a possibility if Type I cells change its membrane potential. Glomus cells not only depolarize, but even more, they are capable of generating action potentials, either spontaneously (Duchen et al., 1988) or evoked by depolarizing currents (López-López et al., 1989). Therefore, the information sensed by Type I cells can be conveyed to CN endings by chemical and/or electrical synapses.

Early evidence of functional coupling at cat CB cells was provided by electrophysiological studies in impaled cells stained with procion navy blue for ulterior recognition, and the dye from these cells spread to others (Baron and Eyzaguirre, 1977). Above study availed the previously described ultrastructural evidence of GJCs between rat glomus cells (McDonald, 1976), as a "plausible explanation" for dye spreading. The species puzzle was solved when lucifer yellow injected into one cat carotid glomus cell spread to other cells (Chou et al., 1998). Finally, Cx43 was identified in rat carotid bodies using Western blots and immunocytochemical methods, clearing up doubts about this issue (Abudara et al., 1999).

Despite the functional evidence, the detection and identification of GJs in the CB was elusive. Controversial evidence was provided using different techniques, even though they were reported by the same group of researchers. Using freeze fracture analysis, Kondo and Yamamoto did not find the characteristic GJC clusters (Kondo, 1981; Kondo and Yamamoto, 1993). However, using freeze substitution after aldehyde-prefixation, they found GJ-like structures between Type I cells and Type I cells and nerve terminals (Kondo and Iwasa, 1996), accounting for the electrocoupling and also enlighten some unexplained data (McQueen and Evrard, 1990). McQueen used selective antagonists to study the role of transmitters in the CB chemotransmission. Although the pharmacological effects were blocked, the response evoked by physiological stimuli still remained. This controversy still persisted because the technique causes artifactual GJ-like structures (Kondo, 2002).

Despite above results, dye spreading from one cell to others occurs because GJCs are able to convey molecules; therefore, current spreading is also possible. This was tested by Eyzaguirre's group (Monti-Bloch et al., 1993), impaling two adjacent Type I cells using independent amplifiers. To obtain the coupling coefficient (Bennett, 1966), they measured the membrane potential of both cells, while current was injected to one of them. They also calculated the coupling resistance (Spray et al., 1981), recording intracellular currents, while clamping the voltage at different potentials or during the application of chemical stimuli. Regarding glomus cells, current spreading from one cell to another (glomic coupling) has two characteristics: (1) coupling is bidirectional: the current spreads from one to another cell, no matter which cell is stimulated; (2) coupling is resistive: the response in the second cell is maintained during the stimulation of the first one (Jiang and Eyzaguirre, 2006; Eyzaguirre, 2007).

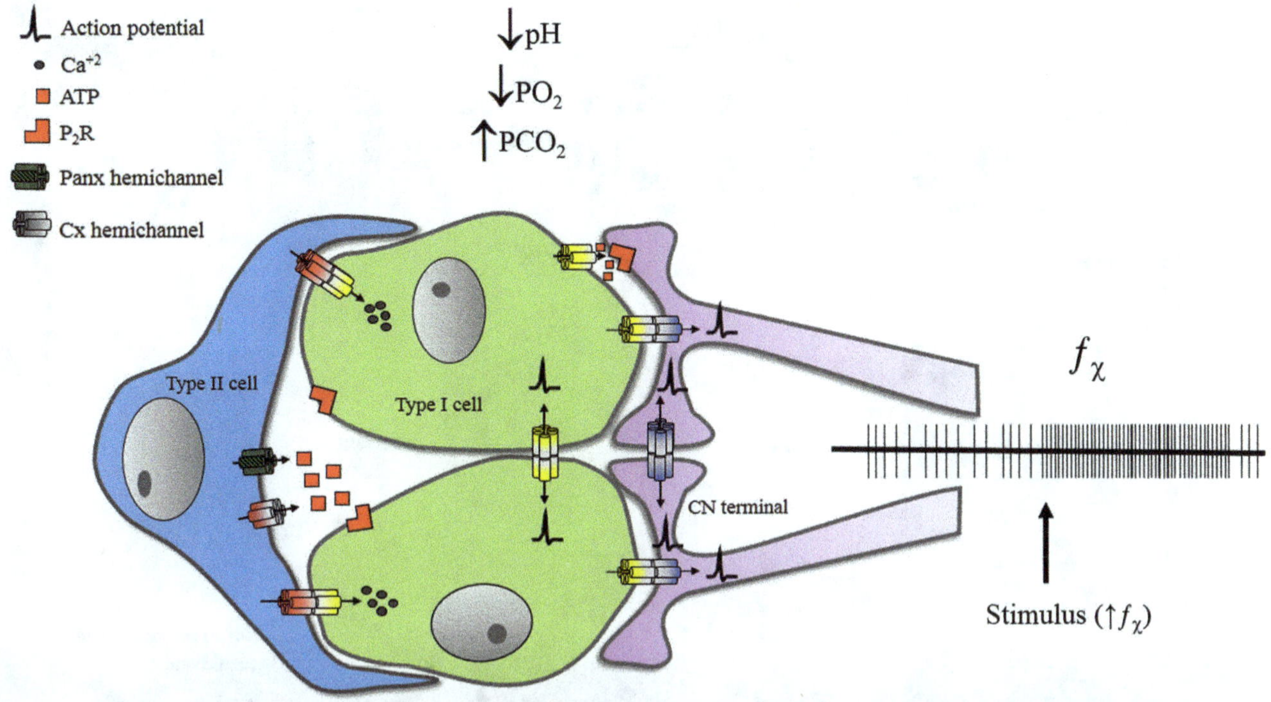

FIGURE 2 | Representation of some pathways proposed for f_χ increase upon stimulation. GJCs may participate on electrical chemosensory transmission from type I cells to CN terminals. Also GJCs may propagate electrical signals between glomus cells and/or between CN terminals. The release of ATP—from glomus and/or sustentacular cells—may also influence the chemosensory discharge of the carotid nerve. Details are discussed in the text. Chemical synapses are not represented.

In eucapnia (normocapnic normoxia), the degree of coupling between glomus cells is variable and it seems to be reduced by CB stimulants such as acid, hypercapnia or hypoxia, in agreement with studies showing that these stimuli close GJC (Peracchia et al., 2003; Peracchia, 2004). Nevertheless, the uncoupling effect of chemosensory stimuli was not uniform, since the majority of coupled glomus cells reduced their coupling, but some are found more coupled (Monti-Bloch et al., 1993; Eyzaguirre and Abudara, 1995, 1999). The explanation for this irregular result considers that glomus cells uncouple for transmitter secretion—just like secretory cells at exocrine glands (for references see Bennett and Spray, 1985; Bennett et al., 1991), hence the uncoupling during stimulation. Also, the enhanced coupling of some cells during stimulation may be compatible with transmitters recharging or production by those cells (Eyzaguirre and Abudara, 1999). We consider that the explanation may be extrapolated for the coupling disparity in basal conditions. Taking into account that chemosensory discharge exists in eupneic condition and in the absence of CO_2 (Eyzaguirre and Lewin, 1961; for a thorough discussion see Zapata, 1997b), it is very likely that some glomus cells were secreting transmitters in those conditions. Therefore, some of glomus cells will be uncoupled and some coupled, secreting and recharging transmitters, respectively.

Other remarkable observation is related to glomus cells depolarization. There seems to be a correlation between depolarization and uncoupling. On the one hand, glomus cells are known to be depolarized during transmitter release (Monti-Bloch et al., 1993). Although the transductional mechanism of chemoreception is not completely understood, there is a consensus that stimuli produce membrane depolarization of glomus cells, leading to increases in $[Ca^{2+}]_i$ and, consequently, transmitter release (see López-López et al., 2001; Weir et al., 2005). Altogether, aforementioned evidence indicates that chemoreceptor stimuli (low PO_2 or pH, high PCO_2 among others) concomitantly depolarize and uncouple glomus cells, stimulating the transmitter secretion and, consecutively, increase CN chemosensory discharge. Since Cx channels are modulated by Ca^{2+} and membrane potential, it is necessary to study how these variables modify the function of GJC and/or hemichannels in glomus cells, and how those possible modifications are relevant for the chemosensory process.

Carotid bodies are involved in the response to acute and chronic hypoxia. Several studies reported that CB responses to physiological and pharmacological stimuli are enhanced after acclimation to chronic hypoxia (Rey et al., 2004; He et al., 2005, 2006). Shortly after Cx43 was described in the CB (Abudara et al., 1999, 2000), the upregulation of this protein by chronic hypoxia was reported (Chen et al., 2002). Interestingly, the augmented response may be related to the increase of transmitter release by the glomus cell (Jackson and Nurse, 1997; Eyzaguirre and Abudara, 1999), which, in turn, can be associated with a decrease in glomus cells coupling (Jiang and Eyzaguirre, 2006). All the evidence presented above are based on the chemical

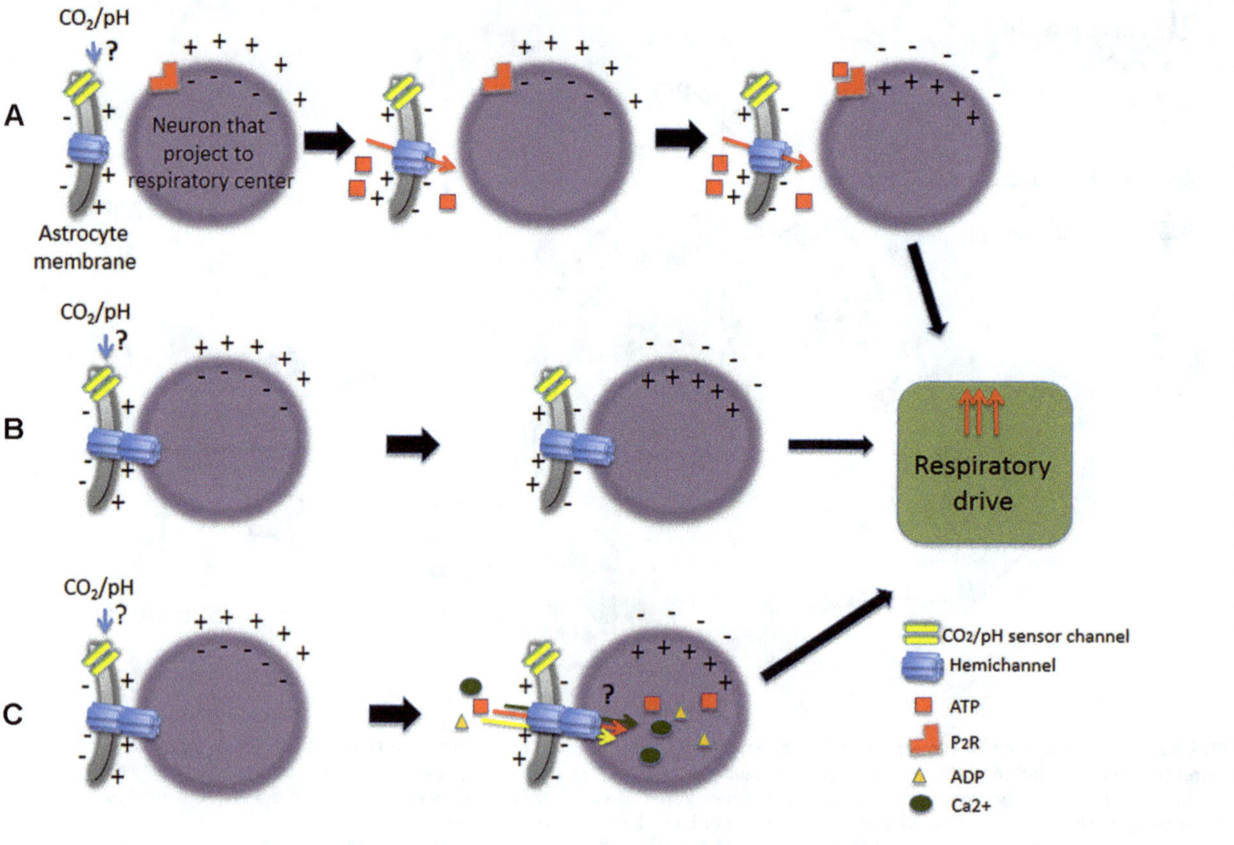

FIGURE 3 | Schematic representation of some of the different pathways proposed upon CO_2-dependent stimulation of astrocytes increasing respiratory drive. Astrocytes could stimulate neurons that project to respiratory center, being either chemosensory neurons or not. **(A)** Chemical stimulation by ATP release. **(B)** Stimulation by electrical coupling. **(C)** Stimulation of neurons by electrical coupling with passage of small molecules.

synapsis between glomus cells and CN endings. The identity of the synaptic transmitter was investigated (see Zapata, 1997a) and solved by Nurse's group using an *in vitro* preparation of co-cultured glomus cell clusters and petrosal neurons. With a cocktail of suramin and hexamethonium, they blocked the hypoxic chemotransmission from glomus cells to neurons (Zhang et al., 2000). We tested the combined cholinergic-purinergic block *in situ* and and *in vitro*, but it did not prevent the hypoxia-induced increases in chemosensory discharge in the CN (Reyes et al., 2007a,b). The fact that chemoreception transmission can be blocked *in vitro*, but not *in situ*, revealed some caveats related to the cell coupling. To our knowledge, there are no results showing the effect of GJC and/or hemichannels blockers in the CB chemosensing process. The lack of GJC blockers in those experiments is relevant because Eyzaguirre's group described dye and electrical coupling between glomus cells and CN endings (Eyzaguirre et al., 2003; Jiang and Eyzaguirre, 2006). Altogether, the evidence suggests that chemotransmission from glomus cells to CN endings may as well include electrical synapses.

It is noteworthy that coupling between glomus cells and CN endings is more complicated than the glomic coupling. First, coupling between glomus cells and CN endings presents a clear rectification. Thus, current from glomus cell spreads to nerve ending as easily as to other glomus cell, but current from nerve ending spreads poorly to glomus cell, thereby the coupling is mostly unidirectional. Also, current transmission is capacitive at the beginning and the end of the stimulus with little or no resistive component during stimulation. Additionally, CN endings are also coupled, and this specific electrical communication is capacitive and bidirectional (Jiang and Eyzaguirre, 2006). Recently, Cx36 was described in the CB, but it is unknown the cell type in which it is expressed (Frinchi et al., 2013). Considering that Cx36 has been described mainly in neuronal cells in the CNS (for review see Condorelli et al., 2000), it may be also present at the nerve endings. Thus, if neurons express mostly Cx36 and glomus cells express mainly Cx43, bidirectional communication between glomus cells and between CN endings, can be explained by the formation of homomeric homotypic GJCs. Furthermore, the formation of homomeric heterotypic GJCs between glomus cells and CN endings may explain the unidirectionality of that coupling (Jiang and Eyzaguirre, 2006). Indeed, the fact that two elements are enough to explain this phenomenon, does not exclude the putative participation of several other Cxs, currently not yet described in the CB system.

After chronic hypoxia, ventilatory or chemosensory discharge responses to different stimuli are augmented. This phenomenon may involve an enhanced release of transmitters by glomus cells, but it may also be clarified by the consideration of electrical synapses and the interaction between cells in the CB. During hypoxia, the glomic coupling is reduced. Conversely, the coupling between glomus cells and CN endings is enhanced, as well as the coupling between CN endings. This boosted coupling may allow the transmission of electrical changes from the glomus cells membrane to CN endings. Also, the enhanced coupling between nerve endings may assure the generation or multiply the action potentials in the CN, depending whether the coupled endings originate from the same neuron or from two independent neurons.

Channels formed by Cxs transmit information in another way. Recent studies show that some Cx hemichannels are permeable to Ca^{2+} (Schalper et al., 2010; Fiori et al., 2012), and it is known that chemosensory stimuli rise glomus cells $[Ca^{2+}]_i$ (Buckler and Vaughan-Jones, 1994a,b; Abudara et al., 2001; Jiang and Eyzaguirre, 2004; Xu et al., 2006; Lowe et al., 2013). Stimulated glomus cell may excite the secretion of a non-stimulated coupled glomus cell as a calcium wave-related second messenger, like AMPc (provided the wave occurs before the aforementioned uncoupling). Also, the same stimulated glomus cell may induce membrane depolarization of the CN endings via Ca^{+2} currents through Cx. Therefore, it could be interesting to test if Cx hemichannels are responsible—at least in part—of this phenomenon.

Bearing in mind that purinegic synapses have been considered as important components of the chemotransmission (Acker and Starlinger, 1984; Alcayaga et al., 2000; Zhang et al., 2000; Xu et al., 2003; Conde and Monteiro, 2004; Reyes et al., 2007a,b; Brown et al., 2011; Lowe et al., 2013; Piskuric and Nurse, 2013), it appears to be relevant that ATP can be released through Cx hemichannels (Kang et al., 2008). In this scenario, a stimulated glomus cell may excite neighboring cells (despite they were coupled or not with the original glomus cell) releasing ATP, which may acutely stimulate the target cell—either glomic, sustentacular or neuron—or have a chronic effect (Lin et al., 2008). However, this hypothesis has yet to be tested.

Finally, Panx-1 has been recently studied in the CB system where Type II cells were found to express this protein (Zhang et al., 2012). Type II cells are in close contact with glomus cells (McDonald and Mitchell, 1975) and some evidence suggests they may be connected via GJCs (Kondo, 2002). Also, Type II cells express metabotropic purinergic receptors, and may differentiate into glomus cells, under the adequate conditions (Pardal et al., 2007). Nonetheless, it is unclear if Type II cells do participate on the chemoreception/transmission processes. Recently, Nurse's group reconstructed *in vitro* a tripartite CB system, using petrosal neurons, glomic cells and sustentacular—Type II—cells. In this preparation, ATP released by Type II cells, via Panx-1, stimulates neurons (Zhang et al., 2012). The three-cell model suggests that ATP released by glomus cells in response to excitatory stimulation, activates inotropic receptors at CN terminals and metabotropic receptors of Type II cells. Consequently, the $[Ca^{+2}]_i$ of Type II cells rises, opening Panx-1 hemichannels, which—in turn—release more ATP to the intercellular medium, thus amplifying the signal.

The functional evidence of Cxs and Panxs at the CB and their plausible participation in chemosensory transmission appears to be consistent, but more investigation on this subject is required. Additionally, it remains to be determined the presence of other types of Cx, as well as its specific location. In vitro preparations, Cxs blockers and Cxs knock-out models may clarify the physiological relevance of the intercellular coupling in the acute chemosensory process and in chronic hypoxia (sustained or intermittent).

Cx- AND Panx CHANNELS IN CENTRAL CHEMORECEPTION

In CNS changes of CO_2/pH are sensed by central chemoreceptors. Its location has been studied using different approaches. In mammals, *in vivo* and *in vitro* findings showed chemosensitive areas diffusely located in the brainstem, including: nucleus of solitary tract (NTS), retrotrapezoide nucleus (RTN), parafacial respiratory group (pFRG), locus coreuleos (LC), raphé nuclei and ventrolateral medulla (VLM; Elam et al., 1981; Loeschcke, 1982; Coates et al., 1993; Wang et al., 1998, 2001; Richerson et al., 2001; Messier et al., 2002; Nattie and Li, 2002a,b; Guyenet, 2008; Li and Nattie, 2008; Gargaglioni et al., 2010; Hodges and Richerson, 2010; Putnam, 2010; Ray et al., 2011; Corcoran et al., 2013; Guyenet et al., 2013). Moreover, areas related to respiratory rhythm generation—as Pre Bötzinger nucleus—also showed chemosensitivity upon exposure to CO_2 (Solomon et al., 2000; Solomon, 2003a). Most of these studies were performed using specific blockers for predominant synapses in each preparation; glutamatergic and GABAergic blockers; or synaptic blockade medium (high Mg^{2+}-low Ca^{2+}). However, none of those experiments included GJC blockers, so it is possible that chemosensory nuclei could be less responsive.

As it is well known, hemichannels participate in diverse functions of CNS (reviewed by Menichella et al., 2003; Kielian, 2008; Kleopa et al., 2010; Abrams and Scherer, 2012; Mika and Prochnow, 2012; Belousov and Fontes, 2013). Since Cxs proteins are expressed in several CNS regions involved in central chemoreception, it is possible that hemichannels may play a role in pH/CO_2 sensing. They could increase the neuronal response in chemosensory nuclei and/or directly sense the hypercapnic stimulus. For many years, pH/CO_2 central chemosensing has been described as a property restricted to neurons, discarding that astrocytes could sense pH/CO_2. Now, if hemichannels are involved in direct chemosensing, it would implied that astrocytes could also be considered as chemoreceptors (**Figure 3**).

At the moment, more than 15 Cxs isoforms are described in the rodent brain (Dermietzel and Spray, 1993; Condorelli et al., 1998), but the first study showing electrotonic coupling between neurons of mammalian brainstem predates all descriptive ones (Llinás et al., 1974). Later, different researchers showed that Cx26, Cx30, Cx32, Cx36 and Cx46 are predominant in the brain with different cellular distribution. While Cx32 and Cx36 are principally expressed in neurons, Cx30 and Cx43 are mainly expressed in astrocytes, and both cell types share Cx26 (Dermietzel et al.,

1989; Yamamoto et al., 1990; Nagy et al., 1997, 1999, 2001; Condorelli et al., 1998; Rash et al., 2001).

Cx26, Cx32 and Cx36 are expressed in rat putative chemosensory nuclei, such as RTN, raphe, LC and Pre Bötzinger (Alvarez-Maubecin et al., 2000; Solomon et al., 2001; Solomon, 2003b). Furthermore, mRNA of Cx36 and Cx43 have been identified in the ventral respiratory group and in XIIn, respectively (Parenti et al., 2000). GJCs were reported in the dorsal aspect of the medulla oblongata showing electric and anatomical coupling in dorsal nucleus of the vagus (DMV) and NTS, during and after at least one exposure to hypercapnic acidosis (Dean et al., 1997, 2002; Huang et al., 1997).

On the other hand, the molecular mechanism of central chemoreception has not been well established yet. Until now, there are several pH-sensitive K^+ channels considered as candidates (Wu et al., 2004; Yuill et al., 2004; Zhang et al., 2006; Lazarenko et al., 2010; Wenker et al., 2010; Huckstepp and Dale, 2011; Hawryluk et al., 2012; Wang et al., 2013b). Also, ATP appears to be involved. It is well known that in peripheral sensory neurons both, ATP receptors—ionotropic P_2X or metabotropic P_2Y—excite afferent fibers. Therefore, ATP contributes significantly to the CB chemotransmission, being released by chemoreceptor cells and thus activating sinus nerve endings (Alcayaga et al., 2000; Rong et al., 2003 and reviewed by Spyer et al., 2004).

Bearing in mind the contribution of ATP in peripheral chemotransmission, its participation has been studied in the central chemoreception. Inspiratory and pre-inspiratory neurons of VLM only expresses the P_2X_2 receptor subunit and its activity was increased by ATP and blocked by suramin (Gourine et al., 2003). In order to accurately measure real time changes of ATP concentration, Gourine group developed a microelectrode biosensor detecting an almost immediate release of ATP upon CO_2 stimulation in rats. Using horizontal slices of medulla oblongata, they detected a marked release of ATP from the most ventral slice (mainly from RTN), upon CO_2-induced acidification of the incubation media. Moreover, blocking ATP receptors at these sites diminishes the chemosensory control of breathing. During hypercapnia, the increase in ATP release occurred 19.5 ± 4.8 ms before the induction of breathing. Based on above evidence, they hypothesized that ATP-mediated afferent transduction may also occur in the central chemoreception (Spyer et al., 2004; Gourine et al., 2005), as is described in peripheral chemoreception (Prasad et al., 2001; Zapata, 2007; Piskuric and Nurse, 2013).

In adult rats, recordings of respiratory activity of phrenic nerve showed that bilateral injections—at RTN level—of a P2 receptor blocker decreased by 30% the ventilatory responses to CO_2. Conversely, the inhibition of P_2Y_1 receptor—at the same level—had no effect on CO_2 responsiveness neither *in vitro* nor *in situ* (Wenker et al., 2012). Taking together, these results indicate that modulation of P_2X_2 receptor function (e.g., during hypercapnia) may contribute to changes in the activity of the VLM respiratory and chemosensory neurons that express those receptors. Interestingly, P_2X_2 and P_2X_3 receptor subunits knockout mice have normal ventilatory response to hypercapnia (Rong et al., 2003 and reviewed by Erlichman et al., 2010).

The CO_2-dependent ATP release persisted in the absence of extracellular Ca^{2+}, i.e., it did not occurred via neuronal exocytosis. This release—presumably from astrocytes in ventral surface of rat brainstems—depends on hemichannels formed by Cx26. Additionally, three different methods showed that HeLa cells expressing Cx26 release ATP in response to CO_2 (whole cell patch-clamp, CO_2-dependent dye uptake and patch clamp "*inside-out* and *outside-out*"). In HeLa cells model, changing PCO_2 from 35–70 mmHg evokes outward currents, increases the current noise, and also causes rapid and large increases of the conductance. The gating of Cx26 hemichannel increased and decreased in response to increases and decreases of PCO_2, respectively. Interestingly, only Cx30 and Cx32 (classified as β Cxs), exhibited sensitivity to changes in PCO_2 (Huckstepp et al., 2010a,b). This evidence indicates that astrocytes (additionally to neurons) could be considered as chemoreceptors in the CNS, and it also suggests that Cxs are sensors for the extracellular CO_2/pH (reviewed by Funk, 2010). Recently, evidence demonstrate CO_2 binding to Cx26, and that this interaction was probably via carbamylation of K125 motif. The authors hypothesized that CO_2 would form a carbamate bridge between the K125 of one subunit and the R104 of the adjacent subunit, therefore opening the Cx26 hemichannel (Meigh et al., 2013).

An alternative hypothesis proposed that astrocytes would be pH-sensitive. This notion derived from *in vitro* studies of RTN, a specific area within VLM. The sensitivity expressed as pH-sensitive currents involved either, Kir4.1-Kir5.1 channels and/or sodium/bicarbonate cotransporter (Wenker et al., 2010). Moreover, removal of pia matter irreversibly eliminates CO_2-evoked ATP release, indicating the importance of structural integrity of the marginal glial layer of the ventral medullary surface. Based on these observations, the marginal glial layer appears to be the likely source of ATP release in response to CO_2/pH (Spyer et al., 2004; Erlichman et al., 2010).

Many putative chemosensory nuclei in the medulla oblongata are ATP-sensitive areas, including RTN, raphé nuclei and LC. As mentioned previously, just a few of them have been studied more thoroughly pointing out the possible involvement of ATP and astrocytes in central chemoreception.

At the LC, the participation of ATP in the central chemosensory mechanism is supported by ATP-induced neuronal depolarization. This depolarization was reduced by 30 mM suramin and abolished by 100 mM suramin. In addition, suramin potentiated the excitatory AMPA effect, but did not alter the inhibitory effect of noradrenaline (Nieber et al., 1997). It remains to be elucidated where ATP is released from, astrocytes, neurons or both. It is unclear if ATP is released as the sole transmitter from purinergic neurons projecting to LC. Also, it is uncertain if ATP is released as co-transmitter with noradrenaline from recurrent axon collaterals—or dendrites—of LC neurons themselves. Finally, the LC responded to CO_2 with synchronic activity maintained in spite of synaptic blockade (Andrzejewski et al., 2001). This may be explained considering the expression of Cx at the LC (Solomon, 2003a) which, as previously mentioned, may be involved in the chemosensory activity.

Early studies in the ventral medulla showed that cells with electrophysiological characteristics of astrocytes depolarized during hypercapnic condition (Fukuda et al., 1978) Many years after, Gourine group (Gourine et al., 2010; Kasymov et al.,

2013) demonstrated that astrocytes from VLM responded to physiological acidity with important increases in intracellular Ca^{2+} and release of ATP. Also, they mimic Ca^{2+} responses evoked by pH, using optogenetic stimulation of astrocytes expressing channelrhodopsin-2. Thus, activating chemoreceptor neurons via ATP-dependent mechanism and triggering robust respiratory response *in vivo*, demonstrated a potential role of brain glial cells in central chemoreception. Cx expressed in astrocytes were related to Ca^{2+} waves, which have been involved in intercellular transmission of information (reviewed by Scemes and Giaume, 2006). Recently, the direct demonstration of Ca^{2+} flux through purified Cx26 hemichannels reconstituted in liposomes, suggested that Ca^{2+} fluxes through hemichannels can be a pathway for Ca^{2+} influx into cells in physiological and pathological conditions (Fiori et al., 2012). Hence, astrocytes could stimulate adjacent neurons by releasing ATP through hemichannels and also by Ca^{2+} waves through GJCs.

In summary, the evidence revisited here indicates that astrocytes may have a preponderant participation in central chemoreception. They respond to CO_2/pH increasing their intracellular Ca^{2+} levels and releasing ATP by mechanisms still unknown that may include Cxs (at least Cx26). Released ATP would excite ATP-sensitive neurons that directly innervate the respiratory controller. Most of other chemosensory areas are ATP-sensitive and express Cxs (potentially forming functional hemichannels). Therefore, ATP and Cxs could be part of a common mechanism in chemosensory nuclei. Considering the evidence, these mechanisms may occur at RTN, but further studies are required to demonstrate the participation of Cxs in other chemosensitive areas.

Finally, despite the knowledge that Panx are expressed in the brain (Panchin et al., 2000; Bruzzone et al., 2003; Baranova et al., 2004), their functional expression in central chemosensory areas has not been studied so far. Panx could be participating in chemosensory processes in a similar way than Cxs do.

Reviewing the abovementioned studies, there is evidence to enlighten the central chemoreception. However, several questions arise about the cellular identity of the chemoreceptor and the signaling pathways involved.

Firstly, is there an overestimation in the number and/or types of chemosensory cells? Considering the absence of GJCs blockers in chemosensory recordings, if a non-chemosensitive cell is coupled to a chemosensitive cell, the first one will also present chemosensory responses to CO_2. This may lead to an overestimation of the chemosensitive cells population. Also, the overestimation may be due to the effect of ATP released from a chemosensitive cell (neuron or astrocyte) exciting neighboring non-chemosensitive cells (neurons and/or astrocytes), which in turn will be considered as chemosensitive cells. As it now appears, the release of ATP in response to CO_2 may involve Cx or Panx hemichannels. Secondly, if astrocytes are also chemoreceptors, do they have different sensitivity to CO_2/pH than that of neurons? If neurons are more sensitive, they will respond to lower changes in CO_2/pH, and then the astrocyte response may increase/potentiate/synchronize the nucleus response. If astrocytes are more sensitive than neurons, they may prime chemosensitive neurons, which directly innervate the respiratory controller. Thirdly, do neurons and astrocytes share a common mechanism of CO_2/pH sensing? Are there multiple mechanisms involved? It seems like Cxs could be sensing CO_2 or pH—as many pH—sensitive K^+ channels, but neurons express both. Therefore, there are many facts still pending to be clarify. Fourthly, are the ATP-sensitive chemosensory areas also sensitive to other transmitters released by astrocytes? It is known that astrocytes release ATP, but they also release adenosine that may as well be involved in the excitation of neighboring cells. Finally, the presence of Cxs and Panx in the chemosensory system may represent an alternate—independent—via to increase the response to hypercapnia.

ACKNOWLEDGMENTS

The authors wish to acknowledge the writing assistance of Ms. Carolina Larraín. This study was supported by FONDECYT 1120214, ANILLO ACT1104 (Retamal, Mauricio Antonio) and Proyecto Interno UDD 23400094 (Reyes, Edison Pablo).

REFERENCES

Abrams, C. K., and Scherer, S. S. (2012). Gap junctions in inherited human disorders of the central nervous system. *Biochim. Biophys. Acta* 1818, 2030–2047. doi: 10.1016/j.bbamem.2011.08.015

Abudara, V., Eyzaguirre, C., and Sáez, J. C. (2000). Short- and long-term regulation of rat carotid body gap junctions by cAMP. Identification of connexin43, a gap junction subunit. *Adv. Exp. Med. Biol.* 475, 359–469. doi: 10.1007/0-306-46825-5_33

Abudara, V., Garcés, G., and Sáez, J. C. (1999). Cells of the carotid body express connexin43 which is up-regulated by cAMP. *Brain Res.* 849, 25–33. doi: 10.1016/s0006-8993(99)01946-0

Abudara, V., Jiang, R. G., and Eyzaguirre, C. (2001). Acidic regulation of junction channels between glomus cells in the rat carotid body. Possible role of [Ca (2+).](i). *Brain Res.* 916, 50–60. doi: 10.1016/s0006-8993(01)02862-1

Acker, H., and Starlinger, H. (1984). Adenosine triphosphate content in the cat carotid body under different arterial O2 and CO2 conditions. *Neurosci. Lett.* 50, 175–179. doi: 10.1016/0304-3940(84)90482-8

Aishah, Z. S., Khairi, M. D. M., Normastura, A. R., Zafarina, Z., and Zilfalil, B. A. (2008). Screening for gap junction protein beta-2 gene mutations in Malays with autosomal recessive, non-syndromic hearing loss, using denaturing high performance liquid chromatography. *J. Laryngol. Otol.* 122, 1284–1288. doi: 10.1017/s0022215108002041

Alcayaga, J., Cerpa, V., Retamal, M. A., Arroyo, J., Iturriaga, R., and Zapata, P. (2000). Adenosine triphosphate-induced peripheral nerve discharges generated from the cat petrosal ganglion in vitro. *Neurosci. Lett.* 282, 185–188. doi: 10.1016/s0304-3940(00)00896-x

Alvarez-Maubecin, V., Garcia-Hernandez, F., Williams, J. T., and Van Bockstaele, E. J. (2000). Functional coupling between neurons and glia. *J. Neurosci.* 20, 4091–4098.

Ambrosi, C., Gassmann, O., Pranskevich, J. N., Boassa, D., Smock, A., Wang, J., et al. (2010). Pannexin1 and Pannexin2 channels show quaternary similarities to connexons and different oligomerization numbers from each other. *J. Biol. Chem.* 285, 24420–24431. doi: 10.1074/jbc.m110.115444

Andrzejewski, M., Mückenhoff, K., Scheid, P., and Ballantyne, D. (2001). Synchronized rhythms in chemosensitive neurons of the locus coeruleus in the absence of chemical synaptic transmission. *Respir. Physiol.* 129, 123–140. doi: 10.1016/s0034-5687(01)00300-0

Baranova, A., Ivanov, D. V., Petrash, N., Pestova, A., Skoblov, M., Kelmanson, I., et al. (2004). The mammalian pannexin family is homologous to the invertebrate innexin gap junction proteins. *Genomics* 83, 706–716. doi: 10.1016/j.ygeno.2003.09.025

Baron, M., and Eyzaguirre, C. (1977). Effects of temperature on some membrane characteristics of carotid body cells. *Am. J. Physiol.* 233, C35–C46.

Baruch, A., Greenbaum, D., Levy, E. T., Nielsen, P. A., Gilula, N. B., Kumar, N. M., et al. (2001). Defining a link between gap junction communication, proteolysis, and cataract formation. *J. Biol. Chem.* 276, 28999–29006. doi: 10.1074/jbc.m103628200

Belousov, A. B., and Fontes, J. D. (2013). Neuronal gap junctions: making and breaking connections during development and injury. *Trends Neurosci.* 36, 227–236. doi: 10.1016/j.tins.2012.11.001

Bennett, M. V. L. (1966). Physiology of electrotonic junctions. *Ann. N. Y. Acad. Sci.* 137, 509–539. doi: 10.1111/j.1749-6632.1966.tb50178.x

Bennett, M. V. L., Barrio, L. C., Bargiello, T. A., Spray, D. C., Hertzberg, E., and Sáez, J. C. (1991). Gap junctions: new tools, new answers, new questions. *Neuron* 6, 305–320. doi: 10.1016/0896-6273(91)90241-q

Bennett, M. V. L., and Spray, D. C. (1985). *Gap Junctions.* New York: Cold Spring Harbor Laboratory.

Beyer, E. C., Paul, D. L., and Goodenough, D. A. (1987). Connexin43: a protein from rat heart homologous to a gap junction protein from liver. *J. Cell Biol.* 105, 2621–2629. doi: 10.1083/jcb.105.6.2621

Billaud, M., Lohman, A. W., Straub, A. C., Looft-Wilson, R., Johnstone, S. R., Araj, C. A., et al. (2011). Pannexin1 regulates α1-adrenergic receptor- mediated vasoconstriction. *Circ. Res.* 109, 80–85. doi: 10.1161/circresaha.110.237594

Brown, S. T., Reyes, E.-P., and Nurse, C. A. (2011). Chronic hypoxia upregulates adenosine 2a receptor expression in chromaffin cells via hypoxia inducible factor-2α: role in modulating secretion. *Biochem. Biophys. Res. Commun.* 412, 466–472. doi: 10.1016/j.bbrc.2011.07.122

Bruzzone, R., Hormuzdi, S. G., Barbe, M. T., Herb, A., and Monyer, H. (2003). Pannexins, a family of gap junction proteins expressed in brain. *Proc. Natl. Acad. Sci. U S A* 100, 13644–13649. doi: 10.1073/pnas.2233464100

Bruzzone, S., Guida, L., Zocchi, E., Franco, L., and De Flora, A. (2001). Connexin 43 hemi channels mediate Ca2+-regulated transmembrane NAD+ fluxes in intact cells. *FASEB J.* 15, 10–12. doi: 10.1096/fj.00-0566fje

Bruzzone, R., White, T. W., and Paul, D. L. (1996). Connections with connexins: the molecular basis of direct intercellular signaling. *Eur. J. Biochem.* 238, 1–27. doi: 10.1111/j.1432-1033.1996.0001q.x

Buckler, K. J., and Vaughan-Jones, R. D. (1994a). Effects of hypercapnia on membrane potential and intracellular calcium in rat carotid body type I cells. *J. Physiol.* 478(Pt. 1), 157–171.

Buckler, K. J., and Vaughan-Jones, R. D. (1994b). Effects of hypoxia on membrane potential and intracellular calcium in rat neonatal carotid body type I cells. *J. Physiol.* 476, 423–428.

Bukauskas, F. F., and Verselis, V. K. (2004). Gap junction channel gating. *Biochim. Biophys. Acta* 1662, 42–60. doi: 10.1016/j.bbamem.2004.01.008

Celetti, S. J., Cowan, K. N., Penuela, S., Shao, Q., Churko, J., and Laird, D. W. (2010). Implications of pannexin 1 and pannexin 3 for keratinocyte differentiation. *J. Cell Sci.* 123, 1363–1372. doi: 10.1242/jcs.056093

Chen, J., He, L., Dinger, B., Stensaas, L., and Fidone, S. J. (2002). Chronic hypoxia upregulates connexin43 expression in rat carotid body and petrosal ganglion. *J. Appl. Physiol.* 92, 1480–1486. doi: 10.1152/japplphysiol.00077.2001

Cherian, P. P., Siller-Jackson, A. J., Gu, S., Wang, X., Bonewald, L. F., Sprague, E., et al. (2005). Mechanical strain opens connexin 43 hemichannels in osteocytes: a novel mechanism for the release of prostaglandin. *Mol. Biol. Cell* 16, 3100–3106. doi: 10.1091/mbc.e04-10-0912

Chou, C.-L., Sham, J. S., Schofield, B., and Shirahata, M. (1998). Electrophysiological and immunocytological demonstration of cell-type specific responses to hypoxia in the adult cat carotid body. *Brain Res.* 789, 229–238. doi: 10.1016/s0006-8993(97)01472-8

Coates, E. L., Li, A., and Nattie, E. E. (1993). Widespread sites of brain stem ventilatory chemoreceptors. *J. Appl. Physiol.* 75, 5–14.

Common, J. E. A., O'Toole, E. A., Leigh, I. M., Thomas, A., Griffiths, W. A. D., Venning, V., et al. (2005). Clinical and genetic heterogeneity of erythrokeratoderma variabilis. *J. Invest. Dermatol.* 125, 920–927. doi: 10.1111/j.0022-202x.2005.23919.x

Conde, S. V., and Monteiro, E. C. (2004). Hypoxia induces adenosine release from the rat carotid body. *J. Neurochem.* 89, 1148–1156. doi: 10.1111/j.1471-4159.2004.02380.x

Condorelli, D., Belluardo, N., Trovato-Salinaro, A., and Mudò, G. (2000). Expression of Cx36 in mammalian neurons. *Brain Res. Brain Res. Rev.* 32, 72–85. doi: 10.1016/s0165-0173(99)00068-5

Condorelli, D. F., Parenti, R., Spinella, F., Trovato Salinaro, A., Belluardo, N., Cardile, V., et al. (1998). Cloning of a new gap junction gene (Cx36) highly expressed in mammalian brain neurons. *Eur. J. Neurosci.* 10, 1202–1208. doi: 10.1046/j.1460-9568.1998.00163.x

Contreras, J. E., Sánchez, H. A., Eugenin, E. A., Speidel, D., Theis, M., Willecke, K., et al. (2002). Metabolic inhibition induces opening of unapposed connexin 43 gap junction hemichannels and reduces gap junctional communication in cortical astrocytes in culture. *Proc. Natl. Acad. Sci. U S A* 99, 495–500. doi: 10.1073/pnas.012589799

Corcoran, A. E., Richerson, G. B., and Harris, M. B. (2013). Serotonergic mechanisms are necessary for central respiratory chemoresponsiveness in situ. *Respir. Physiol. Neurobiol.* 186, 214–220. doi: 10.1016/j.resp.2013.02.015

Corriden, R., and Insel, P. A. (2010). Basal release of ATP: an autocrine-paracrine mechanism for cell regulation. *Sci. Signal.* 3:re1. doi: 10.1126/scisignal.3104re1

Das Sarma, J., Wang, F., and Koval, M. (2002). Targeted gap junction protein constructs reveal connexin-specific differences in oligomerization. *J. Biol. Chem.* 277, 20911–20918. doi: 10.1074/jbc.m111498200

Dean, J. B., Ballantyne, D., Cardone, D. L., Erlichman, J. S., and Solomon, I. C. (2002). Role of gap junctions in CO(2) chemoreception and respiratory control. *Am. J. Physiol. Lung Cell. Mol. Physiol.* 283, L665–L670. doi: 10.1152/ajplung.00142.2002

Dean, J. B., Huang, R. Q., Erlichman, J. S., Southard, T. L., and Hellard, D. T. (1997). Cell-cell coupling occurs in dorsal medullary neurons after minimizing anatomical-coupling artifacts. *Neuroscience* 80, 21–40. doi: 10.1016/s0306-4522(97)00016-x

Dermietzel, R., and Spray, D. C. (1993). Gap junctions in the brain: where, what type, how many and why? *Trends Neurosci.* 16, 186–192. doi: 10.1016/0166-2236(93)90151-b

Dermietzel, R., Traub, O., Hwang, T. K., Beyer, E. C., Bennett, M. V. L., Spray, D. C., et al. (1989). Differential expression of three gap junction proteins in developing and mature brain tissues. *Proc. Natl. Acad. Sci. U S A* 86, 10148–10152. doi: 10.1073/pnas.86.24.10148

DeVries, S. H., and Schwartz, E. A. (1992). Hemi-gap-junction channels in solitary horizontal cells of the catfish retina. *J. Physiol.* 445, 201–230.

Duchen, M. R., Caddy, K. W., Kirby, G. C., Patterson, D. L., Ponte, J., and Biscoe, T. J. (1988). Biophysical studies of the cellular elements of the rabbit carotid body. *Neuroscience* 26, 291–311. doi: 10.1016/0306-4522(88)90146-7

Ebihara, L., Berthoud, V. M., and Beyer, E. C. (1995). Distinct behavior of connexin56 and connexin46 gap junctional channels can be predicted from the behavior of their hemi-gap-junctional channels. *Biophys. J.* 68, 1796–1803. doi: 10.1016/s0006-3495(95)80356-5

Elam, M., Yao, T., Thorén, P., and Svensson, T. H. (1981). Hypercapnia and hypoxia: chemoreceptor-mediated control of locus coeruleus neurons and splanchnic, sympathetic nerves. *Brain Res.* 222, 373–381. doi: 10.1016/0006-8993(81)91040-4

Elfgang, C., Eckert, R., Lichtenberg-Fraté, H., Butterweck, A., Traub, O., Klein, R. A., et al. (1995). Specific permeability and selective formation of gap junction channels in connexin-transfected HeLa cells. *J. Cell Biol.* 129, 805–817. doi: 10.1083/jcb.129.3.805

Erlichman, J. S., Leiter, J. C., and Gourine, A. V. (2010). ATP, glia and central respiratory control. *Respir. Physiol. Neurobiol.* 173, 305–311. doi: 10.1016/j.resp.2010.06.009

Eyzaguirre, C. (2007). Electric synapses in the carotid body-nerve complex. *Respir. Physiol. Neurobiol.* 157, 116–122. doi: 10.1016/j.resp.2007.01.013

Eyzaguirre, C., and Abudara, V. (1995). Possible role of coupling between glomus cells in carotid body chemoreception. *Biol. Signals* 4, 263–270. doi: 10.1159/000109451

Eyzaguirre, C., and Abudara, V. (1999). Carotid body glomus cells: chemical secretion and transmission (modulation?) across cell-nerve ending junctions. *Respir. Physiol.* 115, 135–149. doi: 10.1016/s0034-5687(99)00020-1

Eyzaguirre, C., and Lewin, J. (1961). Effect of different oxygen tensions on the carotid body in vitro. *J. Physiol.* 159, 238–250.

Eyzaguirre, C., Jiang, R. G., and Abudara, V. (2003). "Electric and dye coupling between rat carotid body cells and between these cells and carotid nerve endings," in *Oxygen Sensing: Responses and Adaptation to Hypoxia*, eds S. Lahiri, G. L. Semenza and N. R. Prabhakar (New York: Marcel Dekker), 331–352.

Falk, M. M., Buehler, L. K., Kumar, N. M., and Gilula, N. B. (1997). Cell-free synthesis and assembly of connexins into functional gap junction membrane channels. *EMBO J.* 16, 2703–2716. doi: 10.1093/emboj/16.10.2703

Falk, M. M., Kumar, N. M., and Gilula, N. B. (1994). Membrane insertion of gap junction connexins: polytopic channel forming membrane proteins. *J. Cell Biol.* 127, 343–355. doi: 10.1083/jcb.127.2.343

Fiori, M. C., Figueroa, V., Zoghbi, M. E., Saéz, J. C., Reuss, L., and Altenberg, G. A. (2012). Permeation of calcium through purified connexin 26 hemichannels. *J. Biol. Chem.* 287, 40826–40834. doi: 10.1074/jbc.m112.383281

Frinchi, M., Di Liberto, V., Turimella, S., D'Antoni, F., Theis, M., Belluardo, N., et al. (2013). Connexin36 (Cx36) expression and protein detection in the mouse carotid body and myenteric plexus. *Acta Histochem.* 115, 252–256. doi: 10.1016/j.acthis.2012.07.005

Fukuda, Y., Honda, Y., Schläfke, M. E., and Loeschcke, H. H. (1978). Effect of H+ on the membrane potential of silent cells in the ventral and dorsal surface layers of the rat medulla in vitro. *Pflugers Arch.* 376, 229–235. doi: 10.1007/bf00584955

Funk, G. D. (2010). The "connexin" between astrocytes, ATP and central respiratory chemoreception. *J. Physiol.* 588, 4335–4337. doi: 10.1113/jphysiol.2010.200196

Gargaglioni, L. H., Hartzler, L. K., and Putnam, R. W. (2010). The locus coeruleus and central chemosensitivity. *Respir. Physiol. Neurobiol.* 173, 264–273. doi: 10.1016/j.resp.2010.04.024

Gemel, J., Valiunas, V., Brink, P. R., and Beyer, E. C. (2004). Connexin43 and connexin26 form gap junctions, but not heteromeric channels in co-expressing cells. *J. Cell Sci.* 117, 2469–2480. doi: 10.1242/jcs.01084

George, C. H. (1999). Intracellular Trafficking Pathways in the Assembly of Connexins into Gap Junctions. *J. Biol. Chem.* 274, 8678–8685. doi: 10.1074/jbc.274.13.8678

Gomes, P., Srinivas, S. P., Vereecke, J., and Himpens, B. (2006). Gap junctional intercellular communication in bovine corneal endothelial cells. *Exp. Eye Res.* 83, 1225–1237. doi: 10.1016/j.exer.2006.06.012

Goodenough, D. A., Goliger, J. A., and Paul, D. L. (1996). Connexins, connexons and intercellular communication. *Annu. Rev. Biochem.* 65, 475–502. doi: 10.1146/annurev.biochem.65.1.475

Gourine, A. V., Atkinson, L., Deuchars, J., and Spyer, K. M. (2003). Purinergic signalling in the medullary mechanisms of respiratory control in the rat: respiratory neurones express the P2X2 receptor subunit. *J. Physiol.* 552, 197–211. doi: 10.1113/jphysiol.2003.045294

Gourine, A. V., Kasymov, V., Marina, N., Tang, F., Figueiredo, M. F., Lane, S., et al. (2010). Astrocytes control breathing through pH-dependent release of ATP. *Science* 329, 571–575. doi: 10.1126/science.1190721

Gourine, A. V., Llaudet, E., Dale, N., and Spyer, K. M. (2005). ATP is a mediator of chemosensory transduction in the central nervous system. *Nature* 436, 108–111. doi: 10.1038/nature03690

Guyenet, P. G. (2008). The 2008 Carl Ludwig Lecture: retrotrapezoid nucleus, CO2 homeostasis and breathing automaticity. *J. Appl. Physiol.* 105, 404–416. doi: 10.1152/japplphysiol.90452.2008

Guyenet, P. G., Abbott, S. B. G., and Stornetta, R. L. (2013). The respiratory chemoreception conundrum: light at the end of the tunnel? *Brain Res.* 1511, 126–137. doi: 10.1016/j.brainres.2012.10.028

Hawryluk, J. M., Moreira, T. S., Takakura, A. C., Wenker, I. C., Tzingounis, A. V., and Mulkey, D. K. (2012). KCNQ channels determine serotonergic modulation of ventral surface chemoreceptors and respiratory drive. *J. Neurosci.* 32, 16943–16952. doi: 10.1523/jneurosci.3043-12.2012

He, L., Chen, J., Dinger, B., Stensaas, L., and Fidone, S. J. (2006). Effect of chronic hypoxia on purinergic synaptic transmission in rat carotid body. *J. Appl. Physiol.* 100, 157–162. doi: 10.1152/japplphysiol.00859.2005

He, L., Dinger, B., and Fidone, S. J. (2005). Effect of chronic hypoxia on cholinergic chemotransmission in rat carotid body. *J. Appl. Physiol.* 1297, 614–619. doi: 10.1152/japplphysiol.00714.2004

Hertzbergs, E. L., Disher, R. M., Tiller, A. A., Zhou, Y., and Cook, R. G. (1988). Topology of the Mr 27,000 liver gap junction protein. Cytoplasmic localization of amino- and carboxyl termini and a hydrophilic domain which is protease-hypersensitive. *J. Biol. Chem.* 263, 19105–19111.

Hodges, M. R., and Richerson, G. B. (2010). The role of medullary serotonin (5-HT) neurons in respiratory control: contributions to eupneic ventilation, CO2 chemoreception and thermoregulation. *J. Appl. Physiol.* 108, 1425–1432. doi: 10.1152/japplphysiol.01270.2009

Huang, R. Q., Erlichman, J. S., and Dean, J. B. (1997). Cell-cell coupling between CO2-excited neurons in the dorsal medulla oblongata. *Neuroscience* 80, 41–57. doi: 10.1016/s0306-4522(97)00017-1

Huckstepp, R. T. R., and Dale, N. (2011). CO2-dependent opening of an inwardly rectifying K+ channel. *Pflugers Arch.* 461, 337–344. doi: 10.1007/s00424-010-0916-z

Huckstepp, R. T. R., Eason, R., Sachdev, A., and Dale, N. (2010a). CO2-dependent opening of connexin 26 and related β connexins. *J. Physiol.* 588, 3921–3931. doi: 10.1113/jphysiol.2010.192096

Huckstepp, R. T. R., id Bihi, R., Eason, R., Spyer, K. M., Dicke, N., Willecke, K., et al. (2010b). Connexin hemichannel-mediated CO2-dependent release of ATP in the medulla oblongata contributes to central respiratory chemosensitivity. *J. Physiol.* 588, 3901–3920. doi: 10.1113/jphysiol.2010.192088

Ishikawa, M., Iwamoto, T., Nakamura, T., Doyle, A., Fukumoto, S., and Yamada, Y. (2011). Pannexin 3 functions as an ER Ca(2+) channel, hemichannel and gap junction to promote osteoblast differentiation. *J. Cell Biol.* 193, 1257–1274. doi: 10.1083/jcb.201101050

Jackson, A., and Nurse, C. A. (1997). Dopaminergic properties of cultured rat carotid body chemoreceptors grown in normoxic and hypoxic environments. *J. Neurochem.* 69, 645–654. doi: 10.1046/j.1471-4159.1997.69020645.x

Jiang, R. G., and Eyzaguirre, C. (2004). Effects of hypoxia and putative transmitters on [Ca2+]i of rat glomus cells. *Brain Res.* 995, 285–296. doi: 10.1016/j.brainres.2003.09.075

Jiang, R. G., and Eyzaguirre, C. (2006). Effects of prolonged hypobaric hypoxia on carotid nerve endings and glomus cells. Changes in intercellular coupling. *Brain Res.* 1076, 198–208. doi: 10.1016/j.brainres.2005.08.059

John, S. A. (1999). Connexin-43 hemichannels opened by metabolic inhibition. *J. Biol. Chem.* 274, 236–240. doi: 10.1074/jbc.274.1.236

Kameritsch, P., Khandoga, N., Pohl, U., and Pogoda, K. (2013). Gap junctional communication promotes apoptosis in a connexin-type-dependent manner. *Cell Death Dis.* 4:e584. doi: 10.1038/cddis.2013.105

Kang, J., Kang, N., Lovatt, D., Torres, A., Zhao, Z., Lin, J. H.-C., et al. (2008). Connexin 43 hemichannels are permeable to ATP. *J. Neurosci.* 28, 4702–4711. doi: 10.1523/jneurosci.5048-07.2008

Kasymov, V., Larina, O., Castaldo, C., Marina, N., Patrushev, M., Kasparov, S., et al. (2013). Differential sensitivity of brainstem versus cortical astrocytes to changes in pH reveals functional regional specialization of astroglia. *J. Neurosci.* 33, 435–441. doi: 10.1523/jneurosci.2813-12.2013

Kielian, T. (2008). Glial connexins and gap junctions in CNS inflammation and disease. *J. Neurochem.* 106, 1000–1016. doi: 10.1111/j.1471-4159.2008.05405.x

Kleopa, K. A., Orthmann-Murphy, J., and Sargiannidou, I. (2010). Gap junction disorders of myelinating cells. *Rev. Neurosci.* 21, 397–419. doi: 10.1515/revneuro.2010.21.5.397

Kondo, H. (1981)."Evidence for the secretion of chief cells in the carotid body," in *Arterial Chemoreceptors*, eds C. Belmonte, D. J. Pallot, H. Acker and S. J. Fidone (Leicester: Leicester University Press), 45–53.

Kondo, H. (2002). Are there gap junctions between chief (glomus, type I) cells in the carotid body chemoreceptor? A review. *Microsc. Res. Tech.* 59, 227–233. doi: 10.1002/jemt.10196

Kondo, H., and Iwasa, H. (1996). Re-examination of the carotid body ultrastructure with special attention to intercellular membrane appositions. *Adv. Exp. Med. Biol.* 410, 45–50. doi: 10.1007/978-1-4615-5891-0_4

Kondo, H., and Yamamoto, M. (1993). Multi-unit compartmentation of the carotid body chemoreceptor by perineurial cell sheaths: immunohistochemistry and freeze-fracture study. *Adv. Exp. Med. Biol.* 337, 61–66. doi: 10.1007/978-1-4615-2966-8_9

Kumar, N. M., and Gilula, N. B. (1996). The gap junction communication channel. *Cell* 84, 381–388. doi: 10.1016/s0092-8674(00)81282-9

Lazarenko, R. M., Fortuna, M. G., Shi, Y., Mulkey, D. K., Takakura, A. C., Moreira, T. S., et al. (2010). Anesthetic activation of central respiratory chemoreceptor neurons involves inhibition of a THIK-1-like background K(+) current. *J. Neurosci.* 30, 9324–9334. doi: 10.1523/jneurosci.1956-10.2010

Li, A., and Nattie, E. E. (2008). Serotonin transporter knockout mice have a reduced ventilatory response to hypercapnia (predominantly in males) but not to hypoxia. *J. Physiol.* 586, 2321–2329. doi: 10.1113/jphysiol.2008.152231

Lin, J. H.-C., Lou, N., Kang, N., Takano, T., Hu, F., Han, X., et al. (2008). A central role of connexin 43 in hypoxic preconditioning. *J. Neurosci.* 28, 681–695. doi: 10.1523/jneurosci.3827-07.2008

Lin, J. H.-C., Yang, J., Liu, S., Takano, T., Wang, X., Gao, Q., et al. (2003). Connexin Mediates Gap Junction-Independent Resistance to Cellular Injury. *J. Neurosci.* 23, 430–441.

Llinás, R., Baker, R., and Sotelo, C. (1974). Electrotonic coupling between neurons in cat inferior olive. *J. Neurophysiol.* 37, 560–571.

Loeschcke, H. H. (1982). Central chemosensitivity and the reaction theory. *J. Physiol.* 332, 1–24.

Lohman, A. W., Billaud, M., Straub, A. C., Johnstone, S. R., Best, A. K., Lee, M., et al. (2012). Expression of pannexin isoforms in the systemic murine arterial network. *J. Vasc. Res.* 49, 405–416. doi: 10.1159/000338758

López-López, J., González, C., Ureña, J., and López-Barneo, J. (1989). Low pO2 selectively inhibits K channel activity in chemoreceptor cells of the mammalian carotid body. *J. Gen. Physiol.* 93, 1001–1015. doi: 10.1085/jgp.93.5.1001

López-López, J. R., Pardal, R., and Ortega-Sáenz, P. (2001). Cellular mechanism of oxygen sensing. *Annu. Rev. Physiol.* 63, 259–287. doi: 10.1146/annurev.physiol.63.1.259

Lowe, M., Park, S. J., Nurse, C. A., and Campanucci, V. A. (2013). Purinergic stimulation of carotid body efferent glossopharyngeal neurones increases intracellular Ca2+ and nitric oxide production. *Exp. Physiol.* 98, 1199–1212. doi: 10.1113/expphysiol.2013.072058

Ma, W., Compan, V., Zheng, W., Martin, E., North, R. A., Verkhratsky, A., et al. (2012). Pannexin 1 forms an anion-selective channel. *Pflugers Arch.* 463, 585–592. doi: 10.1007/s00424-012-1077-z

MacVicar, B. A., and Thompson, R. J. (2010). Non-junction functions of pannexin-1 channels. *Trends Neurosci.* 33, 93–102. doi: 10.1016/j.tins.2009.11.007

Martin, P. E. M., Blundell, G., Ahmad, S., Errington, R. J., and Evans, W. H. (2001). Multiple pathways in the trafficking and assembly of connexin 26, 32 and 43 into gap junction intercellular communication channels. *J. Cell Sci.* 114, 3845–3855.

McDonald, D. M. (1976). "Structure and function of reciprocal synapses interconnecting glomus cells and sensory nerve terminals in the rat carotid body," in *Chromaffin, Enterochromaffin and Related Cells*, eds R. Coupland and T. Fujita (Amsterdam, New York: Elsevier Scientific Pub. Co.), 375–394.

McDonald, D. M., and Mitchell, R. A. (1975). The innervation of glomus cells, ganglion cells and blood vessels in the rat carotid body: a quantitative ultrastructural analysis. *J. Neurocytol.* 4, 177–230. doi: 10.1007/bf01098781

McQueen, D. S., and Evrard, Y. (1990). "Use of selective antagonist for studying the role of putative transmitters in chemoreception," in *Arterial Chemoreception*, eds C. Eyzaguirre, S. Fidone, R. Fizgerald, S. Lahiri and D. McDonald (New York: Springer-Verlag New York Inc.), 168–173.

Meigh, L., Greenhalgh, S. A., Rodgers, T. L., Cann, M. J., Roper, D. I., and Dale, N. (2013). CO_2 directly modulates connexin 26 by formation of carbamate bridges between subunits. *Elife* 2, 1–13. doi: 10.7554/elife.01213

Menichella, D. M., Goodenough, D. A., Sirkowski, E., Scherer, S. S., and Paul, D. L. (2003). Connexins are critical for normal myelination in the CNS. *J. Neurosci.* 23, 5963–5973.

Messier, M. L., Li, A., and Nattie, E. E. (2002). Muscimol inhibition of medullary raphé neurons decreases the CO2 response and alters sleep in newborn piglets. *Respir. Physiol. Neurobiol.* 133, 197–214. doi: 10.1016/s1569-9048(02)00168-4

Mika, T., and Prochnow, N. (2012). Functions of connexins and large pore channels on microglial cells: the gates to environment. *Brain Res.* 1487, 16–24. doi: 10.1016/j.brainres.2012.07.020

Milks, L. C., Kumar, N. M., Houghten, R., Unwin, N., and Gilula, N. B. (1988). Topology of the 32-kd liver gap junction protein determined by site-directed antibody localizations. *EMBO J.* 7, 2967–2975.

Monti-Bloch, L., Abudara, V., and Eyzaguirre, C. (1993). Electrical communication between glomus cells of the rat carotid body. *Brain Res.* 622, 119–131. doi: 10.1016/0006-8993(93)90810-a

Musil, L. S., and Goodenough, D. A. (1993). Multisubunit assembly of an integral plasma membrane channel protein, gap junction connexin43, occurs after exit from the ER. *Cell* 74, 1065–1077. doi: 10.1016/0092-8674(93)90728-9

Nagy, J. I., Ochalski, P. A., Li, J., and Hertzbergs, E. L. (1997). Evidence for the co-localization of another connexin with connexin-43 at astrocytic gap junctions in rat brain. *Neuroscience* 78, 533–548. doi: 10.1016/s0306-4522(96)00584-2

Nagy, J. I., Patel, D., Ochalski, P. A., and Stelmack, G. (1999). Connexin30 in rodent, cat and human brain: selective expression in gray matter astrocytes, co-localization with connexin43 at gap junctions and late developmental appearance. *Neuroscience* 88, 447–468. doi: 10.1016/s0306-4522(98)00191-2

Nagy, J. I., Li, X., Rempel, J., Stelmack, G., Patel, D., Staines, W. A., et al. (2001). Connexin26 in adult rodent central nervous system: demonstration at astrocytic gap junctions and colocalization with connexin30 and connexin43. *J. Comp. Neurol.* 441, 302–323. doi: 10.1002/cne.1414

Nattie, E. E., and Li, A. (2002a). Substance P-saporin lesion of neurons with NK1 receptors in one chemoreceptor site in rats decreases ventilation and chemosensitivity. *J. Physiol.* 544, 603–616. doi: 10.1113/jphysiol.2002.020032

Nattie, E. E., and Li, A. (2002b). CO2 dialysis in nucleus tractus solitarius region of rat increases ventilation in sleep and wakefulness. *J. Appl. Physiol.* 92, 2119–2130. doi: 10.1152/japplphysiol.01128.2001

Nieber, K., Poelchen, W., and Illes, P. (1997). Role of ATP in fast excitatory synaptic potentials in locus coeruleus neurones of the rat. *Br. J. Pharmacol.* 122, 423–430. doi: 10.1038/sj.bjp.0701386

Nielsen, M. S., Axelsen, L. N., Sorgen, P. L., Verma, V., Delmar, M., and Holstein-Rathlou, N.-H. (2012). Gap junctions. *Compr. Physiol.* 2, 1981–2035. doi: 10.1002/cphy.c110051

Orellana, J. A., Martínez, A. D., and Retamal, M. A. (2013). Gap junction channels and hemichannels in the CNS: regulation by signaling molecules. *Neuropharmacology* 75, 567–582. doi: 10.1016/j.neuropharm.2013.02.020

Oyamada, M., Oyamada, Y., and Takamatsu, T. (2005). Regulation of connexin expression. *Biochim. Biophys. Acta* 1719, 6–23. doi: 10.1016/j.bbamem.2005.11.002

Panchin, Y., Kelmanson, I., Matz, M., Lukyanov, K., Usman, N., and Lukyanov, S. (2000). A ubiquitous family of putative gap junction molecules. *Curr. Biol.* 10, R473–R474. doi: 10.1016/s0960-9822(00)00576-5

Pardal, R., Ortega-Sáenz, P., Durán, R., and López-Barneo, J. (2007). Glia-like stem cells sustain physiologic neurogenesis in the adult mammalian carotid body. *Cell* 131, 364–377. doi: 10.1016/j.cell.2007.07.043

Parenti, R., Gulisano, M., Zappalà, A., and Cicirata, F. (2000). Expression of connexin36 mRNA in adult rodent brain. *Neuroreport* 11, 1497–1502. doi: 10.1097/00001756-200005150-00027

Peracchia, C. (2004). Chemical gating of gap junction channels; roles of calcium, pH and calmodulin. *Biochim. Biophys. Acta* 1662, 61–80. doi: 10.1016/j.bbamem.2003.10.020

Peracchia, C., Young, K. C., Wang, X. G., Chen, J. T., and Peracchia, L. L. (2003). The voltage gates of connexin channels are sensitive to CO(2). *Cell Commun. Adhes.* 10, 233–237. doi: 10.1080/cac.10.4-6.233.237

Perkins, G., Goodenough, D. A., and Sosinsky, G. E. (1997). Three-dimensional structure of the gap junction connexon. *Biophys. J.* 72, 533–544. doi: 10.1016/s0006-3495(97)78693-4

Piskuric, N. A., and Nurse, C. A. (2013). Expanding role of ATP as a versatile messenger at carotid and aortic body chemoreceptors. *J. Physiol.* 591, 415–422. doi: 10.1113/jphysiol.2012.234377

Prasad, M., Fearon, I. M., Zhang, M., Laing, M., Vollmer, C., and Nurse, C. A. (2001). Expression of P2X2 and P2X3 receptor subunits in rat carotid body afferent neurones: role in chemosensory signalling. *J. Physiol.* 537, 667–677. doi: 10.1113/jphysiol.2001.012836

Putnam, R. W. (2010). CO2 chemoreception in cardiorespiratory control. *J. Appl. Physiol.* 108, 1796–1802. doi: 10.1152/japplphysiol.01169.2009

Quist, A. P., Rhee, S. K., Lin, H., and Lal, R. (2000). Physiological role of gap-junctional hemichannels. Extracellular calcium-dependent isosmotic volume regulation. *J. Cell Biol.* 148, 1063–1074. doi: 10.1083/jcb.148.5.1063

Rackauskas, M., Neverauskas, V., and Skeberdis, V. A. (2010). Diversity and properties of connexin gap junction channels. *Medicina (Kaunas)* 46, 1–12.

Ramachandran, S., Xie, L.-H., John, S. A., Subramaniam, S., and Lal, R. (2007). A novel role for connexin hemichannel in oxidative stress and smoking-induced cell injury. *PLoS One* 2:e712. doi: 10.1371/journal.pone.0000712

Rash, J. E., Yasumura, T., Davidson, K. G. V., Furman, C. S., Dudek, F. E., and Nagy, J. I. (2001). Identification of cells expressing Cx43, Cx30, Cx26, Cx32 and Cx36 in gap junctions of rat brain and spinal cord. *Cell Commun. Adhes.* 8, 315–320. doi: 10.3109/15419060109080745

Ray, R. S., Corcoran, A. E., Brust, R. D., Kim, J. C., Richerson, G. B., Nattie, E., et al. (2011). Impaired respiratory and body temperature control upon acute serotonergic neuron inhibition. *Science* 333, 637–642. doi: 10.1126/science.1205295

Retamal, M. A., Cortés, C. J., Reuss, L., Bennett, M. V. L., and Sáez, J. C. (2006). S-nitrosylation and permeation through connexin 43 hemichannels in astrocytes: induction by oxidant stress and reversal by reducing agents. *Proc. Natl. Acad. Sci. U S A* 103, 4475–4480. doi: 10.1073/pnas.0511118103

Rey, S., Del Rio, R., Alcayaga, J., and Iturriaga, R. (2004). Chronic intermittent hypoxia enhances cat chemosensory and ventilatory responses to hypoxia. *J. Physiol.* 560, 577–586. doi: 10.1113/jphysiol.2004.072033

Reyes, E.-P., Fernández, R., Larraín, C., and Zapata, P. (2007a). Carotid body chemosensory activity and ventilatory chemoreflexes in cats persist after combined cholinergic-purinergic block. *Respir. Physiol. Neurobiol.* 156, 23–32. doi: 10.1016/j.resp.2006.07.006

Reyes, E.-P., Fernández, R., Larraín, C., and Zapata, P. (2007b). Effects of combined cholinergic-purinergic block upon cat carotid body chemoreceptors in vitro. *Respir. Physiol. Neurobiol.* 156, 17–22. doi: 10.1016/j.resp.2006.07.007

Richerson, G. B., Wang, W., Tiwari, J., and Bradley, S. R. (2001). Chemosensitivity of serotonergic neurons in the rostral ventral medulla. *Respir. Physiol.* 129, 175–189. doi: 10.1016/s0034-5687(01)00289-4

Riquelme, M. A., Cea, L. A., Vega, J. L., Boric, M. P., Monyer, H., Bennett, M. V. L., et al. (2013). The ATP required for potentiation of skeletal muscle contraction is released via pannexin hemichannels. *Neuropharmacology* 75, 594–603. doi: 10.1016/j.neuropharm.2013.03.022

Romanov, R. A., Bystrova, M. F., Rogachevskaya, O. A., Sadovnikov, V. B., Shestopalov, V. I., and Kolesnikov, S. S. (2012). The ATP permeability of pannexin 1 channels in a heterologous system and in mammalian taste cells is dispensable. *J. Cell Sci.* 125, 5514–5523. doi: 10.1242/jcs.111062

Rong, W., Gourine, A. V., Cockayne, D. A., Xiang, Z., Ford, A. P. D. W., Spyer, K. M., et al. (2003). Pivotal role of nucleotide P2X2 receptor subunit of the ATP-gated ion channel mediating ventilatory responses to hypoxia. *J. Neurosci.* 23, 11315–11321.

Sáez, J. C., Contreras, J. E., Bukauskas, F. F., Retamal, M. A., and Bennett, M. V. L. (2003). Gap junction hemichannels in astrocytes of the CNS. *Acta Physiol. Scand.* 179, 9–22. doi: 10.1046/j.1365-201x.2003.01196.x

Sáez, J. C., Martínez, A. D., Brañes, M. C., and González, H. E. (1998). Regulation of gap junctions by protein phosphorylation. *Braz. J. Med. Biol. Res.* 31, 593–600. doi: 10.1590/s0100-879x1998000500001

Sáez, J. C., Retamal, M. A., Basilio, D., Bukauskas, F. F., and Bennett, M. V. L. (2005). Connexin-based gap junction hemichannels: gating mechanisms. *Biochim. Biophys. Acta* 1711, 215–224. doi: 10.1016/j.bbamem.2005.01.014

Saito, T., Nishimura, M., Kudo, R., and Yamasaki, H. (2001). Suppressed gap junctional intercellular communication in carcinogenesis of endometrium. *Int. J. Cancer* 93, 317–323. doi: 10.1002/ijc.1350.abs

Scemes, E., and Giaume, C. (2006). Astrocyte calcium waves: what they are and what they do. *Glia* 54, 716–725. doi: 10.1002/glia.20374

Schalper, K. A., Palacios-Prado, N., Retamal, M. A., Shoji, K. F., Martínez, A. D., and Sáez, J. C. (2008). Connexin hemichannel composition determines the FGF-1-induced membrane permeability and free [Ca2+]i responses. *Mol. Biol. Cell* 19, 3501–3513. doi: 10.1091/mbc.e07-12-1240

Schalper, K. A., Sánchez, H. A., Lee, S. C., Altenberg, G. A., Nathanson, M. H., and Sáez, J. C. (2010). Connexin 43 hemichannels mediate the Ca2+ influx induced by extracellular alkalinization. *Am. J. Physiol. Cell Physiol.* 299, C1504–C1515. doi: 10.1152/ajpcell.00015.2010

Solomon, I. C. (2003a). Connexin36 distribution in putative CO2-chemosensitive brainstem regions in rat. *Respir. Physiol. Neurobiol.* 139, 1–20. doi: 10.1016/j.resp.2003.09.004

Solomon, I. C. (2003b). Focal CO2/H+ alters phrenic motor output response to chemical stimulation of cat pre-Botzinger complex in vivo. *J. Appl. Physiol.* 94, 2151–2157. doi: 10.1152/japplphysiol.01192.2002

Solomon, I. C., Edelman, N. H., and O'Neal, M. H. (2000). CO(2)/H(+) chemoreception in the cat pre-Bötzinger complex in vivo. *J. Appl. Physiol.* 88, 1996–2007.

Solomon, I. C., Halat, T. J., El-Maghrabi, R., and O'Neal, M. H. (2001). Differential expression of connexin26 and connexin32 in the pre-Bötzinger complex of neonatal and adult rat. *J. Comp. Neurol.* 440, 12–19. doi: 10.1002/cne.1366

Sosinsky, G. E., Boassa, D., Dermietzel, R., Duffy, H. S., Laird, D. W., MacVicar, B. A., et al. (2011). Pannexin channels are not gap junction hemichannels. *Channels (Austin)* 5, 193–197. doi: 10.4161/chan.5.3.15765

Spray, D. C., Harris, A. L., and Bennett, M. V. L. (1981). Equilibrium properties of a voltage-dependent junctional conductance. *J. Gen. Physiol.* 77, 77–93. doi: 10.1085/jgp.77.1.77

Spyer, K. M., Dale, N., and Gourine, A. V. (2004). ATP is a key mediator of central and peripheral chemosensory transduction. *Exp. Physiol.* 89, 53–59. doi: 10.1113/expphysiol.2003.002659

Stout, C. E., Costantin, J. L., Naus, C. C. G., and Charles, A. C. (2002). Intercellular calcium signaling in astrocytes via ATP release through connexin hemichannels. *J. Biol. Chem.* 277, 10482–10488. doi: 10.1074/jbc.m109902200

Unger, V. M., Kumar, N. M., Gilula, N. B., and Yeager, M. (1999). Expression, two-dimensional crystallization and electron cryo-crystallography of recombinant gap junction membrane channels. *J. Struct. Biol.* 128, 98–105. doi: 10.1006/jsbi.1999.4184

Valladares, D., Almarza, G., Contreras, A., Pavez, M., Buvinic, S., Jaimovich, E., et al. (2013). Electrical Stimuli Are Anti-Apoptotic in Skeletal Muscle via Extracellular ATP. Alteration of This Signal in Mdx Mice Is a Likely Cause of Dystrophy. *PLoS One* 8:e75340. doi: 10.1371/journal.pone.0075340

Vanden, F. A., Bidaux, G., Gordienko, D., Beck, B., Panchin, Y. V., Baranova, A. V., et al. (2006). Functional implications of calcium permeability of the channel formed by pannexin 1. *J. Cell Biol.* 174, 535–546. doi: 10.1083/jcb.200601115

Vinken, M. (2011). Role of connexin-related signalling in hepatic homeostasis and its relevance for liver-based in vitro modelling. *World J. Gastrointest. Pathophysiol.* 2, 82–87. doi: 10.4291/wjgp.v2.i5.82

Wang, S., Benamer, N., Zanella, S., Kumar, N. N., Shi, Y., Bévengut, M., et al. (2013b). TASK-2 channels contribute to pH sensitivity of retrotrapezoid nucleus chemoreceptor neurons. *J. Neurosci.* 33, 16033–16044. doi: 10.1523/JNEUROSCI.2451-13.2013

Wang, N., De Bock, M., Decrock, E., Bol, M., Gadicherla, A., Vinken, M., et al. (2013a). Paracrine signaling through plasma membrane hemichannels. *Biochim. Biophys. Acta* 1828, 35–50. doi: 10.1016/j.bbamem.2012.07.002

Wang, W., Pizzonia, J. H., and Richerson, G. B. (1998). Chemosensitivity of rat medullary raphe neurones in primary tissue culture. *J. Physiol.* 511(Pt. 2), 433–450. doi: 10.1111/j.1469-7793.1998.433bh.x

Wang, W., Tiwari, J. K., Bradley, S. R., Zaykin, R. V., and Richerson, G. B. (2001). Acidosis-stimulated neurons of the medullary raphe are serotonergic. *J. Neurophysiol.* 85, 2224–2235.

Weir, E. K., López-Barneo, J., Buckler, K. J., and Archer, S. L. (2005). Acute oxygen-sensing mechanisms. *N. Engl. J. Med.* 353, 2042–2055. doi: 10.1056/nejmra050002

Wenker, I. C., Kréneisz, O., Nishiyama, A., and Mulkey, D. K. (2010). Astrocytes in the retrotrapezoid nucleus sense H+ by inhibition of a Kir4.1-Kir5.1-like current and may contribute to chemoreception by a purinergic mechanism. *J. Neurophysiol.* 104, 3042–3052. doi: 10.1152/jn.00544.2010

Werner, R., Levine, E., Rabadan-Diehl, C., and Dahl, G. (1989). Formation of hybrid cell-cell channels. *Proc. Natl. Acad. Sci. U S A* 86, 5380–5384. doi: 10.1073/pnas.86.14.5380

Wenker, I. C., Sobrinho, C. R., Takakura, A. C., Moreira, T. S., and Mulkey, D. K. (2012). Regulation of ventral surface CO2/H+-sensitive neurons by purinergic signalling. *J. Physiol.* 590, 2137–2150. doi: 10.1113/jphysiol.2012.229666

Willecke, K., Eiberger, J., Degen, J., Eckardt, D., Romualdi, A., Güldenagel, M., et al. (2002). Structural and functional diversity of connexin genes in the mouse and human genome. *Biol. Chem.* 383, 725–737. doi: 10.1515/bc.2002.076

Wu, J., Xu, H., Shen, W., and Jiang, C. (2004). Expression and coexpression of CO2-sensitive Kir channels in brainstem neurons of rats. *J. Membr. Biol.* 197, 179–191. doi: 10.1007/s00232-004-0652-4

Xu, F., Xu, J., Tse, F. W., and Tse, A. (2006). Adenosine stimulates depolarization and rise in cytoplasmic [Ca2+] in type I cells of rat carotid bodies. *Am. J. Physiol. Cell Physiol.* 290, C1592–C1598. doi: 10.1152/ajpcell.00546.2005

Xu, J., Tse, F. W., and Tse, A. (2003). ATP triggers intracellular Ca2+ release in type II cells of the rat carotid body. *J. Physiol.* 549, 739–747. doi: 10.1113/jphysiol.2003.039735

Yamamoto, T., Ochalski, A., Hertzbergs, E. L., and Nagy, J. I. (1990). LM and EM immunolocalization of the gap junctional protein connexin 43 in rat brain. *Brain Res.* 508, 313–319. doi: 10.1016/0006-8993(90)90415-8

Ye, Z.-C., Wyeth, M. S., Baltan-Tekkok, S., and Ransom, B. R. (2003). Functional hemichannels in astrocytes: a novel mechanism of glutamate release. *J. Neurosci.* 23, 3588–3596.

Yen, M. R., and Saier, M. H. (2007). Gap junctional proteins of animals: the innexin/pannexin superfamily. *Prog. Biophys. Mol. Biol.* 94, 5–14. doi: 10.1016/j.pbiomolbio.2007.03.006

Yuill, K., Ashmole, I., and Stanfield, P. R. (2004). The selectivity filter of the tandem pore potassium channel TASK-1 and its pH-sensitivity and ionic selectivity. *Pflugers Arch.* 448, 63–69. doi: 10.1007/s00424-003-1218-5

Zapata, P. (1997a). "Chemosensory activity in the carotid nerve: effects of pharmacological agents," in *The Carotid Body Chemoreceptors*, ed C. González (Georgetown, TX, USA: Landes Biosciences), 119–146.

Zapata, P. (1997b). "Chemosensory activity in the carotid nerve: effects of physiological variables," in *The Carotid Body Chemoreceptors*, ed C. González (Georgetown, TX, USA: Landes Biosciences), 97–117.

Zapata, P. (2007). Is ATP a suitable co-transmitter in carotid body arterial chemoreceptors? *Respir. Physiol. Neurobiol.* 157, 106–115. doi: 10.1016/j.resp.2007.01.002

Zhang, M., Piskuric, N. A., Vollmer, C., and Nurse, C. A. (2012). P2Y2 receptor activation opens pannexin-1 channels in rat carotid body type II cells: potential role in amplifying the neurotransmitter ATP. *J. Physiol.* 590, 4335–4350. doi: 10.1113/jphysiol.2012.236265

Zhang, Y.-Y., Sackin, H., and Palmer, L. G. (2006). Localization of the pH gate in Kir1.1 channels. *Biophys. J.* 91, 2901–2909. doi: 10.1529/biophysj.106.087700

Zhang, M., Zhong, H., Vollmer, C., and Nurse, C. A. (2000). Co-release of ATP and ACh mediates hypoxic signalling at rat carotid body chemoreceptors. *J. Physiol.* 525(Pt. 1), 143–158. doi: 10.1111/j.1469-7793.2000.t01-1-00143.x

18

Pannexin 1 channels: New actors in the regulation of catecholamine release from adrenal chromaffin cells

Fanny Momboisse[1]*, María José Olivares[1], Ximena Báez-Matus[1], María José Guerra[1], Carolina Flores-Muñoz[1], Juan C. Sáez[1,2], Agustín D. Martínez[1] and Ana M. Cárdenas[1]*

[1] *Centro Interdisciplinario de Neurociencias de Valparaíso, Universidad de Valparaíso, Valparaíso, Chile*
[2] *Departamento de Fisiología, Pontifícia Universidad Católica de Chile, Santiago, Chile*

***Correspondence:**
Fanny Momboisse and Ana M. Cárdenas, Centro Interdisciplinario de Neurociencia de Valparaíso, Universidad de Valparaíso, Avenida Gran Bretaña 1111, Playa Ancha, Valparaíso 2340000, Chile
e-mail: fanny.momboisse@cinv.cl; ana.cardenas@uv.cl

Chromaffin cells of the adrenal gland medulla synthesize and store hormones and peptides, which are released into the blood circulation in response to stress. Among them, adrenaline is critical for the fight-or-flight response. This neurosecretory process is highly regulated and depends on cytosolic [Ca^{2+}]. By forming channels at the plasma membrane, pannexin-1 (Panx1) is a protein involved in many physiological and pathological processes amplifying ATP release and/or Ca^{2+} signals. Here, we show that Panx1 is expressed in the adrenal gland where it plays a role by regulating the release of catecholamines. In fact, inhibitors of Panx1 channels, such as carbenoxolone (Cbx) and probenecid, reduced the secretory activity induced with the nicotinic agonist 1,1-dimethyl-4-phenyl-piperazinium (DMPP, 50 μM) in whole adrenal glands. A similar inhibitory effect was observed in single chromaffin cells using Cbx or ^{10}Panx1 peptide, another Panx1 channel inhibitors. Given that the secretory response depends on cytosolic [Ca^{2+}] and Panx1 channels are permeable to Ca^{2+}, we studied the possible implication of Panx1 channels in the Ca^{2+} signaling occurring during the secretory process. In support of this possibility, Panx1 channel inhibitors significantly reduced the Ca^{2+} signals evoked by DMPP in single chromaffin cells. However, the Ca^{2+} signals induced by caffeine in the absence of extracellular Ca^{2+} was not affected by Panx1 channel inhibitors, suggesting that this mechanism does not involve Ca^{2+} release from the endoplasmic reticulum. Conversely, Panx1 inhibitors significantly blocked the DMPP-induce dye uptake, supporting the idea that Panx1 forms functional channels at the plasma membrane. These findings indicate that Panx1 channels participate in the control the Ca^{2+} signal that triggers the secretory response of adrenal chromaffin cells. This mechanism could have physiological implications during the response to stress.

Keywords: pannexin 1, Ca^{2+} signaling, neurosecretion, catecholamines, chromaffin cells

INTRODUCTION

The release of hormones from neuroendocrine cells is a highly regulated process that has to be adjusted to environmental demands. This regulation is particularly required for the release of catecholamines during stress situations wherein the body reacts to environmental changes (Harvey et al., 1984). The adrenal gland medulla constitutes the major source of catecholamines in the body (Wong, 2003). These glands are innervated by cholinergic terminals of the splanchnic nerve, which activates chromaffin cells in response to stressors. Upon activation with acetylcholine, chromaffin cells secrete to the blood circulation mostly catecholamines (adrenaline, noradrenaline) but also ATP, enkephalins and neuropeptide Y among others peptides and proteins (Aunis and Langley, 1999; Crivellato et al., 2008). The release process depends on a series of events that are finely regulated and occur in a perfect sequence: activation of nicotinic receptors, membrane depolarization, opening of voltage-dependent Ca^{2+} channels, and transient increase of the cytosolic [Ca^{2+}] that ends with the exocytotic release of molecules stored in chromaffin granules (Weiss, 2012). Most of the released transmitters act on autoreceptors present in the chromaffin cell plasma membrane allowing an autocrine feedback of the secretory process (Currie and Fox, 1996; Lim et al., 1997).

Pannexins (Panxs) are a family of three glycoproteins, Panxs 1, 2 and 3, which form pore channels localized at the plasma membrane but also in intracellular compartments such as the endoplasmic reticulum (ER; Penuela et al., 2008). Pannexin-1 (Panx1), the best-characterized member, is involved in various cellular processes in both pathologic and physiologic contexts (Thompson et al., 2008; Orellana et al., 2011; Séror et al., 2011), where it participates in the initiation and/or amplification of ATP release and Ca^{2+} signals (Bao et al., 2004; Vanden Abeele et al., 2006). Most of the described actions of Panx1 channels often take place in tight collaboration with metabotropic or ionotropic purinergic receptors. Furthermore, Panx1 forms a functional association with the purinergic receptor $P2X_7$ ($P2X_7R$) (Locovei et al., 2007). It has been proposed that these types of associations allow an autocrine feedback loop that controls different physiological settings, such as the immune synapse for T-cell activation (Woehrle et al., 2010) or the potentiation of the skeletal muscle contraction (Riquelme et al., 2013). These evidences place Panx1 as an important actor in the control of several signaling processes, mainly through its participation in the regulation of the cytosolic [Ca^{2+}]. Hence, we decided to study the possible role of Panx1 in the process of catecholamine release from the adrenal medulla and its underlying mechanism. We found that Panx1 is expressed at the plasma membrane of adrenal chromaffin cells where it plays an important role controlling the release of catecholamines through the amplification of Ca^{2+} signals triggered by activation of nicotinic receptors.

MATERIALS AND METHODS

CHROMAFFIN CELL CULTURE

Bovine adrenal chromaffin cells were isolated as previously described (Montiel et al., 2003). Cells were plated at a density of 10 × 10^6 cells/ml on collagen-coated glass coverslips (12 or 25 mm, Warner Instrument, Hamden, USA) or on plastic culture dishes (10-cm diameter, Orange scientific, Braine-l'Alleud, Belgium), and kept at 37°C, 5% CO_2 and 100% humidity at least 48 h before the experiments.

INHIBITORS, DYE AND ANTIBODIES

Carbenoxolone (Cbx) and probenecid were purchased from Sigma-Aldrich (Saint Louis, MO, USA) and Molecular Probes Life technologies (Grand Island, USA), respectively. ^{10}Panx1 and ^{10}Panx1 scramble (^{10}Panx1 scrb) peptides were synthesized by Tocris Bioscience (Bristol, UK). A rabbit polyclonal anti-Panx1 serum was a generous gift from Dr. D. Laird (University of Western Ontario, London, Canada). The monoclonal anti-actin antibody and 4′,6-diamidino-2-phenylindole (DAPI) were purchased from Sigma-Aldrich. The α-CY2-labeled secondary anti-rabbit antibody and the HRP-labeled secondary anti-rabbit antibody were purchased from Jackson ImmunoResearch (West Grove, USA) and the HRP-labeled secondary anti-mouse antibody was obtained from Thermoscientific (Waltham, USA) respectively.

RT-PCR AND SINGLE CELL RT-PCR

RT-PCR

Total RNA from bovine adrenal glands or cultured bovine chromaffin cells was isolated using the SV Total RNA Isolation System from Promega (Madison, USA), following the manufacturer's instructions. cDNAs were synthesized from total RNAs using SUPERSCRIPT III First-Strand (Invitrogen). Hot start PCR was performed using 2 ng cDNA in a total volume of 50 μl containing PCR Master Mix (Promega) and specific primers (forward primer 5′ TACTTTGGGGATGCCTGGAG-3′ and reverse primer 5′-GGCGCACTGAAAGACCTC-3′ for dopamine β-hydroxylase (DBH), matched with gene ID NM_180995.2, targeting exon-exon junction with the forward primer, forward primers 5′-CGCAAGAAATCTCCATTGGT-3′ and reverse primer for Panx1, 5′-GGCTTTCCTGTGAACT TTGC-3′, matched with gene ID: NM_001245925.1, targeting exon-exon junction with the reverse primer and forward primer 5′-TTTGTGATGGGTGTGAACCACGAG-3′ and reverse primer 5′-CAACGGATACATTGGGGGTAGGAAC-3′ for GADPH, matched with the gene ID XR 405643.1. PCR reactions were run for 40 cycles and PCR products obtained for DBH (480 pb), Panx1 (396 pb) and GADPH (334 pb) were visualized in 4% agarose gel.

Single cell RT-PCR

The protocol for Single cell RT-PCR was adapted from Phillips and Lipski (2000). Briefly, the cytoplasm of a single cultured cell was picked up using a patch pipette filled with 10 μl of autoclaved diethylpyrocarbonate-treated water. The cell content was expelled by negative pressure into an autoclaved microtube, immediately frozen in liquid nitrogen and keep at −80°C. The reverse transcription was performed as described above. For amplification of the cDNA product, we used a Nested PCR procedure. During the first round of PCR (40 cycles, as described above), Panx1 and DBH external primer pairs (sequence mentioned above) were mixed in the PCR reaction. The 50 μl containing the PCR product was split in two tubes and the second PCR was performed using the internal primers (forward: 5′ TGTGACCCCAACGACTACCT 3′, and reverse primers: 5′ TCGGTCACGTAGCACCAGTA 3′ for DBH, matched with gene ID NM_180995.2, and 5′ TAAGCTGCTTCTCCCCCAGT 3′, and 5′ AGGCACCGTCTCTCAAGTCA 3′ for Panx1, matched with gene ID NM_001245925.1). PCR products (248 pb for DBH and 318 pb for Panx1) were analyzed in 4% agarose gel. For each RT-PCR, negative controls were prepared using water as template.

WESTERN BLOT ANALYSIS

The presence of Panx1 was evaluated in extracts of bovine adrenal glands, cultured chromaffin cells, wild type or Panx1$^{-/-}$ KO mice brain using western blot analysis. Tissue or cell extracts were obtained using a lysis buffer (HaltTM Protease and Phosphatase Inhibitor Cocktail) (Thermoscientific, Waltham, USA) and 0.5 μM EDTA in phosphate buffered saline (PBS) pH 7.4. After 10 s of sonication, extracts were centrifuged 10 min at 14,000 rpm 4°C. The supernatants were collected and proteins were quantified using Qubit® Protein Assay (Molecular Probes, Life Technologies). The amount of 40 μg of proteins of each extract was separated by electrophoresis on a 10% SDS-polyacrylamide gel and proteins were electrotransferred to nitrocellulose membranes. Then, membranes were incubated for 1 h with blocking solution (0.05% Tween-20 with 5% w/v nonfat dry milk) in a Tris-buffered saline (TBS) solution, followed by overnight incubation at 4°C with anti-Panx1 serum diluted (1:1000) in blocking solution. After several washes, membranes were incubated with HRP-labeled secondary anti-rabbit antibody (1:5000) for 1 h and immunoreactive bands were revealed with the ECL Plus system (Amersham GE Lifes sciences, Piscataway, USA). As Actin and Panx1 have similar molecular weight, the membrane were treated with a stripping buffer (glycine 0.2 M, 0.1% SDS, Tween 20 1%, pH 2.2) and then washed, incubated with the blocking solution as described above and incubated with anti-actin antibody (1/500, 4°C overnight in blocking solution). After several washes membranes were incubated with HRP-labeled secondary anti-mouse antibody (1:2500) for 1 h and immunoreactive bands were revealed as described above.

CATECHOLAMINE SECRETION FROM THE ADRENAL GLAND

Fresh bovine adrenal glands were perfused with a Kreb's-Hepes solution (in mM: 140 NaCl, 5.9 KCl, 1.2 $MgCl_2$, 2 $CaCl_2$, 10 Hepes-NaOH, 10 glucose) by means of a peristaltic pump (Variable-speed pump 2, Fisher Scientific, Waltham, USA) at a rate of 4 mL/min. The solution was constantly bubbled with O_2 and the final pH was maintained in the range of 7.4–7.5. After an equilibration time of 1 h, secretion was induced with a 2 min pulse of the nicotinic agonist 1,1-dimethyl- 4-phenyl-piperazinium (DMPP, 50 μM) added into the perfusion stream. A second pulse was applied 30 min later. In order to evaluate the role of Panx1 channels in the catecholamine release, a group of glands was perfused 15 min before and during the second DMPP pulse with Cbx (5 μM) or probenecid (200 μM).

Perfusates were collected in 2 min fractions, 6 min prior to the DMPP pulse to determine the spontaneous secretion of catecholamines, during the pulse, and 14 min after the DMPP pulse. The perfusated samples were collected in tubes containing 0.005 N perchloric acid and kept on ice. The catecholamines released in the background sample were subtracted from those released from the stimulated sample to obtain the net value of secretion. The products generated from the oxidation of noradrenaline and adrenaline by iodine at pH 6 were measured (excitation wavelength 540 nm) (Persky, 1955). Due to the variation in gland shape and size, each gland was used as its own control, thus results were expressed as the percentage of the first stimulation pulse.

IMMUNOFLUORESCENCE AND CONFOCAL IMAGING

Cultured chromaffin cells were first washed with PBS (pH 7.4), and incubated fixed with 4% paraformaldehyde (PFA) in PBS for 15 min at 4°C and permeabilized with a fixative solution containing 0.2% Triton X-100 for 10 min. Then, cells were rinsed several times with PBS, pre-treated with 3% bovine serum albumin (BSA) in PBS for 1 h, and incubated with the anti-Panx1 serum (1:100) for 2 h. After that, cells were washed several times with PBS and incubated for 1 h with α-CY2-labeled secondary anti-rabbit antibody (1:1000). After several washes with PBS, cells were incubated with DAPI (5 μg/ml) for 15 min. Finally, coverslips were washed and mounted with Dako fluorescent mounting medium (Dako, Glostrup, Denmark). Stained cells were visualized with a Nikon C1 Plus laser-scanning confocal microscope, equipped with a 100X objective NA 1.30 and excited with laser line 408 and 488 nm. To determine if Panx1 is localized at the plasma membrane, cultured chromaffin cells were labeled with extracellular biotin. Cells were first washed in PBS (pH 7.4) and incubated 10 min with a solution of Kreb's Hepes containing 90 μM sulfo-NHS-biotin (Thermoscientific Waltham, USA). Then, cells were incubated twice with 15 mM glycin solution (in Kreb's Hepes) for 7 min and then fixed with 4% PFA. The immunofluorescence protocol was similar as described above but using an avidin-Cy3 (1:5000, Jackson ImmunoResearch) with the α-CY2-labeled secondary anti-rabbit antibody. Confocal acquisitions were analyzed and processed using Image-J software (NIH, USA). Biotin or Panx1 fluorescence was measured subtracting the mean fluorescence 1 μm under the cell periphery to the total cell fluorescence intensity. Pearson's correlation was employed to determine the level of colocalization between Panx1 and biotin.

AMPEROMETRIC DETECTION OF EXOCYTOSIS IN CULTURED CHROMAFFIN CELLS

Amperometric recordings were performed as previously described (Ardiles et al., 2006). Carbon fiber electrodes (5 μm diameter, Thornel P-55; Amoco Performance Product, Greenville, USA) held at a potential of 650 mV were used to detect single exocytotic events. A HEKA EPC10 amplifier (HEKA Elektronik, Lambrecht/Pfalz, Germany) controlled by the PatchMaster software (HEKA Elektronik) allowed us to obtain the amperometric signals, which were low-pass filtered at 1 kHz and digitized at 5 Hz. During the recording, cells were maintained in Kreb's-Hepes solution. The exocytosis was evoked by a 10 s pressure ejection of 50 μM DMPP. In order to evaluate the role of Panx1 channels in exocytosis, cells were pre-incubated for 15 min with Cbx (5 μM), or ^{10}Panx1 or ^{10}Panx1 scrb peptides (200 μM) at 37°C. These reagents were maintained in the bath solution during the entire recording. Single exocytotic events were analyzed using a written macro for IGOR (Wavemetrics), obtained from Dr. R. Borges.[1] Only spikes with Imax > 5SD of the noise were counted and analyzed.

CYTOSOLIC [Ca^{2+}] MEASUREMENTS

Variations of cytosolic [Ca^{2+}] were determined using microfluorometry (Cárdenas et al., 2002). Chromaffin cells cultured on glass coverslips were incubated for 40 min at 37°C with 5 mM Indo-1 AM (Molecular Probes, Life technologies) in 0.1% pluronic acid and then washed with Kreb's-Hepes solution (pH 7.4). The coverslips were mounted in a perfusion chamber and placed on the stage of a fluorescence-inverted microscope (Diaphot- 200, Nikon Corp. Tokyo, Japan). The microscope was equipped with two dichroic mirrors: the first one sent excitation light (355 nm) to the cell and the second mirror split the fluorescent light emitted by intracellular Indo-1 (>400 nm) into beams of light centered at 410 and 485 nm. The intensity of the light at each wavelength was measured continuously using two photomultipliers and the digital signal was obtained using HEKA EPC10 amplifier (HEKA Elektronik) controlled by the PatchMaster software (HEKA Elektronik). The cytosolic [Ca^{2+}] was calculated from the F_{410}/F_{485} ratio using the following formula: $[Ca^{2+}] = K_d (R-R_{min})/(R_{max}/R)$. The Ca^{2+} dissociation constant for Indo-1 AM, K_d, was obtained with a calibration curve using various known [Ca^{2+}]. R_{min} was determined incubating the cells with EGTA (10 mM) and R_{max} with a solution containing 10 μM ionomycin with 10 mM $CaCl_2$. Cells were incubated at 37°C with Cbx (5 μM), probenecid (200 μM), ^{10}Panx1 (200 μM) or ^{10}Panx1 scrb peptide (200 μM) 15 min before the experiments. These inhibitors were maintained in the bath solution during the entire recording. The Ca^{2+} signals were evoked by a 10 s pressure ejection of 50 μM DMPP or 50 mM caffeine. In experiments done with DMPP, the cells were kept

[1] http://rborges.webs.ull.es/index.html

in the Kreb's-Hepes solution, but the stimulation with caffeine was done in a Ca^{2+}-free Kreb's-Hepes solution (in mM: 140 NaCl, 5.9 KCl, 1.2 $MgCl_2$, 2 EGTA, 10 Hepes-NaOH, 10 glucose).

DAPI UPTAKE

Chromaffin cells cultured on 12 mm coverslips were washed twice with Krebś Hepes solution and incubated for 15 min in Kreb's Hepes solution with or without Cbx (5 μM) or probenecid (200 μM) at 37°C. Then, the cells were incubated for 10 s with a Kreb's Hepes solution containing 50 μM DAPI (Sigma) in the absence or presence of 50 μM DMPP and the respective Panx1 inhibitors. Then the solution was replaced the same Kreb's Hepes solutions without DMPP for 2 min and the cells were fixed with 4% PFA in PBS for 15 min at 4°C. Finally, coverslips were washed 3 times and mounted with DAKO fluorescent mounting medium (Dako, Glostrup, Denmark). Stained cells were visualized with a Nikon C1 Plus laser-scanning confocal microscope, equipped with a 40X objective (NA = 1.30) and excited with laser lines of 408. Confocal acquisitions were analyzed and processed using Image-J software (NIH, USA). Three independent background fluorescence intensity measurements were averaged and subtracted from the fluorescence intensity of each cell. DAPI uptake, expressed in $AU/\mu m^2$, was evaluated calculating the fluorescence intensity of each chromaffin cell nucleus, divided by the nucleus area.

Data analysis and statistics

Data of catecholamine secretion from adrenal glands correspond to the mean ± SEM of 9 to 10 glands. Amperometric spikes, cytosolic $[Ca^{2+}]$ and DAPI uptake were averaged by individual cells, where "n" refers to the number of tested cells. Data presented correspond to means ± SEM of cell averages from at least three different cultures. The statistical significance of the differences was evaluated using Krukal-Wallis test for nonparametric data, level of $p < 0.05$ was considered statistically significant (*).

Ethics statement

The present work includes the use of bovine adrenal glands obtained from a local slaughterhouse, Frigorific Don Pedro, certificated (Livestock role 04.2.03.0002) by the Agriculture and Livestock Service of the Chilean Government. The slaughterhouse is regularly inspected by a veterinarian of the Chilean Health Service. Transport, processing and elimination of the samples were carried out in strict accordance with the Article 86 of the Sanitary Regulations of the Chilean Government (Supreme decree Nu 977/96). Panx1 knock-out (KO) C57BL/6 mice previously described by Bargiotas et al. (2011) were kindly provided by Dr. Hannah Monyer, University Heidelberg, Germany. These animals were bred in the Animal Facilities of the Pontifícia Universidad Católica de Chile. Wild type C57BL/6 mice were used as control. The use of KO mice was limited to crucial experiments to reduce the number of animals sacrificed. Mouse brain extract were obtained using 9–12 months old male.

All the protocols described in this article were approved by a Committee of Bioethics and Biosafety of the Faculty of Science, University of Valparaíso, directed by Professor Juan Carlos Espinoza, on May, 2, 2011.

RESULTS

Panx1 IS EXPRESSED IN THE ADRENAL GLAND AND PARTICIPATES IN THE SECRETORY RESPONSE INDUCED BY THE ACTIVATION OF NICOTINIC RECEPTORS

Panx1 is expressed in various tissues including neuroendocrine tissues such as the pituitary gland (Li et al., 2011) but until now, its expression in the adrenal gland remains unknown. To investigate Panx1 expression in this tissue, we performed an RT-PCR assay of total RNA obtained from bovine adrenal glands. Bovine brain RNA was used as a positive control. Panx1 transcripts were detected in both tissues (**Figure 1A**). The expression of the protein in the adrenal gland was confirmed by western blot using a specific polyclonal serum against Panx1 (**Figure 2B**). Next, we studied the possible implication of Panx1 expression in the release of catecholamine from intact adrenal glands. To this end, we used two different Panx1 channel inhibitors: Cbx, which at 5 μM blocks Panx1 channels, but not connexin based channels (Bruzzone et al., 2005), and probenecid (200 μM), a Panx1 channel inhibitor (Silverman et al., 2008). To mimic the physiological condition, the glands were stimulated with the nicotinic agonist DMPP. First, the glands were perfused with Krebs's solution for 1 h, then the secretory activity was induced with two 2 min pulses of the nicotinic agonist DMPP (50 μM) applied every 45 min. A group of glands was treated with probenecid or Cbx 15 min before and during the second pulse. In these experiments, the first pulse was used as an internal control. **Figure 1B** shows the catecholamine release after the second DMPP pulse expressed as a percentage of the release induced by the first pulse. In control glands, the secretory response increased up to 144.6% after the second stimulation. Conversely, the treatment with Cbx or probenecid significantly decreased secretory activity of the adrenal gland to 64.1% and 34.9%, respectively. Taken together, these results demonstrate that Panx1 channels regulate the secretory activity of the adrenal gland. Therefore, we decided to study the involved mechanism in cultured chromaffin cells.

Panx1 IS EXPRESSED IN CULTURED CHROMAFFIN CELLS AND CONTROLS THE NUMBER OF SECRETORY EVENTS

We first checked that Panx1 mRNA is expressed in chromaffin cells using single cell RT-PCR. To do so, the cytoplasm of a single chromaffin cell was extracted and we performed the RT-PCR with a nested PCR procedure by using two pairs of primers designed for Panx1. DBH mRNA was used as a positive control. In this assay, we also observed the presence of Panx1 mRNA (**Figure 2A**), confirming its expression in chromaffin cells. In order to verify that Panx1 is expressed at the protein level and how it is distributed in cultured chromaffin cells, we performed western blot and immunofluorescence assays. By western blot, we detected the presence of a band of around 50 kDa in extracts from cultured chromaffin cells and adrenal glands (**Figure 2B**). Brain extracts from WT or $Panx1^{-/-}$ mice were used as positive control and negative control, respectively. An anti-actin antibody was used as a loaded control. Samples processed for

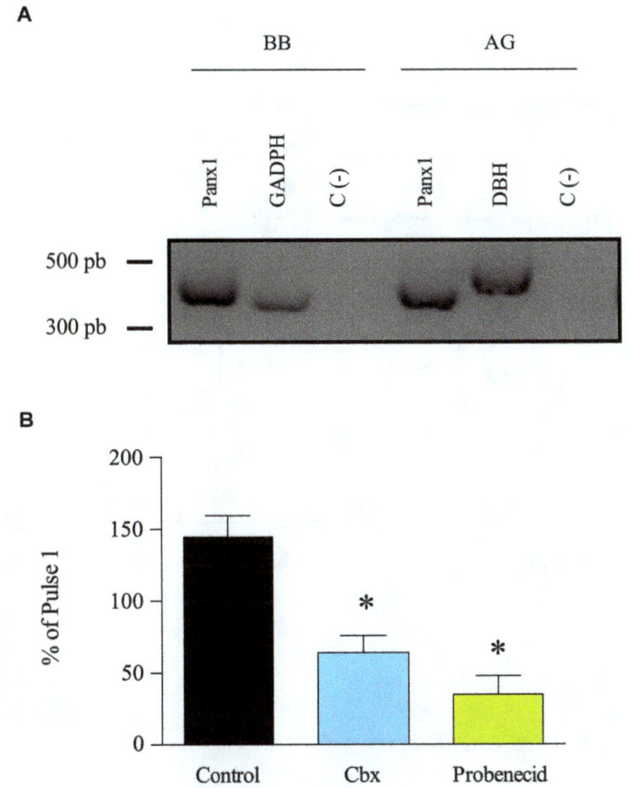

FIGURE 1 | Panx1 is expressed in adrenal gland and participates in the release of catecholamines. (A) Expression of Panx1 was evaluated by RT-PCR using Panx1 primers. The RT-PCR was performed on mRNA extracted from bovine brain (BB) (positive control) and adrenal gland (AG). GADPH and DBH were used as amplification controls; they lead to the generation of 334 pb and 480 pb transcripts, respectively. A negative control without template was also included (C-). The presence of Panx1 transcript (396 pb) was detected in bovine brain as well as in adrenal gland. **(B)** Entire adrenal glands were perfused with Kreb's solution for 1 h, and then the release of catecholamines was induced by a 2 min pulse of the nicotinic agonist DMPP (50 μM). Fifteen minutes before the second pulse, the glands were treated with Cbx (5 μM) or probenecid (200 μM). The drugs were maintained in the perfused solution during the entire sample recollection. Absorbance of the catecholamine oxidation products was measured at 540 nm. Data, expressed as the percentage of catecholamines secreted during the second pulse with respect to the first pulse, are means ± SEM of control glands ($n = 10$), or glands treated with carbenoxolone (Cbx) ($n = 9$) or probenecid ($n = 10$). * $p < 0.05$ compared with control glands (Krukal-Wallis test).

immunofluorescence were visualized using a confocal microscope and we found that cultured chromaffin cells were positively immunolabeled by the anti-Panx1 serum. Panx1 was essentially found at the cell periphery (**Figure 2C**), suggesting that it is mainly localized at the plasma membrane of the chromaffin cells.

Then, we monitored exocytosis in chromaffin cells using amperometry, since this technique provides information about the characteristics of individual release events (González-Jamett et al., 2010). The exocytosis was induced with 50 μM DMPP and Panx1 channels were inhibited with 5 μM Cbx or the mimic peptide ^{10}Panx1 (200 μM), which is a specific Panx1 channel blocker (Pelegrin and Surprenant, 2006). Each inhibitor was added to the culture medium 15 min before the experiment and maintained in the bath solution during the entire recording. The scramble ^{10}Panx1 peptide (^{10}Panx1 scrb) was used as a negative control. All the experiments were performed on uncoupled cells to rule out any possible implication of gap junctional coupling with neighboring cells. **Figure 2D** shows representative amperometric traces of each condition. In non-treated cells (control), a 10 s pulse with 50 μM DMPP induced 29.6 ± 3.4 amperometric spikes in 100 s ($n = 57$). The number of release events was not significantly affected by ^{10}Panx1 scrb (40.3 ± 7.8). However, it was significantly diminished by Cbx or ^{10}Panx1, as compared with non-treated cells (control) or cells treated with the ^{10}Panx1 scrb, respectively (**Figure 2E**). ^{10}Panx1 and Cbx reduced the event number by 81.4% and 51.3% respectively.

Panx1 CHANNELS REGULATE THE KINETIC OF THE SINGLE RELEASE EVENTS

In order to understand how Panx1 channels affect the characteristics of individual release events, we analyzed for each amperometric spike the quantal size (Q), which is proportional to the amount of catecholamines released per event, the time to peak (tP) or rising time that reflects the speed of the expansion of the fusion pore, and the half-width ($t_{1/2}$) that reflects the duration of the exocytotic events (Neco et al., 2008; **Figure 3A**). **Figure 3B** shows examples of amperometric spike of control cells or cells treated with the different Panx1 inhibitors. Q, tP and $t_{1/2}$ values in control cells stimulated with 50 μM DMPP were 1.6 ± 0.1 pC, 6.6 ± 0.6 ms and 14.5 ± 0.8 ms, respectively. As compared with control cells, the cell treatment with ^{10}Panx1 scrb did not affect any of the amperometric parameters. In this last condition Q, tP and $t_{1/2}$ values were 1.6 ± 0.2 pC, 7.3 ± 0.5 ms and 14.4 ± 0.8 ms, respectively. The treatment with Cbx or ^{10}Panx1 also did not significantly change Q values compared with control cells or cells treated with ^{10}Panx1 (**Figure 3C**). However, in the presence of these Panx1 channel inhibitors, tP values were significantly higher (11.6 ± 1.2 and 9.0 ± 0.8 ms for Cbx or ^{10}Panx1 treatment, respectively) (**Figure 3D**). As shown in **Figure 3E**, a similar trend was observed for $t_{1/2}$ values (25.2 ± 2.3 and 23.8 ± 1.7 ms for Cbx or ^{10}Panx1 treatment, respectively). These results indicate that Panx1 channels control the number of release events as well as the kinetics of single fusion events.

Panx1 CHANNELS CONTRIBUTE TO THE Ca^{2+} SIGNALS INDUCED BY ACTIVATION OF NICOTINIC RECEPTORS

Given that the secretory response in chromaffin cells is triggered by a transient increase in the cytosolic [Ca^{2+}] and that Panx1 channels have been implicated in the regulation of Ca^{2+} signals (Vanden Abeele et al., 2006; Pinheiro et al., 2013), we explored the possibility that Panx1 channels affect the secretory activity in chromaffin cells by contributing to the Ca^{2+} signal induced by the activation of nicotinic receptors. Thus, we analyzed the impact of different Panx1 channel inhibitors on the Ca^{2+} signals evoked by the nicotinic agonist DMPP in cells loaded with the Ca^{2+} probe Indo-1. **Figure 4A** shows representative traces of cytosolic [Ca^{2+}] signals in cells treated with the different inhibitors. In control cells, the cytosolic [Ca^{2+}] in resting condition was

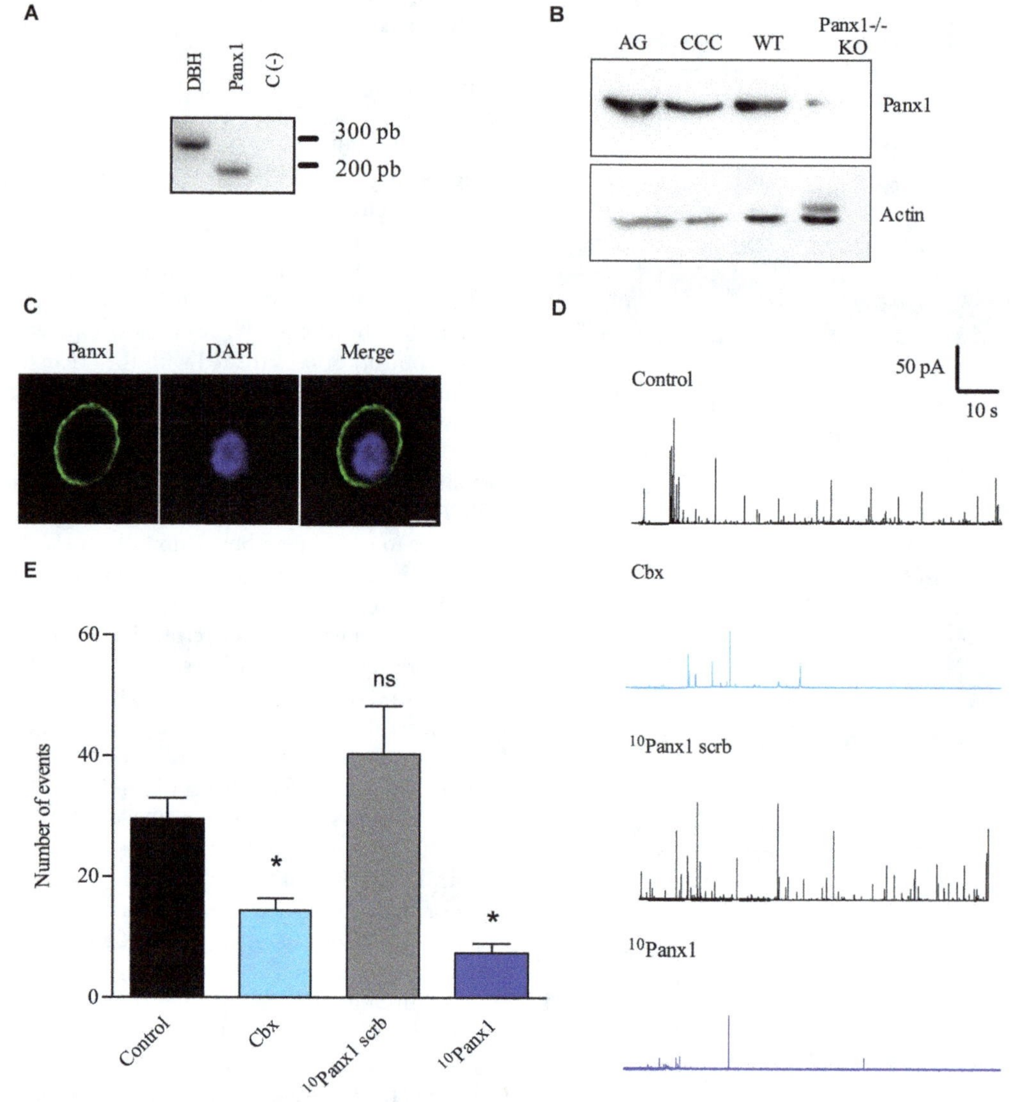

FIGURE 2 | Panx1 is expressed in cultured chromaffin cells and contributes to the exocytotic release of catecholamines. (A) A single cell RT-PCR assay was performed on single chromaffin cell using a nested PCR procedure in order to detect the presence of Panx1 mRNA. After the second PCR, the predicted transcript sizes were observed at 318 pb for Panx1 and 248 pb for DBH (positive control), thus confirming the expression of Panx1 in cultured bovine chromaffin cells. **(B)** Expression of Panx1 at the protein level. Protein extract (40 μg) from adrenal gland (AG) or cultured chromaffin cells (CCC) were used for electrophoresis and western blot analysis using a specific serum against Panx1. A band around 50 kDa was detected confirming the expression of Panx1 in both samples. Brain extracts from WT or Panx1$^{-/-}$ KO mice were used as positive and negative control. Anti-actin antibody was used as a loaded control. **(C)** Localization of Panx1 in cultured chromaffin cells. Confocal immunofluorescence images were obtained by labeling chromaffin cells with a polyclonal anti-Panx1 serum and visualized with an α-CY2-conjugated secondary antibody. Nuclei were visualized with DAPI staining. Scale bar = 10 μm. **(D,E)** The secretory activity was evoked by a 10 s pulse of the nicotinic agonist DMPP (50 μM) and monitored by amperometry. Cells were pre-incubated with the different inhibitors during 15 min at 37°C, which were maintained in the bath solution during the whole experiment. **(D)** Representative amperometry traces from control cells or cells treated with Cbx (5 μM), ^{10}Panx1 scrb peptide (200 μM) or ^{10}Panx1 peptide (200 μM). Note that Panx1 channel inhibitors decreased the number of exocytotic events. **(E)** Data show average values ± SEM of the number of events during 100 s recording of control cells (n = 55), or cells treated with Cbx (n = 20), ^{10}Panx1 scrb peptide (n = 18) or ^{10}Panx1 peptide (n = 25) from three to five different cultures. ns = non-significant, * $p < 0.05$ (Krukal-Wallis test) compre with control cells.

137.5 ± 5.1 nM, cell stimulation with a 10 s pulse of 50 μM DMPP led to a transient increase of the cytosolic [Ca^{2+}] with an amplitude of 1.2 ± 0.2 μM (**Figures 4B,C**). As shown in **Figure 4B**, none of the different Panx1 channel inhibitors affected the resting cytosolic [Ca^{2+}] (125.1 ± 7.4 nM, 194.7 ± 23.1 nM, 145.8 ± 8.1 nM and 145.8 ± 8.1 nM for cells treated with Cbx, probenecid, ^{10}Panx1 scrb and ^{10}Panx1, respectively). However, treatment with Cbx, probenecid or ^{10}Panx1 significantly reduced the amplitude of the Ca^{2+} signal induced by DMPP by ~68%, ~75% and ~85%, respectively, compared

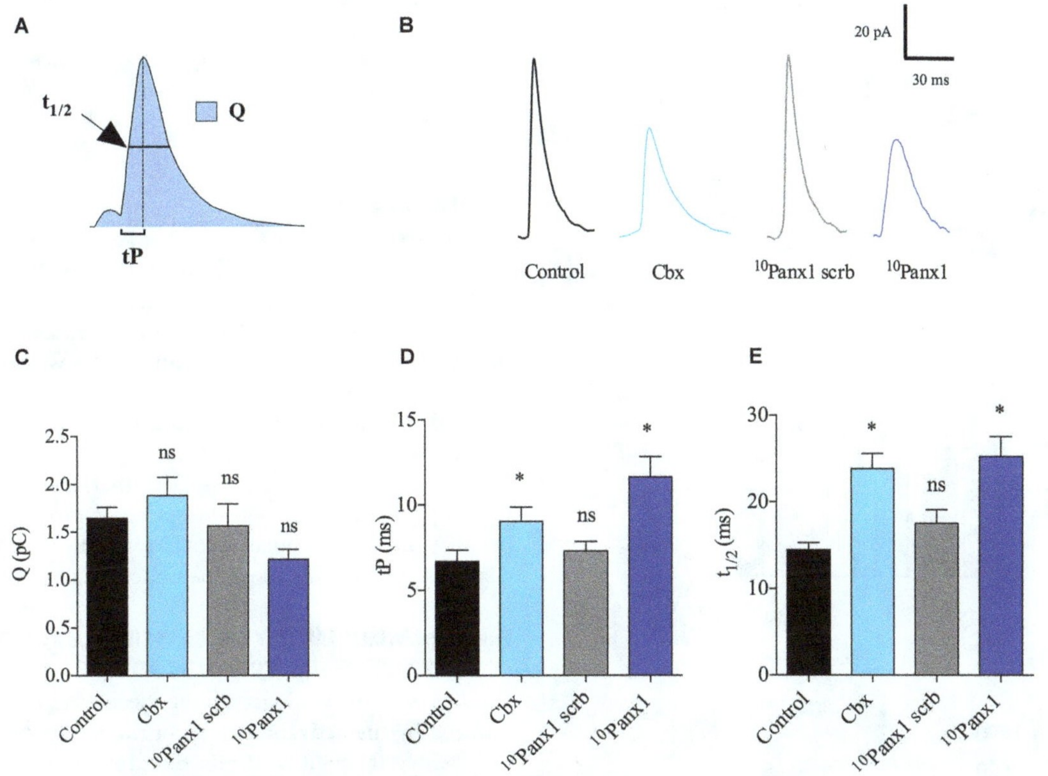

FIGURE 3 | Panx1 channels regulate the kinetics of single exocytotic events. (A) Schema of an amperometric spike with the different parameters analyzed: quantal size (Q), time to peak (tP) and half width ($t_{1/2}$). **(B)** Representative amperometric spikes from control cells or cells treated with 5 μM Cbx, 200 μM ^{10}Panx1 scrb peptide (n = 18) or 200 μM ^{10}Panx1 peptide (n = 25). **(C–E)** Data show average values of Q, tP and $t_{1/2}$ ± SEM from control cells (n = 55), or cells treated with 5 μM Cbx (n = 20), 200 μM ^{10}Panx1 scrb peptide (n = 18) or 200 μM ^{10}Panx1 peptide (n = 25) from three to five different cultures. Parameter values correspond to the mean values of the events from individual cells, thus "n" corresponds to the number of cells analyzed in each group. * $p <$ 0.05 (Mann-Whitney test) compare with control cells. Note that treatment of chromaffin cells with the Panx1 channel inhibitors did not affect Q **(C)**, but slows down the kinetics of single exocytotic events, as observed by an increase of tP **(D)** and $t_{1/2}$ **(E)**. ns = non-significant, * $p < 0.05$ (Krukal-Wallis test).

to control cells (**Figure 4C**). As we found for the secretory response, the strongest inhibition was observed after treatment with ^{10}Panx1.

Panx1 CHANNELS EXPRESSED AT THE PLASMA MEMBRANE CONTRIBUTE TO THE AMPLIFICATION OF DMPP-INDUCED Ca^{2+} SIGNALS

Finally, we explored the mechanism by which Panx1 contributes to the Ca^{2+} response to nicotinic receptor stimulation. One possibility is that the Ca^{2+} signal is mediated by Panx1 channels present at the ER membrane. This idea was supported by two facts: (1) the Ca^{2+} release from ER importantly contributes to the increase of the $[Ca^{2+}]$ in response to nicotinic receptor stimulation (del Barrio et al., 2011); and (2) in other cell types, Panx1 channels present at ER mediate the Ca^{2+} release from intracellular stores (D'hondt et al., 2011). Therefore, we evaluated the effect of Panx1 inhibitors on the Ca^{2+} release from the ER induced by 50 mM caffeine (Cheek et al., 1993; Alonso et al., 1999). These experiments were done in a Ca^{2+}-free medium. Under these conditions, treatment with Cbx (5 μM) or probenecid (200 μM) did not affect the basal $[Ca^{2+}]$ (**Figure 5A**) or the amplitude of the Ca^{2+} signal (**Figure 5B**). Thus, the mechanism by which Panx1 contributed to the Ca^{2+} signals appears to be independent of ER stores.

Another possibility is that Panx1 channels present at the cell periphery contribute to the amplification of the Ca^{2+} response to the nicotinic receptor stimulation. An amplification of Ca^{2+} influx mediated by plasma membrane Panx1 channels has been observed T-cell and platelet (Woehrle et al., 2010; Taylor et al., 2014). Therefore, we evaluated the presence and the functional state of Panx1 channels at the plasma membrane of chromaffin cells. In order to determine if Panx1 is present at this compartment, we co-labeled cells with anti-Panx1 antibody and extracellular biotin, which binds to primary amines of cell surface proteins (Turvy and Blum, 2001). Confocal acquisition showed colocalization of the two markers (**Figure 5C**). The mean score for the Pearson coefficient between biotin and Panx1 was 0.71 ± 0.02 (n = 18), confirming the presence of Panx1 at the plasma membrane. To determine the functional state of plasma membrane Panx1, we performed dye uptake experiments using DAPI. Cultured chromaffin cells were washed and incubated 10 s in a solution containing 50 μM DAPI in the presence (stimulated

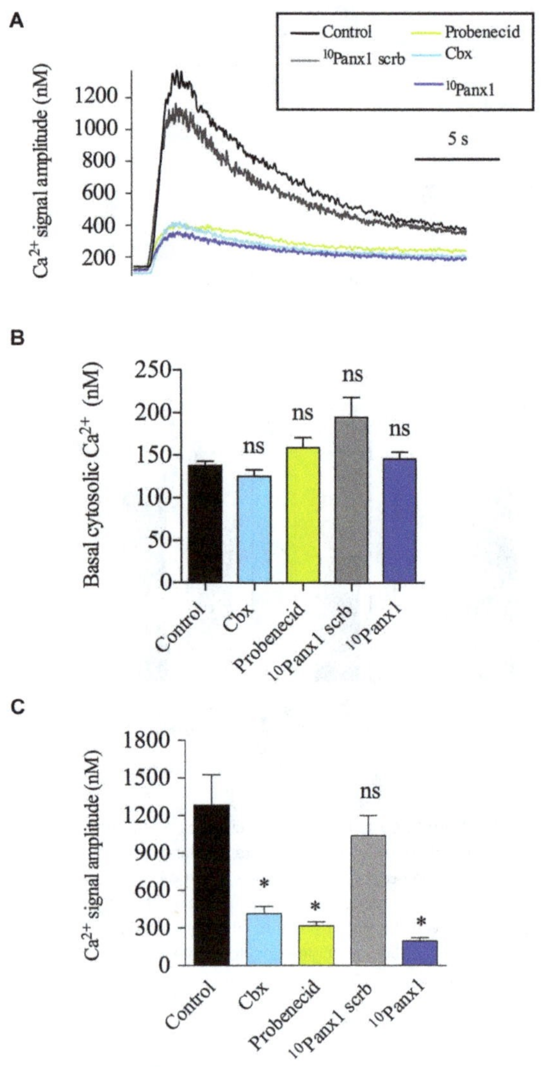

FIGURE 4 | Panx1 channels contribute to the Ca^{2+} signal induced by the activation of nicotinic receptors. Indo-1 loaded chromaffin cells were stimulated with a 10 s pulse of the nicotinic agonist DMPP (50 μM), and the cytosolic $[Ca^{2+}]$ was monitored by microfluorometry. **(A)** Representative traces of $[Ca^{2+}]$ rises induced by DMPP. Note that the evoked $[Ca^{2+}]$ rise was reduced in the presence of Cbx (5 μM), probenecid (200 μM) or ^{10}Panx1 peptide (200 μM). **(B,C)** Data show means ± SEM of basal cytosolic $[Ca^{2+}]$ **(B)** or the maximum amplitude of the Ca^{2+} signals induced by DMPP **(C)** in control cells ($n = 69$), cells treated with probenecid ($n = 40$), ^{10}Panx1 scrb peptide ($n = 37$) or ^{10}Panx1 peptide ($n = 31$). Cbx, probenecid and ^{10}Panx1 peptide treatment significantly reduced the amplitude of the DMPP-induced Ca^{2+} signal. ns = non-significant, * $p < 0.05$ (Krukal-Wallis test) compare with control cells.

cells) or absence (non-stimulated cells) of 50 μM DMPP. Then the cells were incubated in 50 μM DAPI in Kreb's HEPES solution for 2 min and immediately fixed with PFA. Confocal acquisition images and fluorescent analysis showed a low basal DAPI uptake in non-stimulated control cells (152 ± 18 A.U/μm^2), but the nucleus fluorescence intensity significantly increased in cells stimulated with DMPP (332 ± 31 A.U/μm^2). Interestingly, cell treatment with 5 μM Cbx or 200 μM probenecid reduced the dye uptake to values comparable to those observed in non-stimulated cells (105 ± 18 and 101 ± 17 A.U/μm^2 respectively) (**Figures 5D,E**). Taking together, these results strongly suggest that Panx1 channels present at the plasma membrane contribute to the DMPP-induced Ca^{2+} signal in chromaffin cells.

DISCUSSION

The present work demonstrates that Panx1 is expressed in bovine adrenal glands and is involved in the regulation of the catecholamine release evoked by nicotinic receptor activation. At the cellular level, we found that Panx1 is expressed at the plasma membrane of cultured chromaffin cells, where it contributes to the exocytotic release process by regulating the Ca^{2+} signal induced by the activation of nicotinic receptors. These findings define Panx1 channels as new actors in the regulation of the catecholamine release and suggest that these channels play a relevant role in the response to stress. As we discuss below, the peculiarities of the regulation of chromaffin cell activity by Panx1 channels enlighten us about the underlying mechanisms.

Panx1 CHANNELS AMPLIFY THE Ca^{2+} SIGNALS IN CHROMAFFIN CELLS, IMPACTING THE EXOCYTOTIC RELEASE

Until now, most evidence about the nature of the Ca^{2+} signal induced by the activation of nicotinic receptors is explained by Ca^{2+} entry through nicotinic receptors and voltage-dependent Ca^{2+} channels (Arnáiz-Cot et al., 2008). The mechanism of the Ca^{2+} -induced Ca^{2+} -release from the ER is the other important pathway that contributes to the increase of cytosolic Ca^{2+} (del Barrio et al., 2011). As recently demonstrated, depending on the type of nicotinic receptor that is activated, the voltage-dependent Ca^{2+} channels contribute to the Ca^{2+} signals by 15–20%, while the Ca^{2+} -release from the ER contributes over 60% (del Barrio et al., 2011). According to our results using ^{10}Panx1, Panx1 channels contribute ~85% to the Ca^{2+} signal induced by the nicotinic agonist DMPP (**Figure 4C**). Similar results were obtained with the different Panx1 channel inhibitors (**Figure 4C**). These results are in agreement with the fact that Panx1 importantly regulates both the number of exocytotic events in cultured chromaffin cells (**Figures 2D,E**) and the global secretion of catecholamines in perfused glands (**Figure 1B**).

Indeed, the amplitude of the Ca^{2+} signal defines the number of fusion events (Wang et al., 2006; Ardiles et al., 2007). Furthermore, the cytosolic $[Ca^{2+}]$ reached at the release sites can also determine the kinetics of the fusion pore expansion; an intermediate structure formed during the exocytosis processes (Lindau and Alvarez de Toledo, 2003). By using amperometry, the latter is reflected in the rise time and the duration of the single release events (Grabner and Fox, 2006; Ardiles et al., 2007). Thus, the slow-down of the rise time (tP) and the lengthening of the event duration ($t_{1/2}$) observed in the presence of the Panx1 channel inhibitors (**Figure 3**) could be also a consequence of the effect of these agents on the Ca^{2+} signal amplitude. Together these findings reveal that Panx1 channels constitute a new mechanism that importantly contributes to the Ca^{2+} response to the nicotinic receptor activation and impacts the exocytotic release of catecholamines.

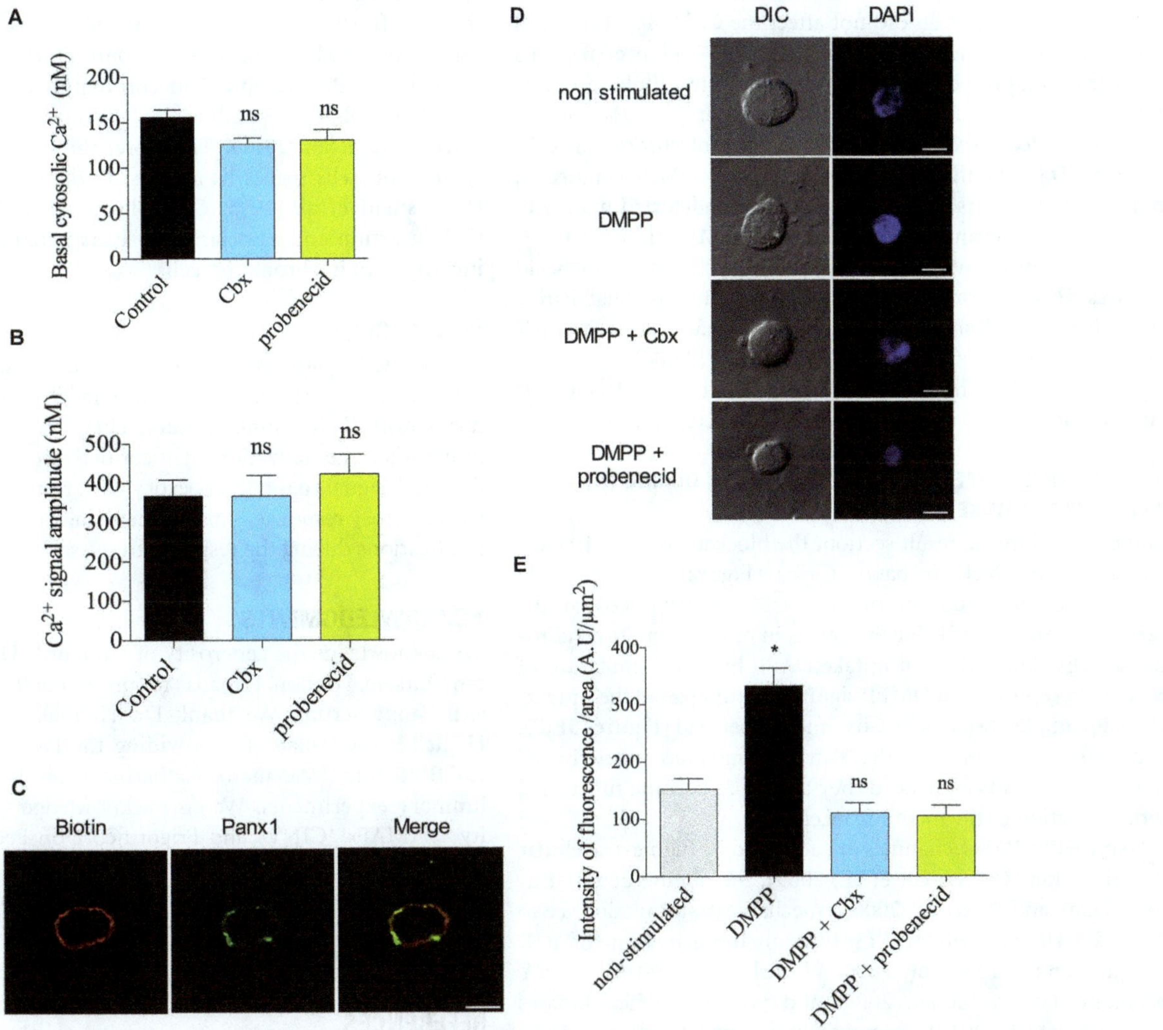

FIGURE 5 | The Panx1 channels do not contribute to the Ca^{2+} signal induced by caffeine but form functional channels at the plasma membrane. (A,B) Indo-1 loaded chromaffin cells were maintained in a Ca^{2+}-free Kreb's Hepes solution (in mM: 140 NaCl, 5.9 KCl, 1.2 $MgCl_2$, 2 EGTA, 10 Hepes-NaOH, 10 glucose) and stimulated with a 10 s pulse of 50 mM caffeine. The cytosolic $[Ca^{2+}]$ was monitored by microfluorometry. Data show means ± SEM of basal cytosolic $[Ca^{2+}]$ **(A)** or the maximum amplitude of the Ca^{2+} signals induced by caffeine **(B)** in control cells ($n = 22$), or cells treated with probenecid ($n = 22$) or Cbx ($n = 21$). ns = non-significant, * $p < 0.05$ (Krukal-Wallis test). **(C)** Colocalization of Panx1 with extracellular biotin at plasma membrane. Confocal immunofluorescence images were obtained by labeling external plasma membrane protein of chromaffin cells with biotin, and Panx1 with a polyclonal anti-Panx1 serum. Biotin labeling was visualized with avidin Cy3 and Panx1 with α-CY2-conjugated secondary antibody. Nuclei were visualized with DAPI staining. Scale bar = 10 µm. **(D,E)** DAPI uptake experiments were performed incubating the cells for 10 s in a Kreb's HEPES solution containing 50 µM DAPI without (unstimulated cells) or with nicotinic agonist receptor DMPP (50 µM). Then cells were incubated with Kreb's HEPES solution containing 50 µM DAPI for 2 min and immediately fixed with 4% PFA (in PBS, pH 7.4). After 3 washes with PBS (pH 7.4), coverslip were mounted for visualization. Treatment with probenecid (200 µM) Cbx (5 µM) were performed 15 min before the experiments and the inhibitors were maintained in the bath solution until cell fixation. **(D)** Confocal imaging and **(E)** quantification of DAPI uptake in unstimulated ($n = 24$) or DMPP stimulated cells, nontreated ($n = 21$) or treated with the Panx channel inhibitors ($n = 24$ and $n = 19$ for probenecid and Cbx respectively). Scale bar = 10 µm. Data shows mean ± SEM of nucleus fluorescence intensity per nucleus area (AU/μm^2). ns = non-significant, * $p < 0.05$ (Krukal-Wallis test) compare with non-stimulated cells.

FUNCTIONAL Panx1 CHANNELS PRESENT AT THE PLASMA MEMBRANE CONTRIBUTE TO THE Ca^{2+} SIGNALS INDUCE BY THE ACTIVATION OF NICOTINIC RECEPTORS

As aforementioned, the Ca^{2+} released from the ER in chromaffin cells importantly contributes to the Ca^{2+} signal induced by the activation of nicotinic receptors (del Barrio et al., 2011). On the other hand, the contribution of Panx1 channels to the Ca^{2+} signals could be also mediated by Panx1 channels present at the ER membrane (D'hondt et al., 2011). In prostate cancer cells Panx1 forms Ca^{2+} channels in the ER mediating the Ca^{2+} release from intracellular stores (Vanden Abeele et al., 2006). In osteoblasts, Panx3 also forms channels in the ER that are activated by PI3K–Akt signaling (Ishikawa et al., 2011). However, in our model, the contribution of Panx1 to the signals in chromaffin cells appears to not come from channels present at the ER because Panx1 inhibition did not affect the Ca^{2+} signal induced by caffeine in the absence of extracellular Ca^{2+} (**Figure 5B**). This idea is also supported by the fact that the extracellular application of ^{10}Panx1 (which theoretically does not cross the plasma membrane) decreased the secretory activity of chromaffin cells (**Figures 2D,E**) and the Ca^{2+} signal induced by DMPP (**Figure 4**). In addition, the facts that Panx1 was inmunodetected mainly at the plasma membrane (**Figure 5C**) and a DMPP-induced DAPI uptake was blocked with the Panx inhibitors Cbx and Probenecid (**Figures 5D,E**), strongly support the idea of functional Panx1 channels at the plasma membrane. Given that Panx1 channels are permeable to Ca^{2+} (Vanden Abeele et al., 2006), a possible mechanism is that the Ca^{2+} entry mediated by Panx1 channels amplifies of the Ca^{2+} signal that induces the secretory process.

Panx1 CHANNELS APPEAR TO BE ACTIVATED JUST DURING THE SECRETORY RESPONSE

As mentioned in the result section, the blockade of Panx1 channels does not affect the basal $[Ca^{2+}]$ (**Figure 4B**); conversely it decreased the amplitude of the Ca^{2+} signal induced by the nicotinic agonist DMPP. Furthermore, in resting condition, chromaffin cells almost did not uptake DAPI, but the stimulation of nicotinic receptors with DMPP significantly increased the uptake, which is completely block by Cbx- and probenecid (**Figures 5E,F**). Thus these findings suggest that Panx1 channels are not activated in a resting condition; instead they appear to become functional upon activation of the nicotinic receptor.

Reportedly, Panx1 channels are activated by high extracellular potassium ions (Silverman et al., 2009), membrane depolarization (Dahl and Locovei, 2006), mechanical stimulation (Bao et al., 2004), extracellular ATP through the activation of P2X (Pelegrin and Surprenant, 2006; Woehrle et al., 2010) or P2Y receptors (Locovei et al., 2006) and possibly by Src kinases (Weilinger et al., 2012). Some of mentioned mechanisms are triggered by the stimulation of nicotinic receptors in chromaffin cells, such as membrane depolarization (Pérez-Alvarez et al., 2012), Src kinase activation (Allen et al., 1996) and ATP release (Rojas et al., 1985). Thus, one or several of these events could activate Panx1 channels after the stimulation of nicotinic receptors.

IS THE ACTIVATION OF Panx1 CHANNELS IN CHROMAFFIN CELLS COUPLED TO PURINERGIC RECEPTORS?

Panx1 channels mediate Ca^{2+} influx to the cells through their association with the $P2X_7R$ (Pelegrin et al., 2008; Iglesias et al., 2009). In this mechanism, the release of ATP through Panx1 channels activates $P2X_7Rs$, which in turn transiently increases the cytosolic $[Ca^{2+}]$ (see review Baroja-Mazo et al., 2013). Hitherto there is no study showing the expression $P2X_7R$ in chromaffin cells. However, by using different agonists and antagonists, Tomé et al. (2007b) suggest the presence of functional purinergic $P2X_7R$ receptor, but only in a fraction (app. 20%) of the cultured bovine chromaffin cells.

Panx1 channels could also be functionally associated with $P2Y_1$ and $P2Y_2$ receptors (Locovei et al., 2006), and chromaffin cells expressed P2Y receptors that mediate Ca^{2+} signals (Tomé et al., 2007a,b). However, and as observed with P2X receptors, they are functionally expressed in only a small fraction of cultured bovine chromaffin cells (Tomé et al., 2007a,b). On the other hand, P2Y receptors in chromaffin cells also inhibit the activity of voltage-dependent Ca^{2+} channels (Ennion et al., 2004; Hernández et al., 2011). Therefore, the dual action of ATP in chromaffin cells makes its actions on the Ca^{2+} signals complex (Reichsman et al., 1995; Carabelli et al., 2001) and therefore their function and association to Panx1 channels needs further investigation in chromaffin cells.

CONCLUSION

Taken together our results reveal a new mechanism that regulates the release of hormones by the adrenal chromaffin cells. In this mechanism the opening of Panx1 channels at the plasma membrane after the activation of nicotinic receptors contributes to the Ca^{2+} signal that triggers exocytosis, resulting in a robust and fast secretory response. This mechanism could have physiological implications during the response to stress.

ACKNOWLEDGMENTS

We acknowledge the generosity of Dr. Laird (University of Western Ontario, London, Canada) for providing the rabbit polyclonal anti-Panx1 serum. We thank Dr. Hannah Monyer, University Heidelberg, Germany for providing the Panx1 knock-out (KO) C57BL/6 mice. We thank Catherine Estay for setting up preliminary experiments. We also acknowledge the confocal facility «LAMAF» (CINV) and Frigorífico Don Pedro for providing bovine adrenal glands. This work was supported by a FONDECYT post-doctoral grant N 3120221 (to Fanny Momboisse), FONDECYT 1110552 (to Ana M. Cárdenas) and P09-022-F from ICM-ECONOMIA, Chile.

REFERENCES

Allen, C. M., Ely, C. M., Juaneza, M. A., and Parsons, S. J. (1996). Activation of Fyn tyrosine kinase upon secretagogue stimulation of bovine chromaffin cells. *J. Neurosci. Res.* 44, 421–429. doi: 10.1002/(sici)1097-4547(19960601)44:5<421::aid-jnr2>3.0.co;2-h

Alonso, M. T., Barrero, M. J., Michelena, P., Carnicero, E., Cuchillo, I., Garcia, A. G., et al. (1999). Ca2+-induced Ca2+ release in chromaffin cells seen from inside the ER with targeted aequorin. *J. Cell Biol.* 144, 241–254. doi: 10.1083/jcb.144.2.241

Ardiles, A. O., Gonzalez-Jamett, A. M., Maripillan, J., Naranjo, D., Caviedes, P., and Cardenas, A. M. (2007). Calcium channel subtypes differentially regulate fusion pore stability and expansion. *J. Neurochem.* 103, 1574–1581. doi: 10.1111/j.1471-4159.2007.04871.x

Ardiles, A. O., Maripillan, J., Lagos, V. L., Toro, R., Mora, I. G., Villarroel, L., et al. (2006). A rapid exocytosis mode in chromaffin cells with a neuronal phenotype. *J. Neurochem.* 99, 29–41. doi: 10.1111/j.1471-4159.2006.04080.x

Arnáiz-Cot, J. J., De Diego, A. M., Hernández-Guijo, J. M., Gandía, L., and García, A. G. (2008). A two-step model for acetylcholine control of exocytosis via nicotinic receptors. *Biochem. Biophys. Res. Commun.* 365, 413–419. doi: 10.1016/j.bbrc.2007.10.151

Aunis, D., and Langley, K. (1999). Physiological aspects of exocytosis in chromaffin cells of the adrenal medulla. *Acta Physiol. Scand.* 167, 89–97. doi: 10.1046/j.1365-201x.1999.00580.x

Bao, L., Locovei, S., and Dahl, G. (2004). Pannexin membrane channels are mechanosensitive conduits for ATP. *FEBS Lett.* 572, 65–68. doi: 10.1016/j.febslet.2004.07.009

Bargiotas, P., Krenz, A., Hormuzdi, S. G., Ridder, D. A., Herb, A., Barakat, W., et al. (2011). Pannexins in ischemia-induced neurodegeneration. *Proc. Natl. Acad. Sci. U S A* 108, 20772–20777. doi: 10.1073/pnas.1018262108

Baroja-Mazo, A., Barbera-Cremades, M., and Pelegrin, P. (2013). The participation of plasma membrane hemichannels to purinergic signaling. *Biochim. Biophys. Acta* 1828, 79–93. doi: 10.1016/j.bbamem.2012.01.002

Bruzzone, R., Barbe, M. T., Jakob, N. J., and Monyer, H. (2005). Pharmacological properties of homomeric and heteromeric pannexin hemichannels expressed in Xenopus oocytes. *J. Neurochem.* 92, 1033–1043. doi: 10.1111/j.1471-4159.2004.02947.x

Carabelli, V., Hernández-Guijo, J. M., Baldelli, P., and Carbone, E. (2001). Direct autocrine inhibition and cAMP-dependent potentiation of single L-type Ca2+ channels in bovine chromaffin cells. *J. Physiol.* 532, 73–90. doi: 10.1111/j.1469-7793.2001.0073g.x

Cárdenas, A. M., Allen, D. D., Arriagada, C., Olivares, A., Bennett, L. B., Caviedes, R., et al. (2002). Establishment and characterization of immortalized neuronal cell lines derived from the spinal cord of normal and trisomy 16 fetal mice, an animal model of down syndrome. *J. Neurosci. Res.* 68, 46–58. doi: 10.1002/jnr.10205

Cheek, T. R., Moreton, R. B., Berridge, M. J., Stauderman, K. A., Murawsky, M. M., and Bootman, M. D. (1993). Quantal Ca2+ release from caffeine-sensitive stores in adrenal chromaffin cells. *J. Biol. Chem.* 268, 27076–27083.

Crivellato, E., Nico, B., and Ribatti, D. (2008). The chromaffin vesicle: advances in understanding the composition of a versatile, multifunctional secretory organelle. *Anat. Rec. (Hoboken)* 291, 1587–1602. doi: 10.1002/ar.20763

Currie, K. P., and Fox, A. P. (1996). ATP serves as a negative feedback inhibitor of voltage-gated Ca2+ channel currents in cultured bovine adrenal chromaffin cells. *Neuron* 16, 1027–1036. doi: 10.1016/s0896-6273(00)80126-9

Dahl, G., and Locovei, S. (2006). Pannexin: to gap or not to gap, is that a question? *IUBMB Life* 58, 409–419. doi: 10.1080/15216540600794526

del Barrio, L., Egea, J., Leon, R., Romero, A., Ruiz, A., Montero, M., et al. (2011). Calcium signalling mediated through alpha7 and non-alpha7 nAChR stimulation is differentially regulated in bovine chromaffin cells to induce catecholamine release. *Br. J. Pharmacol.* 162, 94–110. doi: 10.1111/j.1476-5381.2010.01034.x

D'hondt, C., Ponsaerts, R., De Smedt, H., Vinken, M., De Vuyst, E., De Bock, M., et al. (2011). Pannexin channels in ATP release and beyond: an unexpected rendezvous at the endoplasmic reticulum. *Cell Signal.* 23, 305–316. doi: 10.1016/j.cellsig.2010.07.018

Ennion, S. J., Powell, A. D., and Seward, E. P. (2004). Identification of the P2Y(12) receptor in nucleotide inhibition of exocytosis from bovine chromaffin cells. *Mol. Pharmacol.* 66, 601–611. doi: 10.1124/mol.104.000224

González-Jamett, A. M., Báez-Matus, X., Hevia, M. A., Guerra, M. J., Olivares, M. J., Martínez, A. D., et al. (2010). The association of dynamin with synaptophysin regulates quantal size and duration of exocytotic events in chromaffin cells. *J. Neurosci.* 30, 10683–10691. doi: 10.1523/JNEUROSCI.5210-09.2010

Grabner, C. P., and Fox, A. P. (2006). Stimulus-dependent alterations in quantal neurotransmitter release. *J. Neurophysiol.* 96, 3082–3087. doi: 10.1152/jn.00017.2006

Harvey, S., Phillips, J. G., Rees, A., and Hall, T. R. (1984). Stress and adrenal function. *J. Exp. Zool.* 232, 633–645. doi: 10.1002/jez.1402320332

Hernández, A., Segura-Chama, P., Jiménez, N., García, A. G., Hernández-Guijo, J. M., and Hernández-Cruz, A. (2011). Modulation by endogenously released ATP and opioids of chromaffin cell calcium channels in mouse adrenal slices. *Am. J. Physiol. Cell Physiol.* 300, C610–C623. doi: 10.1152/ajpcell.00380.2010

Iglesias, R., Dahl, G., Qiu, F., Spray, D. C., and Scemes, E. (2009). Pannexin 1: the molecular substrate of astrocyte "hemichannels". *J. Neurosci.* 29, 7092–7097. doi: 10.1523/JNEUROSCI.6062-08.2009

Ishikawa, M., Iwamoto, T., Nakamura, T., Doyle, A., Fukumoto, S., and Yamada, Y. (2011). Pannexin 3 functions as an ER Ca(2+) channel, hemichannel and gap junction to promote osteoblast differentiation. *J. Cell Biol.* 193, 1257–1274. doi: 10.1083/jcb.201101050

Li, S., Bjelobaba, I., Yan, Z., Kucka, M., Tomic, M., and Stojilkovic, S. S. (2011). Expression and roles of pannexins in ATP release in the pituitary gland. *Endocrinology* 152, 2342–2352. doi: 10.1210/en.2010-1216

Lim, W., Kim, S. J., Yan, H. D., and Kim, J. (1997). Ca2+-channel-dependent and -independent inhibition of exocytosis by extracellular ATP in voltage-clamped rat adrenal chromaffin cells. *Pflugers Arch.* 435, 34–42. doi: 10.1007/s004240050481

Lindau, M., and Alvarez de Toledo, G. (2003). The fusion pore. *Biochim. Biophys. Acta* 1641, 167–173. doi: 10.1016/S0167-4889(03)00085-5

Locovei, S., Scemes, E., Qiu, F., Spray, D. C., and Dahl, G. (2007). Pannexin1 is part of the pore forming unit of the P2X(7) receptor death complex. *FEBS Lett.* 581, 483–488. doi: 10.1016/j.febslet.2006.12.056

Locovei, S., Wang, J., and Dahl, G. (2006). Activation of pannexin 1 channels by ATP through P2Y receptors and by cytoplasmic calcium. *FEBS Lett.* 580, 239–244. doi: 10.1016/j.febslet.2005.12.004

Montiel, C., Mendoza, I., Garcia, C. J., Awad, Y., Garcia-Olivares, J., Solis-Garrido, L. M., et al. (2003). Distinct protein kinases regulate Snap-25 expression in chromaffin cells. *J. Neurosci. Res.* 71, 353–364. doi: 10.1002/jnr.10499

Neco, P., Fernandez-Peruchena, C., Navas, S., Gutierrez, L. M., De Toledo, G. A., and Ales, E. (2008). Myosin II contributes to fusion pore expansion during exocytosis. *J. Biol. Chem.* 283, 10949–10957. doi: 10.1074/jbc.M709058200

Orellana, J. A., Froger, N., Ezan, P., Jiang, J. X., Bennett, M. V., Naus, C. C., et al. (2011). ATP and glutamate released via astroglial connexin 43 hemichannels mediate neuronal death through activation of pannexin 1 hemichannels. *J. Neurochem.* 118, 826–840. doi: 10.1111/j.1471-4159.2011.07210.x

Pelegrin, P., Barroso-Gutierrez, C., and Surprenant, A. (2008). P2X7 receptor differentially couples to distinct release pathways for IL-1beta in mouse macrophage. *J. Immunol.* 180, 7147–7157. doi: 10.4049/jimmunol.180.11.7147

Pelegrin, P., and Surprenant, A. (2006). Pannexin-1 mediates large pore formation and interleukin-1beta release by the ATP-gated P2X7 receptor. *EMBO J.* 25, 5071–5082. doi: 10.1038/sj.emboj.7601378

Penuela, S., Celetti, S. J., Bhalla, R., Shao, Q., and Laird, D. W. (2008). Diverse subcellular distribution profiles of pannexin 1 and pannexin 3. *Cell Commun. Adhes.* 15, 133–142. doi: 10.1080/15419060802014115

Pérez-Alvarez, A., Hernández-Vivanco, A., Alonso, Y. G. S., Tabernero, A., Mcintosh, J. M., and Albillos, A. (2012). Pharmacological characterization of native alpha7 nicotinic ACh receptors and their contribution to depolarization-elicited exocytosis in human chromaffin cells. *Br. J. Pharmacol.* 165, 908–921. doi: 10.1111/j.1476-5381.2011.01596.x

Persky, H. (1955). Chemical determination of adrenaline and noradrenaline in body fluids and tissues. *Methods Biochem. Anal.* 2, 57–82. doi: 10.1002/9780470110188.ch2

Phillips, J. K., and Lipski, J. (2000). Single-cell RT-Pcr as a tool to study gene expression in central and peripheral autonomic neurones. *Auton. Neurosci.* 86, 1–12. doi: 10.1016/s1566-0702(00)00245-9

Pinheiro, A. R., Paramos-DE-Carvalho, D., Certal, M., Costa, M. A., Costa, C., Magalhaes-Cardoso, M. T., et al. (2013). Histamine induces ATP release from human subcutaneous fibroblasts, via pannexin-1 hemichannels, leading to Ca2+ mobilization and cell proliferation. *J. Biol. Chem.* 288, 27571–27583. doi: 10.1074/jbc.M113.460865

Reichsman, F., Santos, S., and Westhead, E. W. (1995). Two distinct ATP receptors activate calcium entry and internal calcium release in bovine chromaffin cells. *J. Neurochem.* 65, 2080–2086. doi: 10.1046/j.1471-4159.1995.65052080.x

Riquelme, M. A., Cea, L. A., Vega, J. L., Boric, M. P., Monyer, H., Bennett, M. V., et al. (2013). The ATP required for potentiation of skeletal muscle contraction is released via pannexin hemichannels. *Neuropharmacology* 75, 594–603. doi: 10.1016/j.neuropharm.2013.03.022

Rojas, E., Pollard, H. B., and Heldman, E. (1985). Real-time measurements of acetylcholine-induced release of ATP from bovine medullary chromaffin cells. *FEBS Lett.* 185, 323–327. doi: 10.1016/0014-5793(85)80931-5

Séror, C., Melki, M. T., Subra, F., Raza, S. Q., Bras, M., Saïdi, H., et al. (2011). Extracellular ATP acts on P2Y2 purinergic receptors to facilitate Hiv-1 infection. *J. Exp. Med.* 208, 1823–1834. doi: 10.1084/jem.20101805

Silverman, W. R., De Rivero Vaccari, J. P., Locovei, S., Qiu, F., Carlsson, S. K., Scemes, E., et al. (2009). The pannexin 1 channel activates the inflammasome in neurons and astrocytes. *J. Biol. Chem.* 284, 18143–18151. doi: 10.1074/jbc.M109.004804

Silverman, W., Locovei, S., and Dahl, G. (2008). Probenecid, a gout remedy, inhibits pannexin 1 channels. *Am. J. Physiol. Cell Physiol.* 295, C761–C767. doi: 10.1152/ajpcell.00227.2008

Taylor, K. A., Wright, J. R., Vial, C., Evans, R. J., and Mahaut-Smith, M. P. (2014). Amplification of human platelet activation by surface pannexin-1 channels. *J. Thromb. Haemost.* 12, 987–998. doi: 10.1111/jth.12566

Thompson, R. J., Jackson, M. F., Olah, M. E., Rungta, R. L., Hines, D. J., Beazely, M. A., et al. (2008). Activation of pannexin-1 hemichannels augments aberrant bursting in the hippocampus. *Science* 322, 1555–1559. doi: 10.1126/science.1165209

Tomé, A. R., Castro, E., Santos, R. M., and Rosário, L. M. (2007a). Functional distribution of Ca2+-coupled P2 purinergic receptors among adrenergic and noradrenergic bovine adrenal chromaffin cells. *BMC Neurosci.* 8:39. doi: 10.1186/1471-2202-8-39

Tomé, A. R., Castro, E., Santos, R. M., and Rosário, L. M. (2007b). Selective stimulation of catecholamine release from bovine adrenal chromaffin cells by an ionotropic purinergic receptor sensitive to 2-methylthio ATP. *BMC Neurosci.* 8:41. doi: 10.1186/1471-2202-8-41

Turvy, D. N., and Blum, J. S. (2001). Biotin labeling and quantitation of cell-surface proteins. *Curr. Protoc. Immunol.* Chapter 18, Unit 18.7. doi: 10.1002/0471142735.im1807s36

Vanden Abeele, F., Bidaux, G., Gordienko, D., Beck, B., Panchin, Y. V., Baranova, A. V., et al. (2006). Functional implications of calcium permeability of the channel formed by pannexin 1. *J. Cell Biol.* 174, 535–546. doi: 10.1083/jcb.200601115

Wang, C. T., Bai, J., Chang, P. Y., Chapman, E. R., and Jackson, M. B. (2006). Synaptotagmin-Ca2+ triggers two sequential steps in regulated exocytosis in rat PC12 cells: fusion pore opening and fusion pore dilation. *J. Physiol.* 570, 295–307. doi: 10.1113/jphysiol.2005.097378

Weilinger, N. L., Tang, P. L., and Thompson, R. J. (2012). Anoxia-induced NMDA receptor activation opens pannexin channels via Src family kinases. *J. Neurosci.* 32, 12579–12588. doi: 10.1523/JNEUROSCI.1267-12.2012

Weiss, J. L. (2012). Ca(2+) signaling mechanisms in bovine adrenal chromaffin cells. *Adv. Exp. Med. Biol.* 740, 859–872. doi: 10.1007/978-94-007-2888-2_38

Woehrle, T., Yip, L., Elkhal, A., Sumi, Y., Chen, Y., Yao, Y., et al. (2010). Pannexin-1 hemichannel-mediated ATP release together with P2X1 and P2X4 receptors regulate T-cell activation at the immune synapse. *Blood* 116, 3475–3484. doi: 10.1182/blood-2010-04-277707

Wong, D. L. (2003). Why is the adrenal adrenergic? *Endocr. Pathol.* 14, 25–36. doi: 10.1385/ep:14:1:25

19

Gap junctions and hemichannels composed of connexins: Potential therapeutic targets for neurodegenerative diseases

Hideyuki Takeuchi* and Akio Suzumura

Department of Neuroimmunology, Research Institute of Environmental Medicine, Nagoya University, Nagoya, Japan

***Correspondence:**
Hideyuki Takeuchi, Department of Neuroimmunology, Research Institute of Environmental Medicine, Nagoya University, Furo-cho, Chikusa-ku, Nagoya 464-8601, Japan
e-mail: htake@riem.nagoya-u.ac.jp

Microglia are macrophage-like resident immune cells that contribute to the maintenance of homeostasis in the central nervous system (CNS). Abnormal activation of microglia can cause damage in the CNS, and accumulation of activated microglia is a characteristic pathological observation in neurologic conditions such as trauma, stroke, inflammation, epilepsy, and neurodegenerative diseases. Activated microglia secrete high levels of glutamate, which damages CNS cells and has been implicated as a major cause of neurodegeneration in these conditions. Glutamate-receptor blockers and microglia inhibitors (e.g., minocycline) have been examined as therapeutic candidates for several neurodegenerative diseases; however, these compounds exerted little therapeutic benefit because they either perturbed physiological glutamate signals or suppressed the actions of protective microglia. The ideal therapeutic approach would hamper the deleterious roles of activated microglia without diminishing their protective effects. We recently found that abnormally activated microglia secrete glutamate via gap-junction hemichannels on the cell surface. Moreover, administration of gap-junction inhibitors significantly suppressed excessive microglial glutamate release and improved disease symptoms in animal models of neurologic conditions such as stroke, multiple sclerosis, amyotrophic lateral sclerosis, and Alzheimer's disease. Recent evidence also suggests that neuronal and glial communication via gap junctions amplifies neuroinflammation and neurodegeneration. Elucidation of the precise pathologic roles of gap junctions and hemichannels may lead to a novel therapeutic strategies that can slow and halt the progression of neurodegenerative diseases.

Keywords: glutamate, microglia, neuroinflammation, neurodegeneration, gap junction, hemichannel, connexin

INTRODUCTION

Microglia are macrophage-like immune cells that reside in the central nervous system (CNS), where they play multiple roles: presenting antigen to initiate immunological reactions, directing attack against non-self antigens, debris clearance, support of neuronal circuit development (Kreutzberg, 1996; Kempermann and Neumann, 2003; Block et al., 2007; Takeuchi, 2010; Boche et al., 2013), and so on. Microglia contribute to maintenance of CNS homeostasis, but abnormal activation of these cells often causes damage to surrounding cells and tissues. Microgliosis, the accumulation of activated microglia, is a characteristic pathological feature in many neurologic conditions such as trauma, stroke, inflammation, epilepsy, and neurodegenerative diseases (Cagnin et al., 2001; Eikelenboom et al., 2002; McGeer and McGeer, 2002; Nelson et al., 2002; Orr et al., 2002; Bruijn et al., 2004; Pavese et al., 2006). Activated microglia release massive amounts of glutamate, at much higher levels than astrocytes and neurons (mM vs. μM), and destroy neural cells; these processes have been implicated as a major cause of neuronal damage in neurologic diseases (Piani et al., 1992; Barger and Basile, 2001; Schwartz et al., 2003; Ye et al., 2003; Kipnis et al., 2004; Takeuchi et al., 2005, 2008b; Herman and Jahr, 2007; Liang et al., 2008; Yawata et al., 2008). Therefore, blockade of glutamate signaling and inhibition of microglial activation have been explored as therapeutic candidates for several neurodegenerative diseases. However, glutamate receptor blockers also perturb physiological glutamate signals and cause severe adverse side effects (Parsons et al., 2007). Tetracycline and two of its derivatives (doxycycline and minocycline) have been used as inhibitors of microglial activation, but these compounds exerted little therapeutic benefit, because activated microglia also exert neuroprotective effects such as production of neurotrophic factors and sequestration of neurotoxic substances (Zietlow et al., 1999; Kempermann and Neumann, 2003; Kipnis et al., 2004; Koenigsknecht and Landreth, 2004; Schwab and Schluesener, 2004). Thus, the optimal therapeutic strategy would inhibit the deleterious effects of activated microglia without diminishing their protective roles (Takeuchi, 2010). We recently found that neurotoxic activated microglia secrete glutamate through gap-junction hemichannels. Recent evidence also suggests that neuronal and glial communication by gap junctions amplifies neuroinflammation and neurodegeneration.

Therefore, elucidation of the pathologic roles of gap junctions and hemichannels may provide us with new therapeutic strategies against many neurologic diseases.

MICROGLIA AS THE "ENEMY WITHIN"

Microglia, which originate from bone marrow–derived myeloid cells, account for approximately 10% of cells in the human CNS (Del Rio-Hortega, 1932). Microglia are predominantly observed in gray matter, especially in the olfactory bulb, hippocampus, basal ganglia, and substantia nigra (Lawson et al., 1990). Under healthy physiological conditions, microglia persist in a quiescent state with ramified morphology (resting microglia) and survey the environment of the CNS (Davalos et al., 2005; Nimmerjahn et al., 2005). Under pathological conditions, microglia dramatically change their morphology and adopt an amoeboid appearance in the activated state. Activated microglia express surface molecules such as Fc receptor, CD11b, CD11c, CD14, major histocompatibility complex (MHC) molecules, Toll-like receptors (TLRs), scavenger receptors, and cytokine/chemokine receptors, and they can act as both antigen-presenting cells and immunological effector cells (Suzumura et al., 1987; Rock et al., 2004). In addition to innate immunity, activated microglia also play other beneficial roles, such as neuroprotection mediated by release of neurotrophic factors (Zietlow et al., 1999; Bessis et al., 2007; Liang et al., 2010), maintenance of CNS homeostasis by clearance of cellular debris and toxic substances (Upender and Naegele, 1999; Marin-Teva et al., 2004; Iribarren et al., 2005; Simard et al., 2006; Richard et al., 2008), and guidance of stem-cell migration in neuronal repair and neurogenesis (Aarum et al., 2003; Ziv et al., 2006a,b).

TLRs and scavenger receptors may contribute to diminishing neurotoxicity by sequestering neurotoxic substances such as amyloid β (El Khoury et al., 1996; Coraci et al., 2002; Bamberger et al., 2003; Liu et al., 2005); however, signals downstream of these receptors also enhance microglial neurotoxic effects by producing neurotoxic factors such as cytokines/chemokines, nucleic acids, excitatory amino acids, reactive oxygen species (ROS), and proteases (Kempermann and Neumann, 2003; Takeuchi et al., 2005, 2006; Kawanokuchi et al., 2006; Block et al., 2007). In fact, expression levels of TLRs and scavenger receptors are upregulated in a variety of neurologic diseases (Akiyama and McGeer, 1990; Grewal et al., 1997; El Khoury et al., 1998; Bsibsi et al., 2002; Cho et al., 2005; Carpentier et al., 2008). Therefore, whether activated microglia exert a neurotoxic or a neuroprotective effect may depend on their environment, the spatiotemporal distribution of the microglia themselves, and the type and magnitude of stimuli (Jimenez et al., 2008; Wu et al., 2008; Nakanishi and Wu, 2009; Sawada, 2009). A recently proposed hypothesis suggests, by analogy to macrophage activation, that activated microglia comprise two subpopulations, the neurotoxic (M1) and neuroprotective (M2) species (Mantovani et al., 2002; Henkel et al., 2009; Boche et al., 2013); however, this hypothesis is still open to debate. At least in pathological conditions, deleterious microglial activation is probably involved in the progression of various neurological disorders (Yrjanheikki et al., 1998; Wu et al., 2002; Zhu et al., 2002; Stirling et al., 2004; Boillee et al., 2006; Seabrook et al., 2006). Thus, elucidating the precise mechanism of microglial neurotoxicity is a necessary step toward development of effective therapeutic strategies against neurologic diseases.

GLUTAMATE AS A MAJOR NEUROTOXIC FACTOR FROM MICROGLIA

Glutamate is the most potently neurotoxic factor released from activated microglia. Excessive glutamate induces severe neuronal damage via excitotoxicity (Piani et al., 1992; Barger and Basile, 2001; Takeuchi et al., 2005, 2006). A common misconception is that inflammatory cytokines produced by activated microglia directly induce neuronal damage. In fact, these cytokines have little direct neurotoxic effect (Takeuchi et al., 2006; Takeuchi, 2010). Although tumor necrosis factor-α (TNF-α) and interferon-γ (IFN-γ) are considered to be the most deleterious inflammatory cytokines produced by activated microglia, these cytokines have only weak direct neurotoxic effects because they also enhance neuroprotective cascades involving mitogen-activated protein kinase (MAPK) and expression of nuclear factor κB (NF-κB) (Ghezzi and Mennini, 2001; Kamata et al., 2005). In general, inflammatory cytokines induce neurotoxicity indirectly by stimulating microglia in an autocrine/paracrine manner. These stimuli induce microglia to release high levels of glutamate, resulting in neuronal damage via excitotoxicity. Moreover, a recent paper showed that activated microglial glutamate suppresses astrocytic glutamate transporters, which play a pivotal role in maintenance of the physiological extracellular glutamate level (Takaki et al., 2012); this suppression probably worsens excitotoxic neuronal damage. Although microglia also express glutamate transporters, they seem much less effective at maintaining extracellular glutamate homeostasis than astrocytic glutamate transporters (Liang et al., 2008).

One of the earliest pathologic features of excitotoxicity is formation of neuritic beading, i.e., focal bead-like swelling in dendrites and axons (Takeuchi et al., 2005; Mizuno et al., 2008). Neuritic beading is a common neuropathological hallmark of many neurologic conditions such as ischemia, epilepsy, mechanical pressure, brain tumor, aging, neuroinflammatory diseases, and neurodegenerative diseases such as multiple sclerosis (MS), Alzheimer's disease (AD), Parkinson's disease (PD), and amyotrophic lateral sclerosis (ALS) (Delisle and Carpenter, 1984; Hori and Carpenter, 1994; Takahashi et al., 1997; Trapp et al., 1998; Dickson et al., 1999; Mattila et al., 1999; Swann et al., 2000; Goel et al., 2003; Pavlidis et al., 2003; Saito et al., 2003; Dutta and Trapp, 2007). Recent studies elucidated the detailed role of microglial glutamate in formation of neuritic beading and subsequent neuronal death. Glutamate produced by activated microglia activates neuronal N-methyl-D-aspartate (NMDA) receptor signaling, which promotes Ca^{2+} influx and activates Ca^{2+}/calmodulin-dependent protein kinase (CaMK). CaMK activates neuronal nitric oxide synthase (nNOS) and increases the intracellular concentration of nitric oxide (NO). NO in turn inhibits mitochondrial respiratory chain complex IV, resulting in a rapid reduction in intracellular ATP levels. Ultimately, the loss of intracellular energy pools suppresses dendritic and axonal transport, leading to bead-like accumulation of cytoskeletal and motor proteins along neurites and the formation of neuritic beading. Thus, a low-energy state results in neuronal dysfunction. Persistence of this neuronal dysfunction eventually causes neuronal death [i.e., excitotoxic neuronal death or non-cell-autonomous neuronal death (Lobsiger and Cleveland, 2007)].

Recent studies have revealed the precise mechanism of glutamate production by activated microglia (Takeuchi et al., 2005, 2006) (**Figure 1**). Two pathways are involved in cellular glutamate synthesis (Newsholme and Newsholme, 1989; Newsholme and Calder, 1997; Yudkoff, 1997; Nissim, 1999). One of these pathways is mediated by glutamate dehydrogenase, which converts α-ketoglutarate to glutamate. Most cells use this pathway to maintain cellular homeostasis of glutamate levels. The other pathway is mediated by glutaminase, which produces glutamate from extracellular glutamine brought into the cell via glutamine transporters. Resting microglia maintain their physiological glutamate level via the glutamate dehydrogenase pathway, as in other cell types, and secrete very little glutamate into the extracellular space (**Figure 1**). By contrast, activated microglia produce excessive amounts of glutamate as a result of upregulation of glutaminase, but not glutamate dehydrogenase. Subsequently, activated microglia release massive amounts of glutamate via gap-junction hemichannels. Inflammatory cytokines such as TNF-α and IFN-γ enhance not only glutaminase expression level but also cell-surface localization of hemichannels in microglia (Eugenin et al., 2001; Takeuchi et al., 2006). These two phenomena may act synergistically to release excess glutamate, leading to excitotoxic neuronal damage (**Figure 1**). Moreover, the extracellular glutamine level is critical for microglial glutamate production (Takeuchi et al., 2006). In the CNS, glutamine from astrocytes is essential for glutamate production in neurons (Tsacopoulos and Magistretti, 1996), suggesting that it also plays an important role in microglial glutamate production.

GAP JUNCTIONS IN CNS CELLS

Gap junctions contribute to formation of intercellular channels that directly connect the cytoplasmic compartments of neighboring cells (Yeager and Harris, 2007). These channels pass various small molecules (~1000 Da) and ions, although the charges and shapes of these molecules may affect the rate of transfer through gap junctions (Goldberg et al., 2004). Each gap junction is composed of a pair of hemichannels docked in a head-to-head configuration. Hemichannels are organized as hexagonal cylinders with central pores, and each hemichannel consists of a hexameric cluster of protein subunits called connexins (in vertebrates) or innexins (in invertebrates). Connexins are encoded by a conserved family of genes with at least 21 members in mammals. There are 21 connexin genes in the human genome and 20 connexin genes in the mouse genome; 19 of these proteins have orthologs in both humans and mice (Willecke et al., 2002; Laird, 2006). The connexin isoforms structurally interact in multiple ways. Homomeric hemichannels consist of a single connexin isoform, whereas heteromeric hemichannels contain two or more different connexin isoforms. Likewise, a homotypic gap junction channel is composed of two identical hemichannels, whereas a heterotypic gap junction channel contains two different hemichannels. Thus, the compositions of gap junctions can be classified into four types: homomeric and homotypic; heteromeric and homotypic; homomeric and heterotypic; and heteromeric and heterotypic

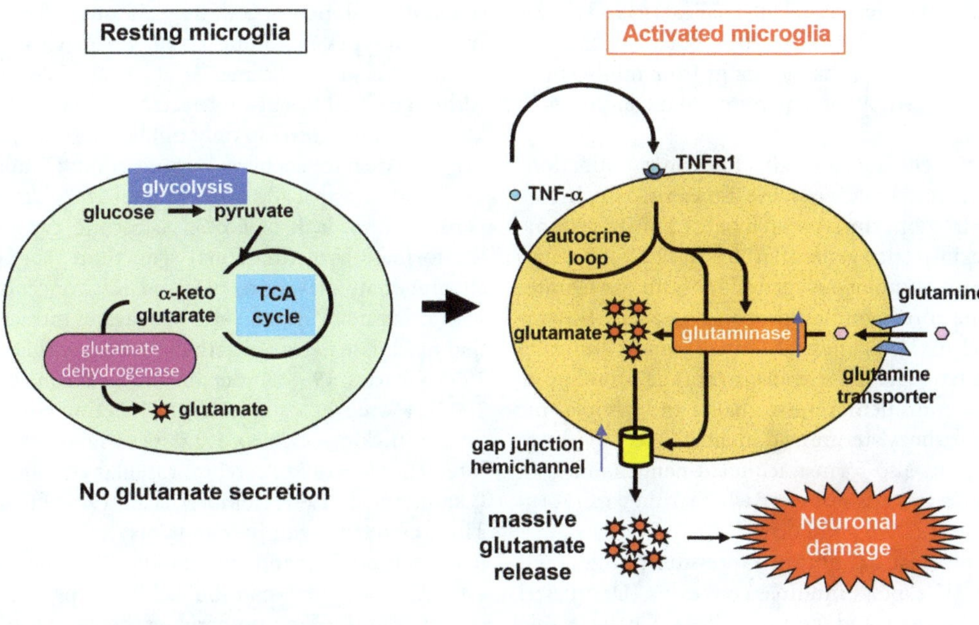

FIGURE 1 | Mechanism of glutamate production and release by activated microglia. Like other types of cells, resting microglia use glutamate dehydrogenase to synthesize glutamate from intracellular α-ketoglutarate in order to maintain a physiologically normal level of glutamate. Under resting conditions, microglia release very little glutamate into the extracellular space. By contrast, under pathological conditions, glutaminase and gap-junction hemichannels are upregulated in activated microglia (e.g., in response to stimulation by TNF-α). Glutaminase synthesizes excess glutamate from extracellular glutamine, which is brought into the cell via glutamine transporters. Subsequently, high levels of glutamate are secreted through gap-junction hemichannels, resulting in eventual neuronal damage.

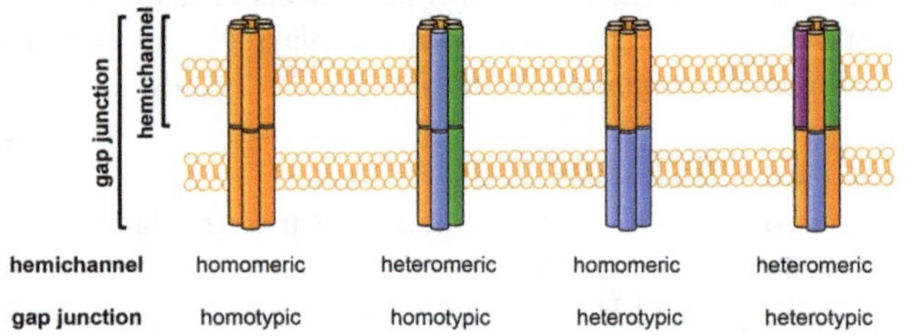

FIGURE 2 | The composition of gap junctions and hemichannels. Each colored column (orange, blue, green, or purple) represents a different connexin isoform. Hemichannels may be homomeric (composed of one connexin isoform) or heteromeric (composed of more than one connexin isoform). Gap-junction channels may be homotypic (formed by identical hemichannels) or heterotypic (formed by different hemichannels).

(**Figure 2**). This heterogeneity of connexin configurations confers complexity to the gap junction/hemichannel system.

Gap junctions allow direct intracellular propagation of second messengers (e.g., Ca^{2+}, IP_3, cAMP, and cGMP), metabolites (e.g., glutamate, glucose, and glutathione), and nucleotides (e.g., ATP, ADP, and RNA) between adjacent cells (Goldberg et al., 1999, 2002; Harris, 2001, 2007; Saez et al., 2003; Valiunas et al., 2005; Laird, 2006). Moreover, recent studies revealed that uncoupled "free" hemichannels facilitate two-way transfer of molecules between the cytosol and extracellular milieu (De Vuyst et al., 2007; Retamal et al., 2007; Laird, 2010). Intracellular communication via gap junctions and hemichannels is regulated by such mechanisms as channel gating via chemicals, pH, and voltage, as well as by changes in connexin transcription, translation, post-translational phosphorylation and ubiquitination, membrane insertion, and hemichannel internalization and degradation (Laird, 2006; Leithe and Rivedal, 2007; Solan and Lampe, 2009). The time courses of these changes range from milliseconds to hours and are influenced by the environmental conditions in cells and tissues.

Whereas vertebrate cells use connexins to form gap junctions and hemichannels, invertebrate cells use innexins, which lack sequence homology to connexins. A search of the human genome identified three innexin-related genes (Barbe et al., 2006). Because of the occurrence of homologous genes in both vertebrates and invertebrates, the corresponding proteins were termed pannexins: pannexin1 (Panx1), pannexin2 (Panx2), and pannexin3 (Panx3). Pannexins have the same transmembrane topology as connexins, with four transmembrane domains and cytoplasmic amino-terminal and carboxyl-terminal domains. Recent evidence indicates that pannexins also form uncoupled hemichannels in mammalian cells; however, it is not clear whether they can form functional gap junctions (Dahl and Locovei, 2006).

Tissues have characteristic connexin expression profiles, and neural cells in the CNS express multiple connexins (Dermietzel et al., 1989, 2000; Bittman and Loturco, 1999; Chang et al., 1999; Nagy and Rash, 2000; Eugenin et al., 2001; Rash et al., 2001; Teubner et al., 2001; Altevogt et al., 2002; Parenti et al., 2002; Rouach et al., 2002; Odermatt et al., 2003; Takeuchi et al., 2006). All neurons express Cx36 and Cx45, whereas other neural connexins are expressed with more specific spatiotemporal profiles (Sohl et al., 2005). Electrical coupling between neurons has been implicated in neuronal synchronization in the CNS (Christie et al., 1989; Bouskila and Dudek, 1993; Wong et al., 1995). Neuronal gap junctions composed of Cx36 and Cx45 are thought to be homomeric and homotypic (Al-Ubaidi et al., 2000; Teubner et al., 2001), and these junctions are important for formation of electrical synapses (Deans et al., 2001; Hormuzdi et al., 2001). Rodent knockout models have shown that other connexins can compensate for the functions of Cx36 and Cx45, despite differences in conformation or permeability (Frank et al., 2010; Zlomuzica et al., 2010). Furthermore, there is accumulating evidence that gap-junction coupling plays a pivotal role in neuronal differentiation. Mice lacking Cx43 exhibit neonatal death and abnormal migration in the neural crest and neocortex (Lo et al., 1999; Xu et al., 2001; Fushiki et al., 2003). Blockade of gap junctions also hampers retinoic acid–induced neuronal differentiation of NT2 and P19 cells (Bani-Yaghoub et al., 1999a,b). Moreover, Cx36-containing gap junctions are important in neuronal remodeling and short-term spatial memory in some mature organisms (Allen et al., 2011; Hartfield et al., 2011). In contrast to neuron–neuron coupling, for which the evidence is convincing, the existence of functional neuron–glia coupling in the CNS is still a matter of debate (Nadarajah et al., 1996; Alvarez-Maubecin et al., 2000; Rash et al., 2001, 2007).

Astrocytes, the main CNS cells coupled via gap junctions, primarily express Cx43 and Cx30 (Dermietzel et al., 1991; Nagy and Rash, 2000). Consistent with this, Cx43/Cx30 double-knockout mice exhibit minimal gap-junction communication between astrocytes (Wallraff et al., 2006; Rouach et al., 2008), suggesting that functional astrocytic gap junctions are composed predominantly of these two connexins. Cx43-deficient astrocytes exhibit reduced gap-junction coupling, although they express other connexin subtypes including Cx30, Cx26, Cx40, Cx45, and Cx46 (Naus et al., 1997; Scemes et al., 1998; Dermietzel et al., 2000). Although Cx30 has been detected exclusively in astrocytes, Cx30 knockout mice develop only mild abnormalities, including hearing loss due to cochlear degeneration (Teubner et al., 2003). Thus, other astrocytic connexin subtypes do not seem to compensate for a lack of Cx43. Astrocytic gap junctions facilitate the formation of functional syncytium that buffers extracellular glutamate elevation, pH, and K^+ concentrations associated with firing neurons, and also propagates intracellular Ca^{2+} waves that modulate neuronal activities (Walz and Hertz, 1983; Jefferys, 1995; Charles, 1998; Anderson and Swanson, 2000; Ransom et al., 2003). Moreover, astrocytic gap-junction communication facilitates trafficking of glucose and its metabolites, thereby mediating interactions between cerebral vascular endothelium and neurons (Giaume et al., 1997; Goldberg et al., 1999; Tabernero et al., 2006). Thus, astrocytic gap junctions play pivotal roles in modulating neuronal activities and maintaining CNS homeostasis. Astrocyte–astrocyte coupling can be achieved by any of the allowed combinations of homomeric or heteromeric hemichannels in homotypic or heterotypic configurations. Cx30 and Cx26 form both heteromeric and heterotypic channels (Nagy et al., 2003; Altevogt and Paul, 2004), whereas Cx43 forms homomeric and homotypic channels (Orthmann-Murphy et al., 2007). A previous report demonstrated that gap-junction coupling in astrocytes results in two distinct subpopulations of cells. Astrocytes expressing glutamate transporters are extensively coupled to each other, whereas strocytes expressing glutamate receptors are not coupled to other astrocytes (Wallraff et al., 2004), suggesting that these cells play a role in buffering extracellular glutamate (Anderson and Swanson, 2000).

Oligodendrocytes primarily express Cx29, Cx32, and Cx47 (Dermietzel et al., 1989; Altevogt et al., 2002; Odermatt et al., 2003). Oligodendrocytic gap junctions facilitate the trafficking of ions and nutrients from somas to myelin layers (Paul, 1995). Mice lacking Cx32 exhibit reduced myelin volume, enhanced excitability in the CNS, and progressive peripheral neuropathies (Anzini et al., 1997; Sutor et al., 2000). Cx32/Cx47 double-knockout mice develop abnormal movements and seizures associated with vacuolated myelin and axonal degeneration in the CNS, whereas Cx47-deficient mice exhibit only minimal CNS abnormalities (Menichella et al., 2003). Cx32 and Cx47 in oligodendrocytes are essential for spatial buffering of K^+ in response to neuronal activity; failure of this function leads to myelin swelling and subsequent axonal degeneration (Menichella et al., 2006). Oligodendrocyte–oligodendrocyte coupling is mediated by gap junctions in homotypic configurations with homomeric or heteromeric hemichannels containing Cx32 or Cx47 (Orthmann-Murphy et al., 2007). Furthermore, oligodendrocytes also couple with astrocytes. Astrocyte–oligodendrocyte coupling may include heterotypic configurations of Cx43–Cx47, Cx30–Cx32, or Cx26–Cx32 (Nagy et al., 2003; Altevogt and Paul, 2004; Orthmann-Murphy et al., 2007). In addition to astrocyte–astrocyte coupling, astrocyte–oligodendrocyte coupling is important in the glial syncytium to facilitate the propagation of Ca^{2+} waves and the buffering of extracellular K^+ and neurotransmitters such as glutamate (Walz and Hertz, 1983; Jefferys, 1995; Charles, 1998; Anderson and Swanson, 2000; Ransom et al., 2003).

Microglia express Cx32, Cx36, and Cx43 (Eugenin et al., 2001; Parenti et al., 2002; Garg et al., 2005; Takeuchi et al., 2006; Kielian, 2008; Talaveron et al., 2014), but form few functional gap junctions under resting conditions. The expression of connexins rises in activated microglia, although it remains unclear whether upregulated expression of connexins leads to enhanced formation of functional gap junctions with microglia and other CNS cells (Eugenin et al., 2001; Garg et al., 2005; Kielian, 2008; Takeuchi, 2010; Wasseff and Scherer, 2014). Recent evidence demonstrates that uncoupled microglial hemichannels play important roles in bidirectional trafficking of small molecules between the cytoplasm and extracellular space (Takeuchi et al., 2011; Eugenin et al., 2012).

The evidence described above might give the false impression that CNS cells express only a narrow range of combinations of homomeric hemichannels and gap junctions. However, the precise configuration of these hemichannels [i.e., homomeric and homotypic; heteromeric and homotypic; homomeric and heterotypic; and heteromeric and heterotypic (**Figure 2**)] has yet to be elucidated. In addition, our recent reverse transcription–PCR analysis using mouse primary cultures indicated that gap junctions/hemichannels in neurons and glial cells may consist of a wider range of combinations of connexins than expected (**Table 1**) (Takeuchi et al., 2014): neurons predominantly express Cx43, Cx50, Cx31, Cx30.3, Cx29, and Cx36; microglia predominantly express Cx46, Cx37, Cx40, Cx33, Cx57, Cx32, Cx31, Cx30.3, Cx47, Cx36, Cx30.2, Cx39, and Cx23; astrocytes predominantly express Cx43, Cx37, Cx57, Cx26, Cx31, Cx30.3, Cx45, Cx30.2, Cx39, and Cx23; and oligodendrocytes predominantly express Cx46, Cx37, Cx40, Cx33, Cx57, Cx32, Cx26, Cx31, Cx30.3, Cx31.1, Cx30, Cx29, Cx36, Cx30.2, Cx39, and Cx23. These findings imply that connexin expression profiles in the CNS are dynamic, both in healthy and pathological states.

Table 1 | mRNA expression levels of mouse connexin in CNS cells.

Mouse		mRNA expression level			
Protein	**Gene**	**Neuron**	**Microglia**	**Astrocyte**	**Oligodendrocyte**
Cx43	*Gja1*	++	±	+++	+
Cx46	*Gja3*	+	+++	+	++
Cx37	*Gja4*	+	+++	++	+++
Cx40	*Gja5*	+	+++	+	+++
Cx33	*Gja6*	+	++	+	++
Cx50	*Gja8*	+++	+	±	+
Cx57	*Gja10*	±	+++	+++	+++
Cx32	*Gjb1*	+	+++	+	+++
Cx26	*Gjb2*	+	+	+++	+++
Cx31	*Gjb3*	++	++	++	++
Cx30.3	*Gjb4*	++	++	++	++
Cx31.1	*Gjb5*	+	+	+	+++
Cx30	*Gjb6*	±	+	+	+++
Cx45	*Gjc1*	−	+	+++	±
Cx47	*Gjc2*	±	+++	+	+
Cx29	*Gjc3*	++	+	±	++
Cx36	*Gjd2*	+++	++	−	++
Cx30.2	*Gjd3*	+	++	++	+++
Cx39	*Gjd4*	±	++	++	+++
Cx23	*Gje1*	±	++	++	+++

−, none; ±, slight; +, low; ++, moderate; +++, high.

GAP JUNCTIONS COMPOSED OF CONNEXINS AS A NOVEL THERAPEUTIC TARGET FOR NEUROLOGIC DISEASES

As mentioned above, glial gap junctions play an important role in maintenance of homeostasis in the CNS under the physiological conditions. These structures, however, also contribute to the initiation and propagation of pathologic conditions (Orellana et al., 2009). Stroke and trauma provide examples that illustrate this mechanism. Ischemia or contusion leads to a rapid decrease in intracellular oxygen levels and subsequent reduction in ATP synthesis, resulting in eventual cell death (Kalogeris et al., 2012). Injured cells contain toxic ions and molecules at high concentrations (e.g., Ca^{2+}, K^+, ROS, and NO). These toxic molecules are propagated from injured cells to healthier cells through gap junctions. Ischemic conditions also induce uncoupled hemichannels to open, leading to paracrine transfer of toxic molecules (Thompson et al., 2006; De Vuyst et al., 2007). These waves of death signals activate astrocytes and microglia, inducing the release of toxic molecules including glutamate, ROS, NO, and pro-inflammatory cytokines and chemokines. This vicious amplification spiral of signaling could worsen neuroinflammation by recruiting leukocytes and increasing the lesion area (Orellana et al., 2009) (**Figure 3**). Moreover, gap junction and hemichannel blockers have exerted therapeutic effects in experimental models of stroke and spinal cord injury (Rawanduzy et al., 1997; Frantseva et al., 2002; De Pina-Benabou et al., 2005; Takeuchi et al., 2008a; Tamura et al., 2011; Huang et al., 2012; Umebayashi et al., 2014).

Abnormal expression of glial connexins has been observed in the inflamed lesions in multiple sclerosis (MS) and an animal model of this disease, experimental autoimmune encephalomyelitis (EAE). In particular, downregulation of oligodendrocytic Cx32 and Cx47 and astrocytic Cx43 have been observed in the active lesions of MS patients and EAE mice (Brand-Schieber et al., 2005; Eugenin et al., 2012; Markoullis et al., 2012). Expression levels of Cx47 and Cx32 were upregulated during remyelination, but downregulated in the relapsing phase, and Cx32 deletion resulted in exacerbates symptoms in EAE, specifically increased demyelination and axonal loss (Markoullis et al., 2012). Whereas mice lacking astrocytic expression of Cx43/Cx30 exhibited white-matter vacuolation and hypomyelination, the severity of EAE in these animals was similar to that in wild-type mice (Lutz et al., 2012). Therefore, oligodendrocytic expression levels of Cx32 and Cx47 appear to be associated with the degree of damage and remyelination, whereas astrocytic expression levels of Cx43 do not. However, recent studies showed that a loss of Cx43 in astrocytes precedes demyelination in the MS-related disorders neuromyelitis optica and Balo's disease (Matsushita et al., 2011; Masaki et al., 2012), suggesting that the temporal expressional pattern of astrocytic Cx43 plays a significant role in the disease process.

Accumulating evidence has also implicated neuroinflammation, including gliosis by activated astrocytes and microglia, in the pathogenesis of such neurodegenerative diseases as HIV encephalopathy, AD, PD, and ALS (Glass et al., 2010; Valcour et al., 2011). Microglial activation followed by astrocytic activation is the earliest pathologic feature in the pre-symptomatic phases of these diseases. Our recent studies have shown that activated microglia release excess glutamate through Cx32 hemichannels, resulting in excitotoxic neuronal death (Takeuchi et al., 2006,

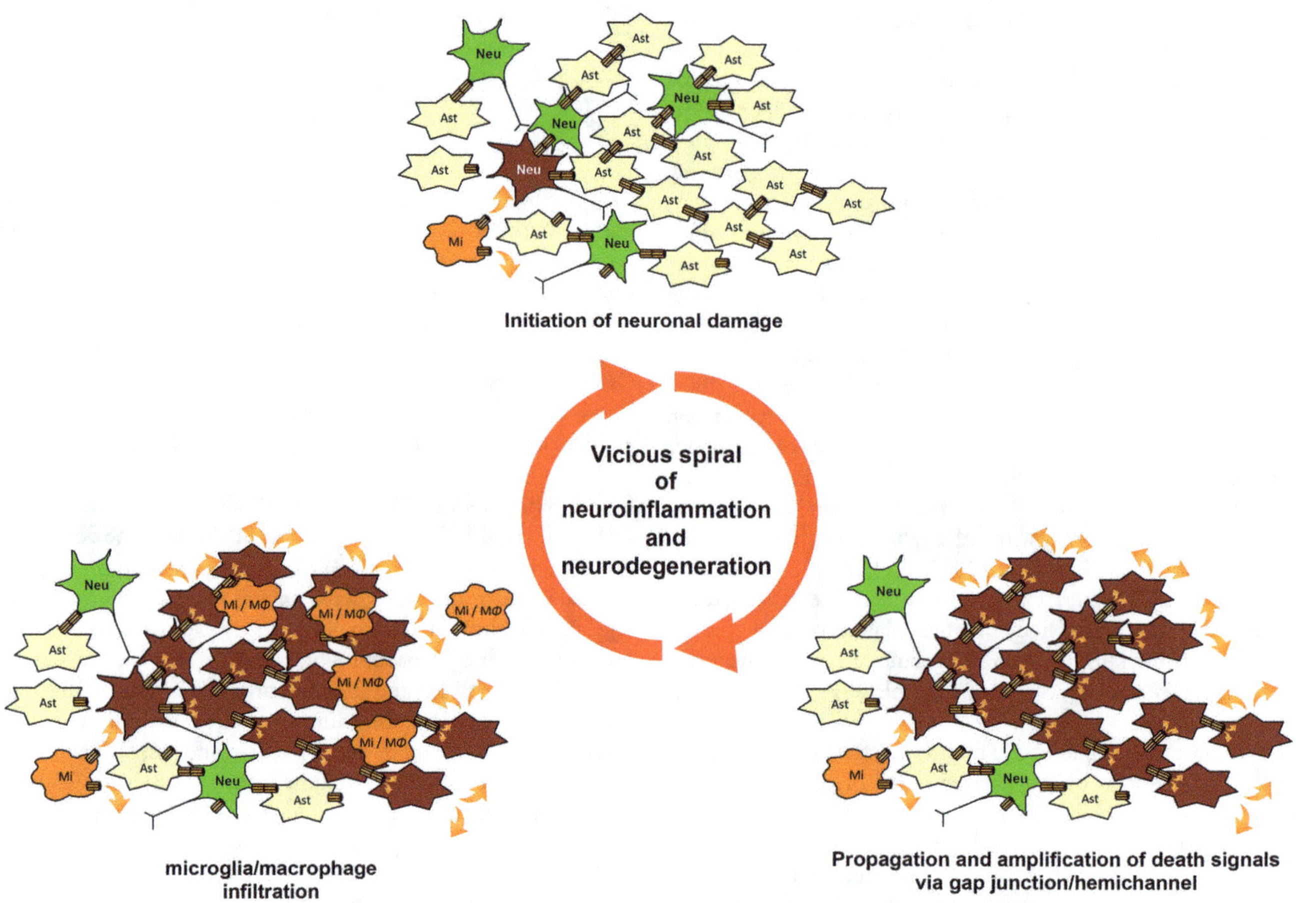

FIGURE 3 | The vicious spiral of neuroinflammation and neurodegeneration mediated by gap junctions and hemichannels. Microglial glutamate release via hemichannels initiates neuronal damage. Then, waves of death signals are propagated and amplified via glial and neuronal gap-junction communication. Chemoattractants from damaged cells induce infiltration by microglia and macrophages. These infiltrating cells initiate further neuronal damage and enlarge the lesion of neuroinflammation and neurodegeneration. This vicious spiral of neuroinflammation and neurodegeneration may be involved in the progression of various neurologic diseases.

2008a; Yawata et al., 2008). Furthermore, microglia-derived glutamate and pro-inflammatory cytokines induce dysfunction of gap junctions and hemichannels in astrocytes (Kielian, 2008), thereby potentially disrupting CNS homeostasis. On the other hand, reactive astrocytes neighboring amyloid β (Aβ) plaques in the brains of AD patients expressed elevated levels of Cx43 and Cx30 (Koulakoff et al., 2012). Aβ peptide induces the release of glutamate and ATP via uncoupled hemichannels in microglia and astrocytes, leading to neuronal death (Orellana et al., 2011). Corroborating this observation, blockade of gap junctions/hemichannels improved memory impairment in a mouse model of AD (Takeuchi et al., 2011). Recent studies also revealed that astrocytic gap junctions/hemichannels are involved in the disease progression of HIV encephalopathy (Eugenin and Berman, 2013; Orellana et al., 2014). PD animal models (MTPT-treated mice and rotenone-treated rats) exhibited upregulation of astrocytic Cx43 expression in affected areas (Rufer et al., 1996; Kawasaki et al., 2009), and a gap junction/hemichannel blocker ameliorated the disease symptoms of a PD mouse model (Suzuki et al., 2014). A recent report revealed that α-synuclein directly binds Cx32, and that overexpression of α-synuclein suppresses the activity of Cx32 in the SH-SY5Y dopaminergic neuroblastoma cell line (Sung et al., 2007). Other studies have shown that microglia and astrocytes are determinants of disease progression in ALS (the non-autonomous neuronal death hypothesis) (Boillee et al., 2006; Yamanaka et al., 2008). Activation of microglia and astrocytes is associated with elevated expression of gap junctions and hemichannels (Cui et al., 2014). In fact, treatment with a gap junction/hemichannel blocker ameliorated disease progression in a mouse model of ALS (Takeuchi et al., 2011). Juvenile neuronal ceroid lipofuscinosis (JNCL) also shows the activation of microglia and astrocytes preceding neuronal loss (Pontikis et al., 2005; Xiong and Kielian, 2013), and treatment with a gap junction/hemichannel blocker attenuated the disease symptoms of a JNCL mouse model (Burkovetskaya et al., 2014). Few reports, however, have focused on the expression profiles and functions of connexins in these diseases. Further studies are needed to elucidate the precise role of glial connexins in the pathogenesis of these diseases.

CONCLUSIONS

A growing body of evidence has demonstrated the pathologic roles of gap junctions and hemichannels in various neurologic diseases. For example, dysfunction and dysregulation of gap junctions and hemichannels in glial cells contribute to neuroinflammation in the CNS, which results in neuronal damage (a situation in which glial cells are "bad neighbors" of neurons) (Block et al., 2007). Despite recent progress in elucidating the pathological roles of gap junctions and hemichannels, many challenges remain, due in part to technical limitations. For instance, few high-quality antibodies against each connexin are available for immunostaining and immunoblotting. Moreover, reagents that are commonly used to block connexin channels are not specific for those channels. In fact, connexin channel blockers such as glycyrrhetinic acid, its derivative carbenoxolone, niflumic acid, and octanol also block pannexin channels. Although the most specific gap junction and hemichannel blockers currently available are mimetic peptides with sequences very similar to that of the extracellular loop of connexins, recent studies showed that mimetic peptides specific for Cx32 (^{32}gap 24 and ^{32}gap 27), Cx43 (^{43}gap 27), or Panx1 (^{10}panx1) non-specifically block both connexins and pannexins (Wang et al., 2007). Although aptamers and siRNA may be used as blockers for specific connexins (Knieps et al., 2007; Xu et al., 2014), they still have a problem of the blood–brain barrier penetration. The heterogeneity of gap-junction and hemichannel configurations (**Figure 2**) and the ability of various connexins to compensate for the loss of other isoforms (e.g., in connexin-knockout studies) also complicate analysis of this system. Although EGFP-tagged connexins have facilitated live-cell imaging, tagging and/or overexpression of connexins in cultured cells often produce abnormally large gap-junction plaques (Lopez et al., 2001; Gaietta et al., 2002; Hunter et al., 2003). Moreover, tagging the amino-termini of connexins results in non-functional channels, whereas tagging the carboxyl-termini alters the properties of the channels (Bukauskas et al., 2000; Contreras et al., 2003). Therefore, future investigations should attempt to elucidate the spatiotemporal expression profiles of connexin isoforms under pathological conditions in the CNS; this work will require development of specific blockers and tracers for each connexin isoform, hemichannel, and gap junction. Understanding the precise pathologic roles of gap junctions and hemichannels may lead to new therapeutic strategies against multiple chronic neurodegenerative diseases.

ACKNOWLEDGMENTS

This work was supported by the Program for Promotion of Fundamental Studies in Health Sciences of the National Institute of Biomedical Innovation (NIBIO); grants from the Ministry of Health, Labour and Welfare of Japan; a grant-in-aid for Scientific Research on Innovative Areas; and a grant-in-aid for the Global Center of Excellence Program from the Ministry of Education, Culture, Sports, Science and Technology of Japan.

REFERENCES

Aarum, J., Sandberg, K., Haeberlein, S. L., and Persson, M. A. (2003). Migration and differentiation of neural precursor cells can be directed by microglia. *Proc. Natl. Acad. Sci. U.S.A.* 100, 15983–15988. doi: 10.1073/pnas.2237050100

Akiyama, H., and McGeer, P. L. (1990). Brain microglia constitutively express beta-2 integrins. *J. Neuroimmunol.* 30, 81–93. doi: 10.1016/0165-5728(90)90055-R

Allen, K., Fuchs, E. C., Jaschonek, H., Bannerman, D. M., and Monyer, H. (2011). Gap junctions between interneurons are required for normal spatial coding in the hippocampus and short-term spatial memory. *J. Neurosci.* 31, 6542–6552. doi: 10.1523/JNEUROSCI.6512-10.2011

Al-Ubaidi, M. R., White, T. W., Ripps, H., Poras, I., Avner, P., Gomes, D., et al. (2000). Functional properties, developmental regulation, and chromosomal localization of murine connexin36, a gap-junctional protein expressed preferentially in retina and brain. *J. Neurosci. Res.* 59, 813–826. doi: 10.1002/(SICI)1097-4547(20000315)59:6<813::AID-JNR14>3.0.CO;2-#

Altevogt, B. M., Kleopa, K. A., Postma, F. R., Scherer, S. S., and Paul, D. L. (2002). Connexin29 is uniquely distributed within myelinating glial cells of the central and peripheral nervous systems. *J. Neurosci.* 22, 6458–6470.

Altevogt, B. M., and Paul, D. L. (2004). Four classes of intercellular channels between glial cells in the CNS. *J. Neurosci.* 24, 4313–4323. doi: 10.1523/JNEUROSCI.3303-03.2004

Alvarez-Maubecin, V., Garcia-Hernandez, F., Williams, J. T., and Van Bockstaele, E. J. (2000). Functional coupling between neurons and glia. *J. Neurosci.* 20, 4091–4098.

Anderson, C. M., and Swanson, R. A. (2000). Astrocyte glutamate transport: review of properties, regulation, and physiological functions. *Glia* 32, 1–14. doi: 10.1002/1098-1136(200010)32:1<1::AID-GLIA10>3.0.CO;2-W

Anzini, P., Neuberg, D. H., Schachner, M., Nelles, E., Willecke, K., Zielasek, J., et al. (1997). Structural abnormalities and deficient maintenance of peripheral nerve myelin in mice lacking the gap junction protein connexin 32. *J. Neurosci.* 17, 4545–4551.

Bamberger, M. E., Harris, M. E., McDonald, D. R., Husemann, J., and Landreth, G. E. (2003). A cell surface receptor complex for fibrillar beta-amyloid mediates microglial activation. *J. Neurosci.* 23, 2665–2674.

Bani-Yaghoub, M., Bechberger, J. F., Underhill, T. M., and Naus, C. C. (1999a). The effects of gap junction blockage on neuronal differentiation of human NTera2/clone D1 cells. *Exp. Neurol.* 156, 16–32. doi: 10.1006/exnr.1998.6950

Bani-Yaghoub, M., Underhill, T. M., and Naus, C. C. (1999b). Gap junction blockage interferes with neuronal and astroglial differentiation of mouse P19 embryonal carcinoma cells. *Dev. Genet.* 24, 69–81. doi: 10.1002/(SICI)1520-6408(1999)24:1/2<69::AID-DVG8>3.0.CO;2-M

Barbe, M. T., Monyer, H., and Bruzzone, R. (2006). Cell-cell communication beyond connexins: the pannexin channels. *Physiology (Bethesda)* 21, 103–114. doi: 10.1152/physiol.00048.2005

Barger, S. W., and Basile, A. S. (2001). Activation of microglia by secreted amyloid precursor protein evokes release of glutamate by cystine exchange and attenuates synaptic function. *J. Neurochem.* 76, 846–854. doi: 10.1046/j.1471-4159.2001.00075.x

Bessis, A., Bechade, C., Bernard, D., and Roumier, A. (2007). Microglial control of neuronal death and synaptic properties. *Glia* 55, 233–238. doi: 10.1002/glia.20459

Bittman, K. S., and Loturco, J. J. (1999). Differential regulation of connexin 26 and 43 in murine neocortical precursors. *Cereb. Cortex* 9, 188–195. doi: 10.1093/cercor/9.2.188

Block, M. L., Zecca, L., and Hong, J. S. (2007). Microglia-mediated neurotoxicity: uncovering the molecular mechanisms. *Nat. Rev. Neurosci.* 8, 57–69. doi: 10.1038/nrn2038

Boche, D., Perry, V. H., and Nicoll, J. A. (2013). Review: activation patterns of microglia and their identification in the human brain. *Neuropathol. Appl. Neurobiol.* 39, 3–18. doi: 10.1111/nan.12011

Boillee, S., Yamanaka, K., Lobsiger, C. S., Copeland, N. G., Jenkins, N. A., Kassiotis, G., et al. (2006). Onset and progression in inherited ALS determined by motor neurons and microglia. *Science* 312, 1389–1392. doi: 10.1126/science.1123511

Bouskila, Y., and Dudek, F. E. (1993). Neuronal synchronization without calcium-dependent synaptic transmission in the hypothalamus. *Proc. Natl. Acad. Sci. U.S.A.* 90, 3207–3210. doi: 10.1073/pnas.90.8.3207

Brand-Schieber, E., Werner, P., Iacobas, D. A., Iacobas, S., Beelitz, M., Lowery, S. L., et al. (2005). Connexin43, the major gap junction protein of astrocytes, is down-regulated in inflamed white matter in an animal model of multiple sclerosis. *J. Neurosci. Res.* 80, 798–808. doi: 10.1002/jnr.20474

Bruijn, L. I., Miller, T. M., and Cleveland, D. W. (2004). Unraveling the mechanisms involved in motor neuron degeneration in ALS. *Annu. Rev. Neurosci.* 27, 723–749. doi: 10.1146/annurev.neuro.27.070203.144244

Bsibsi, M., Ravid, R., Gveric, D., and Van Noort, J. M. (2002). Broad expression of Toll-like receptors in the human central nervous system. *J. Neuropathol. Exp. Neurol.* 61, 1013–1021.

Bukauskas, F. F., Jordan, K., Bukauskiene, A., Bennett, M. V., Lampe, P. D., Laird, D. W., et al. (2000). Clustering of connexin 43-enhanced green fluorescent protein gap junction channels and functional coupling in living cells. *Proc. Natl. Acad. Sci. U.S.A.* 97, 2556–2561. doi: 10.1073/pnas.050588497

Burkovetskaya, M., Karpuk, N., Xiong, J., Bosch, M., Boska, M. D., Takeuchi, H., et al. (2014). Evidence for Aberrant Astrocyte Hemichannel Activity in Juvenile Neuronal Ceroid Lipofuscinosis (JNCL). *PLoS ONE* 9:e95023. doi: 10.1371/journal.pone.0095023

Cagnin, A., Brooks, D. J., Kennedy, A. M., Gunn, R. N., Myers, R., Turkheimer, F. E., et al. (2001). *In-vivo* measurement of activated microglia in dementia. *Lancet* 358, 461–467. doi: 10.1016/S0140-6736(01)05625-2

Carpentier, P. A., Duncan, D. S., and Miller, S. D. (2008). Glial toll-like receptor signaling in central nervous system infection and autoimmunity. *Brain Behav. Immun.* 22, 140–147. doi: 10.1016/j.bbi.2007.08.011

Chang, Q., Gonzalez, M., Pinter, M. J., and Balice-Gordon, R. J. (1999). Gap junctional coupling and patterns of connexin expression among neonatal rat lumbar spinal motor neurons. *J. Neurosci.* 19, 10813–10828.

Charles, A. (1998). Intercellular calcium waves in glia. *Glia* 24, 39–49. doi: 10.1002/(SICI)1098-1136(199809)24:1<39::AID-GLIA5>3.0.CO;2-W

Cho, S., Park, E. M., Febbraio, M., Anrather, J., Park, L., Racchumi, G., et al. (2005). The class B scavenger receptor CD36 mediates free radical production and tissue injury in cerebral ischemia. *J. Neurosci.* 25, 2504–2512. doi: 10.1523/JNEUROSCI.0035-05.2005

Christie, M. J., Williams, J. T., and North, R. A. (1989). Electrical coupling synchronizes subthreshold activity in locus coeruleus neurons *in vitro* from neonatal rats. *J. Neurosci.* 9, 3584–3589.

Contreras, J. E., Saez, J. C., Bukauskas, F. F., and Bennett, M. V. (2003). Gating and regulation of connexin 43 (Cx43) hemichannels. *Proc. Natl. Acad. Sci. U.S.A.* 100, 11388–11393. doi: 10.1073/pnas.1434298100

Coraci, I. S., Husemann, J., Berman, J. W., Hulette, C., Dufour, J. H., Campanella, G. K., et al. (2002). CD36, a class B scavenger receptor, is expressed on microglia in Alzheimer's disease brains and can mediate production of reactive oxygen species in response to beta-amyloid fibrils. *Am. J. Pathol.* 160, 101–112. doi: 10.1016/S0002-9440(10)64354-4

Cui, Y., Masaki, K., Yamasaki, R., Imamura, S., Suzuki, S. O., Hayashi, S., et al. (2014). Extensive dysregulations of oligodendrocytic and astrocytic connexins are associated with disease progression in an amyotrophic lateral sclerosis mouse model. *J. Neuroinflammation* 11:42. doi: 10.1186/1742-2094-11-42

Dahl, G., and Locovei, S. (2006). Pannexin: to gap or not to gap, is that a question? *IUBMB Life* 58, 409–419. doi: 10.1080/15216540600794526

Davalos, D., Grutzendler, J., Yang, G., Kim, J. V., Zuo, Y., Jung, S., et al. (2005). ATP mediates rapid microglial response to local brain injury *in vivo*. *Nat. Neurosci.* 8, 752–758. doi: 10.1038/nn1472

Deans, M. R., Gibson, J. R., Sellitto, C., Connors, B. W., and Paul, D. L. (2001). Synchronous activity of inhibitory networks in neocortex requires electrical synapses containing connexin36. *Neuron* 31, 477–485. doi: 10.1016/S0896-6273(01)00373-7

Delisle, M. B., and Carpenter, S. (1984). Neurofibrillary axonal swellings and amyotrophic lateral sclerosis. *J. Neurol. Sci.* 63, 241–250. doi: 10.1016/0022-510X(84)90199-0

Del Rio-Hortega, P. (1932). *Cytology and Cellular Pathology of the Nervous System.* New York, NY: Penfield Wed.

De Pina-Benabou, M. H., Szostak, V., Kyrozis, A., Rempe, D., Uziel, D., Urban-Maldonado, M., et al. (2005). Blockade of gap junctions *in vivo* provides neuroprotection after perinatal global ischemia. *Stroke* 36, 2232–2237. doi: 10.1161/01.STR.0000182239.75969.d8

Dermietzel, R., Gao, Y., Scemes, E., Vieira, D., Urban, M., Kremer, M., et al. (2000). Connexin43 null mice reveal that astrocytes express multiple connexins. *Brain Res. Brain Res. Rev.* 32, 45–56. doi: 10.1016/S0165-0173(99)00067-3

Dermietzel, R., Hertberg, E. L., Kessler, J. A., and Spray, D. C. (1991). Gap junctions between cultured astrocytes: immunocytochemical, molecular, and electrophysiological analysis. *J. Neurosci.* 11, 1421–1432.

Dermietzel, R., Traub, O., Hwang, T. K., Beyer, E., Bennett, M. V., Spray, D. C., et al. (1989). Differential expression of three gap junction proteins in developing and mature brain tissues. *Proc. Natl. Acad. Sci. U.S.A.* 86, 10148–10152. doi: 10.1073/pnas.86.24.10148

De Vuyst, E., Decrock, E., De Bock, M., Yamasaki, H., Naus, C. C., Evans, W. H., et al. (2007). Connexin hemichannels and gap junction channels are differentially influenced by lipopolysaccharide and basic fibroblast growth factor. *Mol. Biol. Cell* 18, 34–46. doi: 10.1091/mbc.E06-03-0182

Dickson, T. C., King, C. E., McCormack, G. H., and Vickers, J. C. (1999). Neurochemical diversity of dystrophic neurites in the early and late stages of Alzheimer's disease. *Exp. Neurol.* 156, 100–110. doi: 10.1006/exnr.1998.7010

Dutta, R., and Trapp, B. D. (2007). Pathogenesis of axonal and neuronal damage in multiple sclerosis. *Neurology* 68, S22–S31; discussion S43–S54.

Eikelenboom, P., Bate, C., Van Gool, W. A., Hoozemans, J. J., Rozemuller, J. M., Veerhuis, R., et al. (2002). Neuroinflammation in Alzheimer's disease and prion disease. *Glia* 40, 232–239. doi: 10.1002/glia.10146

El Khoury, J., Hickman, S. E., Thomas, C. A., Cao, L., Silverstein, S. C., and Loike, J. D. (1996). Scavenger receptor-mediated adhesion of microglia to beta-amyloid fibrils. *Nature* 382, 716–719. doi: 10.1038/382716a0

El Khoury, J., Hickman, S. E., Thomas, C. A., Loike, J. D., and Silverstein, S. C. (1998). Microglia, scavenger receptors, and the pathogenesis of Alzheimer's disease. *Neurobiol. Aging* 19, S81–S84. doi: 10.1016/S0197-4580(98)00036-0

Eugenin, E. A., Basilio, D., Saez, J. C., Orellana, J. A., Raine, C. S., Bukauskas, F., et al. (2012). The role of gap junction channels during physiologic and pathologic conditions of the human central nervous system. *J. Neuroimmune Pharmacol.* 7, 499–518. doi: 10.1007/s11481-012-9352-5

Eugenin, E. A., and Berman, J. W. (2013). Cytochrome c dysregulation induced by HIV infection of astrocytes results in bystander apoptosis of uninfected astrocytes by an IP3 and calcium-dependent mechanism. *J. Neurochem.* 127, 644–651. doi: 10.1111/jnc.12443

Eugenin, E. A., Eckardt, D., Theis, M., Willecke, K., Bennett, M. V., and Saez, J. C. (2001). Microglia at brain stab wounds express connexin 43 and *in vitro* form functional gap junctions after treatment with interferon-gamma and tumor necrosis factor-alpha. *Proc. Natl. Acad. Sci. U.S.A.* 98, 4190–4195. doi: 10.1073/pnas.051634298

Frank, M., Eiberger, B., Janssen-Bienhold, U., De Sevilla Muller, L. P., Tjarks, A., Kim, J. S., et al. (2010). Neuronal connexin-36 can functionally replace connexin-45 in mouse retina but not in the developing heart. *J. Cell Sci.* 123, 3605–3615. doi: 10.1242/jcs.068668

Frantseva, M. V., Kokarovtseva, L., and Perez Velazquez, J. L. (2002). Ischemia-induced brain damage depends on specific gap-junctional coupling. *J. Cereb. Blood Flow Metab.* 22, 453–462. doi: 10.1097/00004647-200204000-00009

Fushiki, S., Perez Velazquez, J. L., Zhang, L., Bechberger, J. F., Carlen, P. L., and Naus, C. C. (2003). Changes in neuronal migration in neocortex of connexin43 null mutant mice. *J. Neuropathol. Exp. Neurol.* 62, 304–314.

Gaietta, G., Deerinck, T. J., Adams, S. R., Bouwer, J., Tour, O., Laird, D. W., et al. (2002). Multicolor and electron microscopic imaging of connexin trafficking. *Science* 296, 503–507. doi: 10.1126/science.1068793

Garg, S., Md Syed, M., and Kielian, T. (2005). Staphylococcus aureus-derived peptidoglycan induces Cx43 expression and functional gap junction intercellular communication in microglia. *J. Neurochem.* 95, 475–483. doi: 10.1111/j.1471-4159.2005.03384.x

Ghezzi, P., and Mennini, T. (2001). Tumor necrosis factor and motoneuronal degeneration: an open problem. *Neuroimmunomodulation* 9, 178–182. doi: 10.1159/000049024

Giaume, C., Tabernero, A., and Medina, J. M. (1997). Metabolic trafficking through astrocytic gap junctions. *Glia* 21, 114–123. doi: 10.1002/(SICI)1098-1136(199709)21:1<114::AID-GLIA13>3.0.CO;2-V

Glass, C. K., Saijo, K., Winner, B., Marchetto, M. C., and Gage, F. H. (2010). Mechanisms underlying inflammation in neurodegeneration. *Cell* 140, 918–934. doi: 10.1016/j.cell.2010.02.016

Goel, S., Wharton, S. B., Brett, L. P., and Whittle, I. R. (2003). Morphological changes and stress responses in neurons in cerebral cortex infiltrated by diffuse astrocytoma. *Neuropathology* 23, 262–270. doi: 10.1046/j.1440-1789.2003.00510.x

Goldberg, G. S., Lampe, P. D., and Nicholson, B. J. (1999). Selective transfer of endogenous metabolites through gap junctions composed of different connexins. *Nat. Cell Biol.* 1, 457–459. doi: 10.1038/15693

Goldberg, G. S., Moreno, A. P., and Lampe, P. D. (2002). Gap junctions between cells expressing connexin 43 or 32 show inverse permselectivity to adenosine and ATP. *J. Biol. Chem.* 277, 36725–36730. doi: 10.1074/jbc.M109797200

Goldberg, G. S., Valiunas, V., and Brink, P. R. (2004). Selective permeability of gap junction channels. *Biochim. Biophys. Acta* 1662, 96–101. doi: 10.1016/j.bbamem.2003.11.022

Grewal, R. P., Yoshida, T., Finch, C. E., and Morgan, T. E. (1997). Scavenger receptor mRNAs in rat brain microglia are induced by kainic acid lesioning and by cytokines. *Neuroreport* 8, 1077–1081. doi: 10.1097/00001756-199703240-00003

Harris, A. L. (2001). Emerging issues of connexin channels: biophysics fills the gap. *Q. Rev. Biophys.* 34, 325–472. doi: 10.1017/S0033583501003705

Harris, A. L. (2007). Connexin channel permeability to cytoplasmic molecules. *Prog. Biophys. Mol. Biol.* 94, 120–143. doi: 10.1016/j.pbiomolbio.2007.03.011

Hartfield, E. M., Rinaldi, F., Glover, C. P., Wong, L. F., Caldwell, M. A., and Uney, J. B. (2011). Connexin 36 expression regulates neuronal differentiation from neural progenitor cells. *PLoS ONE* 6:e14746. doi: 10.1371/journal.pone.0014746

Henkel, J. S., Beers, D. R., Zhao, W., and Appel, S. H. (2009). Microglia in ALS: the good, the bad, and the resting. *J. Neuroimmune Pharmacol.* 4, 389–398. doi: 10.1007/s11481-009-9171-5

Herman, M. A., and Jahr, C. E. (2007). Extracellular glutamate concentration in hippocampal slice. *J. Neurosci.* 27, 9736–9741. doi: 10.1523/JNEUROSCI.3009-07.2007

Hori, N., and Carpenter, D. O. (1994). Functional and morphological changes induced by transient *in vivo* ischemia. *Exp. Neurol.* 129, 279–289. doi: 10.1006/exnr.1994.1170

Hormuzdi, S. G., Pais, I., Lebeau, F. E., Towers, S. K., Rozov, A., Buhl, E. H., et al. (2001). Impaired electrical signaling disrupts gamma frequency oscillations in connexin 36-deficient mice. *Neuron* 31, 487–495. doi: 10.1016/S0896-6273(01)00387-7

Huang, C., Han, X., Li, X., Lam, E., Peng, W., Lou, N., et al. (2012). Critical role of connexin 43 in secondary expansion of traumatic spinal cord injury. *J. Neurosci.* 32, 3333–3338. doi: 10.1523/JNEUROSCI.1216-11.2012

Hunter, A. W., Jourdan, J., and Gourdie, R. G. (2003). Fusion of GFP to the carboxyl terminus of connexin43 increases gap junction size in HeLa cells. *Cell Commun. Adhes.* 10, 211–214. doi: 10.1080/714040429

Iribarren, P., Chen, K., Hu, J., Gong, W., Cho, E. H., Lockett, S., et al. (2005). CpG-containing oligodeoxynucleotide promotes microglial cell uptake of amyloid beta 1-42 peptide by up-regulating the expression of the G-protein- coupled receptor mFPR2. *FASEB J.* 19, 2032–2034.

Jefferys, J. G. (1995). Nonsynaptic modulation of neuronal activity in the brain: electric currents and extracellular ions. *Physiol. Rev.* 75, 689–723.

Jimenez, S., Baglietto-Vargas, D., Caballero, C., Moreno-Gonzalez, I., Torres, M., Sanchez-Varo, R., et al. (2008). Inflammatory response in the hippocampus of PS1M146L/APP751SL mouse model of Alzheimer's disease: age-dependent switch in the microglial phenotype from alternative to classic. *J. Neurosci.* 28, 11650–11661. doi: 10.1523/JNEUROSCI.3024-08.2008

Kalogeris, T., Baines, C. P., Krenz, M., and Korthuis, R. J. (2012). Cell biology of ischemia/reperfusion injury. *Int. Rev. Cell Mol. Biol.* 298, 229–317. doi: 10.1016/B978-0-12-394309-5.00006-7

Kamata, H., Honda, S., Maeda, S., Chang, L., Hirata, H., and Karin, M. (2005). Reactive oxygen species promote TNFalpha-induced death and sustained JNK activation by inhibiting MAP kinase phosphatases. *Cell* 120, 649–661. doi: 10.1016/j.cell.2004.12.041

Kawanokuchi, J., Mizuno, T., Takeuchi, H., Kato, H., Wang, J., Mitsuma, N., et al. (2006). Production of interferon-gamma by microglia. *Mult. Scler.* 12, 558–564. doi: 10.1177/1352458506070763

Kawasaki, A., Hayashi, T., Nakachi, K., Trosko, J. E., Sugihara, K., Kotake, Y., et al. (2009). Modulation of connexin 43 in rotenone-induced model of Parkinson's disease. *Neuroscience* 160, 61–68. doi: 10.1016/j.neuroscience.2009.01.080

Kempermann, G., and Neumann, H. (2003). Neuroscience. Microglia: the enemy within? *Science* 302, 1689–1690. doi: 10.1126/science.1092864

Kielian, T. (2008). Glial connexins and gap junctions in CNS inflammation and disease. *J. Neurochem.* 106, 1000–1016. doi: 10.1111/j.1471-4159.2008.05405.x

Kipnis, J., Avidan, H., Caspi, R. R., and Schwartz, M. (2004). Dual effect of CD4+CD25+ regulatory T cells in neurodegeneration: a dialogue with microglia. *Proc. Natl. Acad. Sci. U.S.A.* 101(Suppl. 2), 14663–14669. doi: 10.1073/pnas.0404842101

Knieps, M., Herrmann, S., Lehmann, C., Loer, B., Hoch, M., and Famulok, M. (2007). Anti-innexin 2 aptamers specifically inhibit the heterologous interaction of the innexin 2 and innexin 3 carboxyl-termini *in vitro*. *Biol. Chem.* 388, 561–568. doi: 10.1515/BC.2007.074

Koenigsknecht, J., and Landreth, G. (2004). Microglial phagocytosis of fibrillar beta-amyloid through a beta1 integrin-dependent mechanism. *J. Neurosci.* 24, 9838–9846. doi: 10.1523/JNEUROSCI.2557-04.2004

Koulakoff, A., Mei, X., Orellana, J. A., Saez, J. C., and Giaume, C. (2012). Glial connexin expression and function in the context of Alzheimer's disease. *Biochim. Biophys. Acta* 1818, 2048–2057. doi: 10.1016/j.bbamem.2011.10.001

Kreutzberg, G. W. (1996). Microglia: a sensor for pathological events in the CNS. *Trends Neurosci.* 19, 312–318. doi: 10.1016/0166-2236(96)10049-7

Laird, D. W. (2006). Life cycle of connexins in health and disease. *Biochem. J.* 394, 527–543. doi: 10.1042/BJ20051922

Laird, D. W. (2010). The gap junction proteome and its relationship to disease. *Trends Cell Biol.* 20, 92–101. doi: 10.1016/j.tcb.2009.11.001

Lawson, L. J., Perry, V. H., Dri, P., and Gordon, S. (1990). Heterogeneity in the distribution and morphology of microglia in the normal adult mouse brain. *Neuroscience* 39, 151–170. doi: 10.1016/0306-4522(90)90229-W

Leithe, E., and Rivedal, E. (2007). Ubiquitination of gap junction proteins. *J. Membr. Biol.* 217, 43–51. doi: 10.1007/s00232-007-9050-z

Liang, J., Takeuchi, H., Doi, Y., Kawanokuchi, J., Sonobe, Y., Jin, S., et al. (2008). Excitatory amino acid transporter expression by astrocytes is neuroprotective against microglial excitotoxicity. *Brain Res.* 1210, 11–19. doi: 10.1016/j.brainres.2008.03.012

Liang, J., Takeuchi, H., Jin, S., Noda, M., Li, H., Doi, Y., et al. (2010). Glutamate induces neurotrophic factor production from microglia via protein kinase C pathway. *Brain Res.* 1322, 8–23. doi: 10.1016/j.brainres.2010.01.083

Liu, Y., Walter, S., Stagi, M., Cherny, D., Letiembre, M., Schulz-Schaeffer, W., et al. (2005). LPS receptor (CD14): a receptor for phagocytosis of Alzheimer's amyloid peptide. *Brain* 128, 1778–1789. doi: 10.1093/brain/awh531

Lo, C. W., Waldo, K. L., and Kirby, M. L. (1999). Gap junction communication and the modulation of cardiac neural crest cells. *Trends Cardiovasc. Med.* 9, 63–69. doi: 10.1016/S1050-1738(99)00015-8

Lobsiger, C. S., and Cleveland, D. W. (2007). Glial cells as intrinsic components of non-cell-autonomous neurodegenerative disease. *Nat. Neurosci.* 10, 1355–1360. doi: 10.1038/nn1988

Lopez, P., Balicki, D., Buehler, L. K., Falk, M. M., and Chen, S. C. (2001). Distribution and dynamics of gap junction channels revealed in living cells. *Cell Commun. Adhes.* 8, 237–242. doi: 10.3109/15419060109080730

Lutz, S. E., Raine, C. S., and Brosnan, C. F. (2012). Loss of astrocyte connexins 43 and 30 does not significantly alter susceptibility or severity of acute experimental autoimmune encephalomyelitis in mice. *J. Neuroimmunol.* 245, 8–14. doi: 10.1016/j.jneuroim.2012.01.007

Mantovani, A., Sozzani, S., Locati, M., Allavena, P., and Sica, A. (2002). Macrophage polarization: tumor-associated macrophages as a paradigm for polarized M2 mononuclear phagocytes. *Trends Immunol.* 23, 549–555. doi: 10.1016/S1471-4906(02)02302-5

Marin-Teva, J. L., Dusart, I., Colin, C., Gervais, A., Van Rooijen, N., and Mallat, M. (2004). Microglia promote the death of developing Purkinje cells. *Neuron* 41, 535–547. doi: 10.1016/S0896-6273(04)00069-8

Markoullis, K., Sargiannidou, I., Gardner, C., Hadjisavvas, A., Reynolds, R., and Kleopa, K. A. (2012). Disruption of oligodendrocyte gap junctions in experimental autoimmune encephalomyelitis. *Glia* 60, 1053–1066. doi: 10.1002/glia.22334

Masaki, K., Suzuki, S. O., Matsushita, T., Yonekawa, T., Matsuoka, T., Isobe, N., et al. (2012). Extensive loss of connexins in Balo's disease: evidence for an auto-antibody-independent astrocytopathy via impaired astrocyte-oligodendrocyte/myelin interaction. *Acta Neuropathol.* 123, 887–900. doi: 10.1007/s00401-012-0972-x

Matsushita, T., Masaki, K., Suzuki, S., Matsuoka, T., Yonekawa, T., Wu, X. M., et al. (2011). Astrocytopathy in neuromyelitis optica, multiple sclerosis and Balo's disease. *Rinsho Shinkeigaku* 51, 898–900. doi: 10.5692/clinicalneurol.51.898

Mattila, P. M., Rinne, J. O., Helenius, H., and Roytta, M. (1999). Neuritic degeneration in the hippocampus and amygdala in Parkinson's disease in relation to Alzheimer pathology. *Acta Neuropathol.* 98, 157–164. doi: 10.1007/s004010051064

McGeer, P. L., and McGeer, E. G. (2002). Inflammatory processes in amyotrophic lateral sclerosis. *Muscle Nerve* 26, 459–470. doi: 10.1002/mus.10191

Menichella, D. M., Goodenough, D. A., Sirkowski, E., Scherer, S. S., and Paul, D. L. (2003). Connexins are critical for normal myelination in the CNS. *J. Neurosci.* 23, 5963–5973.

Menichella, D. M., Majdan, M., Awatramani, R., Goodenough, D. A., Sirkowski, E., Scherer, S. S., et al. (2006). Genetic and physiological evidence that oligodendrocyte gap junctions contribute to spatial buffering of potassium released during neuronal activity. *J. Neurosci.* 26, 10984–10991. doi: 10.1523/JNEUROSCI.0304-06.2006

Mizuno, T., Zhang, G., Takeuchi, H., Kawanokuchi, J., Wang, J., Sonobe, Y., et al. (2008). Interferon-gamma directly induces neurotoxicity through a neuron specific, calcium-permeable complex of IFN-gamma receptor and AMPA GluR1 receptor. *FASEB J.* 22, 1797–1806. doi: 10.1096/fj.07-099499

Nadarajah, B., Thomaidou, D., Evans, W. H., and Parnavelas, J. G. (1996). Gap junctions in the adult cerebral cortex: regional differences in their distribution and cellular expression of connexins. *J. Comp. Neurol.* 376, 326–342. doi: 10.1002/(SICI)1096-9861(19961209)376:2<326::AID-CNE13>3.0.CO;2-J

Nagy, J. I., Ionescu, A. V., Lynn, B. D., and Rash, J. E. (2003). Coupling of astrocyte connexins Cx26, Cx30, Cx43 to oligodendrocyte Cx29, Cx32, Cx47: implications from normal and connexin32 knockout mice. *Glia* 44, 205–218. doi: 10.1002/glia.10278

Nagy, J. I., and Rash, J. E. (2000). Connexins and gap junctions of astrocytes and oligodendrocytes in the CNS. *Brain Res. Brain Res. Rev.* 32, 29–44. doi: 10.1016/S0165-0173(99)00066-1

Nakanishi, H., and Wu, Z. (2009). Microglia-aging: roles of microglial lysosome- and mitochondria-derived reactive oxygen species in brain aging. *Behav. Brain Res.* 201, 1–7. doi: 10.1016/j.bbr.2009.02.001

Naus, C. C., Bechberger, J. F., Zhang, Y., Venance, L., Yamasaki, H., Juneja, S. C., et al. (1997). Altered gap junctional communication, intercellular signaling, and growth in cultured astrocytes deficient in connexin43. *J. Neurosci. Res.* 49, 528–540. doi: 10.1002/(SICI)1097-4547(19970901)49:5<528::AID-JNR3>3.0.CO;2-D

Nelson, P. T., Soma, L. A., and Lavi, E. (2002). Microglia in diseases of the central nervous system. *Ann. Med.* 34, 491–500. doi: 10.1080/078538902321117698

Newsholme, E. A., and Calder, P. C. (1997). The proposed role of glutamine in some cells of the immune system and speculative consequences for the whole animal. *Nutrition* 13, 728–730. doi: 10.1016/S0899-9007(97)83034-1

Newsholme, P., and Newsholme, E. A. (1989). Rates of utilization of glucose, glutamine and oleate and formation of end-products by mouse peritoneal macrophages in culture. *Biochem. J.* 261, 211–218.

Nimmerjahn, A., Kirchhoff, F., and Helmchen, F. (2005). Resting microglial cells are highly dynamic surveillants of brain parenchyma *in vivo*. *Science* 308, 1314–1318. doi: 10.1126/science.1110647

Nissim, I. (1999). Newer aspects of glutamine/glutamate metabolism: the role of acute pH changes. *Am. J. Physiol.* 277, F493–F497.

Odermatt, B., Wellershaus, K., Wallraff, A., Seifert, G., Degen, J., Euwens, C., et al. (2003). Connexin 47 (Cx47)-deficient mice with enhanced green fluorescent protein reporter gene reveal predominant oligodendrocytic expression of Cx47 and display vacuolized myelin in the CNS. *J. Neurosci.* 23, 4549–4559.

Orellana, J. A., Saez, J. C., Bennett, M. V., Berman, J. W., Morgello, S., and Eugenin, E. A. (2014). HIV increases the release of dickkopf-1 protein from human astrocytes by a Cx43 hemichannel-dependent mechanism. *J. Neurochem.* 128, 752–763. doi: 10.1111/jnc.12492

Orellana, J. A., Saez, P. J., Shoji, K. F., Schalper, K. A., Palacios-Prado, N., Velarde, V., et al. (2009). Modulation of brain hemichannels and gap junction channels by pro-inflammatory agents and their possible role in neurodegeneration. *Antioxid. Redox Signal.* 11, 369–399. doi: 10.1089/ars.2008.2130

Orellana, J. A., Shoji, K. F., Abudara, V., Ezan, P., Amigou, E., Saez, P. J., et al. (2011). Amyloid beta-induced death in neurons involves glial and neuronal hemichannels. *J. Neurosci.* 31, 4962–4977. doi: 10.1523/JNEUROSCI.6417-10.2011

Orr, C. F., Rowe, D. B., and Halliday, G. M. (2002). An inflammatory review of Parkinson's disease. *Prog. Neurobiol.* 68, 325–340. doi: 10.1016/S0301-0082(02)00127-2

Orthmann-Murphy, J. L., Freidin, M., Fischer, E., Scherer, S. S., and Abrams, C. K. (2007). Two distinct heterotypic channels mediate gap junction coupling between astrocyte and oligodendrocyte connexins. *J. Neurosci.* 27, 13949–13957. doi: 10.1523/JNEUROSCI.3395-07.2007

Parenti, R., Campisi, A., Vanella, A., and Cicirata, F. (2002). Immunocytochemical and RT-PCR analysis of connexin36 in cultures of mammalian glial cells. *Arch. Ital. Biol.* 140, 101–108.

Parsons, C. G., Stoffler, A., and Danysz, W. (2007). Memantine: a NMDA receptor antagonist that improves memory by restoration of homeostasis in the glutamatergic system–too little activation is bad, too much is even worse. *Neuropharmacology* 53, 699–723. doi: 10.1016/j.neuropharm.2007.07.013

Paul, D. L. (1995). New functions for gap junctions. *Curr. Opin. Cell Biol.* 7, 665–672. doi: 10.1016/0955-0674(95)80108-1

Pavese, N., Gerhard, A., Tai, Y. F., Ho, A. K., Turkheimer, F., Barker, R. A., et al. (2006). Microglial activation correlates with severity in Huntington disease: a clinical and PET study. *Neurology* 66, 1638–1643. doi: 10.1212/01.wnl.0000222734.56412.17

Pavlidis, M., Stupp, T., Naskar, R., Cengiz, C., and Thanos, S. (2003). Retinal ganglion cells resistant to advanced glaucoma: a postmortem study of human retinas with the carbocyanine dye DiI. *Invest. Ophthalmol. Vis. Sci.* 44, 5196–5205. doi: 10.1167/iovs.03-0614

Piani, D., Spranger, M., Frei, K., Schaffner, A., and Fontana, A. (1992). Macrophage-induced cytotoxicity of N-methyl-D-aspartate receptor positive neurons involves excitatory amino acids rather than reactive oxygen intermediates and cytokines. *Eur. J. Immunol.* 22, 2429–2436. doi: 10.1002/eji.1830220936

Pontikis, C. C., Cotman, S. L., Macdonald, M. E., and Cooper, J. D. (2005). Thalamocortical neuron loss and localized astrocytosis in the Cln3Deltaex7/8 knock-in mouse model of Batten disease. *Neurobiol. Dis.* 20, 823–836. doi: 10.1016/j.nbd.2005.05.018

Ransom, B., Behar, T., and Nedergaard, M. (2003). New roles for astrocytes (stars at last). *Trends Neurosci.* 26, 520–522. doi: 10.1016/j.tins.2003.08.006

Rash, J. E., Olson, C. O., Davidson, K. G., Yasumura, T., Kamasawa, N., and Nagy, J. I. (2007). Identification of connexin36 in gap junctions between neurons in rodent locus coeruleus. *Neuroscience* 147, 938–956. doi: 10.1016/j.neuroscience.2007.04.061

Rash, J. E., Yasumura, T., Dudek, F. E., and Nagy, J. I. (2001). Cell-specific expression of connexins and evidence of restricted gap junctional coupling between glial cells and between neurons. *J. Neurosci.* 21, 1983–2000.

Rawanduzy, A., Hansen, A., Hansen, T. W., and Nedergaard, M. (1997). Effective reduction of infarct volume by gap junction blockade in a rodent model of stroke. *J. Neurosurg.* 87, 916–920. doi: 10.3171/jns.1997.87.6.0916

Retamal, M. A., Schalper, K. A., Shoji, K. F., Bennett, M. V., and Saez, J. C. (2007). Opening of connexin 43 hemichannels is increased by lowering intracellular redox potential. *Proc. Natl. Acad. Sci. U.S.A.* 104, 8322–8327. doi: 10.1073/pnas.0702456104

Richard, K. L., Filali, M., Prefontaine, P., and Rivest, S. (2008). Toll-like receptor 2 acts as a natural innate immune receptor to clear amyloid beta 1–42 and delay the cognitive decline in a mouse model of Alzheimer's disease. *J. Neurosci.* 28, 5784–5793. doi: 10.1523/JNEUROSCI.1146-08.2008

Rock, R. B., Gekker, G., Hu, S., Sheng, W. S., Cheeran, M., Lokensgard, J. R., et al. (2004). Role of microglia in central nervous system infections. *Clin. Microbiol. Rev.* 17, 942–964. doi: 10.1128/CMR.17.4.942-964.2004

Rouach, N., Avignone, E., Meme, W., Koulakoff, A., Venance, L., Blomstrand, F., et al. (2002). Gap junctions and connexin expression in the normal and pathological central nervous system. *Biol. Cell* 94, 457–475. doi: 10.1016/S0248-4900(02)00016-3

Rouach, N., Koulakoff, A., Abudara, V., Willecke, K., and Giaume, C. (2008). Astroglial metabolic networks sustain hippocampal synaptic transmission. *Science* 322, 1551–1555. doi: 10.1126/science.1164022

Rufer, M., Wirth, S. B., Hofer, A., Dermietzel, R., Pastor, A., Kettenmann, H., et al. (1996). Regulation of connexin-43, GFAP, and FGF-2 is not accompanied by changes in astroglial coupling in MPTP-lesioned, FGF-2-treated parkinsonian mice. *J. Neurosci. Res.* 46, 606–617. doi: 10.1002/(SICI)1097-4547(19961201)46:5<606::AID-JNR9>3.0.CO;2-N

Saez, J. C., Berthoud, V. M., Branes, M. C., Martinez, A. D., and Beyer, E. C. (2003). Plasma membrane channels formed by connexins: their regulation and functions. *Physiol. Rev.* 83, 1359–1400.

Saito, Y., Kawashima, A., Ruberu, N. N., Fujiwara, H., Koyama, S., Sawabe, M., et al. (2003). Accumulation of phosphorylated alpha-synuclein in aging human brain. *J. Neuropathol. Exp. Neurol.* 62, 644–654.

Sawada, M. (2009). Neuroprotective and toxic changes in microglia in neurodegenerative disease. *Parkinsonism. Relat. Disord.* 15(Suppl. 1), S39–S41. doi: 10.1016/S1353-8020(09)70011-2

Scemes, E., Dermietzel, R., and Spray, D. C. (1998). Calcium waves between astrocytes from Cx43 knockout mice. *Glia* 24, 65–73. doi: 10.1002/(SICI)1098-1136(199809)24:1<65::AID-GLIA7>3.0.CO;2-#

Schwab, J. M., and Schluesener, H. J. (2004). Microglia rules: insights into microglial-neuronal signaling. *Cell Death Differ.* 11, 1245–1246. doi: 10.1038/sj.cdd.4401487

Schwartz, M., Shaked, I., Fisher, J., Mizrahi, T., and Schori, H. (2003). Protective autoimmunity against the enemy within: fighting glutamate toxicity. *Trends Neurosci.* 26, 297–302. doi: 10.1016/S0166-2236(03)00126-7

Seabrook, T. J., Jiang, L., Maier, M., and Lemere, C. A. (2006). Minocycline affects microglia activation, Abeta deposition, and behavior in APP-tg mice. *Glia* 53, 776–782. doi: 10.1002/glia.20338

Simard, A. R., Soulet, D., Gowing, G., Julien, J. P., and Rivest, S. (2006). Bone marrow-derived microglia play a critical role in restricting senile plaque formation in Alzheimer's disease. *Neuron* 49, 489–502. doi: 10.1016/j.neuron.2006.01.022

Sohl, G., Maxeiner, S., and Willecke, K. (2005). Expression and functions of neuronal gap junctions. *Nat. Rev. Neurosci.* 6, 191–200. doi: 10.1038/nrn1627

Solan, J. L., and Lampe, P. D. (2009). Connexin43 phosphorylation: structural changes and biological effects. *Biochem. J.* 419, 261–272. doi: 10.1042/BJ20082319

Stirling, D. P., Khodarahmi, K., Liu, J., McPhail, L. T., McBride, C. B., Steeves, J. D., et al. (2004). Minocycline treatment reduces delayed oligodendrocyte death, attenuates axonal dieback, and improves functional outcome after spinal cord injury. *J. Neurosci.* 24, 2182–2190. doi: 10.1523/JNEUROSCI.5275-03.2004

Sung, J. Y., Lee, H. J., Jeong, E. I., Oh, Y., Park, J., Kang, K. S., et al. (2007). Alpha-synuclein overexpression reduces gap junctional intercellular communication in dopaminergic neuroblastoma cells. *Neurosci. Lett.* 416, 289–293. doi: 10.1016/j.neulet.2007.02.025

Sutor, B., Schmolke, C., Teubner, B., Schirmer, C., and Willecke, K. (2000). Myelination defects and neuronal hyperexcitability in the neocortex of connexin 32-deficient mice. *Cereb. Cortex* 10, 684–697. doi: 10.1093/cercor/10.7.684

Suzuki, H., Ono, K., and Sawada, M. (2014). Protective effect of INI-0602, a gap junction inhibitor, on dopaminergic neurodegeneration of mice with unilateral 6-hydroxydopamine injection. *J. Neural Transm.* doi: 10.1007/s00702-014-1209-z. [Epub ahead of print].

Suzumura, A., Mezitis, S. G., Gonatas, N. K., and Silberberg, D. H. (1987). MHC antigen expression on bulk isolated macrophage-microglia from newborn mouse brain: induction of Ia antigen expression by gamma-interferon. *J. Neuroimmunol.* 15, 263–278. doi: 10.1016/0165-5728(87)90121-4

Swann, J. W., Al-Noori, S., Jiang, M., and Lee, C. L. (2000). Spine loss and other dendritic abnormalities in epilepsy. *Hippocampus* 10, 617–625. doi: 10.1002/1098-1063(2000)10:5<617::AID-HIPO13>3.0.CO;2-R

Tabernero, A., Medina, J. M., and Giaume, C. (2006). Glucose metabolism and proliferation in glia: role of astrocytic gap junctions. *J. Neurochem.* 99, 1049–1061. doi: 10.1111/j.1471-4159.2006.04088.x

Takahashi, T., Yagishita, S., Amano, N., Yamaoka, K., and Kamei, T. (1997). Amyotrophic lateral sclerosis with numerous axonal spheroids in the corticospinal tract and massive degeneration of the cortex. *Acta Neuropathol.* 94, 294–299. doi: 10.1007/s004010050707

Takaki, J., Fujimori, K., Miura, M., Suzuki, T., Sekino, Y., and Sato, K. (2012). L-glutamate released from activated microglia downregulates astrocytic L-glutamate transporter expression in neuroinflammation: the 'collusion' hypothesis for increased extracellular L-glutamate concentration in neuroinflammation. *J. Neuroinflammation* 9:275. doi: 10.1186/1742-2094-9-275

Takeuchi, H. (2010). Neurotoxicity by microglia: mechanisms and potential therapeutic strategy. *Clin. Exp. Neuroimmunol.* 1, 12–21. doi: 10.1111/j.1759-1961.2009.00001.x

Takeuchi, H., Jin, S., Suzuki, H., Doi, Y., Liang, J., Kawanokuchi, J., et al. (2008a). Blockade of microglial glutamate release protects against ischemic brain injury. *Exp. Neurol.* 214, 144–146. doi: 10.1016/j.expneurol.2008.08.001

Takeuchi, H., Jin, S., Suzuki, H., Doi, Y., Liang, J., Kawanokuchi, J., et al. (2008b). Blockade of microglial glutamate release protects against ischemic brain injury. *Exp. Neurol.* 214, 144–146. doi: 10.1016/j.expneurol.2008.08.001

Takeuchi, H., Jin, S., Wang, J., Zhang, G., Kawanokuchi, J., Kuno, R., et al. (2006). Tumor necrosis factor-alpha induces neurotoxicity via glutamate release from hemichannels of activated microglia in an autocrine manner. *J. Biol. Chem.* 281, 21362–21368. doi: 10.1074/jbc.M600504200

Takeuchi, H., Kawanokuchi, J., Mizuno, T., and Suzumura, A. (2014). Expression profile of connexins in the central nervous system. *Clini. Exp. Neuroimmunol.* doi: 10.1111/cen3.12106. [Epub ahead of print].

Takeuchi, H., Mizoguchi, H., Doi, Y., Jin, S., Noda, M., Liang, J., et al. (2011). Blockade of gap junction hemichannel suppresses disease progression in mouse models of amyotrophic lateral sclerosis and Alzheimer's disease. *PLoS ONE* 6:e21108. doi: 10.1371/journal.pone.0021108

Takeuchi, H., Mizuno, T., Zhang, G., Wang, J., Kawanokuchi, J., Kuno, R., et al. (2005). Neuritic beading induced by activated microglia is an early feature of neuronal dysfunction toward neuronal death by inhibition of mitochondrial respiration and axonal transport. *J. Biol. Chem.* 280, 10444–10454. doi: 10.1074/jbc.M413863200

Talaveron, R., Matarredona, E. R., De La Cruz, R. R., Macias, D., Galvez, V., and Pastor, A. M. (2014). Implanted neural progenitor cells regulate glial reaction to brain injury and establish gap junctions with host glial cells. *Glia* 62, 623–638. doi: 10.1002/glia.22630

Tamura, K., Alessandri, B., Heimann, A., and Kempski, O. (2011). The effect of a gap-junction blocker, carbenoxolone, on ischemic brain injury and cortical spreading depression. *Neuroscience* 194, 262–271. doi: 10.1016/j.neuroscience.2011.07.043

Teubner, B., Michel, V., Pesch, J., Lautermann, J., Cohen-Salmon, M., Sohl, G., et al. (2003). Connexin30 (Gjb6)-deficiency causes severe hearing impairment and lack of endocochlear potential. *Hum. Mol. Genet.* 12, 13–21. doi: 10.1093/hmg/ddg001

Teubner, B., Odermatt, B., Guldenagel, M., Sohl, G., Degen, J., Bukauskas, F., et al. (2001). Functional expression of the new gap junction gene connexin47 transcribed in mouse brain and spinal cord neurons. *J. Neurosci.* 21, 1117–1126.

Thompson, R. J., Zhou, N., and Macvicar, B. A. (2006). Ischemia opens neuronal gap junction hemichannels. *Science* 312, 924–927. doi: 10.1126/science.1126241

Trapp, B. D., Peterson, J., Ransohoff, R. M., Rudick, R., Mork, S., and Bo, L. (1998). Axonal transection in the lesions of multiple sclerosis. *N. Engl. J. Med.* 338, 278–285. doi: 10.1056/NEJM199801293380502

Tsacopoulos, M., and Magistretti, P. J. (1996). Metabolic coupling between glia and neurons. *J. Neurosci.* 16, 877–885.

Umebayashi, D., Natsume, A., Takeuchi, H., Hara, M., Nishimura, Y., Fukuyama, R., et al. (2014). Blockade of gap junction hemichannel protects secondary spinal cord injury from activated microglia-mediated glutamate exitoneurotoxicity. *J. Neurotrauma.* doi: 10.1089/neu.2013.3223. [Epub ahead of print].

Upender, M. B., and Naegele, J. R. (1999). Activation of microglia during developmentally regulated cell death in the cerebral cortex. *Dev. Neurosci.* 21, 491–505. doi: 10.1159/000017416

Valcour, V., Sithinamsuwan, P., Letendre, S., and Ances, B. (2011). Pathogenesis of HIV in the central nervous system. *Curr. HIV/AIDS Rep.* 8, 54–61. doi: 10.1007/s11904-010-0070-4

Valiunas, V., Polosina, Y. Y., Miller, H., Potapova, I. A., Valiuniene, L., Doronin, S., et al. (2005). Connexin-specific cell-to-cell transfer of short interfering RNA by gap junctions. *J. Physiol.* 568, 459–468. doi: 10.1113/jphysiol.2005.090985

Wallraff, A., Kohling, R., Heinemann, U., Theis, M., Willecke, K., and Steinhauser, C. (2006). The impact of astrocytic gap junctional coupling on potassium buffering in the hippocampus. *J. Neurosci.* 26, 5438–5447. doi: 10.1523/JNEUROSCI.0037-06.2006

Wallraff, A., Odermatt, B., Willecke, K., and Steinhauser, C. (2004). Distinct types of astroglial cells in the hippocampus differ in gap junction coupling. *Glia* 48, 36–43. doi: 10.1002/glia.20040

Walz, W., and Hertz, L. (1983). Functional interactions between neurons and astrocytes. II. Potassium homeostasis at the cellular level. *Prog. Neurobiol.* 20, 133–183. doi: 10.1016/0301-0082(83)90013-8

Wang, J., Ma, M., Locovei, S., Keane, R. W., and Dahl, G. (2007). Modulation of membrane channel currents by gap junction protein mimetic peptides: size matters. *Am. J. Physiol. Cell Physiol.* 293, C1112–C1119. doi: 10.1152/ajpcell.00097.2007

Wasseff, S. K., and Scherer, S. S. (2014). Activated microglia do not form functional gap junctions *in vivo*. *J. Neuroimmunol.* 269, 90–93. doi: 10.1016/j.jneuroim.2014.02.005

Willecke, K., Eiberger, J., Degen, J., Eckardt, D., Romualdi, A., Guldenagel, M., et al. (2002). Structural and functional diversity of connexin genes in the mouse and human genome. *Biol. Chem.* 383, 725–737. doi: 10.1515/BC.2002.076

Wong, R. O., Chernjavsky, A., Smith, S. J., and Shatz, C. J. (1995). Early functional neural networks in the developing retina. *Nature* 374, 716–718. doi: 10.1038/374716a0

Wu, D. C., Jackson-Lewis, V., Vila, M., Tieu, K., Teismann, P., Vadseth, C., et al. (2002). Blockade of microglial activation is neuroprotective in the 1-methyl-4-phenyl-1,2,3,6-tetrahydropyridine mouse model of Parkinson disease. *J. Neurosci.* 22, 1763–1771.

Wu, Z., Tokuda, Y., Zhang, X. W., and Nakanishi, H. (2008). Age-dependent responses of glial cells and leptomeninges during systemic inflammation. *Neurobiol. Dis.* 32, 543–551. doi: 10.1016/j.nbd.2008.09.002

Xiong, J., and Kielian, T. (2013). Microglia in juvenile neuronal ceroid lipofuscinosis are primed toward a pro-inflammatory phenotype. *J. Neurochem.* 127, 245–258. doi: 10.1111/jnc.12385

Xu, Q., Cheong, Y. K., He, S. Q., Tiwari, V., Liu, J., Wang, Y., et al. (2014). Suppression of spinal connexin 43 expression attenuates mechanical hypersensitivity in rats after an L5 spinal nerve injury. *Neurosci. Lett.* 566, 194–199. doi: 10.1016/j.neulet.2014.03.004

Xu, X., Li, W. E., Huang, G. Y., Meyer, R., Chen, T., Luo, Y., et al. (2001). Modulation of mouse neural crest cell motility by N-cadherin and connexin 43 gap junctions. *J. Cell Biol.* 154, 217–230. doi: 10.1083/jcb.200105047

Yamanaka, K., Chun, S. J., Boillee, S., Fujimori-Tonou, N., Yamashita, H., Gutmann, D. H., et al. (2008). Astrocytes as determinants of disease progression in inherited amyotrophic lateral sclerosis. *Nat. Neurosci.* 11, 251–253. doi: 10.1038/nn2047

Yawata, I., Takeuchi, H., Doi, Y., Liang, J., Mizuno, T., and Suzumura, A. (2008). Macrophage-induced neurotoxicity is mediated by glutamate and attenuated by glutaminase inhibitors and gap junction inhibitors. *Life Sci.* 82, 1111–1116. doi: 10.1016/j.lfs.2008.03.010

Ye, Z. C., Wyeth, M. S., Baltan-Tekkok, S., and Ransom, B. R. (2003). Functional hemichannels in astrocytes: a novel mechanism of glutamate release. *J. Neurosci.* 23, 3588–3596.

Yeager, M., and Harris, A. L. (2007). Gap junction channel structure in the early 21st century: facts and fantasies. *Curr. Opin. Cell Biol.* 19, 521–528. doi: 10.1016/j.ceb.2007.09.001

Yrjanheikki, J., Keinanen, R., Pellikka, M., Hokfelt, T., and Koistinaho, J. (1998). Tetracyclines inhibit microglial activation and are neuroprotective in global brain ischemia. *Proc. Natl. Acad. Sci. U.S.A.* 95, 15769–15774. doi: 10.1073/pnas.95.26.15769

Yudkoff, M. (1997). Brain metabolism of branched-chain amino acids. *Glia* 21, 92–98. doi: 10.1002/(SICI)1098-1136(199709)21:1<92::AID-GLIA10>3.0.CO;2-W

Zhu, S., Stavrovskaya, I. G., Drozda, M., Kim, B. Y., Ona, V., Li, M., et al. (2002). Minocycline inhibits cytochrome c release and delays progression of amyotrophic lateral sclerosis in mice. *Nature* 417, 74–78. doi: 10.1038/417074a

Zietlow, R., Dunnett, S. B., and Fawcett, J. W. (1999). The effect of microglia on embryonic dopaminergic neuronal survival *in vitro*: diffusible signals from neurons and glia change microglia from neurotoxic to neuroprotective. *Eur. J. Neurosci.* 11, 1657–1667. doi: 10.1046/j.1460-9568.1999.00583.x

Ziv, Y., Avidan, H., Pluchino, S., Martino, G., and Schwartz, M. (2006a). Synergy between immune cells and adult neural stem/progenitor cells promotes functional recovery from spinal cord injury. *Proc. Natl. Acad. Sci. U.S.A.* 103, 13174–13179. doi: 10.1073/pnas.0603747103

Ziv, Y., Ron, N., Butovsky, O., Landa, G., Sudai, E., Greenberg, N., et al. (2006b). Immune cells contribute to the maintenance of neurogenesis and spatial learning abilities in adulthood. *Nat. Neurosci.* 9, 268–275. doi: 10.1038/nn1629

Zlomuzica, A., Reichinnek, S., Maxeiner, S., Both, M., May, E., Worsdorfer, P., et al. (2010). Deletion of connexin45 in mouse neurons disrupts one-trial object recognition and alters kainate-induced gamma-oscillations in the hippocampus. *Physiol. Behav.* 101, 245–253. doi: 10.1016/j.physbeh.2010.05.007

Permissions

All chapters in this book were first published by Frontiers; hereby published with permission under the Creative Commons Attribution License or equivalent. Every chapter published in this book has been scrutinized by our experts. Their significance has been extensively debated. The topics covered herein carry significant findings which will fuel the growth of the discipline. They may even be implemented as practical applications or may be referred to as a beginning point for another development.

The contributors of this book come from diverse backgrounds, making this book a truly international effort. This book will bring forth new frontiers with its revolutionizing research information and detailed analysis of the nascent developments around the world.

We would like to thank all the contributing authors for lending their expertise to make the book truly unique. They have played a crucial role in the development of this book. Without their invaluable contributions this book wouldn't have been possible. They have made vital efforts to compile up to date information on the varied aspects of this subject to make this book a valuable addition to the collection of many professionals and students.

This book was conceptualized with the vision of imparting up-to-date information and advanced data in this field. To ensure the same, a matchless editorial board was set up. Every individual on the board went through rigorous rounds of assessment to prove their worth. After which they invested a large part of their time researching and compiling the most relevant data for our readers.

The editorial board has been involved in producing this book since its inception. They have spent rigorous hours researching and exploring the diverse topics which have resulted in the successful publishing of this book. They have passed on their knowledge of decades through this book. To expedite this challenging task, the publisher supported the team at every step. A small team of assistant editors was also appointed to further simplify the editing procedure and attain best results for the readers.

Apart from the editorial board, the designing team has also invested a significant amount of their time in understanding the subject and creating the most relevant covers. They scrutinized every image to scout for the most suitable representation of the subject and create an appropriate cover for the book.

The publishing team has been an ardent support to the editorial, designing and production team. Their endless efforts to recruit the best for this project, has resulted in the accomplishment of this book. They are a veteran in the field of academics and their pool of knowledge is as vast as their experience in printing. Their expertise and guidance has proved useful at every step. Their uncompromising quality standards have made this book an exceptional effort. Their encouragement from time to time has been an inspiration for everyone.

The publisher and the editorial board hope that this book will prove to be a valuable piece of knowledge for researchers, students, practitioners and scholars across the globe.

List of Contributors

Daniela Boassa, Phuong Nguyen and Junru Hu
National Center for Microscopy and Imaging Research, Center for Research in Biological Systems, University of California, San Diego, La Jolla, CA, USA

Mark H. Ellisman and Gina E. Sosinsky
National Center for Microscopy and Imaging Research, Center for Research in Biological Systems, University of California, San Diego, La Jolla, CA, USA
Department of Neurosciences, University of California, San Diego, La Jolla, CA, USA

Alvaro O. Ardiles, Carolina Flores-Muñoz, Gabriela Toro-Ayala, Ana M. Cárdenas, Adrian G. Palacios, Agustín D. Martínez, Fanny Momboisse, María José Olivares, Ximena Báez-Matus and María José Guerra
Centro Interdisciplinario de Neurociencia de Valparaíso, Facultad de Ciencias, Instituto de Neurociencia, Universidad de Valparaíso, Valparaíso, Chile

Pablo Muñoz
Centro Interdisciplinario de Neurociencia de Valparaíso, Facultad de Ciencias, Instituto de Neurociencia, Universidad de Valparaíso, Valparaíso, Chile
Escuela de Medicina, Facultad de Medicina, Universidad de Valparaíso, Valparaíso, Chile

Marco Fuenzalida
Centro de Neurobiología y Plasticidad Cerebral, Instituto de Fisiología, Universidad de Valparaíso, Valparaíso, Chile

Juan C. Sáez
Departamento de Fisiología, Facultad de Ciencias Biológicas, Pontificia Universidad Católica de Chile, Santiago, Chile
Instituto Milenio, Centro Interdisciplinario de Neurociencia de Valparaíso, Universidad de Valparaíso, Valparaíso, Chile

Paul Castellano and Eliseo A. Eugenin
Public Health Research Institute (PHRI), New Jersey Medical School, Rutgers The State University of New Jersey, Newark, NJ, USA
Department of Microbiology and Molecular Genetics, New Jersey Medical School, Rutgers The State University of New Jersey, Newark, NJ, USA

Angelina Palacios-Muñoz, Alex Vielma, Joaquín Araya, Gonzalo Valdivia, Isaac E. García, Oliver Schmachtenberg and Agustín D. Martínez
Facultad de Ciencias, Centro Interdisciplinario de Neurociencia de Valparaíso, Universidad de Valparaíso, Valparaíso, Chile

Maria J. Escobar and Aland Astudillo
Departamento de Electrónica, Universidad Técnico Federico Santa María, Valparaíso, Chile

José Hurtado and Adrian G. Palacios
Facultad de Ciencias, Centro Interdisciplinario de Neurociencia de Valparaíso, Universidad de Valparaíso, Valparaíso, Chile
Instituto de Sistemas Complejos de Valparaíso, Valparaíso, Chile

Megan Bosch
Department of Pharmacology and Experimental Neuroscience, University of Nebraska Medical Center, Omaha, NE, USA

Tammy Kielian
Department of Pathology and Microbiology, University of Nebraska Medical Center, Omaha, NE, USA

Luis A. Cea and Aníbal A. Vargas
Departamento de Fisiología, Facultad de Ciencias Biológicas, Pontificia Universidad Católica de Chile, Santiago, Chile

Christian A. Verdugo
Facultad de Medicina, Centro de Fisiología Celulare Integrativa, Clínica Alemana Universidad del Desarrollo, Santiago, Chile
Centro de Fisiología Celular e Integrativa, Facultad de Medicina, Clínica Alemana Universidad del Desarrollo, Santiago, Chile

José D. Salas
Centro de Fisiología Celular e Integrativa, Facultad de Medicina, Clínica Alemana Universidad del Desarrollo, Santiago, Chile

Vania A. Figueroa
Departamento de Fisiología, Facultad de Ciencias Biológicas, Pontificia Universidad Católica de Chile, Santiago, Chile
Centro de Fisiología Celular e Integrativa, Facultad de Medicina, Clínica Alemana Universidad del Desarrollo, Santiago, Chile

Oscar Jara and Agustín D. Martínez
Instituto Milenio, Centro Interdisciplinario de Neurociencia de Valparaíso, Universidad de Valparaíso, Valparaíso, Chile

Bruno A. Cisterna
Departamento de Fisiología, Pontificia Universidad Católica de Chile, Santiago, Chile

Christopher Cardozo
Center of Excellence for the Medical Consequences of Spinal Cord Injury, James J. Peters Veterans Affairs Medical Center, Bronx, NY, USA
Departments of Medicine and Rehabilitation Medicine, Icahn School of Medicine at Mount Sinai, New York, USA

Georg Zoidl
Department of Psychology, Faculty of Health, York University, Toronto, ON, Canada
Department of Biology, Faculty of Science, York University, Toronto, ON, Canada

Verónica Abudara and Christian Giaume
Center for Interdisciplinary Research in Biology, Centre National de la Recherche Scientifique, Collège de France, Paris, France

Moises Freitas-Andrade, Maxence Le Vasseur, Jonathan Lelowski, John F. Bechberger, Wun-Chey Sin and Christian C. Naus
Department of Cellular and Physiological Sciences, Faculty of Medicine, Life Sciences Institute, University of British Columbia, Vancouver, BC, Canada

Marijke De Bock, Nan Wang and Luc Leybaert
Department of Basic Medical Sciences - Physiology Group, Faculty of Medicine and Health Sciences, Ghent University, Ghent, Belgium

Geert Bultynck
Laboratory of Molecular and Cellular Signaling, Department of Cellular and Molecular Medicine, KU Leuven, Leuven, Belgium

Nicolás Palacios-Prado and Alberto E. Pereda
Dominick P. Purpura Department of Neuroscience, Albert Einstein College of Medicine, Bronx, NY, USA
Marine Biological Laboratory, Woods Hole, Massachusetts, MA, USA

Wolf Huetteroth
Marine Biological Laboratory, Woods Hole, Massachusetts, MA, USA
Department of Neurobiology, University of Konstanz, Konstanz, Germany

Mauricio A. Retamal
Facultad de Medicina, Centro de Fisiología Celulare Integrativa, Clínica Alemana Universidad del Desarrollo, Santiago, Chile
Departamento de Fisiología, Facultad de Ciencias Biológicas, Pontificia Universidad Católica de Chile, Santiago, Chile
Centro de Fisiología Celular e Integrativa, Facultad de Medicina, Clínica Alemana Universidad del Desarrollo, Santiago, Chile

Julio Alcayaga
Laboratorio de Fisiología Celular, Departamento de Biología, Facultad de Ciencias, Universidad de Chile, Santiago, Chile

Pablo J. Sáez
Departamento de Fisiología, Facultad de Ciencias Biológicas, Pontificia Universidad Católica de Chile, Santiago, Chile

Ricardo Fernández
Facultad de Ciencias Biológicas y Facultad de Medicina, Universidad Andrés Bello, Santiago, Chile

Luis E. León
Facultad de Medicina, Centro de Genética Humana, Clínica Alemana Universidad del Desarrollo, Santiago, Chile

Juan A. Orellana, Gigliola Ramírez and Rommy von Bernhardi
Departamento de Neurología, Escuela de Medicina, Pontificia Universidad Católica de Chile, Santiago, Chile

Dolores Busso and Attilio Rigotti
Departamento de Nutrición, Diabetes y Metabolismo, Escuela de Medicina, Pontificia Universidad Católica de Chile, Santiago, Chile
Centro de Nutrición Molecular y Enfermedades Crónicas, Escuela de Medicina, Pontificia Universidad Católica de Chile, Santiago, Chile

Marlys Campos and Jaime Eugenín
Laboratorio de Sistemas Neurales, Departamento de Biología, Facultad de Química y Biología, Universidad de Santiago de Chile, Santiago, Chile

Helmuth A. Sanchez and Vytas K. Verselis
Dominick P. Purpura Department of Neuroscience, Albert Einstein College of Medicine, Bronx, NY, USA

Giselle Cheung, Oana Chever and Nathalie Rouach
Neuroglial Interactions in Cerebral Physiopathology, Center for Interdisciplinary Research in Biology, Collège de France, CNRS UMR 7241, INSERM U1050, Labex Memolife, PSL Research University, Paris, France

Stefan Kurtenbach, Paige Whyte-Fagundes, Sarah Kurtenbach, Émerson Brazil, Christiane Zoidl and Georg Zoidl
Department of Psychology, Faculty of Health, York University, Toronto, ON, Canada

Lian Gelis and Hanns Hatt
Department of Cell Physiology, Ruhr University Bochum, Bochum, Germany

Valery I. Shestopalov
Department of Ophthalmology, Bascom Palmer Eye Institute, Miller School of Medicine, University of Miami, Miami, FL, USA
Vavilov Institute of General Genetics, Russian Academy of Sciences, Moscow, Russia

Verónica Cerpa, Liliana Corvalán and Mauricio Antonio Retamal
Centro de Fisiología Celular e Integrativa, Facultad de Medicina, Clínica Alemana Universidad del Desarrollo, Santiago, Chile

Edison Pablo Reyes
Centro de Fisiología Celular e Integrativa, Facultad de Medicina, Clínica Alemana Universidad del Desarrollo, Santiago, Chile
Dirección de Investigación, Universidad Autónoma de Chile, Santiago, Chile

Hideyuki Takeuchi and Akio Suzumura
Department of Neuroimmunology, Research Institute of Environmental Medicine, Nagoya University, Nagoya, Japan

Index

Printed in the USA
CPSIA information can be obtained
at www.ICGtesting.com
LVHW071033181223
766281LV00029B/479

9 798887 403786